POWER EXHAUST IN FUSION PLASMAS

Nuclear fusion research is entering a new phase, in which power exhaust will play a vital role. This book presents a comprehensive and up-to-date summary of this emerging field of research in fusion plasmas, focusing on the leading tokamak concept.

Emphasis is placed on rigorous theoretical development, supplemented by numerical simulations, which are used to explain and quantify a range of experimental observations. The text offers a self-contained introduction to power exhaust, and deals in detail with both edge plasma turbulence and edge localized modes, providing the necessary background to understand these important, yet complicated phenomena.

Combining an in-depth overview with an instructive development of concepts, this is an invaluable resource for academic researchers and graduate students in plasma physics.

WOJCIECH FUNDAMENSKI is the leader of the Exhaust Physics Task Force at the Joint European Torus (JET), where he pursues research into edge plasma physics and particle / power exhaust. He is also a Visiting Lecturer in Plasma Physics at Imperial College, London, and a committee member of the Plasma Physics Group of the Institute of Physics.

POWER EXHAUST IN FUSION PLASMAS

WOJCIECH FUNDAMENSKI
Culham Science Centre

CAMBRIDGE
UNIVERSITY PRESS

CAMBRIDGE
UNIVERSITY PRESS

University Printing House, Cambridge CB2 8BS, United Kingdom

Published in the United States of America by Cambridge University Press, New York

Cambridge University Press is part of the University of Cambridge.

It furthers the University's mission by disseminating knowledge in the pursuit of education, learning and research at the highest international levels of excellence.

www.cambridge.org
Information on this title: www.cambridge.org/9781107424210

First published 2010
First paperback edition 2014

A catalogue record for this publication is available from the British Library

ISBN 978-0-521-85171-8 Hardback
ISBN 978-1-107-42421-0 Paperback

dla Oli

Contents

Preface

Power exhaust, by which we mean the safe removal of power from a burning plasma, is an essential requirement for the successful operation of any fusion reactor. Specifically, plasma thermal energy must be conveyed across the first wall without undue damage to plasma facing components (divertor and limiter tiles) by heat load related plasma–surface interactions (ablation, melting, erosion). Unlike other 'technological' problems related to fusion reactor design, e.g. tritium retention in plasma facing materials, neutron damage to structural components or non-inductive current drive, power exhaust is intimately linked to plasma confinement and thus a perennial concern for any fusion reactor. While only a minor issue in existing tokamaks, it will be critical for ITER (the next step plasma-burning experiment) and even more so for DEMO (the demonstration fusion power plant). Even non-burning, superconducting machines, such as EAST, KSTAR, JT60-SA, W7-X, etc. will be forced to tackle this problem due to their long pulse capabilities.

This monograph is an attempt to draw a unified and up-to-date picture of power exhaust in fusion plasmas, focusing primarily on the leading tokamak concept. Emphasis is placed on rigorous theoretical development, supplemented by numerical simulations when appropriate, which are then employed to explain and model a range of experimental observations. The objective is not just to provide the reader with a reliable map of the conquered territory and a guided tour over its many hills and valleys,[1] but also to supply him or her with the tools necessary to embark on independent, and hopefully fruitful, journeys into the uncharted regions, the white spaces on the map, *la terra incognita.* In this respect, the book is aimed both at graduate students of magnetically confined plasmas and at researchers already working in the field wishing to develop a deeper understanding of plasma exhaust physics – a quickly emerging area of fusion research.

[1] This function being well served by regular review articles appearing in topical journals.

Broadly speaking, the text is organized into two parts. The first (Chapters 2 to 4) is dedicated to developing the theoretical framework necessary to describe the equilibrium and stability properties of magnetically confined plasmas, the second (Chapters 5 to 8) deals with plasma transport phenomena necessary to understand power exhaust in real experiments. After a brief examination of charged particle motion, the two basic orderings of plasma dynamics (MHD and drift) are introduced and the corresponding guiding centre kinetic and fluid equations are derived. These are then used to investigate the equilibrium, stability and transport properties of magnetically confined plasmas. Energy transport in the radial, diamagnetic and parallel directions due to collisional (classical and neoclassical) and turbulent (drift-Alfvén and interchange) processes is examined with special emphasis on plasma turbulence in the boundary (edge) plasma and the scrape-off layer (SOL). Next, the relevant experimental results from tokamaks and the modelling approaches typically used to interpret these results are reviewed. Finally, the tools developed hereto are applied collectively to study power exhaust in low and high confinement regime plasmas in tokamaks, in particular to edge / SOL turbulence and edge localized modes (ELMs).

The idea for this book originates with my early inroads into power exhaust on JET and owes much to the difficulties I encountered in finding relevant material in the topical literature. During this period I was, and indeed still am, fortunate enough to benefit from the vibrant scientific environment of the Culham Science Centre. I thus feel highly indebted to my many colleagues and friends at JET and in other labs around the world, without whom this project would certainly not have succeeded. In particular, I would like to thank A. Alonso, P. Andrew, N. Asakura, M. Beurskens, J. Boedo, S. Brezinsek, D. Campbell, C. S. Chang, A. Chankin, J. Connor, G. Corrigan, D. Coster, G. Counsell, T. Eich, S. K. Erents, M. Fenstermacher, O. E. Garcia, B. Gonçalves, P. Helander, T. Hender, C. Hidalgo, G. Huysmann, S. Jachmich, A. Kirk, S. Krashenninikov, A. Kukushkin, B. LaBombard, B. Lipschultz, S. Lisgo, A. Loarte, G. F. Matthews, G. McCracken, W. Morris, D. Moulton, V. Naulin, A. Nielsen, V. Philipps, R. A. Pitts, J. Rapp, R. Schneider, B. Scott, S. Sipilä, P.C. Stangeby, M. Tokar, D. Tskhakaya, S. Wiesen, M. Wischmeyer, G. S. Xu, R. Zagórski and S. Zweben. I would also like to thank UKAEA, EFDA-JET, EPSRC and Imperial College, London, for their support and the many research opportunities which they supplied.

1
Introduction

'Faced with something unusual our thought should not be "What next?" but "Why?". By answering the second of these questions we can answer the first. And this, in brief, is the scientific method.'
Roger Scruton (c. 1990)

By definition, all exothermal reactors, including any fusion reactor one may envisage (tokamak, stellarator, etc.), produce both energy and spent reactants, or ash.[1] In order for the reactor to operate in steady-state, (i) fresh fuel must be added at the rate at which it is consumed, (ii) this fuel must be heated, ideally by the reactions themselves, (iii) fuel must be confined, by whatever means are available, for sufficiently long to allow the exothermic processes to continue, (iv) the energy and ash must be removed from the system at the rate at which they are created, (v) the impurities released from the reactor walls must likewise be removed at the rate at which they are produced, and (vi) the reactor itself, primarily its walls, must not be damaged by all the exhaust processes. Translating the above to a D–T burning tokamak, conditions (i)–(iii) may be labelled loosely as the *ignition* criteria, and conditions (iv)–(vi) as the *exhaust* criteria. Taken together they constitute the criteria of mutual *compatibility* between the burning plasma and first wall materials/components. Since the ignition criteria speak primarily to the central (core) plasma, while the exhaust criteria refer to the boundary (edge) plasma, and since the two regions are coupled by largely self-governing plasma transport processes, it is the exhaust criteria which determine the optimum reactor performance for a given reactor design. In the following, we introduce the basic concepts of fusion reactor operation, including the stability and exhaust limits on reactor performance.

[1] This applies both to chemical reactors, such as a candle or a steam engine, and nuclear reactors, such as a star or a fission power plant. It is equally true for all fusion reactors, irrespective of whether the reacting fuel is confined by *gravity*, as in the Sun, by *magnetic* fields and electric currents, as in a tokamak or stellarator, or by the *inertia* of the ions themselves, as in the violent implosion of a hydrogen ice pellet after it is heated by lasers or heavy ion beams.

1.1 Fusion reactor operating criteria

Let us consider the ignition and exhaust criteria for a *magnetically confined fusion* (MCF) reactor, operated either in steady-state or in successive pulse cycles; although we restrict the discussion to MCF, most of the following remarks apply equally well to *inertially confined fusion* (ICF). There are four fusion reactions of interest for energy production (Krane, 1988),

$$\mathrm{D} + \mathrm{T} \rightarrow \mathrm{He}^4\,(3.52\,\mathrm{MeV}) + \mathrm{n}\,(14.06\,\mathrm{MeV}), \tag{1.1}$$

$$\mathrm{D} + \mathrm{D} \rightarrow \mathrm{T}\,(1.01\,\mathrm{MeV}) + \mathrm{p}\,(3.03\,\mathrm{MeV}), \tag{1.2}$$

$$\mathrm{D} + \mathrm{D} \rightarrow \mathrm{He}^3\,(0.82\,\mathrm{MeV}) + \mathrm{n}\,(2.45\,\mathrm{MeV}), \tag{1.3}$$

$$\mathrm{D} + \mathrm{He}^3 \rightarrow \mathrm{He}^4\,(3.67\,\mathrm{MeV}) + \mathrm{p}\,(14.67\,\mathrm{MeV}), \tag{1.4}$$

where D and T represent the two isotopes of hydrogen: deuterium ($\mathrm{D} \equiv \mathrm{H}^2$) and tritium ($\mathrm{T} \equiv \mathrm{H}^3$). In all four cases, the strong Coulomb repulsion of the positively charged nuclei implies that the fusion cross-sections σ are only significant at ion energies above 10 keV, e.g. at 100 keV, $\sigma_{DT} \sim 5$ barn, $\sigma_{DD} \sim \sigma_{DHe3} \sim 0.01$ barn. In thermonuclear fusion, the supra-thermal particles in the tail of the Maxwellian distribution are responsible for most of the fusion reactions. Since the average reaction rate $\langle \sigma v \rangle$ is largest for (1.1), especially for $T_i < 100$ keV, e.g. at $T_i = 10$ keV, $\langle \sigma v \rangle_{DT} \sim 10^{-22}\,\mathrm{m^3/s}$, while $\langle \sigma v \rangle_{DD} \sim 10^{-24}\,\mathrm{m^3/s}$, a mixture of D and T is the preferred fuel for future fusion reactors, including ITER and DEMO. At keV temperatures, the atoms of hydrogen (for which the ionization potential is only 4 eV), as well as those of most low and medium Z elements, become fully ionized and the neutral gas mixture is transformed into an ion–electron *plasma*.

We now return to our six reactor criteria, the first two of which state that the D–T fuel burned in reaction (1.1) must be replenished, criterion (i), and heated to the operating reactor temperature, criterion (ii). In practice, (i) is achieved either by gas puffing or ice pellet injection, although neither of these methods is capable of delivering the fuel directly to the plasma core, i.e. the hot central region where the thermonuclear burn is active; instead the fresh fuel is deposited (ablated/ionized) in the edge plasma, and only reaches the core by a relatively slow diffusion process. In contrast, the steady flow of power required by (ii) is delivered directly to the core plasma either by external heating, e.g. by neutral beams or radio waves resonant with the gyration frequencies of ions and electrons, or by the charged fusion products, such as the 3.5 MeV alpha particle in (1.1) or the 14.7 MeV proton in (1.4), which are trapped by the magnetic fields.[2]

[2] Neutrons released in fusion reactions do not interact with, and thus cannot heat, the plasma. The same is true for photons released as bremsstrahlung and synchrotron radiation.

Particle and energy confinement of a thermonuclear plasma, criterion (iii), have been the central focus of MCF research over the past 50 years. To appreciate the difficulties posed by this task, recall that energy *break-even*,

$$Q_{DT} \equiv P_{DT}/P_{heat} > 1, \qquad Q_\alpha \equiv P_\alpha/P_{heat} = 0.2Q_{DT}, \tag{1.5}$$

where Q_{DT} is the *energy multiplication factor*, defined as the ratio of fusion and auxiliary heating powers, was only approached recently. Since 80% of the energy released in (1.1) appears as the kinetic energy of neutrons and is thus promptly lost from the plasma, the fusion reactions can only be self-sustaining when $Q_\alpha \gg 1$. Note that P_{DT} and P_α may be evaluated as

$$P_{DT} = 5P_\alpha = \frac{\mathcal{E}_{DT}}{4}\int n^2\langle\sigma v\rangle_{DT}\mathrm{d}\mathbf{x}, \qquad \mathcal{E}_{DT} = 5\mathcal{E}_\alpha = 17.58\,\mathrm{MeV} \tag{1.6}$$

where $\mathcal{E}_{DT}$ is the energy released per fusion reaction, $n = n_D + n_T$ is the particle density, $\langle\sigma v\rangle_{DT}$ is the fusion reaction cross-section and $\int \mathrm{d}\mathbf{x}$ is a volume integral over the plasma.

The slow progress towards $Q_\alpha > 1$ can be ultimately traced to one of the great unsolved problems of classical physics, namely fluid turbulence. Indeed, much of the success of MCF can be ascribed to the basic dimensional scaling: volume/area $\sim$ size, and thus to the building of ever bigger, and more expensive, devices, specifically the toroidal, axis-symmetric, inductively driven *tokamaks*, see Fig. 1.1. It is thus no coincidence that $Q_{DT} \sim 1$ was finally approached in the largest present day

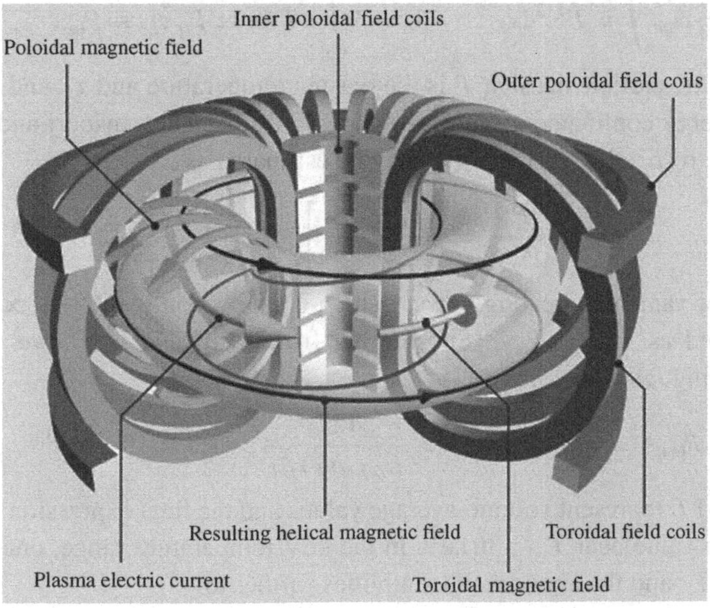

Fig. 1.1. Schematic representation of a *tokamak*. © EFDA-JET.

tokamak, namely the *Joint European Torus* (JET), with $Q_\alpha > 1$ expected in ITER. That future event may be viewed as the watershed beyond which issues related to energy confinement, criterion (iii), will be increasingly overshadowed by those related to power exhaust, criteria (iv)–(vi). This tendency, which first emerged in the technological challenges encountered during the conceptual, and later engineering, design of ITER, is also evident in a new generation of super-conducting, actively cooled machines.

Since both fusion power and ash (He^4) are generated in the plasma centre, their exhaust mechanisms are partly related. Thus, power is removed from an MCF plasma by three channels: (a) by neutrons released in the fusion reaction itself; (b) by photons emitted during bremsstrahlung, synchrotron and (hydrogenic or impurity) line radiation; and (c) by kinetic energy of the ions and electrons which are transported across the magnetic field largely by turbulent plasma motions; in contrast, fusion ash is removed from the core plasma only by turbulent advection. Assuming that bremsstrahlung is the dominant mode of radiation in the hot plasma core, we may approximate the steady-state power balance for a burning fusion plasma as

$$P_{heat} + P_\alpha = (1 + Q_\alpha)P_{heat} \approx P_{br} + P_{tr} = P_{loss}, \tag{1.7}$$

where the left-hand side represents the auxiliary (P_{heat}) and α particle (P_α) heating, and the right-hand side the total losses due to bremsstrahlung (P_{br}) and plasma transport by convection and conduction (P_{tr}),

$$P_{br} = \alpha_{br} \int n^2 T^{1/2} \mathbf{dx}, \qquad W = 3 \int nT\,\mathbf{dx} \equiv P_{tr}\tau_E \equiv P_{loss}\tau_E^*, \tag{1.8}$$

where n is the particle density, T is the plasma temperature and τ_E and τ_E^* are the thermal energy confinement times associated with plasma transport and transport + radiation, respectively. Hence Q_α may be estimated as

$$Q_\alpha \equiv \frac{P_\alpha}{P_{heat}} = \left[\frac{P_{loss}}{P_\alpha} - 1\right]^{-1} = \frac{P_\alpha}{P_{loss} - P_\alpha}. \tag{1.9}$$

The limit of vanishing heating power, which is equivalent to $Q_\alpha \to \infty$, can thus be expressed as $P_\alpha = P_{loss}$. Substituting from (1.6) and (1.8), we obtain the corresponding *ignition criterion*,

$$P_\alpha \geq P_{loss} \qquad \Rightarrow \qquad n\tau_E > \frac{12T}{\mathcal{E}_{DT}\langle\sigma v\rangle_{DT}} > 1.5 \times 10^{20}\,\mathrm{m}^{-3}\,\mathrm{s}, \tag{1.10}$$

where n and T represent volume-average values and the final expression represents a minimum value near $T \approx 30\,\mathrm{keV}$. In the keV temperature range, one finds that $\langle\sigma v\rangle_{DT} \propto T^2$ and the above result simplifies further to

$$nT\tau_E = p\tau_E > 3 \times 10^{21}\,\mathrm{m}^{-3}\,\mathrm{keV\,s} \approx 5\,\mathrm{bar \cdot s}. \tag{1.11}$$

The numerical values in (1.10) and (1.11) assume flat radial profiles of n and T; for peaked profiles, these values are somewhat higher.

Let us next consider a corresponding condition for a *fusion reactor*, in which all power leaving the plasma is converted to electricity with an efficiency η_e and then used to heat the plasma with efficiency η_h. Defining $\eta = \eta_e \eta_h$, for which one expects values in the range 0.2–0.4, the requirement for net energy production may be written as,

$$\eta(P_{fus} + P_{loss}) > P_{loss}, \qquad P_{fus} = P_{DT} + P_{Li}, \tag{1.12}$$

where the additional power P_{Li} refers to the energy released in the breeder blanket by the reaction (1.14), see below. Substituting from (1.6) and (1.8) leads to the celebrated *Lawson's criterion* (Lawson, 1957),

$$n\tau_E > 3T\left(\frac{\eta}{1-\eta}\frac{\mathcal{E}_{DT}}{4}\langle\sigma v\rangle_{DT} - \alpha_{br}T^{1/2}\right)^{-1} \sim 3\times 10^{19}\,\mathrm{m^{-3}s}, \tag{1.13}$$

where the final expression was evaluated near $T \approx 30$ keV and $\eta = 1/3$. The plasma ignition criterion (1.10) is equivalent to (1.13) with $\eta = 0.136$.[3]

Power exhaust channels (a)–(c) lead to three different types of heat loads on the first wall and require three different power removal systems: (a) the *neutron* energy is deposited volumetrically in a neutron-absorbing envelope surrounding the first wall, ideally a *breeder blanket*, employing the reactions,

$$\mathrm{Li}^6 + \mathrm{n} \rightarrow \mathrm{T} + \mathrm{He}^4 + 4.8\,\mathrm{MeV}, \tag{1.14}$$

$$\mathrm{Li}^7 + \mathrm{n} \rightarrow \mathrm{T} + \mathrm{He}^4 + \mathrm{n} - 2.5\,\mathrm{MeV}, \tag{1.15}$$

to breed tritium fuel from solid lithium, (b) the *photon* energy generates a fairly uniform surface heat load on first wall components, and (c) the *plasma* thermal energy is convected and conducted along the magnetic field lines to dedicated heat load bearing tiles. In each case, the power deposited on, or absorbed in, the vessel wall must be removed by an active coolant loop. Moreover, the effective heat load must not exceed some limit imposed by thermo-mechanical constraints. This in turn limits the energy flow crossing the outer boundary of the plasma in each of the three channels, i.e.

$$P_{\perp\sigma}/A_p \equiv q_{\perp\sigma} < q_{\perp\sigma}^{exh}, \qquad \sigma \in \{n, \gamma, tr\}, \tag{1.16}$$

where A_p is the plasma area. In practice, the last of these conditions imposes the most severe constraints on plasma operation, e.g. for ITER, the time-averaged power loads on plasma facing components (PFCs) are limited to $\sim$10 MW/m^2 and

[3] To demonstrate this, it suffices to insert $P_\alpha = P_{loss}$ in (1.12), which yields $\eta = P_\alpha/(P_\alpha + P_{DT} + P_{Li}) = 3.52/(3.52 + 17.58 + 4.8) = 0.136$.

transient energy loads to ~0.5 MJ/m^2 in ~250 μs.[4] Consequently, in the rest of the book we will focus on channel (c) above, i.e. the exhaust of fusion energy (and to a lesser extent, of fusion ash) by plasma transport processes. As motivation for this investigation, we first compare the limits on fusion reactor performance, which for simplicity we assume to be a tokamak, imposed by plasma stability and power exhaust.

1.2 Plasma stability limits on fusion reactor performance

Let us first assess the limits imposed by global (MHD, magneto-hydrodynamic) plasma stability requirements, which will be derived in Section 4.2 and summarized in Table 4.1.[5] Expressing the fusion power density in terms of the toroidal beta,

$$P_{DT}/\mathcal{V} = \tfrac{1}{4}\mathcal{E}_{DT}\langle n^2\langle\sigma v\rangle_{DT}\rangle_a \propto \langle p^2\rangle_a \propto \beta_T^2 B_0^4, \tag{1.17}$$

where β_T is given by (3.19) and $\langle\cdot\rangle_a$ is the average over the plasma volume, $\mathcal{V} = \int_a \mathbf{dx}$, and noting that the toroidal magnetic field (on axis) is limited by technological constraints to roughly $B_0^{max} \sim 5{-}10\,\mathrm{T}$, we find that the MHD pressure limits determine the maximum fusion power density and hence the reactor cost.[6] In order for the burning plasma equilibrium to be MHD stable, the MHD beta limit β_T^{mhd}, as given in Table 4.1, must exceed the minimum beta required for ignition β_T^{ign}, which may be inferred from (1.11).

$$\beta_T^{ign} \propto B_0^{-2}\tau_E^{-1}, \qquad \beta_T^{mhd} \propto \epsilon_a\kappa_a/q_0q_a. \tag{1.18}$$

Here we defined the inverse aspect ratio, $\epsilon_a = a/R_0$, where a and R_0 are the minor and major radii of the torus, the elongation $\kappa_a = A_p/\pi a^2$ where A_p is the cross-sectional area of the plasma, and the safety factors on axis ($r = 0$) and at the edge of the plasma ($r = a$), q_0 and q_a, given by (3.18).

On the other hand, $\beta_T < \beta_T^{mhd}$ amounts to inefficient use of the 'expensive' toroidal magnetic field, which is optimally used only for $\beta_T \approx \beta_T^{mhd}$. Hence, the condition $\beta_T^{ign} \approx \beta_T^{mhd}$ determines the size a_{ign} of the smallest reactor able to ignite for given field B_0, inverse aspect ratio ϵ_a, elongation κ_a, etc.

To evaluate a_{ign}, we need to estimate the energy confinement time τ_E (1.8), e.g. we may assume that radial transport is purely diffusive, so that $\tau_E \approx a^2/\chi_\perp$, where

[4] This value should not be confused with $q_{\perp tr}^{exh}$, which refers to the power flux crossing the last closed flux surface, see Chapter 7.

[5] Here we will anticipate some of the definitions which will be made formally in Chapters 2 to 4.

[6] Since the reactor capital cost is roughly proportional to the plasma volume $\mathcal{V}$, being driven largely by the cost associated with super-conducting poloidal coils, one finds that the cost of electricity it generates scales inversely with $P_{DT}/\mathcal{V}$, i.e. an economical reactor should be as small as possible to generate a desired power output in MWe. Hence, the power density (1.17) may be interpreted as the financial figure of merit for a fusion reactor.

$\chi_\perp$ is the average radial heat diffusivity.[7] Anticipating the results of Chapter 6, we write down the generic expression,

$$B_0\tau_E \propto (q_a\rho_*)^{-x} \propto (q_a\rho_{ti}/a)^{-x} \propto (q_a\sqrt{T}/aB_0)^{-x}, \qquad (1.19)$$

which states that $B_0\tau_E$ scales inversely with the product of the safety factor q_a, or q_* (4.110), and the normalized, toroidal gyro-radius $\rho^* \equiv \rho_{ti}/a$; here $x = 3$ corresponds to the *gyro-Bohm* scaling and $x = 2$ to the *Bohm* scaling.

Inserting (1.19) into (1.11), one finds a scaling of β_T^{ign} with reactor size,

$$\beta_T^{ign} \propto [q_a\sqrt{T}/aB_0]^x/B_0 \propto a^{-x}B_0^{-(x+1)}T^{x/2}q_a^x. \qquad (1.20)$$

Since, the *minimum* beta needed for ignition, β_T^{ign}, decreases sharply with size, whereas the *maximum* beta imposed by MHD stability, β_T^{mhd}, is size independent, we find that ignition is always possible for large enough plasmas.[8] Equating β_T^{ign}, (1.20), and β_T^{mhd}, (1.18), gives the minimum ignition radius,

$$a_{ign} \propto (q_a/B_0)^{1+1/x}T^{1/2}(q_0/\varepsilon_a\kappa_a)^{1/x}, \qquad (1.21)$$

which decreases with toroidal field as $B_0^{-3/2}$ $(x = 2)$ and $B_0^{-4/3}$ $(x = 3)$.

The plasma volume corresponding to (1.21) is found to scale as

$$\mathcal{V}_{ign} \propto a_{ign}^3\kappa_a/\varepsilon_a \propto (q_a/B_0)^{3(1+1/x)}T^{3/2}q_0^{3/x}\varepsilon_a^{-(1+3/x)}\kappa_a^{1-3/x}. \qquad (1.22)$$

Since $\mathcal{V}_{ign}$ increases with q_a, q_0 and T, and decreases with B_0, ϵ_a and κ_a (although the κ_a dependence vanishes for $x = 3$), we would like to minimize (maximize) the former (latter) parameters. This can be done by (i) fixing $q_0 \approx q_0^{mhd} \approx 1$ and $q_a \approx q_a^{mhd} \approx 2 - 3$ at their MHD stability limits, (ii) choosing $T \sim 10$–30 keV, which, although below the maximum of the fusion cross-section, $\langle\sigma v\rangle_{DT}$, minimizes (1.10), and (iii) setting the axial field at $B_0^{max} \sim 5$ T. Moreover, (1.22) strongly favours small aspect ratios ($\epsilon_a \sim 1$) and weakly favours elongated plasma shapes ($\kappa_a > 1$), provided $x < 3$. The upper limit on ϵ_a and lower limit on R_0 are imposed by the requirement for a neutron heat shield on the inner solenoid;[9] the upper limit on κ_a is imposed by an axis-symmetric ($n = 0$) vertical displacement MHD instability, which becomes increasingly acute for elongated poloidal cross-sections. For instance, for ITER, whose aim is to achieve $Q_{DT} = 10$, or $Q_\alpha = 2$, and produce 500 MW of fusion power, the above

[7] This scaling expresses the easily verified fact that larger objects take longer too cool.

[8] However, as will be shown in Section 7.1, power exhaust considerations impose an upper limit on a cost effective reactor, i.e. one with $\beta_T \approx \beta_T^{mhd}$.

[9] Some designs dispense with this requirement by envisioning a replaceable central column, thus allowing a much smaller R_0 and thus a larger inverse aspect ratio ϵ_a.

parameters were carefully optimized with respect to cost and performance to yield

$$a = 2.0\,\text{m}, \quad B_0 = 5.3\,\text{T}, \quad q_a = 3.0, \quad \kappa_a = 1.7, \quad \epsilon_a = 0.33, \tag{1.23}$$

where $q_a = 3$ represents a plasma current of $I_T = 15\,\text{MA}$, see Fig. 8.1.

1.3 Power exhaust limits on fusion reactor performance

The limits on fusion reactor performance imposed by plasma stability, as derived above, should be compared with those imposed by power exhaust (1.16). Let us assume that the radial energy flow at the *last closed flux surface* (LCFS) is limited to some value $q_\perp^{exh}$, which can be written as

$$P_{tr}/A_p < q_\perp^{exh}, \qquad A_p = \int_{LCFS} dS_\perp \approx 4\pi^2 a R_0 \kappa_a \approx 4\pi^2 a^2 \kappa_a/\epsilon_a. \tag{1.24}$$

Defining $0 < f_\alpha \equiv Q_\alpha/(1+Q_\alpha) < 1$, it follows from (1.8) that

$$f_\alpha = \frac{P_\alpha}{P_{loss}} = \frac{\mathcal{E}_\alpha \langle n^2 \langle \sigma v \rangle_{DT} \rangle_a}{12\langle p \rangle_a/\tau_E^*} \propto \frac{\langle p^2 \rangle_a \tau_E^*}{\langle p \rangle_a} \propto \langle p \rangle_a \tau_E^* \propto \beta_T B_0^2 \tau_E^*. \tag{1.25}$$

We next consider the low radiation limit ($P_{br} \ll P_{tr}$) for which $\tau_E^* \approx \tau_E$ and $P_{loss} \approx P_{tr}$. In this case, we can eliminate τ_E in P_{tr} using (1.25),

$$P_{tr}/\mathcal{V} \propto \beta_T B_0^2/\tau_E \propto \beta_T^2 B_0^4/f_\alpha, \qquad P_\alpha/\mathcal{V} \propto \beta_T^2 B_0^4. \tag{1.26}$$

Dividing (1.24) by (1.26) introduces the volume to surface ratio $\mathcal{V}/A_p \approx a/2$, which increases linearly with size. This imposes an upper, *power exhaust* limit a_{exh} on the reactor size for given values of $q_\perp^{exh}$, plasma pressure $p \propto \beta_T B_0^2$, energy confinement τ_E and/or level of ignition f_α, see Fig. 1.2,

$$a < a_{exh} \propto q_\perp^{exh} \tau_E/\beta_T B_0^2 \propto q_\perp^{exh} f_\alpha/\left(\beta_T B_0^2\right)^2. \tag{1.27}$$

For ignited ($f_\alpha = 1$) and marginally MHD stable $\left(\beta_T \approx \beta_T^{mhd}\right)$ plasmas,

$$a_{ign} \le a \le a_{exh}^{mhd} \propto q_\perp^{exh} (q_0 q_*/\epsilon_a \kappa_a)^2 B_0^{-4} \propto q_\perp^{exh} (q_a/\epsilon_a \kappa_a)^2 B_0^{-4}, \tag{1.28}$$

where a_{ign} is given by (1.21) and we set $q_0 = 1$ and $q_* = q_a$. When $q_\perp^{exh}$ is sufficiently large, i.e. when $q_\perp^{exh} > q_\perp^{ign} \propto a_{ign}\left(\beta_T^{mhd} B_0\right)^2$, the maximum β_T power exhaust radius exceeds the ignition radius, $a_{exh}^{mhd} > a_{ign}$, and (1.24) is satisfied automatically in the range of minor radii given by (1.28). This range of optimal a is reduced as $q_\perp^{exh}$ decreases, eventually prohibiting fusion burn at MHD marginal stability when $q_\perp^{exh} < q_\perp^{ign}$ $\left(a_{exh}^{mhd} < a_{ign}\right)$.

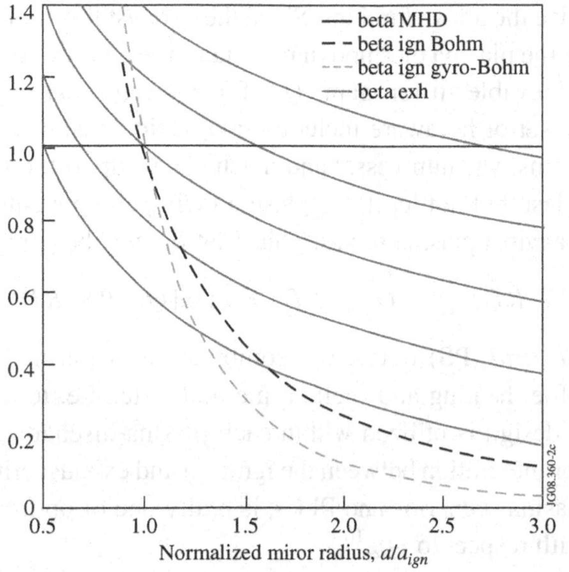

Fig. 1.2. Beta limits (ignition, stability and exhaust) vs. reactor size.

To estimate the minimum ignition radius in that case, we equate the minimum beta needed for ignition β_T^{ign} (1.20), with the maximum beta allowed by the power exhaust limit, which follows from (1.27) with $f_\alpha = 1$,

$$\beta_T^{exh} \propto \left(q_\perp^{exh}/aB_0^4\right)^{1/2}. \tag{1.29}$$

This yields an estimate of the power exhaust limited ignition radius,

$$a_{ign}^{exh} \propto \left(q_\perp^{exh}\right)^{-y/2} q_a^{xy} T^{xy/2} B_0^{(1-x)y}, \qquad y = 1/(x - 1/2). \tag{1.30}$$

Since $\beta_T^{ign} \propto a^{-x}$ decays faster than $\beta_T^{exh} \propto a^{-1/2}$, it is possible to achieve ignition for any value of $q_\perp^{exh}$, by increasing the size of the reactor; this is reflected in the weak, inverse scaling $a_{ign}^{exh} \propto \left(q_\perp^{exh}\right)^{-y/2}$, with the exponent being equal to $-1/3$ for $x = 2$ and $-1/5$ for $x = 3$. In short, power exhaust imposes the minimum reactor size only when $q_\perp^{exh} < q_\perp^{ign}$,

$$a_{min} = \max\left(a_{ign}, a_{ign}^{exh}\right). \tag{1.31}$$

The corresponding toroidal beta, $\beta_T^{min} \propto \left(q_\perp^{exh}/a_{min}B_0^2\right)^{1/2}$, is smaller than β_T^{mhd} thus reducing the fusion power density and the cost effectiveness of the reactor. The determination of $q_\perp^{exh}$ as a function of plasma and field quantities is the chief task of both experimental and numerical power exhaust studies. It is also one of the main incentives for investigating transport processes in the plasma boundary and the ultimate goal of the theoretical development, and the accompanying discussion, in the rest of the book.

Let us summarize the above findings. Since the exhaust limits provide the boundary conditions for the plasma thermodynamic quantities, they effectively determine the maximum achievable fusion gain, Q_α, for a given *reactor design* (RD), by which we mean a set of hardware including magnetic coils, heating, fuelling and current drive systems, vacuum vessel and mechanical support, cooling and pumping systems, and last, but not least, the plasma facing components (PFCs), i.e. the first wall armour against plasma fluxes. This relation may be expressed as

$$Q_\alpha = Q_\alpha(PS, RD), \qquad Q_\alpha^{max}(RD) = \max[Q_\alpha(PS, RD)|PS], \qquad (1.32)$$

where *plasma scenario* (PS) refers to a combination of plasma shape, magnetic field, current profile, heating and fuelling methods, etc., i.e. to the way in which the given reactor design is utilized within each plasma discharge. Thus, the issue of compatibility or integration between the ignition and exhaust criteria, and specifically between plasma scenarios and PFCs, is really one of optimization of the PS for a given RD with respect to Q_α.[10]

Since the fusion power density is roughly proportional to the square of the central fuel ion plasma pressure, (1.6), while the plasma density is limited to roughly the *Greenwald density*, see (7.50), this optimization amounts to maximizing the ion temperature, T_i, and minimizing the effective charge, Z_{eff}, in the centre of the plasma column. In the absence of internal transport barriers, e.g. in the inductive or baseline tokamak plasma scenario, the ion temperature gradient (ITG) is set by the threshold for the ITG drift-wave turbulence (Garbet and Waltz, 1998). Hence, the central ion temperature is a linear function of the edge, or pedestal, temperature, T_{ped}, e.g. in ITER, it is predicted that $T_{ped} \sim 4$ keV is necessary to achieve the desired fusion gain factor, $Q_\alpha \sim 2$ (Doyle *et al.*, 2007).[11] The impact of any given PFC limit on the reactor performance can then be quantified as

$$\zeta(PFC) = 1 - Q_\alpha^{max}(PFC)/Q_\alpha^{max}(\infty), \qquad (1.33)$$

where $Q_\alpha^{max}(PFC)$ is the maximum fusion gain factor for a specified PFC limit, i.e. (1.32) with PFC in place of RD, and $Q_\alpha^{max}(\infty)$ the same factor without any limit on PFC plasma loads, or some previously chosen reference limit value. One can recast (1.33) in terms of density and energy confinement degradation by estimating $Q_\alpha \propto p\tau_E \propto (f_{GW} H_{98})^z$, with $z \sim 2$–3,

$$\zeta(PFC) = 1 - [f_{GW}(PFC)/f_{GW}(\infty)]^z [H_{98}(PFC)/H_{98}(\infty)]^z, \qquad (1.34)$$

[10] It is worth noting that the very terms 'compatibility' and 'integration' reflect the historical disconnection between the tasks of investigating, on the one hand, the plasma equilibrium, stability and transport, and, on the other, its particle and power exhaust properties. Such a disconnection is of course absent in a real plasma where the core and edge regions form an integrated whole.

[11] Whether such high edge plasma temperatures are compatible with the desired lifetime of the divertor and limiter PFCs remains a matter of active research.

where $f_{GW} = \langle n \rangle_a / n_{GW}$ is average density normalized by n_{GW} and $H_{98} = \tau_E / \tau_E^{98,y}$ is the energy confinement time normalized by the ITER98(y,2) scaling; the values required by the ITER reference scenario are $f_{GW} \sim 0.85$ and $H_{98} \sim 1$. Similar impact factors could be obtained for other limits, e.g. tritium retention, dust inventory, etc. Finally, an alternative measure of reactor performance, e.g. price of electricity, neutron fluence, bootstrap fraction, etc., could be used instead of Q_α to defined the impact factor.

In short, terms such as 'integration' and 'compatibility' of plasma scenarios and PFCs are not binary signifiers, but qualitative ones, and refer to the optimization of the former subject to the latter, thus determining the maximum Q_α for a given reactor design. Put another way, PFC limits carry a certain 'cost' in terms of the reduction in Q_α with respect to its unconstrained value, i.e. if arbitrarily high heat loads on divertor and main chamber PFCs were permitted. It will be our task in the following chapters to develop the tools necessary to translate the limits on plasma heat *loads* onto PFCs into limits on plasma *flows* across the last closed flux surface.

1.4 Chapter summary

The remainder of this book is divided into seven chapters, six of comparable lengths, the last somewhat shorter. Their content is briefly summarized below.

Chapter 2 represents a self-contained and rigorous introduction to the theory of magnetized plasma physics. Starting from general considerations of collective behaviour, Section 2.1, the basic plasma parameters are introduced. Next, Section 2.2, the motion of charged particles in both uniform and non-uniform, constant and time-varying magnetic fields is discussed. The notion of the guiding centre and its drift are formulated, as are the two fundamental orderings (MHD and drift) of magnetized plasma dynamics. Angle-action variables and adiabatic invariants for a moving charged particle are derived by introducing and exploiting the Lagrange–Hamilton formalism of classical mechanics. The ensemble of charged particles is then studied in the six-dimensional phase space using the techniques of non-equilibrium statistical mechanics, Section 2.3, culminating in the derivation of the three (MHD, drift and gyro) guiding centre kinetic theories. In Section 2.4, the kinetic equations are integrated over velocity space to yield corresponding fluid equations, in particular the MHD and DHD models.

Chapter 3 addresses the topic of magnetized plasma equilibrium. Beginning with general considerations, it is shown that toroidal topology and the existence of magnetic flux surfaces is a natural requirement for effective plasma confinement, Section 3.1. MHD equilibrium of the plasma is then discussed in magnetic flux co-ordinates, the Grad–Shafranov equation for axis-symmetric equilibria is derived and the origin of the parallel return current in the quasi-neutrality constraint,

$\nabla \cdot \mathbf{J} = 0$, is demonstrated, Section 3.2. Finally, the above formalism is applied to the somewhat idealized case of a large aspect ratio, circular torus, Section 3.3.

Chapter 4 investigates confined plasma stability, or rather the various waves to which the plasma can act as the medium and the various instabilities into which these may develop. Starting with a discussion of Rayleigh–Taylor and Kelvin–Helmholtz instabilities in hydrodynamics, Section 4.1, the bulk of the chapter, Section 4.2, is devoted to ideal MHD waves in a magnetized plasma and instabilities in a magnetically confined plasma. The latter are divided into ideal MHD instabilities, specifically the current-driven (kink) modes and pressure-driven (interchange and ballooning) modes, and resistive MHD instabilities known as tearing modes. The various limits on the current and pressure within the plasma imposed by MHD instabilities are discussed. The final two sections briefly introduce plasma waves and instabilities in the drift ordering, Section 4.3, and in the kinetic approximation, Section 4.4, both of which play an important role in the micro-instabilities which drive plasma turbulence.

Chapter 5 deals with diffusive transport of mass, momentum and energy due to binary Coulomb collisions between charged particles which make up a plasma. Section 5.1 introduces the basic concepts of kinetic theory of gases by deriving the Maxwell–Boltzmann and Fokker–Planck collision operators for binary collisions between neutral particles. These concepts are then extended to binary Coulomb collisions between the ions and electrons, which are characterized by a cubic dependence of the collision rates on the relative particle velocity, Section 5.2. Finally, the resulting transport processes in unmagnetized and magnetized plasmas are discussed in Section 5.3. The latter are divided according to the plasma geometry into collisional transport in cylindrical plasmas, known as classical transport, Section 5.3.2, and in toroidal plasmas, known as neoclassical transport, Section 5.3.3.

Chapter 6 turns our attention to cross-field transport of mass, momentum and energy due to turbulent advection. As before, it begins with an exposition of turbulent flows in hydrodynamics, Section 6.1, including transition from laminar to turbulent state and separate treatments of turbulent flows in two and three dimensions. The basic concepts and theories of turbulent flow analysis are presented: the mean field description and the Reynolds stress, the spectral cascades of energy and enstrophy, the Kolmogorov spectrum and its non-Gaussian variants intermittency modifications, and the statistical description of turbulent flows in terms of probability distribution functions of fluctuating quantities. Special emphasis is placed on the inverse energy cascade and its relation in increased intermittency. These concepts are then extended to magneto-hydrodynamics (MHD), where the discussion is again divided into two- and three-dimensional MHD turbulence, Section 6.2. Finally, the same conceptual framework is applied to drift-hydrodynamics (DHD),

Section 6.3, with the bulk of the discussion dedicated to plasma turbulence in the drift-fluid (DF) and gyro-fluid (GF) approximations, specifically the drift-Alfvén and interchange dynamics; the extension to drift-kinetic (DK) and gyro-kinetic (GK) approximations is touched upon briefly.

Chapter 7 brings the thread of our argument home to our starting point, and applies all the newly developed formalism to the problem of power exhaust in fusion plasmas. To this end, the plasma boundary of magnetically confined plasma, the so-called *scrape-off layer* (SOL), is first described in Section 7.1, including its equilibrium, stability and transport properties. The difference, in all these aspects, between the closed field line, or the *edge* region, and the open field line, or scrape-off layer region, are outlined in some detail. This is supplemented by a discussion of plasma–surface and plasma–neutral interactions, both vital to plasma dynamics in the SOL. The various SOL geometries are compared, with an emphasis on the *divertor* SOL and the phenomenon of divertor plasma *detachment*, and leading edge plasma modelling approaches are compared. In Section 7.2, edge-SOL transport in low confinement (L-mode) plasmas is analyzed. The most recent and relevant experimental observations are reviewed, including observation of radial propagation of plasma filaments (blobs), and are then broadly explained using numerical simulations of edge-SOL turbulence with collisional, (neo)classical dissipation. In Section 7.3, power exhaust in high confinement (H-mode) plasmas is addressed. Starting with a discussion of *edge transport barrier* dynamics and its quasi-periodic relaxation by *edge localized modes* (ELMs), power exhaust during and in between ELMs is treated separately. The onset of Type-I ELMs is shown to depend on the combination of pressure (ballooning) and current (peeling) driven modes, while that of Type-III ELMs to depend on some other non-ideal, resistive MHD or DHD, instability. Once again, a combination of collisional and turbulent effects is evident in the inter-ELM phase of the ELM, while ELM exhaust is characterized by plasma filamentary structures. The fraction of ELM energy reaching the main chamber wall is shown to depend on the combination of radial filament motion (dominated by interchange drive) and parallel losses (initially kinetic free-streaming, later fluid convection) to the divertor targets. The techniques for controlling inter-ELM and ELM power exhaust are then reviewed, including extrinsic impurity seeding, pellet ELM pacing and resonant magnetic perturbation.

Finally, in Chapter 8, the book concludes with a discussion and power exhaust issues in future fusion reactors, which are divided into the next step burning plasma experiment, ITER, and the demonstration power plant, DEMO. To illustrate the issues and apply the physics basis developed in Chapters 1–7, power balance in ITER is analyzed in detail in Section 8.1. The challenges likely to be faced in the design and construction of DEMO are outlined in Section 8.2, including the complete avoidance of ELMs and disruptions, and the need for radiative fractions approaching unity. On the positive note, the opportunities offered by the prospect of advanced fusion fuel cycles and direct energy conversion are highlighted.

Each chapter concludes with a section dedicated to *further reading*, containing a brief selection of recommended texts related to the material discussed in the chapter. Finally, the book is supplemented by two appendices dealing with the Maxwellian distribution and curvilinear co-ordinates.

1.5 Units and notation

Unless otherwise stated, *SI* (*mks*) units are used throughout the text, with the exception of temperatures, which are expressed in eV. This choice seems better suited to many practical problems, in which comparison with experimental data is essential. To convert any SI (mks) expression to Gaussian (cgs) units, simply replace ϵ_0 by $1/4\pi$, μ_0 by $4\pi/c^2$ and $\mathbf{B}$ by $\mathbf{B}/c$. Readers wishing to operate with other unit system may wish to consult the conversion tables in the appendix of Jackson (1975).

A few comments on mathemetical notation are in order. For the sake of brevity, the following shorthand is introduced for differential operators,

$$\partial_t \equiv \frac{\partial}{\partial t}, \qquad \partial_{\mathbf{x}} \equiv \frac{\partial}{\partial \mathbf{x}} \equiv \nabla, \tag{1.35}$$

or more generally for any scalar η and vector $\mathbf{z}$,

$$\mathrm{d}_\eta \equiv \frac{\mathrm{d}}{\mathrm{d}\eta}, \qquad \partial_\eta \equiv \frac{\partial}{\partial \eta}, \qquad \partial_{\mathbf{z}} \equiv \frac{\partial}{\partial \mathbf{z}} \equiv \nabla_{\mathbf{z}}. \tag{1.36}$$

This notation has the advantage of treating the derivative as an operator, such that higher-order derivatives can be denoted by algebraic powers,

$$\mathrm{d}_\eta^k \equiv \frac{\mathrm{d}^k}{\mathrm{d}^k \eta}, \qquad \partial_\eta^k \equiv \frac{\partial^k}{\partial^k \eta}, \qquad \partial_{\mathbf{z}}^k \equiv \frac{\partial^k}{\partial^k \mathbf{z}} \equiv \nabla_{\mathbf{z}}^k. \tag{1.37}$$

The traditional notation will only be used when the denominator contains subscripts, e.g. $\mathrm{d}\psi_T/\mathrm{d}\psi_P$ rather than $\mathrm{d}_{\psi_P}\psi_T$. Otherwise, the shorthand notation will be preferred in both equations and in-line expressions. Finally, special symbols are used for exact derivatives with respect to time ($\dot{\mathcal{A}} \equiv \mathrm{d}_t\mathcal{A}$) and flux surface radius ($\mathcal{A}' \equiv \mathrm{d}\mathcal{A}/\mathrm{d}r_f$), which becomes $\mathcal{A}' \equiv \mathrm{d}_\psi \mathcal{A}$ in symmetry co-ordinates and $\mathcal{A}' \equiv \mathrm{d}_r \mathcal{A}$ in small ϵ tokamak co-ordinates.

The symbol $\perp$ is used in the literature to denote both the direction perpendicular to the magnetic field and that normal to the magnetic flux surface; the two conventions are frequently used side-by-side, with the meaning inferred from the context. To prevent confusion, we use bold font, $\boldsymbol{\perp}$, when referring to the earlier, *gyro-tropic* convention, and regular font, $\perp$, when referring to the latter, *natural magnetic co-ordinates* convention.

To avoid confusion with co-variant components of a vector, denoted by lower case subscripts, e.g. B_ζ, B_θ for co-variant components of **B**, we employ upper case subscripts in the sense of the usual convention in physics, i.e. to denote simple vector components of **B**. Hence, in cylindrical geometry, B_T and B_P represent the axial and poloidal components, while in toroidal geometry they represent the toroidal and poloidal components.

Finally, two notations are used side-by-side for the thermal velocity, namely, $v_{ts}^2 = T_s/m_s$ and $v_{Ts}^2 = 2T_s/m_s$, differing by the factor of 2 appearing in the nominator.

1.6 Further reading

The study of fusion plasmas requires frequent contact with other fields of *theoretical physics*, *applied mathematics* and *engineering*, for which a list of good sources is useful. When in doubt on any of these topics, the reader may wish to consult the following references.

Classical mechanics: Goldstein (1980); Arnold (1989). *Quantum mechanics*: Cohen-Tannoudji *et al.* (1977). *Thermodynamics and statistical physics*: Reif (1965); Callen (1985); Huang (1987). *Electrodynamics*: Jackson (1975). *Nuclear physics*: Krane (1988). *Atomic physics*: Haken and Wolf (1987); Cohen-Tannoudji *et al.* (1992). *Solid state physics*: Kittel (1995). *Chemistry*: Pauling (1960, 1970). *Astrophysics*: Padmanabhan (2000). All other aspects of *theoretical physics*: Landau and Lifschitz (1960); Feynmann *et al.* (1963). *Applied mathematics*: Morse and Feschbach (1953); Smirnov (1964); Boas (1983); Ross (1985); Zwillinger (1989); Arfken and Weber (2001). *Materials*: Budynas (1977); Popov (1978); Ashby and Jones (2005). *Electric circuits*: Dorf (1989). *Thermal circuits*: Wood (1982). *Control systems*: Van de Vegte (1990). *Plasma diagnostics*: Hutchinson (2002). *Nuclear engineering*: Benedict *et al.* (1981); Lamarsh (1983). *MCF*: Miyamoto (1989); Goldston and Rutherford (1995); Wesson (2004). *ICF*: Motz (1979); Hora (1981); Kruer (1988).

2

Magnetized plasma physics

'Physics and other natural sciences are successful because physical phenomena associated with each range of energy and other parameters are explainable to a good, if not perfect, accuracy by an appropriate, self-consistent theory.'

E. H. Lieb (c. 1990)

The physics of plasmas is no exception. We thus begin our journey by introducing the basic concepts, key results and essential theories pertaining to magnetized plasmas, which will underlie the more specialized material found in subsequent chapters. Our discussion will follow the excellent texts by Hazeltine and Meiss (1992) and Hazeltine and Waelbroeck (2004).

2.1 What is a plasma?

The ancient-Greek word $\pi\lambda\alpha\sigma\mu\alpha$ (*plasma*) originates with the verb $\pi\lambda\alpha\sigma\sigma\overline{\omega}$ (*plasso*): *to shape*, *mould*, and signifies a pliant, malleable object (hence plastic, plasticine, etc.). The term *plasma* entered the physics lexicon when used by Tonks and Langmuir to describe the macroscopic behaviour of a glow discharge. More generally, it may be used to denote any *multi-constituent system dominated by collective (as opposed to binary) interactions*. This information-based, systems theory definition includes the traditional electromagnetic plasma of ions and electrons, the gravitational plasma of stars in a galaxy or galaxies in a cluster, and the quark–gluon plasma of elementary particles in a heavy ion collision. It may also be extended beyond the domain of physics, to such fields as finance, semiotics, sociology and cognitive science. One may thus speak of a market plasma in which commodity prices interact via market player transactions, a semantic plasma in which word or sign meanings interact via syntactical references in their respective languages, or a neural plasma in which neurons are inter-linked by multiple pathways in a network. In each case, the trajectory of a single constituent is dominated by the combined effect of many simultaneous interactions, rather than by a series of discrete encounters. This defining characteristic may be quantified in terms of the *plasma parameter*, Λ, which measures the average number of constituents creating the effective force field acting on any given member. A system is said to be in a *plasma state* when $\Lambda \gg 1$, which applies to 99% of the universe, as well as to consciousness, language and society.

2.1.1 Plasma parameter

While the systems theory approach provides valuable insights into the nature of collective behaviour, it is too broad for the purpose of this book. Henceforth, we will restrict the term *plasma* to mean a gas of ions and electrons satisfying the electromagnetic version of the plasma criterion, $\Lambda_p \gg 1$. We expect this criterion to hold due to the long-range effect of the Coulomb potential, $\varphi(r) \propto r^{-1}$. To obtain Λ_p, we begin by noting that, barring relativistic and quantum corrections, an ion–electron gas may be described by *Newton's* and *Maxwell's* equations. Statistical mechanics may then be applied, in analogy with the case of an ideal gas, to define two intensive, thermodynamic variables for each plasma species, $s = (i, e)$, namely: particle density, n_s and kinetic temperature, $T_s = \frac{1}{3} m_s \langle v^2 \rangle_v$, where m_s is the particle mass and $\langle \cdot \rangle_v$ denotes an average over all particle velocities at a given point in space. A two-species plasma may thus be described by six parameters: two densities, n_s, two thermal speeds, $v_{ts} = (T_s/m_s)^{1/2}$ and two magnitudes of the Coulomb interaction, $e_s^2/\epsilon_0 m_s$, e_s being the species charge. Note that global quasi-neutrality requires $n \approx n_i \approx n_e/Z$, where $Z = |e_i/e_e|$, but allows T_i and T_e to differ substantially. Since electrons are much lighter than ions ($m_i/m_e > 10^3$), they are also much more *mobile*, that is

$$v_{te}/v_{ti} \sim (m_i/m_e)^{1/2} \gg 1. \tag{2.1}$$

Consequently, *relativistic effects* may be neglected provided that

$$v_{te}/c < 0.05. \tag{2.2}$$

We can use these six parameters to construct the basic length and time scales of plasma physics, which should be small compared to the system size, L, and the observation time, τ. We thus obtain the *plasma frequency*, ω_{ps}, which describes electrostatic oscillations of species s,

$$\omega_{ps}^2 = n_s e_s^2/\epsilon_0 m_s, \tag{2.3}$$

the *Debye length*, λ_{Ds}, which measures the exponential shielding of a test charge by the rearrangement of free charges,

$$\varphi(r) \propto r^{-1} \exp(-r/\lambda_{Ds}), \tag{2.4}$$

and hence the maximum extent of collective interactions,

$$\lambda_{Ds} = v_{ts}/\omega_{ps} = \left(\epsilon_0 T_s/n_s e_s^2\right)^{1/2}, \tag{2.5}$$

and the classical distance of closest approach in a Coulomb collision, r_{cs},

$$r_{cs} = e_s^2/4\pi\epsilon_0 T_s. \tag{2.6}$$

Comparing this distance with the average inter-particle spacing, $r_{ds} = n_s^{-1/3}$, the *plasma*, or *weak-coupling*, *criterion* may be restated as

$$\Lambda_{ps} = \lambda_{Ds}/r_{cs} = (r_{ds}/r_{cs})^{3/2}/\sqrt{4\pi} = 4\pi n_s \lambda_{Ds}^3 \gg 1. \tag{2.7}$$

It requires that a *Debye sphere* (a sphere of radius λ_{Ds}) contain a large number of particles, such that the combined effect of frequent grazing collisions far exceeds that of rare close encounters; (2.7) is amply satisfied by most ion–electron gases and all fusion plasmas. When it is violated, particles become *strongly coupled* by binary interactions and cease to be a plasma in the sense of systems theory. Strong coupling generally applies to a degenerate quantum gas, e.g. electrons in a metal, *quantum effects* being negligible only when the thermal electron *deBroglie length*, λ_{Be}, is less than r_{de},

$$\lambda_{Be}/r_{de} = n_e^{1/3}\hbar/m_e v_{te} < 1. \tag{2.8}$$

In contrast, (2.8) is satisfied in all fusion plasmas, so that quantum effects may be relegated to nuclear, atomic and molecular interactions, and treated by corresponding transition probabilities and collisional cross-sections.

2.1.2 Magnetization parameter

The presence of an ambient magnetic field **B** introduces additional length and time scales, namely the *gyro-frequency*,

$$\Omega_s = e_s B/m_s, \tag{2.9}$$

and the (thermal) *gyro-radius* of species s,

$$\rho_s(\mathbf{v}) = v_\perp/\Omega_s, \qquad \rho_{ts} = \rho_s(v_{ts}) = v_{ts}/\Omega_s, \tag{2.10}$$

which arise due to gyration of charged particles around the magnetic lines of force. A plasma is said to be *magnetized* when ρ_{ts} is small in relation to the system size or, more stringently, to the (smallest) scale length of the electromagnetic force fields, $L = \min(L_B, L_E)$, $L_B^{-1} = \nabla \ln B$, etc.,

$$\delta_s = \rho_{ts}/L = \omega_{ts}/\Omega_s \ll 1, \qquad \omega_{ts} \equiv v_{ts}/L, \tag{2.11}$$

where ω_{ts} is the local *transit frequency* and δ_s is known as the *magnetization parameter*. The number of particles in a *gyro-sphere* (a sphere of radius ρ_{ts}), namely $n\rho_{ts}^3$, is an alternative, though less frequently used, measure of magnetization. Since $\Omega_e \gg \Omega_i$ and $\rho_{ti} \gg \rho_{te}$ due to the mass difference, (2.11) is equivalent to $\delta_i \ll 1$. This is generally true for fusion plasmas, but can be violated in regions of strong transverse gradients, see Section 7.3.1.

Particle gyration generates plasma currents which in turn produce a plasma magnetic field opposite in direction to the ambient field $\mathbf{B}$, i.e. a *diamagnetic* field. The relative strength of this back-reaction is measured by the ratio β_s of the kinetic ($p_s \equiv n_s T_s$) and magnetic ($B^2/2\mu_0$) pressures,

$$\beta_s = 2\mu_0 p_s/B^2, \qquad \beta = \sum \beta_s = 2\mu_0 p/B^2. \tag{2.12}$$

When the plasma beta, β, is small, as is the case in most fusion plasmas, the ambient magnetic field is merely modulated by plasma dynamics.

It can be shown that the three dimensionless parameters Λ_s (2.7), δ_s (2.11) and β_s (2.12) for each plasma species s, or any combination thereof, are sufficient to completely describe Maxwell's and Newton's equations, which characterize the dynamics of a magnetized plasma. One such combination, widely used in confined plasma research, is the triplet of ρ_s^*, ν_s^* and β_s,

$$\rho_s^* \equiv \rho_{ts}/a, \qquad \nu_s^* \equiv L/\lambda_s \approx L/v_{ts}\tau_s \approx \nu_s/\omega_{ts}, \tag{2.13}$$

where a is the toroidal minor radius of the plasma, $\nu_s = \tau_s^{-1}$ is the collision frequency of species s, λ_s is their collisional mean free path and ν_s^* is known as the *collisionality* parameter, typically defined in terms of the parallel gradient length or the parallel connection length of the magnetic field $L_\parallel$.

2.2 Charged particle motion

All macroscopic phenomena have microscopic origins. This key insight of modern physics pertains to all systems in Nature, including ion–electron plasmas. Consequently, the origin of all plasma behaviour lies in the motions of charged particles and their interactions with ambient and self-generated force fields. Provided (2.2) and (2.8) hold, the trajectory of each plasma particle follows *Newton's equations of motion*,

$$\mathrm{d}_t \mathbf{x} = \mathbf{v}, \qquad \mathrm{d}_t(m_s \mathbf{v}) = m_s \mathbf{a}_s = \mathbf{F}_s + e_s \mathbf{v} \times \mathbf{B}, \tag{2.14}$$

where $\mathbf{F}_s = e_s \mathbf{E} + m_s \mathbf{g}$ is the sum of electric and gravitational forces acting on a particle (the latter represents any generic charge independent force) and the s index on $\mathbf{x}$ and $\mathbf{v}$ was omitted for brevity.

2.2.1 Guiding centre drifts

If the force fields are constant in time and uniform in space, (2.14) may be solved by transforming to a frame moving at the *particle drift* velocity,[1]

$$\mathbf{v}_{GC\perp} \equiv \mathbf{U}_{\perp} \equiv \frac{\mathbf{F}_s \times \mathbf{b}}{e_s B} = \frac{\mathbf{E} \times \mathbf{b}}{B} + \frac{\mathbf{g} \times \mathbf{b}}{\Omega_s} = \mathbf{v}_E + \mathbf{v}_{gs}, \tag{2.15}$$

where $\mathbf{b} = \mathbf{B}/B$ is a unit vector in the direction of $\mathbf{B}$. The *electric drift* $\mathbf{v}_E$ is independent of mass and charge, while the *gravitational drift* $\mathbf{v}_{gs}$ is both mass and charge dependent. The former causes electrons and (all) ions to drift normal to both $\mathbf{E}$ and $\mathbf{b}$, resulting in flow of mass but not charge, while the latter causes opposite charges to drift in opposite directions, normal to both $\mathbf{g}$ and $\mathbf{b}$, producing a net flow of both mass and charge.

At this stage, we write $\mathbf{x}$ and $\mathbf{v}$ as a sum of gyrating and non-gyrating parts, denoted by lower and upper case $\mathbf{r}$ and $\mathbf{u}$, respectively,

$$\mathbf{x} = \mathbf{r} + \mathbf{R}, \qquad \mathbf{v} = \mathbf{u} + \mathbf{U}. \tag{2.16}$$

In addition, we introduce the *gyro-tropic notation*, in which any vector $\mathbf{A}$ is decomposed into projections parallel ($\|$) and perpendicular ($\perp$) to $\mathbf{b}$,

$$\mathbf{A} = \mathbf{A}_{\|} + \mathbf{A}_{\perp} = A_{\|}\mathbf{b} + \mathbf{b} \times (\mathbf{A} \times \mathbf{b}), \qquad A_{\|} = \mathbf{A} \cdot \mathbf{b}. \tag{2.17}$$

Thus, the $\|$ and $\perp$ velocities in the drifting frame are $\mathbf{U}_{\|}$ and $\mathbf{u} \equiv \mathbf{u}_{\perp}$, respectively. The former is unaffected by the magnetic force, $e_s\mathbf{v} \times \mathbf{B}$, whereas the latter describes gyration around the magnetic field lines.

The solution to (2.14) is a combination of free-streaming parallel to $\mathbf{b}$, gyration perpendicular to $\mathbf{b}$ and slow drift normal to both $\mathbf{b}$ and $\mathbf{F}_s$,

$$\mathbf{u}_{\perp} = u_{\perp}(\hat{\mathbf{e}}_2 \cos\gamma - \hat{\mathbf{e}}_3 \sin\gamma) = u_{\perp}\hat{\mathbf{e}}_u, \quad u_{\perp} = \rho_s\Omega_s, \quad \mathbf{u}_{\|} = 0 \tag{2.18}$$

$$\mathbf{U}_{\|} = \mathbf{v}_{\|} = \left(v_{\|0} + F_{\|s}t/m_s\right)\mathbf{b}, \qquad \mathbf{U}_{\perp} = (2.15). \tag{2.19}$$

Here $\gamma = \Omega_s t + \gamma_0$ is the *gyro-phase* and $(\mathbf{b}, \hat{\mathbf{e}}_2, \hat{\mathbf{e}}_3)$ form a right-handed basis. Integration of (2.18)–(2.19) yields a drifting *helical orbit*,

$$\mathbf{r} = \Omega_s^{-1}\mathbf{b} \times \mathbf{u}_{\perp} = \rho_s(\hat{\mathbf{e}}_2 \sin\gamma + \hat{\mathbf{e}}_3 \cos\gamma) = \rho_s\hat{\mathbf{e}}_r, \tag{2.20}$$

$$\mathbf{R} = \left(v_{\|0}t + F_{\|s}t^2/2m_s\right)\mathbf{b} + \mathbf{U}_{\perp}t. \tag{2.21}$$

Note that (2.14) takes on a particularly simple form in the $(\mathbf{b}, \hat{\mathbf{e}}_r, \hat{\mathbf{e}}_u)$ basis,

$$\mathrm{d}_t\mathbf{x} = \mathbf{v}, \qquad \mathrm{d}_t\mathbf{v} = \mathbf{a}_s = \frac{\mathbf{F}_s}{m_s} - \rho_s\Omega_s^2\hat{\mathbf{e}}_r, \tag{2.22}$$

[1] Since a particle drift occurs along with motion of its guiding centre, see below, we denote the drift velocity by the subscript GC.

in which $\mathbf{bb} + \hat{\mathbf{e}}_u\hat{\mathbf{e}}_u + \hat{\mathbf{e}}_r\hat{\mathbf{e}}_r = \mathsf{I}$, $\hat{\mathbf{e}}_u = \partial_\gamma \hat{\mathbf{e}}_r$, $\hat{\mathbf{e}}_r = -\partial_\gamma \hat{\mathbf{e}}_u$ and

$$2\int \hat{\mathbf{e}}_u\hat{\mathbf{e}}_u \mathrm{d}\gamma = 2\int \hat{\mathbf{e}}_r\hat{\mathbf{e}}_r \mathrm{d}\gamma = \hat{\mathbf{e}}_r\hat{\mathbf{e}}_u + \gamma(\hat{\mathbf{e}}_u\hat{\mathbf{e}}_u + \hat{\mathbf{e}}_r\hat{\mathbf{e}}_r) + \text{const.} \tag{2.23}$$

To summarize, the particle gyrates with radius ρ_s and speed $u_\perp = |\mathbf{u}_\perp| = \rho_s\Omega_s$ around its centre of gyration, known as the *guiding centre* (GC). The GC-position, $\mathbf{x}_{GC}$, and velocity, $\mathbf{v}_{GC}$, defined as gyro-averages of $\mathbf{x}$ and $\mathbf{v}$,

$$\mathbf{R} \equiv \mathbf{x}_{GC} \equiv \langle \mathbf{x} \rangle_\gamma \equiv \frac{1}{2\pi}\oint \mathbf{x}\mathrm{d}\gamma, \qquad \mathbf{U} \equiv \mathbf{v}_{GC} \equiv \langle \mathbf{v} \rangle_\gamma = \mathrm{d}_t\mathbf{R}, \tag{2.24}$$

are independent of γ. In magnetized plasmas, the GC-description greatly simplifies charged particle dynamics, as we will see presently.

The above results presuppose stationary and uniform force fields. If the fields are either time-dependent or non-uniform, a moving particle will experience forces which change during the course of its orbit. Provided the force fields vary *adiabatically*, i.e. the forces change little during one gyro-period,

$$\epsilon \equiv \max\left(\rho_s/L, 1/\tau\Omega_s\right) \sim O(\delta_s) \ll 1, \tag{2.25}$$

where $L = \min(L_B, L_E)$ with $L_B^{-1} = \nabla \ln B$, etc. and $\tau \equiv \min(\tau_B, \tau_E)$ with $\tau_B^{-1} = \partial_t \ln B$, etc. are the length and time scales of this variation, then the slow GC-drift may be calculated using a *multiple time scale* expansion (Bogoliubov and Mitropolski, 1961). This technique exploits the large difference in temporal scales, evident in (2.25), to treat the gyro-phase γ and the time t as independent variables. The equations of motion (2.14) are then expanded in the small parameter ϵ, averaged over the rapid gyration using $\langle\cdot\rangle_\gamma = \oint \cdot \mathrm{d}\gamma/2\pi$, and solved order by order in ϵ. Specifically, the gyration radius $\mathbf{r}(\mathbf{R}, \mathbf{U}, t, \gamma)$ and velocity $\mathbf{u}(\mathbf{R}, \mathbf{U}, t, \gamma)$ are expanded in powers of ϵ, denoted by subscripts,

$$\mathbf{x} = \mathbf{R} + \mathbf{r}_0 + \mathbf{r}_1 + \cdots, \qquad \mathbf{v} = \mathbf{U} + \mathbf{u}_0 + \mathbf{u}_1 + \cdots \tag{2.26}$$

The time derivative of the gyro-phase is similarly expanded, $\mathrm{d}_t\gamma = \omega_{-1} + \omega_0 + \cdots$, taking note that, to lowest order, $\mathrm{d}_t\gamma \sim \Omega_s \sim O(\epsilon^{-1})$. The assumed periodicity requires the averages of the gyrating quantities to vanish, $\langle \mathbf{r} \rangle_\gamma = \langle \mathbf{u} \rangle_\gamma = 0$, which sets the *solubility criteria* at each order.

Substituting (2.26) into (2.14), equating terms of same order in ϵ, and satisfying the solubility criteria, yields evolution equations for $\mathbf{U}$ and γ, with $\mathrm{d}_t\mathbf{R} = \mathbf{U}$ being satisfied to all levels of accuracy. Thus, to zeroth order,

$$\omega\partial_\gamma\mathbf{u} - \Omega_s\mathbf{u} \times \mathbf{b} = \frac{e}{m}(\mathbf{E} + \mathbf{U}_0 \times \mathbf{B}), \tag{2.27}$$

where the order-subscript is only retained in $\mathbf{U}$. The average of (2.27) yields the zeroth-order GC-drift, with all terms evaluated at $\mathbf{R}$,

$$\mathbf{U}_0 = \mathbf{U}_{0\parallel} + \mathbf{U}_{0\perp} = U_{0\parallel}\mathbf{b} + \mathbf{v}_E. \tag{2.28}$$

The perpendicular component is just the electric drift, $\mathbf{U}_{0\perp} = \mathbf{v}_E$, as expected for a stationary, uniform force field seen by the particle to lowest order in ϵ. Integrating (2.27) we recover the gyration velocity $\mathbf{u} = \mathbf{u}_\perp = \Omega\partial_\gamma \mathbf{r}$, (2.18) with $\gamma = \Omega_s t + \gamma_0$, and the corresponding gyro-radius $\mathbf{r}$, (2.20).

Gyro-averaging the next-order equation of motion, we find

$$m\mathrm{d}_t \mathbf{U}_0 = e[\mathbf{U}_1 \times \mathbf{B} + E_\parallel \mathbf{b} + \langle \mathbf{u} \times \mathbf{B}(\mathbf{x})\rangle_\gamma], \tag{2.29}$$

where the last term must be evaluated over the *gyro-orbit*, $\mathbf{x} = \mathbf{R} + \mathbf{r}$. This *gyro-average* can be found by Taylor expanding $\mathbf{B}(\mathbf{R} + \mathbf{r})$, with the result

$$e\langle \mathbf{u} \times \mathbf{B}(\mathbf{x})\rangle_\gamma = -\frac{mu_\perp^2}{2B}\nabla B = -\mu\nabla B, \qquad \mu = \frac{mu_\perp^2}{2B}, \tag{2.30}$$

where $\mathbf{m} = -\mu\mathbf{b}$ is the *magnetic moment* of the gyration evaluated at $\mathbf{R}$. Note that $\mathbf{m}$ always opposes the magnetic field, i.e. it is *diamagnetic*, and that μ/m is independent of both mass and charge. The term $-\mu\nabla B$ is known as the *mirror force* for reasons that will become clear shortly.

Combining (2.28)–(2.30), we find

$$m\mathrm{d}_t \mathbf{U}_0 = m\mathrm{d}_t(U_{0\parallel}\mathbf{b} + \mathbf{v}_E) = e[\mathbf{U}_1 \times \mathbf{B} + E_\parallel \mathbf{b} - \mu\nabla B]. \tag{2.31}$$

Dot product with $\mathbf{b}$ yields the evolution of $U_{0\parallel}$, while cross product results in the first-order perpendicular velocity $\mathbf{U}_{1\perp}$,

$$\mathrm{d}_t U_{0\parallel} = \frac{e}{m}E_\parallel + \frac{\mu}{m}\mathbf{b}\cdot\nabla B - \mathbf{b}\cdot\mathrm{d}_t\mathbf{v}_E, \tag{2.32}$$

$$\mathbf{U}_{1\perp} = \frac{\mathbf{b}}{\Omega} \times \left(\mathrm{d}_t\mathbf{U}_0 + \frac{\mu}{m}\nabla B\right) = \frac{\mathbf{b}}{\Omega} \times \left(U_{0\parallel}\mathrm{d}_t\mathbf{b} + \mathrm{d}_t\mathbf{v}_E + \frac{\mu}{m}\nabla B\right). \tag{2.33}$$

The first term in (2.33) is linked to the zeroth-order GC-acceleration, $\mathrm{d}_t\mathbf{U}_0$, the second term to the *mirror force*, $-\mu\nabla B$. The two terms, known as the *inertial* and *magnetic* drifts, can be expressed in the *gravitational form*, $\mathbf{v}_{gs} = \mathbf{g} \times \mathbf{b}/\Omega_s$ (2.15), with $\mathbf{g} = \mathbf{F}_s/m$ replaced by $-\mathrm{d}_t\mathbf{U}_0$ and $-\mu\nabla B/m$, respectively. The $\mathbf{g}$-form offers useful insight into the origin of both drifts, namely the variation of the gyro-radius during the course of gyration.

The inertial drift consists of two parts, made explicit in (2.33), which reflect changes to $\mathbf{U}_{0\parallel}$ and $\mathbf{U}_{0\perp} = \mathbf{v}_E$. Since $\mathbf{b} \times \mathbf{b}\mathrm{d}_t U_{0\parallel} = 0$, the parallel component of the inertial drift depends only on the change in $\mathbf{b}$ perceived by the guiding centre (expressed by a zeroth-order *advective* derivative),

$$\mathrm{d}_t\mathbf{b} = (\partial_t + \mathbf{U}_0\cdot\nabla)\mathbf{b} = \partial_t\mathbf{b} + \mathbf{v}_E\cdot\nabla\mathbf{b} + U_{0\parallel}\mathbf{b}\cdot\nabla\mathbf{b}. \tag{2.34}$$

Parallel motion being generally much faster than any perpendicular drift ($U_{0\parallel} \gg \mathbf{v}_E$), the last term usually dominates. It describes the centripetal acceleration felt by the guiding centre moving along a curved magnetic field line. Introducing the *magnetic curvature* vector, $\boldsymbol{\kappa} \equiv \mathbf{b} \cdot \nabla \mathbf{b}$,[2] this *curvature drift* can also be written in gravitational form with $\mathbf{g} = -\boldsymbol{\kappa} U_{0\parallel}^2$. Using $\mathbf{b} \cdot \nabla \mathbf{b} + \mathbf{b} \times \nabla \times \mathbf{b} = 0$ and Ampere's law, $\nabla \times \mathbf{B} = \mu_0 \mathbf{J}$, we find

$$\boldsymbol{\kappa} \equiv \nabla_\parallel \mathbf{b} = -\mathbf{b} \times (\nabla \times \mathbf{b}) \approx \frac{\nabla_\perp B}{B} + \frac{\mu_0 \mathbf{J} \times \mathbf{b}}{B} \approx \frac{\mu_0}{B^2} \nabla_\perp \left(p + \frac{B^2}{2\mu_0} \right). \quad (2.35)$$

Finally, changes to $\mathbf{U}_{0\perp} = \mathbf{V}_E$ give rise to the *polarization drift*,

$$\mathbf{v}_p = \frac{\mathbf{b}}{\Omega} \times \mathrm{d}_t \mathbf{v}_E = \frac{\mathbf{b}}{\Omega} \times (\partial_t + \mathbf{U}_0 \cdot \nabla) \frac{\mathbf{E} \times \mathbf{b}}{B}, \quad (2.36)$$

where the variation is once again evaluated at the moving guiding centre.

We can now combine (2.28) and (2.33) to write the GC-velocity, $\mathbf{v}_{GC} = \mathbf{U} = \mathbf{U}_0 + \mathbf{U}_1 + O(\epsilon^2)$ to first order in ϵ,

$$\mathbf{U} = U_\parallel \mathbf{b} + \mathbf{v}_E + \frac{\mathbf{g} \times \mathbf{b}}{\Omega_s} + O(\epsilon^2), \qquad \Omega_s^{-1} \sim O(\epsilon), \quad (2.37)$$

$$-\mathbf{g} = (\partial_t \mathbf{b} + \mathbf{v}_E \cdot \nabla \mathbf{b} + \boldsymbol{\kappa} U_\parallel) U_\parallel + \mathrm{d}_t \mathbf{v}_E + \frac{\mu}{m} \nabla B. \quad (2.38)$$

In the typical case of stationary $\mathbf{B}$ and weak, electrostatic $\mathbf{E} = -\nabla \varphi$, the first, second and fourth terms in (2.38) may be omitted, leaving[3]

$$-\mathbf{g} \approx \boldsymbol{\kappa} U_\parallel^2 + \frac{\mu}{m} \nabla B \approx \left(U_\parallel^2 + \frac{u_\perp^2}{2} \right) \frac{\nabla B}{B}. \quad (2.39)$$

The second form applies only to low-β, current free ($\mathbf{J} \times \mathbf{B} \approx 0$) plasmas, for which (2.35) reduces to $\boldsymbol{\kappa} \approx \nabla_\perp \ln B = \nabla \ln B$. For higher β, we find

$$\mathbf{U}_\perp = \mathbf{v}_E + \frac{\mathbf{b}}{\Omega} \times \left(U_\parallel^2 + \frac{u_\perp^2}{2} \right) \frac{\nabla B}{B} + \frac{U_\parallel^2}{\Omega} \frac{(\nabla \times \mathbf{B})_\perp}{B} + O(\epsilon^2). \quad (2.40)$$

The relative strength of the transverse electric field, $E_\perp$, and hence of the electric drift, $v_E = E_\perp / B$, compared to the ion thermal speed, v_{ti}, gives rise to two ordering schemes for magnetized plasma dynamics. These are traditionally referred to by the names of the resulting dynamical equations, irrespective of

[2] The *curvature* of some vector field is defined as the variation of its unit vector along the lines of force. Hence, $\boldsymbol{\kappa} \equiv \mathbf{b} \cdot \nabla \mathbf{b} = \nabla_\parallel \mathbf{b}$ represents the variation of $\mathbf{b}$ along the magnetic field lines.

[3] The first term in this 'negative' gravity is simply the centripetal acceleration of a particle travelling along a curved magnetic field line, and is thus known as the *curvature* drift. The second term is related to the non-uniformity in the the field magnitude and thus the variation of the gyro-radius; it is thus known as the gradient B, or simply ∇B, drift.

whether these are formulated in the kinetic or fluid descriptions. One thus speaks of *magneto-hydrodynamic (MHD) ordering*,

$$v_E/v_{ti} \sim 1, \qquad \omega \sim \omega_{ti} \sim \delta_i \Omega_i, \qquad \delta_i \ll 1, \tag{2.41}$$

and of *drift ordering*,

$$v_E/v_{ti} \sim \delta_i \ll 1, \qquad \omega \sim \delta_i \omega_{ti} \sim \delta_i^2 \Omega_i \ll \omega_{ti}, \tag{2.42}$$

where ω is the typical frequency of dynamical evolution being considered and all expressions are written for ions on account of $\delta_i \gg \delta_e$. In both cases, the remaining GC-drifts are small compared to v_{ti}, $v_{gs}/v_{ti} \sim \delta_i$.

Consider the relative strength of the electrostatic, $\mathbf{E}^\varphi$, and inductive, $\mathbf{E}^A$, contributions to the electric field and hence the electric drift,

$$\mathbf{E} = \mathbf{E}^\varphi + \mathbf{E}^A = -\nabla\varphi - \partial_t \mathbf{A}, \tag{2.43}$$
$$\mathbf{v}_E = \mathbf{v}_E^\varphi + \mathbf{v}_E^A = (\nabla\varphi + \partial_t \mathbf{A}) \times (\mathbf{b}/B), \tag{2.44}$$

where φ is the scalar and $\mathbf{A}$ the vector potential, defined by $\mathbf{B} = \nabla \times \mathbf{A}$. Hence $\mathbf{v}_E^\varphi$ arises due to a perpendicular gradient of φ and $\mathbf{v}_E^A$ due to the time variation of $\mathbf{A}_\perp$ and hence of $B_\parallel = B = |\mathbf{B}|$. The magnitude of these two contributions to $\mathbf{v}_E$ may be estimated as

$$v_E^\varphi = \nabla_\perp \varphi / B \sim \varphi / L_\perp B, \tag{2.45}$$
$$v_E^A = \partial_t A_\perp / B \sim L_\perp \partial_t B / B \sim L_\perp \omega_B \sim L_\perp \omega_{\widetilde{B}} (\widetilde{B}/B). \tag{2.46}$$

Under the MHD ordering, we have $\omega_{\widetilde{B}} = \partial_t \widetilde{B}/\widetilde{B} \sim \omega_{ti}$ so that $v_E^A \sim v_{ti} \widetilde{B}/B$, which is typically larger than v_E^φ, e.g. assuming that $e\varphi$ is comparable to T_e, we find $v_E^\varphi \sim T_e / L_\perp eB \sim \delta_i v_{ti}$ which is a factor $\delta_i / (\widetilde{B}/B)$ smaller than v_E^A. The exception is the important case of so-called *flute-reduced MHD*, which corresponds to a low-beta plasma confined in a strong, externally imposed field, which evolves primarily by perpendicular (flute-like) displacements, see Section 4.2.5. In this case, parallel field fluctuations become

$$\widetilde{B}/B \sim \mu_0 \widetilde{p}/B^2 \sim (\widetilde{p}/p)\beta \sim \delta_i \beta, \qquad v_E^A / v_E^\varphi \sim (\widetilde{B}/B)/\delta_i \sim \beta \ll 1. \tag{2.47}$$

A similar result is found under the drift ordering, (2.42), when the magnetic field is constrained to evolve so slowly that the inductive contribution is negligible, especially in the case of magnetically confined plasmas,

$$\omega_{\widetilde{B}} \sim \delta_i \omega_{ti}, \qquad v_E^A \sim \delta_i v_{ti} (\widetilde{B}/B), \qquad v_E^A / v_E^\varphi \sim \widetilde{B}/B \ll 1. \tag{2.48}$$

In short, for most fusion plasma applications, including power exhaust, one may assume that $\mathbf{v}_E \approx \nabla\varphi \times (\mathbf{b}/B)$ irrespective of the ordering scheme.

The particle kinetic energy, $\mathcal{K} = \frac{1}{2} m v^2$, evolves according with $\mathrm{d}_t \mathcal{K} = \mathbf{v} \cdot \mathbf{F} = e\mathbf{v} \cdot \mathbf{E}$, obtained by taking a dot product of (2.14) with $\mathbf{v}$; the magnetic force, $e\mathbf{v} \times \mathbf{B}$, being normal to $\mathbf{v}$, cannot modify $\mathcal{K}$. Here and below, d_t is taken along the GC-trajectory, so that $\mathrm{d}_t = \partial_t + \mathbf{U}\cdot\nabla$. Expanding $v^2 = \mathbf{v}\cdot\mathbf{v}$, with $\mathbf{v} = \mathbf{U} + \mathbf{u}$ given by (2.26), and gyro-averaging, yields

$$\mathrm{d}_t\mathcal{K} = \frac{m}{2}\mathrm{d}_t\left(U_\parallel^2 + u_\perp^2 + v_E^2\right) = e\mathbf{U}\cdot\mathbf{E} + e\langle\mathbf{u}\cdot\mathbf{E}(\mathbf{x})\rangle_\gamma, \tag{2.49}$$

where, as in (2.29), we find contributions from both GC-motion and gyration. The integral over the gyro-orbit, evaluated analogous to that in (2.29), gives the electromagnetic work, $-\mu\mathbf{b}\cdot\nabla\times\mathbf{E} = \mu\partial_t B$. Hence,

$$\mathrm{d}_t\mathcal{K} = e\mathbf{U}\cdot\mathbf{E} + \mu\partial_t B. \tag{2.50}$$

The rate of change of the total energy, defined as

$$\mathcal{E} = \mathcal{K} + e\varphi = \frac{1}{2}mv^2 + e\varphi = \frac{m}{2}\left(u_\perp^2 + v_E^2 + U_\parallel^2\right) + e\varphi, \tag{2.51}$$

can now be evaluated using (2.50) and (2.43),

$$\mathrm{d}_t\mathcal{E} = \mu\mathrm{d}_t B + e\mathrm{d}_t\varphi + e\mathbf{U}\cdot\mathbf{E} = \mu\partial_t B + e\partial_t\varphi - e\mathbf{U}\cdot\partial_t\mathbf{A}. \tag{2.52}$$

The total energy is evidently conserved in stationary fields. Using (2.32) and (2.33) to eliminate $U_\parallel^2$ and $v_E^2 = \mathbf{v}_E\cdot\mathbf{v}_E$ from (2.51), we find that the magnetic moment is conserved to first order in ϵ,

$$\mathrm{d}_t\mu = \mathrm{d}_t\left(\frac{mu_\perp^2}{2B}\right) = O(\epsilon). \tag{2.53}$$

In Section 2.2.2, we will see that μ is the lowest-order approximation to an *adiabatic invariant*, conserved to all orders in ϵ. We can thus treat $\frac{1}{2}mu_\perp^2 = \mu B$ as an internal energy of the guiding centre,

$$\mathcal{E} = \mathcal{K} + e\varphi = \frac{m}{2}\left(v_E^2 + U_\parallel^2\right) + \mu B + e\varphi, \tag{2.54}$$

and express the parallel GC-velocity as a function of $\mathcal{E}$ and μ,

$$U_\parallel(\mathbf{R},\mathcal{E},\mu) = \pm\sqrt{\frac{2}{m}(\mathcal{E} - \mu B - e\varphi) - v_E^2}, \tag{2.55}$$

where all quantities are now evaluated at the GC-position, $\mathbf{R}$.

Consider the typical case of stationary $\mathbf{B}$ and weak, electrostatic $\mathbf{E} = -\nabla\varphi$, for which the last term in (2.55) may be omitted. To express the GC-velocity (2.37) in terms of the conserved quantities $\mathcal{E}$ and μ, we eliminate ∇B and $\nabla\varphi$ using the gradient of (2.54), $mU_\parallel\nabla U_\parallel = -\mu\nabla B - e\nabla\varphi$, to find

$$\mathbf{U} = U_\parallel\mathbf{b} + \frac{\mathbf{b}}{\Omega}\times\left(\kappa U_\parallel^2 + \frac{\mu}{m}\nabla B + \frac{e}{m}\nabla\varphi\right) = U_\parallel\left[\mathbf{b} + \frac{1}{\Omega}\nabla\times(U_\parallel\mathbf{b})\right]. \tag{2.56}$$

Introducing the *pseudo-magnetic* field, $\mathbf{B}^*$ and vector potential, $\mathbf{A}^*$,

$$\mathbf{B}^* = \nabla \times \mathbf{A}^* = \mathbf{B} + \frac{m}{e}\nabla \times (U_\parallel \mathbf{b}), \qquad \mathbf{A}^* = \mathbf{A} + mU_\parallel \mathbf{b}/e, \tag{2.57}$$

allows (2.56) to be expressed in a particularly simple form, $\mathbf{U} = U_\parallel \mathbf{B}^*/B$, i.e. the GC is seen to follow the pseudo-magnetic field lines. Together, (2.55) and (2.56) give $\mathbf{U}(\mathbf{R}, \mathcal{E}, \mu)$ to order ϵ. More generally,

$$\mathbf{v}_{GC} = \mathbf{U} = U_\parallel \left[\mathbf{b} + \frac{1}{\Omega}\nabla \times (U_\parallel \mathbf{b})\right] - \frac{1}{\Omega}\left(U_\parallel^2 - \frac{u_\perp^2}{2}\right)\frac{(\nabla \times \mathbf{B})_\parallel}{B}, \tag{2.58}$$

where the additional terms represent the parallel drift velocity, $\mathbf{v}_{GC\parallel} = (\mu/m\Omega)$ $(\nabla \times \mathbf{B})_\parallel$, which will be derived in Section 2.3.2, (2.114), and the high-β perpendicular drift, $\left(U_\parallel^2/\Omega\right)(\nabla \times \mathbf{B})_\perp$, already obtained in (2.40).

The origin of the label *mirror force* to denote $-\mu\nabla B$ should now clear. As the guiding centre moves into regions of stronger $\mathbf{B}$ field, the perpendicular energy $\frac{1}{2}mu_\perp^2 = \mu B$ increases to satisfy $\mathrm{d}_t\mu \approx 0$. Since total energy is also conserved ($\mathrm{d}_t\mathcal{E} \approx 0$), parallel energy $\frac{1}{2}mU_\parallel^2$ must be reduced and the guiding centre decelerates. At the *turning point* $\mathbf{R}'$, defined by $U_\parallel(\mathcal{E}, \mu, \mathbf{R}') = 0$, or $\mathcal{E} = \mu B(\mathbf{R}')$, the GC is reflected and the direction of $U_\parallel$ reverses. Defining the velocity pitch angle as $\alpha = \tan^{-1}(u_\perp/U_\parallel)$, we see that

$$\sin^2\alpha = u_\perp^2/v^2 = \mu B/\mathcal{E}, \tag{2.59}$$

and α increases with B to a maximum of $\pi/2$ at the turning point. Since both $\mathcal{E}$ and μ are constants of motion, reflection occurs when

$$\sin^2\alpha(\mathbf{R}) = B(\mathbf{R})/B(\mathbf{R}'), \tag{2.60}$$

which defines the *loss cone* in $U_\parallel - u_\perp$ velocity space; this cone becomes narrower with increasing $B(\mathbf{R}')$, reducing to a ray as $B(\mathbf{R}') \to \infty$.[4]

Consider a gas of charged particles in a (periodic) magnetic well with a minimum and maximum field, B_{min} and B_{max}, at locations $\mathbf{R}_{min}$ and $\mathbf{R}_{max}$ lying on the same field line. Assuming an isotropic velocity distribution at $\mathbf{R}_{min}$, the *fraction* of particles *trapped* in the well follows from (2.60),

$$f_T = 1 - \frac{\alpha(\mathbf{R}_{min})}{\pi/2} = 1 - \frac{2}{\pi}\sin^{-1}\sqrt{\frac{B_{min}}{B_{max}}}. \tag{2.61}$$

In an infinite well, the loss cone closes and all particles are trapped.

[4] Note the analogy with celestial mechanics: only a particle (body) with zero magnetic moment μ (angular momentum ℓ) in the region of vanishing magnetic (gravitational) field, can enter the strong field region; all orbits with finite μ (ℓ) are reflected to satisfy $\mathrm{d}_t\mathcal{E} = \mathrm{d}_t\mu = 0$ ($\mathrm{d}_t\ell = 0$).

Finally, we evaluate the rates of change of $\mathcal{E}$, μ and γ over the exact particle trajectory, i.e. $\mathrm{d}_t = \partial_t + \mathbf{v}\cdot\nabla$, which we will need in Section 2.3.2. The calculation is simplified by adopting the $(\mathbf{b}, \hat{\mathbf{e}}_r, \hat{\mathbf{e}}_u)$ basis, defined by (2.18) and (2.20). Taking dot products of (2.22) with $\mathbf{v}$, $\mathbf{v}_\perp$ and $\hat{\mathbf{e}}_r$, and negotiating some rather lengthy algebra, we obtain

$$\mathrm{d}_t\mathcal{E} = e_s\partial_t\varphi - e_s\mathbf{v}\cdot\partial_t\mathbf{A}, \tag{2.62}$$

$$B\mathrm{d}_t\mu = -\mu\mathrm{d}_t B - m_s v_\parallel \mathbf{v}_\perp\cdot\mathrm{d}_t\mathbf{b} + e_s\mathbf{v}_\perp\cdot\mathbf{E}, \tag{2.63}$$

$$\mathrm{d}_t\gamma = \Omega - \hat{\mathbf{e}}_r\cdot\mathrm{d}_t\hat{\mathbf{e}}_u - \frac{v_\parallel}{v_\perp}\hat{\mathbf{e}}_r\cdot\mathrm{d}_t\mathbf{b} - \frac{e_s}{m_s v_\perp}\hat{\mathbf{e}}_r\cdot\mathbf{E}. \tag{2.64}$$

Here we made use of $\hat{\mathbf{e}}_2\cdot\mathrm{d}_t\hat{\mathbf{e}}_2 = \hat{\mathbf{e}}_3\cdot\mathrm{d}_t\hat{\mathbf{e}}_3 = \hat{\mathbf{e}}_3\cdot\mathrm{d}_t\hat{\mathbf{e}}_2 + \hat{\mathbf{e}}_2\cdot\mathrm{d}_t\hat{\mathbf{e}}_3 = 0$, and

$$\hat{\mathbf{e}}_r\cdot\mathrm{d}_t\mathbf{v}_\parallel = v_\parallel\hat{\mathbf{e}}_r\cdot\mathrm{d}_t\mathbf{b}, \quad \hat{\mathbf{e}}_r\cdot\mathrm{d}_t\mathbf{v}_\perp = v_\perp\hat{\mathbf{e}}_3\cdot\mathrm{d}_t\hat{\mathbf{e}}_2 - v_\perp\mathrm{d}_t\gamma. \tag{2.65}$$

Assuming $\partial_t \sim \delta_s\Omega_s$ and $v_E/v_t \sim \delta_s$, all right-hand side terms in (2.62)–(2.64) are small compared to the gyration term Ω_s. Gyro-averaging with these assumptions, we recover the conservation of $\mathcal{E}$ and μ to order δ_s,

$$\langle\mathrm{d}_t\mathcal{E}\rangle_\gamma \approx e_s\partial_t\varphi - e_s\mathbf{v}_\parallel\cdot\partial_t\mathbf{A}, \quad \langle\mathrm{d}_t\mu\rangle_\gamma \approx -\frac{\mu}{B}\partial_t B, \quad \langle\mathrm{d}_t\gamma\rangle_\gamma \approx \Omega_s. \tag{2.66}$$

2.2.2 Canonical (angle-action) variables

The analysis presented above was based on the expansion of Newton's equations (2.14) in the small parameter ϵ (2.25). This procedure, while formally correct, has several limitations: it becomes unwieldy at higher orders in ϵ; fares poorly in complicated geometries; and is ill-suited to exploiting the symmetry properties that may exist in a given plasma configuration.

To overcome these problems, we resort to the Lagrange–Hamilton formulation of classical mechanics, an elegant and powerful *meta-theory* forming the skeleton of modern physics. Its cornerstone is the *Principle of Least Action*, which states that a dynamical system evolves in such a way as to minimize the *action*, $\mathcal{S} = \int \mathcal{L}(\mathbf{q}, \dot{\mathbf{q}}, t)\mathrm{d}t$, where the *Lagrangian*, $\mathcal{L} = \mathcal{K} - \mathcal{V}$, defined as the difference between the *kinetic* and *potential* energies, is a function of only the generalized co-ordinates $\mathbf{q} = (q_1, q_2, \ldots)$, velocities $\dot{\mathbf{q}} = (\dot{q}_1, \dot{q}_2, \ldots)$ and time. Expressed as a variational principle,

$$\delta\mathcal{S} = \delta\int \mathcal{L}(\mathbf{q}, \dot{\mathbf{q}}, t)\mathrm{d}t = 0, \tag{2.67}$$

it states that the *phase path* taken by the system in the *phase space* formed by $\mathbf{q}$ and $\dot{\mathbf{q}}$ minimizes the *path integral* of $\mathcal{L}$, i.e. the action. Performing the variation we obtain *Lagrange's* equations of motion,

$$\mathrm{d}_t\mathbf{q} = \mathbf{p}, \qquad \mathrm{d}_t\mathbf{p} = \mathbf{F} = \partial_{\mathbf{q}}\mathcal{L}, \tag{2.68}$$

where $\mathbf{p}$ are the generalized momenta and $\mathbf{F}$ the generalized forces. This shows that *dynamical invariants* originate in the *symmetry properties* of the Lagrangian: if $\mathcal{L}$ is independent of co-ordinate q_k, the corresponding momentum p_k becomes a *constant of motion*. For example, consider a closed system of n (self-)interacting particles: $\mathbf{q}_j = \mathbf{x}_j$, $\dot{\mathbf{q}}_j = \mathbf{v}_j$, $\mathcal{L} = \sum \frac{1}{2}mv_j^2 - \mathcal{V}(\mathbf{x}_1, \ldots, \mathbf{x}_n)$. Homogeneity of time ($\partial_t\mathcal{L} = 0$) results in conservation of energy, $\mathcal{E} = \sum \frac{1}{2}mv_j^2 + \mathcal{V}$, homogeneity of space ($\nabla\mathcal{L} = 0$) in conservation of momentum, $\sum m\mathbf{v}_j$, and isotropy of space ($\partial_\phi\mathcal{L} = \partial_\theta\mathcal{L} = 0$) in conservation of angular momentum, $\sum \mathbf{x}_j \times \mathbf{p}_j$, (Landau and Lifschitz, 1960, Vol. I).

To illustrate that Newton's and Lagrange's equations are indeed equivalent, let us consider a single charged particle[5] moving in an electromagnetic field, for which Newton's equations are given by (2.14). In this case, $\mathbf{q} = \mathbf{x}$, $\dot{\mathbf{q}} = \mathbf{v}$ and $\mathcal{L} = \frac{1}{2}m\mathbf{v}^2 + e\mathbf{A}\cdot\mathbf{v} - e\varphi$, where $\mathbf{A}$ is the vector potential, $\mathbf{B} = \nabla \times \mathbf{A}$. Inserting these into (2.68), we recover (2.14), with the generalized momentum $\mathbf{p} = m\mathbf{v} + e\mathbf{A}$. If the magnetic field is symmetric in some q_j, then p_j will be conserved. For instance, in cylindrical geometry (r, θ, z), longitudal symmetry ($\partial_z = 0$) requires $p_z = m\dot{z} + eA_z =$ const, while poloidal symmetry ($\partial_\theta = 0$), results in $p_\theta = mr^2\dot{\theta} + erA_\theta =$ const. Similarly, in toroidal geometry (r, θ, ζ), axis-symmetry ($\partial_\zeta = 0$) implies

$$p_\zeta = mR^2\dot{\zeta} + eRA_\zeta = R(mv_\zeta + eA_\zeta) = \text{const}. \tag{2.69}$$

It is worth noting that the gyro-average of p_ζ, which represents the generalized toroidal momentum of the guiding centre, is likewise conserved,

$$\langle p_\zeta \rangle_B = R(mb_\zeta v_\parallel + eA_\zeta) = mRb_\zeta v_\parallel - e\psi_P \approx \text{const} \approx p_\zeta, \tag{2.70}$$

where we wrote the toroidal GC velocity as $v_\zeta \approx b_\zeta v_\parallel$ and introduced the poloidal flux function, $\psi_P = -RA_\zeta$, see Chapter 3.

Conservation of generalized momenta in response to system symmetries motivates a tranformation from the phase space defined by $\mathbf{q}$ and $\dot{\mathbf{q}}$ to that formed by $\mathbf{q}$ and $\mathbf{p}$. This is accomplished by means of a *Legendre transformation*, and results in *Hamilton's* formulation of classical mechanics. Thus, the *Lagrangian* $\mathcal{L}(\mathbf{q}, \dot{\mathbf{q}}, t)$ is replaced by the *Hamiltonian* $\mathcal{H}(\mathbf{q}, \mathbf{p}, t) = \mathbf{p}\cdot\mathbf{q} - \mathcal{L}$, and (2.68) by *Hamilton's* (or *canonical*) equations of motion,

$$\mathrm{d}_t\mathbf{q} = \partial_{\mathbf{p}}\mathcal{H}, \qquad \mathrm{d}_t\mathbf{p} = -\partial_{\mathbf{q}}\mathcal{H}, \tag{2.71}$$

[5] For simplicity we drop the species index, so that $m = m_s$, $e = e_s$, $x = x_s$, $v = v_s$, etc.

which may also be derived from (2.67) written as $\delta \int \mathbf{p}\cdot d\mathbf{q} - \mathcal{H}dt = 0$. The action $\mathcal{S}$ evolves according to the related *Hamilton–Jacobi* equation,

$$\partial_{\mathbf{q}}\mathcal{S} = \mathbf{p}, \quad \partial_t\mathcal{S} = -\mathcal{H}, \quad d_t\mathcal{S} = \mathcal{L} \quad \Rightarrow \quad \partial_t\mathcal{S} + \mathcal{H}(\mathbf{q}, \partial_{\mathbf{q}}\mathcal{S}) = 0. \tag{2.72}$$

The physical meaning of the Hamiltonian becomes evident when $\mathcal{H}$ is evaluated for a given system: e.g. our self-interacting particles, $\mathcal{H} = \sum m\mathbf{v}_j\cdot\mathbf{v}_j - \mathcal{L} = \sum \frac{1}{2}mv_j^2 + \mathcal{V}$ or a charged particle in an electromagnetic field, $\mathcal{H} = \mathbf{v}\cdot(m\mathbf{v} + e\mathbf{A}) - \mathcal{L} = \frac{1}{2}mv^2 + e\varphi$. In both cases, the Hamiltonian is simply the total energy of the system, $\mathcal{E} = \mathcal{K} + \mathcal{V}$. Conservation of energy may thus be obtained by taking the total time derivative of $\mathcal{H}$, and using (2.71) to eliminate $d_t\mathbf{q}$ and $d_t\mathbf{p}$, which yields $d_t\mathcal{H} = \partial_t\mathcal{H}$. More generally, the evolution of any quantity $\mathcal{A}$ may be expressed as

$$d_t\mathcal{A} = \partial_t\mathcal{A} + \{\mathcal{H}, \mathcal{A}\} = \partial_t\mathcal{A} + \partial_{\mathbf{p}}\mathcal{H}\cdot\partial_{\mathbf{q}}\mathcal{A} - \partial_{\mathbf{q}}\mathcal{H}\cdot\partial_{\mathbf{p}}\mathcal{A}, \tag{2.73}$$

where we introduced the commutator, or *Poisson bracket*, which should be familiar from quantum mechanics. If $\mathcal{A}$ does not explicitly depend on time, it will be a constant of motion ($d_t\mathcal{A} = 0$) provided it commutes with $\mathcal{H}$, that is if $\{\mathcal{H}, \mathcal{A}\} = 0$. It can be shown (*Poisson's theorem*) that a Poisson bracket of two constants of motion is likewise a constant of motion.

The principal advantage of (2.71) over (2.68) is the conjugate symmetry of the *canonical variables* $\mathbf{q}$ and $\mathbf{p}$. This symmetry permits a wide range of *canonical transformations* $\mathbf{q}, \mathbf{p} \rightarrow \mathbf{q}'(\mathbf{q}, \mathbf{p}, t), \mathbf{p}'(\mathbf{q}, \mathbf{p}, t)$, where the new variables satisfy (2.71) with some new $\mathcal{H}'$ as well as the conjugate constraints,

$$\{q'_k, q'_l\} = 0, \qquad \{p'_k, p'_l\} = 0, \qquad \{q'_k, p'_l\} = \delta_{kl}. \tag{2.74}$$

The relations between the new and old quantities may be derived from the *generating function* of the transformation F, defined as $dF = \mathbf{p}\cdot d\mathbf{q} - \mathbf{p}'\cdot d\mathbf{q}' + (\mathcal{H}' - \mathcal{H})dt$. In particular, $\mathcal{H}' = \mathcal{H} + \partial_t F$ so that the Hamiltonian remains constant if F does not explicitly depend on time. In that case, the phase space volume, measured by the integral $\int d\mathbf{q}d\mathbf{p}$, is conserved.

Consider a dynamical system with n degrees of freedom. Its phase space may be described as a $2n$-dimensional *differentiable manifold* and its evolution according with (2.71) may be viewed as a series of infinitesimal canonical transformations within that manifold, with $-\mathcal{S}$ playing the role of the generating function, F. Therefore, provided the system conserves energy ($\partial_t\mathcal{H} = 0$), its flow in phase space is '*incompressible*',

$$\partial_{\mathbf{q}}d_t\mathbf{q} + \partial_{\mathbf{p}}d_t\mathbf{p} = 0 \quad \Leftrightarrow \quad d_t\int d\mathbf{q}d\mathbf{p} = 0. \tag{2.75}$$

This result, expressed above in differential and integral forms, is known as *Liouville's theorem*. It applies to any conjugate pair q_k, p_k, as well as the entire system trajectory $\mathbf{q}, \mathbf{p}$. If the system is dissipative ($\partial_t \mathcal{H} < 0$), its phase space flow is '*compressed*' at the rate of energy loss.

Liouville's theorem may also be expressed in path integral form as

$$\mathrm{d}_t \mathcal{J} = \mathrm{d}_t \oint_{C(t)} \mathbf{p} \cdot \mathrm{d}\mathbf{q} - \mathcal{H}\mathrm{d}t = \mathrm{d}_t \oint_{C(t)} \mathcal{L}\mathrm{d}t = 0, \tag{2.76}$$

where $C(t)$ is any closed path satisfying the dynamical equations (2.71). The integrals $\mathcal{J}_k$, one for each conjugate pair q_k, p_k, are known as *Poincaré invariants* and are evidently conserved.[6]

When n is large, the system is best described in terms of the phase space density of states, $\mathcal{F}(\mathbf{q}, \mathbf{p})$. Since individual trajectories satisfy (2.75), their sum must do likewise, $\mathrm{d}_t \int \mathcal{F}_s \mathrm{d}\mathbf{q}\mathrm{d}\mathbf{p} = 0$, and Liouville's theorem becomes

$$\mathrm{d}_t \mathcal{F} = \partial_t \mathcal{F} + \{\mathcal{H}, \mathcal{F}\} = 0. \tag{2.77}$$

To fully exploit (2.76), we effect a canonical transformation, $(\mathbf{q}, \mathbf{p}) \to (\boldsymbol{\vartheta}, \boldsymbol{\mathcal{I}})$ using $F = \int \mathbf{p} \cdot \mathrm{d}\mathbf{q}$ as the generating function. The quantities $\vartheta_k = \partial_{\mathcal{I}_k} F$ and $\mathcal{I}_k = \mathcal{J}_k / 2\pi$, known as *canonical*, or *angle-action, variables*, form the simplest canonically conjugate pairs for a bounded, conservative system. Their Hamilton's equations (2.71) are found to describe multiple *rotations* with angles, ϑ_k, constant frequencies, ω_k, and constant radii, $\mathcal{I}_k$,

$$\mathrm{d}_t \vartheta_k = \partial_{\mathcal{I}_k} \mathcal{H} = \omega_k(\mathcal{I}_k) = \text{const}, \qquad \mathrm{d}_t \mathcal{I}_k = -\partial_{\vartheta_k} \mathcal{H} = 0. \tag{2.78}$$

For a system with n degrees of freedom, (2.78) define an n-torus, i.e. an n-dimensional hyper-surface formed by *concentric tori* with k-axial radii $\mathcal{I}_k = \text{const}$ and angles ϑ_k; the system evolves on the *n-torus* by constant rotation around each of its axes. Determining all the angle-action variables is thus formally analogous to an eigenvalue decomposition, with the final solution formed by a linear superposition of oscillating modes. Louiville's theorem (2.76) takes on a particularly simple form in the angle-action representation, $\oint \mathcal{L}_k \mathrm{d}\vartheta_k = \mathcal{J}_k = 2\pi \mathcal{I}_k = \text{const}$, showing a formal identity, aside from a numerical constant, between action variables and Poincaré invariants.

Consider a system whose Lagrangian depends on several parameters λ_j (Landau and Lifschitz, 1960). If $\mathrm{d}_t \lambda_j = 0$, all the action variables are conserved and the system evolves according to (2.78). If λ_j vary with time, the action $\mathcal{S}$ is modified by terms involving λ_j, namely $(\partial_{\lambda_j} \mathcal{H})_{q,p} \mathrm{d}\lambda_j = -(\partial_{\lambda_j} \mathcal{L})_{q,\dot{q}} \mathrm{d}\lambda_j$. Effecting a canonical transformation using $F = \mathcal{S}_0(\mathbf{q}, \mathcal{E}; \lambda_j) = \int \mathbf{p}\mathrm{d}\mathbf{q}$, the new Hamiltonian becomes

[6] Their existence can also be deduced from the *complete integral* of (2.72), which contains one *integration constant* c_k for each independent variable q_k: $\mathcal{S} = \mathcal{S}(t, \mathbf{q}, \mathbf{c}) + c_0$. Provided the system is non-degenerate, solving (2.72) by separation of variables yields the required constants of motion, $c_k = \mathcal{J}_k = \text{const}$.

$\mathcal{H}' = \mathcal{H} + \sum \Lambda_j \mathrm{d}_t \lambda_j$ with $\Lambda_j = (\partial_{\lambda_j} \mathcal{S}_0)_{q,I}$. The canonical equations (2.78) are thus replaced by,

$$\mathrm{d}_t \vartheta_k(\lambda_j) = \partial_{\mathcal{I}_k} \mathcal{H}' = \omega_k(\mathcal{I}_k; \lambda_j) + \sum \left(\partial_{\mathcal{I}_k} \Lambda_j\right)_{\vartheta,\lambda_j} \mathrm{d}_t \lambda_j, \tag{2.79}$$

$$\mathrm{d}_t \mathcal{I}_k(\lambda_j) = -\partial_{\vartheta_k} \mathcal{H}' = -\sum \left(\partial_{\vartheta_k} \Lambda_j\right)_{\mathcal{I},\lambda_j} \mathrm{d}_t \lambda_j. \tag{2.80}$$

Let us assume that λ_j vary adiabatically with time, in the sense of (2.25), i.e. that $\mathrm{d}_t \ln \lambda_j = \epsilon \times \min(\omega_k)$, where $\epsilon \ll 1$ and $\omega_k = (\partial_{\mathcal{I}_k} \mathcal{H})_{\lambda_j}$ are the unperturbed angular frequencies. We see from (2.80) that the evolution of the perturbed action variables $\mathcal{I}_k(\lambda_j)$ now involves a product of fast oscillation, represented by the periodic coefficients $(\partial_{\vartheta_k} \Lambda_j)_{\mathcal{I},\lambda_j}$, and slow variation at $\mathrm{d}_t \lambda_j \sim O(\epsilon)$. We can thus average (2.80) over the fast rotation using $\langle \cdot \rangle_{\vartheta_k} = \oint \cdot \mathrm{d}\vartheta_k / 2\pi$, taking $\mathrm{d}_t \lambda_j$ outside the integral, with the result

$$\mathrm{d}_t I_k = \mathrm{d}_t \langle \mathcal{I}_k \rangle \approx -\sum \left\langle \partial_{\lambda_j} \partial_{\vartheta_k} S_0 \right\rangle_{\vartheta_k} \mathrm{d}_t \lambda_j \approx 0. \tag{2.81}$$

The averaged action variables, which represent the *Poincaré invariants*, $\mathcal{J}_k = 2\pi \mathcal{I}_k$ averaged over the fast rotation, are therefore *adiabatically invariant*. Treating ϑ_k as complex variables, it can be shown that $\langle \mathcal{I}_k \rangle$ are indeed conserved to all orders in ϵ as $\exp(-1/\epsilon)$.[7]

We next subject the system to small, sinusoidal perturbations $\lambda_j \sim \exp(\mathrm{i}\omega_{\lambda_j} t)$. If the pitch of the helical field lines covering the n-torus has a 'radial' shear $\mathrm{d}\omega_k / \mathrm{d}\mathcal{I}_k > 0$, then dynamical modes resonant with the perturbations ($\omega_k = m\omega_{\lambda_j}$, where m is an integer) will be excited, in what is known as *parametric resonance*. As a result, the associated toroidal surfaces break up into m *helical islands*, whose *O-points* and *X-points* remain at the 'radial' location of the unperturbed surface, and which are separated from the smooth nested tori by a *separatrix* surface.

What happens when the strength of the perturbation is increased? This question was formally answered by a celebrated Kolmogorov–Arnold–Moser (KAM) theorem, a good account of which may be found in Arnold (1989). The answer is two-fold: (i) the identity of the islands is preserved, even as their 'radial' width increases; and (ii) chaotic volumes appear close to the X-points and expand along with the strength of the perturbation. Eventually, the chaotic region fills the entire volume of the helical island and the unperturbed resonant surface is destroyed, as illustrated in Fig. 2.1. In Chapter 3 we will see that flux surfaces in magnetically confined plasmas are realizations of such abstract canonical tori in three-dimensional space.

Returning to the discussion of charged particle gyration (Section 2.2.1), we recall that all fields were assumed to vary adiabatically ($\epsilon \ll 1$). We could thus proceed

[7] This result is the first 'dividend' of our 'investment' in the Hamilton–Lagrange formalism.

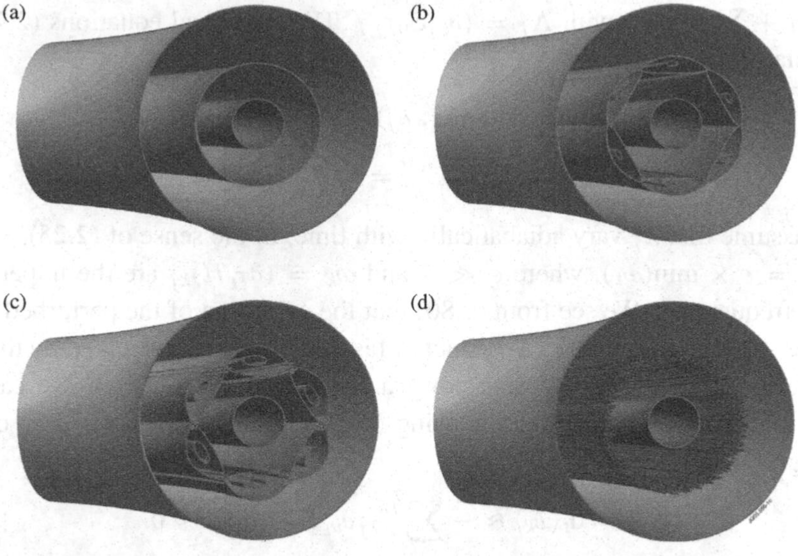

Fig. 2.1. Schematic representation of nested canonical 2-tori, shown for the simple case of two degrees of freedom: (a) unperturbed canonical surfaces; (b) weak perturbation – parametric resonance with a sixth harmonic ($m = 6$); (c) moderate perturbation – appearance of chaotic regions at the X-points; (d) strong perturbation – destruction of the surface and appearance of a chaotic volume.

to calculate the angle-action variables starting with $\mathbf{q} = \mathbf{r}$, $\mathbf{p} = m\mathbf{v} + e\mathbf{A}$ and $\mathcal{L} = \frac{1}{2}m\mathbf{v}^2 + e\mathbf{A}\cdot\mathbf{v} - e\varphi$. However, we anticipate that the gyro-phase γ will play the role of the angle variable ϑ. With this choice, the Poincaré invariant, $\mathcal{J} = \oint_{C(t)} \mathbf{p}\cdot d\mathbf{q}$ is integrated over the exact path of gyration $C(t)$, and the adiabatic invariant, $J = \langle\mathcal{J}\rangle_\gamma = \oint_{C_\odot(t)} \mathbf{p}\cdot d\mathbf{q}$ over the approximate circular path, $C_\odot(t)$,

$$J = \langle\mathcal{J}\rangle_\gamma \approx \frac{1}{\Omega}\oint_{C_\odot} \mathbf{u}_\perp \cdot [m(\mathbf{U}+\mathbf{u}_\perp) + e\mathbf{A}(\mathbf{x})]\mathrm{d}\gamma \approx \frac{2\pi m}{e}\mu + O(\epsilon). \quad (2.82)$$

Here the last term is integrated analogously to (2.49). The magnetic moment μ is thus the first term in the expansion of the adiabatic invariant J in powers of ϵ. The accompanying angle-action variables may be approximated by $I = m\mu/e = \mathcal{K}_\perp/\Omega$ and $\vartheta = \gamma$, with $\mathrm{d}_t I \approx 0$ and $\mathrm{d}_t\vartheta \approx \Omega$.

A systematic application of Lagrange–Hamilton mechanics to charged particle motion offers a complete and self-consistent description of guiding centre dynamics (Littlejohn, 1983; Boozer, 1984). Specifically, averaging $\mathcal{L}$ and $\mathcal{H}$ over the fast gyration, and recalling the definition (2.57), yields the GC-Lagrangian in the extended phase space formed by $(\mathbf{R}, \dot{\mathbf{R}}, U_\parallel, \gamma, t)$,

$$\langle\mathcal{L}\rangle_\gamma = e\mathbf{A}^* \cdot \mathrm{d}_t\mathbf{R} + \frac{m\mu}{e}\mathrm{d}_t\gamma - e\varphi - \mu B - \frac{1}{2}mU_\parallel^2. \quad (2.83)$$

Here the last three terms represent the averaged GC-Hamiltonian, $-\langle\mathcal{H}\rangle_\gamma = -\mathcal{E}$. Applying $\delta\int\langle\mathcal{L}\rangle_\gamma dt = 0$ results in $U_\parallel = \dot{\mathbf{R}}\cdot\mathbf{b}$, $\dot{\mu} = 0$, $\dot{\gamma} = \Omega$, which agree with the previous results. Denoting $\dot{\mathbf{R}}$ by $\mathbf{U}$, one finds

$$e(\mathbf{E}^* + \mathbf{U}\times\mathbf{B}^*) = m\mathbf{b}d_t U_\parallel + \mu\nabla B, \tag{2.84}$$

where $\mathbf{B}^* = \mathbf{B} + \frac{m}{e}\nabla\times(U_\parallel\mathbf{b})$ and $\mathbf{E}^* = \partial_t\mathbf{A}^* - \nabla\varphi = \mathbf{E} - \frac{m}{e}U_\parallel\partial_t\mathbf{b}$ are the *pseudo-magnetic* and *pseudo-electric* fields, respectively.[8] The associated GC-motion, obtained by taking a cross product of (2.84) with $\mathbf{b}$,

$$\mathbf{U} = \frac{1}{B^*_\parallel}\left(U_\parallel\mathbf{B}^* + \mathbf{E}^*\times\mathbf{b} + \frac{\mu}{e}\mathbf{b}\times\nabla B\right), \qquad B^*_\parallel = \mathbf{b}\cdot\mathbf{B}^*, \tag{2.85}$$

satisfies all the requirements of Hamiltonian mechanics: it conserves phase space volume, total energy $\mathcal{E}$ and canonical momenta p_k, reflecting any system symmetries, $\partial_{q_k}\mathcal{L} = 0$. The earlier result (2.37)–(2.39) is recovered by making the substitution $B \to B^*_\parallel$ and $\mathbf{E}\to\mathbf{E}^*$.[9]

If $\mathbf{E}$ and $\mathbf{B}$ do not vary with time so that the energy $\mathcal{E} = \langle\mathcal{H}\rangle_\gamma$ is conserved, then we can use (2.55) to recover the results on page 26. In other words, the GC-velocity (2.56) reduces to $\mathbf{U} = U_\parallel\mathbf{B}^*/B^*_\parallel$, and thus follows the *pseudo-magnetic* field lines. Replacing $B^*_\parallel$ by B, we recover the earlier expression (2.56). Making use of Ampere's law and evaluating the curl at constant $\mathcal{E}$ and μ, the GC-drift can then be recast in a particularly useful form,

$$\mathbf{v}_{GC\perp} = \mathbf{U}_\perp \approx \frac{U_\parallel}{B}\nabla\times\left(\frac{U_\parallel\mathbf{B}}{\Omega}\right) \approx U_\parallel\nabla\left(\frac{U_\parallel}{\Omega}\right)\times\mathbf{b} + \frac{U_\parallel^2}{\Omega B}\mu_0\mathbf{J}. \tag{2.86}$$

2.3 Kinetic description

In Section 2.2, we studied the effect of force fields on the motion of charged particles, treating the fields themselves as given. In reality, this motion produces charge and current densities, $\varrho(\mathbf{x}, t)$ and $\mathbf{J}(\mathbf{x}, t)$, and thus additional fields $\mathbf{E}(\mathbf{x}, t)$ and $\mathbf{B}(\mathbf{x}, t)$, according with *Maxwell's equations*,

$$\nabla\cdot\mathbf{E} = \varrho/\epsilon_0 \qquad \nabla\cdot\mathbf{B} = 0, \tag{2.87}$$

$$\nabla\times\mathbf{E} = -\partial_t\mathbf{B} \qquad \nabla\times\mathbf{B} = \mu_0\mathbf{J} + \partial_t\mathbf{E}/c^2. \tag{2.88}$$

The aim of *plasma theory* is to *close* Maxwell's equations (ensure an equal number of equations and unknowns) by developing *constitutive relations* which express $\varrho(\mathbf{x}, t)$ and $\mathbf{J}(\mathbf{x}, t)$ as functionals of $\mathbf{E}(\mathbf{x}, t)$ and $\mathbf{B}(\mathbf{x}, t)$.

[8] Note that (2.84) is more accurate than the earlier estimate (2.31).

[9] Although both results are accurate to order ϵ, only the latter derives from the GC-Lagrangian, and thus exhibits exact conservation properties of Hamiltonian mechanics.

We first obtain $\varrho(\mathbf{x}, t)$ and $\mathbf{J}(\mathbf{x}, t)$ for a collection of charged particles with trajectories $\mathbf{x}_j(t)$ and $\mathbf{v}_j(t)$ determined by (2.14). The phase space density of particles of species s is given by the sum of delta functions following the individual trajectories, $\mathcal{F}_s(\mathbf{x}, \mathbf{v}, t) = \sum \delta[\mathbf{x} - \mathbf{x}_j(t)]\delta[\mathbf{v} - \mathbf{v}_j(t)]$. The charge and current densities are found by integrating over velocity space,

$$\varrho(\mathbf{x}, t) \equiv \sum \varrho_s \equiv \sum e_s n_s = \sum e_s \int \mathcal{F}_s(\mathbf{x}, \mathbf{v}, t) d\mathbf{v}, \tag{2.89}$$

$$\mathbf{J}(\mathbf{x}, t) \equiv \sum \varrho_s \mathbf{V}_s = \sum e_s \int \mathbf{v}\mathcal{F}_s(\mathbf{x}, \mathbf{v}, t) d\mathbf{v}. \tag{2.90}$$

where we introduced the particle density, $n_s(\mathbf{x}, t)$, the charge density, $\varrho_s(\mathbf{x}, t) = e_s n_s(\mathbf{x}, t)$, and the flow velocity, $\mathbf{V}_s(\mathbf{x}, t)$, of species s.

Liouville's theorem for the phase space density of states (2.77) implies

$$\mathrm{d}_t \mathcal{F}_s = \partial_t \mathcal{F}_s + \{\mathcal{H}, \mathcal{F}_s\} = \partial_t \mathcal{F}_s + \mathbf{v} \cdot \nabla \mathcal{F}_s + \mathbf{a}_s \cdot \partial_{\mathbf{v}} \mathcal{F}_s = 0, \tag{2.91}$$

where the acceleration $\mathbf{a}_s(\mathbf{x}, \mathbf{v}, t)$ follows from (2.14). This exact, microscopic result is known as *Klimontovich's equation*. Given an *omnipotent* computer (memory $\gg 10^{23}$ bytes, speed $\gg 10^{33}$ flops), our task would now be complete. We could simply follow the trajectories with (2.14), updating the fields using (2.87)–(2.91). Such computers could become available by the end of the century, if computing continues to grow at *Moore's law* (doubling every 18 months), and taking $\sim 10^{15}$ flops as the standard for heavy computing in 2008. In the meantime, we need an alternative strategy.

2.3.1 Phase space conservation laws

This strategy is found in performing an *ensemble average* $\langle \cdot \rangle_{ens}$ in the sense of statistical mechanics and separating each quantity into the smooth, averaged part (denoted by a bar) and the spiky, fluctuating part (denoted by a tilde): $\mathcal{A} = \overline{\mathcal{A}} + \widetilde{\mathcal{A}}$ where $\langle \mathcal{A} \rangle_{ens} = \overline{\mathcal{A}}$ and $\langle \widetilde{\mathcal{A}} \rangle_{ens} = 0$. Both parts are defined over the whole phase space $(\mathbf{x}, \mathbf{v})$. Due to its frequent use, the ensemble average of $\mathcal{F}_s$ is denoted simply by $f_s \equiv \overline{\mathcal{F}_s} = \langle \mathcal{F}_s \rangle_{ens}$. The ensemble average of (2.91) now yields the *plasma kinetic equation*,

$$\mathrm{d}_t f_s = \partial_t f_s + \mathbf{v} \cdot \nabla f_s + \overline{\mathbf{a}}_s \cdot \partial_{\mathbf{v}} f_s = \langle \widetilde{\mathbf{a}}_s \cdot \partial_{\mathbf{v}} \widetilde{\mathcal{F}}_s \rangle_{ens} = \mathcal{C}_s(f). \tag{2.92}$$

The two terms involving $\mathbf{a}_s$ arise from $\langle \mathbf{a}_s \cdot \partial_{\mathbf{v}} \mathcal{F}_s \rangle_{ens}$: while the left-hand term contains only statistically independent parts of $\mathbf{a}_s$ and $\mathcal{F}_s$, the right-hand term represents *correlations* between fields and particles, or, ultimately, between individual particle trajectories. In other words, $\langle \widetilde{\mathbf{a}}_s \cdot \partial_{\mathbf{v}} \widetilde{\mathcal{F}}_s \rangle_{ens}$ reflects inter-particle encounters and may thus be expressed in terms of the *collision* operator $\mathcal{C}_s(f) = \sum \mathcal{C}_{ss'}(f_s, f_{s'})$, which in general depends on all plasma species. To close (2.92), we ensemble average (2.87)–(2.90), which now apply to $\overline{\mathbf{E}}, \overline{\mathbf{B}}, \overline{\mathbf{j}}, \overline{\varrho}$ and $\overline{\mathcal{F}}_s = f_s$, and drop the bar henceforth.

In many applications, the effect of correlations (collisions) may be neglected and (2.92) reduces to the *Vlasov equation*,

$$\partial_t f_s + \mathbf{v} \cdot \nabla f_s + \mathbf{a}_s \cdot \partial_{\mathbf{v}} f_s = \partial_t f_s + \nabla \cdot (\mathbf{v} f_s) + \partial_{\mathbf{v}} \cdot (\mathbf{a}_s f_s) = 0, \tag{2.93}$$

where the divergence form follows from $\nabla \cdot \mathbf{v} = 0$ and $\partial_{\mathbf{v}} \cdot \mathbf{a}_s = 0$ for (2.14). All reference to particle discreteness has now been eradicated. Although (2.93) bears a striking resemblance to (2.91), it carries a very different meaning: (2.91) describes the conservation of *discrete particles* as they travel in phase space, (2.93) the continuity of a *kinetic substance* with density $f_s(\mathbf{x}, \mathbf{v}, t)$ as it flows in phase space. Despite these differences, the formal identity between the two equations allows us to use Lagrange–Hamilton mechanics to solve (2.93). The primitive variables $(\mathbf{x}, \mathbf{v})$ can thus be replaced by any canonical $(\mathbf{q}, \mathbf{p})$, such that continuity of $f_s(\mathbf{q}, \mathbf{p}, t)$ implies

$$\mathrm{d}_t f_s = \partial_t f_s + \mathrm{d}_t \mathbf{q} \cdot \partial_{\mathbf{q}} f_s + \mathrm{d}_t \mathbf{p} \cdot \partial_{\mathbf{p}} f_s = 0, \tag{2.94}$$

or more generally, by any generalized, non-Cartesian co-ordinates $\mathbf{z} = \{z^1, \ldots, z^6\}$ obtained from $(\mathbf{x}, \mathbf{v})$ by a smooth, non-singular transformation,

$$\mathrm{d}_t f_s = \partial_t f_s + \mathrm{d}_t \mathbf{z} \cdot \partial_{\mathbf{z}} f_s = 0. \tag{2.95}$$

The ideal choice of $(\mathbf{q}, \mathbf{p})$ are angle-action variables $(\boldsymbol{\vartheta}, \boldsymbol{\mathcal{I}})$, for which

$$\mathrm{d}_t f_s = \partial_t f_s + \mathrm{d}_t \boldsymbol{\mathcal{I}} \cdot \partial_{\boldsymbol{\mathcal{I}}} f_s + \mathrm{d}_t \boldsymbol{\vartheta} \cdot \partial_{\boldsymbol{\vartheta}} f_s = \partial_t f_s + \boldsymbol{\omega}(\boldsymbol{\mathcal{I}}) \cdot \partial_{\boldsymbol{\vartheta}} f_s = 0, \tag{2.96}$$

since $\mathrm{d}_t \boldsymbol{\mathcal{I}} = 0$ and $\mathrm{d}_t \boldsymbol{\vartheta} = \boldsymbol{\omega}(\boldsymbol{\mathcal{I}})$, see (2.78). This is the simplest form of the Vlasov equation and offers the best strategy for solving (2.93), valid to order $\exp(-1/\epsilon)$, namely replacing $\boldsymbol{\mathcal{I}}$ by adiabatic invariants $\mathbf{I} = \langle \boldsymbol{\mathcal{I}} \rangle_{\boldsymbol{\vartheta}}$.

Note that $\mathrm{d}_t f_s = 0$ along individual particle trajectories $\mathbf{x}_j(t), \mathbf{v}_j(t)$, or more generally $\mathbf{q}_j(t), \mathbf{p}_j(t)$, which thus play the role of *characteristics* of the phase space flow, defining surfaces of constant f_s. Consequently, (2.93) is satisfied by any function $g(\mathbf{c})$ of the constants of motion, $c_j(\mathbf{x}, \mathbf{v}, t)$,

$$\mathrm{d}_t g(\mathbf{c}) = \partial_{\mathbf{c}} g \cdot \mathrm{d}_t \mathbf{c} = \sum \partial_{c_j} g \left(\partial_t c_j + \mathbf{v} \cdot \nabla c_j + \mathbf{a}_s \cdot \partial_{\mathbf{v}} c_j\right) = 0. \tag{2.97}$$

Here c_j represent either the exact dynamical invariants $\mathcal{I}_j$ or the adiabatic invariants I_j. An 'infinite' number of these invariants (up to six for each particle) generates an 'infinite' number of conservation laws,

$$\mathrm{d}_t \int G(f_s) \mathrm{d}\mathbf{x} \mathrm{d}\mathbf{v} = 0, \qquad \forall\, G(f_s). \tag{2.98}$$

Setting $G(f_s) = -f_s \ln f_s$ yields conservation of *entropy density*, s_s, requiring that f_s evolve *reversibly*. Conservation of all $G(f_s)$ results in the so-called *shuffling* property of the Vlasov equation, which states that the evolution of f_s proceeds exclusively by rearrangement in phase space of the infinitesimal elements $f_s\mathbf{dxdv}$. Consequently, a configuration in which $f_s(v)$ decreases monotonically ($\partial_v f_s < 0$) is absolutely stable, since the kinetic energy $\int \frac{1}{2}m_s v^2 f_s \mathrm{d}\mathbf{v}$ is minimized. Whereas entropy production and *irreversibility* are introduced by the collision operator, $C_s(f)$, reversible mixing of phase space elements may lead to a collisionless decay of plasma disturbances, known as *Landau damping*. Such *phase mixing* is a fundamental property of all plasmas and one of the hallmarks of collective behaviour.

Since closure of Maxwell's equations (2.87) required only the two lowest velocity moments of f_s, namely n_s and $\mathbf{V}_s$ – a fact which may be traced to (2.67), in which $\mathcal{L}$ is a function of only $\mathbf{q}$ and $\dot{\mathbf{q}}$ – one would expect a low-order *moment formulation*, obtained by taking successive velocity moments of (2.92), to adequately describe most plasma phenomena. *Hydrodynamics* suggests that this moment or *fluid description* is only appropriate for perturbations which are either slow compared to the thermal speed, v_{ts}, or which affect particles with different velocities in the same way, as is true of the electric drift $\mathbf{v}_E$ irrespective of the ordering of v_E/v_{ts}. In either case, particle response is largely independent of $\mathbf{v}$, and particles move collectively in velocity space. The *kinetic flow* in phase space $(\mathbf{x}, \mathbf{v})$ then reduces to the *fluid flow* in co-ordinate space $(\mathbf{x})$, i.e. a standard fluid description.[10]

2.3.2 Guiding centre kinetic theory

In magnetized plasmas, the general kinetic equation (2.92),

$$\partial_t f_s + \mathbf{v}\cdot\nabla f_s + \frac{e_s}{m_s}(\mathbf{E}_\parallel + \mathbf{E}_\perp + \mathbf{v}\times\mathbf{B})\cdot\partial_\mathbf{v} f_s = C_s(f), \qquad (2.99)$$

is greatly simplified by the smallness of the magnetization parameter ($\delta_s \ll 1$), which ensures the dominance of the gyration term $\sim \Omega_s f_s$. Let us consider the size of the other terms in (2.99).[11] We assume that f_s evolves slowly, such that $\partial_t f_s \sim \delta_s\Omega_s f_s$. The advective term in (2.99) may contain contributions from both large, L-scale and small, ρ_{ts}-scale variations of f_s, which we denote by f_{s0} and $f_{s1} = \Delta_s f_{s0}$, respectively; its magnitude is thus found as $\sim (\delta_s + \Delta_s)\Omega_s f_s$. Since

[10] For faster perturbations particle response becomes velocity dependent, so that some particles may resonate with the disturbance, making *kinetic description* unavoidable.

[11] In the rest of the section, the symbol '$\sim$' denotes the asymptotic magnitude and contains '$\ll$' as its subset, i.e. it designates the maximum ordering of a quantity in terms of δ_s.

only the third and last terms affect parallel motion then, by (2.19), they must be comparable to prevent unbalanced acceleration or *runaway*. We assume these terms to have the same asymptotic magnitude as $\partial_t f_s$, i.e. $\sim \delta_s \Omega_s f_s$.

With the above assumptions, we write the asymptotic magnitude of each of the terms in (2.99), normalized by $\Omega_s f_s$, as

$$\delta_s + (\delta_s + \Delta_s) + (\delta_s + v_E/v_{ts} + 1) \sim \delta_s. \tag{2.100}$$

Different orderings of plasma flow, v_E/v_{ts}, and ρ_{ts}-scale variation, Δ_s, produce different versions of the kinetic equation (2.99).[12] The simplest version results from the *drift ordering*, which assumes the electric fields to be sufficiently weak for all drifts to have comparable magnitude (2.42), and omits all ρ_{ti}-scale variations ($\Delta_i = 0$); note that all relations are expressed in terms of ion quantities on account of $\delta_i \gg \delta_e$, $v_{ti} \ll v_{te}$ and $\rho_{ti} \gg \rho_{te}$. The plasma is thus assumed to flow slowly ($v_E \ll v_{ti}$) and vary only gradually in space ($|\nabla \ln f_i|^{-1} \sim L \gg \rho_{ti}$) and time ($\omega \sim \partial_t \ln f_i \ll \Omega_i$). Drift ordering has wide ranging applications, including the study of plasma equilibrium, stability and transport, see Chapter 5. Its main drawback lies in neglecting finite gyro-radius effects, and hence its inability to treat ρ_{ti}-scale perturbations, which are believed to play a crucial role in plasma turbulence.

Allowing for such short-scale disturbances, but assuming these to be weak, gives rise to the *gyro-ordering*: ($v_E/v_{ti} \sim \Delta_i \sim \delta_i$). This approach is dedicated exclusively to the study of ion-related waves, instabilities and turbulence, ρ_{te}-scale disturbances being of limited interest. Finally, the *MHD ordering* (2.41) and $\Delta_i = 0$, assumes electric fields to be sufficiently strong for $\mathbf{v}_E$ to compete with gyration, generating rapid cross-field plasma motion, $v_E \sim v_{ti}$. In the form of MHD-ordered *fluid* equations, or simply MHD, it is widely used to study the global stability and dynamics of magnetized plasmas.

Only one small parameter is common to all three ordering schemes, namely $\delta_i = \rho_{ti}/L \ll 1$. In Section 2.2.1, we developed the GC-approximation by expanding the particle equations of motion in δ_s, averaging over the rapid gyration and then solving the gyro-averaged equations order by order. Starting from (2.99), the same strategy can be applied to construct a guiding centre kinetic theory. As a first step, we select the energy, $\mathcal{E} = \frac{1}{2}mv^2 + e\varphi$, (2.54), the magnetic moment, $\mu = mv_\perp^2/2B$, (2.30), and the gyro-phase, γ, (2.19), as the velocity co-ordinates, motivated by the form (2.96) and the fact that $\mathcal{E}$ and μ are approximate action variables. The exact

[12] Recall that the two orderings, (2.41) and (2.42), are introduced so as to describe plasma phenomena on different temporal scales, i.e. they are in general appropriate to high- and low-frequency dynamics, respectively. The reason for retaining $O(\delta_i)$ terms in the drift orderings is not *rapid spatial* variation but *slow temporal* variation, implied by (2.42), i.e. the plasma remains strongly magnetized ($\delta_i \ll 1$), but is assumed to evolve slowly enough such that $O(\delta_i)$ terms become significant.

particle position, $\mathbf{x}$, is retained as the spatial co-ordinate. The kinetic equation for $f_s(\mathbf{x}, \mathcal{E}, \mu, \gamma, t)$ follows directly from (2.95) with $\mathbf{z} = (\mathbf{x}, \mathcal{E}, \mu, \gamma)$,

$$\partial_t f + \mathrm{d}_t \mathbf{x} \cdot \partial_{\mathbf{x}} f + \mathrm{d}_t \mathcal{E} \partial_{\mathcal{E}} f + \mathrm{d}_t \mu \partial_\mu f + \mathrm{d}_t \gamma \partial_\gamma f = \mathcal{C}(f), \qquad (2.101)$$

where $\mathrm{d}_t = \partial_t + \mathbf{v} \cdot \nabla$ refers to rate of change along the exact particle trajectory and the species index is suppressed for clarity. The advective term takes the usual form $\mathbf{v} \cdot \nabla$ since $\mathrm{d}_t \mathbf{x} = \mathbf{v}$ and $\partial_{\mathbf{x}} = \nabla$. The remaining d_t terms were derived at the end of Section 2.2.1, (2.62)–(2.64).

2.3.2.1 Drift-kinetic equation

We next adopt the *drift ordering* ($v_E/v_t \sim \delta_i$, $\Delta_i = 0$), which requires f to vary gradually in $\mathbf{x}$-space ($L\nabla \sim 1$) and additionally assume gradual variation in $\mathbf{v}$-space, $\mu\partial_\mu \sim \mathcal{E}\partial_{\mathcal{E}} \sim 1$. With these assumptions, all terms in (2.101) are small ($\sim \delta = \delta_i$) compared to the gyration term $\Omega \partial_\gamma f$ which we isolate by combining all lower-order terms into a linear operator $\mathcal{L}$, not to be confused with the Lagrangian, and rewriting (2.101) as

$$\mathcal{L} f + \Omega \partial_\gamma f = \mathcal{C} f, \quad \mathcal{L} = \partial_t + \mathbf{v} \cdot \nabla + \mathrm{d}_t \mathcal{E} \partial_{\mathcal{E}} + \mathrm{d}_t \mu \partial_\mu + (\mathrm{d}_t \gamma - \Omega)\, \partial_\gamma. \quad (2.102)$$

As the next step, we introduce the gyro-average, $\langle \mathcal{A} \rangle_\gamma \equiv \oint \mathcal{A}(\mathbf{x}, \mathcal{E}, \mu, \gamma) \mathrm{d}\gamma / 2\pi$, defined at constant $(\mathbf{x}, \mathcal{E}, \mu)$. We can now separate any quantity $\mathcal{A}$ into the gyro-averaged part $\overline{\mathcal{A}} = \langle \mathcal{A} \rangle_\gamma$ and the γ-dependent part $\widetilde{\mathcal{A}} = \mathcal{A} - \overline{\mathcal{A}}$, where $\partial_\gamma \overline{\mathcal{A}} = 0$ and $\langle \widetilde{\mathcal{A}} \rangle_\gamma = 0$. In particular, $f = \overline{f} + \widetilde{f}$ and the gyration term reduces to $\Omega \partial_\gamma \widetilde{f}$. Since Ω is large, this implies that the γ-dependent part of f must be comparatively small, $\widetilde{f} \sim \delta \overline{f}$. It is convenient to apply the same decomposition to $\mathcal{L} = \overline{\mathcal{L}} + \widetilde{\mathcal{L}}$, calculating $\overline{\mathcal{L}}$ using (2.66),

$$\overline{\mathcal{L}} = \langle \mathcal{L} \rangle_\gamma = \partial_t + \mathbf{v}_\parallel \cdot \nabla + e \left(\partial_t \varphi - \mathbf{v}_\parallel \cdot \partial_t \mathbf{A} \right) \partial_{\mathcal{E}} - \frac{\mu}{B} \partial_t B \partial_\mu, \quad (2.103)$$

$$\widetilde{\mathcal{L}} = \mathbf{v}_\perp \cdot \nabla - e \mathbf{v}_\perp \cdot \partial_t \mathbf{A} \partial_{\mathcal{E}} - \frac{1}{B} \left[\mu \mathbf{v} \cdot \nabla B + \mathbf{v}_\perp \cdot (m v_\parallel \mathrm{d}_t \mathbf{b} - e\mathbf{E}) \right] \partial_\mu, \quad (2.104)$$

where we neglected the ∂_γ term in $\widetilde{\mathcal{L}}$ as it will vanish upon gyro-averaging.

We are now ready to gyro-average the plasma kinetic equation. Since f is required to be 2π-periodic in γ, application of $\langle \cdot \rangle_\gamma$ annihilates all ∂_γ terms. With this observation, gyro-averaging (2.102) yields,[13]

$$\left\langle \mathcal{L} f + \Omega \partial_\gamma \widetilde{f} - \mathcal{C} f \right\rangle_\gamma = \langle (\mathcal{L} - \mathcal{C}) f \rangle_\gamma = 0. \qquad (2.105)$$

This *solubility criterion* is also an abstract form of the *drift-kinetic equation*.

[13] In Chapter 5 we will see that for a Fokker–Planck collision operator, applicable to most plasma correlations, one may safely neglect the γ-dependent part, $\widetilde{\mathcal{C}} \approx 0$, and thus adopt $\mathcal{C} \approx \overline{\mathcal{C}}$.

To solve (2.105), we could apply the *multiple time scale* expansion, $f = f_0 + f_1 + \cdots$, where subscripts denote orders in δ. We will instead employ a recursive technique which greatly simplifies the calculation, while requiring only a single iteration (Hazeltine, 1973).[14] The first recursion consists in subtracting (2.105) from (2.102) and replacing f by $\overline{f}$ in $\mathcal{L}$ and $\mathcal{C}$,

$$(\mathcal{L}-\mathcal{C})\overline{f} + \Omega\partial_\gamma \widetilde{f} - \langle(\mathcal{L}-\mathcal{C})\overline{f}\rangle_\gamma = 0 \quad \Rightarrow \quad \Omega\partial_\gamma \widetilde{f} + \widetilde{\mathcal{L}f} = O(\delta^2). \quad (2.106)$$

The gyro-average of (2.106) vanishes for all $\overline{f}$, as required by (2.105). To estimate $\widetilde{f}$, we integrate (2.106) for Newtonian trajectories (2.22),

$$\widetilde{f} = \mathbf{r}\cdot\mathbf{h} + a_{\widetilde{f}}, \qquad \mathbf{h} = -\nabla\overline{f} - e\mathbf{b}\times\mathbf{U}_\perp\partial_\mu\overline{f} + e\partial_t\mathbf{A}\partial_\mathcal{E}\overline{f}, \quad (2.107)$$

where $\mathbf{r} = \rho\hat{\mathbf{e}}_r$ is the gyro-radius vector (2.20), $\mathbf{U}_\perp = \mathbf{v}_E + \mathbf{v}_g$ is the perpendicular drift velocity (2.15) with $\mathbf{g}$ given by (2.38) and

$$a_{\widetilde{f}} = \frac{v_\parallel \mu}{\Omega}\left(\hat{\mathbf{e}}_r\hat{\mathbf{e}}_u : \nabla\mathbf{b} - \frac{\mathbf{b}}{2}\cdot\nabla\times\mathbf{b}\right)\partial_\mu\overline{f}, \quad (2.108)$$

is an integration constant. For slowly varying fields and nearly isotropic $\overline{f}$, (2.107) reduces to $\widetilde{f} \approx -\mathbf{r}\cdot\nabla\overline{f}$, which depends on γ only through $\mathbf{r}(\gamma) = \rho\hat{\mathbf{e}}_r(\gamma)$. The form appearing in (2.107),

$$f(\mathbf{x},\gamma) \approx \overline{f}(\mathbf{x}) - \mathbf{r}(\gamma)\cdot\nabla\overline{f}(\mathbf{x}), \quad (2.109)$$

suggests that f depends on $\mathbf{x}$ and γ through the GC-position, $\mathbf{x} - \mathbf{r}(\gamma) = \mathbf{R}$. We infer that $f(\mathbf{x},\gamma) = F(\mathbf{R})$, where $F(\mathbf{R},\mathcal{E},\mu,t)$ is a *guiding centre distribution*. Taylor expanding $F(\mathbf{R}) = F(\mathbf{x}-\mathbf{r}(\gamma))$ around $\mathbf{x}$,

$$F(\mathbf{x}-\mathbf{r},\mathcal{E},\mu,t) \approx F(\mathbf{x},\mathcal{E},\mu,t) - \mathbf{r}(\gamma)\cdot\nabla F(\mathbf{x},\mathcal{E},\mu,t), \quad (2.110)$$

shows that the gyro-average of the distribution of particles gives rise to the distribution of guiding centres, i.e. $\overline{f}(\mathbf{x},\mathcal{E},\mu,t) \approx F(\mathbf{x},\mathcal{E},\mu,t)$.[15]

Inserting (2.107) into (2.105) we obtain the desired *drift-kinetic equation*,

$$\langle(\mathcal{L}-\mathcal{C})(\overline{f}+\widetilde{f})\rangle_\gamma = \overline{\mathcal{L}f} + \langle\mathcal{L}(\mathbf{r}\cdot\mathbf{h} + a_{\widetilde{f}})\rangle_\gamma - \mathcal{C}\overline{f} = 0. \quad (2.111)$$

All that remains is to evaluate the gyro-averages.[16] First, a straightforward calculation shows that $\langle\mathcal{L}a_{\widetilde{f}}\rangle = 0$. Second, noting that $\partial_\gamma\mathbf{h} = 0$ and $\partial_\gamma\mathbf{r} = \rho\hat{\mathbf{e}}_u$, and applying the chain rule, we obtain

$$\langle\mathcal{L}(\mathbf{r}\cdot\mathbf{h})\rangle_\gamma = \langle\mathbf{r}\cdot\mathcal{L}\mathbf{h}\rangle_\gamma + \langle\mathcal{L}\mathbf{r}\rangle_\gamma\cdot\mathbf{h}, \quad (2.112)$$

14 The results of the two approaches are of course identical.

15 The fact that $(\partial_\gamma F)_\mathbf{R} = 0$ but $(\partial_\gamma F)_\mathbf{x} \neq 0$ is significant since $\langle\cdot\rangle_\gamma$ is typically defined at fixed $\mathbf{x}$ in *drift-kinetic* theory but at fixed $\mathbf{R}$ in *gyro-kinetic* theory. Note that velocity moments are defined at fixed $\mathbf{x}$ in both theories.

16 The details of the calculation may be found in Hazeltine (1973).

which may be expanded and gyro-averaged to yield,

$$\begin{aligned}\langle \mathbf{r}\cdot\mathcal{L}\mathbf{h}\rangle_\gamma &= \langle \mathbf{r}\mathbf{u}_\perp\rangle_\gamma : \nabla\mathbf{h} + \langle \mathbf{r}\mathrm{d}_t\mu\rangle_\gamma\cdot\partial_\mu\mathbf{h} + \langle \mathbf{r}\mathrm{d}_t\mathcal{E}\rangle_\gamma\cdot\partial_{\mathcal{E}}\mathbf{h}\\ &= \tfrac{1}{2}\mathbf{b}\cdot\nabla\times\mathbf{h} - \mu\mathbf{U}_\perp\cdot\partial_\mu\mathbf{h} - \mu\mathbf{b}\times\partial_\mu\mathbf{A}\cdot\partial_{\mathcal{E}}\mathbf{h}.\end{aligned} \tag{2.113}$$

The same procedure applied to $\langle\mathcal{L}\mathbf{r}\rangle_\gamma$ results in the net GC-drift, $\mathbf{U}$,

$$\begin{aligned}\langle\mathcal{L}\mathbf{r}\rangle_\gamma &= \Omega^{-1}\left(\langle\mathcal{L}\mathbf{b}\times\mathbf{u}_\perp\rangle_\gamma + \mathbf{b}\times\langle\mathcal{L}\mathbf{u}_\perp\rangle_\gamma + \mathbf{b}\times\langle\mathbf{u}_\perp\mathcal{L}\rangle_\gamma\right)\\ &= -\frac{\mu B}{m\Omega}\mathbf{b}(\mathbf{b}\cdot\nabla\times\mathbf{b}) - \mathbf{U}_\perp = -\mathbf{U},\end{aligned} \tag{2.114}$$

where the second and third terms reproduce the familiar perpendicular GC-drift, $\mathbf{v}_{GC\perp}$, while the first term describes the parallel GC-drift, $\mathbf{v}_{GC\parallel}$. Combining (2.111)–(2.114) with $\overline{\mathcal{L}}$ (2.103) and $\mathbf{h}$ (2.107), we find

$$\mathcal{L}_{GC}\overline{f} = \left(\partial_t + \mathrm{d}_t\mathbf{R}\cdot\nabla + \mathrm{d}_t\mathcal{E}_{GC}\partial_{\mathcal{E}} + \mathrm{d}_t\mu_{GC}\partial_\mu\right)\overline{f} = \mathcal{C}\overline{f}, \tag{2.115}$$

where $\mathrm{d}_t\mathbf{R} = \mathbf{U} = \mathbf{v}_\parallel + \mathbf{U}_\perp$ is the net GC-velocity, and

$$\begin{aligned}\mathrm{d}_t\mu_{GC} &= -\frac{v_\parallel}{\Omega}\nabla\cdot\left(\mathbf{b}\times\frac{\partial\mathbf{b}}{\partial t}\right) - (\mathbf{b}\cdot\nabla\times\mathbf{b})\frac{\mathbf{b}}{B}\cdot\frac{\partial\mathbf{A}}{\partial t} + \frac{v_\parallel B}{\Omega}\mathbf{b}\cdot\nabla\left(\frac{v_\parallel}{B}\mathbf{b}\cdot\nabla\times\mathbf{b}\right),\\ \mathrm{d}_t\mathcal{E}_{GC} &= \mu\frac{\partial B}{\partial t} + e\frac{\partial\varphi}{\partial t} - e(\mathbf{v}_\parallel + \mathbf{U}_\perp)\cdot\frac{\partial\mathbf{A}}{\partial t},\end{aligned} \tag{2.116}$$

are the rates of change of μ and $\mathcal{E}$ evaluated at the guiding centre, i.e. $\mathrm{d}_t\mathcal{E}_{GC} \equiv \partial_t\mathcal{E} + \mathbf{U}\cdot\nabla\mathcal{E}$, etc. This is the exact and explicit form of (2.111).

In practice, (2.115) is simplified by neglecting $\mathbf{v}_{GC\parallel}$ and $\mathrm{d}_t\mu_{GC}\partial_\mu$. This is justified by noting that $\mathbf{b}\cdot\nabla\times\mathbf{b} = \mu_0 J_\parallel/B$ contains the parallel current density $J_\parallel$, which can be estimated from the equilibrium force balance $\mathbf{J}\times\mathbf{B}\approx\nabla p$ as $J_\parallel \sim J_\perp \sim p/LB$. Consequently, $\mathrm{d}_t\mu_{GC} \sim v_{D\parallel}/v_t \sim \beta\delta$ with β defined by (2.12). Typically $\beta \ll 1$ and $\partial_\mu \ln\overline{f} \sim \partial_{\mathcal{E}}\ln\overline{f} \sim \delta$, such that

$$\mathrm{d}_t\mu_{GC}\partial_\mu\ln\overline{f} \sim (v_{D\parallel}/v_t)\partial_{\mathcal{E}}\ln\overline{f} \sim \beta\delta^2, \tag{2.117}$$

which leads to a version of (2.115) suitable for most applications,

$$\partial_t F + \mathbf{U}\cdot\nabla F + \left(\mu\partial_t B + e\partial_t\varphi - e\mathbf{v}_\parallel\cdot\partial_t\mathbf{A}\right)\partial_{\mathcal{E}}F = \mathcal{C}F + O(\beta\delta^2). \tag{2.118}$$

Here $\mathbf{U} = \mathbf{v}_\parallel + \mathbf{U}_\perp$ and $\overline{f}$ is replaced by $F(\mathbf{x},\mathcal{E},\mu,t)$ to indicate the GC-aspect of drift-kinetic theory. It is instructive to note that (2.118) follows directly from (2.95) with $\mathbf{z} = (\mathbf{x},\mathcal{E},\mu)$ and $\mathrm{d}_t = \partial_t + \mathbf{U}\cdot\nabla$, i.e. by neglecting the γ-dependence from the outset and evaluating the rates of change along the GC-trajectory. Finally, in terms of $\mathcal{K} = \mathcal{E} - e\varphi$, (2.118) becomes

$$\partial_t F + \mathbf{U}\cdot\nabla F + (\mu\partial_t B + e\mathbf{U}\cdot\mathbf{E})\,\partial_{\mathcal{K}}F = \mathcal{C}F. \tag{2.119}$$

A complete numerical solution of (2.119) for fusion applications would require a memory of $\sim 10^6\delta_i^{-2} \sim 10^{16}$ bytes and speed of $\sim 10^{12}\Omega_i \sim 10^{20}$ flops.[17] By deriving the drift-kinetic equation, i.e. by going from (2.91) to (2.119), we have thus greatly simplified the modelling task.

2.3.2.2 *Gyro-kinetic equation*

Returning to (2.101), we next consider the application of the *gyro-kinetic* ordering ($v_E/v_t \sim \Delta_i \sim \delta_i$). Separating all quantities into the unperturbed and perturbed parts, e.g. $\mathcal{A} = \mathcal{A}_0 + \mathcal{A}_1$, we assume the former to represent equilibrium values ($\partial_t \mathcal{A}_0 = 0$) and the latter to be ordered as $\mathcal{A}_1/\mathcal{A}_0 \sim \Delta_i \sim \delta_i$. Sharp, ρ_{ti}-scale variations are assumed to be purely transverse,

$$L \sim |\nabla \ln \mathcal{A}_0|^{-1} \sim |\nabla_\parallel \ln \mathcal{A}_1|^{-1} \gg |\nabla_\perp \ln \mathcal{A}_1|^{-1} \sim \rho_{ti}. \tag{2.120}$$

Gyro-kinetic theory is most conveniently developed in the phase space defined by GC-co-ordinates $(\mathbf{R}, \mathcal{E}, \mu, \gamma)$, which requires gyro-averages to be performed at fixed $\mathbf{R}$, not at fixed $\mathbf{x}$, cf. (2.30) and (2.49),

$$\langle \mathcal{A}_1 \rangle_\gamma \equiv \frac{1}{2\pi} \oint \mathcal{A}_1(\mathbf{R}+\mathbf{r})\mathrm{d}\gamma, \qquad \mathbf{r} = \rho\hat{\mathbf{e}}_r. \tag{2.121}$$

Since $\mathbf{R}(\mathbf{x}, \mathbf{v}) = \mathbf{x} - \mathbf{r}(\mathbf{R}, \mathbf{v})$, the evaluation of the integral involves a recursive expression for $\mathbf{r}(\mathbf{R}, \mathbf{v}) = \mathbf{b}(\mathbf{R})\times\mathbf{v}_\perp/\Omega(\mathbf{R})$, which greatly complicates higher-order analysis. More importantly, (2.121) introduces *non-local* (finite ρ_{ti}) corrections, which are ultimately responsible for the integro-differential nature of gyro-kinetic theory. *Local* versions of this theory invariably invoke the *eikonal* approximation, $\mathcal{A}_1(\mathbf{x}) = \check{\mathcal{A}}(\mathbf{R})\mathrm{e}^{\mathrm{i}\mathbf{k}_\perp\cdot\mathbf{x}}$, where both the amplitude $\check{\mathcal{A}}$ and the wave number, $\mathbf{k}_\perp$, change gradually with $\mathbf{R}$. With this assumption, the gyro-average, $\langle \mathcal{A}_1 \rangle_\gamma$, becomes a GC-quantity and is found proportional to the zeroth-order Bessel function, J_0,

$$\langle \mathcal{A}_1 \rangle_\gamma(\mathbf{R}) = \check{\mathcal{A}}\mathrm{e}^{\mathrm{i}\mathbf{k}_\perp\cdot\mathbf{R}}\langle \mathrm{e}^{\mathrm{i}\mathbf{k}_\perp\cdot\mathbf{r}}\rangle_\gamma \equiv \hat{\mathcal{A}}\langle \mathrm{e}^{\mathrm{i}\mathbf{k}_\perp\cdot\mathbf{r}}\rangle_\gamma = \hat{\mathcal{A}} J_0(k_\perp\rho). \tag{2.122}$$

Similarly, the eikonal gyro-averages involving $\hat{\mathbf{e}}_u$ and $\hat{\mathbf{e}}_r$ can be expressed in terms of the first-order Bessel function, J_1,

$$\langle \hat{\mathbf{e}}_u \mathrm{e}^{\mathrm{i}\mathbf{k}_\perp\cdot\mathbf{r}}\rangle_\gamma = \mathrm{i}J_1(k_\perp\rho)\frac{\mathbf{k}_\perp \times \mathbf{b}}{k_\perp}, \qquad \langle \hat{\mathbf{e}}_r \mathrm{e}^{\mathrm{i}\mathbf{k}_\perp\cdot\mathbf{r}}\rangle_\gamma = \mathrm{i}J_1(k_\perp\rho)\frac{\mathbf{k}_\perp}{k_\perp}, \tag{2.123}$$

and so on to higher orders; typically one sets $\rho = \rho_{ti}$ in (2.121)–(2.123).

Aside from (2.121), the derivation of the *gyro-kinetic* equation closely resembles the *drift-kinetic* case. Below we follow the accounts of Catto *et al.* (1981) and

[17] Such computational power could become available in 20 years if Moore's law persists.

Hazeltine and Meiss (1992), beginning with the plasma kinetic equation obtained from (2.95) with $\mathbf{z} = (\mathbf{R}, \mathcal{E}, \mu, \gamma)$,

$$\partial_t f + \mathrm{d}_t \mathbf{R} \cdot \partial_{\mathbf{R}} f + \mathrm{d}_t \mathcal{E} \partial_{\mathcal{E}} f + \mathrm{d}_t \mu \partial_\mu f + \mathrm{d}_t \gamma \partial_\gamma f = \mathcal{C} f, \tag{2.124}$$

where, as before, d_t measures the rate of change along the exact particle trajectory. Collisional effects are typically neglected, hence $\mathcal{C} \approx 0$ below.

The coefficients in (2.124) have the form, $\mathrm{d}_t \mathcal{A} = \mathrm{d}_t \mathcal{A}_0 + \mathrm{d}_t \mathcal{A}_1$. The equilibrium parts $\mathrm{d}_t \mathcal{E}_0$, $\mathrm{d}_t \mu_0$ and $\mathrm{d}_t \gamma_0$ follow from (2.62)–(2.64) with $\mathrm{d}_t = \mathbf{v} \cdot \nabla$ and $\mathbf{E}, \mathbf{B} \to \mathbf{E}_0, \mathbf{B}_0$. The perturbed parts are found by detailed application of (2.62)–(2.64), e.g. the energy $\mathrm{d}_t \mathcal{E}_1$ is given by (2.62) with $\mathbf{E}, \mathbf{B} \to \mathbf{E}_1, \mathbf{B}_1$, where for an electromagnetic disturbance $\mathbf{v} \cdot \mathbf{A}_1 \sim \varphi_1$. Likewise,

$$\mathrm{d}_t \mu_1 = \frac{e_s v_\perp}{B_0} \left(\hat{\mathbf{e}}_u \cdot \mathbf{E}_1 - v_\parallel \hat{\mathbf{e}}_r \cdot \mathbf{B}_1 \right), \tag{2.125}$$

$$\mathrm{d}_t \gamma_1 = -\frac{e_s}{m v_\perp} \hat{\mathbf{e}}_r \cdot \mathbf{E}_1 + \Omega_0 \left(\mathbf{b}_0 - \frac{v_\parallel}{v_\perp} \hat{\mathbf{e}}_u \right) \cdot \mathbf{B}_1. \tag{2.126}$$

Finally, $\mathbf{U} = \mathrm{d}_t \mathbf{R}$ follows from $\mathrm{d}_t \mathbf{R}(\mathbf{x}, \mathbf{v}) = \mathbf{v} \cdot \nabla \mathbf{R} + \mathrm{d}_t \mathbf{v} \cdot \partial_{\mathbf{v}} \mathbf{R}$ evaluated using $\mathbf{R}(\mathbf{x}, \mathbf{v}) = \mathbf{x} - \mathbf{r}(\mathbf{R}, \mathbf{v})$. After some lengthy vector algebra, one finds

$$\mathbf{U}_0 = \mathbf{v}_\parallel + \mathbf{v}_{E0} + \mathbf{v} \times (\mathbf{v} \cdot \nabla) \left(\frac{\mathbf{b}_0}{\Omega_0} \right) + \frac{\mathbf{r}}{B_0} (\mathbf{v} \cdot \nabla) B_0, \tag{2.127}$$

$$\mathbf{U}_1 = \mathbf{v}_{E1} - \mathbf{v} \left(\frac{\mathbf{b}_0 \cdot \mathbf{B}_1}{B_0} \right) + \frac{v_\parallel}{B_0} \mathbf{B}_1. \tag{2.128}$$

Examining the magnitude of each term in (2.124) reveals that $\partial_\gamma \ln f \sim \widetilde{f}/\overline{f} \sim \delta_i + \Delta_i$, which permits us to approximate f by $\overline{f} \equiv F(\mathbf{R}, \mathcal{E}, \mu, t)$. Gyro-averaging, we obtain the lowest-order *gyro-kinetic equation*,

$$\partial_t F + \langle \mathbf{U}_0 + \mathbf{U}_1 \rangle_\gamma \cdot \partial_{\mathbf{R}} F + \langle \mathrm{d}_t \mathcal{E}_0 + \mathrm{d}_t \mathcal{E}_1 \rangle_\gamma \partial_{\mathcal{E}} F + \langle \mathrm{d}_t \mu_0 + \mathrm{d}_t \mu_1 \rangle_\gamma \partial_\mu F = 0. \tag{2.129}$$

Evaluating the coefficients, we find $\langle \mathrm{d}_t \mathcal{E}_0 \rangle_\gamma = \langle \mathrm{d}_t \mu_0 \rangle_\gamma = 0$ and

$$\langle \mathbf{U}_0 \rangle_\gamma = \mathbf{v}_\parallel + \mathbf{v}_{E0} + \mathbf{b}_0 \times \left(\mu \nabla B_0 + v_\parallel^2 \kappa \right) / \Omega_0, \tag{2.130}$$

$$\langle \mathrm{d}_t \mathcal{E}_1 \rangle_\gamma = e_s \langle \partial_t \varphi_1 \rangle_\gamma - e_s \langle \mathbf{v} \cdot \partial_t \mathbf{A}_1 \rangle_\gamma, \tag{2.131}$$

$$B_0 \langle \mathrm{d}_t \mu_1 \rangle_\gamma = e_s v_\perp \left\langle \hat{\mathbf{e}}_u \cdot \mathbf{E}_1 \right\rangle_\gamma - e_s v_\perp v_\parallel \left\langle \hat{\mathbf{e}}_r \cdot \mathbf{B}_1 \right\rangle_\gamma, \tag{2.132}$$

$$B_0 \langle \mathbf{U}_1 \rangle_\gamma = \langle \mathbf{E}_1 \rangle_\gamma \times \mathbf{b} + v_\parallel \langle \mathbf{B}_{1\perp} \rangle_\gamma - \langle \hat{\mathbf{e}}_u B_{1\parallel} \rangle_\gamma. \tag{2.133}$$

Since $(\partial_\gamma F)_{\mathbf{R}} = 0$, the GC-distribution $F(\mathbf{R})$ changes only gradually with $\mathbf{R}$, in contrast to the sharp, ρ_{ti}-scale variation of $\mathbf{E}_1$ and $\mathbf{B}_1$.

Since the gyro-kinetic equation, (2.129) with (2.130)–(2.133), is both *non-linear* and *non-local*, its complete numerical solution would require far more computational power,[18] than that of the local, non-linear drift-kinetic equation (2.119). We are thus motivated to examine ways to further simplify (2.129) while retaining the gyro-kinetic ordering (2.120).

As a first step, we thus resort to an *implicit* linearization of (2.129)

$$\partial_t F_1 + \langle \mathbf{U}_0 \rangle_\gamma \cdot \partial_\mathbf{R} F_1 + \langle \mathbf{U}_1 \rangle_\gamma \cdot \partial_\mathbf{R} F_0 + \langle \mathrm{d}_t \mathcal{E}_1 \rangle_\gamma \partial_\mathcal{E} F_0 = 0. \tag{2.134}$$

Note that by omitting high-amplitude, short-scale perturbations ($\Delta_i \ll 1$) we effectively ensure that the plasma remains magnetized ($\delta_i \ll 1$).

The non-local, integral operators arising from (2.121) may be eliminated by assuming the eikonal form $\mathcal{A}_1(\mathbf{x}) \sim \mathrm{e}^{\mathrm{i}(\mathbf{k}\cdot\mathbf{x}-\omega t)}$ and using (2.122)–(2.123) to evaluate the remaining gyro-averages. For instance, (2.131) becomes,

$$\langle \mathrm{d}_t \mathcal{E}_1 \rangle_\gamma = -\mathrm{i}\omega e \left[J_0(k_\perp \rho)(\hat{\varphi} - v_\parallel \hat{A}_\parallel) + J_1(k_\perp \rho) \frac{v_\perp \hat{B}_\parallel}{k_\perp} \right], \tag{2.135}$$

while (2.133) yields $\langle \mathbf{U}_1 \rangle_\gamma = \langle \mathrm{d}_t \mathcal{E}_1 \rangle_\gamma \mathbf{k}_\perp \times \mathbf{b}_0 / eB_0$. This leads to the most popular form of the gyro-kinetic equation, which is both *linear* and *local*,

$$\partial_t F_1 + \mathbf{U}_0 \cdot \partial_\mathbf{R} F_1 + \langle \mathrm{d}_t \mathcal{E}_1 \rangle_\gamma \left(\partial_\mathcal{E} + \frac{\mathbf{k} \times \mathbf{b}}{eB_0} \cdot \nabla \right) F_0 = 0, \tag{2.136}$$

and thus much easier to solve then its non-linear, non-local cousin (2.129).

2.3.2.3 *MHD-kinetic equation*

Finally, we consider the MHD-ordered version of the plasma kinetic equation (2.101). Recall that MHD ordering ($v_E \sim v_{ti}$, $\Delta_i = 0$) assumes strong cross-field flow and neglects finite ρ_{ti} effects. As a result, the derivation of the *MHD-ordered* kinetic equation differs from that of the *drift-ordered* equation in only one tangible point: $\mathcal{E}$, μ and γ must now be defined in the frame moving with the fluid at the centre-of-mass flow velocity $\mathbf{V}$, defined in (2.174); in particular $\mathbf{V}_\perp = \mathbf{v}_E$. Introducing the relative velocity $\mathbf{w} \equiv \mathbf{v} - \mathbf{V} = w_\parallel \mathbf{b} + w_\perp \hat{\mathbf{e}}_u$, we replace (2.53)–(2.54) by

$$\mathcal{E} - e\varphi = \mathcal{K} = \frac{mw^2}{2}, \qquad \mu = \frac{mw_\perp^2}{2B}, \qquad \gamma = \tan^{-1}\left(\frac{w_2}{w_3}\right). \tag{2.137}$$

18 Roughly: memory $\sim 10^7 \delta_i^{-2} \sim 10^{15}$ bytes and speed $\sim 10^{15} \Omega_i \sim 10^{23}$ flops. Assuming Moore's law continues to hold, such computers will become available in 40 years.

A detailed account of MHD-ordered kinetic theory may be found in Rosenbluth and Rostoker (1959). It is convenient to start from (2.95) with $\mathbf{z} = (\mathbf{x}, \mathcal{K}, \mu, \gamma)$, replacing the total energy $\mathcal{E}$ by the kinetic energy $\mathcal{K}$,

$$\partial_t f + (\mathbf{V} + \mathbf{w}) \cdot \nabla f + \mathrm{d}_t \mathcal{K} \partial_{\mathcal{K}} f + \mathrm{d}_t \mu \partial_\mu f + \mathrm{d}_t \gamma \partial_\gamma f = \mathcal{C} f. \tag{2.138}$$

The coefficients may be calculated from (2.137) along the lines of Section 2.2.1. The result is the MHD-ordered version of the (2.62)–(2.64),

$$\mathrm{d}_t \mathcal{K} = -m\mathbf{w} \cdot \left(\mathrm{d}_t \mathbf{V} - \frac{eE_\parallel}{m} \mathbf{b} \right), \qquad \mathrm{d}_t \mu = \partial_t \mu + \mathbf{v} \cdot \nabla \mu - \frac{m\mathbf{w}_\perp \cdot \mathrm{d}_t \mathbf{V}}{B}, \tag{2.139}$$

where $\mathbf{v} = \mathbf{V} + \mathbf{w}$ and $\mathrm{d}_t \mathbf{V} = \partial_t \mathbf{V} + (\mathbf{V} + \mathbf{w}) \cdot \nabla \mathbf{V}$; the angular velocity is easily found as $\mathrm{d}_t \gamma = \Omega[1 + O(\delta_i)]$. Since MHD ordering presupposes $\mathbf{v}_E \cdot \nabla \sim \Omega$, we expect the zeroth-order terms to dominate and can feel justified in omitting $O(\delta_i)$ corrections, i.e. in setting $\rho_{ti} \to 0$. The desired *MHD-ordered kinetic* equation for $F_0 = \overline{f}_0(\mathbf{x}, \mathcal{K}, \mu, t)$ is obtained by gyro-averaging (2.138) and retaining only the lowest-order terms,

$$\partial_t F_0 + (\mathbf{V} + \mathbf{w}_\parallel) \cdot \nabla F_0 + \langle \mathrm{d}_t \mathcal{K} \rangle_\gamma \partial_{\mathcal{K}} F_0 = \mathcal{C} F_0. \tag{2.140}$$

Its evolution depends on a single $\mathbf{v}$-space variable, the kinetic energy $\mathcal{K}$; the ∂_μ-term disappears on account of $\langle \mathrm{d}_t \mu \rangle \sim \delta_i$, while $\mathbf{w}$ acts as a *dependent* variable related to $\mathcal{K}$ by (2.137). The gyro-average of $\dot{\mathcal{K}}$ yields

$$\langle \mathrm{d}_t \mathcal{K} \rangle_\gamma = \frac{e}{m} w_\parallel E_\parallel - w_\parallel \mathbf{b} \cdot \mathrm{d}_t \mathbf{V} - \frac{\mu B}{m} \nabla \cdot \mathbf{V} - \left(w_\parallel^2 - \frac{\mu B}{m} \right) \mathbf{b} \cdot (\mathbf{b} \cdot \nabla) \mathbf{V}. \tag{2.141}$$

In Section 2.4, we'll see that (2.140) may be combined with the zeroth and first velocity moments of (2.92) to achieve rigorous dynamical closure, although the resulting fluid-kinetic hybrid model is rarely used. Instead, MHD ordering is typically implemented using a truncation closure, which leads to the *fluid MHD* model from which the ordering takes its name.

2.4 Fluid description

In Section 2.3 we observed that closure of Maxwell's equations requires only the zeroth and first velocity moments of $f_s(\mathbf{x}, \mathbf{v}, t)$, namely n_s and $\mathbf{V}_s$. The additional information contained in f_s is thus irrelevant, provided that particles respond collectively in $\mathbf{v}$-space, i.e. that plasma motions are too slow to resonate with thermal particles. Under these conditions, fluid description provides a useful shortcut to physical insight, along with a powerful theoretical/computational tool, by collapsing the entire $\mathbf{v}$-space into several thermodynamic variables. We begin by examining these variables and the associated conservation laws before addressing the central problem of fluid theory: how to *close* the infinite sequence of moment equations.

2.4.1 Co-ordinate space conservation laws

The zeroth and first velocity moments, that is the particle density, n_s and flow velocity, $\mathbf{V}_s$, were defined previously in (2.89)–(2.90),

$$n_s(\mathbf{x}, t) = \int f_s d\mathbf{v}, \qquad n_s \mathbf{V}_s(\mathbf{x}, t) = \int f_s \mathbf{v} d\mathbf{v} = n_s \langle \mathbf{v} \rangle_{f_s}. \tag{2.142}$$

Here we introduced the $\mathbf{v}$-space average, $\langle \cdot \rangle_{f_s}$, which we normalize to unity,

$$\langle \mathcal{A} \rangle_{f_s} \equiv \frac{1}{n_s} \int \mathcal{A} f_s(\mathbf{x}, \mathbf{v}, t) d\mathbf{v}, \qquad \langle 1 \rangle_{f_s} = 1. \tag{2.143}$$

Second moments include the *stress tensor*, P_s, *pressure tensor*, p_s, (scalar) pressure, p_s, temperature, $T_s \equiv p_s / n_s$, and energy density, ε_s,

$$\mathsf{P}_s(\mathbf{x}, t) \equiv n_s m_s \langle \mathbf{v}\mathbf{v} \rangle_{f_s}, \quad \varepsilon_s \equiv \tfrac{1}{2}\mathrm{Tr}(\mathsf{P}_s) = \tfrac{3}{2} p_s + \tfrac{1}{2} m_s n_s V_s^2, \tag{2.144}$$

$$\mathsf{p}_s(\mathbf{x}, t) \equiv n_s m_s \langle \mathbf{w}_s \mathbf{w}_s \rangle_{f_s}, \quad p_s \equiv \tfrac{1}{3}\mathrm{Tr}(\mathsf{p}_s) = \tfrac{1}{3} n_s m_s \left\langle w_s^2 \right\rangle_{f_s}, \tag{2.145}$$

where $\mathbf{w}_s \equiv \mathbf{v} - \mathbf{V}_s$ is the relative velocity.

The third moments give rise to the *energy*, Q_s, and *heat*, q_s, *tensors*, and corresponding *energy*, $\mathbf{Q}_s$, and *heat*, $\mathbf{q}_s$, *flux densities*, or *flows*,[19]

$$\mathsf{Q}_s(\mathbf{x}, t) \equiv \tfrac{1}{2} n_s m_s \langle \mathbf{v}\mathbf{v}\mathbf{v} \rangle_{f_s}, \quad \mathsf{q}_s(\mathbf{x}, t) \equiv \tfrac{1}{2} n_s m_s \langle \mathbf{w}_s \mathbf{w}_s \mathbf{w}_s \rangle_{f_s}, \tag{2.146}$$

$$\mathbf{Q}_s(\mathbf{x}, t) \equiv \tfrac{1}{2} n_s m_s \langle v^2 \mathbf{v} \rangle_{f_s}, \quad \mathbf{q}_s(\mathbf{x}, t) \equiv \tfrac{1}{2} n_s m_s \left\langle w_s^2 \mathbf{w}_s \right\rangle_{f_s}. \tag{2.147}$$

The only fourth moments conventionally named are the *energy weighted stress tensor*, Y_s, and the *energy weighted pressure tensor*, y_s,

$$\mathsf{Y}_s(\mathbf{x}, t) \equiv \tfrac{1}{2} n_s m_s \langle v^2 \mathbf{v}\mathbf{v} \rangle_{f_s}, \quad \mathsf{y}_s(\mathbf{x}, t) \equiv \tfrac{1}{2} n_s m_s \langle w_s^2 \mathbf{w}_s \mathbf{w}_s \rangle_{f_s}. \tag{2.148}$$

Here and below, we adopt the convention of denoting *laboratory frame* quantities (stress, energy, etc.) by upper-case symbols and *fluid frame* quantities (pressure, heat, etc.) by lower-case symbols. The two types of quantities are related by a direct expansion of $\langle \mathbf{v}^k \rangle_{f_s}$ with $\mathbf{v} = \mathbf{V}_s + \mathbf{w}_s$,

$$\mathsf{P}_s = \mathsf{p}_s + m_s n_s \mathbf{V}_s \mathbf{V}_s, \qquad \mathbf{Q}_s = \mathbf{q}_s + \mathbf{V}_s \cdot (\varepsilon_s \mathsf{I} + \mathsf{p}_s) \tag{2.149}$$

$$\mathsf{Y}_s = \mathsf{y}_s + \mathbf{V}_s \cdot \left[(\mathbf{Q}_s + \mathbf{q}_s + \mathsf{p}_s \cdot \mathbf{V}_s)\mathsf{I} + \tfrac{1}{2}\mathsf{p}_s \mathbf{V}_s + m_s n_s \langle \mathbf{v}^3 \rangle_{f_s} \right] \tag{2.150}$$

e.g. when $\mathsf{p}_s = p_s \mathsf{I}$ and $V_s \ll v_{ts}$, one finds $\mathbf{Q}_s \approx \mathbf{q}_s + \frac{5}{2} p_s \mathbf{V}_s$.

[19] The term '*flux density*' is often shortened to '*flux*', which can generate confusion, since flux densities have units of $L^{-2}T^{-1}$ while fluxes have units of T^{-1}. We will avoid such elision, preferring the term '*flow*' instead.

Plasma *fluid* equations are most directly obtained from the divergence form of the plasma *kinetic* equation (2.92), to which we add the *source distribution function*, $I_s(\mathbf{x}, \mathbf{v}, t)$, representing atomic or nuclear processes, such as ionization/recombination or nuclear fusion/fission, which can create/destroy particles and add/remove momentum and energy,

$$\partial_t f_s + \nabla \cdot (\mathbf{v} f_s) + \partial_{\mathbf{v}} \cdot (\mathbf{a}_s f_s) = \sum C_{ss'}(f_s, f_{s'}) + I_s. \tag{2.151}$$

We will therefore require velocity moments of both $C_{ss'}(\mathbf{x}, \mathbf{v}, t)$ and $I_s(\mathbf{x}, \mathbf{v}, t)$ which we define in terms of corresponding velocity space integrals,

$$\langle \mathcal{A} \rangle_{C_{ss'}}(\mathbf{x}, t) \equiv \int C_{ss'} \mathcal{A} d\mathbf{v}, \quad \langle \mathcal{A} \rangle_{I_s}(\mathbf{x}, t) \equiv \int I_s \mathcal{A} d\mathbf{v}. \tag{2.152}$$

In addition, we introduce the following notation for most common collisional moments in the laboratory frame: $C_{ss'0} \equiv \langle 1 \rangle_{C_{ss'}}$ and

$$\mathbf{C}_{ss'1} \equiv m_s \langle \mathbf{v} \rangle_{C_{ss'}}, \quad C_{ss'2} \equiv \frac{m_s}{2} \langle v^2 \rangle_{C_{ss'}}, \quad \mathbf{C}_{ss'3} \equiv \frac{m_s}{2} \langle v^2 \mathbf{v} \rangle_{C_{ss'}}, \tag{2.153}$$

and in the fluid frame: $c_{ss'0} \equiv \langle 1 \rangle_{C_{ss'}} = C_{ss'0}$ and

$$\mathbf{c}_{ss'1} \equiv m_s \langle \mathbf{w}_s \rangle_{C_{ss'}}, \quad c_{ss'2} \equiv \frac{m_s}{2} \langle w_s^2 \rangle_{C_{ss'}}, \quad \mathbf{c}_{ss'3} \equiv \frac{m_s}{2} \left\langle w_s^2 \mathbf{w}_s \right\rangle_{C_{ss'}}, \tag{2.154}$$

with similar definitions for $I_{s0}, \mathbf{I}_{s1}, I_{s2}, \ldots$ and $i_{s0}, \mathbf{i}_{s1}, i_{s2}, \ldots$ The lab and fluid frame moments are related via a Galilean transformation, e.g.

$$\mathbf{C}_{ss'1} = \mathbf{c}_{ss'1} + m_s \mathbf{V}_s c_{ss'0}, \tag{2.155}$$

$$C_{ss'2} = c_{ss'2} + \mathbf{V}_s \cdot \mathbf{c}_{ss'1} + \tfrac{1}{2} m_s V_s^2 c_{ss'0}. \tag{2.156}$$

It is customary to assign special symbols to low-order collisional moments. Thus the rate at which species s gain momentum due to collisions with species s' is denoted by $\mathbf{F}_{ss'}$, with the corresponding rate of energy gain in the laboratory frame written as $\mathcal{Q}_{ss'}^L$ and in the fluid frame as $\mathcal{Q}_{ss'}$,

$$\mathcal{Q}_{ss'}^L \equiv C_{ss'2}, \quad \mathcal{Q}_{ss'} \equiv c_{ss'2}, \tag{2.157}$$

$$\mathbf{F}_{ss'} \equiv \mathbf{C}_{ss'1}, \quad \mathcal{Q}_{ss'}^L = \mathcal{Q}_{ss'} + \mathbf{V}_s \cdot \mathbf{F}_{ss'}. \tag{2.158}$$

We define the net friction, $\mathbf{F}_s$, net energy gain, $\mathcal{Q}_s^L$, and net heat gain, $\mathcal{Q}_s$, of species s, and all other moments, $\mathbf{C}_{sk}$, as sums over all species s',

$$\mathbf{F}_s \equiv \sum \mathbf{F}_{ss'}, \quad \mathcal{Q}_s^L \equiv \sum \mathcal{Q}_{ss'}^L, \quad \mathcal{Q}_s \equiv \sum \mathcal{Q}_{ss'}, \quad \mathbf{C}_{sk} \equiv \sum \mathbf{C}_{ss'k}. \tag{2.159}$$

Conservation of particles, momentum and energy in binary collisions,

$$C_{ss'0} = 0, \qquad \mathbf{C}_{ss'1} + \mathbf{C}_{s's1} = 0, \qquad C_{ss'2} + C_{s's2} = 0, \tag{2.160}$$

can now be combined with (2.157)–(2.159) to yield

$$\mathbf{F}_{ss'} + \mathbf{F}_{s's} = 0 = \sum \mathbf{F}_s, \tag{2.161}$$

$$\mathcal{Q}^L_{ss'} + \mathcal{Q}^L_{s's} = 0 = \sum (\mathcal{Q}_s + \mathbf{V}_s \cdot \mathbf{F}_s). \tag{2.162}$$

The first four velocity moments of the *phase space conservation law* (2.151) generate *co-ordinate space conservation laws* for particles, momentum, energy and energy flow, respectively. Multiplying (2.151) in turn by $1, m_s\mathbf{v}, \frac{1}{2}m_s v^2$ and $\frac{1}{2}m_s v^2\mathbf{v}$, integrating over $\mathbf{v}$ and expressing the results in terms of (2.142), (2.144)–(2.148), (2.152)–(2.162), yields the *plasma fluid equations*,[20]

$$\partial_t n_s + \nabla \cdot (n_s \mathbf{V}_s) = I_{s0} \tag{2.163}$$

$$m_s \partial_t (n_s \mathbf{V}_s) + \nabla \cdot \mathsf{P}_s = e_s n_s (\mathbf{E} + \mathbf{V}_s \times \mathbf{B}) + \mathbf{F}_s + \mathbf{I}_{s1} \tag{2.164}$$

$$\partial_t \varepsilon_s + \nabla \cdot \mathbf{Q}_s = (e_s n_s \mathbf{E} + \mathbf{F}_s) \cdot \mathbf{V}_s + \mathcal{Q}_s + I_{s2} \tag{2.165}$$

$$\partial_t \mathbf{Q}_s + \nabla \cdot \mathsf{Y}_s = \tfrac{e_s}{m_s} \left[\varepsilon_s \mathbf{E} + \mathbf{E} \cdot \mathsf{P}_s + \mathbf{Q}_s \times \mathbf{B} \right] + \mathbf{G}_s + \mathbf{I}_{s3} \tag{2.166}$$

All four equations are written in the *divergence form*, in which the *local densities* of particles, n_s, momentum, $m_s n_s \mathbf{V}_s$, energy, ε_s, and energy flux, $\mathbf{Q}_s$, evolve due to divergences of corresponding flows ($n_s\mathbf{V}_s$, P_s, $\mathbf{Q}_s$, Y_s) and due to net volumetric sources, given by the right-hand side terms. These include (i) *atomic/nuclear* sources, $I_{s0}, \mathbf{I}_{s1}, \ldots$, (ii) *collisional* processes, such as inter-species friction, $\mathbf{F}_s$, frictional work, $\mathbf{F}_s \cdot \mathbf{V}_s$, energy exchange, $\mathcal{Q}^L_s$, and energy weighted friction, $\mathbf{G}_s \equiv \mathbf{C}_{s3}$, and (iii) *electromagnetic* effects, such as the Lorentz force on a moving fluid element, $\varrho_s(\mathbf{E} + \mathbf{V}_s \times \mathbf{B})$, the associated work, $\varrho_s \mathbf{E} \cdot \mathbf{V}_s$ and higher-order sources of the energy flow.[21]

Both even and odd moment equations are lowest-order realizations of a general conservation law for the kth moment $\langle \mathbf{v}^k \rangle_{f_s}$, obtained by multiplying (2.151) by a kth rank tensor $\mathbf{v}^k \equiv \mathbf{v}\mathbf{v}\cdots\mathbf{v}$ and integrating over $\mathbf{v}$-space,

$$\partial_t \langle \mathbf{v}^k \rangle_{f_s} + \nabla \cdot \langle \mathbf{v}^{k+1} \rangle_{f_s} = \frac{e_s}{m_s} \{ \mathbf{E} \langle \mathbf{v}^{k-1} \rangle_{f_s} + \langle \mathbf{v}^k \rangle_{f_s} \times \mathbf{B} \}^* + \langle \mathbf{v}^k \rangle_{I_s} + \sum \langle \mathbf{v}^k \rangle_{C_{ss'}}. \tag{2.167}$$

[20] This reduction is analogous to the derivation of equations of hydrodynamics from gas kinetic theory, e.g. Chapman and Cowling (1939); these follow from (2.163)–(2.165) by setting $e_s = 0$.

[21] It is worth noting that *even* moment equations govern the evolution of *scalar* quantities (densities n_s and ε_s) relevant to particle and energy *confinement*, while *odd* moment equations govern the evolution of *vector* quantities (flows $n_s\mathbf{V}_s$ and $\mathbf{Q}_s$) relevant to particle and energy *transport*. Since the former are determined by the divergences of the latter, we expect plasma confinement and transport to be intimately linked.

Here $\{\mathbf{A}\}^*$ denotes a symmetrized form of the tensor $\mathbf{A}$, obtained as a sum of all permutations over one of its indices, which follows from integrating the acceleration term (Hazeltine and Meiss, 1992),

$$\int \partial_{\mathbf{v}} \mathbf{v}^k f_s \mathrm{d}\mathbf{v} = \left\{ \mathsf{I} \langle \mathbf{v}^{k-1} \rangle_{f_s} \right\}^* . \tag{2.168}$$

Tensor contraction, required to obtain scalar quantities from even rank tensors $\mathbf{v}^{2k}$, e.g. $\varepsilon_s \equiv \frac{1}{2}\mathrm{Tr}(\mathsf{P}_s)$ from P_s and (2.165) from (2.167) with $k = 2$, annihilates the gyration term $\Omega_s\{\langle \mathbf{v}^k \rangle_{f_s} \times \mathbf{b}\}^*$ in (2.167). This explains the absence of $\mathbf{B}$ from *even* moment (scalar) equations, cf. (2.163) and (2.165), and indicates that finite $\mathbf{B}$ improves plasma confinement by way of the *odd* moment (vector) equations, i.e. by inhibiting perpendicular transport.

The central problem of plasma fluid theory becomes evident from the order of moments in (2.167): the conservation law for the kth moment introduces the $(k + 1)$th moment, to determine which we need the $(k + 1)$th conservation law, which in turn introduces the $(k + 2)$th moment, and so on. Repeated application of this procedure simply lifts the problem of closure to ever higher rungs of the moment ladder, giving rise to an infinite sequence of moment equations, each one coupled to its immediate neighbours.

There are two general strategies for closing this infinite chain. The first, known as *truncated closure*, consists of either *omitting* the as yet unspecified $(k + 1)$th moments or *prescribing* these via ad-hoc expressions involving lower-order moments. It constitutes a simple but non-rigorous technique which nonetheless, when applied judiciously, has led to some remarkably successful dynamical models. The second strategy, known as *asymptotic* closure, relies on expanding the underlying distribution $f_s(\mathbf{x}, \mathbf{v}, t)$ in some small parameter $\epsilon \ll 1$ and solving the plasma kinetic equation perturbatively to an arbitrary order in ϵ; this solution is then used to estimate the $(k + 1)$th moments, thus closing the dynamical system to a desired level of accuracy.[22] We will discuss such asymptotic fluid closure for a magnetized plasma in Section 2.4.2.

Introducing the *advective derivative*, $\mathrm{d}_{t,s} = \partial_t + \mathbf{V}_s \cdot \nabla$, which measures the rate of change in the fluid frame, (2.163)–(2.165) can be written as

$$\mathrm{d}_{t,s} n_s + n_s \nabla \cdot \mathbf{V}_s = i_{s0}, \tag{2.169}$$

$$m_s n_s \mathrm{d}_{t,s} \mathbf{V}_s + \nabla \cdot \mathsf{p}_s = e_s n_s (\mathbf{E} + \mathbf{V}_s \times \mathbf{B}) + \mathbf{F}_s + \mathbf{i}_{s1}, \tag{2.170}$$

$$\tfrac{3}{2}\mathrm{d}_{t,s} p_s + \tfrac{5}{2} p_s \nabla \cdot \mathbf{V}_s + \boldsymbol{\pi}_s : \nabla \mathbf{V}_s + \nabla \cdot \mathbf{q}_s = \mathcal{Q}_s + i_{s2}, \tag{2.171}$$

[22] Although satisfactory closure can be achieved in just a few iterations (with an estimate of the error being available from the next order in ϵ at the expense of one additional iteration), each successive step entails significant increase in algebraic complexity, which quickly becomes prohibitive. For this reason, the expansion is typically limited to one or two iterations.

Table 2.1. *Number of equations and moments (variables) resulting from frequently encountered fluid closure schemes.*

Quantity	n	$\mathbf{V}$	p	$\boldsymbol{\pi}$	$\mathbf{q}$	$\mathsf{q}-\mathbf{q}$
Moment	$\langle 1\rangle_f$	$\langle \mathbf{v}\rangle_f$	$\langle w^2\rangle_f$	$\langle \mathbf{ww}\rangle_f - \langle w^2\rangle_f$	$\langle \mathbf{w}w^2\rangle_f$	$\langle \mathbf{www}\rangle_f - \langle \mathbf{w}w^2\rangle_f$
5	1	3	1	0	0	0
8	1	3	1	0	3	0
10	1	3	1	5	0	0
13	1	3	1	5	3	0
20	1	3	1	5	3	7

where $\boldsymbol{\pi}_s = \mathsf{p}_s - p_s\mathsf{I}$ is the *viscosity* tensor, containing off-diagonal elements of the pressure tensor and $\nabla\cdot\mathsf{p}_s = \nabla p_s + \nabla\cdot\boldsymbol{\pi}_s$.[23] In deriving (2.169)–(2.171), we made use of the following identities:

$$\tfrac{3}{2}\nabla\cdot(p_s\mathbf{V}_s) + \mathsf{p}_s : \nabla\mathbf{V}_s = \tfrac{5}{2}p_s\nabla\cdot\mathbf{V}_s + \boldsymbol{\pi}_s : \nabla\mathbf{V}_s, \tag{2.172}$$

$$\mathsf{p}_s : \nabla\mathbf{V}_s = \nabla\cdot(\mathbf{V}_s\cdot\mathsf{p}_s) - \mathbf{V}_s\cdot(\nabla\cdot\mathsf{p}_s). \tag{2.173}$$

Fluid conservation laws expressed in the *advective form* (2.169)–(2.171) indicate that aside from atomic/nuclear, collisional and electromagnetic effects previously mentioned, *fluid densities* (n_s, $m_s n_s\mathbf{V}_s$, ε_s) evolve due to (i) fluid compression, $\nabla\cdot\mathbf{V}_s$, including compressional work, (ii) non-uniform pressure/viscous stress and (iii) viscous heating/heat flow.[24] The closure of (2.169)–(2.171) requires that the collisional ($\mathbf{F}_s$, $\mathcal{Q}_s$) and non-Maxwellian ($\boldsymbol{\pi}_s$, $\mathbf{q}_s$) terms be expressed in terms of n_s, $\mathbf{V}_s$ and p_s (see Table 2.1).

In some applications, it is beneficial to transform the dynamical equations into the *centre-of-mass* (CM) frame of reference. We will denote the CM-frame quantities (particle density, $n = n_i$, mass density, ρ, flow velocity, $\mathbf{V}$, charge density, ϱ and current density, $\mathbf{J}$) by omitting the species index,

$$\rho \equiv nm \equiv \sum m_s n_s \approx nm_i, \quad \rho\mathbf{V} \equiv \sum m_s n_s\mathbf{V}_s \approx \rho\mathbf{V}_i, \tag{2.174}$$

$$\varrho = \sum \varrho_s = \sum e_s n_s, \quad \mathbf{J} = \sum \varrho_s\mathbf{V}_s = \sum e_s n_s\mathbf{V}_s, \tag{2.175}$$

where ϱ and $\mathbf{J}$ were previously defined in (2.89)–(2.90). To compute the pressure tensor and the heat flow in the CM-frame, we introduce the CM-frame relative

[23] Note that $\boldsymbol{\pi}_s : \nabla\mathbf{V}_s$ appearing in (2.171) originates in, and is equal to, the symmetric form $\boldsymbol{\pi}_s : \{\nabla\mathbf{V}_s\}^*$, where $\{\nabla\mathbf{V}_s\}^* = \frac{1}{2}\left(\nabla\mathbf{V}_s + \nabla\mathbf{V}_s^\dagger\right)$, see Section 4.1 and Section 6.1.

[24] Note that a pressure gradient accelerates the fluid element due to an imbalance of particles, or their average velocities, entering the element volume from opposite facing sides, i.e. via a *kinematic* rather than a *collisional* mechanism.

velocity, $\mathbf{w} \equiv \mathbf{v} - \mathbf{V}$ and define $\mathbf{p}$ and $\mathbf{q}$ by (2.144) and (2.147) with $\mathbf{w}_s$ replaced by $\mathbf{w}$,

$$\mathsf{p} \equiv \sum n_s m_s \langle \mathbf{w}\mathbf{w} \rangle_{f_s}, \quad \mathbf{q} \equiv \sum \tfrac{1}{2} n_s m_s \langle w^2 \mathbf{w} \rangle_{f_s}. \tag{2.176}$$

The CM-frame scalar pressure, p, temperature, T, and viscosity tensor, π, are defined in a similar fashion, while the CM-frame energy density and energy flow are given by $\varepsilon = \frac{3}{2}p + \frac{1}{2}\rho V^2$ and $\mathbf{Q} = \varepsilon \mathbf{V} + \mathsf{p} \cdot \mathbf{V} + \mathbf{q}$, respectively.

Although similarly defined CM-frame collisional moments cancel exactly due to (2.160), atomic/nuclear reactions can inject momentum and energy into the CM-frame. We define the corresponding CM-frame sources as

$$mI_0 \equiv \sum m_s I_{s0}, \quad \mathbf{I}_1 \equiv \sum \hat{\mathbf{I}}_{s1}, \quad i_2 \equiv \sum \hat{I}_{s2}, \tag{2.177}$$

$$mi_0 \equiv \sum m_s i_{s0}, \quad \mathbf{i}_1 \equiv \sum \hat{\mathbf{i}}_{s1}, \quad i_2 \equiv \sum \hat{i}_{s2}, \tag{2.178}$$

with higher-source moments, $\hat{\mathbf{i}}_{sk}$, and $\mathbf{i}_k$ defined analogously to (2.176).

A derivation similar to that which led to (2.169)–(2.171), concluded by a summation over all species, yields the CM-frame plasma fluid equations,

$$\partial_t \rho + \nabla \cdot \rho \mathbf{V} = mI_0, \tag{2.179}$$

$$\partial_t \rho \mathbf{V} + \nabla \cdot \mathsf{P} = \varrho \mathbf{E} + \mathbf{J} \times \mathbf{B} + \mathbf{I}_1, \tag{2.180}$$

$$\partial_t \varepsilon + \nabla \cdot \mathbf{Q} = \mathbf{J} \cdot \mathbf{E} + I_2, \tag{2.181}$$

here written in divergence form. In advective form, these become

$$\mathrm{d}_t \rho + \rho \nabla \cdot \mathbf{V} = mi_0, \tag{2.182}$$

$$\rho \mathrm{d}_t \mathbf{V} + \nabla \cdot \mathsf{p} = \varrho \mathbf{E} + \mathbf{J} \times \mathbf{B} + \mathbf{i}_1, \tag{2.183}$$

$$\tfrac{3}{2}\mathrm{d}_t p + \tfrac{5}{2} p \nabla \cdot \mathbf{V} + \boldsymbol{\pi} : \nabla \mathbf{V} + \nabla \cdot \mathbf{q} = i_2, \tag{2.184}$$

where the advective derivative, $\mathrm{d}_t = \partial_t + \mathbf{V} \cdot \nabla$, now measures the rate of change in the CM-frame of reference. The above equations differ from the conservation laws for a *neutral gas* only in the appearance of the electromagnetic force on a charged, current carrying fluid element, $\varrho \mathbf{E} + \mathbf{J} \times \mathbf{B}$, and the associated work, $\mathbf{J} \cdot \mathbf{E}$, known as *Joule heating*. Specifically, the magnetic field affects plasma motion only through the perpendicular force $\mathbf{J} \times \mathbf{B}$ in the momentum equation. Since quasi-neutrality requires ϱ to be small, we may conclude that in the lowest approximation, a plasma behaves as a neutral, electrically conducting, magnetized fluid.

2.4.2 Guiding centre fluid theory

In Section 2.3 we saw that the presence of a strong magnetic field ($\delta_s \ll 1$) gives rise to rapid particle gyration, which dominates the plasma *kinetic* equation – recall the discussion of the individual terms appearing in (2.99) and their relative asymptotic magnitudes (2.100). It should come as no surprise that the related *fluid* equations are similarly affected. To illustrate this point, we subject (2.167) to the same type of analysis as that performed for (2.99), which yields asymptotic magnitudes of individual terms in (2.167) normalized by the *gyration term*, $\Omega_s \mathbf{b} \times \langle \mathbf{v}^k \rangle_{f_s}$,

$$\delta_s + (\delta_s + \Delta_s) \sim \left(\delta_s + \frac{v_E}{v_{ts}} + 1 \right) + \delta_s. \tag{2.185}$$

Here, as in (2.99), $\mathbf{E}$ is separated into $\mathbf{E}_\parallel$ and $\mathbf{E}_\perp$ and $\mathbf{I}_s$ is neglected for simplicity. Other than the rearrangement of terms, we effectively recover (2.100). As happens in the kinetic case, different orderings of v_E/v_{ts} and Δ_s give rise to different versions of plasma dynamical equations, with rapid gyration dominating their *drift-ordered* variants.

The divergence term involving the $(k+1)$th moment, which we need to eliminate from (2.167) to close the dynamical system, is found to be $O(\delta_s)$ compared to the leading, kth moment term, which is $O(1)$. This suggests a promising *closure strategy*: (i) estimate the $(k+1)$th moment to order unity, e.g. using the zeroth order, GC-distribution, $\overline{f}_{s0}$; (ii) evaluate the kth moment to $O(\delta_s)$ by isolating the gyration term in (2.167) on the left-hand side and inserting the $(k+1)$th moment from (i),

$$\begin{aligned} \Omega_s \left\{ \langle \mathbf{v}^k \rangle_{f_s} \times \mathbf{b} \right\}^* &= \partial_t \langle \mathbf{v}^k \rangle_{f_s} + \nabla \cdot \langle \mathbf{v}^{k+1} \rangle_{\overline{f}_{s0}} - \frac{e_s}{m_s} \left\{ \mathbf{E} \langle \mathbf{v}^{k-1} \rangle_{f_s} \right\}^* \\ &\quad - \langle \mathbf{v}^k \rangle_{I_s} - \sum \langle \mathbf{v}^k \rangle_{C_{ss'}}; \end{aligned} \tag{2.186}$$

(iii) repeat this procedure to evaluate successively higher moments.

There is one serious deficiency in the above plan: although (2.186) determines the *perpendicular* component of $\langle \mathbf{v}^k \rangle_{f_s}$, it leaves the *parallel* component $\langle \mathbf{v}^k \rangle_{f_s \parallel}$, which represents the homogeneous, non-trivial solution to (2.186), unspecified. Evaluating the latter necessitates knowledge of $f_s(\mathbf{x}, \mathbf{v}, t)$ and thus brings us back to the plasma kinetic equation which the fluid approach promised to circumvent. To make matters worse, (2.186) indicates that the parallel contribution typically dominates, i.e.

$$\langle \mathbf{v}^k \rangle_{f_s \perp} \equiv \{ \langle \mathbf{v}^k \rangle_{f_s} \times \mathbf{b} \}^* \sim O(\delta_s), \quad \langle \mathbf{v}^k \rangle_{f_s} = \langle \mathbf{v}^k \rangle_{f_s \parallel} + O(\delta_s). \tag{2.187}$$

In short, the strategy fails because $\epsilon \sim \delta_s \ll 1$ only constrains *perpendicular* plasma motion leaving *parallel* dynamics largely unaffected. Additional information, the search for which will be the central theme of Chapter 5, is thus needed to adequately close the plasma fluid equations.

Nonetheless, (2.186) represents significant progress. To see why, we write

$$\langle \mathbf{v}^k \rangle_{f_s} = \langle (\mathbf{v}_\parallel + \mathbf{v}_\perp)^k \rangle_{\overline{f}_s + \widetilde{f}_s} = \langle (v_\parallel \mathbf{b} + v_\perp \hat{\mathbf{e}}_u)^k \rangle_{\overline{f}_s + \widetilde{f}_s}, \tag{2.188}$$

where both the gyro-averaged ($\overline{f}_s$) and the γ-dependent ($\widetilde{f}_s$) parts of f_s, as well as the parallel ($\mathbf{v}_\parallel$) and perpendicular ($\mathbf{v}_\perp$) particle velocities, expressed in the ($\mathbf{b}, \hat{\mathbf{e}}_r, \hat{\mathbf{e}}_u$) basis (2.18)–(2.22), are shown explicitly. This form suggests that moments of $\mathbf{v}_\parallel$ depend on $\overline{f}_s$, moments of $\mathbf{v}_\perp$ on $\widetilde{f}_s$ and moments of $\mathbf{v}$ on f_s. Although $\overline{f}_s$ and $\widetilde{f}_s$ are *implicitly* related by the plasma kinetic equation, to express $\widetilde{f}_s$ *explicitly* in terms of $\overline{f}_s$ poses a formidable challenge. These complications are not removed by $\delta_s \ll 1$, which requires $\widetilde{f}_s$ to be small: $\widetilde{f}_s \sim \delta_s \overline{f}_s$ and $f_s = \overline{f}_s + O(\delta_s)$ – recall the derivation of the simplest, *drift-ordered* form of $\widetilde{f}_s$, (2.107). However, ensuring that the gyration term dominates, $\delta_s \ll 1$, allows $\langle \mathbf{v}^k \rangle_{f_s\perp}$ to be computed to $O(\delta_s)$ from the knowledge of f_s to $O(1)$, i.e. from its GC-part, $\overline{f}_s$. Moreover, inserting (2.188) into $\{\langle \mathbf{v}^k \rangle_{f_s} \times \mathbf{b}\}^* = 0$ and (2.187) it can be shown that both $\langle \mathbf{v}^k \rangle_{f_s\parallel}$ and $\langle \mathbf{v}^k \rangle_{f_s\perp}$ can be expressed in terms of $\overline{f}_s$ alone. We thus obtain a fluid theory based entirely on moments of $\overline{f}_s$, i.e. a *GC-fluid theory*, which is the moment equivalent of the *GC-kinetic theory* of Section 2.3.2.

2.4.2.1 *Gyro-tropic tensors*

Let us consider the contracted form of $\langle \mathbf{v}^k \rangle_{f_s\parallel}$, adopting the drift ordering (2.42), such that the right-hand side of (2.186) is $O(\delta_s)$. Solving $\{\langle \mathbf{v}^k \rangle_{f_s} \times \mathbf{b}\}^* = 0$ and exploiting the symmetry properties implied by $\{\cdot\}^*$, one finds that contraction of $\langle \mathbf{v}^k \rangle_{f_s\parallel}$ to a vector yields a parallel flow, while contraction to second rank produces a *gyro-tropic tensor*,

$$\mathsf{A}_{gt} \equiv A_\parallel \mathbf{bb} + A_\perp (\mathsf{I} - \mathbf{bb}), \tag{2.189}$$

whose divergence can be related to the magnetic curvature (2.35) by

$$\nabla \cdot \mathsf{A}_{gt} = \nabla A_\perp + A_\Delta \kappa + \mathbf{B} \nabla_\parallel \left(\frac{A_\Delta}{B} \right), \qquad A_\Delta \equiv A_\parallel - A_\perp. \tag{2.190}$$

For $k = 1$ to 4 we obtain the parallel flow velocity, $\langle \mathbf{v} \rangle_{f_s\parallel} = V_{\parallel s}\mathbf{b}$, the gyro-tropic stress tensor, $n_s m_s \langle \mathbf{v}^2 \rangle_{f_s\parallel} = \mathsf{P}_{gts} = \mathsf{p}_{gts} + O(\delta_s^2)$, the parallel energy flow, $\frac{1}{2} m_s n_s \langle v^2 \mathbf{v} \rangle_{f_s\parallel} = Q_{\parallel s}\mathbf{b}$, and the gyro-tropic energy weighted stress tensor, $n_s m_s \langle v^2 \mathbf{v}^2 \rangle_{f_s\parallel} = \mathbf{Y}_{gts} = \mathbf{y}_{gts} + O\left(\delta_s^2\right)$, respectively. Their scalar components contain velocity moments based entirely on $\overline{f}_s$,

$$V_{\parallel s} = \langle v_\parallel \rangle_{\overline{f}_s}, \qquad Q_{\parallel s} = \tfrac{1}{2} m_s n_s \langle v^2 v_\parallel \rangle_{\overline{f}_s}, \tag{2.191}$$

$$P_{\parallel s} = m_s n_s \left\langle v_\parallel^2 \right\rangle_{\overline{f}_s} \approx p_{\parallel s}, \quad P_{\perp s} = \tfrac{1}{2} m_s n_s \left\langle v_\perp^2 \right\rangle_{\overline{f}_s} \approx p_{\perp s}, \tag{2.192}$$

$$Y_{\parallel s} = m_s n_s \left\langle v^2 v_\parallel^2 \right\rangle_{\overline{f}_s} \approx y_{\parallel s}, \quad Y_{\perp s} = \tfrac{1}{2} m_s n_s \left\langle v^2 v_\perp^2 \right\rangle_{\overline{f}_s} \approx y_{\perp s}. \tag{2.193}$$

In the case of MHD ordering (2.41), the second rank form of $\langle \mathbf{v}^k \rangle_{fs\|}$ is no longer gyro-tropic since the large electric drift $\mathbf{v}_E$ breaks the symmetry between the two gyrating directions via the non-gyro-tropic term $m_s n_s \mathbf{V}_s \mathbf{V}_s$ in the stress tensor P_s, see (2.145). However, this inertial flow is exactly cancelled by the electric field term on the right-hand side of (2.186), which is now large and must be retained while solving $\{\langle \mathbf{v}^k \rangle_{f_s} \times \mathbf{b}\}^* = O(\delta_s)$. Consequently, $n_s m_s \langle \mathbf{v}^2 \rangle_{fs\|} = \mathsf{p}_{gts}$ and $n_s m_s \langle v^2 \mathbf{v}^2 \rangle_{fs\|} = \mathsf{y}_{gts}$, with

$$p_{\|s} = m_s n_s \left\langle w_{\|s}^2 \right\rangle_{\overline{f}_s}, \qquad p_{\perp s} = \tfrac{1}{2} m_s n_s \left\langle w_{\perp s}^2 \right\rangle_{\overline{f}_s}, \tag{2.194}$$

$$y_{\|s} = m_s n_s \left\langle w_s^2 w_{\|s}^2 \right\rangle_{\overline{f}_s}, \qquad y_{\perp s} = \tfrac{1}{2} m_s n_s \left\langle w_s^2 w_{\perp s}^2 \right\rangle_{\overline{f}_s}, \tag{2.195}$$

where $\mathbf{w}_{\|s}$ and $\mathbf{w}_{\perp s}$ are particle velocities relative to $\mathbf{V}_{\|s}$ and $\mathbf{V}_{\perp s}$, respectively. This is the expected result since p_s and y_s are defined in the fluid frame in which the inertial term vanishes and the symmetry between the gyrating directions is restored. In general, second-rank contractions of all fluid frame moments $\left\langle \mathbf{w}_s^k \right\rangle_{fs\|}$, including p_s and y_s, are gyro-tropic irrespective of the ordering scheme for v_E / v_{ts}.

Gyro-tropic tensors are characteristic features of GC-fluid theory. The lowest-order $\overline{f}_s$ which generates such tensors is the bi-Maxwellian distribution f_{2Ms} (A.12), with two distinct temperatures $T_{\|s}$ and $T_{\perp s}$, see Appendix A. In the absence of collisions, $f_s(\mathbf{v})$ is in general non-isotropic so that fluid closure can only be obtained by replacing the isotropic pressure equation by two equations for parallel and perpendicular pressures, known as the *double adiabatic* or *CGL equations* (Chew *et al.*, 1956),

$$\mathrm{d}_{t,s}(T_{\perp s}/B) = 0, \qquad \mathrm{d}_{t,s}\left(T_{\|s} B^2 / n_s^2\right) = 0. \tag{2.196}$$

These equations, which express the conservation of magnetic moment, μ, and energy, $\mathcal{E}$, are derived by multiplying the Vlasov equation (2.93) expressed in terms of integrals of motion, by $m_s w_\perp^2$ and $m_s w_\|^2$ and integrating over $\mathbf{v}$.

In a confined plasma, $\overline{f}_s$ relaxes first to f_{2Ms} (A.12), then to f_{Ms} (A.9), with a single temperature T_s. As a result, all *gyro-tropic* plasma tensors, A_{gt}, become *isotropic*, $A_\|$ and $A_\perp$ converge to a common value A, the anisotropy A_Δ vanishes and the divergence of $\nabla \cdot \mathsf{A}_{gt}$ reduces to ∇A.

The classic example of a gyro-tropic tensor is the *magnetic*, or *Maxwell, stress*, $\mathcal{T}_B$, which follows directly from Ampere's law (2.88),

$$\mathcal{T}_B \equiv -\frac{B^2}{2\mu_0}\mathbf{bb} + \frac{B^2}{2\mu_0}(\mathsf{I} - \mathbf{bb}) = \mathcal{T}_{B\|}\mathbf{bb} + \mathcal{T}_{B\perp}(\mathsf{I} - \mathbf{bb}). \tag{2.197}$$

The Maxwell tensor offers convenient expressions for the Lorentz force on a (quasi-neutral) fluid element $\mathbf{J} \times \mathbf{B} = -\nabla \cdot \mathcal{T}_M$ and the momentum conservation law in the CM-frame, (2.180)

$$\rho \mathrm{d}_t \mathbf{V} + \nabla \cdot (\mathcal{T}_B + \mathsf{p}_s) = \mathbf{i}_1. \tag{2.198}$$

Since $\mathcal{T}_{B\perp} > 0$ ($\mathcal{T}_{B\parallel} < 0$), the magnetic energy density, $B^2/2\mu_0$, increases (decreases) the perpendicular (parallel) stress within the plasma fluid. The former is indicative of *magnetic pressure*, which counteracts compressional deformations, the latter of *magnetic tension*, which counteracts bending and/or twisting deformations.[25]

2.4.2.2 Magnetization law

We next turn to the contracted form of $\langle \mathbf{v}^k \rangle_{f_s\perp}$. Reduced to a vector it yields a perpendicular flow, or *gyro-flow*, of the scalar $\langle v^{k-1} \rangle_{f_s}$. Contracted to second rank it generates a *gyro-stress* tensor whose elements are off-diagonal with respect to (2.189), and represent a *gyro-flow* of the vector $\langle \mathbf{v} v^{k-2} \rangle_{f_s}$. Both originate with particle gyration and thus depend on $\widetilde{f}_s$, but may be expressed in terms of $\overline{f}_s$ using (2.186). Thus, $k = 1$ and $k = 3$ (reduced to a vector) produce gyro-flows of particles and energy,

$$n_s \mathbf{V}_{\perp s} = \frac{\mathbf{b}}{\Omega_s m_s} \times (m_s \partial_t n_s \mathbf{V}_s + \nabla \cdot \mathsf{P}_s - \varrho_s \mathbf{E} - \mathbf{F}_s), \tag{2.199}$$

$$\mathbf{Q}_{\perp s} = \frac{\mathbf{b}}{\Omega_s} \times \left[\partial_t \mathbf{Q}_s + \nabla \cdot \mathsf{Y}_s - \frac{e_s \mathbf{E}}{m_s} \cdot (\varepsilon_s \mathsf{I} + \mathsf{P}_s) - \mathbf{G}_s \right], \tag{2.200}$$

and $k = 2$ and $k = 4$ (reduced to a second-rank tensor) the gyro-stress, $\mathsf{P}_{\perp s} = \{\mathsf{P}_s \times \mathbf{b}\}^*$, and energy weighted gyro-stress, $\mathsf{Y}_{\perp s} = \{\mathsf{Y}_s \times \mathbf{b}\}^*$,

$$\Omega_s \mathsf{P}_{\perp s} = \partial_t \mathsf{P}_s + m_s \nabla \cdot n_s \langle \mathbf{v}^3 \rangle_{f_s} - \varrho_s \{\mathbf{E}\mathbf{V}_s\}^* - \mathbf{C}_{s2}, \tag{2.201}$$

$$\Omega_s \mathsf{Y}_{\perp s} = \partial_t \mathsf{Y}_s + \frac{1}{2} m_s \nabla \cdot n_s \langle v^2 \mathbf{v}^3 \rangle_{f_s} - \frac{e_s}{m_s} \{\mathbf{E}\mathbf{Q}_s\}^* - \mathbf{C}_{s4}. \tag{2.202}$$

Note that for a second-rank tensor $\{\mathsf{A}\}^* = \frac{1}{2}(\mathsf{A} + \mathsf{A}^\dagger)$ where $\mathsf{A}^\dagger$ is a transpose of A. In particular, for two vectors $\mathbf{A}$ and $\mathbf{B}$, $\{\mathbf{AB}\}^* = \frac{1}{2}(\mathbf{AB} + \mathbf{BA})$.

Expanding the divergence in (2.199) and making use of (2.163), we find[26]

$$\mathbf{V}_{\perp s} = \mathbf{V}_E + \frac{\mathbf{b}}{\Omega_s} \times \left(\mathrm{d}_t \mathbf{V}_s + \frac{\nabla \cdot \mathsf{p}_s}{m_s n_s} + \frac{\mathbf{F}_s}{m_s} \right), \tag{2.203}$$

$$= \mathbf{V}_E + \mathbf{V}_{*s} + \mathbf{V}_{ps} + \mathbf{V}_{\pi s} + \mathbf{V}_{\nu s}, \tag{2.204}$$

where $\mathbf{V}_E = \mathbf{E} \times \mathbf{b}/B$ is the *electric* drift, $\mathbf{V}_{*s} = \mathbf{b} \times \nabla p_s / n_s e_s B$ is the (isotropic) *diamagnetic* drift, $\mathbf{V}_{ps} = \mathbf{b} \times m_s \mathrm{d}_t \mathbf{V}_s / e_s B$ is the *polarization* drift, $\mathbf{V}_{\pi s} = \mathbf{b} \times \nabla \cdot$

[25] In this respect a magnetic field line resembles a stretched rubber band, whose lowest energy state is a straight line, and which permits transverse disturbances to propagate along its length with a speed proportional to the square root of the tension-to-linear density ratio. This yields the *shear Alfvén* wave which travels along the field line at the *Alfvén* speed, $V_A^2 = -2\mathcal{T}_{B\parallel}/\rho = B^2/\rho\mu_0$, (2.253), see Section 2.4.2.6.

[26] The total velocity consists of both parallel and perpendicular components, $\mathbf{V}_s = V_{\parallel s}\mathbf{b} + \mathbf{V}_{\perp s}$.

$\pi_s/n_s e_s B$ is the generalized *viscous* drift and $\mathbf{V}_{\nu s} = \mathbf{b} \times \mathbf{F}_s/e_s B$ is the *collisional* drift. Moreover, the gyro-tropic nature of p_s allows us to expand $\nabla \cdot \mathsf{p}_s$ along the lines of (2.190), with the result

$$\mathbf{V}_{\perp s} = \mathbf{V}_E + \mathbf{V}_{ds} + \mathbf{V}_{ps} + \mathbf{V}_{\Delta s} + \mathbf{V}_{\nu s}, \tag{2.205}$$

where $\mathbf{V}_{ds}$ is the proper (anisotropic) *diamagnetic drift*,

$$\mathbf{V}_{ds} \equiv \frac{\mathbf{b} \times \nabla p_{\perp s}}{n_s e_s B} \approx \frac{\mathbf{b} \times \nabla p_s}{n_s e_s B} \equiv \mathbf{V}_{*s}, \tag{2.206}$$

and $\mathbf{V}_{\Delta s} = \mathbf{b} \times p_{\Delta s}\kappa/n_s e_s B \approx \mathbf{V}_{\pi s}$ is the lowest-order gyro-viscous drift.[27]

In the MHD ordering, $\mathbf{V}_E$ clearly dominates, while in the drift ordering, it is augmented by the leading first-order term, that is by $\mathbf{V}_{ds}$,

$$v_E/v_{ts} \sim 1, \quad v_{ds}/v_{ts} \sim \delta_s \Rightarrow \mathbf{V}_{\perp s} = \mathbf{V}_E + O(\delta_s), \tag{2.207}$$

$$v_E/v_{ts} \sim \delta_s, \quad v_{ds}/v_{ts} \sim \delta_s \Rightarrow \mathbf{V}_{\perp s} = \mathbf{V}_E + \mathbf{V}_{ds} + O\left(\delta_s^2\right), \tag{2.208}$$

where we assumed the gyro-flow due to collisional friction to be small. We will return to this and other collisional flows in Chapter 5.

We next insert (2.203) into (2.175), assume quasi-neutrality ($\varrho \approx 0$) and sum over all species to obtain the *gyro-current*, $\mathbf{J}_\perp$, i.e. the flow of charge (in the CM-frame) resulting from particle gyration,

$$\mathbf{J}_\perp \equiv \sum \varrho_s \mathbf{V}_{\perp s} = \mathbf{J}_d + O\left(\delta_i^2\right), \qquad \mathbf{J}_d = \frac{\mathbf{b}}{B} \times \nabla p_\perp, \tag{2.209}$$

where, for simplicity, we assumed the drift ordering, in which the *diamagnetic current*, $\mathbf{J}_d$ dominates.[28] In the above, $p_\perp$ is the total transverse pressure, $p_\perp \approx \sum p_{\perp s} + O\left(\delta_i^2\right)$, which closely approximates the plasma pressure in the CM-frame. The magnetic field, $\mathbf{B}_d = -\frac{1}{2}\beta\mathbf{B}$, induced by $\mathbf{J}_d$, which follows from (2.88), scales linearly with the plasma β (2.12) and is always *diamagnetic*, in analogy to the magnetic moments, $\mathbf{m}_s = -\mu_s\mathbf{b}$, (2.30) of individual particles. Hence the label *diamagnetic drift* assigned to $\mathbf{V}_{ds}$.

Comparison of the fluid velocity (2.203) and the GC-drift (2.28)–(2.33) reveals a number of similarities, e.g. in both cases, the electric and inertial drifts are clearly visible. The remaining terms suggest a link between the pressure gradient on the

[27] The distinction between anisotropic and isotropic diamagnetic drifts ceases to be useful in a confined plasma, in which the velocity distribution and all its even moments ($\mathbf{p}_s, \mathbf{y}_s, \ldots$) are nearly isotropic, see Section 2.4.2.1. In other words, since $\overline{f}_s(\mathbf{v}) = \overline{f}_{Ms}(v; n_s, \mathbf{V}_s, T_s) + O(\delta_s)$ and $p_{\Delta s} \ll p_{\perp s} \approx p_{\parallel s} \approx p$, one finds $\mathbf{V}_{ds} \approx \mathbf{V}_{*s}$. For this reason, the two forms of the diamagnetic flow are used interchangeably in the rest of the book, with the isotropic form ($\mathbf{V}_{*s}, \mathbf{J}_{*s}, \pi_{*s}, \mathbf{q}_{*s}$) preferred, unless the anisotropy of $\overline{f}_s(\mathbf{v})$ is to be retained, in which case the diamagnetic 'd' or gyro-viscous 'gv' labels are retained ($\mathbf{V}_{ds}, \mathbf{J}_{ds}, \pi_{gvs}, \mathbf{q}_{ds}$), see below.

[28] The electric drift does not contribute because $\mathbf{V}_E$ is the same for all plasma species. In analogy with (2.206) we denote the anisotropic and isotropic diamagnetic currents by $\mathbf{J}_d$ and $\mathbf{J}_*$ respectively, where $\mathbf{J}_* = (\mathbf{b}/B) \times \nabla p \approx \mathbf{J}_d$ in a confined plasma, see footnote 27.

one hand and the magnetic drift on the other. This link becomes apparent when we consider the difference between the flow velocity $\mathbf{V}_{\perp s}$ and the average GC-drift $\langle \mathbf{U}_\perp \rangle_{\overline{f}_s}$ with $\mathbf{U}_\perp = \mathbf{v}_E + \mathbf{U}_{1\perp} + O\left(\delta_s^2\right)$ given by (2.33). In performing the $\mathbf{v}$-space average one must be careful to properly treat the parallel velocity $U_\parallel \mathbf{b}$ in the GC-advective derivative, $\mathrm{d}_t = \partial_t + \mathbf{U}_0 \cdot \nabla$, which should not be confused with its fluid counterpart, $\mathrm{d}_t = \partial_t + \mathbf{V} \cdot \nabla$. In the final subtraction, several cancellations ensue (known collectively as the *diamagnetic cancellation*), leaving a single remainder,

$$\mathbf{V}_{\perp s} = \langle \mathbf{U}_\perp \rangle_{\overline{f}_s} - \frac{1}{\varrho_s} \nabla \times \left(\frac{p_{\perp s} \mathbf{b}}{B} \right), \qquad \varrho_s \equiv e_s n_s, \tag{2.210}$$

The term inside the curl represents a single-species *magnetization*, $\mathbf{M}_s$, describing the density of magnetic moments, $\mathbf{m}_s$, (2.30),

$$\mathbf{M}_s \equiv n_s \langle \mathbf{m}_s \rangle_{\overline{f}_s} = -n_s \langle \mu_s \mathbf{b} \rangle_{\overline{f}_s} = -\frac{n_s m_s}{2B} \left\langle w_{\perp s}^2 \right\rangle_{\overline{f}_s} \mathbf{b} = -\frac{p_{\perp s} \mathbf{b}}{B}. \tag{2.211}$$

Equation (2.210) can thus be written in the remarkably simple form,

$$\varrho_s \mathbf{V}_s = \varrho_s \mathbf{V}_{GCs} + \nabla \times \mathbf{M}_s, \qquad \mathbf{V}_{GCs} \equiv \langle \mathbf{v}_{GC} \rangle_{\overline{f}_s} = \langle \mathbf{U} \rangle_{\overline{f}_s}. \tag{2.212}$$

This is the celebrated *magnetization law*, which states that the *plasma fluid flow* consists of the flow of guiding centres and the *magnetization flow* (the curl of the density of magnetic moments). The magnetization law is an exact relation and can be generalized to higher moment gyro-flows as follows,

$$\varrho_s \langle \mathbf{v} v^{k-1} \rangle_{f_s} = \varrho_s \langle \mathbf{U}_\perp v^{k-1} \rangle_{\overline{f}_s} + \nabla \times \mathbf{M}_{k,s}, \tag{2.213}$$

where $\mathbf{M}_{k,s}$ is a kth moment magnetization; (2.213) applies to any intensive thermodynamic variable which is a function of $\langle \mathbf{v}^k \rangle_{f_s}$.

It is worth noting that (2.212) follows directly from *Stoke's law* of vector calculus, which states that the surface integral of a curl is equal to the path integral around the surface boundary. It was first established in magneto-statics to decompose the net current into the movement of free charges $\mathbf{J}_{free}$ and the magnetization current $\nabla \times \mathbf{M}$ due to a gradient of fixed magnetic moments, see Jackson (1975). In plasmas, guiding centres take the role of the former and particle gyration the role of the latter. Performing a summation of (2.212) over all species, we indeed find the magneto-static law, now expressed in terms of plasma variables,

$$\mathbf{J} = \mathbf{J}_{GC} + \nabla \times \mathbf{M}, \qquad \mathbf{M} \equiv \sum \mathbf{M}_s. \tag{2.214}$$

2.4.2.3 Diamagnetic cancellation

At this point, we distinguish between four types of perpendicular flow, see Table 2.2. *Gyro-flow* refers to the total perpendicular flow $\langle \mathbf{v}^k \rangle_{f_s \perp}$, such as $n_s \mathbf{V}_{\perp s}$.

Table 2.2. *Four types of perpendicular flow in a magnetized plasma.*

Label	Definition	Example
gyro-flow	net $\perp$ plasma flow	$\mathbf{V}_{\perp s} = \mathbf{U}_{\perp s} + \nabla \times \mathbf{M}_s = \mathbf{V}_E + \mathbf{V}_d$
GC-flow	$\perp$ flow due to GC-drifts	$\mathbf{U}_{\perp s} = \mathbf{V}_E + \mathbf{U}_{Bs}$
magnetization flow	non-GC $\perp$ flow	$\nabla \times \mathbf{M}_s = \mathbf{V}_{ds} - \mathbf{U}_{Bs}$
diamagnetic flow	$\perp$ flow due to $\nabla p_\perp$	$\mathbf{V}_{ds} = \nabla \times \mathbf{M}_s - \mathbf{U}_{Bs}$

GC-flow is the part of the gyro-flow resulting from the movement of guiding centres, which alone affects the fluid conservation laws. *Magnetization flow* is the solenoidal (divergence free) part of the gyro-flow, which can be written as the curl of some product involving $\mathbf{M}_s$, e.g. $\nabla\times\mathbf{M}_s$ for particles (recall that $\nabla\cdot\nabla\times = 0$). By (2.212), magnetization flow does not involve GC-motion and being divergence free plays no role in fluid conservation laws. Finally, *diamagnetic flow* is the part of the gyro-flow containing gradients of $\langle\mu_s\rangle_{f_s} = p_{\perp s}$. Guiding centres do not experience these gradients, i.e. do not drift in response to $\nabla p_{\perp s}$, which are consequently absent from $n_s\mathbf{V}_{GCs}$. The term *diamagnetic drift* is thus somewhat of a misnomer: $\mathbf{V}_{ds}$ is not a GC-drift, but a purely macroscopic, fluid velocity.[29] Unlike the magnetization flow, it does in general posses a non-vanishing divergence, which then drives a parallel *return flow*. It is worth noting that being normal to the gradients that drive it, diamagnetic flow does not alter these gradients, i.e. it has a non-dissipative character.

It is instructive to compute $\mathbf{V}_{\perp s} = \langle\mathbf{v}_{\perp s}\rangle_{\widetilde{f}_s}$ directly from drift-ordered GC-kinetic theory.[30] We begin by inserting $\widetilde{f}_s$ (2.107) into $\langle\cdot\rangle_{\widetilde{f}_s}$, make use of $\langle\mathbf{A}\cdot\hat{\mathbf{e}}_r\rangle_{f_s} = \frac{1}{2}\mathbf{b}\times\mathbf{A}$ and evaluate by parts the integral containing ∂_μ,

$$\begin{aligned}\mathbf{V}_{\perp s} = \langle\mathbf{v}_\perp\rangle_{\widetilde{f}_s} &= \frac{\mathbf{b}}{2\Omega_s}\times\left\langle v_\perp^2\right\rangle_{\nabla\overline{f}_s + e_s\mathbf{b}\times\mathbf{U}_{\perp s}\partial_\mu\overline{f}_s},\\ &= \frac{\mathbf{b}}{2\Omega_s}\times\left\langle v_\perp^2\right\rangle_{\nabla\overline{f}_s} + \langle v_\parallel\rangle_{\overline{f}_s\partial_\mu(\mu\mathbf{U}_{\perp s}/v_\parallel)}.\end{aligned} \tag{2.215}$$

Computing the above averages with $\mathbf{U}_{\perp s}$ given by (2.40) is laborious but straightforward and, after simplification, leads to the following expression,

$$\begin{aligned}\varrho_s\mathbf{V}_s &= \varrho_s\mathbf{V}_{\parallel s} + \frac{\mathbf{b}}{B}\times\left[\varrho_s\nabla\varphi + (p_{\parallel s} + p_{\perp s})\frac{\nabla B}{B}\right] + \frac{p_{\perp s}}{B}\cdot\frac{\nabla\times\mathbf{B}}{B}\\ &\quad + \frac{p_{\Delta s}}{B}\cdot\frac{(\nabla\times\mathbf{B})_\perp}{B} - \nabla\times\left(\frac{p_{\perp s}\mathbf{b}}{B}\right).\end{aligned} \tag{2.216}$$

[29] Unfortunately, the term is deeply rooted in the literature (e.g. in labels like *drift-waves*) and any attempt to rename it would generate unnecessary confusion.

[30] A good account of the calculation, which continues the development leading up to (2.114), may be found in Mikhailovskii (1974).

The first four terms correspond to the average GC-drift $\langle \mathbf{U}_\perp \rangle_{\overline{f}_s}$ with $\mathbf{U}_{\perp s}$ given by (2.58), and include both high-β and parallel drift contributions,

$$n_s \mathbf{V}^{high\beta}_{GC\perp s} = \frac{p_{\parallel s}}{e_s B} \cdot \frac{(\nabla \times \mathbf{B})_\perp}{B}, \quad n_s \mathbf{V}_{GC\parallel s} = \frac{p_{\perp s}}{e_s B} \cdot \frac{(\nabla \times \mathbf{B})_\parallel}{B}. \tag{2.217}$$

Since the remaining term is just the magnetization flow $\nabla \times \mathbf{M}_s$, we exactly recover (2.212). Similarly, expanding the curl, we obtain (2.208),

$$\mathbf{V}_s = \mathbf{V}_{\parallel s} + \frac{\mathbf{b}}{B} \times \left(\nabla \varphi + \frac{\nabla p_{\perp s}}{\varrho_s} \right) = \mathbf{V}_{\parallel s} + \mathbf{V}_E + \mathbf{V}_{ds}. \tag{2.218}$$

In short, the gyro-flow $n_s \mathbf{V}_{\perp s}$ computed from GC-fluid theory is identical to the GC-kinetic result. Although drift ordering was assumed in each case, the same agreement can be demonstrated for MHD-ordered flows. The effort expanded in the kinetic calculation, which was considerable, bears witness to the power and economy of the fluid approach.

The meaning of the diamagnetic cancellation should now be clear: the average GC-drifts related to the non-uniformity of $\mathbf{B}$ (the three terms in (2.216) in which ∇B and $\nabla \times \mathbf{B}$ appear explicitly) exactly cancel with identical and opposite terms originating from $\nabla \times \mathbf{M}_s$, leaving only the diamagnetic contribution, $n_s \mathbf{V}_{ds}$. In short, $\nabla \times \mathbf{M}_s$ converts the *guiding centre* (*microscopic*) $\nabla \mathbf{B}$ drift, $\mathbf{U}_B$, into the *fluid* (*macroscopic*) $\nabla p_{\perp s}$ flow, $\mathbf{V}_d$, i.e.

$$\mathbf{V}_{\perp s} = \mathbf{U}_{\perp s} + \nabla \times \mathbf{M}_s = (\mathbf{V}_E + \mathbf{U}_{Bs}) + (\mathbf{V}_{ds} - \mathbf{U}_{Bs}) = \mathbf{V}_E + \mathbf{V}_{ds}. \tag{2.219}$$

For a confined plasma, for which $\overline{f}_s = \overline{f}_{Ms} + O(\delta_s)$, (2.216) reduces to

$$\varrho_s \mathbf{U}_{\perp s} = \left(\frac{e_s \nabla \varphi}{T_s} + 2\frac{\nabla B}{B} \right) \times \mathbf{M}_s + M_s \frac{\nabla \times \mathbf{B}}{B}, \tag{2.220}$$

$$\nabla \times \mathbf{M}_s = \left(\frac{\nabla p_s}{p_s} - 2\frac{\nabla B}{B} \right) \times \mathbf{M}_s - M_s \frac{\nabla \times \mathbf{B}}{B}, \tag{2.221}$$

$$\varrho_s \mathbf{V}_{\perp s} = \left(\frac{e_s \nabla \varphi}{T_s} + \frac{\nabla p_s}{p_s} \right) \times \mathbf{M}_s, \quad \mathbf{M}_s = -M_s \mathbf{b} = -\frac{p_s \mathbf{b}}{B}. \tag{2.222}$$

The fact that $\mathbf{M}_s$ appears in all expressions has led to some confusion in the terminology of transverse flows, which we tried to avoid by defining all terms explicitly. The reader is encouraged to verify the above expressions directly by noting that $\partial_\mu f_{Ms} = 0$ and evaluating $\langle \cdot \rangle_{\nabla f_{Ms}}$ in (2.215) using (A.10)–(A.11). Inspection suffices to show that (2.212) is indeed satisfied.[31]

[31] Note that by definition, (2.209), the gyro-current $\mathbf{J}_\perp = \mathbf{J}_d \approx \mathbf{J}_* = B^{-1}\mathbf{b} \times \nabla p$ is normal to both $\mathbf{b}$ and ∇p and therefore flows on *magnetic flux surfaces*, defined by magnetic lines of force, yet perpendicular to the local magnetic field, see Chapter 3. The direction $\mathbf{J}_*/J_*$ defines the *diamagnetic* direction $\hat{\mathbf{e}}_\wedge$ and is frequently referred to as the *ion drift* direction since $\hat{\mathbf{e}}_\wedge = \mathbf{V}_{*i}/V_{*i} = \mathbf{J}_*/J_*$; the *electron drift* direction is defined as $\mathbf{V}_{*e}/V_{*e} = -\hat{\mathbf{e}}_\wedge$.

2.4.2.4 Diamagnetic energy flow

The analysis performed above for $\mathbf{V}_{\perp s}$ is easily extended to higher moment gyro-flows. Thus, the gyro-flow of energy (2.200) may be expressed as

$$\mathbf{Q}_{\perp s} = \mathbf{V}_E \cdot (\varepsilon_s \mathsf{I} + \mathsf{P}_s) + \frac{\mathbf{b}}{\Omega_s} \times (\mathrm{d}_t \mathbf{Q}_s + \nabla \cdot \mathsf{y}_s + \mathbf{V}_s \cdot \nabla \cdot \mathsf{Z}_s). \qquad (2.223)$$

Here we neglected the collisional term (which will be treated in Chapter 5) and combined the compressional and viscous terms to form the tensor Z_s,

$$\mathsf{Z}_s = (\mathbf{q}_s + \mathsf{p}_s \cdot \mathbf{V}_s)\mathsf{I} + \tfrac{1}{2}\mathsf{p}_s \mathbf{V}_s + m_s n_s \left\langle \mathbf{w}_s^3 \right\rangle_{f_s} \qquad (2.224)$$

$$= (\mathbf{q}_s + \mathsf{p}_s \cdot \mathbf{V}_s)\mathsf{I} - \left(\tfrac{5}{2}\mathsf{p}_s + m_s n_s \mathbf{V}_s^2\right) \mathbf{V}_s + m_s n_s \langle \mathbf{v}^3 \rangle_{f_s}. \qquad (2.225)$$

Neglecting $O\left(\delta_s^2\right)$ terms and exploiting the gyro-tropic nature of y_s, we find

$$\mathbf{Q}_{\perp s} = \mathbf{Q}_{Es} + \frac{\mathbf{b}}{\Omega_s} \times (\nabla y_{\perp s} + y_{\Delta s} \boldsymbol{\kappa}), \qquad (2.226)$$

which in the case of MHD and drift ordering, respectively, reduces to

$$v_E/v_{ts} \sim 1 \Rightarrow \quad \mathbf{Q}_{\perp s} = \mathbf{Q}_{Es} + O(\delta_s), \qquad (2.227)$$

$$v_E/v_{ts} \sim \delta_s \Rightarrow \quad \mathbf{Q}_{\perp s} = \mathbf{Q}_{Es} + \mathbf{Q}_{ds} + O\left(\delta_s^2\right), \qquad (2.228)$$

with $\mathbf{Q}_{Es}$ and $\mathbf{Q}_{ds}$ denoting the *electric* and *diamagnetic* energy flows,

$$\mathbf{Q}_{Es} \equiv \mathbf{V}_{Es} \cdot [(\varepsilon_s + p_s)\mathsf{I} + \boldsymbol{\pi}_s + m_s n_s \mathbf{V}_s \mathbf{V}_s], \qquad \mathbf{Q}_{ds} \equiv \frac{\mathbf{b} \times \nabla y_{\perp s}}{\Omega_s}. \qquad (2.229)$$

Approximating $\overline{f}_s$ by f_{2Ms}, (2.195) may be evaluated directly to yield

$$\frac{y_{\perp s}}{p_{\perp s}} = \frac{y_{\parallel s}}{p_{\parallel s}} = \frac{5}{2}\frac{T_s}{m_s}, \qquad \mathbf{b} \times \nabla \cdot \mathbf{y}_s = \frac{5\mathbf{b}}{2m_s} \times (T_s \nabla \cdot \mathsf{p}_s + p_{\perp s} \nabla T_s). \qquad (2.230)$$

In the case of drift ordering (2.42), we can neglect the inertial term, $m_s n_s \mathbf{V}_s \mathbf{V}_s$, and use (2.145) to decompose $\mathbf{Q}_{\perp s}$ into the convective energy flow, $\frac{5}{2} p_s \mathbf{V}_{\perp s}$, and the (non-dissipative) *diamagnetic heat flow*, $\mathbf{q}_{ds}$,

$$\mathbf{Q}_{\perp s} = \frac{5}{2} p_s \left(\mathbf{V}_E + \frac{\mathbf{b} \times \nabla \cdot \mathsf{p}_s}{n_s e_s B} \right) + \frac{5}{2} \frac{p_{\perp s}}{e_s B} \mathbf{b} \times \nabla T_s = \frac{5}{2} p_s \mathbf{V}_{\perp s} + \mathbf{q}_{ds}. \qquad (2.231)$$

For a confined plasma, $f_{2Ms} \to f_{Ms}$, so that $y_{\perp s} = y_{\parallel s} = \frac{5}{2} p_s T_s / m_s$ and[32]

$$\mathbf{Q}_{Es} = \frac{5}{2} p_s \mathbf{V}_E, \quad \mathbf{Q}_{*s} = \frac{5}{2} \frac{\mathbf{b}}{e_s B} \times \nabla(p_s T_s), \quad \mathbf{q}_{*s} = \frac{5}{2} \frac{p_s}{e_s B} \mathbf{b} \times \nabla T_s. \qquad (2.232)$$

[32] Here we replace the anisotropic symbol 'd' by the isotropic '*' in line with the earlier notation.

The energy variant of the *magnetization law*, which states that the net energy flow equals the *advective* flow due to the movement of guiding centres and the *magnetization* flow due to gradients in the density and energy of magnetic moments, is easily obtained by substituting (2.212) into (2.231),

$$e_s\mathbf{Q}_s = e_s\mathbf{Q}_{GCs} + \nabla \times \mathbf{M}_{3s}, \quad \mathbf{Q}_{GCs} = \tfrac{5}{2}p_s\mathbf{V}_{GCs}, \quad \mathbf{M}_{3s} = \tfrac{5}{2}T_s\mathbf{M}_s. \qquad (2.233)$$

2.4.2.5 *Gyro-viscous stress tensor*

The extension of the above analysis from odd to even moments, i.e. from *gyro-flows* to *gyro-stresses*, is complicated by the tensorial nature of the latter. For reasons of economy, we must forgo a detailed exposition,[33] focusing instead on the salient features of the gyro-stress, $\mathsf{P}_{\perp s}$. Analysis of (2.201) reveals this tensor to be symmetric, traceless and purely perpendicular ($\mathbf{b}\cdot\mathsf{P}_{\perp s}\cdot\mathbf{b} = 0$), which points to its origin in the viscosity tensor, $\boldsymbol{\pi}_s \equiv \mathsf{p}_s - \mathsf{p}_{gts}$ composed of off-diagonal elements of p_s; note the extension of our earlier, isotropic definition $\boldsymbol{\pi}_s = \mathsf{p}_s - p_s\mathsf{I}$ to the more general, gyro-tropic case. The stress and pressure tensors may now be written as

$$\mathsf{P}_s = \mathsf{p}_s + m_s n_s \mathbf{V}_s\mathbf{V}_s, \qquad \mathsf{p}_s = p_s\mathsf{I} + (\mathsf{p}_{gts} - p_s\mathsf{I}) + \boldsymbol{\pi}_s, \qquad (2.234)$$

where the first, second and third terms in p_s are $O(1)$, $O(\delta_s)$ and $O\left(\delta_s^2\right)$, respectively; $m_s n_s \mathbf{V}_s\mathbf{V}_s$ is $O(1)$ in MHD ordering and $O\left(\delta_s^2\right)$ in drift ordering. Inverting $\{\mathsf{P}_s \times \mathbf{b}\}^* = \mathsf{P}_{\perp s}$, the stress tensor P_s is found as

$$\mathsf{P}_s = \mathsf{P}_{\parallel s} + \mathsf{P}_s^0 = \mathsf{p}_{gts} - \tfrac{1}{4}\left[\mathbf{b} \times \mathsf{P}_{\perp s} + 3\mathbf{b}\mathbf{b} \times (\mathbf{b}\cdot\mathsf{P}_{\perp s})\right], \qquad (2.235)$$

where $\mathsf{P}_{\parallel s} = \mathsf{p}_{gts}$ is the homogeneous solution to (2.201), given by (2.194), while P_s^0 has the following components in the $(\mathbf{b}, \hat{\mathbf{e}}_2, \hat{\mathbf{e}}_3)$ basis,

$$\mathsf{P}_s^0 = \frac{1}{4}\begin{pmatrix} 0 & 4P_{\perp s13} & 4P_{\perp s12} \\ 4P_{\perp s13} & 4P_{\perp s23} & P_{\perp s33} - P_{\perp s22} \\ 4P_{\perp s12} & P_{\perp s33} - P_{\perp s22} & -2P_{\perp s23} \end{pmatrix}. \qquad (2.236)$$

We defer the discussion of the collisional gyro-stress associated with $\mathbf{C}_{s2}$ to Chapter 5, and concentrate on the part of P_s^0 generated by $\langle \mathbf{v}^3 \rangle_{f_s}$ and $\mathbf{E}\mathbf{V}_s$,

$$\Omega_s\{\mathbf{b} \times \mathsf{P}_{gvs}\}^* = m_s\nabla\cdot n_s\langle \mathbf{v}^3\rangle_{f_s} - \varrho_s\{\mathbf{E}\mathbf{V}_s\}^*, \qquad (2.237)$$

which we denote by P_{gvs}. This tensor represents the so-called *gyro-viscous stress*, which is the second-rank tensorial equivalent of the *diamagnetic flow*, and describes the non-dissipative, non-solenoidal flow of momentum due to gradients in the density and/or energy of magnetic moments. To evaluate $\langle \mathbf{v}^3 \rangle_{f_s}$, we need first-order expressions for $\widetilde{f}_s$ and $\overline{f}_s$. For the former, we take the leading term in

[33] A good account may be found in Braginskii's famous article (Braginskii, 1965).

(2.107), for the latter, the first term in the Taylor expansion of a slowly moving Maxwellian, $f_{Ms}(\mathbf{v} - \mathbf{V}_{\|s})$, with the result

$$f_{1s} = \overline{f}_{1s} + \widetilde{f}_{1s} = \frac{2V_{\|}v_{\|}}{v_{ts}^2} f_{Ms} - \mathbf{r} \cdot \nabla f_{Ms}, \tag{2.238}$$

and $f_s = f_{Ms} + f_{1s} + O\left(\delta_s^2\right)$ – the standard *ansatz* of drift-ordered fluid theory, see Section 2.4.2.7. Performing the averages reveals that $\langle \mathbf{v}^3 \rangle_{f_s}$ is a linear function of T_s, $V_{\|}$, $\mathbf{V}_{\perp s} = \mathbf{V}_E + \mathbf{V}_{*s}$ and $\mathbf{q}_{\perp s} = \mathbf{q}_{*s}$. The desired gyro-stress can be computed by inserting the result into (2.237), and inverting the $\{\cdot\}^*$ operator. The solution is conveniently expressed in terms of the *gyro-viscosity* tensor, $\boldsymbol{\pi}_{*s}$, defined as to exclude the diagonal, inertial term[34]

$$\mathsf{P}_{*s} = \mathsf{p}_{*s} + m_s n_s \mathbf{V}_s \mathbf{V}_s, \qquad \mathsf{p}_{*s} = \boldsymbol{\pi}_{*s} - \tfrac{1}{2} m_s n_s V_s^2 \mathsf{I}, \tag{2.239}$$

$$\mathsf{p}_s = p_s \mathsf{I} + \boldsymbol{\pi}_{*s}, \qquad \boldsymbol{\pi}_{*s} = \boldsymbol{\pi}_{\perp s} + \{\mathbf{b}\boldsymbol{\pi}_{\|s}\}^*, \tag{2.240}$$

where $\boldsymbol{\pi}_{\perp s}$ is the perpendicular gyro-viscosity (tensor),

$$\begin{aligned} \Omega_s \boldsymbol{\pi}_{\perp s} = {} & \tfrac{1}{4} p_s \{(\mathbf{b} \times \nabla)\mathbf{V}_{\perp s} + \nabla(\mathbf{b} \times \mathbf{V}_{\perp s})\}^*, \\ & + \tfrac{1}{10}\{(\mathbf{b} \times \nabla)\mathbf{q}_{\perp s} + \nabla(\mathbf{b} \times \mathbf{q}_{\perp s})\}^*, \end{aligned} \tag{2.241}$$

and $\boldsymbol{\pi}_{\|s}$ is the parallel gyro-viscous flow (vector),

$$\Omega_s \boldsymbol{\pi}_{\|s} = p_s[\mathbf{b} \times \nabla V_{\|s} + \mathbf{b} \cdot \nabla(\mathbf{b} \times \mathbf{V}_{\perp s})] + \tfrac{1}{5}\mathbf{b} \cdot \nabla(\mathbf{b} \times \mathbf{q}_{\perp s}). \tag{2.242}$$

We thus find that gyro-viscous stress P_{gvs} originates with the non-uniformity of the diamagnetic flows of particles and heat.[35]

In summary, $\delta_s \ll 1$ provides a viable strategy for determining *perpendicular* projections of velocity moments, i.e. gyro-flows and gyro-stresses, but offers little help in calculating *parallel* flows and stresses, such as $p_{\|s}$, $q_{\|s}$ or $\boldsymbol{\pi}_{\|s}$, for which additional information about the GC-distribution is required. Invariably, parallel GC-dynamics and the associated parallel plasma transport are the weakest link in the fluid closure chain. In the final two sections, we explore the MHD- and drift-ordered versions of plasma fluid dynamics.

2.4.2.6 MHD-ordered fluid equations: magneto-hydrodynamics

The MHD ordering, (2.41) and (2.207), reduces the burden of closure by ensuring that all gravitational drifts, or more generally all $O(\delta_s)$ terms, can be neglected.

[34] Again we replace the anisotropic symbol 'gv' by the isotropic '*' in line with the earlier notation.

[35] The above results should in general be modified to include magnetic curvature effects, which can have important consequences in toroidal plasmas.

This is formally equivalent to an assumption of zero gyro-radius ($\rho_{ts} = \delta_s = 0$), such that the MHD flow velocity becomes,

$$\mathbf{V} = \mathbf{V}_s = V_\parallel \mathbf{b} + \mathbf{V}_E, \qquad V_\parallel = V_{\parallel i} = V_{\parallel e}. \tag{2.243}$$

Note that $V_\parallel$ is common to both species as required by quasi-neutrality ($n = n_i = n_e/Z$) and weak parallel electric field, $E_\parallel \sim \delta_s E_\perp$. One might naively suspect the gyro-current (2.209) to vanish in the $\delta_s \to 0$ limit. However, since the electric charge can be written as $e_s = m_s v_{ts}/\delta_s B$, one finds that $\mathbf{J}_\perp \sim \delta_s^{-1}\mathbf{V}_\perp$. As a result, small flows, $\mathbf{V}_{*s} \sim O(\delta_s)$ generate large currents, $\mathbf{J}_* \sim O(1)$, which must be retained in MHD-ordered plasma dynamics. Fortunately, Ampere's law, (2.87) can be used to express $\mathbf{J}$ in terms of $\mathbf{B}$, such that the $O(\delta_s)$ contribution to $\mathbf{V}_\perp$ need not be computed explicitly.

Taking the cross product of (2.243) with $\mathbf{B}$, we find

$$\mathbf{E} + \mathbf{V} \times \mathbf{B} = \mathbf{E}' = 0. \tag{2.244}$$

This relation, known as *ideal MHD Ohm's law*, states that the electric field in the fluid frame, $\mathbf{E}'$ (recall the Lorentz transformation), must vanish in MHD flows. Since $\mathbf{E}' = 0$ for $|\mathbf{J}| > 0$ implies vanishing *resistivity*, $\eta \equiv \mathbf{E}'/\mathbf{J} = 0$, we conclude that ideal MHD describes a perfectly conducting fluid.

There exist various strategies of closing the MHD-ordered moment equations, i.e. of expressing $\boldsymbol{\pi}_s$ and $\mathbf{q}_s$ in terms of lower moments. Below we consider two limiting cases. The most complicated (and the only fully rigorous) approach, already mentioned in Section 2.3.2.3, solves the problem of closure by replacing the fluid energy equation with the MHD-ordered plasma kinetic equation, (2.140), thus specifying the CM-frame parallel and transverse pressures, $p_\parallel$ and $p_\perp$. The resulting *kinetic-fluid hybrid*, differs from ideal MHD (developed below), in that it evolves $p_\parallel$, $p_\perp$ and $E_\parallel$ (via quasi-neutrality) in place of p and $\mathbf{J}$. Unfortunately, rigour comes at the price of complexity, so that kinetic-fluid MHD is rarely used in practice.

At the other extreme, the simplest approach adopts the assumption of *local thermodynamic equilibrium* (LTE): $f_s(\mathbf{x}, \mathbf{v}) = f_{Ms}(\mathbf{x}, \mathbf{w}_s)$, see Appendix A, such that both $\boldsymbol{\pi}_s$ and $\mathbf{q}_s$ vanish exactly, and p_s becomes isotropic,

$$\boldsymbol{\pi}_s(f_{Ms}) = 0 = \mathbf{q}_s(f_{Ms}), \qquad \mathsf{p}(f_{Ms}) = p_s \mathsf{I} = n_s T_s \mathsf{I}. \tag{2.245}$$

Hence, the LTE assumption justifies a *truncation* closure in which higher moments are simply omitted. It also motivates the *single fluid* approach, since adiabatic expansion (implied by LTE) guarantees that p_i/p_e remains constant.[36] Applying (2.245) to the CM-frame fluid conservation laws (2.179)-(2.181), yields a set of

[36] It should be noted, however, that the single fluid approach is not a defining feature of MHD.

moment equations involving ρ, $\mathbf{V}$, p, $\mathbf{B}$ and $\mathbf{J}$. Combined with quasi-neutrality, Ampere's and Faraday's laws, (2.87), and the ideal MHD Ohm's law (2.244), these form a closed dynamical system known as *single fluid, ideal MHD*,

$$\mathrm{d}_t\rho + \rho\nabla\cdot\mathbf{V} = mi_0, \qquad \mathbf{E} + \mathbf{V}\times\mathbf{B} = 0, \tag{2.246}$$

$$\rho\mathrm{d}_t\mathbf{V} + \nabla p = \mathbf{J}\times\mathbf{B} + \mathbf{i}_1, \qquad \nabla\times\mathbf{B} = \mu_0\mathbf{J}, \tag{2.247}$$

$$\mathrm{d}_t p + \tfrac{5}{3}p\nabla\cdot\mathbf{V} = \tfrac{2}{3}i_2, \qquad \nabla\times\mathbf{E} + \partial_t\mathbf{B} = 0. \tag{2.248}$$

The above equations imply charge conservation, Coulomb's law and solenoidal magnetic field, i.e. they ensure that $\mathbf{J}$, $\mathbf{E}$ and $\mathbf{B}$ remain divergence free at all times. The resulting flow is fully reversible, adiabatic and isentropic, as required by the absence of viscosity, heat flow and collisional dissipation. This explains the factor of 5/3 in (2.248), which represents the *polytropic exponent* for an adiabatic, ideal gas,

$$\mathrm{d}_t(p\rho^{-5/3}) = 0 = \mathrm{d}_t(Tn^{-2/3}). \tag{2.249}$$

It is noteworthy that setting $\mathbf{B} = 0$ in (2.246)–(2.248), we exactly recover *Euler's equations* describing an adiabatic, neutral fluid. For finite $\mathbf{B}$, (2.246)–(2.248) may be simplified by eliminating $\mathbf{E}$ and $\mathbf{J}$, which, neglecting the volumetric sources, reduces to a set of equations for ρ, p, $\mathbf{V}$ and $\mathbf{B}$,

$$\mathrm{d}_t\rho + \rho\nabla\cdot\mathbf{V} = 0, \qquad \rho\mathrm{d}_t\mathbf{V} + \nabla\left(p + \frac{B^2}{2\mu_0}\right) = \frac{\mathbf{B}}{\mu_0}\cdot\nabla\mathbf{B}, \tag{2.250}$$

$$\mathrm{d}_t p + \tfrac{5}{3}p\nabla\cdot\mathbf{V} = 0, \qquad \mathrm{d}_t\mathbf{B} = (\partial_t + \mathbf{V}\cdot\nabla)\mathbf{B} = \mathbf{B}\cdot\nabla\mathbf{V}. \tag{2.251}$$

The last expression, which may be recast as $\partial_t\mathbf{B} = \nabla\times(\mathbf{V}\times\mathbf{B})$, and which implies (2.244), leads to the characteristic property of ideal MHD, namely *magnetic flux conservation*. This property states that the magnetic flux, Ψ, through any surface $S(t)$ moving with the fluid remains constant,

$$\mathrm{d}_t\int_{S(t)}\mathbf{B}\cdot\mathrm{d}\mathbf{S} = \mathrm{d}_t\Psi = 0, \tag{2.252}$$

The proof can be found in most elementary texts. It follows from the general form of Kelvin's vorticity theorem, which states that $\mathrm{d}_t\int_S \boldsymbol{Q}\cdot\mathrm{d}\mathbf{S} = 0$ for any vector field satisfying $\partial_t\boldsymbol{Q} = \nabla\times(\mathbf{V}\times\boldsymbol{Q})$, see Section 4.1. It is conventional to speak of the flux as being *frozen into* the plasma flow, but one can equally envision the plasma as being frozen into individual magnetic *flux tubes*.[37] The latter become de facto plasma *filaments*, such that mass, energy and magnetic flux are preserved

[37] We define the term *magnetic flux tube* as an infinitesimal tube surrounding a given field line.

within the same volume.[38] As a result, magnetic field lines and fluid elements are inextricably tied and evolve in unison.

As already mentioned, the combination of magnetic tension, $-\mathcal{T}_{B\parallel}$, (2.198) and (ion) inertia gives rise to the basic dynamical mode of ideal MHD: the *shear Alfvén* wave, see Section 4.2. In this mode, disturbances propagate along the field lines at the *Alfvén speed*, V_A, which is just the phase velocity of a transverse wave travelling along a 'magnetic' string,

$$V_A^2 = -\frac{2\mathcal{T}_{B\parallel}}{\rho} = \frac{B^2}{\rho\mu_0}, \qquad V_S^2 = \frac{5}{3}\frac{p}{\rho}, \qquad \frac{V_S^2}{V_A^2} = \frac{5}{3}\frac{\mu_0 p}{B^2} \approx \beta. \tag{2.253}$$

Here V_S is the plasma sound speed and β is given by (2.12).[39] An ideal MHD system evolves on the *Alfvénic time scale*, τ_A, which for low-beta is much shorter than the *sonic time scale*, τ_S,

$$\tau_A \equiv \frac{L_\parallel}{V_A} = \frac{L_\parallel\sqrt{\rho\mu_0}}{B}, \qquad \tau_S \equiv \frac{L_\parallel}{V_S}, \qquad \frac{\tau_A}{\tau_S} \equiv \frac{\omega_S}{\omega_A} = \frac{V_S}{V_A} \approx \sqrt{\beta}. \tag{2.254}$$

Flux conservation forbids changes in magnetic topology, i.e. field lines cannot be broken, flux tubes cannot be pierced, flux surfaces cannot be torn apart. This symmetry is broken by any finite amount of dissipation, typically introduced by replacing (2.244) with a *resistive MHD Ohm's law*,

$$\mathbf{E} + \mathbf{V} \times \mathbf{B} = \mathbf{E}' = \eta\mathbf{J}, \tag{2.255}$$

where η is assumed to be finite and positive, and by replacing the adiabatic value 5/3 with a general *polytropic exponent*, γ, with $5/3 > \gamma > 1$,

$$\mathrm{d}_t p + \gamma p \nabla \cdot \mathbf{V} = 0 \quad \Rightarrow \quad \mathrm{d}_t(p\rho^{-\gamma}) = 0 = \mathrm{d}_t(Tn^{1-\gamma}). \tag{2.256}$$

Single fluid, resistive MHD is given by (2.246)–(2.248) with the energy equation and the ideal Ohm's law replaced by (2.256) and (2.255), respectively. As a result, a diffusive term appears in (2.251),

$$\partial_t \mathbf{B} = \nabla \times (\mathbf{V} \times \mathbf{B}) + D_\eta \nabla^2 \mathbf{B}, \qquad D_\eta \equiv \eta_\parallel/\mu_0, \tag{2.257}$$

where the magnetic diffusivity, D_η, measures the rate at which field lines diffuse relative to the moving plasma. The intimate link between flux and flow is thus severed (the *frozen* flux is allowed to *thaw*), permitting field lines to move relative to the plasma, and the plasma to gradually drift out of magnetic flux tubes.

[38] Likewise, we define the term *plasma filament* as a flute-like perturbation (4.53) of plasma thermodynamic quantities, which is localized in the drift plane and whose characteristic length scales along the field are much longer than those across, e.g. a Gaussian disturbance of the form $\exp(-|\mathbf{k}\cdot\mathbf{x}|^2)$ with $k_\parallel \ll k_\perp$.

[39] Note that $V_A \gg V_S$ in low-beta plasmas, such as those found in large aspect ratio tokamaks, where $\beta < 0.03$, see Section 4.2.

Consequently, one must distinguish between *(magnetic) flux tubes* and *(plasma) filaments*, which are no longer tied by a strictly one-to-one mapping. This has dramatic effects on magnetic topology: besides the ability to bend and twist, field lines can now be broken and reconnected, allowing new topologies to be realized. This process of *magnetic reconnection* occurs on the *resistive time scale*, τ_η, which is typically much slower than either the Alfvénic or sonic time scales, $\tau_\eta \gg \tau_S \gg \tau_A$, with

$$\tau_\eta \equiv \frac{L_\perp^2}{D_\eta} = \frac{\mu_0 L_\perp^2}{\eta_\parallel}, \qquad \text{Lu} \equiv \frac{\tau_\eta}{\tau_A} = \frac{L_\perp^2 V_A}{L_\parallel D_\eta}, \tag{2.258}$$

i.e. the *Lundquist number*, Lu, is typically much larger than one.

2.4.2.7 *Drift-ordered fluid equations: the drift-fluid model*

As noted previously, the MHD ordering is rarely pertinent to confined fusion plasmas, which generally evolve on time scales representative of the drift ordering (2.42). To derive the drift-ordered dynamical equations, it thus becomes necessary to retain all $O(\delta_i)$-terms in (2.169)–(2.171); $O(\delta_e)$ corrections are typically ignored on account of $\delta_e \ll \delta_i$. The new terms, which involve gyro-flows and gyro-stresses and can be viewed as *finite gyro-radius* corrections to the zero gyro-radius description used in MHD, have two important consequences for plasma dynamics:

(i) In what is known as the *drift-fluid correction*, they allow ions and electrons to drift across magnetic field lines, effectively decoupling the evolution of plasma filaments and magnetic flux tubes. In confined plasmas, this leads to the appearance of *drift-waves*, which propagate perpendicular to $\mathbf{B}$ with a velocity $\mathbf{V}_{*i} \approx \mathbf{V}_{di}$ (2.206) and oscillate at the *diamagnetic drift frequency*,

$$\mathbf{k} \cdot \mathbf{V}_{*s} \equiv \omega_{*s} \approx \omega_{ds} \equiv \mathbf{k} \cdot \mathbf{V}_{ds}. \tag{2.259}$$

(ii) In what is known as the *gyro-fluid correction*, they smear out ion related quantities over the extent of a thermal ion gyro-radius,

$$\mathcal{A}_i \to (1 - \rho_{ti}^2 \nabla_\perp^2)\mathcal{A}_i, \qquad \mathcal{A}_i = \{\varphi, n_i, \mathbf{V}_i, p_i, \mathbf{q}_i, \ldots\}, \tag{2.260}$$

which follows from gyro-averaging at fixed $\mathbf{R}$, as in *gyro-kinetic* theory, (2.121). Corrections (i) and (ii) are $O(\delta_i)$ and $O\left(\delta_i^2\right)$, respectively.

It should be stressed that the ultimate reason for retaining finite ρ_i terms is not rapid *spatial* variation but slow *temporal* variation, implied by the drift ordering, (2.42), i.e. the plasma is assumed to remain strongly magnetized ($\delta_i \ll 1$), but to evolve slowly enough such that $O(\delta_i)$ terms become significant. Due to this slowness, the inductive contribution to the perpendicular electric field can be neglected and the electric drift velocity becomes purely electrostatic, recall the discussion in Section 2.2,

$$\mathbf{E}_\perp \approx -\nabla\varphi, \qquad \mathbf{V}_E = \mathbf{E}\times\mathbf{b}/B \approx (\mathbf{b}/B)\times\nabla\varphi. \tag{2.261}$$

In terms of MHD phenomena, this amounts to a quasi-static radial force balance maintained by fast compressional wave dynamics, see Section 4.2.

Let us consider the impact of the drift ordering on the plasma fluid equations, which must now be treated separately for ions and electrons due to the charge dependence in $\mathbf{V}_{*s} \approx \mathbf{V}_{ds}$, (2.206), and $\mathbf{q}_{*s} \approx \mathbf{q}_{ds}$, (2.229). In the drift-fluid approximation, however, one can retain a single conservation law for the plasma density $n = n_i = n_e/Z$,

$$\mathrm{d}_t n + n\nabla\cdot\mathbf{V} = i_0, \qquad \mathbf{V}\approx\mathbf{V}_i = \mathbf{V}_{\parallel i} + \mathbf{V}_E + \mathbf{V}_{*i}, \tag{2.262}$$

where $\mathrm{d}_t = \partial_t + \mathbf{V}\cdot\nabla$ and $\mathbf{V}$ is the CM-frame flow velocity, which, aside from a small (m_e/m_i) correction, is equal to $\mathbf{V}_i$. Taking advantage of the diamagnetic cancellation, we rewrite the above continuity equation as

$$\mathrm{d}_{t,0} n + n\nabla\cdot\mathbf{V}_0 + \nabla p_i\cdot\nabla\times(\mathbf{b}/eB) = i_0, \quad \mathbf{V}_0 = \mathbf{V}_\parallel + \mathbf{V}_E, \tag{2.263}$$

where $\mathbf{V}_0$ is the zeroth-order, or MHD, flow velocity, (2.243), and $\mathrm{d}_{t,0} \equiv \partial_t + \mathbf{V}_0\cdot\nabla$ is the corresponding advective derivative. As a result, $\nabla\cdot n\mathbf{V}_{*i}$ disappears from (2.262) and is replaced by the $\nabla\times(\mathbf{b}/B)$ term. The electron flow velocity $\mathbf{V}_e$ is conveniently expressed in terms of $\mathbf{V}$ and $\mathbf{J}$ as

$$\mathbf{V}_e = \mathbf{V}_{\parallel e} + \mathbf{V}_E + \mathbf{V}_{*e}, \qquad \mathbf{J} = \sum e_s n_s \mathbf{V}_s \approx e n_e(\mathbf{V}-\mathbf{V}_e). \tag{2.264}$$

We next turn to the ion momentum equation, (2.170), in which the gyro-viscous tensor, $\boldsymbol{\pi}_{*i} \approx \boldsymbol{\pi}_{gvi}$, must now be included. This tensor was previously calculated as (2.240)–(2.242) by Taylor expanding a slowly drifting Maxwellian distribution, $f_{Mi}(\mathbf{v}-\mathbf{V})$, with $\mathbf{V}\approx\mathbf{V}_i \sim O(\delta_i)v_{ti}$,

$$f_i = f_{Mi} + \overline{f}_{1i} + \widetilde{f}_{1i} + O\left(\delta_i^2\right) = \left(1 + \frac{2V_\parallel v_\parallel}{v_{ti}^2}\right) f_{Mi} - \mathbf{r}\cdot\nabla f_{Mi} + O\left(\delta_i^2\right). \tag{2.265}$$

Inserting the above expressions for $\mathbf{V}$ and $\boldsymbol{\pi}_{*i}$ into (2.170), we find that the divergence of $\boldsymbol{\pi}_{*i}$ combines with the troublesome ion diamagnetic velocity, $\mathbf{V}_{*i}$, to generate a gyro-correction, χ_g, to the ion pressure, p_i,

$$\rho\mathrm{d}_t\mathbf{V}_{*i} + (\nabla\cdot\boldsymbol{\pi}_{*i})_\perp = -\nabla_\perp\chi_g, \tag{2.266}$$

$$\rho\mathbf{V}_{*i}\cdot\nabla V_{\parallel i} + (\nabla\cdot\boldsymbol{\pi}_{*i})_\parallel = -2\nabla_\parallel\chi_g, \tag{2.267}$$

$$\chi_g = p_i\Omega_\parallel/2\Omega_i, \quad \Omega_\parallel \equiv \mathbf{b}\cdot\nabla\times\mathbf{V}_\perp, \quad \mathbf{V}_\perp = \mathbf{V}_E + \mathbf{V}_{*i}, \tag{2.268}$$

where $2\chi_g/p_i$ is just the parallel *vorticity*, $\Omega_\parallel$, normalized by Ω_i.[40] This reduction, known as the *gyro-viscous cancellation*, is the gyro-stress analogue of the *diamagnetic cancellation* of Section 2.4.2.3. Consequently, $\boldsymbol{\pi}_{*i}$ does not appear explicitly in the resulting ion momentum equations,

$$\rho \mathrm{d}_t \mathbf{V}_E + \nabla_\perp (p - \chi_g) = \mathbf{J} \times \mathbf{B} + \mathbf{i}_{1\perp}, \tag{2.269}$$

$$\rho \partial_t \mathbf{V}_\parallel + (\mathbf{V}_E + \mathbf{V}_\parallel) \cdot \nabla \mathbf{V}_\parallel + \nabla_\parallel (p - 2\chi_g) = \mathbf{i}_{1\parallel}, \tag{2.270}$$

where $\rho \approx m_i n$ and $p = p_i + p_e$ are the CM-frame density and pressure. Combining (2.262) and (2.268), we find that χ_g amounts to a finite gyro-radius correction anticipated in (2.260), i.e. $p_i \rightarrow p_i \left(1 - \rho_{ti}^2 \nabla_\perp^2\right)$.

The ion energy equation can be closed by noting that (2.265) implies (i) $\mathbf{q}_{\parallel i} = 0$ as in (2.245), (ii) $\mathbf{q}_{\perp i} = \mathbf{q}_{*i}$ as in (2.232) and (iii) $\mathsf{p}_i = p_i \mathsf{I} + \boldsymbol{\pi}_{*i}$ as in (2.240); since $\pi_{*i} \sim O\left(\delta_i^2\right)$ and $\mathbf{V} \sim O(\delta_i)$, the viscous heating term $\boldsymbol{\pi}_i : \nabla \mathbf{V} \sim O\left(\delta_i^3\right)$ can be neglected. Once again, the drifting Maxwellian (LTE) assumption justifies a truncation closure, this time in the parallel direction, i.e. as in kinetic-fluid MHD, p_{gti} is calculated by solving the plasma kinetic equation. We thus find the simple result,

$$\tfrac{3}{2} \mathrm{d}_t p_i + \tfrac{5}{2} p_i \nabla \cdot \mathbf{V} + \nabla \cdot \mathbf{q}_{*i} = i_{2i}. \tag{2.271}$$

The diamagnetic flows $\mathbf{V}_{di}$ and $\mathbf{q}_{di}$ again combine into a $\nabla \times (\mathbf{b}/B)$ term,

$$\mathrm{d}_{t,0} p_i + \tfrac{5}{3} p_i \nabla \cdot \mathbf{V}_0 + \tfrac{5}{3} T_i \nabla p_i \cdot \nabla \times (\mathbf{b}/eB) = \tfrac{2}{3} i_{2i}. \tag{2.272}$$

We next consider the electron momentum and energy equations. Making use of (2.264), the former can be replaced by Faraday's and Ampere's laws,

$$\nabla \times \mathbf{E} + \partial_t \mathbf{B} = 0, \qquad \nabla \times \mathbf{B} = \mu_0 \mathbf{J}, \tag{2.273}$$

where the displacement current $\partial_t \mathbf{E}/c^2$ is neglected as usual. To close the energy equation, it is tempting to invoke the LTE assumption for electrons, as we have done for ions, (2.265). Unfortunately, LTE implies $\mathbf{q}_{\parallel s} = 0$, which, due to lower electron mass and higher electron mobility, leads to inconsistencies in the description of parallel electron dynamics. We are thus forced to retain a first-order deviation away from LTE and thus a finite parallel heat flow, $\mathbf{q}_{\parallel e}$, such that the electron energy equation becomes

$$\tfrac{3}{2} \mathrm{d}_{t,e} p_e + \tfrac{5}{2} p_e \nabla \cdot \mathbf{V}_e + \nabla \cdot (\mathbf{q}_{\parallel e} + \mathbf{q}_{*e}) = i_{2e}, \quad \mathrm{d}_{t,e} = \partial_t + \mathbf{V}_e \cdot \nabla. \tag{2.274}$$

[40] Here and elsewhere, upper case Ω is used for both the gyro-frequency and the vorticity; to avoid confusion, the latter will always appear as a component of a vector, e.g. $\Omega_\parallel$ for parallel projection of $\boldsymbol{\Omega} = \nabla \times \mathbf{V}$, while the former will be accompanied only by a species index, e.g. $\Omega_i = eB/m_i$. This unfortunate doubling up of notation is due to a historical clash of conventions between plasma physics and hydrodynamics.

As we will find in Chapter 5, there is no universally valid expression for the parallel heat flux in terms of local values of n_s, $\mathbf{V}_s$ and p_s; in the long mean free path limit ($\nu_e^* \ll 1$), parallel transport becomes essentially non-local and such an expression ceases to be useful. However, it is possible to avoid computing $\mathbf{q}_{\parallel e}$ explicitly. To this end, we introduce the field line average, which for ergodically covered surfaces is equal to the flux surface average,[41]

$$\langle \mathcal{A} \rangle_B \equiv \frac{\oint \mathcal{A} B^{-1} \mathrm{d}l}{\oint B^{-1} \mathrm{d}l}, \qquad \langle \mathbf{B} \cdot \nabla \mathcal{A} \rangle_B = \langle B \nabla_\parallel \mathcal{A} \rangle_B = 0, \tag{2.275}$$

where l measures the distance along the field line, and we take note that the average of $\mathbf{B} \cdot \nabla \mathcal{A}$ must vanish for any single-valued quantity $\mathcal{A}$. Since $\nabla \cdot \mathbf{q}_{\parallel e} = \mathbf{B} \cdot \nabla (q_{\parallel e}/B)$, the field line average of (2.274) yields

$$\left\langle \mathrm{d}_{t,e} p_e + \tfrac{5}{3} p_e \nabla \cdot \mathbf{V}_e + \tfrac{2}{3} \nabla \cdot \mathbf{q}_{*e} - \tfrac{2}{3} i_{2e} \right\rangle_B = 0. \tag{2.276}$$

The above expression assumes *closed* field lines and should not be used when the field lines are *open*, that is when they penetrate a solid surface.

Equations (2.273)–(2.276) provide the basic description of electron dynamics in the drift-fluid model but leave $E_\parallel$ and p_e unspecified. The additional constraints can be obtained from the conservation laws for electron momentum, (2.164), and for energy flow, (2.166),

$$\mathbf{E} + \mathbf{V}_e \times \mathbf{B} \approx \mathbf{E} + (\mathbf{V} - \mathbf{J}/en_e) \times \mathbf{B} \approx (\mathbf{F}_e - \nabla p_e)/en_e, \tag{2.277}$$

$$e\mathbf{q}_e \times \mathbf{B} = m_e \mathbf{G}_e - \tfrac{5}{2} T_e (\mathbf{F}_e + n_e \nabla T_e) = T_e \left(\mathbf{H}_e - \tfrac{5}{2} n_e \nabla T_e \right), \tag{2.278}$$

where we neglected terms of order m_e/m_i, used $\mathsf{p}_e = p_e \mathsf{I}$ to evaluate $\mathbf{Y}_e$ and introduced the normalized *heat weighted friction*, $\mathbf{H}_s$, defined as

$$\mathbf{Y}_e = \tfrac{5}{2} p_e v_{te}^2 \mathsf{I}, \qquad \mathbf{H}_s \equiv \mathbf{G}_s / v_{ts}^2 - \tfrac{5}{2} \mathbf{F}_s. \tag{2.279}$$

The first equation, (2.277), is a drift-ordered version of MHD Ohm's law (2.255), with collisional friction $\mathbf{F}_e$ being proportional to the resistivity, η, and the appearance of an additional, so-called *Hall term*, $(\mathbf{J} \times \mathbf{B} - \nabla p_e)/en_e$, representing adiabatic electron response,

$$\mathbf{E}' = \mathbf{E} + \mathbf{V} \times \mathbf{B} \approx (\mathbf{F}_e + \mathbf{J} \times \mathbf{B} - \nabla p_e)/en_e. \tag{2.280}$$

The second equation, (2.278), has no analogue in MHD, where the LTE assumption ensures zero heat flow. As expected, perpendicular projections of (2.277)–(2.278)

[41] Since the element $\mathrm{d}l/B$ is the specific flux tube volume, the lower integral in (2.275) represents the average flux tube volume covering a given flux surface, see Chapter 3.

reproduce earlier expressions for $\mathbf{V}_{*e}$ and $\mathbf{q}_{*e}$, whereas their parallel projections yield the additional constraints required for closure,

$$en_e E_\parallel = F_{\parallel e} - \nabla_\parallel p_e, \qquad H_{\parallel e} = G_{\parallel e}/v_{te}^2 - \tfrac{5}{2}F_{\parallel e} = \tfrac{5}{2}n_e \nabla_\parallel T_e, \tag{2.281}$$

which in the absence of collisions ($\nu_e^* \ll 1$) reduce to

$$\mathbf{F}_e = \mathbf{G}_e = \mathbf{H}_e = 0, \qquad en_e E_\parallel + \nabla_\parallel p_e = 0, \qquad \nabla_\parallel T_e = 0. \tag{2.282}$$

We thus find that $T_e = p_e/n_e = \text{const}$ along the magnetic field lines (as expected from efficient electron free-streaming) and that the density profile follows the *Maxwell–Boltzmann* distribution of statistical mechanics,[42]

$$\nabla_\parallel (\ln n_e - e\varphi/T_e) = 0 \quad \Rightarrow \quad n/n_0 = n_e/n_{e0} = \exp(e\varphi/T_e). \tag{2.283}$$

The latter is valid provided φ changes *adiabatically* ($\omega_\varphi \equiv \partial_t \ln \varphi \ll \omega_{te}$), such that electrons have ample time to adjust to local variations in $E_\parallel = -\nabla\varphi$ and thus evolve via a succession of quasi-equilibria.[43]

Examining the drift-ordered expressions for $\mathbf{V}_s$, p_s and $\mathbf{q}_s$ used above

$$\mathbf{V}_i = \mathbf{V}_{\parallel i} + \mathbf{V}_E + \mathbf{V}_{*i}, \; \mathbf{q}_i = \mathbf{q}_{\perp i} = \mathbf{q}_{*i}, \quad \mathsf{p}_i = p_i \mathsf{I} + \boldsymbol{\pi}_{*i}, \tag{2.284}$$

$$\mathbf{V}_e = \mathbf{V}_i - \mathbf{J}/en, \qquad \mathbf{q}_e = \mathbf{q}_{\parallel e} + \mathbf{q}_{*e}, \quad \mathsf{p}_e = p_e \mathsf{I}, \tag{2.285}$$

we note the general absence of dissipation from these flows (recall our earlier discussion of $\mathbf{V}_{*s}$, $\mathbf{q}_{*s}$ and $\boldsymbol{\pi}_{*s}$). Dissipative flows, which relax the thermodynamic gradients and thereby increase the local entropy density, appear only with the inclusion of collisional effects, i.e. when $\nu_e^* > 0$.

The set of equations for n, p_i, p_e, $\mathbf{V}_E(\varphi)$, $V_\parallel$ and $J_\parallel(V_{\parallel e})$, (2.262)–(2.278), is known as the *six-field drift-fluid* model.[44] A reduced version, known as the *four-field drift-fluid* model, assumes cold ions and isothermal electrons ($T_i = 0$, $T_e = \text{const}$), and thus follows only n, φ, $V_\parallel$ and $J_\parallel$. Unlike *single-fluid* MHD discussed earlier, the drift-fluid model is an archetypal *two-fluid* description, in which plasma behaviour is characterized by ion inertia (*hydrodynamic ions*) and electron mobility (*adiabatic electrons*). Since $\mathbf{V}_E$ is the same for all species, the MHD ordering (2.41) facilitates a single-fluid description. In contrast, $\mathbf{V}_{*s}$ has opposite signs for ions and electrons and different magnitude for each plasma species, thus motivating a multi-fluid description. However, it must be stressed that the single- or multi-fluid character is not the defining feature of the MHD- vs. drift-ordered dynamics, which are defined solely on the bases of (2.41) and (2.42).

[42] For small perturbations, this leads to a linear relation between $\tilde{n}$ and $\tilde{\varphi}$, i.e. $\tilde{n}/n_0 \approx e\tilde{\varphi}/T_{e0}$.
[43] It is customary to describe the resulting electron response itself as being *adiabatic*.
[44] Note that φ specifies $\mathbf{V}_E$ and $\Omega_\parallel$, while $J_\parallel$ specifies $V_{\parallel e}$, $A_\parallel$ and $\mathbf{B}_\perp$, e.g. see Section 4.2.5.

2.4.2.8 Simplified drift-fluid models

There are two further, widely used simplifications of the drift-fluid model: the *two-field* (n, φ) or *Hasegawa–Wakatani* model, and the *single-field* (φ) or *Hasegawa–Mima* model. The former is the simplest system which captures the *unstable* non-linear drift-wave dynamics, and so the resulting turbulence and transport, while the latter is the simplest system exhibiting non-linear, yet *stable*, drift-wave evolution. The two models follow directly from the four-field model derived in the preceding section under additional simplifications (uniform magnetic field, no parallel ion flow, scalar viscosity, etc.); the reduction is sufficiently simple, and so widely available in the literature, that it can be left as an exercise for the reader. Below we simply present the results, adopting the so-called *Bohm* normalization,[45] see (6.219),

$$t\Omega_i \to t, \qquad \mathbf{x}/\rho_S \to \mathbf{x}, \qquad e\varphi/T_e \to \varphi, \qquad n/n_0 \to n, \tag{2.286}$$

where $\rho_S = c_{Se}/\Omega_i$ is the ion gyro-radius at the cold ion sound speed, $c_{Se}^2 = T_e/m_i$, which is the characteristic spatial scale of drift-wave dynamics.

The *two-field, Hasegawa–Wakatani* model may be obtained from the four-field model by assuming a uniform magnetic field, $\mathbf{B}_0 = B_0\mathbf{b}_0$, and non-uniform background density, $n_0(x)$, neglecting parallel ion dynamics, and expressing the ion viscous stress and ion–electron friction as $\mu_i \nabla \mathbf{V}_E$ and $\eta_\| J_\|$, respectively, where $\eta_\|$ and μ_i are the electrical resistivity and ion viscosity. Using the parallel Ohm's law to eliminate $J_\|$, one finds the evolution equations for $\varphi(\mathbf{x}, t)$ and $n(\mathbf{x}, t)$, in normalized units,

$$\mathrm{d}_t\Omega_E = \mathcal{C}_\|(\varphi - n) + C_\mu \nabla_\perp^2 \Omega_E, \tag{2.287}$$
$$\mathrm{d}_t(n + \ln n_0) = \mathcal{C}_\|(\varphi - n), \tag{2.288}$$

where $\mathrm{d}_t = \partial_t + \mathbf{V}_E \cdot \nabla_\perp$ is the advective derivative, $\mathbf{V}_E = \mathbf{b}_0 \times \nabla_\perp\varphi$ is the perpendicular $\mathbf{E} \times \mathbf{B}$ velocity, $\Omega_E = \nabla_\perp^2\varphi$ is the associated parallel vorticity, and $h_e = n - \varphi$ is a measure of the non-adiabaticity of the parallel electron response. The operator $\mathcal{C}_\|$ and constant C_μ, which measure the strength parallel coupling and viscous dissipation, respectively, are given by,[46]

$$\mathcal{C}_\| = -(T_e/e^2 n_0 \eta_\| \Omega_i)\nabla_\|^2, \qquad C_\mu = \mu_i/m_i\rho_S^2\Omega_i. \tag{2.289}$$

The drift-wave regime is found for $\mathcal{C}_\| \gg 1$ and $C_\mu \ll 1$, whereas for $\mathcal{C}_\| \ll 1$ one recovers the Euler equation for a 2D incompressible fluid. In the former case, the dispersion relation obtained from (2.287)–(2.288),

45 The combination x^2/t thus yields the so-called Bohm diffusivity, $D_B = \rho_S^2\Omega_i = \rho_S c_{Se}$.
46 For clarity, both $\mathcal{C}_\|$ and C_μ are given in original units; both are dimensionless.

$$\omega[1 + k^2/(1 - \mathrm{i}\omega/\mathcal{C}_\parallel)] = -(\mathbf{k} \times \mathbf{b}_0) \cdot \nabla \ln n_0 - \mathrm{i}\mathcal{C}_\mu k^4, \qquad (2.290)$$

may be solved for the real, ω_r, and imaginary, $\omega_i = -\mathrm{i}\gamma$, parts of ω,

$$\omega_r(1 + k^2) \approx \omega_*, \qquad \gamma(1 + k^2) \approx \mathcal{C}_\parallel^{-1} k^2 \omega_r^2 - \mathcal{C}_\mu k^4, \qquad (2.291)$$

where $\omega_* = \mathbf{k} \cdot \mathbf{V}_* = k_\wedge V_*$ and $\mathbf{V}_* = \mathbf{b}_0 \times \nabla \ln n_0$ are the familiar drift frequency and velocity, respectively; as expected the latter is perpendicular to both $\mathbf{b}_0$ and ∇n_0 and thus points in the diamagnetic direction. According to (2.291), both ω_r and γ tend to peak around $k\rho_S \sim 1$, with the latter peak being much more pronounced. This localization in **k**-space indicates that drift-waves are most readily destabilized on the ion gyro-radius scale.

In the collisionless limit ($\eta_\parallel$, $\mu_i \to 0$), the electron response becomes adiabatic ($\mathcal{C}_\parallel \to \infty$, $n \to \varphi$) and the viscous term negligible ($\mathcal{C}_\mu = 0$), so that (2.287)–(2.288) reduce to a *single-field, Hasegawa–Mima* equation,[47]

$$\mathrm{d}_t\,[\Omega_E - \varphi - \ln n_0] = 0. \qquad (2.292)$$

Despite its simplicity, (2.292) is nonetheless both non-linear, due to appearance of φ in the advective derivative, and anisotropic, due to the background density gradient. It supports wave-like solutions governed by a simplified dispersion relation, $\omega(1 + k^2) = \omega_*$. Since an adiabatic electron response is assumed from the outset, the phase shift Δ_1 between n and φ is vanishingly small and the resulting drift-waves are stable; this also follows from (2.290), which yields $\gamma = 0$ for $\mathcal{C}_\parallel \to \infty$ and $\mathcal{C}_\mu = 0$. Finally, since (2.292) is non-dissipative, it exactly conserves two quadratic forms,

$$\mathcal{K} = \frac{1}{2}\int \varphi^2 + V_E^2 \mathbf{dx}, \qquad \mathcal{O} = \frac{1}{2}\int V_E^2 + \Omega_E^2 \mathbf{dx}, \qquad (2.293)$$

corresponding to the generalized energy and enstrophy of the flow.

It is interesting to note that the Hasegawa–Mima equation is identical to the *Charney* equation describing an incompressible, Eulerian fluid forming a thin layer of variable depth, h, and subject to a Coriolis force, $\Omega_c V_\perp$,

$$\mathrm{d}_t \mathbf{V} = -g\nabla h + \Omega_c \mathbf{V} \times \hat{\mathbf{e}}_g, \qquad (2.294)$$

see Table 2.3. This equivalence between drift-waves in a magnetized plasma and *Rossby* waves in a rotating fluid, allows us to understand, and for a classical drift-wave also quantify, the former in terms of the latter.

When $\mathrm{d}_t\varphi \ll \mathrm{d}_t\Omega_E$, i.e. on scales shorter than the gyro-radius, $k\rho_S \gg 1$, and when the background gradient is weak, $\mathrm{d}_t \ln n_0 \ll \mathrm{d}_t\Omega_E$, (2.292) reduces to the

[47] The reader is encouraged to derive this from scratch making the following assumptions: $\mathbf{B}_0 = const$, $n_0 = n_0(x)$, $T_i = 0$, $T_e =$ const, adiabatic electrons (2.283) and drift ordering (2.42).

Table 2.3. *Correspondence of drift- and Rossby waves.*

Equation	Hasegawa–Mima	Charney
state variable	electric potential, φ	fluid depth, h
free energy	density gradient, ∇n_0	fluid depth gradient, ∇h_0
frequency	gyro-frequency, Ω_i	Coriolis frequency, Ω_c
fast velocity	sound speed, $c_{Se}^2 = T_e/m_i$	gravity wave speed, $V_g^2 = gh$
spatial scale	gyro-radius, $\rho_S = c_{Se}/\Omega_i$	geostrophic radius, $\rho_g = V_g/\Omega_c$
slow velocity	electric drift, $\mathbf{V}_E$	geostrophic flow, $\mathbf{V}_\perp$
deflecting force	Lorentz force, $\Omega_i V_E$	Coriolis force, $\Omega_c V_\perp$
dynamical mode	drift-wave	Rossby wave

Euler equation of a 2D incompressible fluid, $d_t\Omega_E = 0$, with φ interpreted as the stream function of the flow. In this case, (2.293) expresses the conservation of energy and enstrophy of a 2D hydrodynamic flow,

$$\mathcal{K} = \frac{1}{2}\int V^2 d\mathbf{x} = \frac{1}{2}\int (\nabla\varphi)^2 d\mathbf{x}, \quad \mathcal{O} = \frac{1}{2}\int \Omega^2 d\mathbf{x} = \frac{1}{2}\int (\nabla^2\varphi)^2 d\mathbf{x}.$$

In Section 6.1.3, we will see that this *simultaneous* conservation of both energy and enstrophy has important consequences for two-dimensional turbulence, such as the flow of energy from small to large scales and the formation of large and relatively long lived, coherent structures.

2.5 The relation between MHD- and drift-ordered dynamics

We conclude the chapter with two poignant remarks from the topical literature on the differences in plasma dynamics resulting from the MHD and drift orderings. The first is from Hazeltine and Meiss (1992):[48]

'The two orderings lead to quite different pictures of plasma evolution. In the MHD-ordered case, electric drifts, with their associated non-linear inertial terms (such as the centrifugal force), dominate the dynamics; in the drift ordering, the electric drift enters only in concert with other slow motions, such as the curvature and gradient-B drifts. While the most violent perturbations may be described by (2.41), the more prevalent ones, at least in modern confinement devices, are consistent with (2.42). Indeed, the replacement of (2.41) by (2.42) as the more realistic ordering is one indicator of progress in plasma confinement: even tokamak disruptions are not fast enough for the MHD ordering.'

And the second from Scott (2001):

'The key concept [in fluid-drift dynamics] is quasi-neutrality: neutral plasma with robust electric field activity. Indeed, one often asks the question: how does, or why should, a

[48] The numbers in brackets refer to equations in this book.

neutral plasma develop electric fields? The understanding of the resolution to that question forms the basis of this subject, and what one gains from that is a physical framework under which the turbulence and transport problems in magnetized plasmas can be addressed. For it is important to understand that they cannot be addressed by the MHD model, because the coupling from pressure to current to electric field is one of the things that are discarded in that model.'

The family of fluid models encompassed by MHD- and drift-ordered plasma dynamics, which collectively may be referred to as *drift-hydrodynamics* (DHD),

$$\text{DHD} \approx \text{MHD} \cup \text{drift} \approx O(1) + O(\delta_i), \tag{2.295}$$

is sufficient for most fusion applications. In subsequent chapters, we will employ both ordering schemes to study plasma equilibrium, stability and transport – the three basic aspects of any dynamical system and the central issues of both plasma energy confinement and power exhaust.

2.6 Further reading

A student approaching plasma physics for the first time is faced with dozens of books, varying in scope, style and complexity, ranging from classic accounts by pioneers in the field to contemporary texts presenting the latest developments. Navigating this imposing, and rapidly growing, body of literature is no small task. In this respect, the outstanding monographs by Hazeltine and Meiss (1992) and Hazeltine and Waelbroeck (2004), to which the foregoing chapter owes a heavy debt, deserve a special mention both for their clarity and their scope. The following texts are also recommended:

Classic texts: Chandrasekhar (1941); Spitzer (1956); Alfvén and Falthammar (1963). *Introductions*: Krall and Trivelpiece (1973); Schmidt (1979); Dolan (1982); Nicholson (1983); Chen (1984); Nishikawa and Wakatani (1994); Goldston and Rutherford (1995); Sitenko and Malnev (1995); Choudhuri (1998). *Advanced introductions*: Gill (1981); Miyamoto (1989); Hazeltine and Meiss (1992); Dendy (1993); Boyd and Sanderson (2003); Hazeltine and Waelbroeck (2004); Wesson (2004). *Plasma kinetic theory*: Shkarofsky *et al.* (1966); Ichimaru (1973); Lifschitz and Pitaevskii (1981); Balescu (1988a). *Magneto-hydrodynamics*: Batemann (1978); Freidberg (1991); Biskamp (1993, 2000, 2003); Goedbloed and Poedts (2004). *Handbooks*: Galeev and Sudan (1983); Huba (2006). *Plasma diagnostics*: Hutchinson (2002).

3
Magnetized plasma equilibrium

'The objective of the controlled thermonuclear fusion program is to heat a gas composed of light elements to a temperature considerably hotter than the centre of the sun and to confine this hot plasma long enough for the resulting nuclear reactions to produce more energy than was consumed.'

Batemann (1978)

Considerably more energy, one might add – in the planned ITER experiment, Q_α (1.5) is expected to reach a maximum value of 2 and a steady-state value of 1. How is it possible to achieve this extraordinary level of plasma energy confinement in a laboratory environment? The theoretical framework developed in the previous chapter provides a simple answer: by embedding the plasma in a strong *ambient* magnetic field, such that the thermal pressure, p, is supported by the magnetic pressure, $B^2/2\mu_0$, see (2.198) and (2.247). Indeed, after no less than half a century of international research, the strategy of *magnetic confinement* appears to offer the most promising route to realizing the long-held dream of constructing a technologically feasible and commercially viable fusion reactor. Yet, it is not immediately obvious what such a reactor should look like, e.g. what is the optimal magnetic geometry? How strong does the ambient magnetic field have to be? Can this field be generated by the plasma itself or are external coils and/or antennas required? etc. In approaching these questions for the first time, or even returning to them after a long spell down a specialist corridor, it is instructive to forego the benefit of hindsight and seek the answers afresh, aided only by theoretical considerations.

In the first instance, let us consider the properties of magnetized plasma *equilibrium*, which is the *sine qua non* of plasma confinement. By equilibrium we actually mean *quasi-equilibrium* or *steady-state*, since global thermodynamic equilibrium precludes the existence of pressure gradients and hence of plasma energy confinement.[1] It should be stressed that the equilibrium field **B** consists of two parts, $\mathbf{B}^{ext}$ and $\mathbf{B}^{int}$, generated in accordance with (2.88) by external, $\mathbf{J}^{ext}$, and internal, $\mathbf{J}^{int}$, current densities,

$$\nabla \times \mathbf{B}^{ext} = \mu_0 \mathbf{J}^{ext}, \qquad \nabla \times \mathbf{B}^{int} = \mu_0 \mathbf{J}^{int}. \tag{3.1}$$

The former, which may be viewed as an ambient field, is under our direct control and can thus be used to modify the confined plasma equilibrium, e.g. change its shape and position. The latter, which represents the self-organized plasma response to external actuators, is beyond our direct control. This lack of control is a matter of serious concern since loss of plasma equilibrium typically leads to a complete collapse, or *disruption*, of the discharge, and results in large mechanical forces and thermal heat loads on plasma facing components. Great care must hence be taken to minimize the likelihood and severity of such events, which in turn requires a good grasp of the mechanisms responsible for internal plasma equilibrium.

3.1 Magnetic geometry and flux co-ordinates

The question of geometry and topology of the confined plasma equilibrium may be answered by making two key assumptions: (i) that the hot fusion plasma is strongly magnetized ($\delta_s \ll 1$), such that fast transverse motions can occur only via electric drifts; and (ii) that it is nearly collisionless ($\nu_s^* \ll 1$), as expected at keV temperatures required for nuclear fusion, such that collisional dissipation may be neglected. With these assumptions, *plasma equilibrium* can be studied using the model of single-fluid ideal MHD, (2.246)–(2.248), and thus becomes synonymous with *MHD equilibrium*, i.e. with the absence of violent ($v_\perp \sim v_E \sim v_{ti}$) bulk plasma motion. Recall that in the MHD model, plasma is frozen into individual magnetic *flux tubes*, and can only escape by parallel losses at the open ends of these flux tubes. In the absence of collisions such losses can only be reduced by the magnetic mirror force (2.60), i.e. by creating a *magnetic well* with $B_{max} \gg B_{min}$. Unfortunately, the *mirror loss fraction*, $f_L = 1 - f_T$ (2.61), scales as the square root of B_{min}/B_{max}, which prevents efficient suppression of end losses. The magnetic mirror density loss time,

$$\tau_L = \frac{\tau_\parallel}{f_L} \approx \frac{L_\parallel}{V_S}\sqrt{\frac{B_{max}}{B_{min}}}, \tag{3.2}$$

is hence determined by the sonic, parallel loss time $\tau_\parallel \sim \tau_S$ and the technological limits related to B_{max}/B_{min}. In short, it is not possible to completely seal the

[1] By considering plasma equilibrium, we exclude other, transient, means of achieving $Q_\alpha > 1$.

magnetic bottle, which largely explains why *magnetic mirror machines*, as such linear devices are known, have failed to achieve $Q_\alpha > 1$.

The only alternative is to bend the magnetic field lines in such a way that they remain confined within some finite volume without ever intersecting a solid boundary defined by the vacuum vessel. In view of the solenoidal aspect of the magnetic field ($\nabla \cdot \mathbf{B} = 0$) its lines of force must ultimately close upon themselves, although this closure may occur *asymptotically* (*ergodically*) in the limit $l \to \infty$, l being the distance along the field line. In order to prevent contact with the solid boundary, such closure must occur on a two-dimensional surface on which the magnetic field cannot vanish (otherwise the plasma would stream out at the sound speed) and its lines of force cannot intersect (in keeping with the frozen flux property of ideal MHD), i.e. on a 2D surface covered by non-vanishing, non-intersecting field lines. If we imagine their direction vectors as *hairs* growing on the surface, such that it is everywhere covered with dense *vector fur*, then we require this fur to be neatly combed over the surface, without forming either bald spots ($\mathbf{B} = 0$) or hairy spikes ($\mathbf{B} \times \hat{\mathbf{e}}_\perp = 0$, where $\hat{\mathbf{e}}_\perp$ is the surface normal). Any surface satisfying these requirements must have the topology of a *torus*, i.e. only a *toroidal surface* can be covered by a non-vanishing vector field.[2] We may conclude that *an equilibrium configuration of a magnetically confined plasma involves an arrangement of field lines into a sequence of nested tori*. The resulting toroidal surfaces are commonly known as *flux surfaces*, since the magnetic field defining the surface can be expressed in terms of *toroidal*, Ψ_T, and *poloidal*, Ψ_P, magnetic fluxes subtended by the surface,

$$\Psi_T \equiv \Psi_\zeta \equiv \int_{S_T} \mathbf{B} \cdot d\mathbf{S}_\zeta = \int_{\mathcal{V}} \mathbf{B} \cdot \nabla\zeta d\mathbf{x}/2\pi = \int_{\mathcal{V}} B^\zeta d\mathbf{x}/2\pi, \tag{3.3}$$

$$\Psi_P \equiv \Psi_\theta \equiv \int_{S_P} \mathbf{B} \cdot d\mathbf{S}_\theta = \int_{\mathcal{V}} \mathbf{B} \cdot \nabla\theta d\mathbf{x}/2\pi = \int_{\mathcal{V}} B^\theta d\mathbf{x}/2\pi. \tag{3.4}$$

Here we designated the *general toroidal co-ordinates* by (r, ζ, θ), with r being the radial ordinate, ζ the toroidal angle and θ the poloidal angle;[3] a brief discussion of curvilinear co-ordinates, including the co- and contra-variant representation of vector components, may be found in Appendix B.

The most general expression for $\mathbf{B}$, and in fact for any solenoidal vector field, in toroidal co-ordinates follows from the so-called *Clebsh representation*, $\mathbf{B} = \nabla\beta \times \nabla\alpha$, which automatically satisfies $\nabla \cdot \mathbf{B} = 0$. The existence of flux surfaces allows us to identify β with some *flux surface label* $F(r)$, that is, any scalar F which takes a unique value on every ergodically covered surface ($\mathbf{B} \cdot \nabla F = 0$), and

[2] This result, known as *Poincaré's theorem*, is both intuitive and fully rigorous.

[3] For axis-symmetric systems, we also introduce the major radius R_0, the distance from the major axis $R = R_0 + r\cos\theta$ and the distance from the equatorial plane $Z = r\sin\theta$.

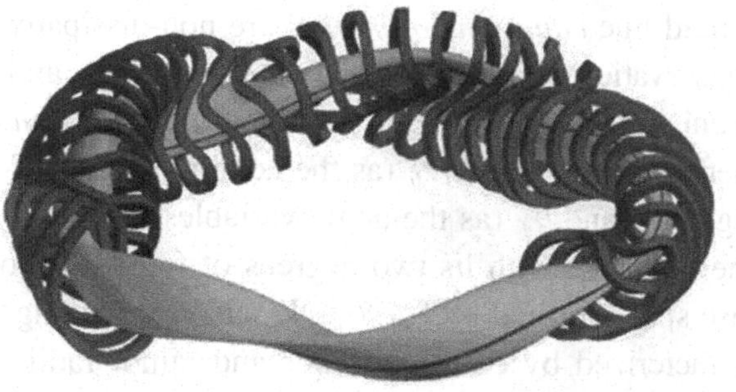

Fig. 3.1. Magnetic field configuration of the Wendelstein-7X stellarator (currently under construction) showing both the external field coils and a typically twisting magnetic flux surface. © Max Planck Institute für Plasmaphysik.

α with some *flux surface potential*, $G(r, \theta, \zeta) = q(r)\theta + \zeta + G_0(r, \theta, \zeta)$, where $G_0(r, \theta, \zeta)$ is a periodic function of both angles and $q(r) = \mathrm{d}\Psi_T/\mathrm{d}\Psi_P$ is known as the *safety factor*,

$$\mathbf{B} = \nabla F \times \nabla G = \nabla F \times \nabla(q\theta - \zeta) + \nabla F \times \nabla G_0(r, \theta, \zeta). \qquad (3.5)$$

The name refers to stability of toroidal plasmas, which requires $q(r) > 1$ on every flux surface, Section 4.2. In a tokamak, see Fig. 1.1, $q > 1$ is generated by a toroidal component of the plasma current driven by an EMF induced with an external transformer winding, whereas in a stellarator, see Fig. 3.1, it is generated by a physical twist of the external coils, which produce both poloidal and toroidal field components and a poloidal flux surface cross-section which changes with toroidal angle. In stellarator research, this helical twist is typically measured by the so-called *rotational transform* $\iota \equiv 2\pi/q$. Whereas, by definition, both q and ι are flux surface labels, the pitch angle of the helical field lines covering the toroidal surface, $\mathrm{d}_\zeta\theta = \mathbf{B}\cdot\nabla\theta/\mathbf{B}\cdot\nabla\zeta = B^\theta/B^\zeta$, changes with position, as does the radial ordinate r.[4]

To progress the above argument, we attack Maxwell's equations with the full arsenal of the Hamilton–Lagrange formalism. We first note that the dynamics of a magnetic field line, defined by $\mathrm{d}_\tau\mathbf{x}\times\mathbf{b} = 0$, where τ is some parameter (an effective time) along the field line, follow from the *principle of least action* (2.67), with the action $\mathcal{S}_B$ defined in terms of the magnetic Lagrangian, $\mathcal{L}_B = \mathbf{A}\cdot\mathrm{d}_\tau\mathbf{x} = A_\parallel \mathrm{d}_\tau l$, $\mathbf{A}$ being the vector potential,

$$\delta\mathcal{S} = 0, \qquad \mathcal{S} = \int \mathcal{L}_B \mathrm{d}\tau = \int A_\parallel(\mathrm{d}_\tau l)\mathrm{d}\tau = \int A_\parallel \mathrm{d}l. \qquad (3.6)$$

[4] At this stage, the *aspect ratio* of the torus (the ratio R_0/a of its major and minor radii), the shape of its poloidal and toroidal cross-sections, and pitch of its field lines remain unspecified.

The resulting field line *equations of motion* are non-dissipative, as required by magnetic flux conservation of ideal MHD, $\mathrm{d}_t\Psi = 0$ (2.252), and can therefore be expressed in canonical form. An obvious choice of *angle-action* variables are the co-variant magnetic fluxes Ψ_ζ and Ψ_θ (as the action variables $\mathcal{I}_1$ and $\mathcal{I}_2$) and the related (flux) angles ζ_f and θ_f (as the angle variables ϑ_1 and ϑ_2).[5] According to Section 2.2.2, the field line with its two degrees of freedom evolves in the four-dimensional phase space formed by $\{\zeta_f, \theta_f, \Psi_\zeta, \Psi_\theta\}$ by rotating around both axis of a 2-torus, characterized by constant major and minor radii, Ψ_ζ and Ψ_θ, and toroidal and poloidal angles, ζ_f and θ_f, respectively. It winds its way around the torus with constant angular velocities, $\omega_\zeta = \mathrm{d}_\tau\zeta_f$ and $\omega_\theta = \mathrm{d}_\tau\theta_f$, such that the pitch angle of its helical trajectory, $\omega_\theta/\omega_\zeta$, is a flux surface label,

$$\frac{1}{q} \equiv \frac{\iota}{2\pi} \equiv \frac{\omega_\theta}{\omega_\zeta} = \frac{\mathrm{d}\theta_f}{\mathrm{d}\zeta_f} = \frac{\mathbf{B}\cdot\nabla\theta_f}{\mathbf{B}\cdot\nabla\zeta_f} = \frac{B^\theta}{B^\zeta} = \text{const.} \tag{3.7}$$

This result is consistent with the earlier definition of the safety factor, $q \equiv \mathrm{d}\Psi_T/\mathrm{d}\Psi_P$, which is now equal simply to the ratio of toroidal and poloidal fluxes, $q = \Psi_T/\Psi_P$. The *radial (flux) ordinate*, r_f can be defined in terms of any monotonic flux surface label, such as Ψ_P and Ψ_T, but not in terms of q, which may have a non-monotonic radial profile.

The toroidal co-ordinates (r_f, θ_f, ζ_f), commonly known as *flux co-ordinates*,[6] provide the most natural representation of a magnetic field possessing toroidal flux surfaces and thus play a central role in magnetized plasma theory (D'haeseleer *et al.*, 1991). Their pre-eminence stems from the constancy of B^θ/B^ζ everywhere on the flux surface (3.7), such that helical field lines in (r, ζ, θ) are transformed into straight lines in (r_f, ζ_f, θ_f), thus greatly simplifying geometrical analysis, see Fig. 3.2, e.g. identifying F with $\psi_P = \Psi_P/2\pi$ and eliminating the periodic function G_0 by an appropriate choice of the poloidal (flux co-ordinate) angle $\theta_f(r, \zeta, \theta)$, we can rewrite (3.5) as,[7]

$$r_f = F = \psi_P = \Psi_P/2\pi, \qquad G(r_f, \theta_f, \zeta_f) = q(r_f)\theta_f - \zeta_f, \tag{3.8}$$

$$\mathbf{B} = \nabla F \times \nabla G = \nabla r_f \times \nabla(q\theta_f - \zeta_f) = \mathbf{B}_T + \mathbf{B}_P, \tag{3.9}$$

$$\mathbf{B}_T = q\nabla\psi_P \times \nabla\theta_f = \nabla\psi_T \times \nabla\theta_f, \quad \mathbf{B}_P = -\nabla\psi_P \times \nabla\zeta_f, \tag{3.10}$$

where $G = q(r_f)\theta_f - \zeta_f$, known as the *field line label*, is constant along the field line and satisfies the periodicity condition, $G\{\theta_f, \zeta_f\} = G\{\theta_f + 2\pi, \zeta_f + 2\pi\}$. The

5 We use ζ and θ in anticipation of the toroidal result, but in fact they represent arbitrary canonical angles.

6 Flux co-ordinates should be treated as a function of position $\mathbf{x}$, i.e. $r_f(\mathbf{x}), \theta_f(\mathbf{x}), \zeta_f(\mathbf{x})$.

7 To avoid confusion with co-variant components of a vector, denoted by lower case subscripts, e.g. B_ζ, B_θ for co-variant components of $\mathbf{B}$, we employ upper case subscripts in the sense of the usual convention in physics, i.e. to denote simple vector components of $\mathbf{B}$. Hence, in cylindrical geometry, B_T and B_P represent the axial and poloidal components, while in toroidal geometry they represent the toroidal and poloidal components (see Chapter 1).

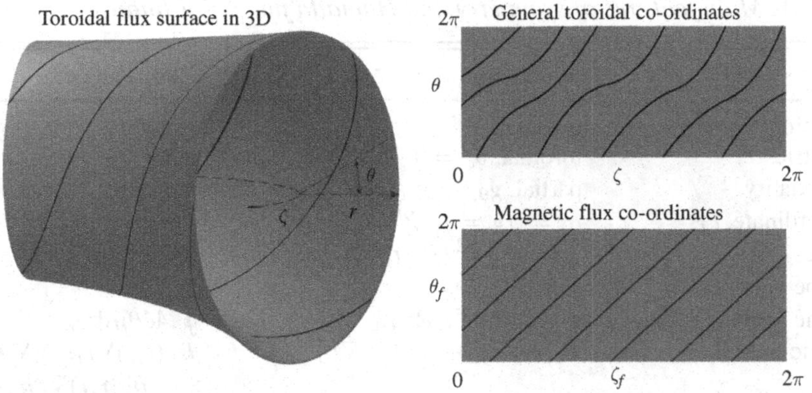

Fig. 3.2. Schematic representation of a three-dimensional toroidal flux surface, which is unfolded in both general toroidal and in magnetic flux co-ordinates.

above choice of F and G is easily verified by substituting (3.9) into (3.3)–(3.4). The magnetic field $\mathbf{B}$, which we decomposed into its toroidal, $\mathbf{B}_T$, and poloidal, $\mathbf{B}_P$, projections, is thus written explicitly in terms of the magnetic fluxes ψ_T and ψ_P in the *contra-variant* basis, $\mathbf{e}^j = \{\nabla r_f, \nabla\theta_f, \nabla\zeta_f\}$. This basis can be viewed as *natural*, since it defines the three principal magnetic directions: (i) *parallel*, $\hat{\mathbf{e}}_\parallel \equiv \mathbf{b} = \mathbf{B}/B$; (ii) *radial*, $\hat{\mathbf{e}}_\perp \propto$ surface normal;[8] and (iii) *diamagnetic*, $\hat{\mathbf{e}}_\wedge \propto \hat{\mathbf{e}}_\parallel \times \hat{\mathbf{e}}_\perp$,

$$\hat{\mathbf{e}}_\parallel \propto \nabla r_f \times \nabla(q\theta_f - \zeta_f), \qquad \hat{\mathbf{e}}_\perp \propto \nabla r_f, \qquad \hat{\mathbf{e}}_\wedge \propto \nabla(q\theta_f - \zeta_f), \tag{3.11}$$

and can be used to define the field line co-ordinates (r_f, l, α), see Fig. 3.2,

$$\hat{\mathbf{e}}_\parallel \propto \nabla l, \qquad \hat{\mathbf{e}}_\perp \propto \nabla r_f, \qquad \hat{\mathbf{e}}_\wedge \propto \nabla\alpha. \tag{3.12}$$

The contra-variant components of $\mathbf{B}$ are found by taking a dot product of (3.9) with $\mathbf{e}^j$ and comparing with (B.1)–(B.2),

$$B^r = 0, \qquad B^\theta = \frac{B^\zeta}{q} = \frac{\psi_P'}{\sqrt{g}} = \frac{1}{\sqrt{g}}, \qquad \mathcal{A}' \equiv \frac{\mathrm{d}\mathcal{A}}{\mathrm{d}r_f} = \frac{\mathrm{d}\mathcal{A}}{\mathrm{d}\psi_P}, \tag{3.13}$$

where we denoted the radial derivative with a prime.

The choice of flux co-ordinates is not unique, since any co-ordinate transformation which preserves the field line label G will yield a new set of flux co-ordinates. The two most common systems are the *symmetry* (r_0, θ_0, ζ_0) and *Hamada* (r_H, θ_H, ζ_H) co-ordinates, whose main features are summarized in Table 3.1. The former are most useful in axis-symmetric systems $(\partial_\zeta = 0)$ and

[8] This usage should not be confused with the gyro-tropic convention (2.17), in which $\boldsymbol{\perp}$ denotes a vector in a plane perpendicular to $\mathbf{b}$, i.e. the $\boldsymbol{\perp}$ $-\wedge$ plane in the new basis (3.11). The two conventions are frequently used side-by-side, with the meaning inferred from the context. To prevent confusion, throughout this book we used bold font, $\boldsymbol{\perp}$, when referring to the gyro-tropic symbol and regular font, $\perp$, when referring to the radial direction.

Table 3.1. *Main features of symmetry and Hamada flux co-ordinates.*

	Symmetry (r_0, θ_0, ζ_0)	Hamada (r_H, θ_H, ζ_H)
application	tokamak	general
symmetries	toroidal, $\partial_\zeta = 0$, $\zeta_0 = -\phi$	none
orthogonality	partial, $g_{\theta\zeta} = g_{r\zeta} = 0$, $g_{\zeta\zeta} = R^2$	none
radial ordinate, r_f	$r_0 = \psi_P = \int_V B^\theta \mathrm{d}\mathbf{x}/(2\pi)^2$	$v_H = \int_V \mathrm{d}\mathbf{x}/(2\pi)^2$
Jacobian $\sqrt{g}$	$\sqrt{g_0} = qR^2/B_\zeta(r_0)$	$\sqrt{g_H} = 1$
flux tube volume $\int B^{-1}\mathrm{d}l$	$\oint R^2 \mathrm{d}\theta_0$	1
field line average $\langle \mathcal{A} \rangle_B$	$\oint \mathcal{A}R^2\mathrm{d}\theta_0 / \oint R^2 \mathrm{d}\theta_0$	$\oint \mathcal{A}\mathrm{d}\theta_H \mathrm{d}\zeta_H$
magnetic field **B**	$B_\zeta(r_0)\nabla\zeta_0 + \nabla\zeta_0 \times \nabla r_0$	$B^\zeta(r_H)\nabla r_H \times \nabla\theta_H$ $+B^\theta(r_H)\nabla\zeta_H \times \nabla r_H$
plasma current **J**	$J^\zeta R^2 \nabla\zeta_0 + \sqrt{g_0}J^\theta\nabla\zeta_0 \times \nabla r_0$	$J^\zeta(r_H)\nabla r_H \times \nabla\theta_H$ $+J^\theta(r_H)\nabla\zeta_H \times \nabla r_H$
parallel current $J_\parallel$	$-(B/\mu_0)B_\zeta' - (B_\zeta/B)p'$	$(J^\theta B - p' B_\zeta/B)/\psi_P'$

are typically used in tokamak research. They exploit this symmetry by setting $\zeta_0 = -\phi$, where ϕ is the simple toroidal angle, which yields orthogonality between ζ_0 and both r_0 and θ_0: $g_{\theta\zeta} = g_{r\zeta} = 0$, $h_\zeta = 1/h^\zeta = R$. As a result, $\mathbf{B}_T$ can be represented in co-variant and $\mathbf{B}_P$ in contra-variant form,

$$\mathbf{B} = \mathbf{B}_T + \mathbf{B}_P = B_\zeta(r_0)\nabla\zeta_0 + \nabla\zeta_0 \times \nabla r_0 = I(\psi_P)\nabla\zeta_0 + \nabla\zeta_0 \times \nabla\psi_P, \quad (3.14)$$

where $I(\psi_P) = B_\zeta(r_0) = RB_T$ is a flux surface label,[9] and we again set $r_0 = \psi_P$, such that $\psi_P' = 1$. The main disadvantage of symmetry co-ordinates is the fact that their Jacobian, $\sqrt{g_0} = qR^2/I(\psi_P)$, and by (3.13), B^θ and B^ζ, vary within the flux surface. Hamada co-ordinates resolve this problem by identifying r_H with the normalized volume enclosed by the flux surface $v_H = \int_V \mathrm{d}\mathbf{x}/(2\pi)^2$, which yields a unit Jacobian, $\sqrt{g_H} = 1$, so that B^θ and B^ζ become flux surface labels. This result greatly simplifies many expressions, such as the flux tube volume $\int B^{-1}\mathrm{d}l$ and field line average $\langle \mathcal{A} \rangle_B$, which reduce to 1 and $\oint \mathcal{A}\mathrm{d}\theta_H \mathrm{d}\zeta_H$, respectively. It also allows both $\mathbf{B}_T$ and $\mathbf{B}_P$ to be represented in purely contra-variant form,

$$\mathbf{B} = \mathbf{B}_T + \mathbf{B}_P = B^\zeta(r_H)\nabla r_H \times \nabla\theta_H + B^\theta(r_H)\nabla\zeta_H \times \nabla r_H. \quad (3.15)$$

Since Hamada co-ordinates lack any inherent symmetries, they are sub-optimal for axis-symmetric devices, such as tokamaks, but are preferred for stellarators, or whenever general applicability is desired.

All field lines covering a flux surface must close upon themselves after m poloidal and n toroidal turns, where m and n are both integers. If the ratio

9 This follows from the radial component of Ampere's law (2.88) with $J^r = 0 = \partial_\zeta$. Care should be taken not to confuse $I(\psi_P) = RB_T$ with the plasma current I, usually given by its toroidal and poloidal components, I_T and I_P.

$m/n = \mathrm{d}\psi_T/\mathrm{d}\psi_P = q$ is *rational*, then both m and n must be finite, i.e. the field line closes upon itself in a *finite* number of turns.[10] The flux surface is then covered by many distinct field lines, which may be distinguished by some field line label, e.g. $G = q(r_f)\theta_f - \zeta_f$. In contrast, if $m/n = q$ is *irrational*, then both m and n must be countably infinite, with $O(\mathcal{N}_\infty) = \aleph_0$. The field line then closes upon itself only asymptotically (as $l, \tau \to \infty$ and $m, n \to \aleph_0$), and densely covers the entire flux surface. Field line identity is thereby lost and $G(r_f, q\theta_f - \zeta_f)$ becomes a flux surface label $G(r_f)$.[11]

The earlier assertion, that the field line average (2.275) reduces to a flux surface average for all ergodic surfaces, should now be self-evident, since

$$\langle \mathcal{A} \rangle_B = \mathrm{d}_\mathcal{V} \int_\mathcal{V} \mathcal{A}\mathrm{d}\mathbf{x}, \quad V \equiv \int_V \mathrm{d}\mathbf{x}, \quad \mathrm{d}\mathbf{x} = \sqrt{g}\mathrm{d}r_f \mathrm{d}\theta_f \mathrm{d}\zeta_f, \tag{3.16}$$

where $\mathcal{V}(r_f)$ is the volume bounded by the flux surface. Note that $\langle F \rangle_B = F$ for any flux surface label $F(r_f)$ and that $\langle \mathbf{B} \cdot \nabla \mathcal{A} \rangle_B = 0$ for any function $\mathcal{A}(\mathbf{x})$, in agreement with (2.275). Finally, making use of flux co-ordinates, (3.16) can be simplified to read

$$\langle \mathcal{A} \rangle_B = (\mathrm{d}_\mathcal{V}\psi_P) \oint \frac{\mathcal{A}}{B^\theta} \mathrm{d}\theta_f \mathrm{d}\zeta_f = \oint \frac{\mathcal{A}}{B^\theta} \mathrm{d}\theta_f \mathrm{d}\zeta_f \Big/ \oint \frac{\mathrm{d}\theta_f \mathrm{d}\zeta_f}{B^\theta} \tag{3.17}$$

Note that $\iota(r_f)$ and $q(r_f)$ can now be defined as flux surface averages,

$$\frac{\iota}{2\pi} = \frac{1}{q} = \left\langle \frac{B^\theta}{B^\zeta} \right\rangle_B = (\mathrm{d}_\mathcal{V}\psi_P) \oint \frac{\mathrm{d}\theta_f \mathrm{d}\zeta_f}{B^\zeta}, \tag{3.18}$$

and that $\oint B^{-1}\mathrm{d}l$ in the numerator of (2.276) represents the average specific volume of a flux tube covering the surface (since $\mathrm{d}l/B = \mathrm{d}\mathcal{V}/\Psi_P$). We also define the volume average values of toroidal, poloidal and total plasma beta,

$$\beta_T \equiv 2\mu_0\langle p \rangle_a / \langle B_T \rangle_a^2 \approx 2\mu_0 \langle p \rangle_a / B_0^2, \tag{3.19}$$

$$\beta_P \equiv 2\mu_0\langle p \rangle_a / \langle B_P \rangle_a^2 \approx 2\mu_0 \langle p \rangle_a / B_P^2(a), \tag{3.20}$$

$$\overline{\beta} \equiv 2\mu_0\langle p \rangle_a / \langle B \rangle_a^2 = \left(\beta_T^{-1} + \beta_P^{-1}\right)^{-1}, \tag{3.21}$$

where the volume average over the entire plasma, $r_f < r_f(a)$, is defined as,

$$\langle \mathcal{A} \rangle_a \equiv \int_{\mathcal{V}(a)} \mathcal{A}\,\mathrm{d}\mathbf{x} / V(a). \tag{3.22}$$

[10] Flux surfaces with rational q are known as *resonant*, those with irrational q as *ergodic*.

[11] Recall that there is a countable infinity of irrational numbers between any two rational numbers, such that the infinity of real numbers $O(\mathcal{R}_\infty) = 2^{\aleph_0}$ greatly exceeds that of rational numbers $O(\mathcal{Q}_\infty) = O(\mathcal{N}_\infty) = \aleph_0$. Hence, ergodic surfaces will dominate the magnetic field structure in the confined plasma volume, with resonant surfaces being comparatively rare, whenever $q(r_f)$ has a finite radial gradient or *magnetic shear*, $s(r_f) \equiv \mathrm{d}\ln q/\mathrm{d}\ln r_f$.

The above discussion of flux surfaces is idealized in several respects:

(i) It neglects the finite gyro-radius of thermal ions ρ_{ti}. In reality, two field lines and/or flux surfaces lose their identity unless they are separated by more than ρ_{ti}. Hence, the ratio $a/\rho_{ti} = 1/\rho^*$ sets the maximum number of distinct flux surfaces, while $2\pi/\rho^*$ determines the maximum number of distinct field lines covering a flux surface. Since a/ρ_{ti} is less than 10^4 in existing experiments, e.g. $\rho^* \sim 10^{-3}$–10^{-4} for typical JET conditions, a flux surface is resonant if $q = m/n$ can be written in decimal notation with less than $|\log_{10}\rho^*| \sim 3$ significant digits and does not overlap with a lower-order resonant surface to the same order of accuracy.[12]

This constraint is made even more rigorous by considering the deviation of an average GC-orbit away from the flux surface. Since the particle energy, $\mathcal{E} = \frac{1}{2}mv^2 + e\varphi$, magnetic moment, $\mu = \frac{1}{2}mv_\perp^2/B$, and toroidal momentum, $p_\zeta = mRv_\zeta - e\psi_P$, are conserved in axis-symmetric toroidal geometry (recall the discussion on page 28), it is easy to show that the GC deviates from the flux surface by a radial distance of order $\rho_P = mv/eB_P$, i.e. by its poloidal gyro-radius. This result, known as *Tamm's theorem*, implies that ions deviate from the flux surface much further than electrons, and that both are confined primarily by the poloidal field.[13] It also suggests that a/ρ_{Pi}, rather than a/ρ_{ti}, determines the number of distinct flux surfaces.

(ii) It omits dissipative effects, e.g. plasma resistivity, $\eta = \eta_\parallel$, and the associated field line mixing with a magnetic diffusivity, $D_\eta = \eta/\mu_0$. For finite η, field lines initially covering the flux surface at r_f, undergo micro-tearing processes which can introduce stochastic mixing and lead to the smearing out of the flux surface over radial extent $\Delta_r \sim (\tau_A D_\eta)^{1/2}$, where the Alfvén time, τ_A, is defined by (2.254).

(iii) It neglects small perturbations to $\mathbf{B}$, denoted as $\widetilde{\mathbf{B}}$, which can arise due to non-uniformities in the structure of the magnetic field[14] or fluctuating currents associated with EM plasma turbulence. The former, although static in real time, are perceived by the field line as periodic perturbations, and can lead to parametric resonance, magnetic island formation and chaotic volume growth at the X-points, in exact analogy with the KAM theorem of classical mechanics. This is illustrated in Fig. 2.1, where the abstract canonical tori represent magnetic flux surfaces expressed in flux co-ordinates. The latter, temporal perturbations, subject the field line to random deflections, leading to a stochastic field line diffusion. The three types of magnetic field line arrangements (resonant, ergodic and chaotic) emerging from the above analysis can be distinguished by their spatial dimensionality, see Table 3.2.

[12] For instance, $q = 2.9$, $q = 3$ and $q = 3.1$ correspond to resonant surfaces with $(m, n) = (29, 10)$, $(3, 1)$ and $(31, 10)$, respectively, $q = 2.999$ and $q = 3.001$ to ergodic surfaces which provide the upper and lower bounds on the radial extent of the $q = 3$ surface (assuming $\rho^* = 10^3$), while $q = 2.999\,999$ and $3.000\,001$ are de facto equivalent to the resonant $q = 3$ surface.

[13] As a corollary, it predicts loss of plasma confinement in regions where $B_P \to 0$, either at O-points, such as the magnetic axis, or X-points, often created with external toroidal coils.

[14] For example, toroidal field ripple or resonant magnetic perturbation in tokamaks.

Table 3.2. *Comparison of three types of magnetic field structures*

Property	Resonant	Ergodic	Chaotic
dimensionality	1D (line)	2D (surface)	3D (volume)
structural labels	$q(r_f)\theta_f - \zeta_f$	r_f	–
$B^r = \mathbf{B}\cdot\nabla r_f$	0	0	finite
field line length, l	finite	∞	∞
safety factor, $q = m/n$	rational	irrational	–

Both (ii) and (iii) above tend to destroy flux surfaces, thus invalidating the flux co-ordinate approach and necessitating a more general representation for **B**. One such representation, which does not presuppose the existence of flux surfaces and is therefore known as *canonical*, can be derived by writing **A** in co-variant form, relating A_θ and A_ζ to the magnetic fluxes ψ_T and ψ_P, and rewriting $A_r\nabla r$ in terms of the radial integral $\int_0^r A_r \mathrm{d}r$,

$$\mathbf{A} = A_\theta\nabla\theta + A_\zeta\nabla\zeta + A_r\nabla r = \psi_T\nabla\theta - \psi_P\nabla\zeta + \nabla\textstyle\int_0^r A_r\mathrm{d}r, \tag{3.23}$$

$$\psi_T = A_\theta - \partial_\theta\textstyle\int_0^r A_r\mathrm{d}r, \qquad \psi_P = -A_\zeta + \partial_\zeta\int_0^r A_r\mathrm{d}r, \tag{3.24}$$

$$\mathbf{B} = \nabla\times\mathbf{A} = \nabla\psi_T\times\nabla\theta - \nabla\psi_P\times\nabla\zeta. \tag{3.25}$$

Inserting this form into (3.6) and equating the parameter τ, with the toroidal angle ζ, we find the canonical form of the field line *action*, $\mathcal{S}_B$,

$$\mathcal{S}_B = \int A_\parallel \mathrm{d}l = \int \psi_T\mathrm{d}\theta - \psi_P\mathrm{d}\zeta = \int\left(\psi_T\mathrm{d}_\zeta\theta - \psi_P\right)\mathrm{d}\zeta. \tag{3.26}$$

Evaluating $\delta\mathcal{S}_B = 0$, yields Hamilton's equations for field line evolution,

$$\mathrm{d}_\zeta\psi_T = -\frac{\partial\psi_P}{\partial\theta}, \qquad \mathrm{d}_\zeta\theta = \frac{\partial\psi_P}{\partial\psi_T}, \tag{3.27}$$

which are formally identical to (2.71) with $\mathcal{H} = \psi_P$, $t = \zeta$, $q = \theta$ and $p = \psi_T$. Expressing (3.25) in terms of (ψ_T, θ, ζ), which are known as *canonical magnetic co-ordinates*, leads to a contra-variant representation for **B**,

$$\mathbf{B} = \nabla\psi_T\times\nabla\theta - \frac{\partial\psi_P}{\partial\theta}\nabla\theta\times\nabla\zeta + \frac{\partial\psi_P}{\partial\psi_T}\nabla\zeta\times\nabla\psi_T. \tag{3.28}$$

Flux co-ordinates (r_f, θ_f, ζ_f) are clearly a special case of canonical co-ordinates (ψ_T, θ, ζ); the relation between these two systems is identical to that between angle-action $(\boldsymbol{\vartheta}, \boldsymbol{\mathcal{I}})$ and generalized $(\mathbf{q}, \mathbf{p})$ variables.

3.2 Plasma current in MHD equilibrium

Having established that efficient plasma confinement can only be achieved in the presence of toroidally nested flux surfaces, we now turn to examine the flow of charge, i.e. the current density, required to satisfy quasi-neutrality and the MHD force balance in toroidal plasmas,

$$\nabla \cdot \mathbf{B} = 0 = \nabla \cdot \mathbf{J}, \qquad \nabla p = \mathbf{J} \times \mathbf{B}, \qquad \mu_0 \mathbf{J} = \nabla \times \mathbf{B}. \tag{3.29}$$

These relations, which are the necessary conditions for MHD equilibrium, follow directly from (2.246)–(2.248) with $\mathrm{d}_t = 0$. The force balance can be rewritten in terms of the *magnetic stress* tensor as (2.198), i.e. $\nabla \cdot (\mathsf{p} + \mathcal{T}_B) = 0$. Expanding $\mathcal{T}_B$ in terms of the total pressure and magnetic tension, transforms (3.29) into a balance of compressive (cp) and tensile (ts) forces,

$$\nabla_\perp \left(p + \frac{B^2}{2\mu_0} \right) = \mathbf{F}_{cp} = \mathbf{F}_{ts} = \frac{\mathbf{B}}{\mu_0} \cdot \nabla \mathbf{B} = \frac{B^2}{\mu_0} \boldsymbol{\kappa}. \tag{3.30}$$

As a result, the sum of plasma and magnetic pressures can only change in the presence, and in the direction, of finite magnetic curvature, $\boldsymbol{\kappa}$.

In toroidal geometry, the magnetic curvature vector may be decomposed into the *normal* and *geodesic* parts, which are its radial and diamagnetic co-variant components, and represent the bending of magnetic field lines normal to, and within, the flux surface, respectively,[15]

$$\boldsymbol{\kappa} = \kappa_r \nabla r_f + \kappa_\zeta \nabla \zeta_f + \kappa_\theta \nabla \theta_f = \kappa_\perp \nabla r_f + \kappa_\wedge (\nabla \zeta_f - q \nabla \theta_f) = \kappa_\perp \hat{\mathbf{e}}_\perp + \kappa_\wedge \hat{\mathbf{e}}_\wedge. \tag{3.31}$$

In regions where the field lines are straight ($\boldsymbol{\kappa} = 0$), the radial gradients of the plasma and magnetic pressures must be identically balanced,

$$\boldsymbol{\kappa} = \mathbf{b} \cdot \nabla \mathbf{b} = \nabla_\parallel \mathbf{b} = 0 \qquad \Rightarrow \qquad p + B^2/2\mu_0 = \text{const.} \tag{3.32}$$

The existence of toroidal equilibrium, requires (3.30) be satisfied in all directions, in particular along both the minor and major radii, ∇r and ∇R. The former, known as the *radial force balance*, is just the requirement of plasma confinement, whereas the latter, known as the *toroidal force balance*, has its origin in the tendency of all toroidal plasmas to expand in major radius R. The two force balances scale differently with the inverse aspect ratio, $\epsilon = r/R_0$, with *net* radial forces being $O(1)$ and toroidal forces being $O(\epsilon)$ or higher, see Section 3.3.

Let us first investigate the dominant, radial force balance. Taking a dot product of (3.29) with $\mathbf{b}$ yields $\mathbf{b} \cdot \nabla p = \nabla_\parallel p = 0$, which states that the pressure gradient is everywhere normal to the flux surface, so that p is a flux surface label. Taking a

[15] Note that $\kappa_\theta = -q\kappa_\zeta$ follows from $\kappa_\parallel = 0$.

cross product of (3.29) with **b** yields the perpendicular current needed to balance the plasma pressure gradient,

$$\mathbf{J}_\perp \equiv \mathbf{b} \times (\mathbf{J} \times \mathbf{b}) = \frac{\mathbf{b}}{B} \times \nabla p = \mathbf{J}_* = J_* \hat{\mathbf{e}}_\wedge, \tag{3.33}$$

which is just the diamagnetic current (2.209) introduced in Section 2.4.2.[16] Since $\mathbf{J}_*$ is normal to both **b** and $\nabla p \propto \hat{\mathbf{e}}_\perp$, it flows within the flux surface in the diamagnetic direction, $\hat{\mathbf{e}}_\wedge$ (3.11). Its lines of force wind their way helically around the torus at right angles to **b**, making q *poloidal* turns for every toroidal turn, in contrast to the **B**-field lines which make q *toroidal* turns for every poloidal turn. Hence, all statements made earlier regarding ergodic, resonant and chaotic regions of **b**, apply equally well to $\mathbf{J}_*$.[17]

Evaluating the divergence of (3.33) explicitly, we find

$$\nabla \cdot \mathbf{J}_* = \frac{2\nabla p}{B^2} \cdot (\mathbf{b} \times \nabla B) = 2\mathbf{b} \cdot \left(\nabla p \times \nabla B^{-1}\right) \approx \frac{2\nabla p}{B} \cdot (\mathbf{b} \times \boldsymbol{\kappa}), \tag{3.34}$$

where the last expression applies for $\epsilon \ll 1$. The above is equal to the compression of the current due to the magnetic drift of guiding centres,

$$\nabla \cdot \mathbf{J}_* = \mathbf{J}_B \cdot \nabla \ln p = \nabla \cdot \mathbf{J}_B, \qquad \mathbf{J}_B = \frac{p}{B}(\nabla \times \mathbf{b} + \mathbf{b} \times \nabla \ln B), \tag{3.35}$$

in line with equivalence of particle and fluid pictures of plasma dynamics. This expression for $\mathbf{J}_B$ follows from averaging the current resulting from combining the guiding centre drift(s) (2.37) with (2.39), over the Maxwellian distribution(s) f_{Ms} using the expression (A.11). In tokamaks, $\nabla B \approx \nabla B_T$ points mainly towards the major axis, cf. (3.89), producing a vertical flow of charge and an outward radial $\mathbf{E} \times \mathbf{B}$ drift. Unchecked, this drift would result in outward movement of the plasma as a whole and consequent loss of equilibrium. To maintain the equilibrium requires a parallel return current which cancels the vertical **E** field.

The divergences in (3.35) vanish exactly in cylindrical geometry, where $\boldsymbol{\kappa} \propto \nabla r$ and $\nabla p \times \nabla B = 0$, see Section 3.3. On the other hand, they do not generally vanish in toroidal geometry, e.g. in a tokamak, B increases towards the major axis ($B \propto R^{-1}$, $\nabla B^{-1} \propto \nabla R$), while pressure increases towards the minor axis ($\nabla p = p' \nabla r_f \propto -\nabla r$). Consequently, (3.34) is zero only at the outer (inner) mid-planes,

[16] Recall that the label *diamagnetic* refers to the field $\mathbf{B}_* = -\frac{1}{2}\beta \mathbf{B}$ induced by $\mathbf{J}_*$, which is proportional to the plasma β (2.12) and always opposes the ambient magnetic field **B**.

[17] To fully appreciate this complementarity, recall the discussion in Section 2.4.2.3, from which it follows that $\mathbf{J}_*$ is the sum of diamagnetic flows $e_s n_s \mathbf{V}_{*s}$ (2.206) of different species, which originate in (i) the radial gradient in the density and/or energy of the gyrating moments (magnetization flow), and (ii) non-uniformities in the magnetic field (GC-flow).

where ∇R and ∇r are (anti-)parallel, and is largest at the top and bottom of the flux surfaces, where $\nabla R \perp \nabla r$. It is instructive to evaluate (3.34) explicitly for small ϵ tokamak, see Section 3.3, which confirms the above picture and predicts a sinusoidal dependence of $\nabla \cdot \mathbf{J}_*$ on the poloidal angle θ, i.e. it vanishes at the inner and outer mid-planes and is largest at top and bottom of the torus.

In order to prevent local accumulation of charge, i.e. to satisfy quasi-neutrality (3.29), the non-solenoidal part of $\mathbf{J}_*$ must be compensated by a parallel *return current*, $\mathbf{J}_\parallel = J_\parallel \mathbf{b}$. This current is specified, up to an additive constant, by $\nabla \cdot (\mathbf{J}_\parallel + \mathbf{J}_*) = 0$, and thus originates in the toroidicity of the confined plasma. Any additional contributions to $J_\parallel$,[18] represented by this constant, must be divergence free, so that the net current $\mathbf{J} = \mathbf{J}_\parallel + \mathbf{J}_*$ is solenoidal everywhere. Once again, calculation of (3.34) for a small ϵ tokamak, see Section 3.3, reveals that $J_\parallel(\theta) \propto \cos\theta$, i.e. it vanishes at the top and bottom of the torus and is largest at the inner and outer mid-planes.

Finally, it is significant that $\mathbf{J}$ is normal to ∇p, i.e. $\mathbf{J} \cdot \nabla r_f = J^r = 0$, and flows within the flux surface at some finite and variable angle, given by $\tan^{-1}(J_*/J_\parallel)$, to the magnetic field lines. The equal flow of opposite charges implied by $J^r = 0$ is known as *ambipolar* transport, and is a stronger constraint than that of global charge conservation, namely $\langle J^r \rangle_B = 0$. In Chapter 5 we will see that $J^r \neq 0$ in the presence of inter-particle collisions.

3.2.1 Hamada co-ordinates

The most general expression for $J_\parallel$ may be obtained in Hamada co-ordinates, in which $B^r = J^r = 0$, $B^\zeta = qB^\theta$ and $\sqrt{g_H} = 1$. Making use of (B.9), (B.10) and (B.13) the MHD equilibrium constraints (3.29) can be written as,

$$\partial_\theta B^\theta + \partial_\zeta B^\zeta = 0 = \partial_\theta J^\theta + \partial_\zeta J^\zeta, \qquad J^\theta B^\zeta - J^\zeta B^\theta = p', \tag{3.36}$$

$$\partial_\theta B_\zeta - \partial_\zeta B_\theta = 0, \qquad \partial_r B_\zeta = \mu_0 J^\theta, \qquad \partial_r B_\theta = \mu_0 J^\zeta. \tag{3.37}$$

The final expression in (3.36), which represents the radial force balance,[19] can be simplified using (3.13) and combined with $\nabla \cdot \mathbf{J} = 0$ to yield,

$$qJ^\theta - J^\zeta = \frac{p'}{\psi_P'} = \frac{\mathrm{d}p}{\mathrm{d}\psi_P}, \qquad (\partial_\theta + q\partial_\zeta)J^\theta = \partial_\zeta\left(\frac{\mathrm{d}p}{\mathrm{d}\psi_P}\right) = 0, \tag{3.38}$$

where we made use of the fact that q, p' and ψ_P' are flux surface labels. Since $\partial_\theta + q\partial_\zeta$ is equivalent to $\mathbf{b} \cdot \nabla = \nabla_\parallel$, the above result implies that $\nabla_\parallel J^\theta = 0 = \nabla_\parallel J^\zeta$,

[18] For instance, any externally driven currents (inductive or non-inductive) in a tokamak.
[19] The other components include terms with $B^r = 0$ and $J^r = 0$, and thus vanish exactly.

i.e., that both J^θ and J^ζ are flux surface labels,[20] which permits $\mathbf{J}$ to be expressed in purely contra-variant form, as shown in Table 3.1. Combining the contra-variant expressions for $\mathbf{J}$ and $\mathbf{B}$, we find

$$\mathbf{J} = \frac{J^\theta}{B^\theta}\mathbf{B} - \frac{\mathrm{d}p}{\mathrm{d}\psi_P}\nabla r_H \times \nabla\theta_H = \frac{1}{\psi_P'}\left(J^\theta\mathbf{B} - p'\nabla r_H \times \nabla\theta_H\right). \quad (3.39)$$

The return current now follows directly from the dot product of $\mathbf{B}$ and $\mathbf{J}$,

$$J_\parallel = \mathbf{b}\cdot\mathbf{J} = \left(J^\theta - \frac{p' b_\zeta}{B}\right)\frac{B}{\psi_P'}, \qquad b_\zeta = B_\zeta/B. \quad (3.40)$$

A similar expression for $J_\parallel$ can also be derived in terms of J^ζ and B_θ,

$$J_\parallel = \left(J^\zeta + \frac{p' b_\theta}{B}\right)\frac{B}{q\psi_P'}, \qquad b_\theta = B_\theta/B. \quad (3.41)$$

In either case, the leading term represents the divergence-free part of the parallel current, which is otherwise not specified and corresponds to the integration constant in the solution of $\nabla\cdot(\mathbf{J}_* + \mathbf{J}_\parallel) = 0$, while the second term is the return current per se required by quasi-neutrality.[21] Indeed, using the foregoing results, it is straightforward to show that

$$\nabla\cdot\left(\mathbf{J}_* + J_\parallel^{ret}\mathbf{b}\right) = 0, \qquad \mathbf{J}_* = \frac{p'}{B}\mathbf{b}\times\nabla r_H, \qquad J_\parallel^{ret} = -\frac{p' b_\zeta}{\psi_P'}. \quad (3.42)$$

As expected from (3.34), $J_\parallel^{ret}$ is linearly proportional to the radial pressure gradient p' and thus plays a vital role in maintaining the MHD equilibrium.[22] As we shall see shortly, $J_\parallel^{ret}$ is also an important element in virtually all aspects of plasma confinement, e.g. it is largely responsible for both neo classical and turbulent transport, see Chapters 5 and 6. As a rule of thumb, whenever parallel return currents are impeded, whether by collisions, inertia, or fluctuations, then radial plasma transport is enhanced.

In Section 2.4.2.7 we formulated a fluid closure constraint for the electron heat flux by applying a field line average (2.275) to the electron energy equation (2.276). The same strategy can be used to constrain the average co-variant field and current components in MHD equilibrium.[23] For instance, applying the flux surface average (3.17) to $B^2 = B_\theta B^\theta + B_\zeta B^\zeta$ yields,[24]

[20] This result is not true in other flux co-ordinate systems, in which $\sqrt{g}$ is non-uniform.

[21] The constraint defining the return current can also be expressed as $\nabla_\parallel(BJ_\parallel) = 2(\mathbf{J}\cdot\nabla)B$ which follows from inserting B_θ and B_ζ obtained from (3.40)–(3.41) into the first relation in (3.37).

[22] MHD equilibria in which $\nabla\cdot\mathbf{J}_* = 0 = J_\parallel^{ret}$, known as *omnigenous*, do exist, but are rare.

[23] The contra-variant components and their radial gradients, are flux surface labels and are thus equal to their flux surface averages. The same is true for q, p, ψ_P and their radial gradients.

[24] For simplicity, we drop the subscript B on the flux surface average.

$$\langle B^2\rangle = B^\theta\langle B_\theta\rangle + B^\zeta\langle B_\zeta\rangle = \psi'_P\langle B_\theta\rangle + q\psi'_P\langle B_\zeta\rangle. \quad (3.43)$$

Similarly, the radial gradients of B_θ and B_ζ are constrained by the flux surface average of the MHD force balance (3.38) and Ampere's law (3.37),

$$q\langle B_\zeta\rangle' + \langle B_\theta\rangle' + \mu_0 \mathrm{d}p/\mathrm{d}\psi_P = 0 = B^\zeta\langle B_\zeta\rangle' + B^\theta\langle B_\theta\rangle' + \mu_0 p'. \quad (3.44)$$

This expression states that plasma energy *confinement* ($p' < 0$) is a direct consequence of plasma *diamagnetizm*, i.e. the tendency of the internal (diamagnetic and return) currents, J^{int}, to generate a field B^{int} opposite in sign to the ambient field, B^{ext}, and thus to reduce the net magnetic pressure, $B^2/2\mu_0 = (B^{int} + B^{ext})^2/2\mu_0$, towards the centre of the plasma, $\langle B^2\rangle' > 0$. The strength of this diamagnetic effect is typically quantified in terms of the *poloidal beta* β_P defined in (3.20). Starting from (3.44), it is easy to show that the typically large ratio $B_T^2(r)/B_P^2(a)$ varies from the edge ($r = a$) to the centre ($r = 0$) of the plasma by roughly $\beta_P - 1$.[25] Finally, flux surface averaging (3.40) and (3.41) yields the final set of constraints,

$$\langle B^2\rangle\langle B_\zeta\rangle' + \mu_0 p'\langle B_\zeta\rangle + \mu_0\psi'_P\langle BJ_\parallel\rangle = 0, \quad (3.45)$$

$$\langle B^2\rangle\langle B_\theta\rangle' + \mu_0 p'\langle B_\theta\rangle - \mu_0 q\psi'_P\langle BJ_\parallel\rangle = 0, \quad (3.46)$$

which relate B_θ and B_ζ to the flux surface average of $BJ_\parallel$.

3.2.2 Symmetry co-ordinates

In axis-symmetric systems, such as tokamaks, the partial orthogonality ($g_{\zeta r} = 0 = g_{\zeta\theta}$) offered by symmetry co-ordinates outweighs the added complexity of a spatially varying Jacobian (recall that $\sqrt{g_0} = qR^2/B_\zeta(r_0) \propto R^2$, which implies that B^θ, B^ζ, J^θ and J^ζ are no longer flux surface labels). The derivation of the parallel current in MHD equilibrium proceeds along similar lines to that presented in Section 3.2.1 for Hamada co-ordinates; an account of the derivation may be found in most texts dealing with ideal MHD theory, e.g. Batemann (1978). Below we briefly summarize these results.

Denoting the co-variant field component of **B** by $B_\zeta(r_0) = RB_T = I(\psi_P)$,[26] the poloidal and toroidal components of Ampere's law can be written as

$$\mu_0 J^\theta = -\frac{I'}{\sqrt{g_0}} = -\frac{II'}{qR^2}, \qquad \mu_0 J^\zeta = \nabla\cdot(R^{-2}\nabla r_0). \quad (3.47)$$

Substituting these into the radial component of the force balance,

25 Hence, the plasma is diamagnetic when $\beta_P > 1$ and paramagnetic when $\beta_P < 1$.

26 The relation between co-variant, contra-variant and simple vector components is explained in Appendix B. In symmetry co-ordinates, $g_{\zeta\zeta} = g^{\zeta\zeta} = R^2$ and $h_\zeta = h^\zeta = R$, so that $A^\zeta = \mathbf{A}\cdot\nabla\zeta_0 = \mathbf{A}\cdot\hat{\mathbf{e}}_\zeta/R = A_T/R$ and $A_\zeta = g_{\zeta\zeta}A^\zeta = R^2 A^\zeta = RA_T$.

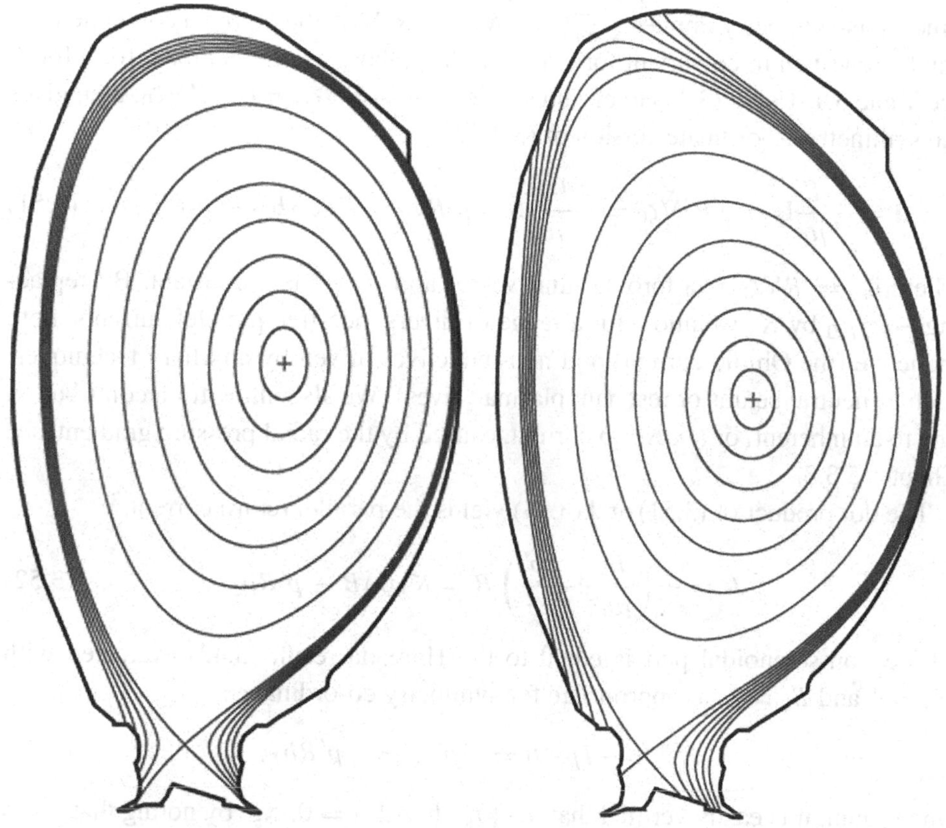

Fig. 3.3. Typical magnetic equilibria, calculated by solving the Grad–Shafranov equation, (3.49), for the JET tokamak with two different plasma shapes; note the shift of the magnetic centre towards the outer mid-plane, a so-called Shafranov shift, and the appearance of the magnetic separatrix, the scrape-off layer (SOL), and a divertor plasma region, see Chapter 7.

$$qJ^\theta - J^\zeta = \frac{p'}{\sqrt{g_0}} = \frac{p'I}{qR^2}, \tag{3.48}$$

yields a second-order elliptic PDE for $r_0(\mathbf{x}) = \psi_P(\mathbf{x}) \equiv \psi(\mathbf{x})$,

$$\Delta^*\psi_P = -II' - \mu_0 R^2 p', \qquad \Delta^* \equiv R^2\nabla\cdot(R^{-2}\nabla), \tag{3.49}$$

commonly known as the *Grad–Shafranov* (G–S) equation. Its solution requires the knowledge of $I(\psi)$ and $p(\psi)$, which act as forcing functions for $\psi(\mathbf{x})$, and the value of ψ on some boundary. A good derivation may be found in Shafranov (1966). A typical solution for an actual tokamak equilibrium is shown in Fig. 3.3.

In exact analogy to the magnetic field, the plasma current can also be written in contra-variant form with J^θ and J^ζ given by (3.47),

$$\mathbf{J} = \sqrt{g_0}(J^\zeta\nabla r_0 \times \nabla\theta_0 + J^\theta\nabla\zeta_0 \times \nabla r_0). \tag{3.50}$$

Since axis-symmetry implies $\sqrt{g_0}\nabla r_0 \times \nabla\theta_0 = R^2\nabla\zeta_0$, the poloidal component $\mathbf{J}_P$ can be rewritten in co-variant form as $J^\zeta R^2\nabla\zeta_0$, thus yielding a mixed form for $\mathbf{J}$, see Table 3.1. Using (3.14) to eliminate $\nabla\zeta_0 \times \nabla r_0 = \mathbf{B}_P = \mathbf{B} - I\nabla\zeta_0$, then gives the symmetry co-ordinate version of (3.39),

$$\mathbf{J} = -\frac{I'}{\mu_0}\mathbf{B} - p'R^2\nabla\zeta_0 = -\frac{I'B}{\mu_0}\mathbf{b} - p'R\hat{\mathbf{e}}_\zeta = K(\psi)B\mathbf{b} - p'R\hat{\mathbf{e}}_\zeta, \tag{3.51}$$

where $\hat{\mathbf{e}}_\zeta = R\nabla\zeta_0$ is a toroidal unit vector and $K(\psi)$ is a constant. By replacing $-I'/\mu_0$ by K, we allow for a range of divergence-free parallel currents, both inductive (the Ohmic current) and non-inductive (driven by auxiliary techniques, such as neutral beams or resonant plasma waves). We also allow for a contribution due to an inherent, or *bootstrap* current, caused by the radial pressure gradient, see Chapter 5.3.3.

The dot product of (3.51) and (3.14) yields the parallel return current,

$$J_\parallel = -\left(\frac{I'}{\mu_0} + \frac{Ip'}{B^2}\right)B = K(\psi)B - p'Rb_T, \tag{3.52}$$

whose non-solenoidal part is equal to the Hamada result (3.42) evaluated with $\psi'_P = 1$ and $B_\zeta = I$, as appropriate for symmetry co-ordinates,

$$J_\parallel^{ret} = -Ip'/B = -p'b_\zeta = -p'Rb_T. \tag{3.53}$$

Once again, it is easily verified that $\nabla\cdot\left(J_\parallel^{ret}\mathbf{b} + \mathbf{J}_*\right) = 0$, e.g. by noting that $J_\parallel^{ret}\mathbf{b}$ cancels with an identical term in the diamagnetic current,

$$\mathbf{J}_* = \frac{\mathbf{b}\times\nabla p}{B} = \frac{\mathbf{b}\times\nabla\psi}{B}p' = -J_\parallel^{ret}\mathbf{b} - p'R\hat{\mathbf{e}}_\zeta = p'R\hat{\mathbf{e}}_\wedge, \tag{3.54}$$

where $\hat{\mathbf{e}}_\wedge = \hat{\mathbf{e}}_\zeta - b_T\mathbf{b}$ and $\nabla\cdot p'R\hat{\mathbf{e}}_\zeta = 0$ on account of axis-symmetry. In deriving (3.54), we used $R^2B^2 = I^2 + |\nabla\psi|^2$ to prove the identity,

$$\mathbf{B}\times\nabla\psi = I\nabla\zeta\times\nabla\psi - |\nabla\psi|^2\nabla\zeta = I\mathbf{B} - R^2B^2\nabla\zeta_0. \tag{3.55}$$

It is instructive to calculate the constant $K(\psi)$ explicitly by dividing (3.51) by B and performing a flux surface average,

$$K(\psi) = (\langle J_\parallel B\rangle + Ip')/\langle B^2\rangle. \tag{3.56}$$

Combining with (3.51), allows $J_\parallel$ to be decomposed into a part with a non-vanishing flux surface average, known as the *Ohmic* current, $J_\parallel^\Omega$, and the remainder, known as the *Pfirsch–Schlüter* (P–S) current, $J_\parallel^{PS}$,

$$J_\parallel = J_\parallel^\Omega + J_\parallel^{PS} = \frac{\langle J_\parallel B\rangle B}{\langle B^2\rangle} - \frac{Ip'}{B}\left(1 - \frac{B^2}{\langle B^2\rangle}\right), \tag{3.57}$$

which clearly satisfies $\langle J_\| B\rangle = \langle J_\|^\Omega B\rangle$ and $\langle J_\|^{PS} B\rangle = 0$. We thus expect the P–S current to vanish, and change sign, near the top and bottom of the torus, where $B^2 \approx \langle B^2\rangle$ and to be largest at the inner and outer mid-planes, where B^2 has its extremal values.

The above analysis may be repeated to calculate the parallel return flows of particles, $n_s V_{\|s}$, and heat, $q_{\|s}$, driven by a finite compressibility (due to toroidal geometry) of their respective gyro-flows, $n_s \mathbf{V}_{\wedge s}$ and $\mathbf{q}_{\wedge s}$. Adopting the drift ordering, the gyro-flow of particles (2.208) can be written as

$$n_s \mathbf{V}_{\wedge s} = n_s(\psi)\omega_s(\psi) R(\hat{\mathbf{e}}_\zeta - b_T \mathbf{b}), \qquad \omega_s \equiv -\varphi' - p_s'/n_s e_s, \tag{3.58}$$

while the gyro-flow of heat is simply the diamagnetic heat flow, $\mathbf{q}_{*s}$ (2.232),

$$\mathbf{q}_{\wedge s} = \mathbf{q}_{*s} = \tfrac{5}{2}\left[p_s(\psi)T_s'(\psi)/e_s B\right] R(\hat{\mathbf{e}}_\zeta - b_T \mathbf{b}). \tag{3.59}$$

In the absence of particle and heat sources, we require $n_s \mathbf{V}_s$ and $\mathbf{q}_s$ to be incompressible,[27] which in turn requires finite parallel return flows,

$$n_s V_{\|s} = n_s(\psi)\omega_s(\psi) R b_T + K_s(\psi) B, \tag{3.60}$$

$$q_{\|s} = -\tfrac{5}{2}\left[p_s(\psi)T_s'(\psi)/e_s\right] R b_T + L_s(\psi) B. \tag{3.61}$$

Combined with (3.58)–(3.59) these yield the net flux surface flows,

$$n_s \mathbf{V}_s = n_s(\psi)\omega_s(\psi) R\hat{\mathbf{e}}_\zeta + K_s(\psi) B\mathbf{b}, \tag{3.62}$$

$$\mathbf{q}_s = -\tfrac{5}{2}\left[p_s(\psi)T_s'(\psi)/e_s B\right] R\hat{\mathbf{e}}_\zeta + L_s(\psi) B\mathbf{b}. \tag{3.63}$$

The first term in (3.62) represents *toroidal* flow of species s, occupying the flux surface characterized by ψ, rotating as a rigid body with a frequency $\omega_s(\psi)$.[28] The second term describes *poloidal* flow, which varies within the flux surface as $nV_\| \propto B \propto R^{-1}$.[29] As expected the summation of charge flows $e_s n_s \mathbf{V}_s$ over all species yields the plasma current (3.51).

Relating the integration constants $K_s(\psi)$ and $L_s(\psi)$ to the flux surface averages, we find the familiar expressions for parallel flows, cf. (3.57),

$$V_{\|s} = V_{\|s}^{PS} + V_{\|s}^{\Omega} = \omega_s R b_T \left(1 - \tfrac{B^2}{\langle B^2\rangle}\right) + \langle V_{\|s} B\rangle \tfrac{B}{\langle B^2\rangle}, \tag{3.64}$$

$$q_{\|s} = q_{\|s}^{PS} + q_{\|s}^{\Omega} = -\tfrac{5}{2}\left(p_s T_s'/e_s\right) R b_T \left(1 - \tfrac{B^2}{\langle B^2\rangle}\right) + \langle q_{\|s} B\rangle \tfrac{B}{\langle B^2\rangle}, \tag{3.65}$$

again divided into the *Pfirsch–Schlüter* and *Ohmic* parts. We will return to these results in Chapter 5.3.3 in the context of neoclassical theory.

[27] This neglects collisional heat exchange Q_s between the various plasma species.
[28] Of course, $\omega_s(\psi)$ may vary with radius, ψ, and with plasma species, s.
[29] This variation originates in the conservation of magnetic flux, $B\mathrm{d}S$ = const, where $\mathrm{d}S$ is the flux tube area, and parallel flow, $n_s V_{\|s}\mathrm{d}S$ = const, so that $V_{\|s} \propto B$.

Finally, we note that by inserting the axis-symmetric expressions for **B** and **J** into the force balance relation (2.35), we obtain the normal, $\kappa_\perp$, and geodesic, $\kappa_\wedge$, components of the magnetic curvature (3.31),

$$\kappa_\perp = -\partial_r \ln R - b_P^2 \Delta^* r, \qquad b_P = B_P/B, \tag{3.66}$$

$$\kappa_\wedge = -q^{-1} b_T^2 \partial_\theta \ln B, \qquad b_T = B_T/B. \tag{3.67}$$

3.3 Large aspect ratio, toroidal equilibrium

Axis-symmetric toroidal equilibrium with finite aspect ratio and arbitrary poloidal cross-section is determined by the solution of the Grad–Shafranov equation (3.49). To illustrate the salient features of the radial and toroidal force balances described by this equation, it is useful to consider MHD equilibria in the large aspect ratio ($\epsilon_a = a/R_0 \ll 1$) limit, where a and R_0 are the minor and major radii of the torus. In this limit, the flux surfaces become circular and concentric to leading order in ϵ_a and resemble a series of thin, nested, cylindrical hoops. Below, we investigate the properties of two small ϵ_a equilibria: the general screw pinch and the small ϵ tokamak.

3.3.1 General screw pinch

The *general screw pinch* is a cylindrical plasma with poloidal and axial fields (B_P, B_T) and currents (J_P, J_T), both of which are flux surface labels, i.e. $B_P = B_P(r)$,[30] etc. It is best described in cylindrical co-ordinates (r, θ, z), in which the field line element projections are related by

$$dl/B = dz/B_T = r d\theta/B_P, \qquad dl^2 = dz^2 + r^2 d\theta. \tag{3.68}$$

Field and current are linked by (3.29), which in cylindrical co-ordinates, with d_r denoted by a prime, yields

$$r^{-1}\partial_\theta B_P + \partial_z B_T = 0, \qquad J_P B_T - J_T B_P = p', \tag{3.69}$$

$$\mu_0 J_P = -B_T', \qquad \mu_0 J_T = r^{-1}(r B_P)'. \tag{3.70}$$

The axial and poloidal fields are related to the total currents by,

$$B_T(r) = B_0 - \mu_0 I_P(r)/2\pi R_0, \qquad B_P(r) = \mu_0 I_T(r)/2\pi r, \tag{3.71}$$

where $I_T(r)$ and $I_P(r)$ are the (integrated) axial and poloidal currents inside the cylinder $r =$ const, $B_0 = B_T(0)$ is the (axial) field at $r = 0$ and $2\pi R_0 = L_T$ is the axial length of the cylinder.

[30] Once again, recall that B_P and B_T are the simple vector components of **B**, whereas B_ζ, B_θ refer to its co-variant components, see comments on notation, page 15.

The MHD force balance follows directly from (3.69)–(3.70),

$$[p + (B_P^2 + B_T^2)/2\mu_0]' + B_P^2/\mu_0 r = 0. \tag{3.72}$$

The second term above implicitly gives the magnetic curvature $\boldsymbol{\kappa}$. To calculate it directly, we write down the magnetic field unit vector,

$$\mathbf{b} = b_T\hat{\mathbf{e}}_z + b_P\hat{\mathbf{e}}_\theta, \qquad b_T = B_T/B, \qquad b_P = B_P/B, \tag{3.73}$$

the gradient operator in cylindrical co-ordinates,

$$\nabla = b_r\hat{\mathbf{e}}_r + b_T\hat{\mathbf{e}}_z + r^{-1}b_P\hat{\mathbf{e}}_\theta, \tag{3.74}$$

and its projection along the field line,

$$\nabla_\parallel = \mathbf{b}\cdot\nabla = b_T\partial_z + r^{-1}b_P\partial_\theta. \tag{3.75}$$

Finally, we apply it to $\mathbf{b}$ to find the magnetic curvature,

$$\boldsymbol{\kappa} = \nabla_\parallel\mathbf{b} = (b_T\partial_z + r^{-1}b_P\partial_\theta)(b_T\hat{\mathbf{e}}_z + b_P\hat{\mathbf{e}}_\theta) = -b_P^2\hat{\mathbf{e}}_r/r. \tag{3.76}$$

Hence, curvature is purely poloidal, being largest for $b_P = 1$ and smallest for $b_P = 0$.[31] Since $b_P = b_P(r)$ is a flux surface label, so is $\kappa_r = \kappa_r(r)$. The *general screw pinch* allows for an arbitrarily tight poloidal winding of the field lines around the cylindrical axis, such that b_P can reach both of these limiting values; typically, $b_T \sim b_P \sim O(1)$ is assumed.

One consequence of (3.76) is that all diamagnetic flows must be solenoidal, e.g. the divergence of the diamagnetic current (3.34) vanishes since ∇p, ∇B and $\boldsymbol{\kappa}$ are all parallel to $\hat{\mathbf{e}}_r$, so that $\nabla p \times \nabla B = 0 = \nabla p \times \boldsymbol{\kappa}$. Hence, there is no source for parallel return currents, such that $J_\parallel^{PS} = 0$.

The *screw pinch* safety factor, $q(r)$ may be derived by directly calculating the rotational transform ι. The latter is easily found by making use of (3.68),

$$\iota(r) = \int_0^{L_T} (\mathrm{d}_z\theta)\mathrm{d}z = \int_0^{2\pi R_0} (B_P/B_T r)\mathrm{d}z = 2\pi R_0 B_P/rB_T. \tag{3.77}$$

$$q(r) = 2\pi/\iota(r) = rB_T/R_0B_P = \epsilon B_T/B_P, \quad s(r) = \frac{\mathrm{d}\ln q}{\mathrm{d}\ln r} = rq'/q. \tag{3.78}$$

The same expression may be obtained from $q(r) = \mathrm{d}\psi_T/\mathrm{d}\psi_P$, and direct evaluation of axial and poloidal fluxes Ψ_T and Ψ_P,

$$\psi_T(r) = \textstyle\int \mathbf{B}\cdot\mathrm{d}\mathbf{S}_z/2\pi = \int B_T(r)r\mathrm{d}r, \tag{3.79}$$

$$\psi_P(r) = \textstyle\int \mathbf{B}\cdot\mathrm{d}\mathbf{S}_\theta/2\pi = \int B_P(r)R_0\mathrm{d}r, \tag{3.80}$$

[31] The former occurs when the field is purely poloidal, which is known as the *Z-pinch*, and yields $|\kappa_r| = -1/r$, while the latter occurs when the field is purely axial, which is known as the *θ-pinch*, and yields $\kappa_r = 0$. Note that curvature $\boldsymbol{\kappa}$ is said to be poloidal when it points along the minor radius of the torus.

For $\epsilon_a = a/R_0 \ll 1$, (3.78) implies $b_P/b_T = r/qR$, which with $b_P \sim b_T$ gives the screw pinch ordering $q \sim O(\epsilon_a)$. In contrast, the tokamak ordering $q \sim O(1)$, implies $b_T^2 \approx 1$ and $b_P^2 \sim O\left(\epsilon_a^2\right)$, such that $\kappa_r = -r/q^2R_0^2$.

The tokamak ordering, see (3.85), assumes dominant axial (toroidal) field, $B_P/B_T \sim O(\epsilon_a)$, and hence $q \sim O(1)$, i.e. the field lines have a very loose poloidal winding, as measured by the rotational transform ι, such that their unit vector is predominantly toroidal; this is in contrast to the general screw pinch ordering which allows for a tighter poloidal winding of the field lines around the minor axis, and hence $B_P/B_T \sim O(1)$. There are two common tokamak orderings for plasma beta: $\beta \sim O\left(\epsilon_a^2\right)$ (Ohmically heated tokamak) or $\beta \sim O(\epsilon_a)$ (high-beta tokamak).

The screw pinch safety factor at the plasma edge, $q_a = q(a)$, is equal to the *cylindrical safety factor*, $q_* \equiv aB_0/R_0\overline{B}_P$ with $\overline{B}_P = \mu_0 I_T/2\pi a$,[32]

$$q(r) = 2B_0/\mu_0 R_0 \langle J_T\rangle_r \qquad \Rightarrow \qquad q_a = 2B_0/\mu_0 R_0\langle J_T\rangle_a \equiv q_*. \tag{3.81}$$

Here we made use of (3.71) and (3.78) and defined $\langle J_T\rangle_r = I_T(r)/A_p$ as the average current density within the cylinder $r = \text{const}$,

$$\langle J_T\rangle_r = 2\pi\int_0^r J_T r\mathrm{d}r/\pi a^2 = I_T(r)/\pi r^2, \tag{3.82}$$

where $I_T = I_T(a)$ is the total current in the plasma. Hence, a hollow $q(r)$ profile, $q(r)' > 0$, implies a peaked current profile, $J_T(r)' < 0$.

The volume averaged axial (toroidal) plasma beta (3.19) now becomes

$$\beta_T = \frac{2\mu_0\langle p\rangle_a}{\left\langle B_T^2\right\rangle_a} \approx \frac{4\mu_0}{a^2B_0^2}\int_0^a pr\mathrm{d}r, \tag{3.83}$$

where we estimated $\left\langle B_T^2\right\rangle_a$ by the axial field B_0. The volume average poloidal beta is found by making the approximation $\left\langle B_P^2\right\rangle_a \approx B_P^2(a) = (\mu_0 I_T/2\pi a)^2 = \overline{B}_P^2$, with the last expression given by (3.71),

$$\beta_P = \frac{2\mu_0\langle p\rangle_a}{\left\langle B_P^2\right\rangle_a} \approx \frac{4\mu_0}{a^2B_P^2(a)}\int_0^a pr\mathrm{d}r \approx \frac{16\pi^2}{\mu_0 I_T^2}\int_0^a pr\mathrm{d}r. \tag{3.84}$$

The two are related by $\beta_T/\beta_P = (b_P/b_T)^2 = (\epsilon_a/q_a)^2 = (\epsilon_a/q_*)^2$.

[32] More generally q_* is defined with πa^2 replaced by the cross-sectional area $A_p = \kappa\pi a^2$ where κ is the vertical elongation (not to be confused with the curvature).

3.3.2 Cylindrical tokamak

Aside from its obvious application to a cylindrical plasma column, the general screw pinch also forms the basis of an idealization, often invoked in the literature and variously known as the *cylindrical* or *straight* tokamak, which involves straightening a small ϵ tokamak (see below) into a circular cylinder and imposing axial symmetry after a length $L_T = 2\pi R_0$, so that $dz = 2\pi R_0 d\zeta$; in this approximation, a cylindrical tokamak follows naturally from a general screw pinch, upon subjecting the latter to the tokamak ordering (3.85). Both these systems neglect the effect of toroidal *curvature*, but the cylindrical tokamak retains the effect of toroidal *topology*. Since magnetic curvature is purely poloidal in cylindrical geometry, see (3.76), i.e. $\boldsymbol{\kappa} = \kappa_r(r)\hat{\mathbf{e}}_r = -b_P^2\hat{\mathbf{e}}_r/r$ is a flux surface label, while it is predominantly toroidal in toroidal geometry, see (3.97), i.e. $\boldsymbol{\kappa} \approx \kappa_R\hat{\mathbf{e}}_R \approx -\hat{\mathbf{e}}_R/R_0$ is not a flux surface label. This is the primary difference between the cylindrical tokamak and the small ϵ tokamak (see below). Note the curvature is said to be toroidal when $\boldsymbol{\kappa}$ points along the major radius of the torus. Hence, all results of Section 3.3.1 apply directly to the cylindrical tokamak.

3.3.3 Large aspect ratio (small ϵ) tokamak

Finally, we repeat the above analysis for a large aspect ratio (small ϵ) tokamak, i.e. a toroidal plasma consisting of axis-symmetric, circular and nearly concentric flux surfaces described by $r \approx$ const, whose poloidal and toroidal fields (B_P, B_T) and currents (J_P, J_T), are not flux surface labels. A small ϵ tokamak resembles a general screw pinch whose axis is bent into a hoop of radius R_0, and whose field lines are wound loosely around the minor axis, such that the magnetic field is predominantly toroidal,

$$B_P/B_T \sim O(\epsilon_a), \qquad q \sim O(1). \tag{3.85}$$

The above relations are known as the *tokamak ordering*, and are usually supplemented by two orderings for plasma beta: $\beta_T \sim O\left(\epsilon_a^2\right)$, the *Ohmically heated* tokamak, or $\beta_T \sim O(\epsilon_a)$, the *high-beta* tokamak.

A small ϵ tokamak is most easily described in orthogonal toroidal co-ordinates (r, θ, ζ), where θ and ζ are the geometrical poloidal and toroidal angles, and r and $R = R_0 + r\cos\theta$ are the radial distances from the minor and major axes. The field line element projections are then related by

$$dl/B = Rd\zeta/B_T = rd\theta/B_P, \qquad dl^2 = dr^2 + R^2d\zeta^2 + r^2d\theta^2. \tag{3.86}$$

Note that the outer side, $\theta = [-\pi/2, \pi/2]$, of a torus has larger area than the inner side, $\theta = [\pi/2, 3\pi/2]$, which for $r =$ const are easily found as,

$$S_{in} = 2\pi^2 R_0 r - 4\pi r^2, \qquad S_{out} = 2\pi^2 R_0 r + 4\pi r^2, \tag{3.87}$$

$$S = S_{in} + S_{out} = 4\pi^2 R_0 r, \quad \Delta S = S_{out} - S_{in} = 8\pi r^2 = 2\epsilon S/\pi. \tag{3.88}$$

The fact that **B** and **J** are *not* flux surface labels, i.e. depend on both r and θ, is largely a consequence of the fact that the *external*, or vacuum, toroidal field scales inversely with R, as is easily demonstrated by integrating Ampere's law in (2.273) over a toroidal ring $R = \text{const}$,

$$B_T^{ext}(r, \theta) = B_0 R_0 / R(r, \theta), \qquad B_0 = B_T^{ext}(0) = \mu_0 I_c / 2\pi R_0, \tag{3.89}$$

where B_0 is the (external) toroidal field on the minor axis and I_c is the current flowing in external (poloidal) coils. As always, the *net* field and current are related by (3.29), which in toroidal co-ordinates, yields

$$r^{-1}\partial_\theta B_P + R^{-1}\partial_\zeta B_T = 0, \qquad J_P B_T - J_T B_P = p', \tag{3.90}$$

$$\mu_0 J_P = -R^{-1}(R B_T)', \qquad \mu_0 J_T = r^{-1}(r B_P)'. \tag{3.91}$$

The lower relations may be integrated radially to yield,

$$B_T(r, \theta) = B_T^{ext}(r, \theta) - \mu_0 I_P(r)/2\pi R, \qquad B_P(r) = \mu_0 I_T(r)/2\pi r, \tag{3.92}$$

where $I_T(r)$ and $I_P(r)$ are the toroidal and poloidal currents inside the annulus $r =$ const. Since $\epsilon = r/R_0 \ll 1$, the toroidal field can be approximated as $B_T/B_0 \approx 1 - \epsilon \cos\theta$, where $B_0 = B_T(r = 0)$ is the toroidal field on axis. The MHD force balance follows directly from the above,

$$\left[p + \left(B_P^2 + B_T^2\right)/2\mu_0\right]' + B_P^2/\mu_0 r + B_T^2/\mu_0 R = 0. \tag{3.93}$$

To calculate $\boldsymbol{\kappa}$, we once again write the magnetic field unit vector,

$$\mathbf{b} = b_T \hat{\mathbf{e}}_\zeta + b_P \hat{\mathbf{e}}_\theta, \qquad b_T = B_T/B, \qquad b_P = B_P/B, \tag{3.94}$$

the gradient operator in toroidal co-ordinates,

$$\nabla = b_r \hat{\mathbf{e}}_r + R^{-1} b_T \hat{\mathbf{e}}_\zeta + r^{-1} b_P \hat{\mathbf{e}}_\theta, \tag{3.95}$$

and its projection along the field line,

$$\nabla_\parallel = \mathbf{b} \cdot \nabla = R^{-1} b_T \partial_\zeta + r^{-1} b_P \partial_\theta. \tag{3.96}$$

Finally, we apply it to **b** to find the magnetic curvature,

$$\kappa = \frac{b_T^2}{R}(\partial_\zeta \hat{\mathbf{e}}_\zeta) + \frac{b_P^2}{r}(\partial_\theta \hat{\mathbf{e}}_\theta) = -b_T^2 \frac{\hat{\mathbf{e}}_R}{R} - b_P^2 \frac{\hat{\mathbf{e}}_r}{r} + O(\epsilon^2). \tag{3.97}$$

which has both poloidal and toroidal contributions. Their ratio $|\kappa_r/\kappa_R|$ is large in the screw pinch ordering, and small in the tokamak ordering,

$$b_P/b_T \sim O(1) \Rightarrow |\kappa_r/\kappa_R| \sim (R/r)(b_P/b_T)^2 \sim O(\epsilon^{-1}) \gg 1, \tag{3.98}$$
$$b_P/b_T \sim O(\epsilon) \Rightarrow |\kappa_r/\kappa_R| \sim (R/r)(b_P/b_T)^2 \sim O(\epsilon) \ll 1. \tag{3.99}$$

This result confirms our earlier assertion that curvature in a small ϵ tokamak is predominantly toroidal, and to lowest order may be approximated as $\kappa = -\hat{\mathbf{e}}_R/R + O(\epsilon)$. It may also be derived by evaluating (3.66)–(3.67) in the small ϵ limit, taking advantage of the fact that in this limit flux surfaces have near-circular poloidal cross-sections,

$$\kappa_\perp \approx \kappa_r \approx -\partial_r \ln R - (r/R_0 q)^2/r \approx -b_T^2/r - \cos\theta/R, \tag{3.100}$$
$$\kappa_\wedge = \kappa_\zeta \approx -\kappa_\theta/q \approx q^{-1}\partial_\theta \ln R \approx -r\sin\theta/qR = -b_P \sin\theta. \tag{3.101}$$

To see that the above are indeed identical to (3.97), it suffices to expand $\kappa \approx -\nabla R/R - b_P^2 \nabla r/r$ with $R = R_0 + r\cos\theta$.

The cylindrical expressions for $q(r)$, $s(r)$, β_T, β_P remain valid to $O(\epsilon)$ for the small ϵ tokamak, e.g. the *toroidal* $q(r)$ may be obtained using (3.86),

$$\iota(r) = \int_0^{2\pi} (\mathrm{d}_\zeta \theta) R \mathrm{d}\zeta = \int_0^{2\pi} (RB_P/rB_T)\mathrm{d}\zeta = 2\pi R_0 B_P/rB_T. \tag{3.102}$$

$$q(r) = 2\pi/\iota(r) = rB_T/R_0 B_P, \qquad s(r) = \mathrm{d}\ln q/\mathrm{d}\ln r = (r/q)q'. \tag{3.103}$$

The appearance of R_0 in place of R is equivalent to surface averaging with

$$\langle \mathcal{A} \rangle_B = \frac{1}{S}\int_0^{2\pi} \mathcal{A} \mathrm{d}S = \frac{1}{\pi R_0}\int_0^{\pi} \mathcal{A}(R_0 + r\cos\theta)\mathrm{d}\theta, \tag{3.104}$$

which yields $\langle R \rangle_B = R_0$ and $\langle r \rangle_B = r$. Applied to (3.97), in which we estimate $b_T^2 \approx 1$ and $b_P^2 \approx (r/qR_0)^2 \sim O(\epsilon^2)$, this gives the average curvature,

$$\langle \kappa \rangle_B = -\frac{\hat{\mathbf{e}}_R}{R_0} + O(\epsilon^2) \approx \langle \kappa_R \rangle_B \hat{\mathbf{e}}_R, \qquad \langle \kappa_R \rangle_B = -\frac{1}{R_0}, \tag{3.105}$$

so that R_0 may be interpreted as the average radius of curvature.

Let us next illustrate the radial and toroidal force balances for a small ϵ tokamak. Inserting the $B^2 \approx B_T^2 \propto R^{-2}$ variation of the magnetic pressure into the net compressional force on a plasma fluid element, $\mathbf{F}_{cp}$, defined in (3.30), and exploiting the fact that $p = p(r)$ is a flux surface label, we find

$$\mathbf{F}_{cp} = -\nabla\left(\frac{B^2}{2\mu_0} + p\right) = -\partial_R\left(\frac{B^2}{2\mu_0}\right)\nabla R - \partial_r p \nabla r. \tag{3.106}$$

Since plasma confinement requires that $p' \equiv \partial_r p < 0$, the projection of $\mathbf{F}_{cp}$ along R is always positive, and thus describes a net outward force,

$$F_{cp}^R = \mathbf{F}_{cp} \cdot \nabla R = B^2/\mu_0 R - p' \cos\theta > 0, \tag{3.107}$$

where $\cos\theta = \nabla r \cdot \nabla R$. The two components of F_{cp}^R are known as the *hoop force* and the *tyre tube force*, respectively, the former originating in the R^{-2} variation of magnetic pressure and the latter in the radial plasma pressure gradient and the R variation of toroidal flux surface area.[33]

It is instructive to flux surface average (3.107) using (3.104), which yields

$$\left\langle F_{cp}^R \right\rangle_B \approx \frac{B_0^2}{R_0 \mu_0} + \frac{r}{2R_0}|p'| + O(\epsilon^2). \tag{3.108}$$

Estimating $|p'|$ as p_0/a, where p_0 is the plasma pressure on the minor axis, and defining β_0 as (2.12) evaluated with p_0 and B_0, we find

$$\left\langle F_{cp}^R \right\rangle_B \approx \left(B_0^2/\mu_0 R_0\right)(1 + \beta_0/4) + O(\epsilon^2). \tag{3.109}$$

Since $\epsilon \ll 1$ implies $\beta_0 \sim O(\epsilon^2) \ll 1$, the tyre tube force (the second term above) is typically much smaller than the hoop force (the first term). However, the hoop force is typically balanced to $O(1)$ by the inward magnetic tension force, i.e. by the magnetic curvature term in (3.30).

To demonstrate this, we first use the estimate of κ for a small ϵ tokamak, (3.97), whose projection along ∇R gives the desired toroidal curvature,

$$\kappa_R \approx -b_T^2/R - b_P^2 \cos\theta/r, \tag{3.110}$$

to show that the projection of the magnetic tension force, $\mathbf{F}_{tn} = B^2\kappa/\mu_0$, defined in (3.30), along ∇R, is always negative (inward pointing),

$$F_{tn}^R = \kappa_R B^2/\mu_0 = -B_T^2/R - B_P^2 \cos\theta/r < 0. \tag{3.111}$$

Hence, its flux surface average is dominated by the toroidal curvature,

$$\left\langle F_{tn}^R \right\rangle_B = -B_0^2/\mu_0 R_0 + O(\epsilon^2) \tag{3.112}$$

and exactly balances the hoop force term in (3.109), leaving the sum,

$$\left\langle F_{cp}^R + F_{tn}^R \right\rangle_B = \left(B_0^2/\mu_0 R_0\right)\beta_0/4 + O(\epsilon^2). \tag{3.113}$$

The remaining net outward force, which is just the tyre tube component of $\mathbf{F}_{cp}$, (3.107), shows that a purely concentric, toroidal equilibrium is only possible in the limit of vanishing plasma beta, $\beta_T \to 0$. When β_T is finite, radial pressure gradients tend to shift the centre of inward flux surfaces (smaller r) towards the

33 Note that the outer side of the torus has larger area than the inner side, see (3.87).

outer mid-plane (in the direction of ∇R), such that the circular flux surfaces are no longer co-axial.

This so-called *Shafranov shift* augments (reduces) the poloidal field on the outboard (inboard) side of the torus. Its magnitude may be calculated by expanding (3.49) in ϵ and adopting the (r, θ, ζ) co-ordinates. The solution of the resulting toroidal force balance yields the desired outward shift, Δ,

$$\frac{\Delta}{b} = \frac{b}{2R_0}\left[\left(\beta_P + \frac{l_i - 1}{2}\right)\left(1 - \frac{a^2}{b^2}\right) + \ln\frac{b}{a}\right] - \frac{B_Z}{B_P(b)}, \tag{3.114}$$

where a and b are the radial positions of the plasma boundary and the conducting wall, respectively, $B_P(b) = \mu_0 I_T(a)/2\pi b$ is the poloidal field at $r = b$, B_Z is an externally applied vertical field, β_P is given by (3.84) and l_i is the *internal inductance* of the plasma, defined as

$$l_i[B_P(r)] = \frac{2}{a^2 B_P^2(a)}\int_0^a B_P^2 r\,\mathrm{d}r, \qquad l_i^{max} = l_i[B_P(a)\delta(r-a)] = 1. \tag{3.115}$$

Note that Δ vanishes in the presence of a tight-fitting conducting wall $b = a$, and scales linearly with β_P and l_i in the no-wall limit. In practice, the vertical field B_Z is used to counteract the outward shift.[34] Assuming a small, positive $\Delta/a \sim O(\epsilon_a)$, the (normal) magnetic curvature becomes

$$\kappa_\perp(r,\theta) = -\frac{b_P^2}{r} - \frac{\cos\theta}{R_0} + \frac{r\cos^2\theta}{R_0^2} + \frac{\Delta\sin^2\theta}{rR_0} + O(\epsilon^2), \tag{3.116}$$

where the first term is the poloidal curvature, the second is the $O(1)$ toroidal curvature, the third is the $O(\epsilon)$ toroidal correction, and the last is the curvature caused by the Shafranov shift. Upon flux surface averaging, the second term disappears, while $\cos^2\theta$ and $\sin^2\theta$ are replaced by $1/2$,

$$\langle\kappa_\perp\rangle_B \approx -\frac{r}{q^2R_0^2}\left(1 - \frac{q^2}{2}\right) + \frac{\Delta}{2rR_0} \approx -\frac{r}{q^2R_0^2}\left(1 - q^2\right). \tag{3.117}$$

In the second expression, we assumed that $\Delta/r \sim r/R_0$, so that the two stabilizing contributions are comparable. Hence, the unfavourable poloidal curvature is overcome by the average $O(\epsilon)$ toroidal corrections, making the average curvature favourable, provided that $q > 1$. Since this is necessary for stability against the $m = 1$ internal kink mode, and imposed by sawtooth oscillations, we may conclude that the small ϵ tokamak is also stable against local interchange modes, see Section 4.2.8.

[34] Note that if the plasma pressure was to be promptly reduced, the vertical field would tend to push the plasma in the direction of the inner wall, i.e. $-\nabla R$. This is observed in tokamaks during certain plasma disruptions (so called ideal-limit disruptions, in which the thermal quench is particularly fast), and most of the plasma wall contact occurs on the inner wall.

Finally, let us calculate the *Pfirsch–Schlüter* current for a small ϵ tokamak. Substituting $1 - B^2/\langle B^2\rangle = 2\epsilon\cos\theta + O(\epsilon^2)$ into (3.57), we find

$$J_{\parallel}^{PS} = -\frac{I}{B}\frac{\mathrm{d}p}{\mathrm{d}\psi}\left(1 - \frac{B^2}{\langle B^2\rangle}\right) \approx -\frac{2p'}{B_P}\cos\theta \approx -\frac{2p'}{B}\frac{q}{\epsilon}\cos\theta, \qquad (3.118)$$

so that $\langle J_{\parallel}^{PS} B\rangle \approx 0$ to the above accuracy. This result can also be derived directly by integrating the divergence of the diamagnetic current, assuming $\mathbf{b} \times \nabla B$ points downwards, so that $(\hat{\mathbf{e}}_r, \hat{\mathbf{e}}_\theta, \hat{\mathbf{e}}_\zeta)$ form a right-handed basis,

$$\nabla \cdot \mathbf{J}_* \approx \frac{2p'\hat{\mathbf{e}}_r}{B_0 R_0} \cdot (\mathbf{b} \times \hat{\mathbf{e}}_R) \approx \frac{2p'\mathbf{b}}{B_0 R_0} \cdot (\hat{\mathbf{e}}_r \times \hat{\mathbf{e}}_R) = -\frac{2p'}{B_0 R_0}\sin\theta. \qquad (3.119)$$

As a concluding remark, we note that from (3.57), $J_T \sim J_{\parallel}^{\Omega} \sim B_P/\mu_0 r$ and $J_{\parallel}^{PS} \sim \epsilon p/r B_P$, so that the ratio of P–S and Ohmic currents, $J_{\parallel}^{PS}/J_{\parallel}^{\Omega} \sim \epsilon\beta_P$, vanishes in the cylindrical limit ($\epsilon \to 0$), consistent with the origin of the former in the toroidal curvature of the tokamak plasma.

3.4 Further reading

An excellent treatment of MHD equilibrium may be found in Freidberg (1991). Good accounts are also given in Batemann (1978); Miyamoto (1989); Whyte (1989); Nishikawa and Wakatani (1994); Goedbloed and Poedts (2004) and Wesson (2004, Chapter 3). Magnetic geometry and flux co-ordinates are comprehensively reviewed in D'haeseleer *et al.* (1991).

4

Magnetized plasma stability

'Let the nature of a fluid be assumed to be such that of its parts, which lie evenly and are continous, that which is under lesser pressure is driven along by that under greater pressure.'

Archimedes (c. 500 BC)

Dynamical equilibrium does not guarantee dynamical stability. In general, an equilibrium is said to be stable if the system remains bounded, i.e. confined to its neighbourhood, after being subjected to a small perturbation. There are many different classifications of stability, e.g. *linear* stability implies small perturbations, while *non-linear* stability allows for perturbations of arbitrary size. The former is equivalent to *spectral* stability, which occurs when all eigenvalues of the linearized dynamical operator have real parts which are positive or zero. The most general formulation of linear stability is the *energy principle*, which states that an equilibrium point is stable when it represents a minimum of the potential energy of the system. Most dynamical systems have both stable and unstable equilibria, e.g. the lowest and highest position of a pendulum. Having identified an equilibrium point, one should next investigate its stability properties and, if the point proves unstable, the physical mechanism responsible for the instability.[1]

4.1 Hydrodynamic waves and instabilities

By way of introduction to plasma instabilities, let us consider the stability properties of a stratified neutral fluid in the presence of a gravitational field $\mathbf{g} = g\hat{\mathbf{e}}_g$. Note that MHD differs from hydrodynamics only by the appearance of the Lorentz force term $\mathbf{J} \times \mathbf{B}$ in the momentum conservation equation (2.247) and by an additional

[1] The term *instability* is generally understood to refer to this physical mechanism, although it is also used to refer to the unstable evolution of the dynamical system.

evolution equation for $\mathbf{B}$, including the resistivity η. Moreover, the presence of $\mathbf{B}$ provides a preferred direction. When $\mathbf{B} \to 0$, or $\beta \gg 1$, isotropy is restored and MHD reduces to hydrodynamics.

The fluid is assumed to conserve mass, momentum and energy,

$$\mathrm{d}_t\rho + \rho\nabla\cdot\mathbf{V} = 0, \tag{4.1}$$

$$\rho\mathrm{d}_t\mathbf{V} + \nabla p + \nabla\cdot\boldsymbol{\pi} = \rho\mathbf{g}, \tag{4.2}$$

$$\mathrm{d}_t p + \gamma p\nabla\cdot\mathbf{V} + (\gamma-1)[\boldsymbol{\pi} : \nabla\mathbf{V} + \nabla\cdot\mathbf{q}] = 0, \tag{4.3}$$

where $\mathrm{d}_t = \partial_t + \mathbf{V}\cdot\nabla$ is the convective time derivative, $\gamma = c_p/c_v$ is the ratio of heat capacities at constant pressure and volume, $\boldsymbol{\pi}$ is the viscous stress and $\mathbf{q}$ is the heat flow. In Section 5.1.2 we will see that $\boldsymbol{\pi}$ and $\mathbf{q}$ can be related to the velocity and temperature gradients, respectively, as

$$\boldsymbol{\pi} = -\mu\mathsf{W}, \qquad \mathbf{q} = -\kappa\nabla T, \qquad T = p/n = pm/\rho, \tag{4.4}$$

$$\mathsf{W} = 2\{\nabla\mathbf{V}\}^* - \tfrac{2}{3}(\nabla\cdot\mathbf{V})\mathsf{I}, \quad W_{ij} = \partial_{x_j}V_i + \partial_{x_i}V_j - \tfrac{2}{3}\delta_{ij}\nabla\cdot\mathbf{V}, \tag{4.5}$$

$$\nabla\cdot\mathsf{W} = \nabla^2\mathbf{V} + \tfrac{1}{3}\nabla(\nabla\cdot\mathbf{V}), \qquad \{\nabla\mathbf{V}\}^* = \tfrac{1}{2}(\nabla\mathbf{V} + \nabla\mathbf{V}^\dagger), \tag{4.6}$$

$$\mathsf{W} : \nabla\mathbf{V} = \mathsf{W} : \{\nabla\mathbf{V}\}^* = 2\{\nabla\mathbf{V}\}^* : \{\nabla\mathbf{V}\}^* - \tfrac{2}{3}(\nabla\cdot\mathbf{V})^2, \tag{4.7}$$

where μ is the dynamic viscosity, κ is the heat conductivity and W is the (traceless) *rate-of-strain tensor*;[2] it is also customary to introduce the kinematic viscosity, $\nu = \mu/\rho$, and heat diffusivity, $\chi = \kappa/n = \kappa m/\rho$.

The equilibrium conditions of hydrostatics are obtained by setting $\mathrm{d}_t = 0$ and $\mathbf{V} = 0$ in (4.1)–(4.3), which yields

$$\nabla p = \rho\mathbf{g}, \qquad \kappa\nabla T = \text{const.} \tag{4.8}$$

If the flow is non-dissipative ($\nu = 0 = \chi$), then (4.1)–(4.3) reduce to

$$\mathrm{d}_t\rho + \rho\nabla\cdot\mathbf{V} = 0 = \mathrm{d}_t T + \gamma T\nabla\cdot\mathbf{V}, \tag{4.9}$$

$$\mathrm{d}_t\mathbf{V} + \rho^{-1}\nabla p = \mathbf{g}, \tag{4.10}$$

where (4.9) yields the polytropic relation, $p \propto \rho^\gamma$, or $T \propto \rho^{\gamma-1}$.[3] Together with (4.10) this relation allows the propagation of compressible (acoustic, sound) waves with a phase velocity equal to the sound speed, V_S,

$$\omega^2 = (kV_S)^2 \quad\Rightarrow\quad v_{ph} \equiv \frac{\omega}{k} = V_S = \sqrt{\frac{\gamma p}{\rho}} = \sqrt{\frac{\gamma T}{m}}. \tag{4.11}$$

[2] Note that since $\mathsf{W}^\dagger = \mathsf{W}$, $\mathsf{W}^\dagger : \nabla\mathbf{V}^\dagger = \mathsf{W} : \nabla\mathbf{V}$ and $\{\nabla\mathbf{V}\}^{*\dagger} = \{\nabla\mathbf{V}\}^*$, then $\mathsf{W} : \nabla\mathbf{V} = \frac{1}{2}(\mathsf{W} + \mathsf{W}^\dagger) : \nabla\mathbf{V} = \frac{1}{2}(\mathsf{W} : \nabla\mathbf{V} + \mathsf{W}^\dagger : \nabla\mathbf{V}^\dagger) = \mathsf{W} : \{\nabla\mathbf{V}\}^*$, as mentioned following (2.171).

[3] Combining the polytropic relation with (4.8), leads to the so-called byrotropic equilibrium $\rho(z) = \rho(0)(1 - (1-\gamma^{-1})z/L_g)^{1/(\gamma-1)}$, where z is the vertical distance and $L_g = p(0)/\rho(0)g$ is a characteristic length scale. When $\gamma = 1$ one finds $\rho(z) = \rho(0)\exp(-z/L_g)$. A version of (4.1)–(4.3) for a layer of thickness $L \ll L_g$ is known as the *Boussinesq* approximation.

Acoustic waves are longitudinal disturbances of the form ρ_1, p_1, $\mathbf{V}_1 \cdot \mathbf{k} \propto \exp(\mathrm{i}\mathbf{k} \cdot \mathbf{x} - \mathrm{i}\omega t)$, where $\mathbf{k}$ is the wave vector and ω is the wave frequency. Since their group velocity, $v_{gr} \equiv \partial_{\mathbf{k}}\omega = V_S$ is independent of k, acoustic waves are non-dispersive – a fact which can be ultimately traced to the absence of an intrinsic length scale in hydrodynamics. Acoustic waves, which are the only waves possible in a uniform neutral fluid, are the primary means of establishing and restoring the hydrostatic equilibrium (4.8).

If the flow is incompressible ($\nabla \cdot \mathbf{V} = 0$), which pertains when the characteristic flow velocity is small compared to the sound speed, i.e. when the Mach number, $M_S = V/V_S \ll 1$,[4] then (4.1) and (4.3) reduce to

$$\mathrm{d}_t\rho = 0, \qquad \mathrm{d}_t T = (\gamma - 1)[2\nu m\{\nabla\mathbf{V}\}^* : \{\nabla\mathbf{V}\}^* + \chi\nabla^2 T]. \tag{4.12}$$

Moreover, the *Navier–Stokes* equation (4.2) can then be replaced by,

$$\mathrm{d}_t\boldsymbol{\Omega} = \boldsymbol{\Omega} \cdot \nabla\mathbf{V} + \nu\nabla^2\boldsymbol{\Omega}, \qquad \boldsymbol{\Omega} \equiv \nabla \times \mathbf{V}, \tag{4.13}$$

obtained by taking the curl of (4.2), which eliminates the pressure gradient;[5] note that $\mathbf{V}$ may be computed a posteriori from $\nabla^2\mathbf{V} = -\nabla \times \boldsymbol{\Omega}$. Since $\boldsymbol{\Omega}$ is just the *vorticity*, (4.13) is known as the *vorticity* equation in the fluid frame of reference; it may also be written in the laboratory frame as

$$\partial_t\boldsymbol{\Omega} = \nabla \times (\mathbf{V} \times \boldsymbol{\Omega}) + \nu\nabla^2\boldsymbol{\Omega}. \tag{4.14}$$

The above equations imply that, for incompressible flows, vorticity can only be created/destroyed by vortex stretching $\boldsymbol{\Omega} \cdot \nabla\mathbf{V}$ and viscous dissipation $\nu\nabla^2\boldsymbol{\Omega}$. In the absence of viscosity, the total *circulation* of the flow must remain constant and the vorticity is effectively *frozen into* the flow,

$$\mathrm{d}_t \int_S \boldsymbol{\Omega} \cdot \mathrm{d}\mathbf{S} = \mathrm{d}_t \int_S \nabla \times \mathbf{V} \cdot \mathrm{d}\mathbf{S} = \mathrm{d}_t \oint_C \mathbf{V} \cdot \mathrm{d}\mathbf{l} = 0, \tag{4.15}$$

a result known as *Kelvin's vorticity theorem*. There is a clear resemblance between the vorticity, $\Omega = \nabla \times \mathbf{V}$, the current density, $\mu_0\mathbf{J} = \nabla \times \mathbf{B}$ and the magnetic field, $\mathbf{B} = \nabla \times \mathbf{A}$. This similarity is based on a general version of Kelvin's theorem, which states that $\mathrm{d}_t \int_S \boldsymbol{Q} \cdot \mathrm{d}\mathbf{S} = 0$ for any vector field satisfying $\partial_t\boldsymbol{Q} = \nabla \times (\mathbf{V} \times \boldsymbol{Q})$. Hence, the conservation of magnetic flux in ideal MHD, and of circulation in ideal HD, stems from same vector relation. For future reference we also write down the MHD version of (4.13),

$$\mathrm{d}_t\boldsymbol{\Omega} = \boldsymbol{\Omega} \cdot \nabla\mathbf{V} + (\mathbf{B} \cdot \nabla\mathbf{J} - \mathbf{J} \cdot \nabla\mathbf{B})/\rho + \nu\nabla^2\boldsymbol{\Omega}, \tag{4.16}$$

[4] Since the compressible term $T\nabla \cdot \mathbf{V}$ in (4.3) is smaller than $\mathbf{V} \cdot \nabla T$ by a factor of order M_S^2.

[5] Here we used the fact that $\mathbf{V} \cdot \nabla\mathbf{V} = \frac{1}{2}\nabla(\mathbf{V} \cdot \mathbf{V}) - \mathbf{V} \times \nabla \times \mathbf{V} = \frac{1}{2}\nabla V^2 - \mathbf{V} \times \boldsymbol{\Omega}$ and that $\nabla \times (\mathbf{V} \times \boldsymbol{\Omega}) = \mathbf{V} \cdot \nabla\boldsymbol{\Omega} + \boldsymbol{\Omega} \cdot \nabla\mathbf{V}$ for incompressible flows ($\nabla \cdot \mathbf{V} = 0$).

which shows that the Lorentz force, $\mathbf{J} \times \mathbf{B}$, provides a source of vorticity.

As a consequence of (4.15), an initially irrotational flow ($\mathbf{\Omega} = 0$) must remain so at all times. The velocity can then be written as the gradient of some potential ($\mathbf{V} = -\nabla\phi$) and the incompressibility condition, $\nabla \cdot \mathbf{V} = 0$, may be recast in the form of Laplace's equation, $\nabla^2\phi = 0$.

Finally, if the flow is both incompressible and non-dissipative, then

$$d_t\rho = 0 = d_t T, \qquad d_t\mathbf{V} + \rho^{-1}\nabla p = \mathbf{g}, \qquad d_t\mathbf{\Omega} = \mathbf{\Omega} \cdot \nabla\mathbf{V}. \tag{4.17}$$

Consider two fluids having different densities and zero relative velocity. This system has two possible hydrostatic equilibria: (i) the heavy fluid is above ($\rho_a > \rho_b$), and (ii) below ($\rho_a < \rho_b$), the lighter fluid. Of these, only (ii) is stable, since it clearly represents the lowest energy state of the system. In contrast, the gravitational potential energy stored in (i) can be converted into kinetic energy, driving the heavier fluid down, and the lighter up, in some complicated 3D motion. The physical mechanism leading to interchange of the lighter and heavier fluids is the *bouyancy* force, $\mathbf{g}\cdot\nabla\rho$, experienced by elements of the lighter fluid, and the process is known as an *interchange*, or *Rayleigh–Taylor* (R–T), instability. It also applies to a stratified fluid in which the density increases gradually with height ($\hat{\mathbf{e}}_g \cdot \nabla\rho < 0$), e.g. in a fluid heated from below, provided the fluid expands with increasing temperature, which is the case for nearly all fluids,[6] e.g. for an ideal gas, in which $p \propto \rho T$, one finds $\partial \ln \rho / \partial \ln T = -1$ and $\alpha_\rho = 1/T > 0$.

The instability occurs when the $\hat{\mathbf{e}}_g \cdot \nabla T$ is super-adiabatic, i.e. when the specific entropy, $s = (c_p/\gamma)\ln(p\rho^{-\gamma})$, decreases with height,

$$-N^2 \equiv \mathbf{g} \cdot (\nabla \ln T - (1 - 1/\gamma)\nabla \ln p) = \mathbf{g} \cdot \nabla \ln(p\rho^{-\gamma})/\gamma > 0, \tag{4.18}$$

where N is the *Brunt–Vaisala* frequency of internal surface waves, see below. Combined with equilibrium (4.8) and the ideal gas law, this gives the *Schwarzschild instability condition*,

$$\hat{\mathbf{e}}_g \cdot \nabla T > (1 - 1/\gamma)mg. \tag{4.19}$$

The above variant of the R–T instability, in which $\nabla\rho$ is caused by ∇T, is known after *Rayleigh* and *Benard* (R–B). In the presence of constant heating, (4.3) requires constant energy flow from the bottom to the top of the system. When (4.19) is satisfied, the system becomes R–B unstable, leading to upward flow of warm, light fluid, and downward flow of cold, heavy fluid. The resulting thermally driven flow is known as R–B, or *thermal*, convection.

If the light and heavy fluids have a finite relative velocity along their horizontal interface, $\hat{\mathbf{e}}_g \times (\mathbf{V}_a - \mathbf{V}_b) \neq 0$, then the kinetic energy of this flow can amplify

[6] That is the thermal expansion coefficient $\alpha_\rho \equiv -(\partial_T \ln \rho)_p$ is typically positive.

vertical perturbations. In the non-linear phase of this process, the surface ripples evolve into complicated vortices, destroy the smooth interface and effectively mix the two fluids. This mechanism, known as the *Kelvin–Helmholtz* (K–H) instability, applies equally well to a sheared flow, e.g. when the horizontal velocity changes with height, $\hat{\mathbf{e}}_g \times \nabla\mathbf{V} \neq 0$.

The linear stability criteria for both R–T and K–H instabilities in ideal, incompressible fluids, can be derived by linearizing (4.17) around the hydrostatic equilibrium (4.8), and assuming small, vertical perturbations of the two-fluid interface, $\xi \propto \exp(ikx - i\omega t)$, where x is the distance and k the wave number along the surface. Since the initial flows are irrotational ($\mathbf{\Omega} = 0$) everywhere, except at the interface, which is thus a thin vorticity sheet, the perturbed velocity may be written as $\mathbf{V}_1 = \nabla\phi_1$. Denoting quantities above and below the interface by subscripts a and b, respectively, and the vertical distance away from the surface as z, we choose ϕ_{1a} and ϕ_{1b} so that $\mathbf{V}_{1a}$ and $\mathbf{V}_{1b}$ decrease exponentially with $|z|$:[7] $\phi_{1a} \propto \exp(ikx + kz - i\omega t)$ and $\phi_{1b} \propto \exp(ikx - kz - i\omega t)$. Inserting these quantities into (4.17), introducing the effect of *surface tension*, $\mathcal{T}_S$,[8] by adding the term $(\mathcal{T}_S/\rho)\partial_x^2\xi$ on the left-hand side of (4.10), and matching the conditions at the interface, yields the phase velocity of *surface gravity* waves (Choudhuri, 1998),

$$\frac{\omega}{k} = \frac{\rho_b V_b + \rho_a V_a}{\rho_b + \rho_a} \pm \sqrt{\frac{g}{k}\left[\frac{\rho_b - \rho_a}{\rho_b + \rho_a} + \frac{k^2 \mathcal{T}_S}{g(\rho_b + \rho_a)}\right] - \frac{\rho_a \rho_b (V_b - V_a)^2}{(\rho_b + \rho_a)^2}}. \tag{4.20}$$

Since all perturbed quantities are assumed to vary with time as $\exp(-i\omega t)$, the real part of ω represents the frequency of sinusoidal oscillations, while the imaginary part is the rate of exponential growth or decay, i.e. when $\omega = \mathrm{Re}(\omega) \neq 0$, the wave propagates along the surface with velocity (4.20),[9] when $\mathrm{Im}(\omega) < 0$, it is quickly damped, and when $\mathrm{Im}(\omega) > 0$ it quickly grows without bound. The last case is just the instability we seek.

The condition for R–T instability follows from (4.20) with $V_a = V_b$,

$$\rho_b + k^2 \mathcal{T}_S/g < \rho_a, \tag{4.21}$$

which agrees with the intuitive result ($\rho_b < \rho_a$) for $\mathcal{T}_S = 0$ and shows that surface tension has a strong restoring effect on small-scale (large k) perturbations. To appreciate this effect, imagine that an R–T unstable system is contained in a vertical cylinder. As the diameter d of the cylinder is reduced, $k_{min} = 2\pi/d$ increases

[7] This assumes infinite 'height' and 'depth' to our two-fluid system. Introduction of finite height and depth does not alter the basic results as presented below.

[8] By definition $\mathcal{T}_S$ counteracts any bending deformations of the two-fluid interface.

[9] These waves are dispersive, since their phase velocity increases as the square root of the wavelength, i.e. $\omega/k \propto k^{-1/2}$.

until, for a sufficiently small d, $k_{min}^2 > g(\rho_a - \rho_b)/\mathcal{T}_S$ is achieved and the R–T instability is suppressed.[10]

Similarly, we obtain the condition for the K–H instability,

$$g\left[\frac{\rho_b - \rho_a}{k} + \frac{k\mathcal{T}_S}{g}\right] < \frac{\rho_a \rho_b (V_b - V_a)^2}{(\rho_b + \rho_a)}, \tag{4.22}$$

which indicates that an R–T stable system ($\rho_b > \rho_a$) becomes K–H unstable for sufficiently large velocity discontinuity $|V_a - V_b|$. Although short scales (large k) are more unstable, they are also more effectively stabilized by surface tension, so that the K–H instability is most likely to occur at some intermediate k, determined from (4.22).

In a stratified fluid the *surface* gravity waves are replaced by *internal* gravity waves, and the stabilizing effect of surface tension by that of viscosity and heat diffusivity. The dispersion relation for internal gravity waves may be derived by linearizing (4.1)–(4.3) about the equilibrium point (4.8). In the absence of velocity shear, this yields

$$(k_{\perp g} N/k)^2 = \omega^2 + \mathrm{i}\omega k^2(\nu + \chi) - \nu\chi k^4, \tag{4.23}$$

where $k_{\perp g} = |\mathbf{k} \times \hat{\mathbf{e}}_g|$ and N is the *Brunt–Vaisala* frequency (4.18), which in the incompressible limit ($\gamma \to \infty$) reduces to $N^2 = \mathbf{g}\cdot\nabla \ln \rho = -\alpha_\rho \mathbf{g}\cdot\nabla T = \alpha_\rho g \partial_z T$. The system becomes (R–T) unstable when $\omega^2 < 0$ or

$$(k_{\perp g} N/k)^2 < k^4(\nu^2 + \chi^2 - 2\nu\chi). \tag{4.24}$$

Since the right-hand side is always positive, the diffusive terms have a stabilizing effect, predominantly at small scales (large k), allowing an incompressible fluid to support a finite temperature gradient, $-\partial_z T$, and a compressible fluid to support a super-adiabatic gradient, in excess of (4.19).[11]

The relationship between waves and instabilities, as outlined above, is common to both fluid and plasma dynamics. Magnetized plasma equilibria are thus subject to a range of instabilities, most of which can be associated with corresponding plasma waves. Below, we briefly examine the MHD, drift- and (gyro-)kinetic waves and instabilities of most relevance to power exhausts in confined plasmas. For further reading on these topics, which have been actively studied for many years, see Section 4.5.

[10] Indeed, since the volume to surface ratio scales linearly with d, one would expect surface tension to dominate at sufficiently small scales.

[11] In an incompressible ($\nabla \cdot \mathbf{V} = 0, \gamma \to \infty$), non-dissipative ($\nu = 0 = \chi$), fluid, (4.23) and (4.24) become $(\omega k/k_{\perp g} N)^2 = 0$ and $\hat{\mathbf{e}}_g \cdot \nabla T > 0$, or $\partial_z T < 0$, respectively.

4.2 MHD waves and instabilities

The underlying reason for fluid, and specifically magneto-hydrodynamic, instabilities has been succinctly summarized by Biskamp (2003):

'A fluid, in particular a plasma, becomes unstable when the gradient of velocity, pressure, or magnetic field exceeds a certain threshold, which occurs, roughly speaking, when the convective transport of momentum, heat, or magnetic flux is more efficient than the corresponding diffusive transport by viscosity, thermal conduction, or resistivity. There are hence three types of instabilities, which play a fundamental role in macroscopic plasma dynamics: the Kelvin–Helmholtz instability driven by a velocity shear; the Rayleigh–Taylor instability caused by the buoyancy force in a stratified system; and current-driven MHD instabilities in a magnetized plasma.'

4.2.1 Ideal MHD waves in a uniform plasma

Before investigating MHD instabilities, all of which are related to plasma non-uniformity, it is worthwhile to review the basic MHD waves in a uniform magnetized plasma. For this purpose we subject an MHD equilibrium to small perturbations of the form $\exp(\mathrm{i}\mathbf{k}\cdot\mathbf{x}-\mathrm{i}\omega t)$,[12] which is equivalent to Fourier transforming (2.246)–(2.248) in space and time[13] and neglecting products of perturbed quantities, such as $\mathbf{V}_1\times\mathbf{B}_1$,

$$\omega\rho_1=\rho_0\mathbf{k}\cdot\mathbf{V}_1,\qquad \mathbf{E}_1+\mathbf{V}_1\times\mathbf{B}_0=0, \tag{4.25}$$

$$\omega\rho_0\mathbf{V}_1=\mathbf{k}p_1+\mathbf{B}_0\times\mathbf{J}_1/\mathrm{i}\qquad \mathrm{i}\mathbf{k}\times\mathbf{B}_1=\mu_0\mathbf{J}_1, \tag{4.26}$$

$$\omega p_1=\gamma p_0\mathbf{k}\cdot\mathbf{V}_1,\qquad \mathbf{k}\times\mathbf{E}_1-\omega\mathbf{B}_1=0. \tag{4.27}$$

In the above, we also neglected the source terms and replaced the ideal gas ratio of specific heats 5/3 by some general value γ. Note that $\nabla\cdot\mathbf{B}=0$ implies $\mathbf{k}\cdot\mathbf{B}_1=0$, which also follows from (4.27). After simplification, (4.25)–(4.27) reduce to a single equation for the perturbed velocity, $\mathbf{V}_1$,

$$\omega^2\rho_0\mathbf{V}_1=\left[\frac{\mathbf{B}_0}{\mu_0}\times(\mathbf{k}\times\mathbf{B}_0)+\gamma p_0\mathbf{k}\right]\mathbf{k}\cdot\mathbf{V}_1-\mathbf{k}\cdot\frac{\mathbf{B}_0}{\mu_0}(\mathbf{k}\times\mathbf{V}_1)\times\mathbf{B}_0. \tag{4.28}$$

This velocity can have both longitudinal ($\mathbf{k}\cdot\mathbf{V}_1$) and transverse ($\mathbf{k}\times\mathbf{V}_1$) components, thus allowing both compressional and shear deformations. We next choose a Cartesian co-ordinate system (x, y, z) in which the z-axis is aligned with the magnetic field ($\hat{\mathbf{e}}_z=\mathbf{b}_0=\mathbf{B}_0/B$). Without loss of generality, we can assume the wave to propagate in the y–z plane, so that $\mathbf{k}=k_\parallel\hat{\mathbf{e}}_z+k_\perp\hat{\mathbf{e}}_y$. The vector equation (4.28)

[12] We denote equilibrium and perturbed quantities by subscripts 0 and 1, respectively.
[13] Fourier transform reduces the differential operators to algebraic ones: $\partial_t\to-\mathrm{i}\omega$ and $\nabla\to\mathrm{i}\mathbf{k}\cdot$. Since equilibrium is assumed to be stationary, $\mathbf{V}_0=0$, then $\mathrm{d}_t=\partial_t+\mathbf{V}\cdot\nabla\to-\mathrm{i}\omega+\mathrm{i}\mathbf{V}_1\cdot\mathbf{k}\cdot$.

then yields three algebraic equations for the velocity components V_{1x}, V_{1y}, V_{1z}. The eigenmodes of this system may be found by requiring the determinant of the matrix to vanish, which yields the following dispersion relation,

$$\left[\omega^2 - (k_\parallel V_A)^2\right]\left[\omega^4 - (\omega k V_{max})^2 + (k k_\parallel V_S V_A)^2\right] = 0, \tag{4.29}$$

where V_S is the sound speed (4.11), V_A is the Alfvén speed (2.253), $V_{max}^2 = V_S^2 + V_A^2$ and $k^2 = k_\parallel^2 + k_\perp^2$. The three roots of this equation, namely,

$$\omega_A^2 = k_\parallel^2 V_A^2, \qquad \omega_\pm^2 = \tfrac{1}{2} k^2 V_{max}^2 \left(1 \pm \sqrt{1 - \alpha^2}\right), \tag{4.30}$$

$$\alpha^2 = 4(k_\parallel / k)^2 \left(V_A V_S / V_{max}^2\right)^2, \tag{4.31}$$

represent the three possible MHD waves in a uniform plasma. Since the discriminant in (4.30) is positive definite,[14] all three waves are stable.

The first root (ω_A) corresponds to the *shear Alfvén* wave, which travels along the field lines ($\mathbf{k} \parallel \mathbf{b}$) at a phase velocity equal to the Alfvén speed, $\omega_A / k_\parallel = \pm V_A$.[15] It consists of purely transverse, and hence incompressible ($\mathbf{V}_1 \cdot \mathbf{k} = 0$), plasma motions and magnetic field displacements ($\mathbf{B}_1 \propto \mathbf{V}_1$). The origin of the shear Alfvén wave is the combination of ion inertia and field line tension, $\mathcal{T}_{B\parallel}$, as discussed in Sections 2.4.2.1 and 2.4.2.6.

The second (ω_+) and third (ω_-) roots of (4.30) represent the *fast* and *slow magneto-sonic* waves, which as the names suggest are a combination of Alfvén and sound waves. They can propagate both along and across $\mathbf{B}_0$,

$$\mathbf{k} \perp \mathbf{b}_0 \;\Rightarrow\; \omega_+^2 = k_\perp^2 V_{max}^2, \quad \omega_-^2 = 0, \tag{4.32}$$

$$\mathbf{k} \parallel \mathbf{b}_0 \;\Rightarrow\; \tfrac{1}{2} < \omega_+^2 / k_\parallel^2 V_{max}^2 < 1,\ 0 < \omega_-^2 / k_\parallel^2 V_{max}^2 < \frac{1}{2}. \tag{4.33}$$

The fast wave includes in-phase oscillations of plasma pressure and magnetic stress, $p + \mathcal{T}_B$, with $\mathcal{T}_B$ given by (2.197), while the slow wave includes out-of-phase oscillations; the perturbations to p and $\mathcal{T}_{B\perp}$ are compressional, while those to $\mathcal{T}_{B\parallel}$ are tensional, i.e. involve transverse shearing. Recall that $\mathcal{T}_{B\perp} > 0$ is indicative of magnetic pressure, which counteracts compressional deformations, while $\mathcal{T}_{B\parallel} < 0$ represents magnetic tension, which counteracts bending and/or twisting deformations. Hence (2.247) can be written as $\rho d_t \mathbf{V} + \nabla \cdot (p + \mathcal{T}_B) = 0$, see (2.198). In the limits of small and large plasma pressure, compared to the magnetic pressure, i.e. when $\beta \approx (V_S / V_A)^2 \ll 1$ or $\gg 1$, the fast wave reduces to the Alfvén and acoustic waves, respectively, while the slow wave vanishes entirely,

[14] Since $k_\parallel / k \le 1$ and $(a^2 + b^2)^2 / a^2 b^2 = 2 + a^2/b^2 + b^2/a^2 \ge 4$, hence $\alpha^2 \le 1$.

[15] The Alfvén waves are non-dispersive, with the same group velocity $v_{gr} = \partial_k \omega_A = \pm V_A$ for all k.

$$\beta \ll 1 \quad \Rightarrow \quad \omega_+^2 \approx k^2 V_A^2,\ \omega_-^2 \approx 0, \tag{4.34}$$

$$\beta \gg 1 \quad \Rightarrow \quad \omega_+^2 \approx k^2 V_S^2,\ \omega_-^2 \approx 0. \tag{4.35}$$

The low-β perpendicular wave (4.32), involves purely longitudinal deformations and is hence known as the *compressional Alfvén* wave.

4.2.2 MHD waves and instabilities in a stratified plasma

We next consider the effect of finite $\mathbf{B}_0$ on the stability of a plasma in a gravitiation field, i.e. on the MHD version of the R–T and K–H instabilities considered in Section 4.1. In ideal MHD ($\nu = \chi = \eta = 0$), (4.23) becomes

$$(k_{\perp g} N/k)^2 + k_\parallel^2 V_A^2 = \omega^2, \qquad \mathbf{k}_{\perp g} \equiv \hat{\mathbf{e}}_g \times (\mathbf{k} \times \hat{\mathbf{e}}_g) \tag{4.36}$$

which shows that the magnetic field has a stabilizing effect on the R–T instability provided that $k_\parallel = \mathbf{k} \cdot \mathbf{b}$ does not vanish. The effect is largest for $k_\parallel = k$, when vertical motions ($\mathbf{V}_1 \cdot \hat{\mathbf{e}}_g$) involve bending deformations of the magnetic field lines.[16] As a result, the R–T instability in a magnetized plasma proceeds by internal gravity waves perpendicular to both $\mathbf{g}$ and $\mathbf{b}$, i.e. by $\mathbf{B}$-field aligned ($\mathbf{k} \propto \mathbf{g} \times \mathbf{b}$, $k_\parallel \approx 0$, $k = k_\perp$), or *flute-like*, perturbations[17] for which the stabilizing effect of magnetic tension $\mathcal{T}_{B\parallel}$, which plays a role analogous to the surface tension $\mathcal{T}_S$ discussed in Section 4.1, is minimized.

Similarly, neglecting surface tension and assuming $\rho_a = \rho_b$ for simplicity, one finds the dispersion relation for the K–H instability across a vorticity sheet in a non-dissipative magnetized plasma,

$$\frac{\omega}{k} = \frac{1}{2}(V_b + V_a) \pm \sqrt{V_A^2 - \frac{1}{4}(V_b - V_a)^2}, \tag{4.37}$$

which is the ideal MHD version of (4.20). Once again magnetic field has a stabilizing effect, with the K–H instability suppressed entirely when $V_A > |V_a - V_b|/2$, i.e. when the kinetic energy of the flow discontinuity can be transported away from the disturbance by the shear Alfvén wave.

4.2.3 Ideal MHD waves and instabilities in a confined plasma

We are now ready to consider the stability properties of a magnetically confined plasma equilibrium, i.e. a non-uniform, toroidal plasma supporting a radial

[16] Such deformations excite shear Alfvén waves, which transport the free energy released by the bouyancy force away from the point of disturbance. For $k_{\perp g} = k_\parallel = k$, the phase velocity becomes $\omega^2/k^2 = (N/k)^2 + V_A^2$ and the waves are R–T unstable when $-N^2 > (kV_A)^2$.

[17] The name refers to the flutes in a Grecian column, which field aligned perturbations in a cylindrical plasma column resemble, see below.

pressure gradient. This subject has received considerable attention since the beginnings of fusion research. As a result, proper treatment of MHD stability, our understanding of which is now at a very advanced stage, requires a dedicated monograph. Fortunately for the reader, several excellent texts on the subject are already in existence, see Section 4.5. Below we restrict the discussion to physical mechanisms driving MHD instabilities, mathematical techniques employed to calculate stability criteria and main analytical and numerical results for axis-symmetric geometry. We begin with (2.247), replacing ∇p by the divergence of the pressure tensor, $\nabla \cdot \mathsf{p} = \nabla p + \nabla \cdot \boldsymbol{\pi}$,

$$\mathbf{f} \equiv \rho \mathrm{d}_t \mathbf{V} + \nabla \cdot \boldsymbol{\pi} = \mathbf{J} \times \mathbf{B} - \nabla p, \qquad \mathrm{d}_t = \partial_t + \mathbf{V} \cdot \nabla. \tag{4.38}$$

Here we denote the left-hand side, which represents the plasma inertia and/or the net force acting on a fluid element, by $\mathbf{f}$; note that $\mathbf{f} = 0$ in equilibrium, i.e. in the absence of flow, $\mathbf{V} = 0 = \boldsymbol{\pi}$. Let us consider $\mathbf{f}$ as an operator acting on a small displacement away from equilibrium $\boldsymbol{\xi}(\mathbf{x})$,[18]

$$\mathbf{f}(\boldsymbol{\xi}) = \mathbf{J}_1 \times \mathbf{B}_0 + \mathbf{J}_0 \times \mathbf{B}_1 - \nabla p_1. \tag{4.39}$$

Here we linearized $\mathbf{f}$, neglecting products of small quantities; the perturbed pressure, p_1, current, $\mathbf{J}_1$, and field, $\mathbf{B}_1$, are found by linearizing (2.246)–(2.248) and integrating over time (for p_1 and $\mathbf{B}_1$ only),

$$\mu_0 \mathbf{J}_1 = \nabla \times \mathbf{B}_1, \qquad \mathbf{B}_1 = \nabla \times (\boldsymbol{\xi} \times \mathbf{B}_0), \qquad p_1 = -\gamma p_0 \nabla \cdot \boldsymbol{\xi} - \boldsymbol{\xi} \cdot \nabla p_0. \tag{4.40}$$

Inserting (4.39) into (4.38) gives the explicit form of the linear force operator,

$$\mathbf{f}(\boldsymbol{\xi}) = \mu_0^{-1}(\nabla \times \mathbf{B}_0) \times \mathbf{B}_1 + \mu_0^{-1}(\nabla \times \mathbf{B}_1) \times \mathbf{B}_0 + \nabla(\gamma p_0 \nabla \cdot \boldsymbol{\xi} + \boldsymbol{\xi} \cdot \nabla p_0). \tag{4.41}$$

This operator has several important properties: (i) it is self-adjoint,

$$\int \boldsymbol{\eta}^* \cdot \mathbf{f}(\boldsymbol{\xi}) \mathrm{d}\mathbf{x} = \int \boldsymbol{\xi}^* \cdot \mathbf{f}(\boldsymbol{\eta}) \mathrm{d}\mathbf{x}, \tag{4.42}$$

for any two displacements $\boldsymbol{\xi}$ and $\boldsymbol{\eta}$; (ii) the squares of its eigenvalues ω^2 are purely real for all discrete normal modes $\boldsymbol{\xi}(\mathbf{x}, t) = \boldsymbol{\xi}_\omega(\mathbf{x}) \exp(-\mathrm{i}\omega t)$, such that its eigenvalues ω are either real or purely imaginary;[19] (iii) as a result, all discrete normal modes are orthogonal,

$$\left(\omega_n^2 - \omega_m^2\right) \int \rho \boldsymbol{\xi}_m^* \cdot \boldsymbol{\xi}_n \mathrm{d}\mathbf{x} = 0; \tag{4.43}$$

(iv) $\mathbf{f}(\boldsymbol{\xi})$ permits both discrete and continuum eigenvalues, although the latter are only allowed in the stable domain, i.e. for $\omega^2 \geq 0$, and represent wave continua;

[18] Note that $\boldsymbol{\xi} = 0$ in equilibrium and that $\mathbf{V}_1 = \partial_t \boldsymbol{\xi}$. The adjoint of $\boldsymbol{\xi}$ is denoted as $\boldsymbol{\xi}^*$.

[19] This pertains only to ideal MHD. Dissipation introduces complex parts to ω^2 and ω.

(v) consequently, the transition from unstable to stable domains must occur via the *marginally stable* state, defined by $\omega = 0 = \mathbf{f}(\boldsymbol{\xi}_\omega)$. The above properties suggest that *normal mode analysis* is well suited for investigating MHD spectral stability of confined plasmas and, indeed, it is the most common technique used for this purpose.

An alternative approach relies on variational techniques and the *energy principle*, which must be satisfied for MHD linear stability, namely

$$\delta^2 W[\boldsymbol{\xi}(\mathbf{x})] > 0 \tag{4.44}$$

where $\delta^2 W$ is the second variation of the potential (free) energy;[20] the first variation δW vanishes in equilibrium. This variation is equal to the work done in displacing the plasma element against the force $\mathbf{f}$, given by (4.39),

$$\delta^2 W[\boldsymbol{\xi}] = -\frac{1}{2}\int \boldsymbol{\xi}^* \cdot \mathbf{f}(\boldsymbol{\xi})\mathbf{dx}. \tag{4.45}$$

Inserting (4.41) into (4.45) and simplifying yields the following expression,

$$\begin{aligned}\delta^2 W[\boldsymbol{\xi}] = \frac{1}{2}\int_p \gamma p_0 |\nabla\cdot\boldsymbol{\xi}|^2 + (\boldsymbol{\xi}\cdot\nabla p_0)(\nabla\cdot\boldsymbol{\xi}^*) - \mathbf{J}_0\cdot(\mathbf{B}_1\times\boldsymbol{\xi}^*) + \frac{B_1^2}{\mu_0}\mathbf{dx}\\ +\frac{1}{2}\int_s\left(p_1 + \mathbf{B}_0\cdot\frac{\mathbf{B}_1}{\mu_0}\right)\boldsymbol{\xi}\cdot\mathbf{dS} + \frac{1}{2}\int_v \frac{B_1^2}{\mu_0}\mathbf{dx},\end{aligned} \tag{4.46}$$

where the subscripts p, v and s denote integrals over the *plasma* volume, $\delta^2 W_p$, the *vacuum* envelope, which is assumed to surround the plasma, $\delta^2 W_v$, and the plasma–vacuum *surface*, $\delta^2 W_s$, respectively. The surface integral, which follows from application of Gauss's theorem, vanishes for purely tangential displacements, e.g. for a perfectly conducting boundary. This suggests a natural division of MHD instabilities into *internal*, or *fixed boundary*, modes, which do not deform the plasma–vacuum surface, and *external*, or *free boundary*, modes, which include such deformations. The *fixed boundary* form of the MHD energy principle may be obtained by neglecting the surface and vacuum terms, separating $\boldsymbol{\xi}$, $\mathbf{B}_1$ and $\mathbf{J}_1$ into parallel and perpendicular components, and effecting several cancellations involving $\xi_\parallel$,

[20] The relation between the two approaches is precisely the same as that between the Schrödinger and Heisenberg pictures of quantum mechanics. The former relies on normal mode analysis in a Hilbert space (wave functions and Hamiltonian formalism), the other on variational techniques and functionals (the principle of least action and the Lagrangian formalism). Both approaches have been widely applied to the study of equilibria, stability and dynamics of confined plasmas.

$$\delta^2 W_p[\boldsymbol{\xi}] = \frac{1}{2}\int_p \gamma p_0|\nabla\cdot\boldsymbol{\xi}|^2 - 2(\boldsymbol{\xi}_\perp\cdot\nabla p_0)(\boldsymbol{\xi}_\perp^*\cdot\boldsymbol{\kappa}) - \mathbf{J}_{\|0}\cdot(\mathbf{B}_{1\perp}\times\boldsymbol{\xi}_\perp^*)$$
$$+\frac{B_{1\perp}^2}{\mu_0} + \frac{B_0^2}{\mu_0}|\nabla\cdot\boldsymbol{\xi}_\perp + 2\boldsymbol{\xi}_\perp\cdot\boldsymbol{\kappa}|^2 d\mathbf{x} > 0. \tag{4.47}$$

Let us consider the energy associated with each of the terms in (4.47):
1st term ⇒ *fluid compression* ⇒ acoustic waves,
2nd term ⇒ flux tube *interchange* ⇒ R–T instability,
3rd term ⇒ flux tube *twisting* ⇒ *kink* instability,[21]
4th term ⇒ *magnetic tension*, or flux tube *bending* ⇒ shear Alfvén waves,
5th term ⇒ *magnetic compression* ⇒ compressional Alfvén waves.
Only the kink and interchange terms above[22] can lower the plasma free energy, thus destabilizing the equilibrium. In contrast, the compressional and tensile terms, which perturb the total stress tensor, $\mathsf{p}+\mathcal{T}_B$, see (2.197), always increase this energy thus providing a stabilizing effect. As a result, the most stringent stability criteria are imposed by those perturbations which satisfy $\nabla\cdot(\mathsf{p}_1+\mathcal{T}_{B1}) = 0$. In other words, an equilibrium is MHD stable, if and only if, it is stable against incompressible displacements.

Let us now return to (4.38) to derive the primary dynamical relation for plasma fluid instabilities. Taking a parallel projection of the curl of (4.38) yields the *general plasma vorticity* equation,

$$\mathbf{b}\cdot(\nabla\times\mathbf{f} - 2\boldsymbol{\kappa}\times\mathbf{f}) = \mathbf{b}\cdot\left[B^2\nabla(J_\|/B) + 2\boldsymbol{\kappa}\times\nabla p\right], \tag{4.48}$$

where **f** now denotes the fluid inertia, i.e. the left-hand side of (4.38). This relation, also known as the *shear Alfvén law*, is of fundamental importance in the theory of confined plasma stability.[23] It is worth noting that $\nabla(J_\|/B)$ and ∇p in (4.48) originate in the gradients of parallel and perpendicular currents. To illustrate this fact, (4.48) may be recast term by term into a charge conservation equation, $\nabla\cdot\mathbf{J} = 0$, with $\mathbf{J} = \mathbf{J}_\| + \mathbf{J}_* + \mathbf{J}_p$, or

$$\nabla\cdot\mathbf{J}_p = -\nabla\cdot\mathbf{J}_\| - \nabla\cdot\mathbf{J}_*, \tag{4.49}$$

expressing the total current continuity (quasi-neutrality) as the sum of polarization (2.36), parallel return (3.53) and diamagnetic currents (3.33),

$$\mathbf{J}_p = (\rho/B)\mathbf{b}\times(\partial_t + \mathbf{V}_E\cdot\nabla)\mathbf{V}_E, \qquad \mathbf{J}_* = \mathbf{b}\times\nabla p/B. \tag{4.50}$$

21 The name originates in the kinking (buckling or twisting into a spiral) of a current filament.

22 These two terms are known collectively as the *Newcomb* terms. The interchange term typically changes sign as one travels poloidally around the flux surface, i.e. it is positive (destabilizing) on the outer side of the torus and negative (stabilizing) on the inner side. However, global stability is determined by the volume, i.e. flux surface, average of all the terms in (4.47).

23 Although derived here for ideal MHD, its validity does not depend on the ordering scheme.

Here it is useful to recall (3.34) and the discussion in Section 3.2. In steady-state ($d_t = 0$), we recover the equilibrium relation, $\nabla \cdot (\mathbf{J}_\parallel + \mathbf{J}_*) = 0$, whereas transiently the net compression of $\mathbf{J}_\parallel + \mathbf{J}_*$ gives rise to the polarization current, and hence to perpendicular plasma motion. Depending on the ordering of v_E, this motion can either be fast ($v_E/v_{ti} \sim 1$), as in the MHD ordering assumed here, or slow ($v_E/v_{ti} \sim \delta_i$), as in the drift ordering. While global equilibrium can be maintained by a parallel return current, polarization currents are always generated on small scales. They are the origin of so-called micro-instabilities (e.g. drift-waves, interchange modes, etc.) which are responsible for plasma turbulence. Due to the ever-present interchange forcing, micro-instabilities exhibit a ballooning character, i.e. are most active in regions of unfavourable magnetic curvature, see Section 4.2.8 and Section 6.3. In short, the two terms on the right-hand side of (4.48), whose energies are given by the third and fourth and second terms in (4.47), represent vorticity sources due to gradients of parallel current and scalar pressure (or diamagnetic current) and give rise to instabilities involving flux tube bending/twisting and interchange, respectively.

Let us first consider the pressure gradient term in (4.48). This term, which can also be written as $2\mathbf{b} \times \boldsymbol{\kappa} \cdot \nabla p$, vanishes when $\boldsymbol{\kappa} \times \nabla p = 0$, i.e. when $\boldsymbol{\kappa} \parallel \nabla p$, and is largest when $(\mathbf{b} \times \boldsymbol{\kappa}) \parallel \nabla p$. It vanishes for the equilibrium pressure gradient ∇p_0 at the inboard and outboard mid-planes of the torus, since there $\boldsymbol{\kappa} \parallel \nabla p_0 \parallel \nabla R$, but generates a vorticity source of $2\kappa_n k_\wedge p_1$ due to diamagnetic (poloidal) pressure variation. Here we assume an eikonal of $S = \mathbf{k} \cdot \mathbf{x}$, and decompose $\mathbf{k}_\perp$ into diamagnetic and radial components,

$$\boldsymbol{\xi} \propto \exp(i\mathbf{k} \cdot \mathbf{x}), \qquad \mathbf{k} = k_\parallel \mathbf{b} + k_\perp \hat{\mathbf{e}}_\perp = k_\parallel \mathbf{b} + k_\wedge \hat{\mathbf{e}}_\wedge + k_\perp \hat{\mathbf{e}}_\perp. \tag{4.51}$$

Substituting for p_1 with $p_1 \approx -\boldsymbol{\xi} \cdot \nabla p_0$, (4.40), we see that such poloidal variation is caused by flute-like radial displacements, generating a vorticity source of $-2k_\wedge \xi_\perp \kappa_n p_0'$. We thus expect the condition for interchange instability to be related to the sign of the product $\kappa_n p_0'$ or, more generally, of $\boldsymbol{\kappa} \cdot \nabla p_0$. This is confirmed by the second term in (4.47), which indicates that $\boldsymbol{\xi}_\perp$ lowers the potential energy of the plasma, $\delta^2 W_p[\boldsymbol{\xi}]$, and hence drives the interchange instability, when $(\boldsymbol{\xi}_\perp^* \cdot \boldsymbol{\kappa})(\boldsymbol{\xi}_\perp \cdot \nabla p_0) > 0$. This corresponds to the outboard side of the torus, where $\boldsymbol{\kappa} \cdot \nabla p_0 > 0$, which defines the region of *unfavourable*, or *bad*, magnetic curvature. Similarly, $\boldsymbol{\kappa}$ is said to be *favourable*, or *good*, when $\boldsymbol{\kappa} \cdot \nabla p_0 < 0$, which occurs on the inboard side of the torus. Based on this preliminary analysis, we may conclude that the magnetic curvature $\boldsymbol{\kappa}$ plays the role of negative gravity, $-\mathbf{g}$, and subjects the plasma fluid to a buoyancy force proportional to $-\boldsymbol{\kappa} \cdot \nabla p_0$.

By expressing the grad-B and curvature GC-drifts in gravitational form, recall (2.37)–(2.38), this 'negative' gravity is seen to represent the centripetal acceleration of a charged particle (fluid element) moving along a curved magnetic field line (flux tube). The buoyancy force is easily understood in the particle picture: following a flute-like perturbation $\xi_\perp \propto \exp(ik_\wedge\theta)$ of the flux surface $r = \text{const}$, the curvature GC-drift, $\mathbf{v}_\kappa \propto e_s\boldsymbol{\kappa} \times \mathbf{b} \propto e_s\hat{\mathbf{e}}_\wedge$, which has opposite sign for ions and electrons, leads to a poloidal electric field, and thus to a radial (outward) $\mathbf{E}\times\mathbf{B}$ drift. It is worth stressing that the plasma remains quasi-neutral at all times. In effect, the radial plasma motion is required to prevent the separation (polarization) of charge that would otherwise ensue due to an unchecked curvature drift, i.e. $\mathbf{v}_\kappa$ ($\mathbf{J}_*$) leads to V_E ($\mathbf{J}_p$) necessary to ensure that (4.50) is satisfied. Hence, the frequently cited explanation of the radial plasma motion as being due to *charge polarization* is not, strictly speaking, correct.

Let us next examine the current gradient term (4.48), which, when linearized, can be decomposed into two parts,

$$B^2\mathbf{b}\cdot\nabla(J_\parallel/B) \approx B_0\mathbf{B}_1\cdot\nabla(J_{\parallel 0}/B_0) + B_0\mathbf{B}_0\cdot\nabla(J_{\parallel 1}/B_0), \tag{4.52}$$

corresponding to the Newcomb terms in (4.47). Here, the first part, which contains gradients of the *equilibrium* parallel current, corresponds to term (iii) in (4.47), associated with *twisting* deformations and the kink instability. Similarly, the second part, which contains gradients of the *perturbed* parallel current, corresponds to term (iv) in (4.47), associated with *bending* deformations (note that $\mu_0 J_{\parallel 1} = \nabla \times \mathbf{B}_{\perp 1}$ is largest when $\mathbf{b}_1 \perp \mathbf{b}_0$) and the familiar stabilizing effect of magnetic tension. Hence, the displacement $\boldsymbol{\xi}(\mathbf{x})$ is most likely to grow when it exhibits a *flute-like* character,[24]

$$\mathbf{b}_0\cdot\nabla = \nabla_{\parallel 0} \approx \nabla_\parallel \approx 0, \qquad \mathbf{b}_0 = \mathbf{B}_0/B_0, \tag{4.53}$$

so that field line bending is minimized. The inhomogeneous form of (4.53), namely $\nabla_\parallel f = \mathcal{A}/B$, is known as the *magnetic differential equation* and has a solution if, and only if, it satisfies the *Newcomb* solubility conditions,

$$\nabla_\parallel f = \mathcal{A}/B \quad \Rightarrow \quad \oint(\mathcal{A}/B)\mathrm{d}l = 0 = (\mathcal{A}/B^\theta)_{mn}. \tag{4.54}$$

To analyze the spectral MHD stability of the plasma, we express $\boldsymbol{\xi}(\mathbf{x})$ in flux co-ordinates (r_f, θ_f, ζ_f), and Fourier expand $\boldsymbol{\xi}(r_f, \theta_f, \zeta_f)$ in terms of the poloidal and toroidal harmonics with mode numbers m and n, and Fourier components $\xi_{mn}(r_f)$, which we assume to be flux surface labels,

$$\boldsymbol{\xi}(r_f, \theta_f, \zeta_f) = \sum_{m,n}\xi_{mn}(r_f)\exp(im\theta_f - in\zeta_f). \tag{4.55}$$

Combining (4.55) and (4.53) yields a criterion for flute-like perturbations,

$$B^\theta(\partial_{\theta_f} + q\partial_{\zeta_f})\boldsymbol{\xi} \approx 0, \qquad \Leftrightarrow \qquad \xi_{mn}(r_f) \approx 0, \quad \forall\, m/n \neq q, \tag{4.56}$$

[24] Or more accurately, the normal mode amplitude $\xi_\omega(\mathbf{x})$, as discussed earlier.

which states that $\xi_{mn}(r_f)$ must vanish on all *ergodic* flux surfaces, so that ξ becomes localized to *resonant* flux surfaces on which $\xi_{mn}(r_f)$ are field line labels.[25] Since $\mathbf{k} = (m - qn)\nabla\theta_f$, (4.56) may also be written as

$$k_\parallel \approx (m - qn)\psi_p'/\sqrt{g}B \ll k_\wedge \approx k, \qquad k_\perp \approx 0, \tag{4.57}$$

which is clearly satisfied near resonant surfaces, $q \approx m/n$. Note that $k_\parallel \ll k_\wedge$ implies that the frequency of the magneto-sonic waves, ω_+ (4.32), greatly exceeds that of the shear Alfvén waves, ω_A (4.30), such that the former equilibrate on the Alfvén time scales considered, and can thus be neglected in MHD stability analysis. This equilibration establishes a quasi-static perpendicular force balance which leads to a purely electrostatic perpendicular electric field, $E_\perp = -\nabla_\perp\varphi$. This relation is employed in both flute-reduced MHD, see Section 4.2.5, and in DHD turbulence, Section 6.3.

4.2.4 Ideal MHD waves and instabilities in a general screw pinch

Although the above equations contain all the essential elements of MHD stability, they are too complicated for analytic treatment in all but a few simple systems. Below we examine two such simple systems, the *general screw pinch*, see Section 3.3.1, and the *cylindrical* tokamak, see Section 3.3.2; both of which neglect the effect of toroidal *curvature*, so that $\boldsymbol{\kappa} = \kappa_r(r)\hat{\mathbf{e}}_\perp = -b_P^2\hat{\mathbf{e}}_\perp/r$ is purely poloidal, see (3.76), and hence a flux surface label. Likewise, both systems exhibit θ and z (or ζ) symmetry, so that the displacement $\boldsymbol{\xi}$ is easily Fourier transformed in θ and z, and requires only a single radial function, $\boldsymbol{\xi}(\mathbf{x}) = \xi_{mn}(r)\exp(im\theta - in\zeta) = \xi_{mn}(r)\exp(ik_P r\theta - ik_T z)$. As a result of these symmetries, the Fourier components $\xi_{mn}(r)$ in (4.55) become decoupled, i.e. evolve separately, allowing the energy variation (4.46) and the full eigenmode equation for $\xi_{mn}(r)$ to be evaluated explicitly.[26] In short, the model retains the essential elements of both kink and interchange modes[27] while dispensing with most of the algebraic complexities related to toroidal geometry. It is thus very useful for introducing both instabilities.

Writing the displacement $\boldsymbol{\xi}(\mathbf{x})$ as $\xi_\parallel\mathbf{b} + \xi_\perp\hat{\mathbf{e}}_\perp + \xi_\wedge\hat{\mathbf{e}}_\wedge$, where

$$\mathbf{b} = b_P\hat{\mathbf{e}}_\theta + b_T\hat{\mathbf{e}}_z, \qquad \hat{\mathbf{e}}_\wedge = b_T\hat{\mathbf{e}}_\theta - b_P\hat{\mathbf{e}}_z, \tag{4.58}$$

$$\xi_\parallel = \xi_P b_P + \xi_T b_T, \qquad \xi_\wedge = \xi_P b_T - \xi_T b_P, \tag{4.59}$$

the plasma energy variation may be calculated as (Freidberg, 1991),

[25] The approximate sign in (4.53) and (4.56) refers to the impossibility of ideal flute perturbations in the presence of finite magnetic shear, $s = d\ln q/d\ln r_f = r_f q'/q$.

[26] A detailed derivation may be found in Appendices B and C of Freidberg (1991).

[27] Although it is far less accurate for the latter, which are directly related to magnetic curvature.

$$\frac{\delta^2 W_p[\boldsymbol{\xi}]}{2\pi^2 R_0/\mu_0} = \int_0^a f\xi_\perp'^2 + g\xi_\perp^2 \mathrm{d}r + \left[\frac{k_\| k_\|^*}{k^2}\right]_a B_a^2 \xi_{\perp a}^2, \quad ' = \mathrm{d}_r, \tag{4.60}$$

where $k_\|$ and $k_\wedge$ are the parallel and diamagnetic components of $\mathbf{k}$,

$$k_\| = k_T b_T + k_P b_P, \qquad k_\wedge = k_T b_P - k_P b_T, \tag{4.61}$$

$$k_\|^* = k_T b_T - k_P b_P, \qquad k^2 = k_T^2 + k_P^2, \tag{4.62}$$

k_T and k_P are its axial (toroidal) and poloidal components,

$$k_T = -n/R_0, \qquad k_P = m/r, \tag{4.63}$$

and $f(r)$ and $g(r)$ are functions of radial position, given by

$$f(r) = r\left(\frac{k_\| B}{k}\right)^2 = \frac{r k_T^2 B_T^2}{k^2}\left(1 - \frac{m}{qn}\right)^2 = \frac{B_P^2}{rk^2}(qn - m)^2, \tag{4.64}$$

$$g(r) = 2\frac{k_T}{k}\mu_0 p_0' + \left(k^2 r^2 - 1 + 2\frac{k_T^2 k_\|^*}{k^2 k_\|}\right)\frac{f(r)}{r^2}. \tag{4.65}$$

The vacuum energy variation is found as

$$\frac{\delta^2 W_v[\boldsymbol{\xi}]}{2\pi^2 R_0/\mu_0} = \left[\frac{\Lambda_w a^2 k_\|^2}{|m|}\right]_a B_a^2 \xi_{\perp a}^2 \tag{4.66}$$

where Λ_w is a geometrical function of the minor radius, a, and the conducting wall radius, b, involving modified Bessel functions taking the argument $k_T a$ and $k_T b$. For $k_T b \ll 1$, this function may be estimated as

$$\Lambda_w \approx \frac{1 + (a/b)^{2|m|}}{1 - (a/b)^{2|m|}}, \qquad k_T b \ll 1, \qquad a < b, \tag{4.67}$$

which for $a/b \ll 1$ reduces to $1 + 2(a/b)^{2|m|} \approx 1$, while for $a/b \to 1$ becomes singular as $[(1 - a/b)|m|]^{-1} \to \infty$; in this limit, a finite $\xi_{\perp a}$ leads to an infinite increase of $\delta^2 W_v$ and is thus prohibited, i.e. $\xi_\perp(a) \to 0$. The above relations exactly specify the MHD energy for a general screw pinch, and thus combined with the energy principle (4.46), its MHD stability properties.

Imposing the condition that $\xi_\perp(r)$ minimize $\delta^2 W_p$ (4.60), where we neglect the vacuum energy $\delta^2 W_v$ (4.66) for simplicity, allows us to derive the corresponding eigenmode equation for $\xi_\perp(r)$. Setting the variation of (4.60) with respect to $\xi_\perp(r)$ to zero, yields an *Euler–Lagrange equation*,

$$[f(r)\xi_\perp']' - g(r)\xi_\perp = 0, \tag{4.68}$$

with $f(r)$ and $g(r)$ defined above. This MHD dispersion relation is the cylindrical version of (4.29), and describes all MHD waves and instabilities in a marginally stable screw pinch equilibrium.

We next write down the corresponding full eigenmode problem for $\boldsymbol{\xi}(\mathbf{x})$,

$$\mathbf{f}[\boldsymbol{\xi}(\mathbf{x})] + \rho\omega^2\boldsymbol{\xi}(\mathbf{x}) = 0 \tag{4.69}$$

where $\mathbf{f}[\boldsymbol{\xi}(\mathbf{x})]$ is given by (4.41). For a general screw pinch this problem takes the form of a second-order ODE for the radial displacement $\xi_\perp$,

$$[\mathcal{F}(r)(r\xi_\perp)']' - \mathcal{G}(r)r\xi_\perp = 0, \tag{4.70}$$

where $\mathcal{F}(r)$ and $\mathcal{G}(r)$ are functions of frequency, wave number and radius,

$$\mathcal{F}(r) = \frac{\rho V_{max}^2}{r}\frac{\left(\omega^2-\omega_A^2\right)\left(\omega^2-\omega_H^2\right)}{\left(\omega^2-\omega_+^2\right)\left(\omega^2-\omega_-^2\right)}, \tag{4.71}$$

$$\begin{aligned}\mathcal{G}(r) = {} & -\frac{\rho\left(\omega^2-\omega_A^2\right)}{r} + \left(\frac{4k_T^2V_A^2B_P^2}{\mu_0 r^3}\right)\frac{\left(\omega^2-\omega_S^2\right)}{\left(\omega^2-\omega_+^2\right)\left(\omega^2-\omega_-^2\right)} \\ & + \left[\frac{B_P^2}{\mu_0 r^2} - \left(\frac{2k_T G B_P}{\mu_0 r^2}\right)\frac{V_{max}^2\left(\omega^2-\omega_H^2\right)}{\left(\omega^2-\omega_+^2\right)\left(\omega^2-\omega_-^2\right)}\right]'.\end{aligned} \tag{4.72}$$

Here, the Alfvénic and magneto-sonic frequencies, ω_A and $\omega_\pm$, defined in (4.30), are supplemented by the sonic and hybrid frequencies, ω_S and ω_H,

$$\omega_S^2 = k_\parallel^2V_S^2, \qquad \omega_H^2 = k_\parallel^2V_H^2, \qquad V_H^2 = V_A^2V_S^2/V_{max}^2, \tag{4.73}$$

with V_A given by (2.253), V_S by (4.11) and $V_{max}^2 = V_A^2 + V_S^2$; finally, $G = k_PB_T - k_TB_P$. The eigenmode equation (4.70) is a more general version of the marginally stable dispersion relation (4.68), and describes all possible MHD waves and instabilities in a general screw pinch. Indeed, it reduces to the latter result when the marginal stability condition ($\omega = 0$) is imposed on (4.71)–(4.72). The agreement between the two results confirms our earlier assertion that a marginally stable eigenmode of the MHD force operator $\mathbf{f}[\xi(\mathbf{x})]$ (4.41) must minimize the free energy of the plasma.[28]

4.2.5 Flute-reduced MHD

The results of Section 4.2.3 indicate that toroidal plasmas are most unstable to flute-like perturbations $\boldsymbol{\xi} \propto \exp(i\mathbf{k}\cdot\mathbf{x})$, which satisfy (4.53). These are in general radially localized to resonant flux surfaces, although in large aspect ratio (small ϵ) tokamaks, where all displacements satisfy (4.53),

$$\mathbf{b}_0\cdot\nabla \sim k_\parallel/k_\wedge \approx r/R_0q = \epsilon/q \sim O(\epsilon) \ll 1, \tag{4.74}$$

[28] This is a direct consequence of the self-adjointness of $\mathbf{f}$, see (4.42).

they also apply on ergodic flux surfaces. To show this, we first note that in cylindrical geometry, the poloidal and axial components of **k** are $k_P = m/r$ and $k_T = -n/R_0$, while parallel and diamagnetic components become[29]

$$k_{\parallel} R_0 = m/q - n, \qquad k_{\wedge} r \approx m, \tag{4.75}$$

so that the ratio $k_{\parallel}/k_{\wedge} \sim O(\epsilon/q)$ is small everywhere when $\epsilon \ll 1$ and $q \sim O(1)$. Of course, the same is not true for all perturbations when ϵ is not small, although it is always satisfied close to resonant flux surfaces, $q \approx m/n$. Hence, it is useful to derive a version of the MHD (and more generally of the DHD) model in which flute-like displacements are assumed from the outset. Such a model, known as *flute-reduced MHD*, or simply *reduced MHD*, may be derived by imposing the flute-ordering (4.74) on the MHD equations. This is accomplished by applying a multi-scale expansion in the small parameter $\epsilon \approx qk_{\parallel}/k \ll 1$, similar to the one used in Section 2.2, see Section 4.2.5, to the MHD (2.246)–(2.248), or more generally to the DHD (2.182)–(2.184), equations and retaining only leading order terms. The resulting equations are a useful reference point for the study of stability and dynamics of toroidal plasmas.[30] Below we outline the derivation, broadly following the original account of Strauss (1976), yet making explicit the multiple-scale expansion.

Multi-scale expansion involves separation of *slow*, $O(1)$, and *fast*, $O(\epsilon^{-1})$, spatial variations into (independent) co-ordinates $\mathbf{X}$ and $\mathbf{x}$, and *slow*, $O(\epsilon^2)$, and *fast*, $O(1)$, temporal variations into co-ordinates T and t,

$$\mathbf{x} \rightarrow \epsilon^{-1}\mathbf{x} + \mathbf{X}, \qquad t \rightarrow t + \epsilon^2 T. \tag{4.76}$$

Hence, the net gradient ∇ can be expressed as the sum of slow, $\nabla_0 = \partial_{\mathbf{X}}$, and fast, $\nabla_1 = \partial_{\mathbf{x}} = \nabla_{\perp} + O(\epsilon)$, gradients,[31] and the net time derivative as the sum of slow, ∂_T, and fast, ∂_t, time derivatives,

$$\nabla \rightarrow \epsilon^{-1}\nabla_{\perp} + \nabla_0, \qquad \partial_t \rightarrow \partial_t + \epsilon^2 \partial_T. \tag{4.77}$$

Similarly, any quantity $\mathcal{A}$ is expanded in powers of ϵ into an equilibrium part $\mathcal{A}_0$ and the flute-perturbed parts $\mathcal{A}_1, \mathcal{A}_2, \ldots$, satisfying (4.74), as follows:

$$\mathcal{A} = \mathcal{A}_0(\mathbf{X}, T) + \sum_{n=1}^{\infty} \epsilon^n \mathcal{A}_n(\mathbf{x}, \mathbf{X}, t, T). \tag{4.78}$$

29 Here we assume that the perturbation is localized to the flux surface $r = const$ and thus exclude the radial component of **k**. Hence $\mathbf{k}_{\perp} = \mathbf{b} \times (\mathbf{k} \times \mathbf{b})$ reduces to the projection in the diamagnetic direction $\mathbf{k}_{\wedge} = k_{\wedge}\hat{\mathbf{e}}_{\wedge}$. More generally, $k_{\wedge}$ should be replaced by $k_{\perp} = |\mathbf{k}_{\wedge}|$.

30 Here we consider only the MHD model, with the DHD model treated in Section 6.3.1.

31 Note that $\nabla_1 \approx \nabla_{\perp}$, because by (4.74), $\mathbf{b}_0 \cdot \nabla_1 \approx 0$, where $\mathbf{b}_0 = \mathbf{B}_0/B_0$.

Hence the 'fast' variables $\mathbf{x}$ and t measure the variation of the flute-perturbed quantities, and the 'slow' variables $\mathbf{X}$ and T that of equilibrium quantities.

First, we assume the existence of an MHD equilibrium for $\mathbf{B}_0$, $\mathbf{J}_0$ and p_0,

$$\mathbf{J}_0 \times \mathbf{B}_0 = \nabla_0 p_0, \qquad \nabla_0 \times \mathbf{B}_0 = 0, \qquad \mu_0 \mathbf{J}_0 = \nabla_0 \times \mathbf{B}_0. \tag{4.79}$$

We next note that in a tokamak, parallel projection of Ohm's law expresses the balance between inductive EMF and collisional resistivity, so that the equilibrium electric field is purely parallel,

$$-\partial_t A_{0\|} = E^A_{0\|} = E_{0\|} = \eta_\| J_{0\|}, \qquad \mathbf{E}_{0\perp} = -\nabla_\perp \varphi_1 - \partial_t \mathbf{A}_1 = 0, \tag{4.80}$$

and the first-order potentials must vanish. Flute-reduction then allows us to express the electromagnetic (*vector*) field in terms of two *scalar* fields ψ_1 and φ_2. Due to the flute-ordering, (4.74), the perturbed electric field is mainly transverse and electrostatic,[32]

$$\mathbf{E}_1 \approx \mathbf{E}_{\perp 1} \approx -\nabla_\perp \varphi_2. \tag{4.81}$$

The perturbed magnetic field, which follows from $\nabla \cdot \mathbf{B} = 0$ and (4.74), implies $\nabla_\perp \cdot \mathbf{B}_1 \approx 0$ to leading order. Hence, $\mathbf{B}_1$ may be written as

$$\mathbf{B}_1 = \nabla_\perp \times \mathbf{A}_2 = \nabla_\perp \times (\mathbf{A}_{2\perp} + \mathbf{b}_0 A_{2\|}) = B_{1\|}\mathbf{b}_0 + \nabla_\perp \psi_1 \times \mathbf{b}_0, \tag{4.82}$$

where $\psi_1 = A_{2\|}$ denotes the parallel vector potential. As a result, the parallel gradient, $\nabla_\| \equiv (\mathbf{B}/B_0) \cdot \nabla$, can be simplified as

$$\nabla_\| = \breve{\mathbf{B}}_0 \cdot \nabla_0 + \breve{\mathbf{B}}_1 \cdot \nabla_\perp = \mathbf{b}_0 \cdot \nabla_0 - B_0^{-1}\{\psi_1, \ \} \approx \mathbf{b}_0 \cdot \nabla_0 - \{\breve{\psi}_1, \ \}, \tag{4.83}$$

where the breve denotes division by B_0 and $\{\mathcal{A}, \ \}$ is the Poisson bracket,

$$\breve{\mathcal{A}} = \mathcal{A}/B_0, \qquad \{\mathcal{A}, \ \} \equiv \mathbf{b}_0 \cdot \nabla_\perp \mathcal{A} \times \nabla_\perp = \mathbf{b}_0 \times \nabla_\perp \mathcal{A} \cdot \nabla_\perp. \tag{4.84}$$

The first term in the parallel gradient represents the variation along the equilibrium field, while the second term represents variation due to field line bending and perpendicular gradients. By nature of (4.74), only transverse deflection $\mathbf{B}_{1\perp} = \nabla_\perp \psi_1 \times \mathbf{b}_0$ contributes to the latter.

The total current follows from Ampere's law (2.88),

$$\mu_0 \mathbf{J} = \mu_0 \mathbf{J}_0 - \mathbf{b}_0 \nabla_\perp^2 \psi_1 - \mathbf{b}_0 \times \nabla_\perp B_{1\|}, \tag{4.85}$$

which indicates $O(\epsilon)$ corrections to parallel and perpendicular currents,

$$\mu_0 J_{1\|} = -\nabla_\perp^2 \psi_1, \qquad \mu_0 \mathbf{J}_{1\perp} = -\mathbf{b}_0 \times \nabla_\perp B_{1\|}. \tag{4.86}$$

[32] Recall the discussion of inductive vs. electrostatic contributions in Section 2.2.1. In the MHD picture, this implies a quasi-static force balance maintained by compressional waves.

Combined with (2.247) this yields a force balance relation between perturbed magnetic tension and perturbed plasma pressure,

$$B_{1\|}/\mu_0 + \breve{p}_1 = 0. \tag{4.87}$$

The evolution equation for ψ_1 follows from the parallel component of the resistive Ohm's law (2.255), with the assumption $D_\eta = \eta_\|/\mu_0 \sim O(\epsilon^2)$,

$$\partial_t \psi_1 + \nabla_\| \varphi_2 = -\eta_\| J_{1\|} = D_\eta \nabla_\perp^2 \psi_1. \tag{4.88}$$

The perpendicular projection of (2.255) yields, after some lengthy algebra, the perturbed transverse velocities, i.e. the first- and second-order GC-drifts,

$$\mathbf{V}_{1\perp} = \mathbf{V}_E \approx \breve{\mathbf{b}}_0 \times \nabla_\perp \varphi_2 \approx \mathbf{b}_0 \times \nabla_\perp \breve{\varphi}_2, \tag{4.89}$$
$$\mathbf{V}_{2\perp} \approx \breve{\mathbf{b}}_0 \times \left(\partial_t \mathbf{A}_2 + \nabla_0 \varphi_2 + \nabla_\perp \varphi_3 - V_{1\|} \nabla_\perp \psi_1 - \breve{B}_{1\|} \nabla_\perp \varphi_2\right) + D_\eta \nabla_\perp B_{1\|}.$$

Of those, only $\mathbf{V}_{1\perp}$ enters the lowest-order advective derivative,

$$\mathrm{d}_t \approx \partial_t + \mathbf{V}_E \cdot \nabla_\perp \approx \partial_t + \mathbf{b}_0 \times \nabla_\perp \breve{\varphi}_2 \cdot \nabla_\perp = \partial_t + \{\breve{\varphi}_2, \}. \tag{4.90}$$

Since $\nabla \cdot \mathbf{V}_E = O(\epsilon)$, (2.246) implies that, to lowest order, the flute-perturbed flow is incompressible. This allows us to decouple ρ_1 from the other perturbed variables, and thus to treat it as a passive scalar in the dynamics.[33]

The evolution equation for $V_{1\|} = \mathbf{b}_0 \cdot \mathbf{V}_1$ can be found by inserting the above expressions into the parallel projection of (2.183),

$$\rho_0 \mathrm{d}_t V_{1\|} \approx -\nabla_\| p_1 + \mathbf{b}_0 \cdot \nabla_\perp \breve{\psi}_1 \times \nabla_0 p_0 + \mu_\perp \nabla_\perp^2 V_{1\|}, \tag{4.91}$$

where $\mu_\perp$ is the leading order viscosity. Likewise, the evolution equation for $\mathbf{V}_E$, expressed in terms of the *parallel vorticity*,

$$\Omega_{1\|} = \mathbf{b}_0 \cdot \nabla_\perp \times \mathbf{V}_E \approx \nabla_\perp^2 \breve{\varphi}_2, \tag{4.92}$$

may be obtained by imposing the flute-ordering on the vorticity equation (4.48). The result is commonly known as the *flute-reduced shear Alfvén law*,

$$\rho_0 \mathrm{d}_t \Omega_{1\|} = B_0 \nabla_\| J_{1\|} + 2\mathbf{b}_0 \times \boldsymbol{\kappa}_0 \cdot \nabla_\perp p_1 + \mu_\perp \nabla_\perp^2 \Omega_{1\|}. \tag{4.93}$$

Similarly, the evolution equation for p_1 can be found from (2.184),

$$\mathrm{d}_t p_1 + \mathbf{V}_E \cdot \nabla_0 p_0 + \gamma p_0 (\nabla_0 \cdot \mathbf{V}_1 + \nabla_\perp \cdot \mathbf{V}_2) = \chi_\perp \nabla_\perp^2 p_1, \tag{4.94}$$

where $\chi_\perp$ is the leading order heat diffusivity. Evaluating the divergences and simplifying terms leaves the following expression,

$$\mathrm{d}_t p_1 + \mathbf{V}_E \cdot \nabla_0 p_0 = \wp_0 \rho_0 V_{A0}^2 (\nabla \cdot \mathbf{V})_2 + [\chi_\perp/(1 + \gamma \beta_0/2)] \nabla_\perp^2 p_1, \tag{4.95}$$
$$(\nabla \cdot \mathbf{V})_2 \approx 2\mathbf{b}_0 \cdot \boldsymbol{\kappa}_0 \times \nabla_0 \breve{\varphi}_2 - \nabla_\| V_{1\|} + V_{1\|} \mathbf{b}_0 \cdot \nabla_0 \ln B_0 + \left(D_\eta / B_0^2\right) \nabla_\perp^2 p_1,$$

[33] This explains why the equilibrium density, ρ_0, is commonly set to unity.

where $(\nabla \cdot \mathbf{V})_2$ is the net flow compressibility to order ϵ^2 and $\wp_0$ is a measure of plasma pressure, related to plasma beta,

$$\wp_0 = \frac{V_{S0}^2}{V_{A0}^2 + V_{S0}^2} = \frac{\gamma \beta_0}{2 + \gamma \beta_0}, \qquad \beta_0 = \frac{2\mu_0 p_0}{B_0^2} = \frac{2}{\gamma} \frac{V_{S0}^2}{V_{A0}^2}. \tag{4.96}$$

Here the Alfvén and sound speeds are evaluated with equilibrium values. The above evolution equations for p_1, $\mathbf{V}_E(\varphi_1, \Omega_{1\|})$, $V_{1\|}$ and $J_{1\|}(\psi_1)$ constitute the (four-field) *flute-reduced MHD* model. This MHD model is a complete subset of the four-field drift-fluid model derived in Section 2.4.2.7. The relation between the flute-reduced MHD and DHD models will be discussed in Section 6.3 in the context of the resulting turbulent flows.

In the low-beta ($\beta_0 \sim \wp_0 \ll 1$) limit, the flow compressibility term in (4.95) may be neglected, and $V_{1\|}$ decouples from the other dynamical variables, i.e. it can also be treated as a passive scalar. This leaves a set of three evolution equations for ψ_1 (4.88) and φ_1 (4.93) and p_1,

$$\mathrm{d}_t p_1 + \mathbf{V}_E \cdot \nabla_0 p_0 = \chi_\perp \nabla_\perp^2 p_1, \tag{4.97}$$

which are collectively known as *low-beta, flute-reduced MHD*.

For a small ϵ tokamak further simplifications are possible, by writing the low-beta, flute-reduced MHD equations in orthogonal toroidal co-ordinates (r, θ, ζ),[34] already introduced in Section 3.3. In these co-ordinates, the magnetic curvature of a small ϵ tokamak is almost purely toroidal and points towards the major axis,[35] see (3.97),

$$\kappa_0 = \mathbf{b}_0 \cdot \nabla \mathbf{b}_0 = -\hat{\mathbf{e}}_R / R_0 + O(\epsilon), \tag{4.98}$$

the Poisson bracket (2.73) reduces to gradients in the poloidal plane, and $\nabla_\|$ can be divided into toroidal and poloidal gradients,

$$\{\mathcal{A}, \ \} \equiv \hat{\mathbf{e}}_\zeta \cdot \nabla_\perp \mathcal{A} \times \nabla_\perp, \qquad R_0 \nabla_\| = \partial_\zeta - a\{\psi_1, \ \}. \tag{4.99}$$

With these simplifications, (4.93) and (4.97) can be written as,

$$\rho_0 \mathrm{d}_t \Omega_{\|1} = B_0 \nabla_\| J_{1\|} - 2\{R, p_1\}/R_0 + \mu_\perp \nabla_\perp^2 \Omega_{\|1}, \tag{4.100}$$

$$\mathrm{d}_t p_1 + \{\breve{\varphi}_2, p_0\} = \chi_\perp \nabla_\perp^2 p_1, \tag{4.101}$$

[34] Recall that θ and ζ are the toroidal and poloidal angles, and r and $R = R_0 + r \cos\theta$ are the radial distances from the minor and major axes.

[35] This is not the case for a medium or low aspect ratio tokamak, for which the poloidal curvature, which points towards the minor axis, and is given by $\kappa_r = -b_P^2 \hat{\mathbf{e}}_r / r$, becomes significant.

which together with (4.88) are known simply as *reduced MHD*. Neglecting dissipation, yields the ideal MHD version of these equations,

$$\rho_0 \mathrm{d}_t \Omega_{\|1} = B_0 \nabla_\| J_{1\|} - 2\{R, p_1\}/R_0, \tag{4.102}$$

$$\mathrm{d}_t p_1 + \{\breve{\varphi}_2, p_0\} = 0, \qquad \partial_t \psi_1 + \nabla_\| \varphi_2 = 0. \tag{4.103}$$

These can be simplified even further for a *cylindrical* tokamak, for which the curvature is not toroidal but poloidal, i.e. given by (3.76) not by (4.98). Linearizing and Fourier transforming (4.102)–(4.103), along the lines of (4.25)–(4.27) and (4.55), and imposing the flute ordering (4.74), yields the spectral equations of ideal, reduced MHD for a *cylindrical* tokamak,

$$\rho_0 \mathrm{d}_t \Omega_{\|1} = B_0 \mathrm{i}[k_\| J_{1\|} + (r/R_0) k_\wedge J_0' \psi_1] + \mathrm{i} k_\wedge \kappa_\perp p_1, \tag{4.104}$$

$$\mathrm{d}_t p_1 = \mathrm{i} k_\wedge p_0' \breve{\varphi}_2, \qquad \partial_t \psi_1 = \mathrm{i} k_\| \varphi_2, \qquad \kappa_\perp = -b_P^2/r. \tag{4.105}$$

To obtain a flute-reduced version of the screw pinch eigenmode equation (4.68), we may (i) impose the tokamak ordering (3.85), which implies the flute-ordering $k_\|/k \sim O(\epsilon)$, on $f(r)$ and $g(r)$ defined in (4.65), (ii) investigate (4.68) near a resonant flux-surface on which $k_\| \approx 0$, or (iii) make use of the spectral flute-reduced, low-beta ideal MHD equations for a cylindrical tokamak derived above. Following the third approach, we Fourier transform (4.104)–(4.105) in time and combine into a single ODE for $\breve{\varphi}_2 = \varphi_2/B_0$.[36] This yields the cylindrical tokamak MHD dispersion relation,

$$[\mathcal{A}(r) r^3 (\breve{\varphi}_2/r)']' - \mathcal{A}(r)(m^2 - 1)\breve{\varphi}_2 - r^2 (k_\wedge V_A)^2 \beta_0' \kappa_\perp \breve{\varphi}_2 \approx 0. \tag{4.106}$$

Here $\beta_0 = 2\mu_0 p_0/B_0^2$ is the zeroth-order plasma beta, $V_A = B_0^2/\mu_0 \rho_0$ is the Alfvén speed and $\mathcal{A}(r) \equiv \omega^2 - \omega_A^2(r)$ is the shear Alfvén dispersion relation. The three terms in (4.106) give rise to (i) shear Alfvén waves, (ii) the kink instability and (iii) the interchange instability, respectively. Below, we examine each of these in both cylindrical and toroidal geometry.

4.2.6 Non-homogeneous shear Alfvén waves

Shear Alfvén waves in a cylindrical tokamak are given by the dispersion relation $\mathcal{A}(r) = 0$, which is consistent with (4.30), i.e. $\omega_A^2 = k_\|^2 V_A^2$, except that V_A now changes with radius. Since $k_\|$, as given by (4.75), vanishes on resonant flux surfaces, i.e. when $q(r) = m/n$, the dispersion relation gives rise to a band of allowed frequencies from $\omega = 0$ to $\omega_{max}^2 = (k_\| V_A)_{max}^2$; this band, or bands, are generally known as the Alfvén continuum. Due to strong coupling between modes in

[36] And hence for the velocity $\mathbf{V}_1 = \nabla\zeta \times \nabla\breve{\varphi}_2 \approx k_\wedge \breve{\varphi}_2 \hat{\mathbf{e}}_\perp$ and displacement $\boldsymbol{\xi} = \int \mathbf{V}_1 \mathrm{d}t \approx \xi_\perp \hat{\mathbf{e}}_\perp$.

this continuum, the modes are quickly damped, making them highly attractive for plasma heating applications.

The cylindrical results are accurate to $O(\epsilon)$ for the small ϵ tokamak since the dominant, toroidal curvature is a flux surface label to the same level of accuracy, see (4.98). For smaller aspect ratios, one has to revert to the general reduced MHD model, (4.102)–(4.103), in which the curvature is no longer a flux surface label but varies poloidally according to (3.97). This has the effect of coupling different (m, n) modes, and hence different resonant surfaces, creating forbidden and allowed bands in the Alfvén continuum.[37]

In the gaps between any two allowed bands, there appear discrete modes known as *toroidal Alfvén eignemodes* (TAEs). These are only weakly coupled to the Alfvén continuum and are damped much slower than other shear Alfvén waves.[38] For this reason, TAEs are easily destabilized by supra-thermal particles, see Section 4.4, and are thus a particular concern for burning plasmas, in which they can be excited by the radial gradient in α-particle pressure or by fast particles generated with external antennas in order to drive the toroidal current. While only weakly coupled to the continuum, they are strongly coupled to other TAEs by toroidicity and magnetic shear, allowing energy transfer between the modes. Since these modes are typically driven at large scale (low-n) by MHD instabilites and are ultimately damped at small scales (high-n) by collisional processes, they can produce a cascade of energy from low-n to high-n TAEs known as an *Alfvén cascade* – a common phenomenon in tokamaks with auxiliary heating in regions of *reversed* magnetic shear, i.e. where $q' < 0$. By exciting ever higher number modes, an Alfvén cascade redistributes energy radially in a quasi-diffusive processes. It is also an instance of a *direct* energy cascade from large to small scales, which is a hallmark of the inertial range in fluid turbulence, see Section 6.1. An Alfvén cascade can thus be viewed as an aspect of transient MHD turbulence in toroidal plasmas, see Section 6.2.

4.2.7 Current-driven ideal MHD instabilities: kink modes

We next turn to tokamak MHD instabilities driven by the equilibrium parallel current and its gradient, i.e. to kink modes. Since curvature plays only a minor role in this case, kink modes are well described by the cylindrical tokamak model, see Section 4.2.4, for which the potential energy variation and the associated dispersion relation are given by (4.60)–(4.67),

37 Similar to the allowed and forbidden bands for electron wave functions in a crystal lattice.

38 In fact, TAEs can be damped via several mechanisms, the dominant ones being coupling to the Alfvén continuum and ion Landau damping.

$$\frac{\delta^2 W_p[\xi]}{2\pi^2 R_0/\mu_0} = \int_0^a f\xi_\perp'^2 + g\xi_\perp^2 dr + \left[\frac{k_\| k_\|^*}{k^2} + \frac{\Lambda_w a^2 k_\|^2}{|m|}\right]_a B_a^2 \xi_{\perp a}^2, \tag{4.107}$$

where $f(r)$ and $g(r)$ are given by (4.64)–(4.65) with $p_0' = 0$. The second term can be neglected for internal (fixed boundary) modes, $\xi_{\perp a} = 0$, which leads to the explicit form of the plasma energy,

$$\delta^2 W_p \propto \int_0^a \left[\left(r\xi_\perp'\right)^2 + (m^2-1)\xi_\perp^2\right]\left(1 - \frac{m}{qn}\right)^2 r dr. \tag{4.108}$$

The associated dispersion relation is again given by (4.68) or, in flute-reduced form, by (4.106).[39] The factor $(1 - m/qn)^2$ implies that the stabilizing line-bending term vanishes on resonant surfaces, $q = m/n$, making the eigenmode problem singular. However, the pre-factor $m^2 - 1$, whose sign determines kink mode stability, indicates that only the $m = 0$ and $m = 1$ modes are potentially unstable. In fact, both these modes require $\mathbf{V}_1 = O(\epsilon^2)$, and are thus linearly stable. Higher-order analysis reveals that the $m = 1$ mode can be weakly destabilized at order $O(\epsilon^4)$, by either pressure or current gradients, provided there exists a $q = 1$ surface inside the plasma and $q(r)$ increases monotonically with radius,

$$q_0 \equiv q(r=0) < 1 < q(r=a) \equiv q_a, \qquad q' > 0. \tag{4.109}$$

The second condition ($q' > 0$) is equivalent to the current profile being peaked towards the centre of the plasma: recall that the safety factor $q(r)$ is related to the toroidal current density by (3.81),[40] i.e.

$$q(r) = rB_T(r)/R_0 B_P(r) = 2B_0/\mu_0 R_0 \langle J_T \rangle_r(r), \tag{4.110}$$

where $B_0 = B_T(r=0)$ is the (toroidal) field on the minor axis and $\langle J_T \rangle_r = 2\pi \int_0^r J_T r dr/\pi r^2 = I_T(r)/\pi r^2$ is the average toroidal current density within the annulus $r =$ const. Hence, a hollow $q(r)$ profile, $q(r)' > 0$, implies a peaked current profile, $J_T(r)' < 0$. It is worth noting that since the plasma resistivity decreases with temperature, $\eta_\| \propto \sigma_{\|e}^{-1} \propto T_e^{-3/2}$, see Section 5.2, the current profile in an Ohmic discharge has a tendency to peak, and the $q(r)$ profile to become hollow, in the hotter, central part of the plasma.

The effect of toroidicity and radial pressure gradient on low-n internal kink modes can be investigated using the ballooning formalism developed in Section 4.2.8. Since toroidicity introduces a coupling between different (m, n)

39 This is easily demonstrated by integrating (4.106), with $\beta_0' = 0$, radially from 0 to a and applying the energy principle (4.47). Alternatively, (4.60) may be evaluated for $p_0' = 0$ to $O(\epsilon^2)$, revealing an integrand proportional to $(n/m - 1/q)^2[r^2\xi_\perp'^2 + (m^2-1)\xi_\perp^2]r$.

40 Note that $q(r)$ can also be expressed as $-(r/a)B_0/\psi_0'$, where $\psi_0(r) = A_{\|0}(r)$ is the lowest-order parallel vector potential.

modes, the (1, 1) mode can be stabilized by the (2, 1) mode. In this case, the lowest value of q_0 is set by the (1, 2) mode which is always unstable. Since the strength of the coupling is proportional to the magnetic shear s, which for realistic pressure and current profiles is too low near the axis to stabilize the (1, 1) mode, one finds that $q_0 > 1$ is again required for stability.

The $m = 1$ kink mode, which corresponds to a lateral displacement of the entire flux surface, is at the root of many *core plasma* oscillations in real tokamaks. For instance, its non-linear evolution, together with dissipative effects, is responsible for the *sawtooth* oscillation which periodically destroys the $q = 1$ surface, thus maintaining $q(r) > 1$ everywhere in the plasma;[41] in fact, combined with the just mentioned tendency of q_0 to slowly decrease with time, this effectively clamps q_0 at unity during the discharge. Similarly, the $m = 1$ mode can be destabilized by supra-thermal particles to yield the *fishbone* oscillation which is often observed with auxiliary heating. To summarize, all internal kink modes in a small ϵ tokamak are stable, except the $m = 1$ mode, which becomes weakly unstable when $q_0 < 1$.

We next consider *external* kink modes, which involve the displacement of the plasma–vacuum boundary. Detailed analysis of (4.46), including the surface and vacuum integrals (Freidberg, 1991), reveals that the $m = 1$ mode becomes unstable whenever the edge safety factor falls below unity. To ensure stability against this mode requires $q_a > 1$, thus setting an upper limit on the toroidal current in a small ϵ tokamak, independent of the current and pressure profiles, as well as the location of a conducting wall. This follows from the definition of $q(r)$, (4.110), which implies that

$$q_a = 2B_0 A_p/\mu_0 R_0 I_T, \qquad I_T = I_T(a) = 2\pi \int_0^a J_T r \mathrm{d}r \tag{4.111}$$

where I_T is the total current inside the annulus $r = a$, $A_p = \kappa\pi a^2$ is the cross-sectional, poloidal area of the plasma, and κ is the vertical elongation. Hence, $q_a > 1$ imposes an upper limit on the plasma current $I_T < 2A_p B_0/\mu_0 R_0$. This limit can also be expressed in terms of the average current density, $\langle J_T \rangle_a = I_T/A_p < 2B_0/\mu_0 R_0$.

In general, stability against any (m, n) external kink mode requires that $q_a > m/n$. Conversely, an $(m \geq 2, n)$ external kink mode is unstable when

$$(m - j)/n < q_a < m/n, \tag{4.112}$$

where $j > 0$ is a function of the current density, $J(r)$, in the vicinity of the plasma boundary, i.e. of J_a, J'_a, and so on. The most pessimistic condition ($j = 1$) is

[41] In between the sawtooth crashes, the volume inside the $q = 1$ surface is subject to a range of benign MHD activity, in particular core localized TAEs, also known as *tornado modes* due to their distinctive signature on spectral amplitude diagrams, Section 4.2.6.

found for finite J_a, e.g. for $n = 1$ this implies instability to some $(m, 1)$ mode for all values of q_a. However, since this condition assumes an unphysical current discontinuity at $r = a$, it overestimates the drive of the instability; by considering $J(r)$ profiles which decay smoothly to zero at $r = a$, more optimistic stability criteria are obtained. In particular, it is found that for high-m modes, the external kink instability should be confined to a narrow radial layer around $r = a$.[42]

Stability properties of high-m external kink, or *peeling*, modes have been derived using asymptotic analysis of (4.46) for $m \gg 1$ (Freidberg, 1991). For finite current at $r = a$, this analysis predicts an instability condition given by (4.112) with $j = J_a/\langle J\rangle_a$, which is the ratio of the current density on the plasma boundary to its volume average value. This corresponds to a narrow q_a instability band around m/n, whose width is linearly proportional to j. As $J_a \to 0$, the width of this band vanishes, and the (m, n) mode is stabilized for all values of q_a, expect $q_a = m/n$.

The case of zero boundary current ($J_a = 0$) requires inclusion of higher m-order corrections in the asymptotic expansion. The result is once again (4.112), but with $j = \exp(2m\langle J\rangle_a/aJ'_a)$,[43] which states that the width of the instability band decreases exponentially for increasing m and decreasing J'_a, normalized by the average gradient of $J(r)$, i.e. $2\langle J\rangle_a/a$. When $J_a = 0 = J'_a$, the width of this band vanishes, and all high-m modes are stabilized.

The high-m (peeling) mode results presented above overestimate the stability boundaries of low-m external kink modes. These have been calculated numerically in Wesson (2004) by considering a family of cylindrical tokamak equilibria with quasi-parabolic current profiles,

$$J_T(r) = J_0(1-u)^\nu, \qquad u = (r/a)^2, \qquad J_0 = (I_T/\pi a^2)(\nu+1), \tag{4.113}$$

for which the $q(r)$ profile increases monotonically as

$$q(r) = q_a u/[1-(1-u)^{\nu+1}], \qquad q_a/q_0 = \nu + 1. \tag{4.114}$$

Here q_a/q_0 and ν both measure the degree of current profile peaking, e.g. $J_T(r)$ is flat for $\nu = 0$ and $q_a/q_0 = 1$. Significantly, both J_a and J'_a vanish for $\nu > 1$ or $q_a/q_0 > 2$. The results are represented graphically in Figure 6.10.4 in Wesson (2004)[44] and differ from the asymptotic results only for low-m modes, which can become unstable even for $J_a = 0 = J'_a$. This occurs only in finite instability bands in q_a, along the lines of (4.112), with j dependent on both q_a/q_0 and the mode numbers (m, n). The broadest of these bands belongs to the (2, 1) mode, which

[42] Since this layer is removed, or 'peeled off', in the non-linear phase of the instability, high-m external kink modes are often referred to as *peeling* modes.

[43] The factor inside the exponential is negative since $|J_a|' < 0$ while $\langle |J| \rangle_r > 0$.

[44] Which is also reproduced in Freidberg (1991, Figure 9.35) and in Biskamp (1993, Figure 4.2).

closes entirely only for $q_a/q_0 > 3.5$, beyond which point the equilibrium becomes stable to all external kink modes.

4.2.8 Pressure-driven ideal MHD instabilities: ballooning modes

We next consider pressure-driven tokamak instabilities. These are most pronounced for flute-like perturbations (4.53), especially those radially localized to the vicinity of resonant flux surfaces, $q = m/n$. The instability criterion for interchange modes in the cylindrical tokamak may be derived in one of two ways: (i) by retaining finite $\beta' = \beta_0'$ terms in the dispersion relation (4.106);[45] or (ii) by Taylor expanding (4.68) around a resonant flux surface $r = r_\star$, so that $k_\parallel = k_{\parallel\star}(r - r_\star) \equiv k_{\parallel\star}x$, etc. and $f(r)$ and $g(r)$ become

$$f \approx \left[r(k_\parallel B)'^2/k^2\right]_\star x^2, \qquad g \approx \left[2\mu_0 p'(k_T/k)^2\right]_\star . \tag{4.115}$$

Both approaches lead to the eigenvalue problem,[46]

$$\left(x^2\xi_\perp'\right)' + D_S\xi_\perp = 0, \qquad D_S = -\left[2\mu_0 p' k_T^2/r(k_\parallel B)'^2\right]_\star > 0, \tag{4.116}$$

which is easily solved using power law trial functions,

$$\xi_\perp = c_+x^{y_+} + c_-x^{y_-}, \qquad 2y_\pm = -1 \pm \sqrt{1-4D_S}. \tag{4.117}$$

When $D_S < 1/4$, so that the discriminant is positive, both exponents are real and the system is stable. On the other hand, an (internal) interchange mode is necessarily destabilized at $r = r_\star$ when $D_S(r_\star) \equiv D_{S\star} > 1/4$ or

$$D_{S\star} = -r_\star\left(2\mu_0 p'/B_T^2\right)_\star / \left(rq'/q\right)_\star^2 = -(r\beta'/s^2)_\star > 1/4, \tag{4.118}$$

with all quantities evaluated at $r = r_\star$. This result, which represents a sufficient condition for instability, is known as *Suydam's criterion*. It predicts that an interchange mode localized at the resonant flux surface $r = r_\star$ becomes unstable when $-4(r\beta')_\star$ exceeds the square of the average magnetic shear, $s_\star = (rq'/q)_\star$. The stabilizing effect of finite shear reflects the work done in bending magnetic field lines when they are radially displaced (interchanged) at a flux surface with $q_\star' \neq 0$; the corresponding energy is represented by the fourth term in (4.47).

To highlight the origin of the interchange instability in the combination of radial pressure gradient and (normal) magnetic curvature, the Suydam's criterion (4.118) may be rewritten as

$$\kappa_\perp\beta' > (s/2R_0q)^2, \qquad \kappa_\perp = \kappa_r = -(r/qR_0)^2/r. \tag{4.119}$$

[45] In the rest of the section we drop the equilibrium subscript '0' on β_0 and p_0.
[46] Note that $k_\parallel = b_Tk_T + b_Pk_P \approx 0$ at $r \approx r_\star$ and $k_P = m/r$ implies $k_T = -(mB_P/rB_T)_\star$.

Since $\kappa_\perp$ and β' are both negative everywhere in a cylindrical tokamak, their product is positive and hence destabilizing. However, the level of magnetic shear required for stability against kink modes, which can be roughly expressed as $q_0 \geq 1$ and $q_a \geq 2-3$, see Section 4.2.7, is sufficient to counteract the interchange drive over most of the plasma volume.[47]

To illustrate the stabilizing effect of magnetic shear, we again turn to the family of cylindrical tokamak equilibria (4.113), for which the two competing terms in (4.118) may be evaluated directly. Adopting the tokamak ordering (3.85), allows us to write $B_T(r) \approx B_0$ and $-p' \approx J_T B_P$. Making use of (3.71) to compute $B_P(r)$, and (4.111) to express the constant J_0 in (4.113) as $\mu_0 J_0 = 2B_0/q_0 R_0$, the driving term in (4.118) is easily evaluated as

$$-r\beta' = (4/q_0 q_a)\epsilon^2(1-u)^\nu[1-(1-u)^{\nu+1}], \tag{4.120}$$

whereas the stabilizing term follows from differentiating (4.114),

$$\frac{s^2}{4} = \left[1 - \frac{(\nu+1)u(1-u^\nu)}{1-(1-u)^{\nu+1}}\right]^2. \tag{4.121}$$

Let us compare these terms for $q_0 = 1$, $q_a > 2$. In this case, the driving term vanishes for both $r = 0$ and $r = a$, reaching a maximum value of order $\epsilon^2/q_a \ll 1$ for some intermediate r, while the stabilizing term increases monotonically from zero at $r = 0$ to unity at $r = a$. Hence, the former dominates only in the vicinity of the axis. The width of this interchange-unstable region, in which (4.118) is satisfied, may be estimated by Taylor expanding (4.120) and (4.121) around $r = 0$, with the result $r_c/a \approx 4\epsilon/\nu$, e.g. $r/a < 0.2$ for $q_a = 3$ and $\epsilon = 0.1$. This is also illustrated in Fig. 4.1, in which the radial profile of the Suydam parameter, D_S, is plotted, along with profiles of J_T, B_P, q, s and β, for typical values of $\epsilon = 0.3$ and $q_a = 3-5$.

We expect the pressure gradient to be strongly flattened in this region, down to the level required for marginal stability. Setting $\beta' = -\min\left(|\beta'_{eq}|, |\beta'_{int}|\right)$, where $-r\beta'_{int} = s^2/4$, and integrating β' from $r = 0$ to $r = a$ assuming $\beta_a = 0$, yields an estimate for the central beta in a cylindrical tokamak,

$$\beta_0 \approx 1.55\epsilon^2(f_{int}/q_a) < 4\%, \qquad f_{int} \approx 1-\epsilon, \qquad \epsilon < 0.5. \tag{4.122}$$

Here the factor f_{int} represents the reduction in β_0 due to internal interchange modes. It becomes significant only for $\epsilon > 0.1$, e.g. with $q_a = 3$, $\beta_0 \approx 3\%$ for $\epsilon = 0.3$ and $\beta_0 = 6\%$ for $\epsilon = 0.5$.[48]

47 Recall that $1/q_a$ measures the plasma current I_T, see (4.111).

48 It is interesting to note that a comparable beta limit may be derived by replacing $s_\star$ and $\beta'_\star$ in (4.118) by their volume average values, $\langle s\rangle_a \approx (q_a - q_0)(q_a - q_0) \approx 1/2$, and $\langle\beta'\rangle_a \approx (\beta_a - \beta_0)/a \approx -\beta_0/a$, which yields $\beta_0 < 1/16 \approx 0.06$.

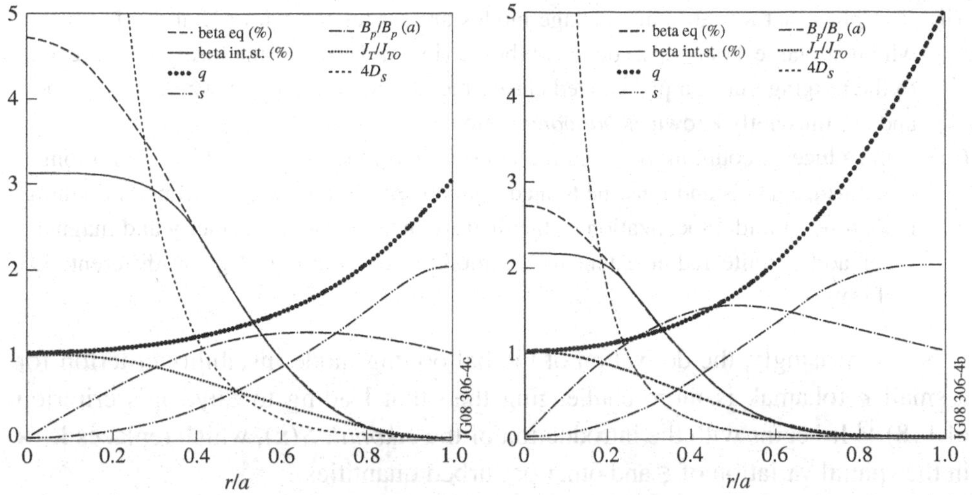

Fig. 4.1. Radial profiles of toroidal current J_T, poloidal field B_P, safety factor q, magnetic shear s, equilibrium beta β_{eq}, marginally interchange-stable beta β_{int} and the Suydam parameter $4D_S$ (4.118), for a cylindrical tokamak with a family of current profiles given by (4.113) with $\epsilon = 0.3$ and $q_a = 3$ (left) and $q_a = 5$ (right).

The volume average toroidal beta (3.19) is found by integrating $r^2\beta'$,

$$\beta_T = \frac{2}{a^2}\int_0^a \beta r \mathrm{d}r \approx \frac{4\mu_0}{a^2 B_0^2}\int_0^a pr\mathrm{d}r \approx (\epsilon/q_a)^2 \beta_P, \qquad (4.123)$$

where the poloidal beta β_P, (3.20), can be evaluated as

$$\beta_P \approx \frac{4\mu_0}{a^2 B_P^2(a)}\int_0^a pr\mathrm{d}r \approx (1-\epsilon^2)\left(1 - 1/q_a^2\right), \qquad \epsilon \leq 0.5,\; q_a \geq 2, \; (4.124)$$

e.g. β_T is limited to roughly one per cent for $\epsilon \approx 0.3$ and $q_a \approx 3$.

Unfortunately, the cylindrical analysis presented above, in which magnetic curvature was assumed to be purely poloidal (3.76), does not accurately describe pressure-driven instabilities in a real tokamak, where toroidal curvature dominates, see (3.97). Toroidal analysis is complicated by the fact that $\kappa_\perp \approx -\cos\theta/R$, (3.100), varies within the flux surface, being unfavourable (favourable) on the outboard (inboard) side of the torus. This poloidal variation has three important consequences:

(i) Since $R_{in} < R_{out}$ and thus $|\kappa_\perp^{in}| > |\kappa_\perp^{out}|$, it provides a net stabilizing effect on radially localized interchange modes with $q > 1$; this follows from the flux surface average of the magnetic curvature $\langle\kappa_\perp\rangle_B \approx -\left(r/q^2 R_0^2\right)\left(1-q^2\right)$, see (3.117), which is positive for $q > 1$.

(ii) It implies that unstable interchange modes are associated with radial displacements which are largest (smallest) on the outboard (inboard) side of the torus; these are akin to the bulging out of a pressurized elastic membrane in the place where it is thinnest and are thus aptly known as *ballooning* modes.

(iii) It introduces a coupling between the Fourier components, ξ_{mn}, which are no longer flux surface labels and must be treated collectively. In the case of strong ballooning, i.e. strong poloidal localization of the distrurbance, toroidal symmetry and magnetic shear add a finite radial extent to the mode, thus coupling ξ_{mn} on different flux surfaces.

Not surprisingly, the derivation of the ballooning mode instability criterion for a small ϵ tokamak is more challenging than that leading to Suydam's criterion (4.118).[49] It begins with the introduction of the *eikonal*, $S(\mathbf{x})$, which replaces $\mathbf{k}\cdot\mathbf{x}$ in the spatial variation of $\boldsymbol{\xi}$ and other perturbed quantities,

$$\boldsymbol{\xi} \approx \boldsymbol{\xi}_\perp = \boldsymbol{\eta}_\perp \exp(\mathrm{i}S), \qquad \mathbf{k} = \mathbf{k}_\perp = \nabla_\perp S, \qquad k_\parallel = \nabla_\parallel S = 0, \tag{4.125}$$

where flute-like displacements are assumed; scale separation is introduced by assuming slow variation of $\boldsymbol{\eta}_\perp$, and fast variation of S, with $\mathbf{x}$.[50] Substituting (4.125) into (4.47), and assuming that $\boldsymbol{\eta}_\perp$ are incompressible, yields

$$\delta^2 W_p \approx \frac{1}{2\mu_0}\int_p |\mathbf{Q}_\perp|^2 + B^2|\mathrm{i}\mathbf{k}_\perp\cdot\boldsymbol{\eta}_\perp + \nabla\cdot\boldsymbol{\eta}_\perp + 2\boldsymbol{\kappa}\cdot\boldsymbol{\eta}_\perp|^2$$
$$-2\mu_0(\boldsymbol{\eta}_\perp\cdot\nabla p)\left(\boldsymbol{\eta}_\perp{}^*\cdot\boldsymbol{\kappa}\right) - \mu_0 J_\parallel\left(\boldsymbol{\eta}_\perp{}^*\times\mathbf{b}\right)\cdot\mathbf{Q}_\perp \mathrm{d}\mathbf{x}, \tag{4.126}$$

where $\mathbf{Q}_\perp = \nabla\times(\boldsymbol{\eta}_\perp\times\mathbf{B})_\perp$ represents field line bending. The displacement $\boldsymbol{\eta}_\perp$ is next expanded in the small parameter ϵ/qn, i.e. roughly $k_\parallel/k_\perp$, whose powers are indicated by subscripts. The zeroth-order contribution to (4.126) comes from the only term containing the fast variation, such that

$$\mathbf{k}_\perp\cdot\boldsymbol{\eta}_{\perp 0} = 0, \qquad \boldsymbol{\eta}_{\perp 0} = (X/B)\mathbf{b}\times\mathbf{k}_\perp, \tag{4.127}$$

where the amplitude of the ballooning displacement, X, varies slowly with $\mathbf{x}$. The first-order contribution comes from the magnetic compression term, which implies that most unstable $\boldsymbol{\eta}_{\perp 1}$ do not compress the magnetic field. Finally, the kink term makes no contribution to (4.126) to second order in ϵ/qn, as can be shown using the identity $\mathbf{Q}_\perp = \mathbf{b}\times\mathbf{k}_\perp\nabla_\parallel X$. Combining these results we find that the plasma potential energy for interchange modes *localized* on a resonant flux surface is determined

[49] A detailed derivation, while relevant to power exhaust in tokamaks, is beyond the scope of this book. Fortunately, good accounts may be found in most texts on MHD stability theory and several review articles, e.g. Connor *et al.* (1998); Connor and Hastie (1999). Below, we present a simplified discussion following Freidberg (1991).

[50] Here $\boldsymbol{\eta}_\perp$ is the envelope and S the phase of the eigenmode.

entirely by competition between the interchange drive and field line bending,

$$\delta^2 W_p \approx \frac{1}{2\mu_0}\int_p k_\perp^2|\nabla_\parallel X|^2 - 2\mu_0(\mathbf{b}\times\mathbf{k}_\perp\cdot\nabla p)(\mathbf{b}\times\mathbf{k}_\perp\cdot\boldsymbol{\kappa})|X/B|^2\mathrm{d}\mathbf{x}. \quad (4.128)$$

Since only the parallel gradient of X appears above, the MHD instability criterion, $\delta^2 W_p < 0$, takes the form of a one-dimensional ODE, describing the parallel (poloidal) variation of X within the flux surface.

To better understand (4.128), it is helpful to recast it in terms of natural field line co-ordinates (r_f, α, l), as defined in (3.12); as with symmetry co-ordinates, the radial ordinate r_f is typically identified with the normalized poloidal flux $\psi \equiv \psi_P$, which gives a simple Jacobian of $\jmath = \sqrt{g} = B^{-1}$. In this co-ordinate system, $\boldsymbol{\kappa}$ and $\mathbf{k}$ are given by (3.31) and (4.51),

$$\boldsymbol{\kappa} = \kappa_\perp\hat{\mathbf{e}}_\perp + \kappa_\wedge\hat{\mathbf{e}}_\wedge, \qquad \mathbf{k}_\perp = k_\perp\hat{\mathbf{e}}_\perp + k_\wedge\hat{\mathbf{e}}_\wedge. \quad (4.129)$$

The latter may be recast in eikonal form (4.125) with

$$k_\perp = (\hat{\mathbf{e}}_\perp\cdot\nabla\psi)\partial_\psi S, \qquad k_\wedge = (\hat{\mathbf{e}}_\wedge\cdot\nabla\alpha)\partial_\alpha S, \qquad k_\parallel = \partial_l S = 0. \quad (4.130)$$

Substituting into (4.128) and noting that $\mathbf{b}\times\mathbf{k}_\perp = k_\wedge\hat{\mathbf{e}}_\perp - k_\perp\hat{\mathbf{e}}_\wedge$, we find

$$\delta^2 W_p \approx \frac{1}{2\mu_0}\int_p k_\perp^2|\nabla_\parallel X|^2 - 2\mu_0 p' k_\wedge(k_\wedge\kappa_\perp - k_\perp\kappa_\wedge)|X/B|^2\mathrm{d}\mathbf{x}, \quad (4.131)$$

where $k_\perp^2 = k_\perp^2 + k_\wedge^2$. When $\eta_{\perp 0}$ is localized radially to some resonant flux surface, then $k_\perp \approx 0$; in that case, $\mathbf{k}_\perp \approx k_\wedge\hat{\mathbf{e}}_\wedge$ is tangential to, and $\mathbf{b}\times\mathbf{k}_\perp \approx k_\wedge\hat{\mathbf{e}}_\perp$ is normal to, the flux surface, and (4.131) becomes

$$\delta^2 W_p \approx \frac{1}{2\mu_0}\int_p k_\wedge^2\left(|\nabla_\parallel X|^2 - \beta'\kappa_\perp X^2\right)\mathrm{d}\mathbf{x}, \qquad \beta' \approx 2\mu_0 p'/B^2. \quad (4.132)$$

Here we recover the product $k_\wedge^2\beta'\kappa_\perp$, which already appeared in the cylindrical tokamak MHD dispersion relation (4.106). In the more general case, both radial variation, $k_\perp$, and geodesic curvature, $\kappa_\wedge$, must be retained.

Inserting the volume element $\mathrm{d}\mathbf{x} = B^{-1}\mathrm{d}\psi\mathrm{d}\alpha\mathrm{d}l$ into (4.131), we find

$$\delta^2 W_p = \frac{1}{2\mu_0}\int_p W(\psi,\alpha)\mathrm{d}\psi\mathrm{d}\alpha, \quad (4.133)$$

$$W(\psi,\alpha) \approx \oint\left[k_\perp^2(\partial_l X)^2 - \beta' k_\wedge(k_\wedge\kappa_\perp - k_\perp\kappa_\wedge)X^2\right]\frac{\mathrm{d}l}{B}, \quad (4.134)$$

where wave numbers are given by (4.130) and the integral is evaluated over the length of the field line.[51] For an interchange perturbation of a thin flux tube lying

[51] When divided by $\oint B^{-1}\mathrm{d}l$, it represents a flux surface average of the expression in square brackets.

on some resonant flux surface $\psi = \psi_\star$, $X(\psi, \alpha, l)$ reduces to $X(\psi - \psi_\star, \alpha - \alpha_\star)$ and the field line bending term, involving $\partial_l X$, disappears from (4.134). Likewise, the average geodesic curvature $\langle \kappa_\wedge \rangle_B$ is equal to zero; this is easily shown by expressing the term involving $\kappa_\wedge$ as a magnetic differential equation (4.54), $\kappa_\wedge p'/B = \partial_l(J_\|/B)$, and noting that it vanishes when integrated over the field line, i.e. $\langle \kappa_\wedge \rangle_B = 0$.

The remaining term involves the average normal curvature $\langle \kappa_\perp \rangle_B$, which alone determines the instability criterion for local interchange modes,

$$W(\psi, \alpha) \approx -\oint k_\wedge^2 \beta' \kappa_\perp X^2 \frac{\mathrm{d}l}{B} < 0, \qquad \Leftrightarrow \qquad \langle \kappa_\perp \rangle_B < 0, \tag{4.135}$$

i.e. local instability ensues when the *average* curvature is unfavourable.[52] For a small ϵ tokamak, for which $\langle \kappa_\perp \rangle_B \propto -(1 - q^2)$, see (3.117), this corresponds to the region $q < 1$. Since $q < 1$ also destabilizes the $m = 1$ internal kink mode leading to semi-periodic sawtooth oscillations which restore $q_0 > 1$, we expect tokamak discharges to be stable against local interchange modes.

The above analysis neglects the effect magnetic shear by assuming exact flute perturbations, $k_\| = \nabla_\| S = 0$. In regions of finite magnetic shear ($q' \neq 0 \neq s$), these are incompatible with the periodicity constraint,

$$S(\theta_f, \zeta_f) = S(\theta_f + 2\pi m, \zeta_f + 2\pi n), \tag{4.136}$$

where m and n are integers, which expresses the requirement that all physical quantities be single valued, i.e. (4.136) implies $k_\| \neq 0$ whenever $s \neq 0$.

Let us demonstrate this for a small ϵ tokamak with circular flux surfaces. For this purpose, we adopt the simple toroidal co-ordinates (r, θ, ζ) for which

$$\mathrm{d}\mathbf{x} = R\mathrm{d}\zeta r \mathrm{d}\theta \mathrm{d}r, \qquad \mathrm{d}l/B = r\mathrm{d}\theta/B_P, \qquad \mathrm{d}\psi = RB_P \mathrm{d}r. \tag{4.137}$$

Assuming axis-symmetry we can integrate over ζ to yield a volume element $\mathrm{d}\mathbf{x} = 2\pi Rr\mathrm{d}\theta\mathrm{d}r$. Substituting into (4.131), or (4.133)–(4.134), we find

$$\delta^2 W_p = \frac{\pi}{\mu_0} \int_0^a W(r)\mathrm{d}r, \tag{4.138}$$

$$W(r) \approx \int_0^{2\pi} \left[k_\perp^2 |\nabla_\| X|^2 - \beta' k_\wedge (k_\wedge \kappa_\perp - k_\perp \kappa_\wedge) X^2\right] Rr\mathrm{d}\theta. \tag{4.139}$$

Note that axis-symmetry reduces the dimensionality of the problem, with field line labels, $W(\psi, \alpha)$, replaced by flux surface labels, $W(r)$. Similarly, the flux surface average becomes a single integral over the poloidal angle, θ.

52 As before, this is a sufficient condition for instability. Conversely, $\langle \kappa_\perp \rangle_B > 0$ is a necessary condition for, but does not guarantee, stability against local interchange modes.

Axis-symmetry permits a Fourier transform with respect to ζ, so that $\partial_\zeta S = -n$ and $S = S_n(r, \theta) - n\zeta$. The condition $k_\| = 0$ now reads

$$k_\| = \nabla_\| S = \partial_l S = \partial_\theta S + q(r)\partial_\zeta S = \partial_\theta S_n - q(r)n = 0, \tag{4.140}$$

where $q(r)$ is given by (3.78). Integrating with respect to θ, we find

$$S_n(r, \theta) = q(r)n(\theta - \theta_\star), \qquad S(r, \theta, \zeta) = n\left[q(r)(\theta - \theta_\star) - \zeta\right], \tag{4.141}$$

where $\theta_\star$ is a constant of integration. We next expand $S(r, \theta, \zeta)$ around a resonant surface $r = r_\star$ and introduce new variables $x = r - r_\star$ and $y = \theta - \theta_\star$,

$$S(x, y, \zeta) = n\left[q_\star y + xq'_\star y - \zeta\right], \tag{4.142}$$

where $q_\star$ and $q'_\star$ are evaluated at $x = 0$. Since $q_\star n = m$ on a resonant surface, the first term is always 2π-periodic in θ, while the last term is necessarily 2π-periodic in ζ, i.e. the sum of these terms satisfies the periodicity constraint (4.136). In contrast, $xq'_\star$, being a product of two independent real numbers is unlikely to be an integer in the vicinity of a resonant surface,[53] i.e. the second term is not 2π-periodic in θ.[54] Hence, finite magnetic shear and poloidal periodicity (4.136) forbid exact flute perturbations!

This dilemma, i.e. the apparent incompatibility between magnetic shear and poloidal periodicity, is resolved by means of the *ballooning expansion*,

$$\boldsymbol{\xi}_n(r, \theta) = \sum \boldsymbol{\xi}_n^j(r, \theta + 2\pi j), \qquad j = 0, \pm 1, \pm 2, \ldots, \tag{4.143}$$

where $\boldsymbol{\xi}_n$ is by definition 2π-periodic in θ, provided the *quasi-modes*, $\boldsymbol{\xi}_n^j$, are bounded in the limits $\theta \to \pm\infty$. The method is formally identical to the WKB theory, best known as the quasi-classical approach in quantum mechanics, see Morse and Feschbach (1953), but is frequently used in problems of wave propagation, e.g. its application to plasma waves may be found in Stix (1992) and Swanson (2003).[55] Since the linearized MHD force operator, $\mathbf{f}(\boldsymbol{\xi}_n)$ (4.41), is likewise 2π-periodic, i.e. satisfies (4.136), it follows that

$$\mathbf{f}(\boldsymbol{\xi}_n) = 0 \quad \Leftrightarrow \quad \mathbf{f}\left(\boldsymbol{\xi}_n^j\right) = 0, \ \forall j. \tag{4.144}$$

Hence, the foregoing analysis remains valid even in the presence of finite magnetic shear, provided that X, S and $\boldsymbol{\xi}$ are interpreted as the *quasi-mode* amplitude, X^j, eikonal, S^j, and displacement, $\boldsymbol{\xi}_n^j$, and poloidal integration in (4.139) is extended

[53] Recall that the neighbourhood of a resonant, or rational, surface is densely populated by ergodic, or irrational, surfaces, see Section 3.1.

[54] Note that $k_\|/k_\perp \approx \epsilon/qn \to 0$ for finite ϵ and q implies $n \to \infty$, so that the $nxq'_\star$ remains finite even when $x \to 0$ and gives the value of $k_\|$ needed to satisfy the periodicity constraint.

[55] Although, the development of the WKB theory can be traced to the original work of Rayleigh (1912), it was first applied to toroidal MHD stability by J. W. Connor and J. B. Taylor (1987).

from a single poloidal circuit to an infinite number of circuits. In short, the ballooning formalism resolves the conflict between magnetic shear and periodicity by constructing a periodic $\xi_n = \eta_\perp \exp(\mathrm{i}S_n)$ from non-periodic $\xi_n^j = \eta_\perp^j \exp\left(\mathrm{i}S_n^j\right)$; note that violation of (4.136) by $S_n^j(\theta)$ is physically admissible provided that it is satisfied by $S_n(\theta)$.

For a cylindrical tokamak, see Section 3.3.1, integration of (4.142) yields

$$k_\perp = nq_\star' y, \qquad k_\wedge \approx \frac{n}{R_0}\left(\frac{b_T^2}{b_P} + b_P\right) \approx \frac{n}{R_0 b_P} \approx \frac{nq_\star}{r}, \tag{4.145}$$

where we neglected the small term $xq_\star'$ in $k_\wedge$ and assumed $b_P \approx \epsilon/q \ll b_T \approx 1$. Since the curvature (3.76) is purely normal, i.e. $\kappa_\perp = -b_P^2/r$ and $\kappa_\wedge = 0$, the potential energy $\delta^2 W_p$ is given by (4.138) with[56]

$$W(r) \approx \int_{-\infty}^{\infty} \left[\left(k_\perp^2 + k_\wedge^2\right)(\partial_\theta X)^2 + \beta' k_\wedge^2 X^2\right]\left(b_P^2/r\right) \mathrm{d}\theta. \tag{4.146}$$

Minimization of this energy yields the eigenmode equation,

$$\mathrm{d}_y(y^2 \mathrm{d}_y X) + D_{S\star} X = 0, \qquad D_S = -2\mu_0 p' q^2 / r B_T^2 q'^2 \approx -r\beta'/s^2, \tag{4.147}$$

where D_S is just the Suydam parameter (4.116). Indeed, the instability criterion for interchange modes, $D_{S\star} \equiv D_S(r_\star) > 1/4$, which follows from (4.147) is identical to Suydam's criterion (4.118).

The similarity between the eigenmode problems (4.116) and (4.147), suggests that x and y, or r and θ, are conjugate variables, in the sense of the Fourier transform.[57] Recall that conjugate Fourier variables x and k satisfy the '*Heisenberg*' uncertainty $\Delta x \Delta k > 1$, which states that a wave packet can be either well localized in real space, or in wave space, but not in both. In the case of ballooning modes, with $k \to nq'y$, radial localization implies poloidal uniformity, and vice versa, so that a poloidally asymmetrical mode, i.e. localized to the outboard side of the torus, must have finite radial extent. It also implies that only high-n modes can be radially localized, while low-n modes must be extended radially.

In toroidal geometry, Suydam's criterion is modified by effects of geodesic curvature and poloidal extent of the modes, to yield an analogous *Mercier's criterion*, $D_M > 1/4$. For a small ϵ tokamak, see Section 3.3.3, this criterion may be derived by setting $k_\perp$ and $k_\wedge$ as (4.145), only with R_0 replaced by R, and taking $\kappa_\perp$ and $\kappa_\wedge$ from (3.100)–(3.101) or, including the effect of the Shafranov shift, from (3.117).

[56] The integral may also be evaluated over the field line, with $r\mathrm{d}\theta = b_P \mathrm{d}l$.

[57] The integral transform relating r and θ is known as the *ballooning* transform. It is worth noting that (4.147), which expresses an eigenmode problem in θ, or y, is effectively a ballooning transform of (4.116), which involves r, or x. The fact that both give the same solution is reassuring and provides an important check of the ballooning formalism.

Inserting these into (4.138)–(4.139) yields the ballooning mode potential energy, $\delta^2 W_p$ (4.138), with

$$W(r) \approx \frac{1}{2\pi} \int_{-\infty}^{\infty} f(\partial_\theta X)^2 + gX^2 \mathrm{d}\theta, \tag{4.148}$$

and the corresponding eigenmode problem,

$$\partial_\theta (f \partial_\theta X) - gX = 0, \tag{4.149}$$

where f and g are given by

$$f = \left(k_\perp^2 + k_\wedge^2\right)/\hat{q}, \qquad g = -\hat{q}R^2\beta' k_\wedge (k_\wedge \kappa_\perp - k_\perp \kappa_\wedge), \tag{4.150}$$

and $\hat{q}$ is the local, and hence poloidally varying, safety factor,

$$\hat{q}(r,\theta) = rB_T/RB_P \approx rB_0R_0/R^2B_P(r), \qquad \langle\hat{q}\rangle_B = q. \tag{4.151}$$

The above may be used to derive the desired interchange stability boundary. A detailed calculation, including the Shafranov shift, yields *Mercier's* instability criterion for a small ϵ tokamak (Ware and Haas, 1966),

$$D_M = D_S(1-q^2) = -r\beta'(1-q^2)/s^2 > 1/4. \tag{4.152}$$

In effect, Mercier's criterion corresponds to replacing the normal curvature in Suydam's criterion by its poloidally averaged value,

$$\langle\kappa_\perp\rangle_B \beta' > (s/2R_0q)^2, \qquad \langle\kappa_\perp\rangle_B \approx -\left(r/q^2R_0^2\right)\left(1-q^2\right), \tag{4.153}$$

which in toroidal geometry is positive, i.e. *favourable*, when $q > 1$. Since $q_0 > 1$ is required for stability against $m = 1$ internal kink modes, and is in practice imposed by sawtooth oscillations, while the small ϵ approximation is always valid near the minor axis ($r = 0$), we may conclude that Mercier's criterion is typically satisfied in tokamaks.[58]

It should be stressed that the above derivation is not valid for non-circular flux surfaces, for which flux co-ordinates must be used. For instance, adopting $(r_f, \theta_f, \zeta_f) = (\psi, \chi, \zeta)$, it is possible to derive an eigen problem analogous to (4.149), which follows from (4.149) with the substitutions,

$$\theta \to \chi, \qquad p' \to rB_P p', \qquad \hat{q}(\psi,\chi) \to \jmath B_T/R \approx \jmath B_0R_0/R^2,$$

where $\jmath$ is the Jacobian of the flux co-ordinate system. Although the corresponding Mercier's criterion is more complicated, e.g. see Freidberg (1991, Chapter 10), it

58 Alternatively, we expect interchange modes to be only active within the volume defined by the $q = 1$ flux surface, e.g. driving low-n MHD modes, such as the TAE (tornado) modes mentioned earlier, Section 4.2.6, which cascade to ever smaller scales (higher-n) in a direct energy cascade, initiating MHD turbulence and enhancing transport within this volume.

contains little new physics beyond that expressed by its simpler version (4.152). It is notable, however, for capturing the effect of non-circular plasma shape, as expressed by the lowest-order moments of the poloidal cross-section: the vertical elongation, $\kappa = A_p/\pi a^2$, and the outward triangularity, δ. Using the flux co-ordinate version of Mercier's criterion it can be shown that a combination of elongation ($\kappa > 1$) and triangularity ($\delta > \epsilon\kappa^2/2$) can stabilize interchange modes even for $q < 1$. In general, elongated, *D-shaped* cross-sections are more stable against pressure-driven MHD modes than circular cross-sections for the same values of ϵ_a, $q(\psi)$, $p(\psi)$, etc. because of an extended favourable curvature region in the D-shape. Note that such plasma shapes are formed naturally in the limit of small aspect ratio ($\epsilon \approx 1$), in which the plasma resembles an apple with its core removed.

Let us next consider perturbations $X(\theta)$ which have a pronounced ballooning character, i.e. poloidal asymmetry with $X(0) > X(\pi)$, and thus finite radial extent. Once again we return to our small ϵ tokamak, but this time assume that $a\beta' \sim O(\epsilon)$ in some thin layer near $r = r_\star$.[59] Under these assumptions, the coefficients in (4.148)–(4.149) may be evaluated as,

$$f = 1 + \Lambda^2, \qquad g = -\alpha(\Lambda \sin\theta + \cos\theta) \tag{4.154}$$

$$\Lambda(\theta) = s(\theta - \theta_\star) - \alpha(\sin\theta - \sin\theta_\star) \approx s\theta - \alpha\sin\theta, \tag{4.155}$$

where $s = rq'/q$ is the magnetic shear and α is the *ballooning parameter*[60]

$$\alpha \equiv -2\mu_0 r^2 p'/R_0 B_P^2 = -q^2 R_0 \beta', \qquad ' = \mathrm{d}_r. \tag{4.156}$$

Note that the Suydam parameter may be expressed in terms of s and α as

$$D_S = \epsilon\alpha/q^2 s^2 = (r/a)(\epsilon_a\alpha/q^2 s^2), \qquad \epsilon = r/R_0, \qquad \epsilon_a = a/R_0. \tag{4.157}$$

The eigenproblem may be solved numerically, see Freidberg (1991, Fig. 10.11), or analytically, for sufficiently simple ballooning trial functions, e.g. selecting $X(\theta) = 1 + \cos\theta$ for $-\pi < \theta < \pi$, and zero otherwise, one finds

$$W(r) \approx 1.39s^2 - 2.17s\alpha + \alpha^2 - \alpha + 0.5, \tag{4.158}$$

which yields the marginal stability condition,

$$W(r) = 0, \qquad \Rightarrow \qquad s \approx 0.78\alpha \pm \sqrt{0.72\alpha - 0.36 - 0.11\alpha^2}. \tag{4.159}$$

This curve, which has a double-valued solution for $\alpha > 0.55$, defines the ballooning instability region, which is bounded by the upper and lower values of s,

[59] Recall that $a\beta' \sim O(\epsilon^2)$ in the low-beta tokamak ordering. However, we may expect $a\beta' \sim O(\epsilon)$ to hold in the edge plasma region near $r = a$, see Section 7.1.

[60] Not to be confused with the field line variable α denoting the diamagnetic angle.

i.e. $s_-(\alpha)$ and $s_+(\alpha)$. It is perhaps surprising to find stability for $s < s_-$, since magnetic shear always exerts a stabilizing influence. The explanation lies in the poloidal variation of *local* magnetic shear, $\hat{s}(\theta)$,

$$\hat{s}(\theta) = r\hat{q}'/\hat{q} = \partial_\theta \Lambda = s - \alpha \cos\theta = s - \alpha[X(\theta) - 1], \tag{4.160}$$

in which the α-term represents a modulation of $\hat{s}(\theta)$ by the radial pressure gradient. While this term does not alter the average magnetic shear, $\langle \hat{s} \rangle_B = s$, it provides a stabilizing effect in the region of unfavourable curvature. Thus, for $s \ll \alpha$, the local value $\hat{D}_S$ is reduced on the outer mid-plane,

$$\hat{D}_S(\theta) \approx \epsilon/q^2\alpha\cos^2\theta = (r/a)[\epsilon_a/q^2\alpha(X-1)^2]. \tag{4.161}$$

In contrast to D_S, which increases with α and decreases with s, see (4.157), $\hat{D}_S$ *decreases* with α and is *independent* of s, so that it remains small even for vanishing average shear, i.e. when $D_S \gg 1$. The local version of Suydam's criterion, $\hat{D}_S(\theta) > 1/4$, is satisfied in the poloidal region defined by $\cos^2\theta < 4\epsilon/q^2\alpha$, so that shear-stabilization is strongest at the outer mid-plane.

Numerical solution of (4.154)–(4.155) confirms the above conclusions, although it indicates that the instability region extends towards the origin of an $s-\alpha$ diagram along a narrow corridor, which separates the stable region into two, topologically distinct halves. The *first stability* region, in which average magnetic shear provides the stabilizing influence, corresponds to low values of α and high values of s, with the boundary defined roughly by

$$\alpha/s = q^3\beta'/q'\epsilon \approx 0.6. \tag{4.162}$$

The *second stability* region, in which pressure-induced modulation of local magnetic shear provides the stabilizing influence, corresponds to high values of α and low values of s.[61] The intermediate region separating these two, in which α and s are comparable, is the domain of ballooning instabilities.

The first stability boundary (4.162) sets an upper limit on the plasma stored energy, expressed by the toroidal beta (3.19), for given current and field profiles. To calculate this beta limit, we again adopt the family of current profiles (4.113), for which s is given by (4.121), and assume the pressure gradient to be marginally stable for ballooning modes at all radial locations.[62] Such marginally stable β' profiles are given by

[61] Since access to the second stability region offers an enticing possibility of tokamak operation at high β_T, its theoretical 'discovery' was received with some excitement. As we will see later on, Nature proved less generous than initially expected due to the combined effect of peeling and ballooning modes which prevent access to the second stability regime.

[62] Note that such marginal stability effectively assumes that the narrow layer of strong pressure gradient is successively localized at different radial locations.

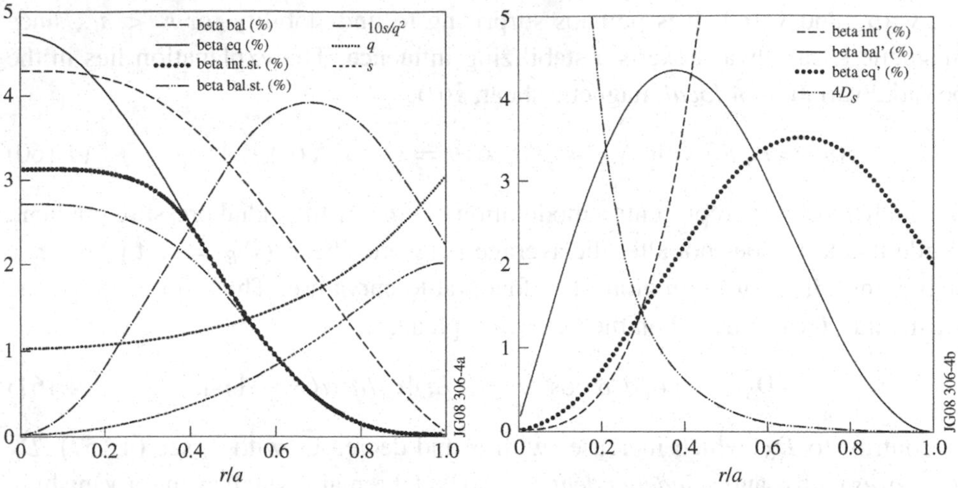

Fig. 4.2. Radial profiles of various field, current and plasma quantities for a (circular) small ϵ tokamak with a family of current profiles given by (4.113) with $\epsilon_a = 0.3$ and $q_a = 3$. The left frame shows $q(r)$ (4.114), $s = rq'/q$ (4.121), $s/q^2 \propto \beta'$, enhanced by a factor of ten for legibility, which is proportional to β' for ballooning marginal-stability ($\alpha \propto s$), and equilibrium and ballooning marginally stable pressure profiles, (4.164). The right frame shows the corresponding radial gradients of β, including the interchange marginally stable β' in a cylindrical tokamak and the associated Suydam parameter D_S (4.118).

$$-r\beta'_{int} = s^2/4, \qquad -r\beta'_{bal} \approx 0.6\epsilon s/q^2 = 0.6(r/a)(\epsilon_a s/q^2), \tag{4.163}$$

where β'_{int} refers to a cylindrical tokamak, since in toroidal geometry interchange modes are stabilized for $q > 1$. The corresponding radial profiles of various field, current and plasma quantities, including equilibrium and marginally stable pressure gradients, are shown in Fig. 4.2 for $\epsilon_a = 0.3$ and $q_a = 35$. Numerical integration of β'_{bal} then yields

$$\beta_T < 0.33\left(\epsilon_a/q_a^{1.2}\right)\left(1 - 1/q_a^{1.2}\right), \qquad q_0 = 1, \tag{4.164}$$

which limits β_T to below $\approx 2\%$ for $q_a \geq 3$ and $\epsilon_a \leq 0.3$.[63]

The above result is largely confirmed by numerical studies of ballooning stability for arbitrary ϵ_a, which yield the so-called *Sykes beta limit*,

$$\beta_T < 0.22(\epsilon_a\kappa/q_0 q_*) \approx 0.044 I_0/aB_0, \qquad q_0 \geq 1, \tag{4.165}$$

where the cylindrical safety factor q_* is given by (4.110). The product I_0/aB_0 is commonly used to define the *normalized toroidal beta*,

[63] Note that the corresponding values of poloidal beta, $\beta_P = (q_a/\epsilon_a)^2\beta_T$, are comparable to unity in the range $q_a \sim 3$ and $\epsilon_a \sim 0.3$, and become larger than unity for $q_a \gg 3$ and $\epsilon_a \ll 0.3$. This follows from the assumption $a\beta' \sim O(\epsilon_a)$ adopted earlier.

$$\beta_N \equiv \beta_T[\%]a[\mathrm{m}]B_0[\mathrm{T}]/I_0[\mathrm{MA}], \tag{4.166}$$

so that (4.164) implies $\beta_N < 4\%$ for $q_a \geq 3$ and $\epsilon_a \leq 0.3$. We observe that (4.164) can now be expressed in terms of β_N and the internal inductance l_i (3.115), evaluated for the family of current profiles (4.113),

$$\beta_N < 4l_i\,\%, \qquad l_i \approx 0.73 + 0.33\ln q_a. \tag{4.167}$$

This relation, which is commonly known as the *no-wall beta limit*, has been confirmed experimentally on many machines under a range of conditions.

Returning to Fig. 4.2, we note that $\beta'_{eq} > \beta'_{bal}$ in the inner half of the plasma, indicating ballooning instability. As before, we repeat the radial integration using $\beta' = -\min\left(|\beta'_{eq}|, |\beta'_{bal}|\right)$ to ensure that the pressure profile satisfies both equilibrium and stability requirements. The result is similar to (4.123)–(4.124) and can be expressed as $\beta_T \approx (\epsilon_a/q_a)^2\beta_P$ with

$$\beta_P \approx \left(1 - \epsilon_a^{1.5}\right)\left(1 - 1/q_a^{1.5}\right), \qquad \epsilon_a \leq 0.5,\; q_a \geq 2. \tag{4.168}$$

As expected, the resulting β_T is reduced compared to (4.164) by a factor roughly proportional to ϵ_a/q_a. This reduction reflects the assumed family of current profiles (4.113), which ensure vanishing edge current density and its derivative(s) at the expense of reduced radial pressure gradients. In contrast, the marginally stable pressure profile (4.165) assumes that the current profile has been optimized to achieve the maximum pressure gradients.

Comparing (4.168) with the cylindrical result (4.124) suggests that pressure-driven instabilities are only moderately affected by finite toroidicity. Indeed, while toroidal curvature stabilizes local interchange modes, it destabilizes non-local, i.e. global, ballooning modes, so that toroidal beta is only moderately reduced for typical values of ϵ_a and q_a.

In the same context, it is worth remarking on the conflicting stability requirements imposed on the $q(r)$ profile by current- and pressure-driven modes. The former are destabilized by parallel current density and its gradient, and thus require high values of q for stability, e.g. $q_0 > 1$ and $q_a > 2-3$. The latter are driven by radial pressure gradients and unfavourable curvature, and are stabilized by magnetic shear. The interplay between these effects determines the marginally stable β'_{bal} profiles given by (4.163), e.g. the appearance of q^2 in the denominator of β'_{bal} suggests that ballooning stability, in contrast to kink stability, requires small values of q. A compromise between these conflicting requirements determines the shape of the β' profile in Fig. 4.2, which peaks in the mid-radius region, where s/q^2 is largest. It also suggests that the combined effect of pressure- and current-driven modes is likely to be more destabilizing than either of these effects acting separately.

This is particularly true for external modes, of which the most unstable are the low-n ballooning–kink modes, driven by a combination of pressure and current gradients.[64] To illustrate this effect, it is useful to write down the surface energy (4.46) for a small ϵ tokamak with all the current flowing on the boundary surface at $r = a$ (Freidberg and Haas, 1973),

$$\frac{\delta^2 W_s}{W_0} = -\frac{1}{2\pi}\int_0^{2\pi} g_a(\theta)\xi^2 \mathrm{d}\theta, \qquad W_0 = 2\pi^2\epsilon^2 B_0^2 R_0/\mu_0, \tag{4.169}$$

$$g_a(\theta) = (\pi/2q_*)^2(1+\cos\theta)/2 + (\beta_T/\epsilon)\cos\theta, \tag{4.170}$$

which was evaluated at the obtained *equilibrium* limit, $B_P \to 0$, as

$$\epsilon\beta_P = \beta_T q_*^2/\epsilon \le (\pi/4)^2, \qquad q_a \to \infty. \tag{4.171}$$

The two terms in (4.170) represent the kink and ballooning terms, respectively, both of which are destabilizing on the outboard side of the torus. The ballooning term is small in the low-β tokamak ordering, $\beta_T \sim \epsilon^2$, but is comparable to the kink term in the high-β ordering, $\beta_T \sim \epsilon$.

Normal mode analysis of (4.169)–(4.170) leads to a rather lengthy expression for $\delta^2 W$, containing poloidal integrals of the form

$$G_{lm} = \frac{1}{2\pi}\int_0^{\pi}(1+\cos\theta)\cos[(l-m)\theta]\mathrm{d}\theta = \delta_{l-m-1}, \tag{4.172}$$

where m and l are both integers. The marginal stability boundary is determined by the external $n = 1$ mode and can be approximated as

$$\beta_T/\epsilon \approx \min\left(\max[0, (q_* - 1.65)/3], (\pi/4q_*)^2\right) \le 0.22. \tag{4.173}$$

This indicates that $q_* = 2A_p B_0/\mu_0 I_0 R_0$, rather than q_a, determines external mode stability, which partially explains the appearance of q_* in (4.165) and the normalization used in β_N (4.166). It should be stressed that q_a and q_* are linked by a one-to-one mapping, which is a complicated function of κ_a, δ_a, ϵ_a, β_T as well as $q(r)$ and $p(r)$. For $q_* < 1$, the plasma is unstable for any value of β_T; for $1 < q_* < 1.65$, maximum β_T increases roughly linearly with q_*; finally, for $q_* > 1.65$ it decreases as q_*^{-2} due to (4.171).[65]

The combined effect of internal and external, current- and pressure-driven, modes has been studied numerically for a range of plasma shapes, aspect ratios

64 This result once again assumes the high-β tokamak ordering $\left(a\beta_T' \sim \epsilon\right)$, since under the low-$\beta$ ordering $\left(a\beta_T' \sim \epsilon^2\right)$ the pressure gradient has a negligible effect on external modes.

65 Note that maximum β_T is comparable to that predicted by the Sykes limit at $q_* = 1$.

and current profiles (Troyon *et al.*, 1984).[66] The optimized $\beta(r)$ and $q(r)$ profiles were found to be fairly flat over most of the plasma, becoming increasingly steep in the edge region. The associated *Troyon beta limit*,

$$\beta_T < 0.14(\epsilon\kappa_a/q_0 q_*) \quad \Leftrightarrow \quad \beta_N < 2.8\%, \tag{4.174}$$

is remarkably similar, albeit with a smaller pre-factor, to the Sykes limit (4.165), which is based only on ideal ballooning modes! Although triangularity δ does not explicitly appear in either (4.165) or (4.174), it is found to stabilize high-n ballooning modes, i.e. ensure $D_M < 1/4$, in vertically elongated cross-sections. While (4.174) suggests that vertical elongation κ_a likewise exerts a stabilizing influence, simulations indicate that $q_*/\kappa_a \to$ const for high values of κ_a and that $\kappa_a > 2$ tends to destabilize low-n external kinks, the suppression of which requires $q_a > 2{-}3$. Indeed, β_T/ϵ reaches a maximum value of $\sim 15\%$ for $q_a = 2(q_* \approx 1.5)$ and $\kappa_a \approx 1.6$. Hence, the minimum value of q_* is set by external kink limits $(q_a > 2-3)$, so that MHD stability is optimized for moderate vertical elongation, $1.5 < \kappa_a < 2$, and outward triangularity, $0.3 < \delta_a < 0.5$. In the design of ITER, these shaping factors were carefully optimized at $\kappa_a \approx 1.7$ and $\delta_a \approx 0.5$, see (1.23).

For comparison, the various ideal MHD beta limits discussed above are plotted together in Fig. 4.3. Note that both the Sykes (4.165) and Troyon (4.174) limits

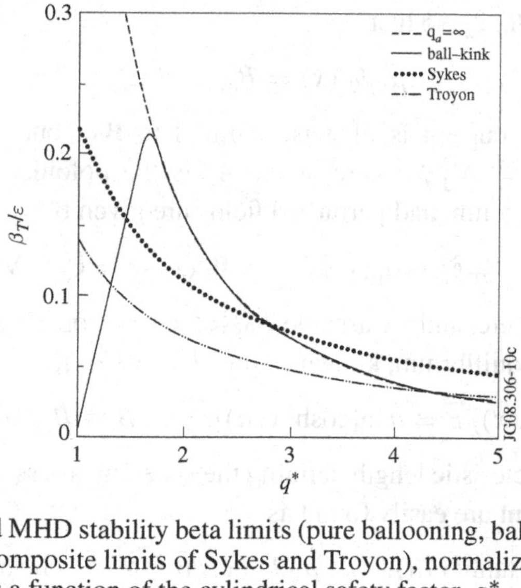

Fig. 4.3. Selected MHD stability beta limits (pure ballooning, ballooning–kink, as well as the two composite limits of Sykes and Troyon), normalized by the inverse aspect ratio, ϵ, as a function of the cylindrical safety factor, q^*.

[66] The study involved optimization of the pressure profiles against high-n interchange and ballooning modes, low-n external ballooning–kink modes and the $n = 1$ internal kink mode.

exhibit weaker dependence on q_*, namely $\beta_T \propto 1/q_* \propto I_0/B_0$, than the sharp boundary model (4.171), which predicts $\beta_T \propto 1/q_*^2$.

4.2.9 Resistive MHD instabilities: tearing modes

Recalling the quote at the opening of Section 4.2, one might expect the presence of parallel electrical resistivity, $\eta = \eta_\| = 1/\sigma_{\|e}$, and other dissipative processes, such as heat conductivity and viscosity, to exert a stabilizing effect on MHD instabilities; indeed, this effect does play a role for pressure-driven instabilities, e.g. resistive-ballooning modes, see Section 6.3.1. However, finite dissipation, especially η, relaxes the *frozen-in flux* constraint (2.252) allowing the topology of the magnetic field to be modified. This process of magnetic field line *tearing* and *reconnection*, see Section 2.4.2.6, opens the door to states of lower potential energy and a new class of instabilities, known as *tearing modes*. More generally, we speak of resistive MHD instabilities, of which tearing modes are a subset, see below. The brief discussion of tearing modes presented below is guided by the excellent accounts of resistive MHD instabilities given in Biskamp (1993) and Biskamp (2000).

In order to illustrate the basic mechanism of a tearing mode, let us consider the stability of a plane current sheet in Cartesian, or slab, geometry (x, y, z), which approximates a flattened-out magnetic flux surface, with $(\hat{\mathbf{e}}_x, \hat{\mathbf{e}}_y, \hat{\mathbf{e}}_z)$ representative of $(\hat{\mathbf{e}}_\perp, \hat{\mathbf{e}}_\wedge, \hat{\mathbf{e}}_\|)$. The equilibrium current is assumed to flow parallel to the dominant (axial) field, $\mathbf{B}_{0z} = B_{0z}\hat{\mathbf{e}}_z$, so that

$$\mathbf{J}_0 = J_{0z}(x)\hat{\mathbf{e}}_z, \qquad \mu_0 J_{0z}(x) = B'_{0y} = \psi''_0, \qquad ' = \partial_x, \tag{4.175}$$

while the perturbed current is likewise parallel to $\mathbf{B}_{0z}$, but depends on both x and y, $\mu_0 J_{1z}(x, y) = \nabla_\perp^2 \psi_1$; here $\psi = A_\|$ is the poloidal flux function. The corresponding equilibrium and perturbed fields are given by

$$\mathbf{B}_0 = B_{0z}\hat{\mathbf{e}}_z + \psi'_0(x)\hat{\mathbf{e}}_y, \qquad \mathbf{B}_1(x, y) = \hat{\mathbf{e}}_z \times \nabla\psi_1, \tag{4.176}$$

where $\mathbf{B} = \mathbf{B}_0 + \mathbf{B}_1$, etc. and we assume $B_{0y} \ll B_{0z} \approx$ const.[67] Below we consider a simple, low-beta equilibrium, known as the *Harris sheet*,

$$\psi_0(x)/\overline{B} = a\ln[\cosh(x/a)], \qquad \overline{B} \equiv B_{0y}(a), \tag{4.177}$$

where a is the characteristic length defining the sheet thickness; the related poloidal field and axial current are easily found as

$$B_{0y}(x)/\overline{B} = \tanh(x/a), \qquad \mu_0 J_{0z}(x)/\overline{B} = a^{-1}\text{sech}^2(x/a). \tag{4.178}$$

[67] This simplification is justified since the dominant field plays no role in the resulting dynamics, aside from defining the drift plane $x-y$ for $\mathbf{E} \times \mathbf{B}$ motions. The resulting flows are effectively two-dimensional within this drift plane and thus correspond to the reduced MHD picture, Section 4.2.5. Consistent with this picture, (4.176) is identical to (4.82).

We assume the sheet evolves in accordance with low-beta ($\beta \ll 1$), incompressible ($M_S \ll 1$), inviscid ($\mu_\perp = 0$) and flute-reduced MHD,

$$\rho_0 \mathrm{d}_t \Omega_{z1} = \mathbf{B} \cdot \nabla J, \qquad \mathrm{d}_t \psi = \eta_\parallel J_{1z} = D_\eta \nabla_\perp^2 \psi, \tag{4.179}$$

$$\Omega_{z1} = \hat{\mathbf{e}}_z \cdot \nabla \times \mathbf{V}_E = \nabla_\perp^2 \breve{\varphi}_2, \qquad \breve{\varphi}_2 = \varphi_2 / B_{0z}, \tag{4.180}$$

where Ω_{z1} is the vorticity of $\mathbf{E} \times \mathbf{B}$ motions in the $x{-}y$ plane, $\mathrm{d}_t = \partial_t + \mathbf{V}_E \cdot \nabla$ is the advective derivative and $D_\eta = \eta_\parallel/\mu_0$ is the magnetic diffusivity, see Section 4.2.5. We next perturb the Harris sheet $\psi_0(x)$ with $\psi_1(x, y)$, which has a specified sinusoidal dependence in the poloidal direction,[68]

$$\psi(x, y) = \psi_0(x) + \psi_1(x, y) \approx \overline{B} x^2/2a + \psi_{1k}(x) \cos(k_\wedge y). \tag{4.181}$$

Note that $\psi_1(x, y)$ corresponds to a parallel current, $\mu_0 J_{1z} = -k_\wedge^2 \psi_{1k} \sin(k_\wedge y)$, and normal magnetic field, $B_{1x}\hat{\mathbf{e}}_x = k_\wedge \psi_{1k} \sin(k_\wedge y)\hat{\mathbf{e}}_x$.

The plot of $\psi(x, y)$ reveals two topologically distinct regions, separated by the line $\psi = \psi_{1k}$. The current sheet is thus broken up into a series of current filaments, or *magnetic islands*, whose flux surfaces are closed in the $x{-}y$ plane. This pattern is identical to that produced in the break-up of a resonant canonical surface of a Hamiltonian system undergoing parametric resonance,[69] see Fig. 2.1 and Section 2.2.2. In toroidal geometry, the neighbourhood of a resonant flux surface is similarly transformed into a series of magnetic islands which are closed in the poloidal ($r_f{-}\theta_f$) plane.

The number of islands created is determined by the poloidal mode number m of the resonant surface. Each island consists of a new subset of locally nested flux surfaces, which are co-axial about the centre of the island and carry most of the current flowing in the original sheet. The observed filamentation, or condensation, of the parallel current is indicative of the tearing and reconnection of poloidal magnetic flux along the resonant surface, which must have occurred to form the magnetic islands. These are separated from each other, and from the open, weakly undulating flux surfaces, by the *separatrix* surface, $\psi = \psi_{1k}$. The original resonant surface is thus replaced by a series of X and O neutral points, which alternate with a spacing of $\Delta y = \pi/k_\wedge$. The width of the island, i.e. the x-distance between the two branches of the separatrix at the O-point, can be estimated as

$$w = \Delta x \approx 4\sqrt{\psi_{1k}/\psi_0''} \qquad \Rightarrow \qquad w/a \approx 4\sqrt{\psi_{1k}/a\overline{B}}. \tag{4.182}$$

68 Recall that a resonant surface is composed of an infinite number of separate field lines, distinguished by field line labels $G(r_f, \theta_f)$. In contrast, an ergodic surface is densely covered by a single field line, and thus requires a single surface label $F(r_f)$. Hence, only a resonant flux surface can support the poloidal gradients implied by the perturbation (4.181).

69 For example, a simple pendulum whose pivot point is subjected to a vertical oscillation which is harmonically resonant with the natural frequency of the pendulum.

In order to study the spectral stability of (4.179), we linearize it to find

$$-i\omega\rho_0\Omega_{z1} = \mathbf{B}_0 \cdot \nabla J_1 + \mathbf{B}_1 \cdot \nabla J_0, \tag{4.183}$$

$$-i\omega\psi_1 = \mathbf{B}_0 \cdot \nabla\breve{\varphi}_2 + D_\eta \nabla_\perp^2 \psi_1. \tag{4.184}$$

Assuming D_η is small, it is natural to divide the domain into the interior, resistive region and the exterior, ideal region; the former can then be treated by one-dimensional resistive MHD, the latter by two-dimensional ideal MHD, and the two solutions can be merged by requiring smooth transition at some intermediate x-location, e.g. in the asymptotic limit of the two approximations.[70] In the inner region, whose thickness, δ_η, is comparable to the island width w, (4.183)–(4.184) can be approximated as

$$-i\omega\rho_0\breve{\varphi}_2'' \approx ik_\wedge x B_{0y}' \psi_{1k}'' - ik_\wedge \mu_0 J_{0z}' \psi_{1k}, \tag{4.185}$$

$$-i\omega\psi_{1k} \approx ik_\wedge x B_{0y}' \breve{\varphi}_2 + D_\eta \psi_{1k}'', \tag{4.186}$$

where only the dominant, radial gradients of φ_2 and ψ_{1k} were retained. In the outer, ideal region, $\mathbf{B} \cdot \nabla\mathbf{J}$ is dominant, and ψ_{1k} is governed by

$$\psi_{1k}'' - \left[k_\wedge^2 + \mu_0 J_{0z}'/B_{0y}(x)\right]\psi_{1k} \approx 0, \tag{4.187}$$

which can be integrated analytically for the Harris sheet to yield,

$$\psi_{1k} = \overline{B}e^{-k_\wedge|x|}[1 + \tanh(x/a)/k_\wedge a], \qquad \psi_{1k}(\pm\infty) \to 0. \tag{4.188}$$

The inner and outer solutions are merged by requiring that the change in the logarithmic derivative of $\psi_{1k}(x)$, evaluated at $x = 0$,

$$\Delta' \equiv (\ln\psi_{1k})'(0) = \left[\psi_{1k}'(0_+) - \psi_{1k}'(0_-)\right]/\psi_{1k}(0), \tag{4.189}$$

be identical in both cases.[71] The definition of $\psi_{1k}'(0_\pm)$ clearly differs in the interior and exterior regions, with the former implying the outermost boundary of the resistive region and the latter the innermost boundary of the ideal region. The exterior case reflects the discontinuity in ψ_{1k}' which occurs in the solution of (4.187) at $x = 0$, e.g. for the Harris sheet we find

$$\Delta' a = 2[(k_\wedge a)^{-1} - k_\wedge a], \tag{4.190}$$

which implies that $\Delta' > 0$ for $k_\wedge a < 1$, i.e. for wavelengths longer than the sheet thickness, a, and $\Delta' < 0$ for $k_\wedge a > 1$.

In order to relate Δ' to the growth rate of the mode, $\gamma = -\text{Im}\,\omega$, a solution of (4.185)–(4.186) is necessary. Here we assume the mode to be ideally stable, such

[70] Of course, this is just the usual approach of boundary layer theory.

[71] That is, Δ' is the inverse radial gradient length of ψ_{1k}. Note that ψ_{1k}'' can be expressed in terms of Δ' as $\psi_{1k}\Delta'/\max(\delta_\eta, w)$.

that its growth rate is sufficiently slow to ensure ample time for resistive diffusion of ψ_{1k} across the layer, i.e.

$$\Delta'^{-1} \gg \delta_\eta \sim \sqrt{D_\eta \gamma} \quad \Rightarrow \quad \psi_{1k} \approx \psi_{1k}(0). \tag{4.191}$$

The above also assumes a linear phase of the instability, when the island thickness, w (4.182), is smaller than δ_η, thus ensuring the constancy of ψ_{1k} across both the island and the resistive layer. Indeed, the constancy of ψ_1 is part of the definition of *tearing* modes, which are assumed to be ideal MHD stable and thus due entirely to the presence of finite resistivity. When this condition is violated, one speaks of resistive MHD modes, e.g. a resistive kink mode, for which $\psi(0) \to 0$, or a resistive interchange mode, for which $w \to 0$ and $\Delta' \to \infty$, etc. It can then be shown (Furth *et al.*, 1963) that ψ_{1k} is unstable if, and only if, $\Delta' > 0$, i.e. for wavelengths larger than the thickness of the sheet, $k_\wedge a < 1$. In that case, the perturbation grows exponentially with time at a rate given by

$$\gamma \approx 0.5(\Delta' a)^{4/5}(k_\wedge a)^{2/5}\left(aB'_{0y}/\overline{B}\right)^{2/5}\tau_A^{-2/5}\tau_\eta^{-3/5}, \tag{4.192}$$

where the resistive (2.258) and Alfvén (2.254) times are given by

$$\tau_\eta = \mu_0 a^2/\eta_\parallel = a^2/D_\eta, \qquad \tau_A = a/V_A = a\sqrt{\rho_0\mu_0}/\overline{B}. \tag{4.193}$$

The layer thickness then follows from the dispersion relation, $\Delta' \approx 2\gamma\delta_\eta/D_\eta$:

$$\delta_\eta/a \approx (\Delta' a)^{1/5}(k_\wedge a)^{-2/5}\left(aB'_{0y}/\overline{B}\right)^{-2/5}(\tau_A/\tau_\eta)^{2/5}. \tag{4.194}$$

Note that $\overline{B}$ in the definition of τ_A (4.193) cancels with that appearing in (4.192) and (4.194), so that both γ and δ_η depend only on B'_{0y}. When J_{0z} is driven by a toroidal EMF, as is the case in a tokamak, then $B'_{0y} \propto J_{0z} \propto \eta^{-1}$, and (4.192) and (4.194) scale as $\gamma \propto \eta^{1/5}$ and $\delta_\eta \propto \eta^{4/5}$.

Inserting (4.190) into (4.192), one finds that for small $k_\wedge$ the growth rate scales as $\gamma \propto k_\wedge^{-2/5}$ so that modes with largest wavelengths grow most rapidly. As the island size increases, eventually w exceeds δ_η and resistive diffusion can no longer ensure $\psi_1 \approx$ const across the island; the net effect is a reduction of Δ' and γ. Replacing the assumption $\psi_{1k} \approx \psi_{1k}(0)$ by $\Delta'^{-1} \approx \delta_\eta$, one finds that γ reaches a maximum value given by,

$$\delta_\eta/a \approx (\Delta' a)^{-1} \sim k_\wedge a \quad \Rightarrow \quad \gamma_{max} \approx 0.5(\tau_\eta\tau_A)^{-1/2} \tag{4.195}$$

at some intermediate wave number, $k_\wedge a \approx (\tau_A/\tau_\eta)^{1/4}$.[72]

[72] There is a close analogy to the effect of surface tension on the K–H instability, see (4.22); in both cases, the optimum wave number is determined by competition between diffusive and inertial effects, and is a quatric root of the diffusion coefficient.

As the island grows further, it is w, rather than δ_η, which dominates in

$$\mu_0 J_{1z} \approx \psi''_{1k} \approx \psi_{1k}\Delta'/\max(\delta_\eta, w), \tag{4.196}$$

so that the non-linear evolution of the island size becomes limited by the diffusive broadening of the current perturbation, ψ_1 (Rutherford, 1973). Inserting (4.196) into (4.179) shows that ψ_1 evolves according with $d_t\psi_{1k} = \eta_\parallel J_{1z} \approx D_\eta \psi_{1k}\Delta'/w$, or, in terms of the island width w,

$$d_t w \approx D_\eta \Delta', \qquad w_{cr} < w \ll w_{sat}, \tag{4.197}$$

where $w_{cr} \approx \delta_\eta$ is the critical size and w_{sat} is the saturated size of the island. We thus find that $d_t w$ is independent of both w and t, and the island size increases *linearly*, rather than *exponentially*, with time. The saturated size is determined by the condition $d_t w_{sat} = 0$ or

$$\Delta'(w_{sat}) \approx [\psi'_{1k}(w_{sat}/2) - \psi'_{1k}(-w_{sat}/2)]/\psi_{1k}(0) = 0. \tag{4.198}$$

For a Harris sheet, w_{sat} is comparable to a, but more generally, it is a function of both current and resistivity profiles.

It is straightforward to extend the slab analysis presented above to include poloidal curvature, i.e. to extend it to the cylindrical tokamak. In the outer, ideal region, one finds the following evolution equation for ψ_1,

$$\nabla^2\psi_1 - \left(\mu_0 J'_{T0} k_P / k_\parallel B_0\right)\psi_1 \approx 0, \tag{4.199}$$

where $k_\parallel$ and k_P are given by (4.63), in particular,

$$\nabla_\parallel \sim k_\parallel = k_P b_P + k_T b_T = k_P b_P (1 - qn/m) \tag{4.200}$$

and the prime now denotes a radial derivative. With these definitions, (4.199) may be expressed in a form similar to (4.187),

$$r^{-1}\left(r\psi'_1\right)' - [(m/r)^2 + \left(\mu_0 J'_{T0}/B_P\right)(1 - qn/m)^{-1}]\psi_1 \approx 0. \tag{4.201}$$

The singularity in (4.187), which occurred due to a vanishing poloidal field $B_{0y}(0)$, now takes place only on resonant flux surfaces, on which $k_\parallel$ and the associated line bending term approach zero. In the vicinity of the resonant surface, the factor $1 - qn/m$ may be approximated by the first term in its Taylor expansion, which is just the change in the magnetic shear,

$$1 - qn/m \approx -(q'/q)_\star x = s_\star - s, \qquad x = r - r_\star. \tag{4.202}$$

Since the above is equivalent to $b_\wedge/b_P$, (4.201) can also be rewritten as

$$r^{-1}\left(r\psi'_1\right)' - \left[k_P^2 + \left(\mu_0 J'_{T0}/B_\wedge\right)\right]\psi_1 \approx 0, \tag{4.203}$$

where $B_\wedge = -B_P(s_\star - s)$, which is a toroidal version of (4.187). On account of the stabilizing poloidal field line bending term, $k_P^2\psi_1 = (m/r)^2\psi_1$, we expect only the lowest m modes to become tearing unstable.

For arbitrary current profile, $J_{T0}(r)$, and edge safety factor, q_a, (4.201) must be solved numerically on either side of the singularity at $r_\star$. As before, this singularity means that the drive for the instability comes predominantly from the radial gradient of the parallel current in the vicinity of the resonant surface, i.e. $J'_{T0}(r_\star)$.[73] It also leads to a jump in $\Delta' = (\ln\psi_1)'$ across the resonant surface, which must be matched with the resistive MHD solution in the inner, diffusive layer. Due to the one-dimensional nature of this layer, the slab results (4.192)–(4.194) remain valid provided a is again interpreted as the minor radius and the following substitutions are made:[74]

$$k_\wedge \to k_P = m/r, \qquad B'_{0y}/\overline{B} \to q'/q = s/r, \qquad \overline{B} \to B_P. \tag{4.204}$$

In the non-linear phase, the evolution equation (4.197) becomes

$$\mathrm{d}_t w \approx D_\eta[\Delta'(w) - \alpha w], \tag{4.205}$$

where the stabilizing term αw is related to local plasma quantities. Hence, the saturated island width w_{sat} is determined by $\Delta'(w_{sat}) = \alpha w_{sat}$.

A more complete analysis reveals several additional effects, which must be included in order to explain the observed behaviour of magnetic islands in tokamaks: (i) spatial variation of $\eta_\parallel \propto T_e^{-3/2}$, which enters in the line bending term $\mathbf{B}\cdot\nabla\mathbf{J}$ in (4.179), and introduces non-linear effects in $\mathrm{d}_t\psi_1 = \eta_\parallel J_{1z}$; (ii) parallel variation of electron and ion temperatures along the field lines, due to finite parallel heat diffusivities, $\chi_{\parallel s} \propto T_s^{5/2}/n_s$; (iii) *neoclassical* effects associated with trapped vs. passing particles, which occur due to magnetic mirror forces in toroidal geometry, see Section 5.3.3; (iv) polarization currents created by plasma turbulence; and (v) additional perturbations to p, J and $\eta_\parallel$ profiles within the island, e.g. due to pellet injection, sawteeth, fishbones or edge localized modes, asymmetric (error) fields, etc.

Let us first consider the neoclassical effects. Aside from corrections to $\eta_\parallel$ and $\chi_\parallel$, these influence island evolution by the so-called *bootstrap current*,

$$J_T^{bs} \approx \left\langle J_\parallel^{bs}\right\rangle_B \sim -\sqrt{\epsilon}\,p'/B_P, \tag{4.206}$$

i.e. a toroidal current generated by a radial gradient of the plasma pressure. The flattening of the pressure gradient within the island reduces J_T^{bs}, which leads to a

[73] Hence, it is possible to construct current profiles which are stable against tearing modes.
[74] Since $k_\parallel = k_P b_P + k_T b_T \approx 0$, one can alternatively choose $k_\wedge \to k_T = -n/R$ and $\overline{B} \to B_T$.

corresponding change in Δ', given by (Sauter *et al.*, 1997)

$$\Delta'(w) \approx 16\left(\mu_0 J_T^{bs}/B_P\right)(q/q'w). \tag{4.207}$$

The resulting instability is known as a *neoclassical tearing mode* (NTM).[75] Although NTMs are relatively benign, with the exception of the (2,1) mode, which can lead to a complete loss of equilibrium and a termination of the discharge, i.e. a *plasma disruption*, they tend to reduce the plasma stored energy by flattening the pressure profile within the island, e.g. the (2,1), (3,2) and (4,3) NTMs typically reduce β_T by 50%, 20–30% and 10–15%, respectively. For a nice discussion of NTM physics, see Sauter *et al.* (1999).

The dominant *ideal MHD* instabilities discussed in Sections 4.2.7 and 4.2.8 are likewise aggravated by parallel electrical resistivity. Foremost among the resulting resistive MHD modes is the $m = 1$ internal kink mode, which is largely responsible for the sawtooth oscillation, and low-m external kink modes. When the latter are ideal MHD stable, they may be treated by the tearing mode analysis presented above. One thus finds a continuous transition from external kink modes for $q_a < m/n$, when the low-m resonant surface lies outside the plasma ($r_\star > a$), to internal tearing modes for $q_a > m/n$, when it crosses the edge of the plasma ($r_\star < a$), see Wesson (2004, Fig. 6.9.2). The presence of a perfectly conducting wall at radius $r = b \geq a$ reduces the growth rate of these low-m tearing modes provided the resonant surface is located close to the edge of the plasma, e.g. for $b = a$, the (2,1) mode is stabilized when $r_\star > 0.95a$.

More generally, the wall has a finite resistivity, η_w, and finite thickness, δ_w. Its effect on tearing mode stability may be quantified by replacing the factor Λ_w (4.67), which enters the vacuum potential energy $\delta^2 W_v$ and the edge boundary condition, $\left(\psi_1'/\psi_1\right)_a = -(m/a)\Lambda_w$, with (Batemann, 1978)

$$\Lambda_w \approx \frac{1 + f_w(a/b)^{2|m|}}{1 - f_w(a/b)^{2|m|}}, \qquad k_T b \ll 1, \qquad a \leq b. \tag{4.208}$$

Here f_w is the resistive wall correction factor,

$$f_w \approx (1 + \mathrm{i}m/\omega\tau_w)^{-1}, \qquad \tau_w = b\delta_w\mu_w/2\eta_w, \qquad \delta_w \ll b/m, \tag{4.209}$$

where τ_w is the resistive time constant of the wall and $\omega = \omega_r + \mathrm{i}\gamma$ is the frequency of the mode.[76] This factor approaches unity for $\omega_r\tau_w \gg 1$, when magnetic oscillations are much faster than τ_w, and are thus shielded from the bulk of the wall by eddy currents induced in the surface, so that the wall behaves as a perfect

[75] It should be clear from the discussion in Section 2.5 that neoclassical effects lie beyond the scope of the MHD model. Hence the appearance of the ion gyro-radius in NTM theory.

[76] In practice, ω_r may be interpreted as the toroidal rotation of the plasma and γ as the growth rate of the mode, see discussion of plasma rotation and mode locking below.

conductor. In contrast, f_w approaches zero for $\omega_r \tau_w \ll 1$, i.e. for sufficiently slow oscillations, which can diffuse into the bulk of the wall up to a depth $\sqrt{\eta_w/\mu_w\omega}$. In this case $f_w \to \gamma\tau_w/(m+\gamma\tau_w)$ and the presence of the wall can affect the growth rate, but not the stability boundary of the mode, i.e. $f_w = 0$ for $\gamma = 0$ and $f_w \to 1$ for $\gamma\tau_w \gg 1$.

When the mode grows to large amplitudes, interaction with a resistive wall can decelerate plasma rotation and lock the mode to asymmetric *error fields*;[77] such *locked modes* usually terminate the plasma discharge in a disruption and can lead to substantial local heat loads on in-vessel components. To study the effect of tearing mode locking, it is first necessary to examine the rotation of the plasma in the laboratory frame of reference, in which the external coils, the static error fields and the resistive wall are all at rest. Assuming the plasma rotates as a rigid body, the evolution of the mode frequency ω may be calculated from momentum conservation,

$$\mathrm{d}_t\omega = \mathbf{k}\cdot\mathrm{d}_t\mathbf{v} = \frac{m\int_w rF_P \mathrm{d}\mathcal{V}}{\int_p \rho r^2 \mathrm{d}\mathcal{V}} - \frac{n\int_w F_T \mathrm{d}\mathcal{V}}{R\int_p \rho \mathrm{d}\mathcal{V}}, \tag{4.210}$$

where the top integrals, representing the poloidal and toroidal forces on the plasma, and the equal and opposing forces on the vessel wall, are evaluated over the wall volume, while the bottom integrals, representing plasma inertia, are evaluated over the plasma volume. For a small ϵ tokamak, the force on the wall is dominantly poloidal, so that $F \approx F_P = J_{T1}B_{r1}$ and $F_T \approx (k_T/k_P)F$. With these approximations, (4.210) may rewritten as

$$\mathrm{d}_t\omega \approx -\frac{J}{m}\int_0^{2\pi}\int_b^{\infty} Fr^2 \mathrm{d}r\mathrm{d}\theta, \qquad J = \frac{m^2}{\int_p \rho r^3 \mathrm{d}r} + \frac{n^2}{R^2\int_p \rho r \mathrm{d}r}. \tag{4.211}$$

Since $r \ll R$, we may expect the first term in J, and hence the poloidal plasma rotation to dominate. However, the $1/R$ variation of B_T and the non-circular shape of typical poloidal cross-sections, means that poloidal rotation is damped far more effectively than toroidal rotation. As a result, the plasma may be assumed to spin toroidally with angular frequency ω.

Substituting for J_{T1} and B_{r1} from the tearing mode analysis, i.e. from the solution of (4.201) with the boundary condition as given above (4.208), leads to the following closed form for $\mathrm{d}_t\omega$ (Nave and Wesson, 1990),

[77] The name refers to errors in the alignment of external coils or in their internal winding. Low-n error fields may be corrected with additional control coils. High-n error fields such as the toroidal field ripple caused by a finite number of poloidal coils, are inherent in any tokamak design and can only be reduced by adding more coils or inserting ferromagnetic elements.

$$\mathrm{d}_t\omega \approx -J\left(\psi_{1a}^2/\mu_0\right)(a/b)^{2|m|}\omega\tau_w/g_w, \tag{4.212}$$

$$g_w = 1 + \left[(1-(a/b)^{2|m|})\omega\tau_w/m\right]^2. \tag{4.213}$$

We observe that toroidal deceleration, $-\mathrm{d}_t\omega$, increases as the square of the tearing mode amplitude and is a strong function of the frequency itself, scaling as $(\omega\tau_w)^{-1}$ for $\omega\tau_w \gg 1$, and as $\omega\tau_w$ for $\omega\tau_w \ll 1$. In the former case, ω is only weakly damped, while in the latter, it decays exponentially with time. Hence, plasma rotation decreases sharply once its angular frequency becomes comparable to the resistive time constant of the wall, provided the amplitude of the mode is sufficiently large and the wall is sufficiently close.

In the presence of an error field, ψ_E, with toroidal phase given by $n_{EF}(\zeta - \zeta_0)$, the phase of the mode locks to that of the error field, and the plasma comes to rest at $\zeta = \zeta_0 + (\pi j/2n_E)$, where j is an integer. The size of the island can then increase even further until it saturates when

$$\Delta'(w_{sat}) = \Delta_0'(w) + 2(m/r_\star)(\psi_E/\psi)_\star \approx 0, \tag{4.214}$$

where $\Delta_0'(w)$ represents the driving forces in the absence of error fields.

During the non-linear evolution of all tearing instabilities, chaotic volumes appear around the X-points, gradually increase with the amplitude of the mode, and eventually destroy the separatrix surface, filling the entire neighbourhood of the resonant surface, including the island region, with chaotic field lines; once again the process, known as *ergodization*,[78] has an exact analogy in the canonical tori describing a Hamiltonian system undergoing parametric resonance, see Fig. 2.1. As the tearing instability continues to grow, islands originating from different resonant surfaces eventually begin to overlap and the entire plasma volume is filled by chaotic field lines. This occurs when the *Chirikov parameter*, defined as the square of the unperturbed magnetic island width, w, divided by its separation from a neighbouring island, $K = (w/s)^2$, approaches unity (Chirikov, 1959). At that time, the nested flux surfaces required for plasma confinement are destroyed and plasma stored energy is quickly removed by parallel losses along the chaotic field lines.

The limits imposed by ideal and resistive MHD instabilities on the equilibrium current and pressure profiles may be summarized as, see Table 4.1,[79]

$$q_0 > 1, \qquad q_a > 2\text{–}3, \qquad \beta_T < 0.14\epsilon\kappa/q_0 q_*, \qquad \beta_N < (3\text{–}4)l_i\%. \tag{4.215}$$

[78] A better term would be *chaotization*, to distinguish from the label *ergodic* flux surface.
[79] Their impact on fusion reactor performance was already discussed in Section 1.2.

Table 4.1. *Current and pressure limits imposed by MHD stability.*

Quantities affected	Instability (mode)	Limit (criterion)
central current density $J_T(0)$	$n = m = 1$ internal kink	$q_0 > 1$
	high-n internal ballooning	$q_0 \geq 1$
total current $I_T(a)$, edge	$n = m = 1$ external kink	$q_a > 1$
current density $J_T(a)$, $J_T'(a)$	low-n external kink – ballooning	$q_a > 2$
	$m \geq 2$ external kink	$q_a/q_0 \geq 2 - 3$
	neoclassical tearing modes	$q_a/q_0 \geq 2 - 3$
toroidal beta β_T, plasma	high-n ballooning	$\beta_T < 0.22\epsilon\kappa/q_0 q_*$
pressure, stored energy, etc.	low-n external kink – ballooning	$\beta_T < 0.14\epsilon\kappa/q_0 q_*$
	combined external modes	$\beta_N < 4l_i\%$
	neoclassical tearing modes	$\beta_N < (3-4)l_i\%$

4.3 Drift-waves and instabilities

The phenomena described in Sections 4.2.1–4.2.8 were derived assuming the MHD ordering (2.41), in which all finite gyro-radius effects, i.e. all terms of order δ_i, are omitted. Inclusion of these effects leads to a new class of waves and instabilities, typically designated by the label '*drift*' on account of their origin in the drift ordering (2.42), since they disappear in the MHD limit $\delta_i \to 0$. Below, we present a brief account of drift-waves and instabilities, also known as *micro-instabilities* due to the small scales (high $k_\perp$) involved, making use of the *drift-fluid model* derived in Section 2.4.2.7.

The simplest, and quintessential, drift-wave can be studied using the electrostatic ($\partial_t \mathbf{B} = 0 = \nabla \times \mathbf{E}$, $\mathbf{E} = -\nabla\varphi$, $\nabla_\parallel \approx \mathbf{b}_0 \cdot \nabla$) version of this model. Neglecting collisional dissipation $(\nu_e^* \to 0)$, volumetric sources ($\mathbf{i}_n = 0$) and equilibrium electromagnetic field non-uniformity $(B_0' = 0 = \varphi_0')$, the corresponding drift-fluid model equations can be evaluated to order δ_i as

$$\mathrm{d}_{t,0} n + n\nabla_\parallel V_\parallel \approx 0, \qquad \rho \mathrm{d}_t V_\parallel + \nabla_\parallel (p_i + p_e) \approx 0, \tag{4.216}$$
$$\mathrm{d}_{t,0} p_i + \tfrac{5}{3} p_i \nabla_\parallel V_\parallel \approx 0, \qquad \nabla_\parallel p_e \approx e n_e \nabla_\parallel \varphi, \tag{4.217}$$

and $\nabla_\parallel T_e \approx 0$, where $\mathbf{V}_0 = V_\parallel \mathbf{b} + \mathbf{V}_E \approx \mathbf{V} - \mathbf{V}_{*i}$ is the MHD-ordered flow velocity and $\mathrm{d}_{t,0} = \partial_t + \mathbf{V}_0 \cdot \nabla$ is the associated advective derivative.[80] Adopting the slab geometry, which we used previously for tearing mode analysis, Section 4.2.9, and subjecting the system to small perturbations of the form $n_1(x)\exp(\mathrm{i}k_\wedge y + \mathrm{i}k_\parallel z - \mathrm{i}\omega t)$, (4.216)–(4.217) linearizes to

$$\omega n_1/n_0 \approx k_\parallel V_{\parallel 1} + \omega_{*n} e\varphi_2/T_{e0}, \quad \omega V_{\parallel 1} \approx k_\parallel (p_{e1} + p_{i1})/m_i n_0, \tag{4.218}$$

[80] Note that $\nabla_\parallel V_\parallel$ appears in place of $\nabla \cdot \mathbf{V}_0$ since $\nabla \cdot \mathbf{V}_E = 0$ for $\mathbf{B} = \text{const}$.

$$\omega p_{i1}/n_0 \approx \tfrac{5}{3}T_{i1}k_{\|}V_{\|1} - \omega_{*i}e\varphi_2,\ n_1/n_0 \approx e\varphi_2/T_{e0}, \tag{4.219}$$

$$p_{i1} \approx n_0 T_{i1} + n_1 T_{i0}, \quad p_{e1} \approx n_{e1}T_{e0}, \tag{4.220}$$

and $T_{e1} \approx 0$, where ω_{*n}, ω_{*e} and ω_{*i} are drift frequencies based on the equilibrium density, electron pressure and ion pressure gradients, respectively,[81]

$$\omega_{*s} \equiv \mathbf{k}\cdot\mathbf{V}_{*s} = \frac{k_\wedge p'_{s0}}{n_{s0}e_s B_0} = \frac{k_\wedge T_{s0}}{e_s B_0 L_{\perp s}}, \quad L_{\perp s}^{-1} = p'_{s0}/p_{s0}. \tag{4.221}$$

$$\omega_{*n} = \omega_{*e}[T'_{e0} = 0] = -\frac{k_\wedge T_{e0}}{eB_0 L_{\perp n}}, \quad L_{\perp n}^{-1} = n'_0/n_0. \tag{4.222}$$

where $L_{\perp n}$ and $L_{\perp s}$ are the radial gradient lengths of plasma density and pressure $p_s = n_s T_s$ of species s. It should be emphasized that ω_{*s} is proportional to the magnetization parameter of species s, defined as $\delta_s = \rho_{ts}/L_{\perp s}$,

$$\omega_{*s} = \frac{k_\wedge v_{ts}^2}{L_{\perp s}\Omega_s} = \frac{k_\wedge v_{ts}\rho_{ts}}{L_{\perp s}} = k_\wedge v_{ts}\delta_s = (k_\wedge \rho_{ts})\omega_{ts}, \quad v_{ts}^2 = \frac{T_{s0}}{m_s}. \tag{4.223}$$

As expected, the drift frequency vanishes ($\omega_{*s} \to 0$) in the MHD ordering ($\delta_s \to 0$), in which case only MHD and EM waves survive, e.g. for a finite gyro-radius, ω_{*s} vanishes in the absence of radial gradients in thermodynamic quantities (n'_{s0}, $T'_{s0} \to 0$), which implies $L_{\perp s} \to \infty$ and $\delta_s \to 0$.

The classical, or *electron drift-wave*,[82] occurs in the frequency range

$$k_{\|}v_{ti} \ll \omega \sim \omega_{*n} \ll k_{\|}v_{te}, \tag{4.224}$$

$$k_\wedge \rho_S \sim 1, \quad \omega_{*n} \sim \omega_{tS}, \tag{4.225}$$

in which oscillations are fast compared to the ion parallel transit time and slow compared to the electron parallel transit time and comparable to the sound wave transit frequency, $\omega_{tS} = c_{Se}/L_\perp$, where c_{Se} is the cold ion sound speed, $c_{Se}^2 = p_e/\rho_i = ZT_e/m_i$, defined in (4.227); the spatial scales are comparable to the sound speed gyro-radius, $\rho_S = c_{Se}/\Omega_i$. This frequency range requires $k_{\|}$ to be finite ($k_{\|} \gg \omega/v_{te}$), so that perturbations must be slightly misaligned with respect to the magnetic field. In the absence of such misalignment, i.e. for ideal flute perturbations ($k_{\|} = 0$), the parallel gradient length and the electron parallel transit time become infinite, thus violating the adiabatic condition (2.283) or (4.217). Consequently, ions respond acoustically, or *hydrodynamically*, while electrons respond quasi-statically, or *adiabatically*, in accordance with (2.283).

81 The drift frequency ω_{*s} appears due to the advective term in the linearized derivative $\mathrm{d}_{t,0}\mathcal{A} \approx -i\omega\mathcal{A}_0 - ik_\wedge(\varphi_2/B)\mathcal{A}'_0$. The definition $\omega_{*s} \equiv \mathbf{k}\cdot\mathbf{V}_{*s}$ suggests that it may be interpreted as the compressibility of the diamagnetic flow velocity, $\mathbf{V}_{*s}$ (2.206).

82 Used by itself, the term '*drift-wave*' (DW) is understood to imply '*electron* drift-wave' (EDW). Since the source of free energy is n'_0, it is also known as the *density gradient* (DG) mode.

Indeed, adiabatic electron response is *the* defining feature of drift-waves, which consequently cannot propagate unless electrons are warm. Such response is absent from MHD waves, in which $E_\parallel = 0 = \nabla_\parallel \varphi$ is implied by ideal MHD Ohm's law (2.244).

Setting $T_i = p_i = \omega_{*i} = 0$ and $T'_{e0} = 0$ in (4.218)–(4.219) and simplifying, yields the electron drift-wave dispersion relation,

$$\omega(\omega - \omega_{*n}) = k_\parallel^2 c_{Se}^2 \qquad \Rightarrow \qquad \omega \approx \omega_{*n} + k_\parallel^2 c_{Se}^2/\omega_{*n}, \tag{4.226}$$

$$c_{Se}^2 \equiv p_e/\rho_i = ZT_e/m_i = \tfrac{3}{5} V_S^2 [T_i = 0]. \tag{4.227}$$

For parallel disturbances ($k = k_\parallel$, $k_\wedge = 0$), the above reduces to $\omega^2 = k_\parallel^2 c_{Se}^2$, indicating a parallel sound wave; for diamagnetic (poloidal) disturbances ($k \approx k_\wedge \gg k_\parallel$), (4.226) has two solutions: one static, $\omega = 0$, and the other purely oscillatory, $\omega_r = \mathrm{Re}(\omega) \approx \omega_{*n}$, which represents a drift-wave travelling along the magnetic flux surface in the direction $\mathbf{k}/k = (k_\wedge \hat{\mathbf{e}}_\wedge + k_\parallel \mathbf{b})/k$. Indeed, the spectrum of mainly electrostatic ($\widetilde{B}/B_0 \ll \delta_i \sim \widetilde{n}/n_0 \sim e\widetilde{\varphi}/T_{e0}$) fluctuations observed in magnetically confined plasmas contains a pronounced peak near ω_{*n}, and exhibits a linear Maxwell–Boltzmann relation, $\widetilde{n}/n_0 \sim e\widetilde{\varphi}/T_{e0}$, (4.219), characteristic of the adiabatic electron response and hence drift-wave turbulence, see Section 6.3.

The physical mechanism responsible for drift-wave propagation should now be clear:[83] since n_1 and φ_2 are coupled by the adiabatic electron response, (4.219), a poloidal density disturbance leads to a poloidal electric field $E_{\wedge 1} = -k_\wedge \varphi_2$, which gives rise to radial plasma motion with a velocity $V_{E1} = V_{\perp 1} = -k_\wedge \varphi_2/B_0$. In the continuity equation (4.218), this $\mathbf{E} \times \mathbf{B}$ advection, represented by the term $V_{\perp 1}/L_{\perp n} = \omega_{*n} e\varphi_2/T_{e0}$, competes with parallel acoustic response, $k_\parallel V_{\parallel 1} = k_\parallel^2 e\varphi_2/\omega m_i$, which provides the restoring effect needed for oscillatory motion. The absence of $\mathbf{V}_*$ from the continuity (4.218) and pressure (4.219) equations, indicates that the drift-wave does not originate in the equilibrium diamagnetic flow, but rather, in the transient compressibility of that flow.

In (4.226), the radial density gradient n'_0 defines the frequency of oscillation ω_{*n}, but in the absence of dissipation cannot provide the free energy required to destabilize the wave. This energy becomes available to the wave if, and only if, there exists a phase shift, Δ_1, between n_1 and φ_2, i.e. a temporal delay in the adiabatic electron response (4.219),

$$n_1/n_0 \approx (e\varphi_2/T_{e0})(1 - \mathrm{i}\Delta_1). \tag{4.228}$$

[83] There is a rough analogy between drift-waves, *internal gravity* waves in a stratified fluid, see Section 4.1, and *Rossby* waves in a rotating fluid (Landau and Lifschitz, 1960, Vol.VI), Table 2.3. All three waves propagate perpendicular to the equilibrium density gradient and hence do not reduce this gradient (do not relax the system towards equilibrium) and so are non-dissipative.

This conclusion follows directly from the corresponding dispersion relation,

$$(\omega^2 - k_\parallel^2 c_{Se}^2)(1 - \mathrm{i}\Delta_1) \approx \omega_{*n}\omega \;\Rightarrow\; \mathrm{Im}(\omega) \approx (\omega_{*n} + k_\parallel^2 c_{Se}^2/\omega_{*n})\Delta_1, \quad (4.229)$$

which indicates instability for $\Delta_1 > 0$, i.e. when the electron density *lags* behind the electric potential.[84] Thus any process which inhibits parallel electron motion will tend to destabilize the electron drift-wave. There are many such processes, foremost among which is collisional dissipation,[85] e.g. the electrical resistivity, $\eta_\parallel$, which enters the Ohm's law (2.281) via $F_{\parallel e}$.

It is worth noting that adiabatic electron response provides an *oscillating* parallel return current, $J_{\parallel 1}$, analogous to the *equilibrium* parallel return current, $J_{\parallel 0}$, (3.45). Whereas the latter is necessary for MHD equilibrium, the former prevents the drift-wave from becoming unstable. As we will see in Section 5.3.3, a similar analogy can be drawn with the parallel collisional friction which lies at the origin of radial transport in neoclassical theory.

Adding $\eta_\parallel J_{\parallel 1}$ to the right-hand side of the electron momentum equation in (4.217), yields a phase lag of $\Delta_1 = \eta_\parallel J_{\parallel 1}/k_\parallel \varphi_2$. The oscillating parallel return current may be calculated from charge conservation $\nabla \cdot (J_{\parallel 1}\mathbf{b} + \mathbf{J}_{\perp 1}) = 0$, with $\mathbf{J}_\perp$ determined from the ion momentum equation (2.269),

$$\mathbf{J}_{\perp 1} \approx \mathbf{J}_{p1} \approx (\mathbf{b}/B) \times \rho_0 \mathrm{d}_t \mathbf{V}_{E1} \approx -\mathbf{k}_\perp (\rho_0/B_0^2)(\omega + \omega_{*n})\varphi_2. \quad (4.230)$$

Here we neglected the diamagnetic term, containing the cross product $\mathbf{b} \times \nabla p_1$, which has a vanishing divergence in a uniform magnetic field. For the assumed oscillations, $\mathbf{k}_\perp = k_\wedge \hat{\mathbf{e}}_\wedge$, we find the following expression for Δ_1,

$$\Delta_1 = 2\omega_{*n}\frac{\rho_0\eta_\parallel}{B_0^2}\left(\frac{k_\wedge}{k_\parallel}\right)^2 = \frac{2\omega_{*n}D_\eta}{V_A^2}\left(\frac{k_\wedge}{k_\parallel}\right)^2 = \frac{m_e}{m_i}\frac{2\omega_{*n}\nu_e}{\Omega_i^2}\left(\frac{k_\wedge}{k_\parallel}\right)^2, \quad (4.231)$$

where $\nu_e = \nu_{ei}$ is the electron–ion collision frequency, see Section 5.2.[86] Not surprisingly, we find that Δ_1 increases linearly with the background density gradient, ω_{*n}, and with the collision frequency, ν_e. Moreover, as with MHD instabilities, finite $k_\parallel$ (required by finite magnetic shear, s) has a stabilizing effect (reduces the growth rate) on drift-wave instabilities.

In the absence of non-adiabatic processes, the free energy required to destabilize the drift-wave can only come from additional electron or ion temperature gradients. Although both T_{e0}' and T_{i0}' can provide this energy, the former does not appear in (4.218)–(4.219) suggesting more complicated dynamics, involving electron kinetics and toroidal geometry. Moreover, we expect such *electron temperature gradient*

[84] Note that because of causality n cannot *lead* φ, so that Δ_1 is either positive or zero.

[85] Secondary processes include finite electron inertia, magnetic trapping, inductive breaking (displacement current) or even non-linear scattering from other drift-waves, see Section 6.3.

[86] Since $m_e/m_i \ll 1$, we expect Δ_1, $J_{\parallel 1}/J_{\parallel 0} \ll 1$, in keeping with the electrostatic assumption.

(ETG) modes to be more readily damped, e.g. by collisionless (Landau) damping, see Section 4.4, or by magnetic mirror forces in toroidal geometry, see Section 2.2. This expectation is confirmed by numerical simulations of ETG modes, which indicate that these are governed by the response of the electron population trapped by the mirror forces associated with the $1/R$ toroidal field variation, Section 5.3.3; the resulting instability, known as the *trapped electron mode* (TEM), is largely responsible for radial electron heat transport in tokamaks. For a short introduction to turbulent transport in tokamaks see Garbet (2006).

In contrast, T'_{i0} enters the ion pressure equation (4.219) via ω_{*i} and hence gives rise to a robust drift instability, known as the *ion temperature gradient* (ITG) mode. The dispersion relation for the ITG mode may be derived from (4.218)–(4.219) by retaining the finite ion temperature and its gradient. Noting that the perturbed ion pressure has contributions due to perturbed density and temperature (4.220), we find the following relation,

$$\begin{aligned}\omega(\omega - \omega_{*n}) &\approx k_\parallel^2 c_{Se}^2[\tfrac{5}{3}\tau + 1 + \tau(\omega_{*n}/\omega)(\eta_i - \tfrac{2}{3})] \\ &= k_\parallel^2[c_S^2 + v_{ti}^2(\omega_{*n}/\omega)(\eta_i - \tfrac{2}{3})], \end{aligned} \tag{4.232}$$

where η_s is the ratio of density and temperature radial gradient lengths and τ is the ratio of ion and electron pressures,[87]

$$\eta_s \equiv \frac{\mathrm{d}\ln T_s}{\mathrm{d}\ln n_s} = \frac{T'_s/T_s}{n'_s/n_s} = \frac{L_n}{L_{T_s}}, \qquad \tau \equiv \frac{p_i}{p_e} = \frac{T_i}{ZT_e}. \tag{4.233}$$

The first term on the right-hand side of (4.232) represents the warm ion sound speed, $c_S^2 = c_{Se}^2(\frac{5}{3}\tau + 1) = (\frac{5}{3}T_i + ZT_e)/m_i \approx V_s^2$, and hence parallel acoustic response. The second term, which involves the ion thermal speed, $c_{Se}^2\tau = T_i/m_i = v_{ti}^2$, represents the ITG drive resulting in diamagnetic plasma flow at the ion drift frequency, $\omega_{*i} = -\tau\omega_{*n}(1 + \eta_i)$.

For parallel disturbances, we find the warm ion sound wave, $\omega^2 \approx k_\parallel^2 c_S^2$, while for purely diamagnetic disturbances we recover the electron drift-wave, $\omega \approx \omega_{*n}$. For arbitrary $k_\parallel$, (4.232) yields a modified sound wave, described by $\omega(k_\parallel)$, which can become unstable when $\eta_i > \eta_{ic} = 2/3$. In fact, instability is only found for a narrow range of $k_\parallel$, namely $0 < k_\parallel < k_{\parallel c} \sim \omega_{*n}/c_{Se}$. The critical wave number, $k_{\parallel c}$, is an increasing function of $\eta_i - \eta_{ic}$, vanishing for $\eta_i = \eta_{ic}$, and becoming comparable to ω_{*n}/c_{Se} for $\eta_i > 2\eta_{ic}$. Since magnetic shear s tends to increase $k_\parallel$, it exerts a stabilizing influence.

More generally, one finds that $\eta_{ic} \approx 1$ with the exact value determined by magnetic geometry, dissipative processes, etc. When parallel electron response is

87 The choice of symbols for η_s and τ, (4.233), are historical. These should not be confused with the resistivity, $\eta_{\parallel s}$ (sometimes denoted as η_s), and the collision time, τ_s, see Section 5.

treated kinetically, this η_i-threshold is replaced by a pure ITG-threshold, $1/L_{Tic}$, or R/L_{Tic} in toroidal plasmas; a similar, ETG-threshold, R/L_{Tec}, is predicted for trapped electron modes.[88] As a result, L_{Ti} and L_{Te} are typically clamped at L_{Tic} and L_{Tec} everywhere in the plasma, irrespective of the heating scheme, input power, magnetic field, toroidal current, fuelling rate, etc.[89] This means that central temperatures $T_i(0)$ and $T_e(0)$ are largely determined by their boundary values $T_i(a)$ and $T_e(a)$, which in turn are determined by transport/exhaust processes in the edge of the plasma.

In the presence of an equilibrium radial electric field, $E_{\perp 0} = -\varphi_1'$, the above analysis remains valid in the frame of reference moving with $\mathbf{V}_{E0} = -(\varphi_1'/B)\hat{\mathbf{e}}_\wedge$. As a result, ω becomes Doppler-shifted by the electric drift frequency $\omega_E = \mathbf{k} \cdot \mathbf{V}_{E0} = k_\wedge V_{E0}$, which entails the substitution $\omega \to \omega - \omega_E$ in (4.216)–(4.232). In other words, the frequency range is determined by the relation $\mathbf{k} \cdot \mathbf{V}_{\perp s} = \omega$, where $\mathbf{V}_{\perp s}$ is given by (2.208),

$$\mathbf{k} \cdot \mathbf{V}_{\perp s} \sim \mathbf{k} \cdot (\mathbf{V}_{E0} + \mathbf{V}_{*s}) \quad \Rightarrow \quad \omega \sim \omega_E + \omega_{*s}, \tag{4.234}$$

so that the characteristic frequencies of MHD(drift)-ordered plasma dynamics, ω_E (ω_{*s}), correspond to the $\mathbf{E} \times \mathbf{B}$ (diamagnetic) flows due to the radial gradients in the electric potential (plasma pressure).

To obtain an electromagnetic version of the drift-wave, one needs to additionally include the linearized version of Ampere's and Faraday's laws,

$$i\mathbf{k}_{\perp 1} \times \mathbf{B}_1 \approx \mu_0 J_{\parallel 1}\mathbf{b}, \qquad \mathbf{k}_{\perp 1} \times \mathbf{E}_1 \approx \omega \mathbf{B}_1. \tag{4.235}$$

Hence, an oscillating return current $J_{\parallel 1}$ gives rise to a perturbed magnetic field, $B_{\perp 1} = k_\wedge A_{\parallel 1} = \mu_0 J_{\parallel 1}/k_\wedge$, which is conveniently expressed in terms of the parallel component of the vector potential $\psi_1 = A_{\parallel 1} = \mu_0 J_{\parallel 1}/k_\wedge^2$. This linear relation is identical to that obtained in flute-reduced MHD, see Section 4.2.5, and when combined with (4.218)–(4.220) gives rise to the *drift-Alfvén* wave, characterized by the dispersion relation (Scott, 2001),

$$\omega(\omega - \omega_{*n}) = k_\parallel^2 \left[c_S^2 + V_A^2 + v_{ti}^2(\omega_{*n}/\omega)(\eta_i - \tfrac{2}{3})\right]. \tag{4.236}$$

In exact analogy to the transition from ideal to resistive MHD, the inclusion of non-adiabatic effects, e.g. finite resistivity $\eta_\parallel$, in the parallel electron response transforms the drift-Alfvén wave into a *drift-tearing* instability.

In toroidal geometry, magnetic non-uniformity substantially modifies drift-wave dynamics, e.g. toroidal curvature introduces a finite compressibility of diamagnetic flows, while the polarization drift gives rise to a parallel vorticity correction,

88 Both the ITG- and ETG-thresholds have been detected in tokamak plasmas (Garbet, 2006).
89 On account of this robustness against external factors, the resulting profiles are said to be *stiff*.

χ_g to the plasma pressure (2.268). These second-order corrections usually provide the dominant drive for drift-wave instabilities. Similarly, kinetic effects related to particle trapping have a pronounced effect on parallel response to electrostatic perturbations, so that drift instabilities are generally divided into *trapped particle* and *passing particle* modes, see Section 5.3.3. A short introduction to the rich zoology of drift-waves and instabilities which ensue may be found in Wesson (2004, Chapter 8).

To conclude, we may now answer the question raised in the quote on page 72: the electric field in a quasi-neutral plasma originates in the net compression of diamagnetic and parallel return currents, which due to charge conservation, $\nabla \cdot (\mathbf{J}_{\parallel} + \mathbf{J}_{*} + \mathbf{J}_{p}) = 0$, gives rise to a net polarization current. The physical mechanisms behind $\nabla \cdot (\mathbf{J}_{\parallel} + \mathbf{J}_{*})$ are radial thermodynamic gradients, magnetic curvature and all dissipative (non-adiabatic) processes inhibiting parallel electron and ion motion. We may add that turbulent flows originating from the non-linear evolution of drift-wave instabilities are observed, both experimentally as well as numerically, to dominate the radial transport of mass, momentum and energy in magnetically confined plasmas, see Section 6.3. Specifically, ion and electron heat transport is typically dominated by ITG and ETG (trapped electron, TEM) modes, respectively.

4.4 Kinetic waves and instabilities

As already mentioned in Section 2.3.1, the study of fast perturbations requires a *kinetic* description, since plasma response becomes velocity dependent and some particles may resonate with the disturbance. To investigate this resonance between waves and particles, we first note that transverse ($\mathbf{E} \perp \mathbf{k}$) electromagnetic waves, which travel at the speed of light, cannot resonate with non-relativistic particles. Therefore, we concentrate on longitudinal ($\mathbf{E} \parallel \mathbf{k}$), electrostatic ($\nabla \times \mathbf{E} = 0, \mathbf{E} = -\nabla\varphi$) oscillations.[90]

For simplicity, we assume an unmagnetized plasma ($\mathbf{B}_0 = 0$) for which (2.87) and (2.93) reduce to the *Vlasov–Poisson* system,

$$\partial_t f_s + \mathbf{v} \cdot \nabla f_s + \frac{e_s \mathbf{E}}{m_s} \cdot \partial_{\mathbf{v}} f_s = 0, \qquad \epsilon_0 \nabla \cdot \mathbf{E} + \sum_s e_s \int f_s d\mathbf{v} = 0. \quad (4.237)$$

Consider the response of (4.237) to plane wave disturbances, $e^{i(\mathbf{k}\cdot\mathbf{x}-\omega t)}$, for which the wave-particle resonance occurs at $\mathbf{k} \cdot \mathbf{v} = \omega$. Since $v_{ti} \ll v_{te}$, the ions may be treated as a stationary background ($f_{i0} = \text{const}$) when considering perturbations ($f_e = f$) fast enough to resonate with thermal electrons. Following Landau

[90] We broadly follow the elegant introduction given in Hazeltine and Waelbroeck (2004).

(1946), who formulated a general method of solving (4.237), we apply the Fourier transform, $\mathcal{A}_{\mathbf{k}} = \int_{-\infty}^{\infty} \mathcal{A} e^{i\mathbf{k}\cdot\mathbf{x}} d\mathbf{x}$,

$$(\partial_t + i\mathbf{k}\cdot\mathbf{v})\, f_{\mathbf{k}}(\mathbf{v}, \mathbf{t}) - \frac{e\mathbf{E}_{\mathbf{k}}}{m_e}\cdot\partial_{\mathbf{v}} f_0 = 0, \quad i\epsilon_0 \mathbf{k}\cdot\mathbf{E}_{\mathbf{k}} + e\int_{-\infty}^{\infty} f_{\mathbf{k}}(\mathbf{v}, t)d\mathbf{v} = 0,$$

followed by a Laplace transform, $\mathcal{A}_{\mathbf{k},\omega} = \int_0^{\infty} \mathcal{A}_{\mathbf{k}} e^{i\omega t} dt$,

$$(\omega - \mathbf{k}\cdot\mathbf{v})\, f_{\mathbf{k},\omega}(\mathbf{v},\omega) - i f(\mathbf{v}, 0) - \frac{ie\mathbf{E}_{\mathbf{k},\omega}}{m_e}\cdot\partial_{\mathbf{v}} f_0 = 0, \tag{4.238}$$

$$i\epsilon_0 \mathbf{k}\cdot\mathbf{E}_{\mathbf{k},\omega} + e\int_{-\infty}^{\infty} f_{\mathbf{k},\omega}(\mathbf{v},\omega)d\mathbf{v} = 0,$$

where $f_0(\mathbf{v})$ is the background electron distribution and $f(\mathbf{v}, 0)$ is the initial perturbation. We can solve for $f_{\mathbf{k},\omega}$ by eliminating $\mathbf{E}_{\mathbf{k},\omega}$,

$$f_{\mathbf{k},\omega}(\mathbf{v},\omega) = \frac{i}{\omega - \mathbf{k}\cdot\mathbf{v}}\left(f(\mathbf{v}, 0) + \frac{e\mathbf{E}_{\mathbf{k},\omega}}{m_e}\cdot\partial_{\mathbf{v}} f_0\right). \tag{4.239}$$

The solution $f = f_e(\mathbf{x}, t)$ is the inverse transform of (4.239). The first term gives $e^{i\mathbf{k}\cdot(\mathbf{x}-\mathbf{v}t)}$, which describes ballistic or free-streaming propagation of $f(\mathbf{v}, 0)$, while the second term represents the response of the background electrons to the perturbed field $\mathbf{E}_{\mathbf{k},\omega}$. Both terms exhibit a *resonance singularity* when the wave and particle velocities are equal, $\mathbf{k}\cdot\mathbf{v} = \omega$. Using (4.239) to eliminate $f_{\mathbf{k},\omega}$, the perturbed electric field becomes

$$\epsilon_0 \mathbf{k}\cdot\mathbf{E}_{\mathbf{k},\omega}(\mathbf{v},\omega) + \frac{e}{\mathcal{K}_e(\omega,\mathbf{k})}\int \frac{f(\mathbf{v}, 0)}{\omega - \mathbf{k}\cdot\mathbf{v}} d\mathbf{v} = 0, \tag{4.240}$$

$$\mathcal{K}_e(\omega,\mathbf{k}) \equiv 1 + \chi_e(\omega,\mathbf{k}) \equiv 1 + \frac{\omega_{pe}^2}{kn_0}\int \frac{\partial_{\mathbf{v}} f_0}{\omega - \mathbf{k}\cdot\mathbf{v}}\cdot d\mathbf{v}.$$

The *electron dielectric function*, $\mathcal{K}_e(\omega,\mathbf{k})$, and *electric susceptibility*, $\chi_e(\omega,\mathbf{k})$, describe the Debye shielding of $\mathbf{E}_{\mathbf{k},\omega}$ by mobile electrons, and determine the *eigenmodes*, $f_{\mathbf{k},\omega} e^{i(\mathbf{k}_m\cdot\mathbf{x} - \omega_m t)}$, as the roots of the *dispersion relation*, $\mathcal{K}_e(\omega_m, \mathbf{k}_m) = 0$. The resulting oscillations are summarized below:

- *Cold electrons* ($v_{te} = 0$): evaluating $\mathcal{K}_e(\omega,\mathbf{k}) = 0$ with $f_0(\mathbf{v}) = n_0\delta(\mathbf{v})$ yields a stationary plasma oscillation, $\omega^2 = \omega_{pe}^2$ with $v_{gr} = d_k\omega = 0$.
- *Warm electrons* ($v_{te} \ll v_{ph} = \omega/k$): expanding $\mathcal{K}_e(\omega,\mathbf{k})$ in $v/v_{ph} \ll 1$ yields a *Langmuir wave*, $\omega^2 = \omega_{pe}^2 + k^2 v_{te}^2$, which propagates with the group velocity, $v_{gr} = d_k\omega \sim k v_{te}^2/\omega_{pe} \ll v_{te}$. Thus, thermal motion transforms the stationary plasma oscillation ($v_{gr} = 0$) into a wave ($v_{gr} > 0$).
- *Hot electrons* ($v_{te} \sim v_{ph}$): the resonance at $\mathbf{k}\cdot\mathbf{v} = \omega$ now takes effect. It is best handled by treating both ω and $\mathbf{v}$ as complex variables, and evaluating inversion integrals along chosen contours in the complex plane. For example, the Laplace inversion,

$\mathbf{E_k}(t) = \int_C \mathbf{E_k}(\omega)e^{-i\omega t}d\omega/2\pi$ should be evaluated over a contour C lying above all singularities of $\mathbf{E_k}(\omega)$.

In the last case, we write $\omega = \omega_r + i\gamma$ and $\mathbf{v} = \mathbf{v}_r + i\mathbf{v}_i$, where ω_r is the oscillating frequency and $\gamma = \omega_i$ the growth rate of the perturbation. The upper half of the complex ω-plane ($\gamma > 0$) thus corresponds to unstable modes, while the lower half ($\gamma < 0$) to stable, damped modes. To evaluate the contour integral, we need the singularities of $\mathbf{E_k}(\omega)$ in (4.240). These consist of the eigenmodes ω_m, which lead to unstable-damped pairs with symmetric $\mathrm{Im}(\omega_m) = \pm\gamma_m$, and the *wave-particle resonance*, $\mathbf{k} \cdot \mathbf{v} = \omega$, which produces a discontinuity of $\mathcal{K}_e(\omega, \mathbf{k})$ on the real ($\gamma = 0$) axis. The latter may be evaluated using *Plemelj's formula* of complex analysis,

$$\mathcal{K}_e(\omega \pm i0, \mathbf{k}) = 1 + \frac{\omega_{pe}^2}{kn_0} P \int \frac{\partial_{\mathbf{v}} f_0}{\omega - \mathbf{k} \cdot \mathbf{v}} \cdot d\mathbf{v} \mp i\pi \frac{\omega_{pe}^2}{k^2 n_0} (\partial_{\mathbf{v}} f_0)_{\omega/k}, \tag{4.241}$$

where P denotes the principal value of the integral and the last term represents the residue of the discontinuity. We can use (4.241) to define $\kappa_e(\omega, \mathbf{k})$ as the analytic continuation of $\mathcal{K}_e(\omega, \mathbf{k})$ below the real axis and solve the new dispersion relation $\kappa_e(\omega, \mathbf{k}) = 0$ by Taylor expansion around ω_0, the approximate real root of $\mathcal{K}_e(\omega, \mathbf{k})$,

$$\omega = \omega_0 + \omega_1 \approx \omega_0 - \mathcal{K}_e(\omega + i0, \mathbf{k}) \left(\partial_\omega \mathcal{K}_e\right)_{\omega_0}^{-1}. \tag{4.242}$$

Using (4.240) to evaluate $(\partial_\omega \mathcal{K}_e)_{\omega_0} \approx 2/\omega_{pe}$, and (4.241) to determine $\mathcal{K}_e(\omega + i0, \mathbf{k})$, we find the imaginary root of $\mathcal{K}_e(\omega, \mathbf{k})$ as $\omega_1 = i\gamma_L$ where

$$\gamma_L = \frac{\pi}{2} \frac{\omega_{pe}^3}{k^3 n_0} \mathbf{k} \cdot (\partial_{\mathbf{v}} f_0)_{\omega_{pe}/k}, \tag{4.243}$$

is the *Landau damping* rate. In other words, small perturbations are damped provided $\mathbf{k} \cdot \partial_{\mathbf{v}} f_0 < 0$, evaluated at the resonant velocity, $v_{ph} = \omega_{pe}/k$.

Such collisionless, or *Landau*, damping is an archetypal plasma phenomenon, originating in *collective* particle interactions.[91] It is best understood by adapting the frame of reference moving with the disturbance $(f_{\mathbf{k}}, \mathbf{E_k})$ at $\mathbf{v} = v_{ph}\mathbf{k}/k$. The resonant particles are stationary in this frame and tend to amplify $\mathbf{E_k}$, while $\mathbf{k} \cdot \partial_{\mathbf{v}} f_0$ stretches and shears (in phase space) the ballistically propagating disturbance $f_{\mathbf{k}}$,[92] thereby reducing $\mathbf{E_k}$ at a rate proportional to $\partial_v f_0$. The shuffling property of the Vlasov equation ensures net decay of $(f_{\mathbf{k}}, \mathbf{E_k})$ whenever $\partial_v f_0 < 0$, e.g. for a Maxwellian distribution $f_M(\mathbf{v})$, (A.1). In that case most near-resonant particles

[91] Recall that we assumed $C_{ss'} = 0$ from the outset, so that the resulting plasma dynamics are clearly independent of inter-particle collisions. As a result, Landau damping is a quite general plasma phenomenon, occurring in all types of plasmas mentioned in Section 2.1.

[92] This is an instance of phase mixing discussed in Section 2.3.1.

move slower than v_{ph}, and plasma energy $\sum_s \int \frac{1}{2} m_s v_s^2 f_s \mathrm{d}\mathbf{v}\mathrm{d}\mathbf{x}$ grows at the expense of wave energy $\int \frac{1}{2}\epsilon_0 |\mathbf{E}|^2 \mathrm{d}\mathbf{v}\mathrm{d}\mathbf{x}$. When $\partial_v f_0 > 0$, i.e. when the population density is *inverted* at the resonance, the wave may, but need not, gain free energy from the plasma. If this occurs, $f_{\mathbf{k}}$ grows at the rate γ_L, which is the same everywhere in velocity space and takes the form of an unstable eigenmode, $f_{\mathbf{k}} \mathrm{e}^{\mathrm{i}(\mathbf{k}\cdot\mathbf{x}-\omega_0 t)} \mathrm{e}^{\gamma_L t}$. Since the simplest example of population inversion is the case of two counter-streaming beams, kinetic modes arising from $\partial_v f_0 > 0$ are generally known as *two-stream instabilities*.

Unlike perturbation analysis, Landau's approach can harness the full power of complex analysis. For example, it can be used to show that $\partial_v f_0 > 0$ is a necessary, but not a sufficient condition for the onset of the two-stream instability. The exact result can be derived using the *Nyquist theorems*, which relate the number of zeros of an analytic function to the contour turns around the origin. When applied to the dispersion relation (4.240), they yield the *Penrose instability criterion* (Penrose, 1960),

$$\int \frac{f_0(v) - f_0(v_{min})}{(v - v_{min})^2} \mathrm{d}v > 0, \tag{4.244}$$

where v_{min} is the velocity at which $f_0(v)$ has a minimum. Thus the well in $f_0(v)$ must exceed a certain threshold for instability to grow.

Landau damping can occur at any wave-particle resonance, a common example being the ion acoustic wave. To see this, we include the ion response to $\mathbf{E}_{\mathbf{k}} \| \mathbf{k}$ in (4.240), which yields the *plasma dielectric function*,

$$\mathcal{K}(\omega, \mathbf{k}) = 1 + \sum_s \chi_s(\omega, \mathbf{k}) = 1 + \sum_s \frac{\omega_{ps}^2}{k n_0} \int \frac{\partial_{\mathbf{v}} f_{s0}}{\omega - \mathbf{k}\cdot\mathbf{v}} \cdot \mathrm{d}\mathbf{v}. \tag{4.245}$$

As before, the dispersion relation $\mathcal{K}(\omega, \mathbf{k}) = 0$ determines the eigenmodes of the oscillation. For a Maxwellian $f_{s0}(v)$, $\chi_s(\omega, \mathbf{k})$ may be written as

$$\chi_s(\omega, k) = 2\zeta_{ps}^2 \left[1 + \zeta_s \mathcal{Z}(\zeta_s)\right], \qquad \zeta_s = v_{ph}/v_{ts} = \omega/k v_{ts}, \qquad \zeta_{ps} = \zeta_s(\omega_p),$$

where $\mathcal{Z}(\zeta)$ is known as the *plasma dispersion function*,

$$\mathcal{Z}(\zeta) = \frac{1}{\sqrt{\pi}} \int_{-\infty}^{\infty} \frac{\mathrm{e}^{-\xi^2}}{\xi - \zeta} \mathrm{d}\xi = \mathrm{i}\sqrt{\pi}\mathrm{e}^{-\zeta^2} \operatorname{erfc}(-\mathrm{i}\zeta), \qquad \operatorname{Im}(\zeta) > 0. \tag{4.246}$$

Using $\mathcal{Z}'(\zeta) = -2[1 + \zeta \mathcal{Z}(\zeta)]$, the dispersion relation becomes,

$$2k^2\lambda_{De}^2 - \mathcal{Z}'(\zeta_e) - \mathcal{Z}'(\zeta_i)/\tau = 0, \tag{4.247}$$

with τ given by (4.233). For a disturbance with $v_{ti} \ll v_{ph} \ll v_{te}$, hence $\zeta_i \gg 1$ and $\zeta_e \ll 1$, the real part of (4.247) gives the ion acoustic mode,

$$\frac{\omega_r^2}{k^2} \approx \frac{(p_i + p_e)/n_i m_i}{1 + k^2 \lambda_{De}^2} + \frac{3}{2} v_{ti}^2 \approx v_{ti}^2 \left[\frac{1 + \tau^{-1}}{1 + k^2 \lambda_{De}^2} + \frac{3}{2} \right], \tag{4.248}$$

while the imaginary part gives the desired Landau damping rate,

$$\gamma_L \approx \frac{\omega_r}{(1 + k^2 \lambda_{De}^2)^{3/2}} \left[\sqrt{\frac{m_e}{m_i}} + \tau^{-3/2} \exp\left(-\frac{2/\tau}{1 + k^2 \lambda_{De}^2} - \frac{3}{2} \right) \right], \tag{4.249}$$

where the electron contribution is given by the first, and the ion by the second term. In fusion plasmas $\tau \sim 1$ and the two terms are comparable.

Landau damping is also active in magnetized plasmas ($\delta_i \ll 1$). The plane waves $e^{i(\mathbf{k} \cdot \mathbf{x} - \omega t)}$ must then resonate with helical trajectories (2.19)–(2.21) which introduces several important modifications:

- The resonance $\omega = \mathbf{k} \cdot \mathbf{v}$ is replaced by a series of resonances $\omega_l = k_\| v_\| + l\Omega_s$, where l is an integer and $k_\| v_\| = (\mathbf{k} \cdot \mathbf{b})(\mathbf{v} \cdot \mathbf{b})$. The resonance occurs when the frequency, Doppler-shifted by the parallel velocity, is a harmonic of the gyration frequency. This fact is exploited in plasma heating methods based on launching EM waves with frequencies close to the ion, electron or hybrid resonances, which fall in the RF range.
- Since particle gyration tends to average out perturbations smaller than the gyro-radius – recall our development of gyro-kinetic theory – this leads to the appearance of lth-order Bessel functions, $J_l(k_\perp v_\perp / \Omega_s)$, which we have already encountered in (2.122)–(2.123).
- As a result, *Landau damping* caused by phase mixing in ($\mathbf{x}$, $\mathbf{v}$)-space is transformed into *cyclotron damping* caused by gyration phase mixing.

4.5 Further reading

MHD has been the bread and butter of fusion research since its earliest beginnings. As such, the field has produced an imposing body of literature, much of which is highly specialized; the highlights are compiled in the bibliography of Wesson (2004, Chapter 6). The reader may also wish to consult the following texts: *Hydrodynamic stability*: Lamb (1932); Landau and Lifschitz (1960, Vol. V), Streeter *et al.* (1962); Batchelor (1967); Milne-Thomson (1968). *Ideal MHD stability*: Batemann (1978); Miyamoto (1989); Whyte (1989); Freidberg (1991); Nishikawa and Wakatani (1994); Goedbloed and Poedts (2004). *Resistive MHD*: Biskamp (2000). *Non-linear MHD*: Biskamp (1993). *Drift and kinetic instabilities*: Krall and Trivelpiece (1973); Mikhailovskii (1974, 1992, 1998); Hasegawa (1975). *Plasma waves*: Stix (1992); Swanson (2003).

5

Collisional transport in magnetized plasmas

'Collisions or encounters in a gas of low density are mainly between pairs of molecules, whereas in a solid or liquid each molecule is usually near or in contact with several neighbours. The legitimate neglect of all but binary encounters in a gas is one of the important simplifications that have enabled the theory of gases to attain its present high development.'

Chapman and Cowling (1939)

In the previous two chapters, we investigated the equilibrium and stability properties of magnetized plasmas. In the next two chapters, we will assume that a stable equilibrium has been achieved and will consider the relatively slow transport processes, associated with inter-particle collisions, Chapter 5, and plasma turbulence, Chapter 6, which redistribute mass, momentum and energy within the plasma and are responsible for their exhaust via the plasma transport channel. To emphasize the slowness of these processes, we will henceforth adopt the so-called *transport ordering*,

$$\partial_t \sim \delta_s^2 \omega_t \sim \delta_s^3 \Omega_s, \qquad \nu_s/\Omega_s = \rho_{ts}/\lambda_s \sim \delta_s \ll 1, \qquad V_E \sim \delta_i v_{ti}, \tag{5.1}$$

where $\omega_t = v_t/L_\parallel$ is a characteristic transit frequency and $\delta_s = \rho_{ts}/L_\perp$ is the magnetization parameter (2.11). The transport ordering also assumes that the ratio of collision and gyro-frequencies, ν_s/Ω_s, or equivalently, of the gyro-radius and the collisional mean free path, ρ_{ts}/λ_s, is of the same order of smallness as δ_s, so that particle gyration is unaffected by collisions.

An additional expansion in the *Knudsen number*, $\Delta_s = \lambda_s/L_\parallel = 1/\nu_s^*$, defined as the inverse of the plasma collisionality, ν_s^* (2.13), is employed in the collisional regime ($\Delta_s \ll 1$), but it is otherwise independent of the transport ordering (5.1), which is assumed to hold in all collisionality regimes.[1]

[1] Before proceeding further, the reader may wish to review the material covered in Chapter 2, especially Sections 2.1, 2.2.2, 2.3.1, 2.4.1 and 2.4.2.

5.1 Collisional transport in a neutral gas

By way of introduction to collisional transport in a plasma, let us first consider collisional processes in a neutral gas. For this purpose we introduce the concept of a *differential scattering cross-section*, $d\sigma_{ss'}$, between a *test* particle s and *field* particles s', defined as the differential scattering rate $dN_s(\vartheta)$ normalized to the incident flow of test particles, $n_s u_\infty$,

$$d\sigma_{ss'} \equiv \frac{dN_s(\vartheta)}{n_s u_\infty} = 2\pi b db = 2\pi b |d_\vartheta b| d\vartheta = \frac{b}{\sin\vartheta} |d_\vartheta b| d\varpi. \qquad (5.2)$$

Here ϑ is the angle through which the test particle is deflected in the CM frame,[2] $dN_s = 2\pi b db$ is the number of particles scattered per unit time into the angular interval $d\vartheta$ to $\vartheta + d\vartheta$, $b(\vartheta)$ is the impact parameter measuring the inter-particle distance perpendicular to their relative velocity at infinity $\mathbf{u}_\infty$ and n_s is the (test) particle density in the incident beam. The total cross-section $\sigma_{ss'} = \int d\sigma_{ss'}$ is simply the integral of (5.2) over the solid angle $d\varpi = 2\pi \sin\vartheta d\vartheta$ and represents the effective area of the scattering, or field, particles in the gas target, experienced by the scattered, or test, particles in the incident beam. For interactions between identical rigid spheres of radius a,[3] it is easy to show that $b(\vartheta) = a\cos(\vartheta/2)$ and (5.2) reduces to

$$d\sigma_{ss'} = \tfrac{1}{2}\pi a^2 \sin\vartheta d\vartheta = \tfrac{1}{4}a^2 d\varpi, \qquad \sigma_{ss'} = \pi a^2, \qquad (5.3)$$

i.e. $\sigma_{ss'}$ is just the cross-sectional area 'seen' by the incident beam.

5.1.1 Maxwell–Boltzmann collision operator

The *collision frequency* between field and test particles, $\nu_{ss'}$, and the *mean free path* of test particles between collisions, $\lambda_{ss'}$, can be estimated as

$$\nu_{ss'} \approx n_{s'} u_\infty \sigma_{ss'}, \qquad \lambda_{ss'} \approx u_\infty/\nu_{ss'} \approx (n_{s'}\sigma_{ss'})^{-1}. \qquad (5.4)$$

As expected,[4] $\nu_{ss'}$ is proportional to the density of field particles, the relative speed and the scattering cross-section, while the mean free path is independent of u_∞ and scales inversely with both $n_{s'}$ and $\sigma_{ss'}$.

The differential form of (5.4), namely $\nu_{ss'} \approx f_{s'} d\mathbf{v}' u d\sigma_{ss'}$ where $\mathbf{u} = \mathbf{v} - \mathbf{v}'$ is the relative velocity[5] and $d\sigma_{ss'}$ is given by (5.2), can be used to construct the *Maxwell–Boltzmann* collision operator,

[2] To obtain $d\sigma$ in the laboratory frame of reference, ϑ must be related to the lab-frame scattering angles, θ_s and $\theta_{s'}$. For elastic collisions, this is found from conservation of momentum and energy, with the result $\tan\theta_s = m_s \sin\vartheta/(m_s + m_{s'}\cos\vartheta)$ and $\theta_{s'} = (\pi - \vartheta)/2$.

[3] That is, interacting via an isotropic, infinite-well potential, $\phi(r) = -\infty, r < a; \phi(r) = 0, r > a$.

[4] And easily confirmed by taking a run through a forest with one's eyes closed.

[5] In the rest of the chapter, we drop the indices s and s' on all velocity variables, adopting instead the following notation: $\mathbf{v} \equiv \mathbf{v}_s$, $\mathbf{v}' \equiv \mathbf{v}_{s'}$, $\mathbf{u} \equiv \mathbf{u}_{ss'} = \mathbf{v}_s - \mathbf{v}_{s'} = \mathbf{v} - \mathbf{v}'$.

$$\mathcal{C}_{ss'}(f_s, f_{s'}) = \int \left(f_s^+ f_{s'}^+ - f_s f_{s'}\right) u \mathrm{d}\sigma_{ss'} \mathrm{d}\mathbf{v}', \tag{5.5}$$

in which the superscript '+' refers to instances after the collision and its absence to instances before the collision, so that

$$f_s = f_s(\mathbf{x}, \mathbf{v}, t), \;\; f_{s'} = f_{s'}(\mathbf{x}, \mathbf{v}', t), \tag{5.6}$$

$$f_s^+ = f_s(\mathbf{x}, \mathbf{v}^+, t), \;\; f_{s'}^+ = f_{s'}(\mathbf{x}, \mathbf{v}'^+, t). \tag{5.7}$$

The new velocities, $\mathbf{v}^+$ and $\mathbf{v}'^+$, can be expressed in terms of $\mathbf{v}$ and $\mathbf{v}'$ using the conservation of momentum and energy in elastic collisions. In effect, (5.5) states that the number of particles in a phase space element $\mathrm{d}\mathbf{x}\mathrm{d}\mathbf{v}$ increases (decreases) due to particles scattering into (out of) this element. It is easily shown, e.g. Huang (1987), that the operator (5.5) vanishes only for a Maxwellian distribution (A.1) and increases the entropy of the system for any other f_s, i.e. that it satisfies *Boltzmann's H-theorem.*

The products $f_s f_{s'}$ appearing in (5.5) are in fact the leading terms in the expression for the 2-particle distribution function $f_{ss'}^{(2)}$ obtained by integrating the phase space density $\mathcal{F}(\mathbf{x}_j, \mathbf{v}_j, t | j = 1, \ldots, N)$, defined as an explicit function of all particle co-ordinates,[6] over *all but two* particles,

$$f_{ss'}^{(2)}(\mathbf{x}_1, \mathbf{v}_1; \mathbf{x}_2, \mathbf{v}_2; t) = \int \mathcal{F}(\mathbf{x}_j, \mathbf{v}_j, t) \mathrm{d}^{N-2}\mathbf{x}_j \mathrm{d}^{N-2}\mathbf{v}_j. \tag{5.8}$$

The integral may be evaluated to yield $f_{ss'}^{(2)} = f_s f_{s'} + g_{ss'}$, where $g_{ss'}$ is the correlation between f_s and $f_{s'}$, which, as already discussed in Section 2.3.1, is a direct consequence of a binary collision.[7] Thus, by neglecting $g_{ss'}$ in (5.5), we effectively assume the two distributions to be statistically independent, i.e. invoke the assumption of *molecular chaos.*

The above procedure may be generalized into a *cluster expansion*,

$$f_{ab}^{(2)} = f_a f_b + g_{ab}, \qquad f_{abc}^{(3)} = f_a f_b f_c + f_a g_{ab} + f_b g_{bc} + f_c g_{cb} + g_{abc}, \tag{5.9}$$

where $f^{(k)}$ is the k-particle distribution function defined as the phase space average over all but k particles. Successive application of (5.9) leads to an infinite sequence of equations for $f^{(k)}$, $k = 1, \ldots, N$, known as the *BBGKY hierarchy*; at each level the k-particle correlation, $g^{(k)}$, is expressed in terms of the $k+1$-particle correlation, $g^{(k+1)}$, thus preventing a closure of this infinite chain. Fortunately, in a dilute gas, one is justified in truncating the hierarchy at the level of binary interactions,

[6] This form differs from our definition of $\mathcal{F}(\mathbf{x}, \mathbf{v}, t)$ in Section 2.3 which involved a sum of delta functions centred on $\mathbf{x}_j$, $\mathbf{v}_j$ and was thus a function of only $\mathbf{x}$ and $\mathbf{v}$.

[7] Indeed, the two particles are only correlated within the range of the intermolecular force, $r < r_0$. Beyond this range, the 2-particle distribution is clearly a product of 1-particle distributions. The ratio $g_{ss'}/f_s f_s' \sim r_0/\lambda_{ss'}$ is therefore small for a dilute gas, justifying the form (5.5).

and thus solving only the kinetic equation for the 1-particle distribution function, $f_s^{(1)} = f_s$.

The resulting integro-differential equation (2.151) with $C_{ss'}$ given by (5.5),

$$\mathrm{d}_t f_s = \partial_t f_s + \nabla \cdot (\mathbf{v} f_s) + \partial_\mathbf{v} \cdot (\mathbf{a}_s f_s) = \sum C_{ss'}(f_s, f_{s'}), \tag{5.10}$$

where $\mathrm{d}_t = \partial_t + \mathbf{v} \cdot \nabla + \mathbf{a}_s \cdot \partial_\mathbf{v}$ is the total derivative and $\mathbf{a}_s$ represents the acceleration due to external forces, is in general too complicated to solve analytically.[8] The same may be said of the infinite chain of moment equations obtained by multiplying (5.10) by powers of the test particle velocity $\mathbf{v}$ and integrating over $\mathbf{v}$, see Section 2.4.2. This leads to a general moment equation for a molecular quantity, $\mathcal{A}_s(\mathbf{v}) = \mathbf{v}^k$,[9]

$$\partial_t n_s \langle \mathcal{A}_s \rangle_{f_s} + \nabla \cdot n_s \langle \mathbf{v} \mathcal{A}_s \rangle_{f_s} - n_s \langle \partial_\mathbf{v} \cdot \mathbf{a}_s \mathcal{A}_s \rangle_{f_s} = \sum \langle \mathcal{A}_s \rangle_{C_{ss'}}, \tag{5.11}$$

where the right-hand side includes integrals over both $\mathbf{v}$ and $\mathbf{v}'$,

$$\langle \mathcal{A}_s \rangle_{C_{ss'}} \equiv \int C_{ss'} \mathcal{A}_s \mathrm{d}\mathbf{v} = \iint \left(f_s^+ f_{s'}^+ - f_s f_{s'} \right) u \mathrm{d}\sigma_{ss'} \mathrm{d}\mathbf{v}' \mathcal{A}_s \mathrm{d}\mathbf{v}. \tag{5.12}$$

Since Newton's equations are invariant under time reversal, we may interchange the quantities before and after the collision, with the result

$$\langle \mathcal{A}_s \rangle_{C_{ss'}} = \iint (\mathcal{A}_s^+ - \mathcal{A}_s) f_s f_{s'} u \mathrm{d}\sigma_{ss'} \mathrm{d}\mathbf{v}' \mathrm{d}\mathbf{v}, \tag{5.13}$$

where the difference in the brackets is the change of $\mathcal{A}_s$ in a binary collision. Conservation of momentum and energy in elastic scattering yields the intuitive result $\mathbf{F}_{ss'} = -\mathbf{F}_{s's}$ and $\mathcal{Q}_{ss'} = -\mathcal{Q}_{s's}$, see (2.160)–(2.162).

For *like-particle* collisions, we may also interchange the indices s and s', both of which refer to the same particle species, to find

$$\langle \mathcal{A}_s \rangle_{C_{ss'}} = \frac{1}{2} \int \left(\mathcal{A}_s^+ + \mathcal{A}_{s'}^+ - \mathcal{A}_s - \mathcal{A}_{s'} \right) f_s f_{s'} u \mathrm{d}\sigma_{ss'} \mathrm{d}\mathbf{v}' \mathrm{d}\mathbf{v}. \tag{5.14}$$

In this case, the four functions (5.6)–(5.7) appearing in (5.12), or f_s and $f_{s'}$ appearing in (5.13), reflect a single distribution function $f(\mathbf{x}, \mathbf{v}, t)$. The bracketed quantity in (5.14), and hence the entire integral, are identically zero for $\mathcal{A}_s = 1$, $\mathbf{v}$ and v^2, expressing the obvious fact that a gas cannot gain/lose momentum or energy due to collisions between its constituents.[10]

[8] Although it may be progressed numerically for any f_s and $f_{s'}$.

[9] It should be clear that (5.11) evaluated for $\mathcal{A} = 1$, $\mathbf{v}$ and v^2 yields the familiar *Navier–Stokes* (*5-moment*) equations of hydrodynamics, (4.1)–(4.3), e.g. Chapman and Cowling (1939). Inclusion of higher moments yields the various fluid equation sets listed in Table 2.1.

[10] The same cannot be said for unlike-particle collisions, which create a net friction force between different gases and tend to relax their temperatures towards a common value, (5.33)–(5.34).

5.1.2 Chapman–Enskog expansion

In order to tackle (5.10), it is useful to assume that the gas is strongly collisional in the sense $\Delta_s = \lambda_{ss'}/L \ll 1$, where L is the gradient length of the lowest-order moments of f_s, i.e. to adopt the usual assumption of hydrodynamics. One may then expand f_s and $f_{s'}$ in powers of Δ_s and retain only the zeroth-order (Maxwellian) and first-order (perturbed) terms, $f_s \approx f_{s0} + f_{s1}$, where $f_{s0} = f_{Ms}$ is the Maxwellian distribution (A.1),[11]

$$\mathcal{C}_{ss'}(f_s) = \int \left(f_{s'0}^+ f_{s1}^+ + f_{s0}^+ f_{s'1}^+ - f_{s'0} f_{s1} - f_{s0} f_{s'1}\right) u \mathrm{d}\sigma_{ss'} \mathrm{d}\mathbf{v}'. \tag{5.15}$$

As an illustration, let us evaluate only the third term above,[12] which yields

$$- f_{s1} \int f_{s'0} u \mathrm{d}\sigma_{ss'} \mathrm{d}\mathbf{v}' \approx \nu_{ss'} f_{s1}. \tag{5.16}$$

Inserting this term into the kinetic equation (5.10) in place of $\mathcal{C}_{ss'}$,

$$\mathrm{d}_t f_s \approx -\nu_{ss'} f_{s1} = -\nu_{ss'}(f_s - f_{Ms}) \qquad \Rightarrow \qquad f_{s1} \approx -\tau_{ss'} \mathrm{d}_t f_{Ms}, \tag{5.17}$$

allows us to estimate f_{s1} from the total derivative $\mathrm{d}_t = \partial_t + \mathbf{v} \cdot \nabla + \mathbf{a}_s \cdot \partial_{\mathbf{v}}$ of f_{Ms}. Evaluating the derivatives and simplifying yields

$$\frac{f_{s1}}{f_{Ms}} \approx -\tau_{ss'} \left[\left(\frac{\nabla T_s}{T_s} \cdot \mathbf{w} \right) \left(\frac{w^2}{v_{Ts}^2} - \frac{5}{2} \right) + \frac{2\{\nabla \mathbf{V}\}^*}{v_{Ts}^2} : \left(\mathbf{ww} - \frac{w^2}{3} \mathsf{I} \right) \right], \tag{5.18}$$

where $\mathbf{w} = \mathbf{v} - \mathbf{V}_s$ is the relative velocity and $\{\nabla \mathbf{V}_s\}^* = \frac{1}{2}\left(\nabla \mathbf{V}_s + \nabla \mathbf{V}_s^\dagger\right)$ is the symmetrized velocity gradient, i.e. $\{\nabla V\}_{ij}^* = \frac{1}{2}(\partial_{x_j} V_i + \partial_{x_i} V_j)$. Apparently, f_{s1} is only non-zero in the presence of velocity and temperature gradients. The associated transport of momentum and heat can now be evaluated directly using the definitions (2.144) and (2.147), with the result

$$\boldsymbol{\pi}_s = -\mu_s \left[2\{\nabla \mathbf{V}_s\}^* - \tfrac{2}{3}(\nabla \cdot \mathbf{V}_s)\mathsf{I} \right] = -\mu_s \mathsf{W}_s, \qquad \mathbf{q}_s = -\kappa_s \nabla T_s, \tag{5.19}$$

where W_s is the (traceless) *rate-of-strain tensor* (4.5), μ_s is the *dynamic viscosity* and κ_s the *heat conductivity*; the two coefficients are defined as

$$\mu_s = \frac{m_s^2 \tau_{ss'}}{T_s} \int w_1^2 w_2^2 f_{Ms} \mathrm{d}\mathbf{v} = \tau_{ss'} p_s, \tag{5.20}$$

[11] This approach was already used in (2.238) and (2.265) to calculate the plasma viscosity tensor $\boldsymbol{\pi}_s$. It is worth noting the resemblance of this expansion to the GC-closure employed for perpendicular transport in Section 2.4.2, which relied on the expansion of $f_s(\mathbf{v})$ in the small parameter $\epsilon \sim \delta_s$. Both schemes are instances of asymptotic closure discussed in Section 2.4.1.

[12] This is justified by noting that $\langle \mathcal{A}_s \rangle_{\mathcal{C}_{ss'}}$ may be expressed in terms of the change of $\mathcal{A}_s$ in a binary collision weighted only by the product $f_{s'} f_s$ (5.13), i.e. the third term in (5.15).

$$\kappa_s = \frac{m_s \tau_{ss'}}{6T_s} \int w^4 \left(\frac{w^2}{v_{Ts}^2} - \frac{5}{2} \right) f_{Ms} \mathrm{d}\mathbf{v} = \frac{5}{2} \tau_{ss'} p_s / m_s = \frac{5}{2} \mu_s / m_s. \tag{5.21}$$

To evaluate $\tau_{ss'}$, we note that the collision rate density for Maxwellian test and field particles undergoing 'solid-sphere' scattering is given by

$$C_{ss'}(f_{Ms}) = \int f_{Ms} f_{Ms'} u \mathrm{d}\sigma_{ss'} \mathrm{d}\mathbf{v}' \approx n_s n_{s'} \sigma_{ss'} u_T, \tag{5.22}$$

where $u_T^2 = v_{Ts}^2 + v_{Ts'}^2$ is the relative thermal speed of the two species. Noting that $\lambda_{ss'} = n_{s'} u_T / C_{ss'} \approx (n_s \sigma_{ss'})^{-1}$ and that $\lambda_{ss'} = \tau_{ss'} u_T$, we are justified in using the expressions (5.4) for $\tau_{ss'} = \nu_{ss'}^{-1}$ and $\lambda_{ss'}$ with the substitution $u_\infty \to u_T$. For a single-species gas ($s = s'$), inserting this expression into (5.20) leads to a simple estimate for μ_s and κ_s,

$$\nu_{ss} \approx v_{ts} n_s a^2, \qquad \tfrac{2}{5} m_s \kappa_s \approx \mu_s \approx n_s T_s / \nu_{ss} \approx m_s v_{ts} / a^2. \tag{5.23}$$

The fact that μ_s and κ_s, whether in (5.20) or (5.23), are linearly proportional to $\tau_{ss'}$, and hence inversely proportional to $\nu_{ss'}$, indicates that collisions *inhibit* net transport of momentum and heat. This important conclusion is easily understood in terms of a diffusive, or *random-walk*, process with characteristic time step $\tau_{ss'}$ and length step $\lambda_{ss'}$. As is evident from (5.4), enhanced collisionality (higher $n_{s'} \sigma_{ss'}$) linearly reduces both these quantities, so that the net diffusivity, $D_{ss'} \sim \lambda_{ss'}^2 / \tau_{ss'} \sim u_T \lambda_{ss'}$, is likewise reduced.

The expansion of f_s in the small parameter $\Delta_s = \lambda_s / L$, generally known as the *Chapman–Enskog* expansion (Chapman, 1916; Enskog, 1917), forms the basis of a rigorous technique of calculating collisional flows in a dilute, yet collision dominated ($\Delta_s \ll 1$), gas, i.e. one which is close to local thermodynamic equilibrium (LTE), so that $f_s = f_{Ms} + O(\Delta_s)$. In other words, it provides an exact asymptotic closure of the fluid equations (5.11) by calculating $\mathbf{F}_s$, $\mathcal{Q}_s$, $\boldsymbol{\pi}_s$ and $\mathbf{q}_s$ in terms of lower-order moments (n_s, $\mathbf{V}_s$, T_s) and their spatial gradients. The technique, which is comprehensively described in Chapman and Cowling (1939), begins with the formal expansion[13]

$$f_s = f_{Ms}(n_s, \mathbf{V}_s, T_s) \left[1 + \mathbf{c}_{1s} \cdot \mathbf{w} + \mathsf{C}_{2s} : \mathbf{w}^2 + \mathsf{C}_{3s} : \mathbf{w}^3 \right] + O\left(\Delta_s^4\right), \tag{5.24}$$

where $\mathbf{c}_{1s} \sim O(\Delta_s)$, $\mathsf{C}_{2s} \sim O\left(\Delta_s^2\right)$ and $\mathsf{C}_{3s} \sim O\left(\Delta_s^3\right)$ are velocity independent tensors. These can be related to the viscous stress and heat flow tensors as

$$\mathbf{c}_{1s} = -\mathbf{q}_s / n_s v_{ts}^2, \qquad \mathsf{C}_{2s} = -\boldsymbol{\pi}_s / 2 p_s v_{ts}^2, \qquad \mathsf{C}_{3s} = \mathsf{q}_s / 6 p_s v_{ts}^4, \tag{5.25}$$

so that (5.24) may be written explicitly in terms of the stresses $\boldsymbol{\pi}_s$ and q_s,

[13] For simplicity, we assume the case of a single-species gas ($s' = s$).

$$\frac{f_s}{f_{Ms}} = 1 - \frac{m_s n_s}{2p_s^2}\boldsymbol{\pi}_s : \mathbf{w}^2 + \frac{m_s^2 n_s^2}{6p_s^3}\left(\mathsf{q}_s : \mathbf{w}^2 - 6v_{ts}^2\mathbf{q}_s\right)\cdot\mathbf{w} + O\left(\Delta_s^4\right). \quad (5.26)$$

As expected, we find that $f_s = f_{Ms}$ if, and only if, $\boldsymbol{\pi}_s = 0 = \mathsf{q}_s$.[14] Inserting (5.26) in the velocity space averages in (5.11), yields the conservation equation for mass ($\mathcal{A}_s = 1$), momentum ($\mathcal{A}_s = \mathbf{v}$), stress tensor ($\mathcal{A}_s = \mathbf{w}^2$) and heat flow tensor ($\mathcal{A}_s = \mathbf{w}^3$). The conservation equations for energy and heat flow follow from the contraction of the last two equations or from direct evaluation of (5.11) with $\mathcal{A}_s = w^2$ and $\mathcal{A}_s = \mathbf{w}w^2$, respectively. Significantly, the form (5.26) implies that the energy weighted stress tensor, $\mathsf{y}_s = m_s n_s\langle\mathbf{w}^4\rangle_{f_s}$, which appears in the conservation equation for $\mathsf{q}_s = m_s n_s\langle\mathbf{w}^3\rangle_{f_s}$, can be expressed entirely in terms of the pressure tensor,

$$\mathsf{y}_s = v_{ts}^2(p_s\mathsf{II} - \mathsf{I}\boldsymbol{\pi}_s) + O\left(\Delta_s^4\right), \quad (5.27)$$

where $\mathsf{II} = \delta_{ij}\delta_{kl} + \delta_{ik}\delta_{jl} + \delta_{il}\delta_{jk}$ and $\mathsf{I}\boldsymbol{\pi}_s$ is evaluated as

$$\mathsf{I}\boldsymbol{\pi}_s = \delta_{ij}\pi_{kl} + \delta_{ik}\pi_{jl} + \delta_{il}\pi_{jk} + \delta_{jk}\pi_{il} + \delta_{jl}\pi_{ik} + \delta_{kl}\pi_{ij} + O\left(\Delta_s^4\right). \quad (5.28)$$

Consequently, the chain of moment equations is closed at the level of $\mathsf{y}_s(n_s, \mathsf{P}_s)$, resulting in 20 PDEs for 20 components of fluid variables, or moments, as summarized in Table 2.1. This *20-moment approximation* can be alternatively derived using the so-called *Grad's even-moment closure* (Grad, 1949), which begins from the assumption (5.27), expressed in the form

$$\mathsf{y}_s = P_{ij}P_{kl} + P_{ik}P_{jl} + P_{il}P_{jk} + O\left(\Delta_s^4\right), \quad (5.29)$$

without making any specific assumption about the form of f_s. Although the Chapman–Enskog and Grad techniques produce slightly different evolution equations for the heat flux tensor q_s, the two sets of 20-moment equations can be shown to be identical through order Δ_s^3, i.e. to differ only at $O\left(\Delta_s^4\right)$. The equivalence of the two methods was not recognized until some time after Grad's original paper, which has resulted in some confusion in the transport theory literature. At present, both forms of transport equations, each having certain advantages for particular problems and numerical techniques, are widely used, and it is understood that their results must be identical.[15]

Since the seven off-diagonal components of $\mathsf{q}_s \sim O\left(\Delta_s^3\right)$ are small compared to $\mathbf{q}_s \sim O\left(\Delta_s\right)$ or $\boldsymbol{\pi}_s \sim O\left(\Delta_s^2\right)$, they may be neglected with little loss in accuracy, so that $\mathsf{q}_s = \frac{2}{5}\mathsf{I}\mathbf{q}_s$ and (5.26) becomes

[14] Despite a clear similarity between (5.18) and (5.26), the latter is more accurate, due to the inclusion of the third-order, heat flux tensor terms. Moreover, the viscous and heat fluxes in (5.26) have not yet been explicitly related to gradients of velocity and temperature.

[15] A good comparison of the two methods for a neutral gas may be found in Gombosi (1994).

$$\frac{f_s}{f_{Ms}} = 1 - \frac{m_s n_s}{2p_s^2}\boldsymbol{\pi}_s : \mathbf{w}^2 + \frac{m_s^2 n_s^2}{5p_s^3}\left(w^2 - 5v_{ts}^2\right)\mathbf{q}_s \cdot \mathbf{w} + O\left(\Delta_s^4\right). \tag{5.30}$$

The resulting *13-moment approximation* contains five equations for the components of the viscous stress and another three for those of the heat flow, e.g. see (Gombosi, 1994, p. 211). Retaining only the leading-order terms in these equations yields a balance between the collisional change to $\boldsymbol{\pi}_s$, or $\mathbf{q}_s$, and the spatial gradients of $\mathbf{V}_s$, or T_s, respectively,

$$\langle \mathbf{w}^2 \rangle_{C_{ss'}} \approx -p_s \mathsf{W}_s, \qquad \langle \mathbf{w} w^2 \rangle_{C_{ss'}} \approx \frac{5}{2}\frac{p_s}{m_s}\nabla T_s. \tag{5.31}$$

The collisional terms can now be calculated using (5.14), which requires explicit knowledge of $d\sigma_{ss'}$, however, and entails substantial algebraic complexity. Alternatively, they may be computed, with little loss in accuracy, using the relaxation time, or *BGK*, approximation (Bhatnagar *et al.*, 1954). This approach relies on the insight (5.17), that f_s relaxes to a Maxwellian f_{Ms} with a characteristic collision time, $\tau_{ss'} = \nu_{ss'}^{-1}$, so that

$$C_{ss'} = \nu_{ss'}(f_s - f_{Ms}) \quad \Rightarrow \quad \langle \mathcal{A}_s \rangle_{C_{ss'}} = \nu_{ss'} \int (f_s - f_{Ms}) \mathcal{A}_s d\mathbf{v}. \tag{5.32}$$

Inserting (5.30) for f_s, leaves a comparatively simple integral involving products of f_{Ms} and $\mathcal{A}_s(\mathbf{v})$. Evaluating the integrals for $\mathcal{A}_s = \mathbf{w}^2$ and $\mathbf{w}w^2$ yields the intuitive expressions, $\langle \mathbf{w}^2 \rangle_{C_{ss'}} = -\nu_{ss'}\boldsymbol{\pi}_s$ and $\langle \mathbf{w}w^2 \rangle_{C_{ss'}} = -\nu_{ss'}\mathbf{q}_s$. As required by (5.30), we find that both $\boldsymbol{\pi}_s$ and $\mathbf{q}_s$ decay towards zero as $f_s \to f_{Ms}$, and that all three processes occur on the same time scale.[16] Finally, inserting these into (5.31), we recover the earlier result (5.19)–(5.20), indicating a linear relation between thermodynamic gradients and collisional flows with the viscosity, μ_s, and heat conductivity, κ_s, being the constants of proportionality.[17] This result can now be seen as the lowest-order estimate to a more complicated dependence, described by the evolution equations for $\boldsymbol{\pi}_s$ and $\mathbf{q}_s$, or more generally for q_s.

For a multi-species gas, f_{Ms} appearing in (5.32) must be replaced by $f_{s(s')M}$, i.e. a Maxwellian that would result in the sole presence of collisions between species s and s'. It is given by f_M (A.1) characterized by a CM velocity, $\mathbf{V} = \mathbf{V}_{s(s')}$, and temperature, $T = T_{s(s')}$, where

$$\mathbf{V} = \mathbf{V}_{s(s')} = \mathbf{V}_{s'(s)} = (m_s \mathbf{V}_s + m_{s'} \mathbf{V}_{s'})/(m_s + m_{s'}), \tag{5.33}$$

[16] In a multi-species gas, the collisional relaxation of momentum and energy occurs on different time scales. We will return to this theme in the next section, when investigating a plasma.

[17] This linear relationship between thermodynamic forces A_j and the resulting flows Γ_i, can be generalized as $\Gamma_i = -\sum \mathcal{L}_{ij} A_j = -\boldsymbol{\mathcal{L}} : \mathbf{A}$, where $\boldsymbol{\mathcal{L}}$ is known as the transport matrix and its components $\mathcal{L}_{ij}$ as transport coefficients. The determination of $\mathcal{L}_{ij}$ is the central aim of *local transport theory*, which applies to fluids close to local thermodynamic equilibrium (LTE).

$$T = T_{s(s')} = T_s + \frac{2m_{ss'}(T_s - T_{s'})}{m_s + m_{s'}} + \frac{m_s m_{ss'}|\mathbf{V}_s - \mathbf{V}_{s'}|^2}{3(m_s + m_{s'})}, \tag{5.34}$$

and $m_{ss'}$ is the *reduced mass*,

$$m_{ss'} \equiv m_s m_{s'}/(m_s + m_{s'}) = m_{s's}. \tag{5.35}$$

This can then be used to calculate the friction force, $\mathbf{F}_{ss'}$ (2.158), and energy exchange, $\mathcal{Q}^L_{ss'}$, or heat exchange, $\mathcal{Q}_{ss'}$ (2.157), between the two gases,

$$\mathbf{F}_{ss'} = \mathbf{C}_{ss'1} = m_s\langle \mathbf{v}\rangle_{\mathcal{C}_{ss'}} = m_s n_s \nu_{ss'}(\mathbf{V}_{s'} - \mathbf{V}_s), \tag{5.36}$$

$$\mathcal{Q}_{ss'} = \mathrm{C}_{ss'2} = \frac{m_s}{2}\langle w^2\rangle_{\mathcal{C}_{ss'}} = \frac{3m_s n_s \nu_{ss'}}{m_s + m_{s'}}(T_s - T_{s'}). \tag{5.37}$$

Thus inter-particle collisions tend to relax the velocities and temperatures of the two gases towards common values given by (5.33)–(5.34).

5.1.3 Fokker–Planck collision operator

The Maxwell–Boltzmann collision operator (5.5) is valid for arbitrary scattering angles $0 < \vartheta < \pi$. In many circumstances, e.g. for long-range interactions or when the test particles are much heavier than the field particles, small-angle scattering ($\vartheta \ll 1$) via grazing collisions predominates, and $f_s^+ = f_s(\mathbf{x}, \mathbf{v}^+, t)$ and $f_{s'}^+ = f_{s'}(\mathbf{x}, \mathbf{v}'^+, t)$ may be expanded in the change in the test and field particle velocities, $\Delta\mathbf{v} = \mathbf{v}^+ - \mathbf{v}$ and $\Delta\mathbf{v}' = \mathbf{v}'^+ - \mathbf{v}'$,

$$f_s^+ = f_s + \partial_{\mathbf{v}} f_s \cdot \Delta\mathbf{v} + \tfrac{1}{2}\partial^2_{\mathbf{v}} f_s : \Delta\mathbf{v}\Delta\mathbf{v} + \cdots, \tag{5.38}$$

$$f_{s'}^+ = f_{s'} + \partial_{\mathbf{v}'} f_{s'} \cdot \Delta\mathbf{v}' + \tfrac{1}{2}\partial^2_{\mathbf{v}'} f_{s'} : \Delta\mathbf{v}'\Delta\mathbf{v}' + \cdots. \tag{5.39}$$

Inserting (5.38)–(5.39) into (5.5) and simplifying, we find

$$\mathcal{C}_{ss'}(f_s, f_{s'}) = \int (\mathbf{X}_{ss'} \cdot \Delta\mathbf{u} - \mathsf{Y}_{ss'} : \Delta\mathbf{u}\Delta\mathbf{u}) u \,\mathrm{d}\sigma_{ss'} \mathrm{d}\mathbf{v}', \tag{5.40}$$

where $\Delta\mathbf{u} = \mathbf{u}^+ - \mathbf{u}$ is the change in the relative velocity, $\mathbf{u} = \mathbf{v} - \mathbf{v}'$, as a result of a binary collision, and $\mathbf{X}_{ss'}$ and $\mathsf{Y}_{ss'}$ are given by

$$\mathbf{X}_{ss'} = \frac{m_{ss'}}{m_s} f_{s'}\partial_{\mathbf{v}} f_s - \frac{m_{ss'}}{m_{s'}} f_s \partial_{\mathbf{v}'} f_{s'}, \tag{5.41}$$

$$\mathsf{Y}_{ss'} = \frac{m^2_{ss'}}{m_s m_{s'}}(\partial_{\mathbf{v}} f_s)(\partial_{\mathbf{v}'} f_{s'}) - \frac{1}{2}\frac{m^2_{ss'}}{m^2_s} f_{s'}\partial^2_{\mathbf{v}} f_s - \frac{1}{2}\frac{m^2_{ss'}}{m^2_{s'}} f_s \partial^2_{\mathbf{v}'} f_{s'}. \tag{5.42}$$

To simplify further, we introduce the nth-order *transport cross-section*, $\sigma^{(n)}_{ss'}$,

$$\sigma^{(n)}_{ss'}(u) = \int (1 - \cos^n \vartheta)\mathrm{d}\sigma_{ss'}, \qquad n = 1, 2, \ldots, \tag{5.43}$$

which is evaluated by writing $d\sigma_{ss'} = (d_{\varpi}\sigma_{ss'})d\varpi$ and integrating over the complete solid angle 4π. Performing this integration over (5.40) yields

$$\mathcal{C}_{ss'} = \int \left[\mathbf{X} \cdot \sigma^{(1)} \mathbf{u} + \mathsf{Y} : \left\{ \frac{1}{2} \sigma^{(2)} (u^2 \mathsf{I} - 3\mathbf{u}^2) + 2\sigma^{(1)} \mathbf{u}^2 \right\} \right] u d\mathbf{v}', \tag{5.44}$$

where for clarity we dropped the double subscript ss' appearing on all terms.

Combining (5.41)–(5.44) leads to a lengthy expression, which after some manipulation can be reduced to a second-order differential operator,

$$\mathcal{C}_{ss'}(f_s) = -\partial_{\mathbf{v}} \cdot (\mathbf{G}_{ss'} + \mathbf{H}_{ss'}) f_s + \partial_{\mathbf{v}}^2 : (\mathsf{K}_{ss'} + \mathsf{L}_{ss'}) f_s + \mathcal{O}_{ss'} f_s, \tag{5.45}$$

whose coefficients involve velocity integrals of $\sigma^{(n)}$ and $\mathbf{u}$ weighted by $f_{s'}$,

$$\mathbf{G}_{ss'} = \frac{m_{ss'}}{m_s} \int f_{s'} \sigma^{(1)} u \mathbf{u} d\mathbf{v}', \qquad \mathsf{K}_{ss'} = \frac{m_{ss'}^2}{m_s^2} \int f_{s'} \sigma^{(1)} u \mathbf{u}^2 d\mathbf{v}',$$

$$\mathbf{H}_{ss'} = \frac{2m_{s'}}{m_s + m_{s'}} \partial_{\mathbf{v}} \cdot \int f_{s'} \left[\sigma^{(1)} u \mathbf{u}^2 + \frac{1}{4} \sigma^{(2)} u (u^2 \mathsf{I} - 3\mathbf{u}^2) \right] d\mathbf{v}',$$

$$\mathsf{L}_{ss'} = \frac{m_{ss'}^2}{4m_s^2} \int f_{s'} \sigma^{(2)} u \left(u^2 \mathsf{I} - 3\mathbf{u}^2 \right) d\mathbf{v}', \tag{5.46}$$

$$\mathcal{O}_{ss'} = \partial_{\mathbf{v}} \cdot \left[\frac{m_s}{m_{ss'}} \mathbf{G}_{ss'} + \frac{m_s + m_{s'}}{2m_{s'}} \partial_{\mathbf{v}} \cdot \mathbf{H}_{ss'} \right].$$

Equation (5.45) is the most general form of the *Fokker–Planck* collision operator, describing the cumulative effect of multiple small-angle deflections. The vectors $\mathbf{G}_{ss'}$ and $\mathbf{H}_{ss'}$ represent the *dynamical friction* between field and test particles, while the tensors $\mathsf{K}_{ss'}$ and $\mathsf{L}_{ss'}$ represent *velocity space diffusion.* The two terms are often denoted as

$$\mathbf{A}_{ss'} = \mathbf{G}_{ss'} + \mathbf{H}_{ss'} = \langle \Delta \mathbf{v} \rangle / \Delta t, \quad \mathsf{D}_{ss'} = \mathsf{K}_{ss'} + \mathsf{L}_{ss'} = \langle \Delta \mathbf{v} \Delta \mathbf{v} \rangle / \Delta t, \tag{5.47}$$

allowing (5.45) to be written in the more connotative form,

$$\mathcal{C}_{ss'}(f_s) = -\partial_{\mathbf{v}} \cdot \left(\frac{\langle \Delta \mathbf{v} \rangle}{\Delta t} f_s \right) + \frac{1}{2} \partial_{\mathbf{v}}^2 : \left(\frac{\langle \Delta \mathbf{v} \Delta \mathbf{v} \rangle}{\Delta t} f_s \right) + \mathcal{O}_{ss'} f_s. \tag{5.48}$$

This form, aside from the scalar $\mathcal{O}_{ss'}$, which represents the change of f_s in the absence of $\partial_{\mathbf{v}} f_s$, may also be derived by assuming that velocity evolution is a Markov process, i.e. that $\mathbf{v}$ evolves via a sequence of small and statistically independent changes, $\Delta \mathbf{v}$. This allows one to write $f_s(\mathbf{v}, t)$ as

$$f_s(\mathbf{v}, t) = \int f_s(\mathbf{v} - \Delta \mathbf{v}, t - \Delta t) P(\mathbf{v} - \Delta \mathbf{v}; \Delta \mathbf{v}, \Delta t) d\Delta \mathbf{v}, \tag{5.49}$$

where $P(\mathbf{v}; \Delta \mathbf{v}, \Delta t)$ is the probability that a test particle will change its velocity by $\Delta \mathbf{v}$ during the time interval Δt (Chandrasekhar, 1943). Taylor expanding $P f_s$ and retaining only second-order terms, we recover (5.48).

In the next section, we will evaluate the Fokker–Planck operator for the Coulomb potential and apply it to charged particle collisions in a plasma.

5.2 Charged particle collisions in a plasma

Coulomb collisions between charged particles play a crucial role in a variety of transport, relaxation and dissipative phenomena in magnetically confined plasmas. They are characterized by an inverse-square dependence of the collision cross-section on the relative particle kinetic energy,

$$d\sigma_{ss'} = \frac{\pi\alpha_{ss'}^2}{m_{ss'}^2 u_\infty^4}\frac{\cos(\vartheta/2)d\vartheta}{\sin^3(\vartheta/2)} = \left(\frac{\alpha_{ss'}}{2m_{ss'}u_\infty^2\sin^2(\vartheta/2)}\right)^2 d\varpi, \tag{5.50}$$

where $\varphi(r) = \alpha_{ss'}/r$ is the Coulomb potential.[18]

Recalling the quote at the beginning of the chapter, we expect any dilute gas to be dominated by binary encounters. When considering collisions in a plasma, it is still possible to retain only binary encounters, but it now becomes necessary to introduce the collective effect of all other field particles, which exponentially shield the Coulomb potential at long distances (2.4), thus limiting the range of the interaction to roughly a Debye length, λ_D (2.5). Since by definition (2.7), a plasma has many particles in a Debye sphere ($\Lambda_p \gg 1$), we expect grazing collisions to dominate the motion of a test particle. Hence, the Fokker–Planck collision operator (5.48) should accurately describe the evolution of f_s.

5.2.1 Coulomb collision operator

In light of the above comments, it is straight forward, if somewhat tedious, to calculate the Fokker–Planck coefficients (5.46) appropriate to Coulomb interactions. One begins by inserting (5.50) into (5.43) to obtain the Coulomb transport cross-sections $\sigma_{ss'}^{(1)}$ and $\sigma_{ss'}^{(2)}$. The next step involves evaluating the angular integrals between $\vartheta_{min}(b_{max})$ and $\vartheta_{max}(b_{min})$. The minimum impact parameter can be approximated as the larger of the *deBroglie* length, λ_{Bs}, (2.8) and the classical distance of closest approach, r_{cs} (2.6),

$$b_{min} \approx \max\left(\frac{\hbar}{2m_{ss'}\langle u\rangle_{ss'}}, \frac{e_s e_{s'}}{4\pi\epsilon_0^2 m_{ss'}\langle u\rangle_{ss'}^2}\right), \tag{5.51}$$

[18] The derivation of the Coulomb collision cross-section, also known as the *Rutherford formula*, may be found in most texts on classical mechanics, e.g. Landau and Lifschitz (1960, Vol. I) or Goldstein (1980). Note that the Coulomb cross-section (5.50) is independent of the sign of $\alpha_{ss'}$ and is thus equal for attractive and repulsive collisions.

where $\langle u \rangle_{ss'} \sim u_T$ denotes an average over both f_s and $f_{s'}$. The maximum impact parameter may be approximated as the effective Debye length,

$$b_{max} \approx \lambda_D^{eff} = \left(\frac{\epsilon_0 T_s}{\sum_a n_a e_a^2} \right)^{1/2} \tag{5.52}$$

with the summation performed over all species a whose thermal velocity exceeds the average relative velocity, $v_{Ta} > \langle u \rangle_{ss'}$. The ratio $\Lambda_{ss'} = b_{max}/b_{min}$ is just the plasma parameter $\Lambda_p \gg 1$ defined by (2.7). Its logarithm,

$$\ln \Lambda_{ss'} \equiv \ln (b_{max}/b_{min}) \sim 20, \tag{5.53}$$

generally known as the *Coulomb logarithm*, enters the final expressions for $\sigma_{ss'}^{(n)}$. As expected, $\ln \Lambda_{ss'}$ has only a weak dependence on n_s, T_s and the combination of s and s', with typical values in the range 15–25.[19]

5.2.1.1 *Fokker–Planck coefficients and Rosenbluth potentials*

Inserting $\sigma_{ss'}^{(1)}$ and $\sigma_{ss'}^{(2)}$ into (5.46), one obtains the Fokker–Planck coefficients $\mathbf{A}_{ss'}$ and $\mathsf{D}_{ss'}$ as integrals of $\mathbf{u}$ and $f_{s'}$ over the field particle velocity $\mathbf{v}'$,

$$\mathbf{A}_{ss'}(\mathbf{v}) = -2\frac{\gamma_{ss'}}{m_s^2}\left(1 + \frac{m_s}{m_{s'}}\right) \int f_{s'} \frac{\mathbf{u}}{u^3} d\mathbf{v}', \tag{5.54}$$

$$\mathsf{D}_{ss'}(\mathbf{v}) = 2\frac{\gamma_{ss'}}{m_s^2} \int f_{s'} \mathsf{U} d\mathbf{v}', \qquad \mathsf{U} \equiv \frac{\mathsf{I}}{u} - \frac{\mathbf{uu}}{u^3}, \tag{5.55}$$

where the charge related constant $\gamma_{ss'}$ is defined as

$$\gamma_{ss'} \equiv e_s^2 e_{s'}^2 \ln \Lambda_{ss'}/8\pi\epsilon_0^2 = \gamma_{s's}. \tag{5.56}$$

Note that the friction force, $\mathbf{F}_{ss'}$, and heat exchange, $\mathcal{Q}_{ss'}$, defined in Section 2.4.1, involve the averages of $\mathbf{A}_{ss'}$ and $\mathsf{D}_{ss'}$ over the test particle velocity $\mathbf{v}$,

$$\mathbf{F}_{ss'} = m_s n_s \langle \mathbf{A}_{ss'} \rangle_{f_s}, \quad \mathcal{Q}_{ss'} = m_s n_s \langle \mathbf{w}_s \cdot \mathbf{A}_{ss'} + \tfrac{1}{2} \mathrm{Tr}\, \mathsf{D}_{ss'} \rangle_{f_s}, \tag{5.57}$$

where $\mathbf{w} = \mathbf{v} - \mathbf{V}_s$ and the velocity space average is defined by (2.143).

The Fokker–Planck coefficients (5.54)–(5.55) can be conveniently expressed as velocity gradients of the scalar functions $g_{s'}(\mathbf{v})$ and $h_{s'}(\mathbf{v})$, known as *Rosenbluth* potentials (Rosenbluth *et al.*, 1957; Trubnikov, 1958),

$$\mathbf{A}_{ss'}(\mathbf{v}) = 2\frac{\gamma_{ss'}}{m_s^2}\left(1 + \frac{m_s}{m_{s'}}\right) \partial_{\mathbf{v}} h_{s'}, \qquad \mathsf{D}_{ss'}(\mathbf{v}) = 2\frac{\gamma_{ss'}}{m_s^2} \partial_{\mathbf{v}}^2 g_{s'}, \tag{5.58}$$

$$h_{s'}(\mathbf{v}) = \int f_{s'} u^{-1} d\mathbf{v}', \qquad g_{s'}(\mathbf{v}) = \int f_{s'} u d\mathbf{v}'. \tag{5.59}$$

[19] Useful approximations to $\ln \Lambda_{ss'}$ may be found in the *NRL Plasma Formulary* (Huba, 2006).

It is easy to confirm this result by noting that the velocity gradients of u and u^{-1} give rise to the expressions appearing in (5.54)–(5.55),

$$\partial_{\mathbf{v}} u^{-1} = -\mathbf{u}/u^3, \qquad \partial_{\mathbf{v}}^2 u = \mathsf{U}, \tag{5.60}$$

$$\partial_v^2 u^{-1} = -4\pi\,\delta(\mathbf{u}), \qquad \partial_v^2 u = 2/u. \tag{5.61}$$

These relations allow $g_{s'}$, $h_{s'}$ and $f_{s'}$ to be related as

$$\partial_v^2 g_{s'} = 2h_{s'}, \qquad \partial_v^2 h_{s'} = -4\pi f_{s'}, \tag{5.62}$$

which provides a simple analogy to electrostatics, i.e. $h_{s'}$ is the net velocity space 'potential' at $\mathbf{v}$ due to a collection of point 'charges' $f_{s'}\mathrm{d}\mathbf{v}'$ each contributing according to the inverse distance in velocity space, i.e. as u^{-1}. Consequently, (5.62) are analogous to the Poisson equation of electrostatics.

When the field particle distribution is isotropic, i.e. $f_{s'}(\mathbf{v}') = f_{s'}(v')$, the Fokker–Planck coefficients $\mathbf{A}_{ss'}$ and $\mathsf{D}_{ss'}$ (5.58) can be simplified to read,

$$\mathbf{A}_{ss'}(\mathbf{v}) = -\nu_{ss'}^t \mathbf{v}, \qquad \mathsf{D}_{ss'}(\mathbf{v}) = D_{\parallel ss'}^t \frac{\mathbf{v}\mathbf{v}}{v^2} + D_{\perp ss'}^t \left(\mathsf{I} - \frac{\mathbf{v}\mathbf{v}}{v^2}\right), \tag{5.63}$$

where $\nu_{ss'}^t$ is the slowing down rate (or frequency), while $D_{\parallel ss'}^t$ and $D_{\perp ss'}^t$ are the parallel and transverse velocity space diffusion coefficients,[20]

$$\nu_{ss'}^t(v) = -\frac{2\gamma_{ss'}}{m_s^2}\left(1 + \frac{m_s}{m_{s'}}\right) v^{-1}\mathrm{d}_v h_{s'}(v), \tag{5.64}$$

$$D_{\parallel ss'}^t(v) = \frac{2\gamma_{ss'}}{m_s^2}\mathrm{d}_v^2 g_{s'}(v), \qquad D_{\perp ss'}^t(v) = \frac{2\gamma_{ss'}}{m_s^2} v^{-1}\mathrm{d}_v g_{s'}(v). \tag{5.65}$$

We also define the associated dispersion and deflection rates,

$$\nu_{\parallel ss'}^t(v) = v^{-2} D_{\parallel ss'}^t(v), \qquad \nu_{\perp ss'}^t(v) = 2v^{-2} D_{\perp ss'}^t(v), \tag{5.66}$$

and use conservation of energy to derive the energy loss rate, $\nu_{\varepsilon ss'}^t$,

$$\nu_{\varepsilon ss'}^t = 2\nu_{ss'}^t - \nu_{\perp ss'}^t - \nu_{\parallel ss'}^t. \tag{5.67}$$

The Rosenbluth potentials are easily evaluated for Maxwellian field particles with zero flow velocity, $\mathbf{V}_{s'} = 0$, and finite temperature, $T_{s'}$,

$$g_{s'M}(x) = \frac{n_{s'} v_{Ts'}}{2x}\left[(1 + 2x^2)\Phi + x\mathrm{d}_x\Phi\right], \qquad h_{s'M}(x) = \frac{n_{s'}}{v_{Ts'}}\frac{\Phi}{x}, \tag{5.68}$$

[20] The symbols $\parallel$ and $\perp$ refer to the directions of the initial particle velocity and should not be confused with the gyro-tropic notation relating to the direction of the magnetic field. We introduce the superscript t on the relaxations rates to emphasize that they refer to a test particle and are thus functions of its velocity $\mathbf{v}$. Later on, we will introduce average relaxation rates and distinguish these from the test particle values by the absence of the superscript t.

where $\Phi(x)$ is the *error* function, $x \equiv v/v_{Ts'}$ is the test particle speed normalized by the field particle thermal speed, $v_{Ts'}^2 \equiv 2T_{s'}/m_{s'}$, and $n_{s'}$, $\mathbf{V}_{s'}$ and $T_{s'}$ are the field particle fluid variables, see (2.142) and (2.145),

$$n_{s'} = \int f_{s'} \mathrm{d}\mathbf{v}', \quad n_{s'}\mathbf{V}_{s'} = \int \mathbf{v} f_{s'} \mathrm{d}\mathbf{v}', \quad \tfrac{3}{2} n_{s'} T_{s'} = \int \tfrac{1}{2} m_{s'} |\mathbf{v} - \mathbf{V}_s|^2 f_{s'} \mathrm{d}\mathbf{v}'.$$

The three relaxation rates (5.64)–(5.66) are then found as

$$\nu_{ss'}^t(x) = 4\frac{\gamma_{ss'} n_{s'}}{m_s^2 v_{Ts'}^3}\left(1 + \frac{m_s}{m_{s'}}\right)\frac{\Psi(x)}{x}, \tag{5.69}$$

$$\nu_{\parallel ss'}^t(x) \equiv \frac{D_{\parallel ss'}}{v^2} = \frac{4\gamma_{ss'} n_{s'}}{m_s^2 v_{Ts'}^3}\frac{\Psi(x)}{x^3}, \tag{5.70}$$

$$\nu_{\perp ss'}^t(x) \equiv \frac{2D_{\perp ss'}}{v^2} = \frac{4\gamma_{ss'} n_{s'}}{m_s^2 v_{Ts'}^3}\frac{\Phi(x) - \Psi(x)}{x^3}. \tag{5.71}$$

The *error*, $\Phi(x)$, and *Chandrasekhar*, $\Psi(x)$, functions are defined by

$$\Phi(x) = \frac{2}{\sqrt{\pi}}\int_0^x \mathrm{e}^{-\xi^2}\mathrm{d}\xi, \qquad \Psi(x) = \frac{1}{2x^2}\left[\Phi - x \mathrm{d}_x \Phi\right], \tag{5.72}$$

where $\mathrm{d}_x\Phi = 2\pi^{-1/2}\mathrm{e}^{-x^2}$; they approach the following asymptotic limits,

$$\lim_{x\to 0}\Phi(x) = \lim_{x\to 0}\frac{2x}{\sqrt{\pi}} = 0, \qquad \lim_{x\to\infty}\Phi(x) = 1, \tag{5.73}$$

$$\lim_{x\to 0}\Psi(x) = \lim_{x\to 0}\frac{2x}{3\sqrt{\pi}} = 0, \qquad \lim_{x\to\infty}\Psi(x) = \lim_{x\to\infty}\frac{1}{2x^2} = 0.$$

and take on the following values at $x = 1$: $\Phi(1) \approx 0.8$, $\Psi(1) \approx 0.2$. The resulting dependence of (5.69)–(5.71) is plotted in Fig. 5.1.

5.2.1.2 *Landau–Boltzmann collision operator*

Inserting the Fokker–Planck coefficients (5.58) or, for isotropic field particles, the three relaxation rates (5.64)–(5.66), into (5.45) yields a Fokker–Planck collision operator appropriate for Coulomb collisions. It is significant that $\mathcal{O}_{ss'}$ in (5.46) vanishes for a Coulomb potential, so that (5.45) can then be written in a purely advective-diffusive (conservative) form,

$$\mathcal{C}_{ss'} = -\partial_{\mathbf{v}} \cdot \mathbf{A}_{ss'} f_s + \tfrac{1}{2}\partial_{\mathbf{v}}^2 : \mathsf{D}_{ss'} f_s = -\partial_{\mathbf{v}} \cdot \left(\mathbf{A}_{ss'} f_s - \tfrac{1}{2}\partial_{\mathbf{v}} \cdot \mathsf{D}_{ss'} f_s\right). \tag{5.74}$$

Note that the quantity being differentiated in (5.74), which can be denoted as $\Gamma_{ss'}$ is the phase space flow of the test particle kinetic substance, i.e. its 1-particle distribution function f_s, due to collisions with species s'. Thus, (5.74) represents the velocity space divergence of this flow, $-\partial_{\mathbf{v}} \cdot \Gamma_{ss'}$, and thus conserves particle number. Since Coulomb collisions are elastic, (5.74) also conserves the momentum and energy, thus satisfying (2.160)–(2.162).

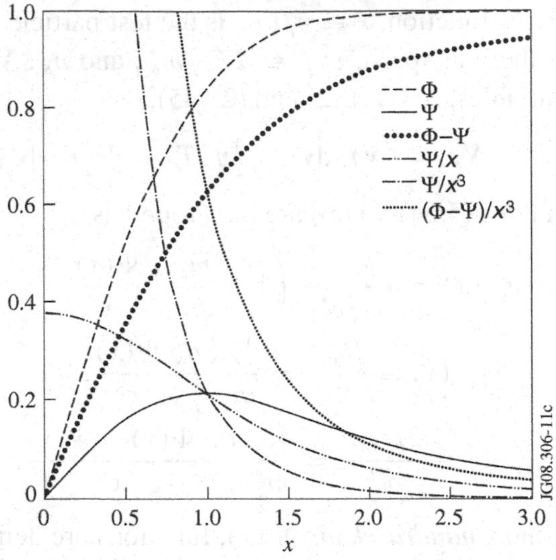

Fig. 5.1. Velocity dependence of selected slowing down functions.

Combining (5.74) with (5.54)–(5.55) and noting that $\partial_{\mathbf{v}} \cdot \mathbf{U} = -\partial_{\mathbf{v}'} \cdot \mathbf{U} = -2\mathbf{u}/u^3$, yields the so-called *Landau–Boltzmann* collision operator,

$$C_{ss'} = \partial_{\mathbf{v}} \cdot \frac{\gamma_{ss'}}{m_s} \int f_s f_{s'} \boldsymbol{\chi}_{ss'} \cdot \mathbf{U} d\mathbf{v}' \equiv \partial_{\mathbf{v}} \cdot \boldsymbol{\Gamma}_{ss'}, \qquad (5.75)$$

$$\boldsymbol{\chi}_{ss'} \equiv \frac{1}{m_s} \partial_{\mathbf{v}} \ln f_s - \frac{1}{m_{s'}} \partial_{\mathbf{v}'} \ln f_{s'}, \quad f_s f_{s'} \boldsymbol{\chi}_{ss'} = \frac{f_{s'}}{m_s} \partial_{\mathbf{v}} f_s - \frac{f_s}{m_{s'}} \partial_{\mathbf{v}'} f_{s'}.$$

In fact, (5.75) pre-dates the Fokker–Planck form of $C_{ss'}$ (5.74), and was the earliest form of the Coulomb collision operator ever derived (Landau, 1936).[21] It retains some advantages over (5.74), e.g. its symmetry leads directly to the conservation of mass, momentum and energy (2.160)–(2.162).

Decomposing f_s and $f_{s'}$ into Maxwellian and perturbed parts, e.g. $f_s = f_{Ms}(1 + \hat{f}_s) + O(\epsilon^2)$, where $\hat{f}_s \equiv f_{s1}/f_{Ms}$, and substituting into (5.75), we obtain the *linearized* collision operator, $C^l_{ss'}$, given by (5.75) with

$$f_s f_{s'} \boldsymbol{\chi}_{ss'} = \frac{1}{m_s} \partial_{\mathbf{v}} \hat{f}_s - \frac{1}{m_{s'}} \partial_{\mathbf{v}'} \hat{f}_{s'} + \left(\frac{\mathbf{v}'}{T_{s'}} - \frac{\mathbf{v}}{T_s} \right) (\hat{f}_s + \hat{f}_{s'}) + O(\epsilon^2). \qquad (5.76)$$

When the test particles are much lighter than the field particle, e.g. for electrons scattering off ions, the latter are largely unaffected by the collisions and remain roughly stationary, $\mathbf{V}_i \approx 0$. In that case, $\mathbf{A}_{ei}$ is small and D_{ei} may be evaluated with $f_i = n_i \delta(\mathbf{v}')$ to yield

[21] The derivation employed the scattering angle expansion (5.38)–(5.39), combined with a cut-off of the angular integral at the minimum and maximum angles given by (5.51) and (5.52).

$$\mathsf{D}_{ei} = \frac{\gamma_{ei} n_i}{m_e^2} \mathbf{U}(\mathbf{v}), \qquad \mathbf{U}(\mathbf{v}) = \frac{\mathsf{I}}{v} - \frac{\mathbf{vv}}{v^3}. \tag{5.77}$$

Inserting into (5.74), we find the *Lorentz* collision operator,[22]

$$\begin{aligned} C_{ei}^0(f_e) &= \frac{\gamma_{ei} n_i}{m_e^2} \partial_{\mathbf{v}} \cdot \mathbf{U}(\mathbf{v}) \cdot \partial_{\mathbf{v}} f_e, & (5.78) \\ &= \frac{\gamma_{ei} n_i}{m_e^2 v^3} \left[\partial_\xi (1 - \xi^2) \partial_\xi + (1 - \xi^2)^{-1} \partial_\phi^2 \right] f_e, & (5.79) \end{aligned}$$

where the second expression follows from writing $\mathbf{v}$ in terms of the speed v, the parallel pitch angle cosine, $\xi = v_\parallel / v = \cos\vartheta = \cos\theta$, and the azimuthal angle in the $\perp$ plane, ϕ.[23] In short, in the case of a large mass ratio, m_i/m_e, the collision operator reduces from integro-differential to purely differential form and f_e evolves mainly due to velocity space diffusion.[24] This allows the kinetic equation to be solved in terms of the eigenvalues of the Lorentz operator, i.e. the *spherical harmonics*, $Y_{lm}(\theta, \phi)$, or for azimuthal symmetry, the *Legendre* polynomials, $P_l(\cos\theta)$.

To take into account the relatively slow ion flow velocity, it is necessary to replace $\mathbf{U}(\mathbf{v})$ in (5.78) by $\mathbf{U}(\mathbf{v} - \mathbf{V}_i)$, or to replace $f_e(\mathbf{v})$ by $f_e(\mathbf{v} - \mathbf{V}_i)$. Both methods produce the following linearized operator

$$\begin{aligned} C_{ei}^l(f_e) &= \frac{\gamma_{ei} n_i}{m_e^2} \partial_{\mathbf{v}} \cdot \left[\mathbf{U}(\mathbf{v}) \cdot \partial_{\mathbf{v}} f_e - 4 \frac{\mathbf{V}_i}{v_{Te}^2 v} f_{Me} \right], & (5.80) \\ &= C_{ei}^0 + \frac{\gamma_{ei} n_i}{m_e^2 v^3} \frac{4 \mathbf{V}_i \cdot \mathbf{v}}{v_{Te}^2} f_{Me}, & (5.81) \end{aligned}$$

where C_{ei}^0 is the Lorentz operator given by (5.78).

5.2.1.3 Balescu–Lenard collision operator

The most accurate derivation of $C_{ss'}$ was carried out by Balescu (1960) and Lenard (1960). This more general, *Balescu–Lenard* operator may be obtained from (5.75) by effecting the substitution

$$\mathbf{U} \cdot \ln \Lambda_{ss'} \quad \Rightarrow \quad \frac{1}{\pi} \int \frac{\mathbf{kk}\delta(\mathbf{k} \cdot \Delta\mathbf{v})}{k^4 |\mathcal{K}(\mathbf{k} \cdot \mathbf{v}, \mathbf{k})|^2} d\mathbf{k}, \tag{5.82}$$

where $\mathcal{K}(\mathbf{k} \cdot \mathbf{v}, \mathbf{k})$ is the dielectric function defined in Section 4.4. The derivation, a good account of which may be found in Hazeltine and Waelbroeck (2004, App. A), relies on calculating the 2-particle correlation $g_{ss'}$. In the large $\ln \Lambda_{ss'}$ limit,

[22] The superscript 0 refers to the assumption that the field particles (ions) remain stationary.
[23] The choice of the parallel direction (axis) is of course arbitrary in an unmagnetized plasma.
[24] Note that the above analysis remains equally valid for scattering of light ions by heavy ions, provided that their mass ratio $m_i/m_i' = A/A'$ is much larger than unity.

$\mathcal{K}(\mathbf{k}\cdot\mathbf{v},\mathbf{k})$ reduces to $\mathcal{K}(0,\mathbf{k}) \approx 1 + (k\lambda_D)^{-2}$ and the *Balescu–Lenard* operator gives the earlier result (5.75).

5.2.2 Test particle dynamics in a plasma

The Fokker–Planck form of $\mathcal{C}_{ss'}$ (5.74) can be used to describe the dynamics of a test particle, s, launched into a collection of field particles, s'. For this purpose, we assume a uniform beam (or an ensemble) of test particles all moving with an initial velocity, $\mathbf{V}_s(0) = \mathbf{V}_{s0}$, and hence an initial distribution, $f_s(\mathbf{v}, 0) = n_s\delta(\mathbf{v} - \mathbf{V}_{s0})$. To calculate the slowing down rate we assume f_s broadens isotropically in velocity space so that its first moment may be obtained by neglecting this broadening altogether, i.e. using $f_s(\mathbf{v}, t) = n_s\delta(\mathbf{w}(t))$, where $\mathbf{w}(t) = \mathbf{v} - \mathbf{V}_s(t)$. Inserting this expression into (5.74), multiplying by $\mathbf{v}$ and integrating over $\mathbf{v}$ yields the deceleration,

$$\mathrm{d}_t\langle\Delta\mathbf{v}\rangle \equiv \mathrm{d}_t\langle\mathbf{v}\rangle_{f_s} = \mathrm{d}_t\mathbf{V}_s = \mathbf{A}_{ss'}(\mathbf{V}_s). \tag{5.83}$$

Similarly, the rates of lateral deflection and parallel dispersion of the test particles in the incident beam are obtained by considering velocity space broadening of $f_s(\mathbf{v}, t)$. It can be shown that f_s broadens diffusively in accordance with the second term in (5.74), evolving as a Gaussian wave packet,

$$f_s(\mathbf{w}, t) = (2\pi)^{-3/2} n_s \exp(-\tfrac{1}{2}\mathbf{w}\cdot\mathsf{M}_{ss'}^{-1}\cdot\mathbf{w})(\det\mathsf{M}_{ss'})^{-1/2}, \tag{5.84}$$

where $\mathsf{M}_{ss'} = \int_0^t \mathsf{D}_{ss'}(\mathbf{V}_s(t'))\mathrm{d}t'$ is the dispersion tensor (Chandrasekhar, 1943). Using this result one finds the desired deflection and dispersion rates,

$$\mathrm{d}_t\langle(\Delta v_\perp)^2\rangle \equiv \mathrm{d}_t\langle w_\perp^2\rangle_{f_s} = \mathrm{d}_t(2T_\perp/m_s) = \hat{\mathbf{e}}_\perp\cdot\mathsf{D}_{ss'}\cdot\hat{\mathbf{e}}_\perp, \tag{5.85}$$

$$\mathrm{d}_t\langle(\Delta v_\parallel)^2\rangle \equiv \mathrm{d}_t\langle w_\parallel^2\rangle_{f_s} = \mathrm{d}_t(T_\parallel/m_s) = \hat{\mathbf{e}}_\parallel\cdot\mathsf{D}_{ss'}\cdot\hat{\mathbf{e}}_\parallel, \tag{5.86}$$

where $\hat{\mathbf{e}}_\parallel$ and $\hat{\mathbf{e}}_\perp$ are unit vectors parallel and perpendicular to the initial test particle velocity $\mathbf{V}_0$. Finally, the *energy loss* rate is found as

$$\mathrm{d}_t\langle v^2\rangle_{f_s} \equiv \mathrm{d}_t\langle|\mathbf{V}_s + \mathbf{w}|^2\rangle_{f_s} = -\nu_{\varepsilon ss'}^t v^2. \tag{5.87}$$

For isotropic field particles, i.e. $f_{s'}(\mathbf{v}') = f_{s'}(v')$, we can use $\mathbf{A}_{ss'}$ and $\mathsf{D}_{ss'}$ (5.63) evaluated at $v = V_s$. Hence, (5.83)–(5.87) reduce to

$$\mathrm{d}_t V_s = -\nu_{ss'}^t(V_s)V_s, \qquad \mathbf{V}_s = V_s\hat{\mathbf{e}}_\parallel, \tag{5.88}$$

$$\mathrm{d}_t\langle w_\perp^2\rangle_{f_s} = \mathrm{d}_t(2T_\perp/m_s) = 2D_{\perp ss'}^t(V_s) = \nu_{\perp ss'}^t(V_s)V_s^2, \tag{5.89}$$

$$\mathrm{d}_t\langle w_\parallel^2\rangle_{f_s} = \mathrm{d}_t(T_\parallel/m_s) = D_{\parallel ss'}^t(V_s) = \nu_{\parallel ss'}^t(V_s)V_s^2, \tag{5.90}$$

$$\mathrm{d}_t\langle v^2\rangle_{f_s} \equiv \mathrm{d}_t(2\varepsilon_s/m_s) = -\nu_{\varepsilon ss'}^t V_s^2, \tag{5.91}$$

where the relaxation rates are given by (5.64)–(5.67), or in the case of Maxwellian field particles by (5.69)–(5.71).[25]

Referring to (5.73) and Fig. 5.1, we observe that in the high energy limit, the rate of test particle slowing down decreases as the cube of its velocity. If the particle is accelerated by a constant force, e.g. by parallel induction, this can lead to a *runaway effect* when the inductive field exceeds a critical value, typically a few per cent of the *Dreicer field*, $E_D = m_e v_{te} \nu_e / e$.[26]

The slowing down expressions (5.88)–(5.91) are also used in determining plasma heating due to ion and electron cyclotron resonant waves, neutral beams and fusion reaction products, i.e. primarily alpha particles.

5.2.3 Collisional momentum exchange

When the test particles are likewise Maxwellian, and the thermal speed of the heavier species is comparable to or smaller than that of the lighter species, then for each combination of s and s', the four frequencies (5.69)–(5.71) reduce to a single characteristic collision rate $\nu_{ss'}$.[27] This rate effectively represents the time required to change the direction of the test particles by a right angle from their initial direction of motion. Hence, the term *collision rate* tacitly assumes *momentum exchange*, unless the exchange of some other quantity, e.g. energy, is explicitly stated.

In analogy with the slowing down rate of a single test particle, given in (5.64), we define $\nu_{ss'}$ as the rate of change of the fluid momentum due to successive binary collisions between species s and s'. This quantity appears naturally when evaluating the friction force $\mathbf{F}_{ss'}$ (2.158), see (5.36),

$$\mathbf{F}_{ss'} \equiv m_s \langle \mathbf{v} \rangle_{C_{ss'}} \equiv \int m_s \mathbf{v} C_{ss'}(f_s, f_{s'}) \mathrm{d}\mathbf{v} = -m_s n_s \nu_{ss'} (\mathbf{V}_s - \mathbf{V}_{s'}) . \tag{5.92}$$

Recall that $\mathbf{F}_{ss'}$ enters the momentum equation for species s (2.170),

$$m_s n_s \mathrm{d}_{t,s} \mathbf{V}_s + \nabla \cdot \mathsf{p}_s = e_s n_s (\mathbf{E} + \mathbf{V}_s \times \mathbf{B}) + \textstyle\sum_{s'} \mathbf{F}_{ss'}, \tag{5.93}$$

where n_s, $\mathbf{V}_s$ and T_s are again defined by (2.142) and (2.144),

$$n_s = \int f_s \mathrm{d}\mathbf{v}, \qquad n_s \mathbf{V}_s = \int \mathbf{v} f_s \mathrm{d}\mathbf{v}, \qquad \tfrac{3}{2} n_s T_s = \int \tfrac{1}{2} m_s |\mathbf{v} - \mathbf{V}_s|^2 f_s \mathrm{d}\mathbf{v}.$$

[25] It is worth noting that the problem of test particle moving in a Maxwellian field particle background was solved half a century ago by Spitzer (1940) and Chandrasekhar (1941, 1942).

[26] The runaway effect, which is manifest primarily in *runaway electrons* on account of their smaller mass, is observed in some plasma 'disruptions'. Such runaway electron beams can then impact on plasma facing components, posing serious challenges to first wall design.

[27] This reduction has led to some differences in the definition of this basic collisional rate in the topical literature; the expressions in Hinton (1983) are recommended as a good benchmark.

Hence, $\nu_{ss'}$ measures the deceleration of $\mathbf{V}_s$ due to collisions with species s',

$$\mathrm{d}_{t,s}\mathbf{V}_s = -\sum_{s'} \nu_{ss'} (\mathbf{V}_s - \mathbf{V}_{s'}), \qquad \mathrm{d}_{t,s} \equiv \partial_t + \mathbf{V}_s \cdot \nabla, \tag{5.94}$$

where we neglect all other terms in (5.93).

To evaluate $\mathbf{F}_{ss'}$ we first combine (5.57) and (5.58), and integrate by parts

$$\mathbf{F}_{ss'} = 2\frac{\gamma_{ss'}}{m_s}\left(1 + \frac{m_s}{m_{s'}}\right)\int h_{s'}(\mathbf{v})\partial_{\mathbf{v}} f_s \mathrm{d}\mathbf{v}. \tag{5.95}$$

We next assume that both test and field particles are Maxwellian,

$$f_s(\mathbf{v}) = f_{Ms}(\mathbf{w}_s; n_s, \mathbf{V}_s, T_s), \quad f_{s'}(\mathbf{v}') = f_{Ms'}(\mathbf{w}_{s'}; n_{s'}, \mathbf{V}_{s'}, T_{s'}), \tag{5.96}$$

and that the relative flow velocity is small compared to the test particle thermal speed, $|\mathbf{V}_s - \mathbf{V}_{s'}| \ll v_{Ts}$. Inserting the above into the definition of $h_{s'}$ (5.59), and transforming variables, we find that $h_{s'}(\mathbf{v}) = h_{s'M}(|\mathbf{v} - \mathbf{V}_{s'}|/v_{Ts'})$, (5.68). Expanding f_s in the small ratio $|\mathbf{V}_s - \mathbf{V}_{s'}|/v_{Ts}$,

$$f_s(\mathbf{v}) = f_{Ms}(\mathbf{v}; n_s, T_s)\left[1 + 2\frac{(\mathbf{V}_s - \mathbf{V}_{s'}) \cdot \mathbf{v}}{v_{Ts}^2}\right], \tag{5.97}$$

and integrating (5.95), we find that $\mathbf{F}_{ss'}$ is proportional to the difference of the two flow velocities, i.e. we recover the form (5.92), with $\nu_{ss'}$ given by,

$$\nu_{ss'} = \frac{8\gamma_{ss'}n_{s'}}{3\sqrt{\pi}m_{ss'}m_s(v_{Ts}^2 + v_{Ts'}^2)^{3/2}}. \tag{5.98}$$

Alternatively, the desired result may be obtained by direct calculation of (5.92) from some approximation of the Coulomb collision operator $C_{ss'}$. The reader is encouraged to check that it is not sufficient to simply insert $v = v_{Ts}$ into (5.69), as the non-linearities in the integral generate errors of order unity; Spitzer (1956) derived a general expression for $\nu_{ss'}$ but included only the thermal velocity of the field particles in his collision rate.[28]

Substituting $\gamma_{ss'}$ from (5.56) into (5.98) gives the following form

$$\nu_{ss'} = \frac{\sqrt{2}n_{s'}Z_s^2 Z_{s'}^2 e^4 \ln \Lambda_{ss'}}{12\pi^{3/2}\epsilon_0^2 m_{ss'}m_s[(T_s/m_s) + (T_{s'}/m_{s'})]^{3/2}}, \tag{5.99}$$

where $Z_s = e_s/e$ and $Z_{s'} = e_{s'}/e$ are the charge states of species s and s', respectively. Writing the reduced mass in (5.35) explicitly, yields

$$\nu_{ss'} = \frac{\sqrt{2}n_{s'}Z_s^2 Z_{s'}^2 e^4 \ln \Lambda_{ss'}}{12\pi^{3/2}\epsilon_0^2 m_s^{1/2} T_s^{3/2}} \frac{1 + m_s/m_{s'}}{[1 + (T_{s'}/m_{s'})/(T_s/m_s)]^{3/2}}. \tag{5.100}$$

28 The more complete expression may be found in most sources, e.g. Hinton (1983).

It follows that $n_s m_s \nu_{ss'} = n_{s'} m_{s'} \nu_{s's}$, consistent with momentum conservation in Coulomb collisions, which requires that $\mathbf{F}_{ss'} + \mathbf{F}_{s's} = 0$. The ratio $\nu_{ss'}/\nu_{s's} = (n_s/n_{s'})(m_{s'}/m_s)$ then implies that light particles are subject to much stronger scattering, and hence deceleration, than the heavy particles.

Let us write down the various combinations of test, s, and field, s', particle species explicitly for a so-called *simple plasma*, i.e. for a plasma with a single ion species, satisfying the quasi-neutrality condition, $n_e = Zn_i$. With multiple ion species, it can be shown that Z in subsequent expressions should be replaced by the *effective charge state*, Z_{eff},

$$n_e Z_{eff} \equiv \sum n_i Z_i^2. \tag{5.101}$$

Selecting $s = e$ and $s' = i$, we obtain the *electron–ion collision rate*, ν_{ei}, by inserting $Z_e = 1$, $Z_i = Z$, $m_i \gg m_e$ and thus $m_{ei} \approx m_e$ into (5.99),

$$\nu_{ei} = \frac{\sqrt{2} n_i Z^2 e^4 \ln \Lambda_{ei}}{12\pi^{3/2}\epsilon_0^2 m_e^{1/2} T_e^{3/2}} = \frac{\sqrt{2} n_e Z e^4 \ln \Lambda_{ei}}{12\pi^{3/2}\epsilon_0^2 m_e^{1/2} T_e^{3/2}}. \tag{5.102}$$

In the above expression we have assumed $T_i/T_e \ll m_i/m_e$, and thus excluded the case when ions are much hotter than electrons such that their thermal speeds become comparable ($v_{Ti} \sim v_{Te}$). The complementary *ion–electron collision rate*, ν_{ie}, is found by setting $s = i$, $s' = e$, and thus inserting $Z_e = 1$, $Z_i = Z$, $m_i \gg m_e$ and $m_{ei} = m_e$ into (5.99),

$$\nu_{ie} = \frac{\sqrt{2} m_e^{1/2} n_e Z^2 e^4 \ln \Lambda_{ie}}{12\pi^{3/2}\epsilon_0^2 m_i T_e^{3/2}} = \left(\frac{m_e}{m_i}\right) Z \nu_{ei}. \tag{5.103}$$

The two rates differ by the product of the ratio of electron and ion masses, which is much smaller than one, and by the ratio of electron and ion densities, which is equal to Z, so that $\nu_{ie}/\nu_{ei} \ll 1$.[29] In other words, the lighter electrons are quickly deflected by collisions with the heavier ions, whereas the course of ions is only gradually modified by collision with electrons.

Let us next consider like-particle collisions. Selecting $s = s' = e$, for which $m_{ee} = m_e/2$ and $Z_e = 1$, we obtain the *electron–electron collision rate*, ν_{ee},

$$\nu_{ee} = \frac{n_e e^4 \ln \Lambda_{ee}}{12\pi^{3/2}\epsilon_0^2 m_e^{1/2} T_e^{3/2}} = \frac{\nu_{ei}}{\sqrt{2}Z} \frac{\ln \Lambda_{ee}}{\ln \Lambda_{ei}} \approx \frac{\nu_{ei}}{\sqrt{2}Z}, \tag{5.104}$$

which differs from ν_{ei} only by the ratio of ion to electron densities ($n_i/n_e = 1/Z$), and the factor of $1/\sqrt{2}$, which comes from the reduced mass and the sum of

[29] Note that by virtue of quasi-neutrality (5.102) and (5.103) satisfy the momentum conservation constraint $m_e n_e \nu_{ei} = m_i n_i \nu_{ie}$.

thermal velocities in (5.98). Thus electrons are deflected by other electrons at a rate comparable to their deflection by ions.

Similarly, the *ion–ion collision rate*, ν_{ii}, in a simple plasma is found by setting $s = s' = i$, and thus inserting $m_{ii} = m_i/2$ and $Z_i = Z$ into (5.99),

$$\nu_{ii} = \frac{n_i Z^4 e^4 \ln \Lambda_{ii}}{12\pi^{3/2}\epsilon_0^2 m_i^{1/2} T_i^{3/2}} \approx \left(\frac{m_e}{m_i}\right)^{1/2} \left(\frac{T_e}{T_i}\right)^{3/2} \frac{Z^2}{\sqrt{2}} \nu_{ei}. \tag{5.105}$$

Thus ions are deflected by other ions much more effectively than by electrons, although the rate of ion–ion collisions is slower than that of electron–electron collisions by the square root for the mass ratio, $(m_i/m_e)^{1/2}$.

The relations between the four collision rates,[30]

$$\nu_{ie} : \nu_{ii} : \nu_{ee} : \nu_{ei} \approx \left(\frac{m_e}{m_i}\right) Z : \left(\frac{m_e}{m_i}\right)^{1/2} \left(\frac{T_e}{T_i}\right)^{3/2} \frac{Z^2}{\sqrt{2}} : \frac{1}{\sqrt{2}Z} : 1, \tag{5.106}$$

clearly express the more efficient scattering of the lighter electrons than that of the heavier ions. The rates are ordered by the square root of the mass ratio, with $e{-}e$ and $e{-}i$ collisions being fastest, $i{-}i$ collisions slower by $(m_i/m_e)^{1/2}$ and i–e collisions slower still by the same ratio.

Finally, we define the *electron* and *ion collision rates* simply as

$$\nu_e \equiv \nu_{ei} = \frac{\sqrt{2} n_i Z^2 e^4 \ln \Lambda_{ei}}{12\pi^{3/2}\epsilon_0^2 m_e^{1/2} T_e^{3/2}}, \qquad \nu_i \equiv \nu_{ii} = \frac{n_i Z^4 e^4 \ln \Lambda_{ii}}{12\pi^{3/2}\epsilon_0^2 m_i^{1/2} T_i^{3/2}}. \tag{5.107}$$

5.2.4 Collisional energy (heat) exchange

The rate of collisional exchange of thermal energy (heat) due to Coulomb collisions can be calculated in a similar manner. The general expression for Maxwellian field and test particles was first derived by Spitzer (1940).

The energy conservation equation for particle species s (2.171),

$$\tfrac{3}{2} \mathrm{d}_{t,s} p_s + \tfrac{5}{2}\, p_s \nabla \cdot \mathbf{V}_s + \boldsymbol{\pi}_s : \nabla \mathbf{V}_s + \nabla \cdot \mathbf{q}_s = \sum \mathcal{Q}_{ss'}, \tag{5.108}$$

includes the *collisional heat exchange* between species s and s', $\mathcal{Q}_{ss'}$ (2.157),

$$\mathcal{Q}_{ss'} \equiv \frac{1}{2} m_s \int |\mathbf{v} - \mathbf{V}_s|^2 C_{ss'}(f_s, f_{s'}) \mathrm{d}\mathbf{v}. \tag{5.109}$$

This quantity has already been expressed in terms of the Fokker–Planck coefficients as (5.57). Combining with (5.58) and integrating by parts yields

$$\mathcal{Q}_{ss'} = 2\frac{\gamma_{ss'}}{m_s} \int \left[\left(1 + \frac{m_s}{m_{s'}}\right) (\mathbf{v} - \mathbf{V}_s) \cdot \partial_{\mathbf{v}} h_{s'} + h_{s'} \right] f_s(\mathbf{v}) \mathrm{d}\mathbf{v}. \tag{5.110}$$

[30] The approximate sign reflects small differences in the Coulomb logarithms, see Huba (2006).

The integral does not depend on the relative flow velocity, $\mathbf{V}_s - \mathbf{V}_{s'}$, and can therefore be evaluated by assuming Maxwellian test and field particles with $\mathbf{V}_s = \mathbf{V}_{s'} = 0$. Performing the integration, we obtain

$$\mathcal{Q}_{ss'} = \frac{8}{\sqrt{\pi}} \frac{\gamma_{ss'} n_s n_{s'} (T_{s'} - T_s)}{m_s m_{s'} (v_{Ts}^2 + v_{Ts'}^2)^{3/2}} = 3\left(\frac{m_s}{m_s + m_{s'}}\right) \nu_{ss'} n_s (T_{s'} - T_s).$$

It is customary to define the *thermal equilibration rate*, $\nu_{ss'}^{\varepsilon}$, as the rate at which T_s and $T_{s'}$ converge to a common value due to collisions,

$$\mathrm{d}_t T_s \equiv \textstyle\sum_{s'} \nu_{ss'}^{\varepsilon} (T_{s'} - T_s). \tag{5.111}$$

Inserting (5.111) into (5.108) and neglecting all other terms on the left-hand side, gives the required rate,

$$\nu_{ss'}^{\varepsilon} = \frac{16}{3\sqrt{\pi}} \frac{\gamma_{ss'} n_{s'}}{m_s m_{s'} (v_{Ts}^2 + v_{Ts'}^2)^{3/2}} = 2\left(\frac{m_s}{m_s + m_{s'}}\right) \nu_{ss'}. \tag{5.112}$$

We conclude that for a mass ratio of order unity the momentum and heat exchange rates are comparable, while for disparate masses the lighter particles gain/lose energy much slower than they gain/lose momentum due to collisions with the heavier particles.[31]

For like-particle collisions ($s = s'$), ν_{ss}^{ε} does not represent the rate of heat exchange between two Maxwellian fluids since, by (5.111), $\mathcal{Q}_{ss} = 0$, and no heat is exchanged. Rather, it measures the rate at which species s approach local thermodynamic equilibrium, i.e. the rate at which $f_s(v) = f_{s'}(v) = f(v)$ relaxes to a Maxwellian, $f_{Ms}(v)$. In this sense, the label *thermalization rate* is better suited to ν_{ss}^{ε}.[32]

Let us again consider all combinations of s and s' for a simple plasma. The *electron–ion* and *ion–electron equilibration rates* are thus found as

$$\nu_{ei}^{\varepsilon} = 2\left(\frac{m_e}{m_i}\right) \nu_{ei} = \frac{\sqrt{2} m_e^{1/2} n_e Z e^4 \ln \Lambda_{ei}}{6\pi^{3/2} \epsilon_0^2 m_i T_e^{3/2}}, \tag{5.113}$$

$$\nu_{ie}^{\varepsilon} = 2\nu_{ie} = 2\left(\frac{m_e}{m_i}\right) Z \nu_{ei} = Z \nu_{ei}^{\varepsilon}. \tag{5.114}$$

Similarly, the *electron–electron* and *ion–ion thermalization rates* are found to be identical to the momentum relaxation rates given previously,

$$\nu_{ee}^{\varepsilon} = \nu_{ee} = \nu_e, \qquad \nu_{ii}^{\varepsilon} = \nu_{ii} = \nu_i. \tag{5.115}$$

[31] Also note that $\mathcal{Q}_{ss'} + \mathcal{Q}_{s's} = 0$, consistent with energy conservation in Coulomb collisions.
[32] It is worth noting that the former is identical to the momentum exchange rate, ν_{ss}.

The ratios between the four thermal equilibration rates can be related to those between the four collision or momentum exchange rates as follows:

$$\nu_{ie}^{\varepsilon} : \nu_{ii}^{\varepsilon} : \nu_{ee}^{\varepsilon} : \nu_{ei}^{\varepsilon} \approx 2\nu_{ie} : \nu_{ii} : \nu_{ee} : 2\left(\frac{m_e}{m_i}\right)\nu_{ei}. \tag{5.116}$$

The only significant difference in the relative magnitude of the heat and momentum exchange rates occurs for electrons colliding with ions, which differ by twice the mass ratio. Combining (5.116) with (5.106) we find

$$\nu_{ie}^{\varepsilon} : \nu_{ii}^{\varepsilon} : \nu_{ee}^{\varepsilon} : \nu_{ei}^{\varepsilon} \approx Z : \left(\frac{m_i}{m_e}\right)^{1/2}\left(\frac{T_e}{T_i}\right)^{3/2}\frac{Z^2}{2^{3/2}} : \left(\frac{m_i}{m_e}\right)\frac{1}{2^{3/2}Z} : 1, \tag{5.117}$$

which confirms the intuitive result that electrons thermalize, or approach local thermodynamic equilibrium, much faster than ions, and that ions thermalize much faster than the rate of thermal equipartition between the ions and electrons. Perhaps less intuitively, the rates of these three processes differ roughly by the square root of the ion-to-electron mass ratio.

5.3 Collisional transport in a plasma

We are now ready to quantify collisional transport in a plasma. This topic is comprehensively summarized in review articles by Braginskii (1965) and Hinton (1983), as well as research monographs by Balescu (1988a) and Helander and Sigmar (2002). Below, we briefly present the results, dividing the discussion into unmagnetized and magnetized plasmas.

5.3.1 Collisional transport in an unmagnetized plasma

To apply the Chapman–Enskog expansion, developed in Section 5.1 for a neutral gas, to the case of an unmagnetized plasma, requires two modifications: (i) $C_{ss'}$ must be chosen as the Coulomb collision operator, in either the L–B (5.75) or F–P (5.74) forms; and (ii) the summation over s' must involve both ions and electrons, so that for a simple plasma one obtains

$$C_e(f_e) = C_{ee}(f_e) + C_{ei}(f_e, f_i), \qquad C_i(f_i) = C_{ii}(f_i) + C_{ie}(f_i, f_e). \tag{5.118}$$

The relative magnitude of the four terms can be inferred from the collision rates (5.106). In particular, $C_{ee} \sim C_{ei}$ in a low-Z plasma so that electron scattering by both ions and other electrons must be considered, while $C_{ii} \gg C_{ie}$, so that ion self-scattering dominates and collisions with electrons may be neglected.[33] Otherwise, the procedure is much the same as for a neutral gas and relies on solving the plasma kinetic equation order by order, at each stage using the kth order distribution function f_{ks} to calculate the next higher-order correction $f_{(k+1)s}$. Thus to lowest order, we may insert $f_{s0} = f_{Ms}$ on the left-hand side of (2.101), from which the slow temporal evolution and both terms relating to particle gyration may be neglected. On the right-hand side, we may replace $\mathcal{C}_s(f_s)$ by the linearized operator $\mathcal{C}_s^l(f_{s1})$ to obtain an implicit linear PDE for $\hat{f}_s = f_{s1}/f_{Ms}$,

$$(\mathbf{v} \cdot \nabla + \mathbf{a}_s \cdot \partial_{\mathbf{v}}) f_{Ms} = \mathcal{C}_s^l(f_{s1}). \tag{5.119}$$

This type of equation is generally known as a *Spitzer problem*. For instance, when $\mathbf{V}_s = \mathbf{V}_{s'} = 0$, the left-hand side of (5.119) can be evaluated as

$$(\mathbf{v} \cdot \nabla + e_s \mathbf{E} \cdot \mathbf{v} \partial_{\mathcal{E}}) f_{Ms} = f_{Ms} \mathbf{v} \cdot (\nabla \ln f_{Ms} - e_s \mathbf{E}/T_s). \tag{5.120}$$

The expression in brackets can then be computed with the help of (A.10),

$$\nabla \ln f_{Ms} - e_s \mathbf{E}/T_s = \mathbf{A}_{s1} + \left(x^2 - \tfrac{5}{2}\right) \mathbf{A}_{s2}, \quad x \equiv v/v_{Ts}, \tag{5.121}$$

and thus expressed in terms of the thermodynamic forces $\mathbf{A}_{s1}$ and $\mathbf{A}_{s2}$,

$$\mathbf{A}_{s1} = \nabla \ln p_s - e_s \mathbf{E}/T_s, \qquad \mathbf{A}_{s2} = \nabla \ln T_s. \tag{5.122}$$

The right-hand side of (5.119) determines the perturbation $\hat{f}_s = \sum \hat{f}_{ss'}$ in response to these forces, which can then be used to calculate the resulting flows. The perturbations $\hat{f}_{ss'}$ are generally found to be inversely proportional to $\nu_{ss'}$ and directly proportional to a linear combination of $\mathbf{A}_{sl}$, i.e. $\hat{f}_{ss'} \propto \nu_{ss'}^{-1} \sum g_l(x) \mathbf{A}_{sl}$. The solution of the electron Spitzer problem is complicated by the presence of like-particle collisions in $\mathcal{C}_e$, however, which can only be neglected in the Lorentz limit ($Z \to \infty$). In this limit, $\hat{f}_{ei}$ can be calculated by replacing $\mathcal{C}_{ei}^l$ in (5.119) by the Lorentz operator (5.80),

$$\nu_{ei} x^{-3} \partial_\xi (1 - \xi^2) \partial_\xi \hat{f}_{ei} = v_{Te} x \xi \left[A_{e1} + \left(x^2 - \tfrac{5}{2}\right) A_{e2}\right], \tag{5.123}$$

where we moved $\mathcal{C}_{ei}^l(\hat{f}_{ei})$ to the left-hand side, since it contains the unknown function $\hat{f}_{ei}$ which we are trying to determine. The forces A_{e1} and A_{e2} are just the projections of (5.122) along the same axis used to define the pitch angle cosine ξ. The solution follows upon double integration of (5.123),

$$\hat{f}_{ei} = -(v_{Te}/2\nu_{ei}) x^4 \xi \left[A_{e1} + \left(x^2 - \tfrac{5}{2}\right) A_{e2}\right], \tag{5.124}$$

[33] It is easy to show that dynamical friction dominates in ion–electron scattering, so that to a good approximation ions experience only an effective drag due to collisions with electrons. To include this effect, it is sufficient to replace the electric field $\mathbf{E}$ appearing in (2.92) by $\mathbf{E} + \mathbf{F}_{ie}/e_i n_i$.

and indicates that the normalized perturbation $\hat{f}_{ei}$ increases monotonically with x as the square or even the cube of the electron energy. Since the number of energetic electrons in a Maxwellian distribution, $f_{Me}(v)$, falls off exponentially with energy for $x \gg 1$, the actual perturbation, $f_{ei1} = \hat{f}_{ei} f_{Me}$, is a product of increasing and decreasing functions of x, and therefore peaks at some intermediate value, $x_c \sim 1$ (typically, $x_c \approx 3$–5).[34]

The above perturbation can then be used to calculate the flows of electrons, $\Gamma_{e1} \equiv nV_e$, and electron heat, $\Gamma_{e2} \equiv q_e/T_e$, both of which can be written in the form, $\mathbf{\Gamma} = -\boldsymbol{\mathcal{L}} \cdot \mathbf{A}$, or $\Gamma_k = -\sum \mathcal{L}_{kl} A_l$, where $\mathcal{L}_{kl} \sim p_e/m_e \nu_{ei}$. We thus obtain the desired relation between thermodynamic forces, $\mathbf{A}$, and flows, $\mathbf{\Gamma}$, which are linked by the components $\mathcal{L}_{kl}$ of the transport matrix $\boldsymbol{\mathcal{L}}$.

Let us consider the results of the application of this method to a simple plasma with arbitrary ion charge (Spitzer, 1956). There are several approaches of solving the associated electron and ion Spitzer problems in the collisional limit ($\Delta_s \ll 1$), all of which amount to a Chapman–Enskog expansion in the parallel direction (Helander and Sigmar, 2002):

(i) a numerical solution, e.g. using finite difference discretization;

(ii) a variational (trial function) technique, which exploits the self-adjoint property of the linearized collision operator, $C_s^l = C_{ss}^l + \sum C_{ss'}$,

$$\int \hat{g} C_s^l (f_{Ms} \hat{h}) \mathrm{d}\mathbf{v} = \int \hat{h} C_s^l (f_{Ms} \hat{g}) \mathrm{d}\mathbf{v}; \tag{5.125}$$

(iii) an approximation of the like-particle collision operator by

$$C_{ss}^l(f_s) = \nu_{\perp ss}^t \left[C_{ss}^0(f_s) + 2\frac{\mathbf{v} \cdot \mathbf{V}^*}{v_{Ts}^2} f_{Ms} \right], \qquad \mathbf{V}^* \equiv \frac{3}{2} \frac{\langle \mathbf{v} \nu_{\perp ss}^t \rangle_{f_s}}{\langle x^2 \nu_{\perp ss}^t \rangle_{f_{Ms}}},$$

where $\nu_{\perp ss}^t$ is the deflection rate and C_{ss}^0 is the Lorentz operator;

(iv) an expansion of f_s in orthogonal polynomials, e.g. the *Sonine* (generalized *Laguerre*) polynomials with an index $m = 3/2$,

$$L_k^{(3/2)}(x) \equiv \frac{\mathrm{e}^x}{k! x^{3/2}} \mathrm{d}_x^k \left(x^{k+3/2} \mathrm{e}^{-x} \right), \qquad \hat{f}_s = \frac{v_\parallel}{v_{ts}^2} \sum u_{sk} L_k^{(3/2)} \left(x_s^2 \right),$$

where u_{sk} are coefficients with a dimension of velocity, e.g. $u_{s0} = V_s$, $u_{s1} = -\frac{2}{5} q_s/p_s$; such an expansion transforms the integro-differential Spitzer problem into a set of coupled algebraic equations,

$$\sum \left(\nu_{ss'}/v_{ts}^2 \right) \left(M_{ss'}^{kl} u_{sl} + N_{ss'}^{kl} u_{s'l} \right) = A_{s1}\delta_{k0} - \frac{5}{2} A_{s2} \delta_{k1}, \tag{5.126}$$

where $M_{ss'}^{kl}$ and $N_{ss'}^{kl}$ are the matrix elements of the collision operator.

[34] As a consequence of $x_c \sim 3$–5, the heat flow is dominated by supra-thermal electrons.

Table 5.1. *Classical electron coefficients as a function of ion charge Z.*

Z	r_1	r_2	r_3	r_4	$L_{11} = r_1$	$L_{12} = L_{21} = r_1 r_2$	$L_{22} = r_1 r_2^2 + r_3$
1	1.975	0.703	3.197	4.664	1.975	1.389	4.174
2	2.320	0.908	4.916	3.957	2.320	2.107	2.910
4	2.665	1.091	6.972	3.603	2.665	2.910	10.15
16	3.132	1.346	10.63	3.338	3.132	4.216	16.31
∞	3.395	1.5	13.58	3.25	3.395	5.093	21.22

The earliest solution (Spitzer, 1956) was in fact obtained using technique (i). It was subsequently confirmed by the analytic approaches (ii)–(iv), most notably the Sonine polynomial expansion (Braginskii, 1965).

Having calculated $\hat{f}_i$ and $\hat{f}_e$, the next task is the evaluation of moments of f_{1i} and f_{1e}. The results are conveniently expressed in terms of electron transport coefficients, r_k and L_{kl}, which can be used to construct the transport matrix, $\mathcal{L}$, and thus to relate the forces, $\mathbf{A}_k$, and flows, $\mathbf{\Gamma}_l$, see below. These are tabulated in Table 5.1 as a function of the ion charge Z (Hinton, 1983). They exhibit a weak dependence on Z due to the changing balance between dynamical friction and velocity space diffusion, the latter becoming increasingly dominant at higher Z, e.g. r_3 may be approximated (to within 2% accuracy) over the range $1 \le Z \le 16$ as $r_3 = 3.153 + 2.7 \ln Z$, while r_4 is given exactly by the formula $r_4 = 3.25 + \sqrt{2}/Z$.

One thus obtains the *friction force* between the electrons and ions,

$$\mathbf{F}_e = \mathbf{F}_{ei} = -\mathbf{F}_{ie} = -\mathbf{F}_i = \mathbf{F}_V + \mathbf{F}_{eT}, \qquad \mathbf{F}_{ee} = \mathbf{F}_{ii} = 0 \tag{5.127}$$

where $\mathbf{F}_V$ and $\mathbf{F}_{eT}$ are *frictional* and *thermal* forces on the electron fluid,

$$\mathbf{F}_V = en_e\mathbf{J}/\sigma_e = -e^2 n_e^2 \Delta\mathbf{V}/\sigma_e = -r_1^{-1} m_e n_e \nu_e \Delta\mathbf{V}, \tag{5.128}$$

$$\mathbf{F}_{eT} = -r_2 n_e \nabla T_e, \qquad \mathbf{J} = -en\Delta\mathbf{V} = \sigma_e \mathbf{E} + \cdots. \tag{5.129}$$

Here $\sigma_e = r_1 e^2 n_e / m_e \nu_e$ is the *electrical conductivity* due to the combined effect of $e{-}e$ and $e{-}i$ collisions[35] and $\Delta\mathbf{V} = \mathbf{V}_e - \mathbf{V}_i$ is the relative flow velocity of electron and ion fluids. The electrical conductivity, which represents the flow of charge due to an electric field in the absence of density and temperature gradients, $\mathbf{J} = \sigma_e \mathbf{E}$, is closely related to the particle diffusivity,

$$D = r_1 v_{te}^2/\nu_e = \sigma_e T_e / e^2 n_e, \qquad \mathbf{V} = -D\nabla \ln n_e + \cdots, \tag{5.130}$$

which measures the CM flow of mass due to a density gradient for uniform temperature and electric potential. The former corresponds to $\mathbf{A}_{s1} = e_s\mathbf{E}/T_s$, $\mathbf{A}_{s2} = 0$,

[35] It is frequently replaced by its inverse $\eta = 1/\sigma_e$, known as the the electrical *resistivity*, which enters into resistive Ohm's law (2.255) and determines the magnetic diffusivity, $D_\eta = \eta/\mu_0$.

the latter to $\mathbf{A}_{s1} = \nabla \ln n_e$, $\mathbf{A}_{s2} = 0$, i.e. σ_e and D measure the response to the two components of $\mathbf{A}_{s1}$, which explains the same pre-factor r_1 appearing in both terms.

In a similar fashion, one finds the heat exchange terms,

$$\mathcal{Q}_i = \tfrac{3}{2}\nu_{ei}^{\epsilon} n_i (T_e - T_i), \qquad \mathcal{Q}_e + \mathcal{Q}_i + \mathbf{F}_e \cdot \Delta\mathbf{V} = 0, \tag{5.131}$$

where ν_{ei}^{ϵ} is given by (5.113); the viscous stresses,[36]

$$\boldsymbol{\pi}_s = -\mu_s \left[2\{\nabla \mathbf{V}_s\}^* - \tfrac{2}{3}(\nabla \cdot \mathbf{V}_s)\mathsf{I}\right] = -\mu_s \mathsf{W}_s, \tag{5.132}$$

$$\mu_i = 0.96 p_i/\nu_i, \qquad \mu_e = 0.73 p_e/\nu_e, \tag{5.133}$$

where W_s is the rate-of-strain tensor given by (4.5); and the heat flows,

$$\mathbf{q}_i = -\kappa_i \nabla T_i, \qquad \mathbf{q}_e = \mathbf{q}_{eT} + \mathbf{q}_{eV} = -\kappa_e \nabla T_e + r_2 p_e \Delta\mathbf{V}, \tag{5.134}$$

$$\kappa_i = n_i \chi_i = 3.91 p_i/m_i \nu_i, \qquad \kappa_e = n_e \chi_e = r_3 p_e/m_e \nu_e, \tag{5.135}$$

where $\mathbf{q}_{eT}$ and $\mathbf{q}_{eV}$ are the electron *thermal gradient* and *frictional* heat flows, respectively. The *thermoelectric* flows, $\mathbf{F}_{eT}$ and $\mathbf{q}_{eV}$, indicate that an electric current can be driven by an electron temperature gradient and that electron heat flow can be driven by an electric current. The fact that the numerical constants in both terms are identical, reflects the so-called *Onsager symmetry*, which states that the transport matrix $\mathcal{L}$ must be symmetrical.

5.3.2 Collisional transport in a cylindrical plasma

To repeat the above analysis for a magnetized, cylindrical plasma, we must again solve the ion and electron Spitzer problems (5.119), but this time retain the Lorentz force, $e_s \mathbf{v} \times \mathbf{B}$, in the acceleration $\mathbf{a}_s$ appearing in (2.99).[37] Transforming the kinetic equation to the fluid frame of reference and evaluating the total derivative of $f_{s0} = f_{Ms}$ in analogy with (5.119)–(5.122), we obtain the 'thermodynamic force' form for the two Spitzer problems,

$$\left(C_{ii} - \frac{e}{m_i}\mathbf{v} \times \mathbf{B} \cdot \partial_{\mathbf{v}}\right) f_{i1} = \left[\left(x_i^2 - \frac{5}{2}\right)\mathbf{v} \cdot \nabla \ln T_i + \left(\mathbf{vv} - \frac{1}{3}v^2\mathsf{I}\right) : \frac{\mathsf{W}_i}{v_{Ti}^2}\right] f_{Mi},$$

$$\left(C_{ee} + C_{ei}^0 + \frac{e}{m_e}\mathbf{v} \times \mathbf{B} \cdot \partial_{\mathbf{v}}\right) f_{e1} \tag{5.136}$$

$$= \left[\left(x_e^2 - \frac{5}{2}\right)\mathbf{v} \cdot \nabla \ln T_e + \mathbf{v} \cdot \left(\frac{\mathbf{F}_e}{p_e} + \frac{\nu_e \Delta\mathbf{V}}{v_{te}^2}\right) + \left(\mathbf{vv} - \frac{1}{3}v^2\mathsf{I}\right) : \frac{\mathsf{W}_e}{v_{Te}^2}\right] f_{Me}.$$

[36] Note that $\pi_e \ll \pi_i$ due to the large mass ratio $m_i/m_e \gg 1$.

[37] Once again, we restrict the discussion to a simple plasma with singly charged ions. Since we are ultimately interested in the slow evolution of 'equilibrium' quantities, we assume the existence of magnetic flux surfaces and hence of the three magnetic directions $\|$, $\wedge$ and $\perp$. The reader may wish to review the material covered in Sections 2.3.2.1 and 2.4.2.1–2.4.2.5.

Here we neglected the divergence of the viscous stress, $\nabla \cdot \boldsymbol{\pi}_s$, which is smaller by $O(\delta_s, \nu_s/\Omega_s)$ from the other terms appearing inside the square brackets. Aside from the thermodynamic forces, $\mathbf{A}_{1e}, \mathbf{A}_{2e}, \mathbf{A}_{2i}$, represented by the first term (two terms) inside the square brackets in the ion (electron) equation, we also find additional driving terms due to gradients in the flow velocities $\mathbf{V}_s$, i.e. the rate-of-strain tensors, W_s. The absence of the adiabatic force, $\mathbf{A}_{1i}$, can be explained by the already mentioned transformation, $\mathbf{E} \to \mathbf{E} - \mathbf{F}_e/en_e$, which removes the friction force from the ion equation. In effect, this states that only electrons participate in the plasma adiabatic response.

The four techniques (i)–(iv) described on page 186 are easily extended to the solution of these two problems. Moreover, the assumed transport ordering (5.1), allows us to replace the plasma kinetic equation (2.99)–(2.102) for the particle distribution $f_s(\mathbf{x}, \mathbf{v})$ by the drift-kinetic equation (2.111)–(2.118) for the guiding centre distribution, $\overline{f}_s(\mathbf{x} - \mathbf{r}, \mathbf{v}) = \overline{f}_s(\mathbf{R}, \mathcal{E}, \mu)$. The γ-dependent correction to the Maxwellian, $\widetilde{f}_s \approx -\mathbf{r} \cdot \nabla f_{Ms}$, see (2.107), can then be integrated to yield the gyro-flows and gyro-stresses, e.g.

$$\begin{pmatrix} n_s \mathbf{V}_{\wedge s} \\ \boldsymbol{\pi}_{\wedge s} \\ \mathbf{q}_{\wedge s} \end{pmatrix} = -\int \begin{pmatrix} \mathbf{v} \\ m_s(\mathbf{w}_s\mathbf{w}_s - \frac{1}{3}\mathsf{I}w_s^2) \\ (\frac{1}{2}m_s w_s^2 - \frac{5}{2}T_s)\mathbf{w}_s \end{pmatrix} \mathbf{r} \cdot \nabla f_{Ms} d\mathbf{v}. \tag{5.137}$$

We can also make use of $\widetilde{f}_s$ to evaluate the associated collisional moments $\mathbf{C}_{si}$. By inserting these moments into (2.199)–(2.202), e.g. into

$$\mathbf{V}_{\perp s} = \mathbf{V}_E + \mathbf{V}_{*s} + \frac{\mathbf{b}}{\Omega_s} \times \left(\mathrm{d}_{t,s}\mathbf{V}_s + \frac{\nabla \cdot \boldsymbol{\pi}_s}{m_s n_s} + \frac{\mathbf{F}_s}{n_s m_s} \right), \tag{5.138}$$

and retaining only $O(\delta_s^2)$ terms, we obtain the desired *collisional flows* and *stresses*, which we denote with a superscript c, i.e. $\mathbf{V}^c_{\perp s}, \mathsf{P}^c_{\perp s}, \mathbf{Q}^c_{\perp s}$. Such perpendicular collisional flows, which include both radial and diamagnetic components, are collectively referred to as *classical collisional flows* and the process itself as *classical collisional transport*.

We are particularly interested in the radial components of these flows and stresses ($\mathbf{V}^c_{\perp s}, \mathsf{P}^c_{\perp s}, \mathbf{Q}^c_{\perp s}, \ldots$), which alone affect the confinement properties of a plasma with nested flux surfaces and are directly responsible for particle and power exhaust.[38] To understand the origin of such flows, consider the radial particle flow $V^c_{\perp s}$, defined as the radial projection of $\mathbf{V}^c_{\perp s}$,

[38] For now, we restrict the discussion to a laminar plasma, i.e. assume the absence of turbulent advection. In most applications, the term classical flow is restricted to the radial component of the perpendicular collisional flow. The more general usage only becomes necessary when (i) flux surfaces are destroyed, (ii) when considering transport on open field lines, or (iii) when considering strong field aligned perturbations, i.e. plasma filaments.

$$V^c_{\perp s} = \hat{\mathbf{e}}_\perp \cdot \mathbf{V}^c_{\perp s} \approx -\hat{\mathbf{e}}_\perp \cdot \frac{\mathbf{b} \times \mathbf{F}_s}{n_s m_s \Omega_s} = \frac{F_{\wedge s}}{n_s m_s \Omega_s}. \quad (5.139)$$

The final expression indicates that a *radial* collisional flow of species s originates in the *diamagnetic* component of the friction force $\mathbf{F}_s$, i.e. in the collisional drag between counter-streaming ion and electron gyro-flows! On the microscopic level, it represents a gravitational drift of guiding centres, $\mathbf{v}^c_{\perp g}$ (2.15), due to the diamagnetic component of the acceleration, $\hat{\mathbf{e}}_\wedge \cdot \mathbf{g}_s = F_{\wedge s}/m_s$, in response to the frictional force, $F_{\wedge s}$, experienced by the gyrating particles. Hence, this drift may be included by effecting a substitution $\mathbf{E} \rightarrow \mathbf{E} + \mathbf{F}_s/e_s n_s$, which gives rise to an equivalent $\mathbf{E} \times \mathbf{B}$ drift.[39]

There are two important conclusions to be drawn from (5.139) and the properties of $\mathbf{F}_{ss'}$, e.g. (5.92): (i) since $\mathbf{F}_{ee} = \mathbf{F}_{ii} = 0$, like-particle collisions evidently do not contribute to radial transport of mass, which is therefore determined solely by unlike-particle collisions;[40] (ii) since $\mathbf{F}_e + \mathbf{F}_i = 0$, one finds $V^c_{\perp i} = V^c_{\perp e}$ and $J^c_\perp = 0$, i.e. there is no net outflow of charge and the flow is said to be *ambipolar*. It is significant that the above conclusions are independent of the details of the Coulomb collision operator $\mathcal{C}_{ss'}$.

In going from (5.138) to (5.139), we retained only the friction force, $\mathbf{F}_s$, and neglected the inertial and viscous forces, $m_s \mathrm{d}_{t,s} \mathbf{V}_s + \nabla \cdot \boldsymbol{\pi}_s / n_s$, which can be shown to be smaller than $\mathbf{F}_s$ by the ratio $(m_i/m_e)^{1/2}\delta_i^2$. Hence,

$$V^{visc}_{\perp i}/V^{fric}_{\perp i} \sim (m_i/m_e)^{1/2}\delta_i^2 \ll 1, \qquad \Leftarrow \qquad \delta_i \ll (m_e/m_i)^{1/4} \sim 0.1,$$

which states that the viscous and inertial classical flows can be neglected provided that δ_i is sufficiently small.[41] However, these flows must be retained when calculating the net radial current, which to lowest order is given by

$$J_\perp = \sum e_s n_s V_{\perp s} = -B^{-1} \sum m_s n_s \mathrm{d}_{t,s} V_{\wedge s} + (\nabla \cdot \boldsymbol{\pi}_s)_\wedge. \quad (5.140)$$

This current generates a radial electric field in accordance with Maxwell's displacement current relation, $J_\perp = -\epsilon_0 \partial_t E_\perp$, which then evolves in response to the diamagnetic inertial and viscous flows, see Section 5.3.2.3.

To obtain the γ-dependent perturbation $\widetilde{f}_{1s}$, which can then be used to calculate $\mathbf{F}_s$ and higher collisional moments, we insert (A.10) into (2.107),

$$\hat{f}_s = \frac{\widetilde{f}_{1s}}{f_{Ms}} = -\mathbf{r} \cdot \nabla_\perp \ln f_{Ms} = -\mathbf{r} \cdot \left[\nabla_\perp \ln p_s + \frac{e_s \mathbf{E}_\perp}{T_s} + \left(x_s^2 - \frac{5}{2} \right) \nabla_\perp \ln T_s \right].$$

[39] Note that the drift $\mathbf{v}^c_{\perp g}$ and flow $\mathbf{V}^c_{\perp s}$ provide $O(\delta_s^2)$ terms appearing in (2.37) and (2.208).

[40] The same is not true for diffusion of momentum or heat as measured by $\mathsf{P}^c_{\perp s}$ and $\mathbf{Q}^c_{\perp s}$.

[41] This condition is satisfied in the core plasma, where $\delta_i \sim \rho_{ti}/a = \rho_{i*} \sim 10^{-3}$, but can be violated in regions of steep radial gradients, e.g. in the so-called *transport barriers*, either in the *interior* of the plasma (ITB) or near the edge (ETB), in which $L_\perp \ll a$ and $\delta_i \sim 0.01 - 0.1$.

The collisional friction between ions and electrons is then found as

$$\mathbf{F}_{\perp ei} = \frac{n_e \nu_e}{\Omega_e}\left[\nabla_\perp(p_i + p_e) - \frac{3}{2}\mathbf{b}\times\nabla_\perp T_e\right], \tag{5.141}$$

which combined with (5.139) gives the radial classical flow,

$$V_\perp^c = -\frac{F_{\wedge ei}}{n_e m_e \Omega_e} = \frac{\nu_e}{n_e m_e \Omega_e^2}\left[\nabla_\perp(p_i + p_e) - \frac{3}{2}\nabla_\perp T_e\right]. \tag{5.142}$$

When the temperatures are uniform, this yields the well-known *Fick's law*,

$$V_\perp^c = -D_\perp^c \nabla_\perp \ln n_e, \quad D_\perp^c = \frac{\nu_e(T_i + T_e)}{m_e \Omega_e^2} \approx 2\nu_e \rho_{te}^2 = 2D_{\perp e}, \tag{5.143}$$

where $D_\perp^c$ is the classical diffusivity and $\rho_{te} = v_{te}/\Omega_e$ is the electron thermal gyroradius. Based on this simple example, we may conclude that collisional flows tend to relax the gradients which drive them and are thus dissipative in nature.[42] This is confirmed by the fact that the entropy production rate (A.5) is linearly proportional to the collision frequency, $\Theta_s \sim \mathbf{\Gamma}_s \cdot \mathbf{A}_s \propto \nu_s$.

It is instructive to once again calculate the perturbation to the electron distribution $\hat{f}_{ei}$ due to collisions with stationary ions with infinite charge ($Z \to \infty$). In this limit, electron scattering is dominated by velocity space diffusion and $C_{ei}(f_{e1})$ can be approximated by the Lorentz operator (5.80). The plasma kinetic equation can then be solved for $\hat{f}_{ei}$, with the result[43]

$$\hat{f}_{ei} \equiv \hat{f}_\parallel + \hat{f}_\wedge + \hat{f}_\perp = -\left(\frac{\mathbf{v}_\parallel}{\nu_e} - \frac{\mathbf{b}\times\mathbf{v}}{\Omega_e} + \frac{\nu_e \mathbf{v}_\perp}{\Omega_e^2}\right)\cdot\left[\mathbf{A}_{e1} + \left(x_e^2 - \frac{5}{2}\right)\mathbf{A}_{e2}\right].$$

Evaluating the integrals in the definitions of $\mathbf{V}_e$ and $\mathbf{q}_e$, then yields

$$V_{\parallel e} = -(r_1/\nu_e m_e)(\nabla_\parallel p_e/n_e + eE_\parallel + r_2\nabla_\parallel T_e),$$
$$\mathbf{V}_{\wedge e} = \mathbf{V}_E + \mathbf{V}_{*e}, \qquad \mathbf{V}_{\perp e} = -(\nu_e/m_e\Omega_e^2)(\nabla_\perp p_e/n_e + e\mathbf{E}_\perp - \tfrac{3}{2}\nabla_\perp T_e),$$
$$q_{\parallel e} = r_2 p_e V_{\parallel e} - r_3(p_e/\nu_e m_e)\nabla_\parallel T_e, \qquad \mathbf{q}_{\wedge e} = \tfrac{5}{2}(p_e/\Omega_e m_e)\mathbf{b}\times\nabla T_e,$$
$$\mathbf{q}_{\perp e} = -\tfrac{3}{2}(p_e\nu_e/\Omega_e)\mathbf{b}\times\mathbf{V}_{e\wedge} - r_4(p_e\nu_e/m_e\Omega_e^2)\nabla_\perp T_e,$$

and the friction force on the electrons due to collisions with ions, $\mathbf{F}_e$,

$$\mathbf{F}_e/n_e = -r_2\nabla_\parallel T_e - \nu_e\left[m_e(r_1^{-1}\mathbf{V}_{\parallel e} + \mathbf{V}_{\wedge e}) + \tfrac{3}{2}\Omega_e^{-1}\mathbf{b}\times\nabla T_e\right]. \tag{5.144}$$

[42] This should be contrasted to gyro-flows which are normal to the thermodynamic forces, i.e. the gradients which drive them, and are therefore non-dissipative in nature.

[43] The solution requires the assumption of strong magnetization, such that $|\nu_e/\Omega_e| \ll 1$. Recall that $\Omega_s = e_s B/m_s$, so that $\Omega_i = ZeB/m_i$ is positive while $\Omega_e = -eB/m_e$ is negative.

These flows can also be recast in terms of thermodynamic forces as

$$n_e \mathbf{V}_{\parallel e} = -(p_e/\nu_e m_e)(L_{11}\mathbf{A}_{e1} + L_{12}\mathbf{A}_{e2})_{\parallel}, \tag{5.145}$$
$$\mathbf{q}_{\parallel e}/T_e = -(p_e/\nu_e m_e)(L_{21}\mathbf{A}_{e1} + L_{22}\mathbf{A}_{e2})_{\parallel}, \tag{5.146}$$
$$n_e \mathbf{V}_{\wedge e} = -(p_e/\Omega_e m_e)(\mathbf{b} \times \mathbf{A}_{e1}), \tag{5.147}$$
$$\mathbf{q}_{\wedge e}/T_e = -(p_e/\Omega_e m_e)(\mathbf{b} \times \tfrac{5}{2}\mathbf{A}_{e2}), \tag{5.148}$$
$$n_e \mathbf{V}_{\perp e} = -(p_e \nu_e/\Omega_e^2 m_e)(\mathbf{A}_{e1} - \tfrac{3}{2}\mathbf{A}_{e1})_{\perp}, \tag{5.149}$$
$$\mathbf{q}_{\perp e}/T_e = -(p_e \nu_e/\Omega_e^2 m_e)(-\tfrac{3}{2}\mathbf{A}_{e1} + r_4\mathbf{A}_{e1})_{\perp}, \tag{5.150}$$

where the transport coefficients L_{kl} are given in Table 5.1.

We next observe that the parallel, diamagnetic and radial perturbations $(\hat{f}_{\parallel}, \hat{f}_{\wedge}, \hat{f}_{\perp})$, and the resulting flows, $\mathbf{\Gamma}_e = (n_e\mathbf{V}_e, \mathbf{q}_e/T_e)$, are ordered by the ratio of collision frequency, ν_e, and the gyro-frequency, Ω_e,

$$\hat{f}_{\parallel} : \hat{f}_{\wedge} : \hat{f}_{\perp} \approx \mathbf{\Gamma}_{\parallel e} : \mathbf{\Gamma}_{\wedge e} : \mathbf{\Gamma}_{\perp e} \approx 1 : \nu_e/\Omega_e : (\nu_e/\Omega_e)^2. \tag{5.151}$$

For the plasma to be considered magnetized, this ratio must be small ($|\nu_e/\Omega_e| \sim \delta_e \nu_* \ll 1$), allowing an electron to gyrate many times before suffering a collision, i.e. before being deflected by 90°. Consequently, we expect collisional transport to be fastest, $O(1)$, in the $\parallel$ direction, moderate, $O(\delta_s)$, in the $\wedge$ direction, and slowest, $O(\delta_s^2)$, in the $\perp$ direction, thus naturally leading to a strong scale separation in the three magnetic directions, see Section 7.

In summary, the knowledge of $C_{ss'}$ combined with the smallness of δ_s can be exploited to close the fluid equations in both the diamagnetic and radial directions, leaving only the cumbersome parallel closure. Here different strategies are required in different limits of Δ_s; we will consider the resulting transport equations below. In the hydrodynamic limit of small Δ_s, it should come as no surprise to find that parallel transport is effectively given by the unmagnetized, collisional case discussed in the previous section.

5.3.2.1 Strongly collisional limit ($\Delta_s \ll 1$): Braginskii equations

The complete solution of the ion and electron Spitzer problems for a strongly collisional ($\Delta_s \ll 1$), magnetized ($\delta_s \ll 1$), hydrogenic ($Z = 1$) plasma, was first obtained by Braginskii (1965), who employed a Sonine polynomial expansion to calculate the desired perturbations, $\hat{f}_{1i}$ and $\hat{f}_{1e}$. The technique itself, although important for quantitative studies, introduces little new physics beyond what was already discussed. Therefore, we present the results without further delay, referring the interested reader either to the original article or to several good accounts of the derivation (Hinton, 1983; Balescu, 1988a; Helander and Sigmar, 2002). The resulting set of two-fluid equations, (2.169)–(2.171) closed by the collisional flows and stresses $\boldsymbol{\pi}_s$, $\mathbf{q}_s$, $\mathbf{F}_s$, $\mathcal{Q}_s$ as given below, are widely known as *Braginskii equations*.[44] Although they were derived for hydrogenic ions, we present the results for arbitrary Z, making use of transport coefficients in Table 5.1.

One thus finds that the friction force between electrons and ions can again be written as (5.127), but with $\mathbf{F}_V$ and $\mathbf{F}_{eT}$ now given by

$$\mathbf{F}_V = -m_e n_e \nu_e (r_1^{-1}\Delta\mathbf{V}_\parallel + \Delta\mathbf{V}_\perp) = en_e(\mathbf{J}_\parallel/\sigma_{e\parallel} + \mathbf{J}_\perp/\sigma_{e\perp}), \qquad (5.152)$$

$$\mathbf{F}_{eT} = -r_2 n_e \nabla_\parallel T_e + \tfrac{3}{2}(n_e\nu_e/\Omega_e)\mathbf{b}\times\nabla T_e, \qquad (5.153)$$

where $\sigma_{e\parallel}$ and $\sigma_{e\perp}$ are the parallel and radial components of the electrical conductivity tensor $\boldsymbol{\sigma}_e$, defined by $\mathbf{J} = \boldsymbol{\sigma}_e\cdot\mathbf{E}+\cdots$,

$$\sigma_{e\parallel} = \sigma_e = r_1\sigma_{e\perp}, \qquad \sigma_{e\perp} = e^2 n_e/m_e\nu_e. \qquad (5.154)$$

Note that the first terms in (5.152)–(5.154) are the same as for an unmagnetized plasma, while the second terms originate in plasma magnetization and represent collisional gyro-flows. The relation (5.130) remains valid, provided it is recast in terms of diffusivity and conductivity tensors, $e\mathbf{D}/T_e = \boldsymbol{\sigma}_e/en_e$, where $\mathbf{J} = \boldsymbol{\sigma}_e\cdot \mathbf{E}+\cdots$ and $\Delta\mathbf{V} = -\mathbf{D}\cdot\nabla\ln n_e+\cdots$.

Similarly, the heat exchange terms are again expressed as (5.131), but with $\mathbf{F}_e = \mathbf{F}_{eT}+\mathbf{F}_V$ given by (5.152)–(5.153). Comparing the resulting magnetic diffusivity, (2.257), with the collisional electron diffusivity, (5.143),

$$D_{\perp e}/D_\eta = D_{\perp e}\sigma_e\mu_0 = (r_1/2)2\mu_0 p_e/B^2 \approx \beta_e, \qquad (5.155)$$

we observe that the former is much larger in a low-beta plasma, i.e. magnetic flux diffuses more readily than mass, breaking the correspondence between magnetic flux-tubes and plasma filaments present under the MHD ordering.

The components of the viscous stress tensor, $\boldsymbol{\pi}_s$, can be calculated as[45]

$$\pi_{22} = -\tfrac{1}{2}\mu_\parallel(W_{22}+W_{33}) - \tfrac{1}{2}\mu_\perp(W_{22}-W_{33}) - \mu_\wedge W_{23}, \qquad (5.156)$$

$$\pi_{33} = -\tfrac{1}{2}\mu_\parallel(W_{22}+W_{33}) + \tfrac{1}{2}\mu_\perp(W_{22}-W_{33}) + \mu_\wedge W_{23}, \qquad (5.157)$$

$$\pi_{11} = -\mu_\parallel W_{11}, \qquad \pi_{23} = \pi_{32} = -\mu_\perp W_{23} + \tfrac{1}{2}\mu_\wedge(W_{22}-W_{33}), \qquad (5.158)$$

$$\pi_{21} = \pi_{12} = -4\mu_\perp W_{21} - 2\mu_\wedge W_{31}, \qquad (5.159)$$

$$\pi_{31} = \pi_{13} = -4\mu_\perp W_{31} + 2\mu_\wedge W_{21}, \qquad (5.160)$$

44 It is interesting to note that Braginskii equations do not invoke the MHD ordering (2.41) and are thus, strictly speaking, a branch of DHD, and not MHD, see the discussion in Section 2.5. Despite this fact, they are sometimes described as *two-fluid MHD* or *non-ideal MHD*.

45 Here the indices (1, 2, 3) refer to the basis $(\mathbf{b}, \hat{\mathbf{e}}_2, \hat{\mathbf{e}}_3)$ or $(\mathbf{b}, \hat{\mathbf{e}}_\wedge, \hat{\mathbf{e}}_\perp)$ defined in Section 2.2.

where the parallel, diamagnetic and radial viscosities are given by

$$\mu_{\parallel i} = 0.96 p_i/\nu_i, \quad \mu_{\wedge i} = p_i/2\Omega_i, \quad \mu_{\perp i} = 0.3 p_i \nu_i/\Omega_i^2, \tag{5.161}$$

$$\mu_{\parallel e} = 0.73 p_e/\nu_e, \quad \mu_{\wedge e} = p_e/2\Omega_e \quad \mu_{\perp e} = 0.51 p_e \nu_e/\Omega_e^2. \tag{5.162}$$

Since the viscous stress constitutes only a second-order correction to f_{Ms}, so that $\pi_e \sim (m_e/m_i)\pi_i$ is negligible under most circumstances, the electron viscosities are evaluated only for a hydrogenic plasma. It is worth noting that Braginskii (1965) obtained the viscosities (5.161)–(5.162) by assuming the MHD ordering (2.41), and that these values differ somewhat from those obtained based on the drift ordering (2.42), see Claassen *et al.* (2000). The other elements of the collisional closure, namely $\mathbf{q}_s$, $\mathbf{F}_s$ and $\mathcal{Q}_s$ are independent of the assumed plasma ordering.

Similarly, the thermal gradient heat flow for species s can be written as

$$\mathbf{q}_{Ts} = -\kappa_{\parallel s} \nabla_\parallel T_s - \kappa_{\wedge s} \mathbf{b} \times \nabla_\perp T_s - \kappa_{\perp s} \nabla_\perp T_s, \tag{5.163}$$

where the heat conductivities for ions and electrons are found as

$$\kappa_{\parallel i} = 3.91 p_i/m_i \nu_i, \quad \kappa_{\wedge i} = \tfrac{5}{2} p_i/m_i \Omega_i, \quad \kappa_{\perp i} = 2 p_i \nu_i/m_i \Omega_i^2, \tag{5.164}$$

$$\kappa_{\parallel e} = r_3 p_e/m_e \nu_e, \quad \kappa_{\wedge e} = -\tfrac{5}{2} p_e/m_e \Omega_e, \quad \kappa_{\perp e} = r_4 p_e \nu_e/m_e \Omega_e^2. \tag{5.165}$$

As before, the ion heat flow is driven purely by the ion temperature gradient, $\mathbf{q}_i = \mathbf{q}_{Ti}$, while the electron heat flow consists of both thermal gradient and frictional contributions, $\mathbf{q}_e = \mathbf{q}_{Te} + \mathbf{q}_{Ve}$. The latter is now found as

$$\mathbf{q}_{Ve} = r_2 p_e \Delta \mathbf{V}_\parallel - \tfrac{3}{2}(p_e \nu_e/\Omega_e)\mathbf{b} \times \Delta \mathbf{V}_\perp, \tag{5.166}$$

and indicates that electron heat flow can be driven by both parallel and radial currents (recall that $\mathbf{J} = -en\Delta\mathbf{V}$). Once again, we observe the equality of the (normalized) thermoelectric terms, $\mathbf{F}_{Te}$ (5.153) and $\mathbf{q}_{Ve}$ (5.166), as required by *Onsager symmetry* ($\mathcal{L}_{kl} = \mathcal{L}_{lk}$). We also note that consistent with (5.151), the parallel, diamagnetic and radial components of viscosities and heat conductivities are ordered by the small ratio $\nu_s/\Omega_s \sim \delta_s \ll 1$,

$$\mu_{\parallel s} : \mu_{\wedge s} : \mu_{\perp s} \approx \kappa_{\parallel s} : \kappa_{\wedge s} : \kappa_{\perp s} \approx 1 : \nu_s/\Omega_s : (\nu_s/\Omega_s)^2. \tag{5.167}$$

Examination of (5.161)–(5.162) and (5.164)–(5.165) reveals the different effect of Coulomb collisions on transport processes in these three directions: (i) as already mentioned in Section 5.3.1, parallel transport scales inversely with ν_s, and is thus *impeded* by collisions; (ii) diamagnetic transport is independent of ν_s, and is thus *unaffected* by collisions; and (iii) radial transport scales linearly with ν_s, and is thus *enhanced* by collisions. Point (ii) above should be clear from our discussion of gyro-flows and gyro-stresses in Section 2.4.2, while point (iii) is easily understood in terms of a random walk of guiding centres with radial step, ρ_{ts}, and time,

$\tau_s = 1/\nu_s$. Since $\rho_{ts} \propto T_s^{1/2}$ and $\nu_s \propto T_s^{-3/2}$, the radial diffusivity, $\rho_{ts}^2 \nu_s$, decreases with temperature as $T_s^{-1/2}$, in contrast to the parallel diffusivity, which increases as $T_s^{5/2}$.[46] Inserting (5.107) into (5.161)–(5.162) and (5.164)–(5.165), one finds

$$\mu_{\|s} \propto T_s^{5/2} m_s^{1/2}, \qquad \mu_{\wedge s} \propto p_s m_s, \qquad \mu_{\perp s} \propto n_s^2 T_s^{-1/2} m_s^{3/2}, \tag{5.168}$$

$$\kappa_{\|s} \propto T_s^{5/2} m_s^{-1/2}, \qquad \kappa_{\wedge s} \propto p_s, \qquad \kappa_{\perp s} \propto n_s^2 T_s^{-1/2} m_s^{1/2}. \tag{5.169}$$

Note that the increasing density dependence (from left to right) is consistent with (5.167) and $\nu_s \propto n_s T_s^{-3/2} m_s^{-1/2}$, and that the square dependence in both $\mu_{\perp s}$ and $\kappa_{\perp s}$ reflects the binary nature of Coulomb collisions.

Writing down the ratios of terms in (5.168)–(5.169) for ions and electrons,

$$\mu_{\|i}/\mu_{\|e} \gg 1, \qquad \mu_{\wedge i}/\mu_{\wedge e} \gg 1, \qquad \mu_{\perp i}/\mu_{\perp e} \gg 1, \tag{5.170}$$

$$\kappa_{\|i}/\kappa_{\|e} \ll 1, \qquad \kappa_{\wedge i}/\kappa_{\wedge e} \sim 1, \qquad \kappa_{\perp i}/\kappa_{\perp e} \gg 1, \tag{5.171}$$

allows us to make several important observations: (i) the ion viscosity is always much larger than the electron viscosity ($\boldsymbol{\mu}_i \gg \boldsymbol{\mu}_e$), suggesting that the latter can be neglected in practically all applications; (ii) the ion and electron heat conductivities exhibit a very different relative size in the three magnetic directions: (iia) the electron heat conductivity is much larger in the parallel direction, due to the higher electron mobility; (iib) the two are comparable in the diamagnetic direction, in which the collision frequency plays no role; and (iic) the ion heat conductivity is much larger in the radial direction due to the larger ion gyro-radius, i.e. since $\rho_{ts} \propto m_s^{1/2}$ and $\nu_s \propto m_s^{-1/2}$, the radial diffusivity, $\rho_{ts}^2 \nu_s$, increases with mass as $m_s^{1/2}$. We may conclude that parallel heat flow is dominated by electrons,[47] diamagnetic heat flow involves both species, and radial heat flow is dominated by ions.

5.3.2.2 *Weakly collisional limit ($\Delta_s \sim 1$): ZNC equations*

The collisional flows derived in the previous section are valid only in the strongly collisional limit ($\Delta_s \ll 1$). This is particularly true of the parallel flows ($\mathbf{q}_{\|s}, \boldsymbol{\pi}_{\|s}$), which are proportional to parallel gradients with coefficients $\mu_{\|s}, \kappa_{\|s} m_s \sim p_s/\nu_s$, and hence imply infinite flows for finite gradients, i.e. $\mu_{\|s}, \kappa_{\|s} \to \infty$ as $\nu_s \to 0$.[48] This unphysical prediction reveals the dangers of applying the Braginskii equations beyond their domain of validity. It also indicates the need for alternative transport equations valid for all degrees of collisionality, which reduce to the Braginskii

[46] Recall, however, that particle diffusion can only occur as a result of unlike-particle collisions. The same is not true for diffusion of momentum or heat, as measured by $\boldsymbol{\mu}_s$ and $\boldsymbol{\kappa}_s$.

[47] Assuming comparable parallel gradients of ion and electron temperatures, i.e. $\nabla_\| T_e \sim \nabla_\| T_i$.

[48] In reality, parallel flows must of course be limited by parallel kinetic free-streaming at $v_\| \sim v_{ts}$.

equations when $\Delta_s \ll 1$ and to the double adiabatic, or CGL, equations (2.196) when $\Delta_s \gg 1$.

Such *parallel* transport equations can indeed be derived by multiplying the drift-kinetic equation by $w_{\|s}^k w_{\perp s}^l$ and integrating over velocity space. This yields the *gyro-tropic moment* equation (Zawaideh *et al.*, 1986, 1988),

$$\begin{aligned} &\mathrm{d}_t n_s \langle w_\|^k w_\perp^l \rangle_{f_s} + \nabla_\| n_s \langle w_\|^{k+1} w_\perp^l \rangle_{f_s} - \left(\frac{e_s E_\|}{m_s^2} - \frac{\mathrm{d}_t V_{\|s}}{m_s} \right) k n_s \langle w_\|^{k-1} w_\perp^l \rangle_{f_s} \\ &+ (\nabla_\| \ln B) \frac{k}{2} n_s \langle w_\|^{k-1} w_\perp^{l+2} \rangle_{f_s} + \left[(k+1) \nabla_\| V_{\|s} - \frac{l+2}{2} \mathrm{d}_t \ln B \right] n_s \langle w_\|^k w_\perp^l \rangle_{f_s} \\ &- (\nabla_\| \ln B) \frac{l+2}{2} n_s \langle w_\|^{k+1} w_\perp^l \rangle_{f_s} = n_s \langle w_\|^k w_\perp^l \rangle_{C_s}, \end{aligned} \tag{5.172}$$

where we dropped the species index inside the averages and assumed that $\mathbf{V}_s = V_{\|s}\mathbf{b}$ so that $\mathrm{d}_t = \partial_t + V_{\|s}\nabla_\|$. Recalling the definitions (2.191)–(2.195), it is possible to write down (5.172) for $k, l \leq 2$,

$$\begin{aligned} \mathrm{d}_t \rho_s + \rho_s \nabla_\| V_{\|s} - \rho_s V_{\|s} \nabla_\| \ln B &= 0, \\ \rho_s \mathrm{d}_t V_{\|s} + \nabla_\| p_{\|s} - (p_{\|s} - p_{\perp s}) \nabla_\| \ln B &= e_s n_s E_\| + F_{\|s}, \\ \mathrm{d}_t p_{\|s} + 3 p_{\|s} \nabla_\| V_{\|s} + \nabla_\| q_{\|s} + (2 q_{\perp s} - q_{\|s} - p_{\|s} V_{\|s}) \nabla_\| \ln B &= \mathcal{Q}_{\|s}, \\ \mathrm{d}_t p_{\perp s} + p_{\perp s} \nabla_\| V_{\|s} + \nabla_\| q_{\perp s} - 2(q_{\perp s} + p_{\perp s} V_{\|s}) \nabla_\| \ln B &= \mathcal{Q}_{\perp s}, \end{aligned} \tag{5.173}$$

and to complement these with evolution equations for $q_{\|s}$ and $q_{\perp s}$,

$$\begin{aligned} &\mathrm{d}_t q_{\|s} - 3 \left(\frac{e_s E_\|}{m_s} - \mathrm{d}_t V_{\|s} \right) p_{\|s} + (4 \nabla_\| V_{\|s} - V_{\|s} \nabla_\| \ln B) q_{\|s} + \nabla_\| \rho_s \langle w_\|^4 \rangle_{f_s} \\ &+ (3 \rho_s \langle w_\|^2 w_\perp^2 \rangle_{f_s} - \rho_s \langle w_\|^4 \rangle_{f_s}) \nabla_\| \ln B = \rho_s \langle w_\|^3 \rangle_{C_s}, \end{aligned} \tag{5.174}$$

$$\begin{aligned} &\mathrm{d}_t q_{\perp s} - \left(\frac{e_s E_\|}{m_s} - \mathrm{d}_t V_{\|s} \right) p_{\perp s} + 2 (\nabla_\| V_{\|s} - V_{\|s} \nabla_\| \ln B) q_{\perp s} + \frac{1}{2} \nabla_\| \rho_s \langle w_\|^2 w_\perp^2 \rangle_{f_s} \\ &+ (\frac{1}{4} \rho_s \langle w_\perp^4 \rangle_{f_s} - \rho_s \langle w_\|^2 w_\perp^2 \rangle_{f_s}) \nabla_\| \ln B = \frac{1}{2} \rho_s \langle w_\| w_\perp^2 \rangle_{C_s}. \end{aligned} \tag{5.175}$$

The even, fourth-order moments are evaluated assuming $f_s = f_{2Ms}$,

$$\langle w_\|^4 \rangle_{f_s} = 3 \left(\frac{T_{\|s}}{m_s} \right)^2, \quad \langle w_\|^2 w_\perp^2 \rangle_{f_s} = 2 \frac{T_{\|s} T_{\perp s}}{m_s^2}, \quad \langle w_\perp^4 \rangle_{f_s} = 8 \left(\frac{T_{\perp s}}{m_s} \right)^2.$$

The next step involves the calculation of collisional terms in (5.173)–(5.175). While the Braginskii result may be used for the friction force, $F_{\|s}$, the second- and third-order moments give rise to gyro-tropic expressions. These may be estimated using the BGK approach with respect to f_{2Ms},

$$\mathcal{Q}_{\parallel s} = 2\nu_s^{(1)}(p_{\perp s} - p_{\parallel s}) + \sum \nu_{ss'}^{(2)}(p_{\parallel s'} - p_{\parallel s}),$$
$$\mathcal{Q}_{\perp s} = -\nu_s^{(1)}(p_{\perp s} - p_{\parallel s}) + \sum \nu_{ss'}^{(2)}(p_{\perp s'} - p_{\perp s}), \quad (5.176)$$
$$\rho_s \langle w_\parallel^3 \rangle_{C_s} = \nu_s^{(3)} q_{\parallel s}, \qquad \tfrac{1}{2}\rho_s \langle w_\parallel w_\perp^2 \rangle_{C_s} = \nu_s^{(3)} q_{\perp s},$$

where the collision frequencies are closely related to ν_s and $\nu_{ss'}^{\epsilon}$,

$$\nu_s^{(1)} = \tfrac{5}{2}\nu_s(T_s^*), \qquad \nu_{ss'}^{(2)} = \nu_{ss'}^{\varepsilon}, \qquad \nu_s^{(3)} = C_q \nu_s. \quad (5.177)$$

Here C_q is a constant of order unity, to be determined by comparison with the Braginskii expression in the collisional limit, and $\nu_s(T_s^*)$ is given by ν_s (5.107) evaluated at the effective temperature T_s^* (Ichimaru, 1992),

$$T_s^* = T_{\parallel s}\left[\frac{15}{4\xi^2}\left(-3 + (\xi + 3)\frac{g(\xi^{1/2})}{\xi^{1/2}}\right)\right]^{-2/3}, \quad \xi \equiv \frac{T_{\perp s} - T_{\parallel s}}{T_{\parallel s}}, \quad (5.178)$$

where $g(\sqrt{x}) = \{\tan^{-1}\sqrt{x}, x > 0; \tanh^{-1}\sqrt{-x}, x < 0\}$.

By forming linear combinations of gyro-tropic moment equations, one obtains evolution equations for isotropic and anisotropic quantities,[49]

$$\mathcal{A} = \frac{1}{3}(\mathcal{A}_\parallel + 2\mathcal{A}_\perp), \qquad \delta\mathcal{A} = \frac{2}{3}(\mathcal{A}_\parallel - \mathcal{A}_\perp) = \frac{2}{3}\mathcal{A}_\Delta. \quad (5.179)$$

No information is lost in making this transformation and both equation sets are valid for any degree of anisotropy, $f_s(w_\parallel, w_\perp)$. However, the evolution equations for isotropic quantities, which dominate in the collisional limit,

$$\Delta_s \ll 1, \qquad \Rightarrow \qquad \mathcal{A}_\parallel \approx \mathcal{A}_\perp \approx \mathcal{A}, \qquad \delta\mathcal{A}/\mathcal{A} \ll 1, \quad (5.180)$$

provide a natural transition to the Braginskii equations.

To illustrate this transition, we evaluate the parallel viscous stress for gradual magnetic field variation, $L_B \gg L_V$, where $L_{\mathcal{A}} = (\nabla_\parallel \ln \mathcal{A})^{-1}$,

$$\pi_{\parallel s} = -\mu_{\parallel s}\nabla_\parallel V_{\parallel s}, \qquad \mu_{\parallel s} \approx \left(\tfrac{4}{9}p_s/\nu_s^{(1)}\right)\left(1 + \tfrac{7}{9}|\nabla_\parallel V_{\parallel s}|/\nu_s^{(1)}\right)^{-1}. \quad (5.181)$$

This expression provides an insight into the nature of the parallel viscosity $\mu_{\parallel s}$, namely the *anisotropy* of the pressure tensor, i.e. $\pi_{\parallel s} = \delta p_s$. In the collisional limit, we recover the Braginskii expression $\mu_{\parallel s} \sim p_s/\nu_s$, while in the opposite limit, we find $\mu_s \approx \frac{4}{7}p_s/|\nabla_\parallel V_{\parallel s}|$ so that the $\pi_{\parallel s}$ saturates at the free-streaming value, $\pi_{\parallel s}^{FS} \approx -\frac{4}{7}p_s$. This result may also be written as a harmonic average of the Braginskii and free-streaming expressions,

$$\frac{\pi_{\parallel s}}{\pi_{\parallel s}^{Br}} = \frac{\mu_{\parallel s}}{\mu_{\parallel s}^{Br}} = \left(1 + \frac{\pi_{\parallel s}^{Br}}{\pi_{\parallel s}^{FS}}\right)^{-1} = \left(1 + \frac{C_s^{\pi}}{\alpha_s^{\pi}}\left|\frac{\nabla_\parallel V_{\parallel s}}{\nu_s}\right|\right)^{-1}, \quad (5.182)$$

[49] For a complete set of moment equations see Fundamenski (2005, p. R188–189).

where $C_i^\pi = 0.96$, $C_e^\pi = 0.73$ are the Braginskii coefficients given by (5.161)–(5.162), while $\alpha_s^\pi = 4/7$ is the *flux limiting factor* for parallel viscosity.

A similar expression may be derived for $q_{\|s}$ and $\kappa_{\|s}$,

$$\frac{q_{\|s}}{q_{\|s}^{Br}} = \frac{\kappa_{\|s}}{\kappa_{\|s}^{Br}} = \left(1 + \frac{q_{\|s}^{Br}}{q_{\|s}^{FS}}\right)^{-1} = \left(1 + \frac{C_s^q}{\alpha_s^q}\Delta_s\right)^{-1}, \tag{5.183}$$

where $C_i^q = 3.91$, $C_e^q = r_3$ are determined from (5.164)–(5.165), while α_s^q is the flux limiting factor for parallel heat flow, also known as the *heat flux limit*. Kinetic calculations indicate a wide range of α_s^q, from 0.03 to above unity, depending on a number of factors; typically $\alpha_s^q \sim 0.3-1$ are found in the important case of parallel transport on open magnetic field lines.

The topic of *heat flux limits* on open field lines has been recently reviewed in Fundamenski (2005). There it was recommended to apply the expression (5.183) with caution, since under some circumstances, the heat flow may actually exceed the Braginskii value, e.g. in the case of strong parallel temperature gradients, when hot electrons stream into a cold plasma region. This recommendation is now strongly supported by recent 1D kinetic simulations of parallel transport (in the SOL, see Section 7.1), which indicate strong parallel variation of flux limiting factors (Tskhakaya *et al.*, 2008).

5.3.2.3 Radial flows in a cylindrical plasma

To see the consequences of the Braginskii equations, let us write down the radial component of classical collisional flows in a cylindrical plasma. Denoting the radial derivative by a prime, we thus find the radial velocities,

$$\begin{pmatrix} V_{\perp e} \\ q_{\perp e}/p_e \\ q_{\perp i}/p_i \end{pmatrix} = -\begin{pmatrix} D_{\perp e} & -\frac{3}{2}D_{\perp e} & 0 \\ -\frac{3}{2}D_{\perp e} & \chi_{\perp e} & 0 \\ 0 & 0 & \chi_{\perp i} \end{pmatrix} \begin{pmatrix} (p_e + p_i)'/p_e \\ T_e'/T_e \\ T_i'/T_i \end{pmatrix}, \tag{5.184}$$

where $D_{\perp e} = T_e\nu_e/m_e\Omega_e^2 = \rho_{te}^2\nu_e$ is the classical particle diffusivity, arising due to unlike $(e-i)$ collisions, while $\chi_{\perp e} = \kappa_{\perp e}/n_e = r_4 D_{\perp e}$ and $\chi_{\perp i} = \kappa_{\perp i}/n_i$, are the electron and ion heat diffusivities.[50]

Similarly, we obtain the parallel and diamagnetic projections of the ion viscous stress, i.e. the collisional flows of axial and poloidal momentum,

$$(\nabla\cdot\boldsymbol{\pi}_i)_\| = -4r^{-1}(r\mu_{\perp i}V_{\|i}')', \qquad (\nabla\cdot\boldsymbol{\pi}_i)_\wedge = -r^{-1}(r\mu_{\perp i}V_{\wedge i}')', \tag{5.185}$$

which are driven by the radial gradients of axial and poloidal velocities. The latter can be used to determine the evolution of the radial electric field, $\partial_t E_\perp = -J_\perp/\epsilon_0$,

[50] We wrote the flows in the transport matrix symmetric form to emphasize the Onsager symmetry.

generated by the non-ambipolar outflow of charge (5.140). The radial electric field in the frame moving with the ion diamagnetic velocity, $V_{*i} = -p_i'/e_i n_i B$, can written as $\Delta E_\perp = -\varphi' - p_i'/e_i n_i$, which, aside from the definition of the radial co-ordinate, and hence the meaning of the radial derivative implied by $'$, is just (diamagnetic, toroidal) rotation velocity, ω_i, defined in (3.58). As a result, we find that $\Delta E_\perp = \omega_i \psi' \approx \omega_i B_P R_0$ evolves according to the diffusion equation,

$$\mathrm{d}_t\omega_i = \partial_t\omega_i + \frac{V_\perp}{r}(r\omega_i)' \approx \frac{(r\mu_{\perp i}\omega_i')'}{n_i m_i r}, \tag{5.186}$$

with the viscosity $\mu_{\perp i} \sim \rho_{ti}^2 \nu_i$ playing the role of momentum diffusivity.[51]

It is possible to repeat the above analysis for a multi-species plasma,[52] in which, according to (5.100), the frictional force on lowest-Z ions i is dominated by collisions with the highest-Z impurity ions z, i.e. $\mathbf{F}_i \sim \mathbf{F}_{iz} \gg \mathbf{F}_{ie}$. When $Z_z \equiv e_z/e \gg 1$, C_{iz} may be approximated by the Lorentz collision operator and the solution to the $i-z$ Spitzer problem may be obtained from the $e-i$ Spitzer problem by effecting the substitution $i \to e$, $z \to i$, so that

$$\mathbf{F}_{iz} = \mathbf{F}_{iz}^V + \mathbf{F}_{iz}^T = -m_i n_i \nu_{iz}(\mathbf{V}_i - \mathbf{V}_z) + \tfrac{3}{2} n_i(\nu_{iz}/\Omega_i)\mathbf{b}\times\nabla T_i. \tag{5.187}$$

The radial hydrogenic ion flow implied by (5.187) is therefore found as

$$V_{\perp i} = -D_{\perp i}\left(\frac{p_i'}{p_i} - \frac{T_z}{Z_z T_i}\frac{p_z'}{p_z} - \frac{3}{2}\frac{T_i'}{T_i}\right), \qquad D_{\perp i} = \frac{v_{ti}^2 \nu_{iz}}{\Omega_i^2} = \rho_{ti}^2 \nu_{iz}, \tag{5.188}$$

while the electron flow is still given by (5.184). The pressure gradient terms (5.188) are more accurate than $(p_i + p_z)'/p_i$, which might be expected from (5.184). Indeed, the driving term $(p_e + p_i)'/p_e$, which involves a simple sum of ion and electron pressures, follows from $(p_e'/p_e) - (T_i/ZT_e)(p_i'/p_i)$ and the quasi-neutrality condition, $n_e = Zn_i$. Since $D_{\perp i}/D_{\perp e} \gg 1$, the ion flow is now much larger, $V_{\perp i} \gg V_{\perp e}$, which together with the quasi-neutrality condition, $\sum e_s n_s V_{\perp s} \approx 0$, implies opposite flows of hydrogenic and impurity ions, i.e. $n_i V_{\perp i} \approx -Z_z n_z V_{\perp z}$. Hence, an *outflow* of low-Z ions (hydrogen fuel, helium ash) gives rise to an *inflow* of high-Z impurities,[53] i.e. there is a tendency for impurities to accumulate in the core of the plasma! To quantify this effect consider the steady ($V_{\perp i} = 0 = V_{\perp z}$) radial profile of impurity concentration for equal ion and impurity temperatures,

$$n_z/n_i \propto n_i(r)^{Z_z-1} T_i(r)^{Z_z/2-1} \propto p_i(r)^{Z_z-1} T_i(r)^{-Z_z/2}. \tag{5.189}$$

51 This is easily understood by noting that $\Delta E_\perp$ corresponds to a poloidal flow $V_{\wedge i} = -\Delta E_\perp/B$.
52 In tokamaks, the effective ion charge, Z_{eff} (5.101), is typically in the range 1.5–3, due to the presence of both intrinsic (eroded) and extrinsic (seeded) impurities, see Chapter 7.
53 These opposing main ion and impurity flows are a direct consequence of conservation of momentum in $i-z$ collisions, which implies equal and opposite friction forces, $\mathbf{F}_{iz} = -\mathbf{F}_{zi}$.

Since pressure and temperature profiles decrease monotonically with radius in a confined plasma, n_z/n_i peaks close to the axis for high-Z impurities.

5.3.3 Collisional transport in a toroidal plasma

The results of the previous section neglect the effect of toroidal curvature and are therefore accurate only for a *cylindrical* plasma. Broadly speaking, the inclusion of finite toroidicity increases the net radial collisional flow by a factor of q^2, or $(q/\epsilon)^2$, above the *classical* values. The rigorous theory of collisional transport in a *toroidal* plasma, known as *neoclassical* transport, is one of the great achievements of magnetized plasma physics.[54] Below we merely introduce this important topic, loosely following the excellent account of neoclassical theory in Helander and Sigmar (2002).

5.3.3.1 Guiding centre orbits in axis-symmetric, toroidal geometry

Neoclassical theory begins with the fact that toroidal curvature, and the associated $B_T \propto 1/R$ variation on any given flux surface, creates a magnetic well on the low field (outboard) side of the torus. Recalling the discussion on page 26, we expect the related mirror forces to generate two types of charged particle populations: (i) passing particles with $\alpha < \alpha(B_{min})$ and (ii) trapped particles with $\alpha > \alpha(B_{min})$ or, written another way,

$$\text{passing } (\circ) \quad \Rightarrow \quad 0 \leq \lambda \leq \lambda_c \equiv B_0/B_{max}, \tag{5.190}$$

$$\text{trapped } (\triangleright) \quad \Rightarrow \lambda_c \leq \lambda \leq \lambda_{max} \equiv B_0/B_{min}, \tag{5.191}$$

where $\lambda = \mu B_0/\mathcal{E} = (v_\perp^2/v^2)(B_0/B)$ is the ratio of the magnetic moment $\mu = \frac{1}{2}mv_\perp^2/B$ and the total energy $\mathcal{E} = \frac{1}{2}mv^2 + e\varphi$, normalized by some reference magnetic field B_0, e.g. the RMS field $\langle B^2\rangle_B^{1/2}$. Since GC-orbits deviate from the flux surface by a radial distance of order $\rho_P = v_\perp/\Omega_P$, see discussion on pages 28 and 82, then provided $\rho_P/r \ll 1$, their radial position may be described by any flux surface label, e.g. the poloidal flux $\psi = \psi_P$. Moreover, under the same conditions, the electric potential φ is an approximate flux surface label, so that the particle speed remains roughly constant.[55] As a result, the three *exact* constants of motion $(\mathcal{E}, \mu, p_\zeta)$ may be replaced by the three *approximate* constants of motion (v, λ, ψ), which can then be used to characterize every trapped orbit. Due to the bi-directionality of circulating orbits, an additional constant of motion, $\sigma \equiv v_\parallel/|v_\parallel| = \pm 1$, representing the sign of the parallel GC-velocity, (2.55), i.e.

[54] It was developed by a number of authors following the ideas of I. E. Tamm and A. D. Sakharov; see Helander and Sigmar (2002, p. 117) for a short historical background.

[55] This approximate constancy of φ and v breaks down in the presence of rapid toroidal rotation.

$$v_\zeta / b_\zeta = v_T / b_T = v_\parallel = \sigma v \sqrt{1 - \lambda B / B_0}, \tag{5.192}$$

is required to characterize every passing orbit.[56]

Assuming isotropic velocity distribution at the bottom of the magnetic well, i.e. at the outer mid-plane ($\theta = 0$), the fraction of particles trapped by the well is simply the fraction of velocity space covered by (5.191). Since the velocity space element may be written in terms of $(v, \lambda, \psi, \sigma)$ as

$$\mathrm{d}\mathbf{v} = 2\pi v_\perp \mathrm{d}v_\perp \mathrm{d}v_\parallel = \sum_\sigma \frac{\pi v^3 B}{|v_\parallel| B_0} \mathrm{d}v \mathrm{d}\lambda = \sum_\sigma \frac{2\pi B}{m^2 |v_\parallel| B_0} \mathcal{E} \mathrm{d}\mathcal{E} \mathrm{d}\lambda, \tag{5.193}$$

with $v_\parallel$ given by (5.192), the trapped particle fraction may be estimated as

$$f_\triangleright = \int_{\lambda_c}^{\lambda_{max}} \frac{\mathrm{d}\lambda}{\sqrt{1 - \lambda B / B_0}} \Big/ \int_0^{\lambda_{max}} \frac{\mathrm{d}\lambda}{\sqrt{1 - \lambda B / B_0}}. \tag{5.194}$$

To evaluate this fraction, and to illustrate the shape of the resulting orbits, we once again turn to the small ϵ tokamak, see Section 3.3, for which the magnetic field may be written as $B/B_0 \approx 1 - \epsilon \cos\theta$, so that λ_{max} and λ_c may be replaced by $1+\epsilon$ and $1-\epsilon$, respectively. With these substitutions, (5.194) becomes $f_\triangleright \sim \sqrt{\epsilon} \ll 1$, indicating that only a small fraction of particles are trapped by the magnetic well.

Let us next estimate the maximum radial deflection, $\Delta r = r_{max} - r_{min}$, of both passing and trapped orbits.[57] From the definition of ψ, Δr is related to $\Delta\psi$ by $\Delta r \approx \Delta\psi / R B_P$, while conservation of $p_\zeta \approx m R v_\parallel - e\psi$ implies $m R \Delta v_\parallel \approx e \Delta\psi$. Combining these relations, we find

$$\Delta r \approx \Delta\psi / R B_P \approx \Delta v_\parallel / \Omega_P, \qquad \Omega_P \equiv e B_P / m. \tag{5.195}$$

Using (5.192) to estimate $\Delta v_\parallel$ for a small ϵ tokamak, it then follows that $\Delta v_\parallel / v \sim \sigma\epsilon$ for the vast majority of passing orbits (for which $1 - \lambda \gg \epsilon$) and $\Delta v_\parallel / v \sim 2\sigma_0\sqrt{\epsilon}$ for all trapped orbits, where $\sigma_0 \equiv \sigma(\theta = 0)$. As a result,

$$\text{passing } (\circ) \quad \Rightarrow \quad \Delta r_\circ \sim \sigma\epsilon v / \Omega_P, \tag{5.196}$$

$$\text{trapped } (\triangleright) \quad \Rightarrow \quad \Delta r_\triangleright \sim 2\sigma_0 \sqrt{\epsilon} v / \Omega_P = 2\sigma_0 \sqrt{\epsilon} \rho_P, \tag{5.197}$$

and since $v \approx v_\perp$ for trapped particles, we find that trapped orbits correspond to much larger radial excursions, $\Delta r_\triangleright / \Delta r_\circ \sim 1/\sqrt{\epsilon} \gg 1$.

To understand this difference, it is helpful to visualize the poloidal projection of both passing and trapped orbits, e.g. see Helander and Sigmar (2002, Fig. 7.7).

56 Note that σ is not conserved for trapped orbits, since it changes sign with each reflection. The two directions ($\sigma = \pm 1$) are typically denoted as co-current and counter-current, depending on their relation to the direction of toroidal plasma current.

57 According to Tamm's theorem, see page 82, this deflection is comparable to the poloidal gyro-radius $\rho_P \sim v_\perp / \Omega_P$ and is therefore much larger for ions than for electrons.

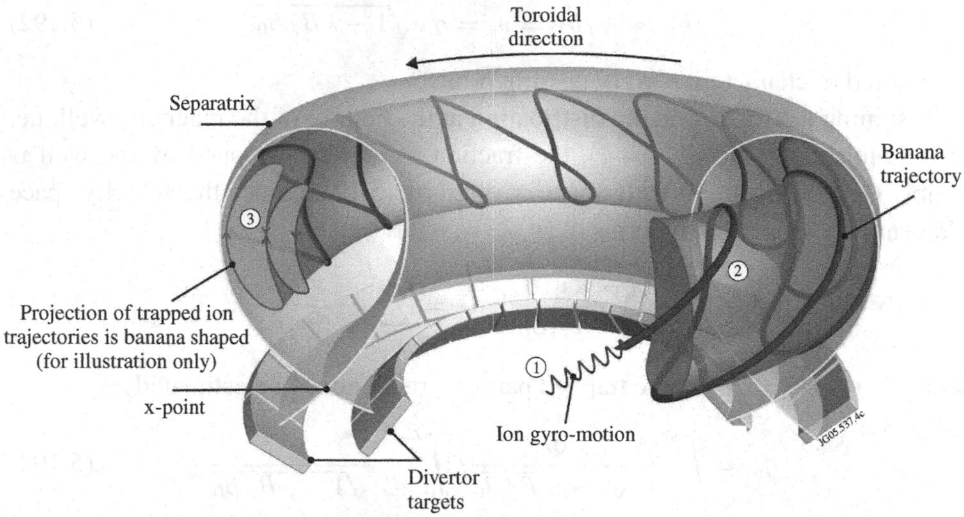

Fig. 5.2. Illustration of a banana orbit in a *tokamak*. © EFDA-JET

Choosing σ_0 such that $\Delta r < 0$, and assuming all orbits to pass through the same point at the outer mid-plane ($\psi = \psi_0$, $\theta = 0$), we find that the former resemble *circles* with radii $r - \Delta r_{\circ}/2$, and the latter inward pointing crescents with radii of curvature $r - \Delta r_{\triangleright}/2$ and widths $\Delta r_{\triangleright}$.[58] Let us denote the position of the tips of each crescent, or *banana* as it is commonly known, which correspond to the turning points of each trapped orbit, by the angle $\theta_{\triangleright} = [\pi, 0]$, i.e. $v_{\parallel}(\theta_{\triangleright}) \equiv 0$; such a banana orbit slowly drifts (precesses) toroidally, as shown in Fig. 5.2. Its width, $\Delta r_{\triangleright}$, increases monotonically with $\theta_{\triangleright}$, so that the thickest banana orbits occur for $\lambda = 1 - \epsilon$, when the tips extend all the way to the inner mid-plane ($\theta_{\triangleright} = \pi$), and the thinnest for $\lambda = 1 + \epsilon$, when the tips meet at the outer mid-plane ($\theta_{\triangleright} \to 0$) and the banana width vanishes. Finally, in the vicinity of the magnetic axis, i.e. within the region $r < \sqrt{\epsilon}\rho_P$, trapped orbits become kidney, or *potato*, shaped and their radial width becomes comparable to r.

The above results can also be derived by integrating the first-order GC-drift, previously derived as (2.40), along the zeroth-order GC-trajectory, i.e. along the circular flux surface. For a small ϵ tokamak, this drift reduces to

$$\mathbf{v}_{GC\perp} = \mathbf{U}_{\perp} \approx \frac{\mathbf{b}}{\Omega} \times (v_{\parallel}^2 + \frac{1}{2}v_{\perp}^2)\frac{\nabla B}{B} \approx -\frac{v_{\parallel}^2 + v^2}{2\Omega}\frac{\mathbf{e}_Z}{R}, \tag{5.198}$$

indicating a vertical drift of ions (in the direction of the vector $\mathbf{B} \times \nabla B$) and electrons (in the opposite direction). The direction of $\mathbf{B} \times \nabla B$ is determined solely by the toroidal direction of $\mathbf{B}$, since ∇B always points towards the major axis, i.e.

[58] For opposite sign of σ_0, the radii of both passing and trapped orbits become $r + \Delta r_{\circ,\triangleright}/2$.

$\nabla B/|\nabla B| = -\mathbf{e}_R$. The toroidal direction of $\mathbf{B}$, which leads to downward $\mathbf{B} \times \nabla B$ (frequently denoted as $\mathbf{B} \times \nabla B \downarrow$), is generally known as the *normal*, or *forward*, field direction, and corresponds to a downward drift of ions and upward drift of electrons. The opposite direction of $\mathbf{B}$, for which $\mathbf{B} \times \nabla B$ points upwards (denoted as $\mathbf{B} \times \nabla B \uparrow$) is known as the *reversed* field direction. Evidently, both drift directions are reversed upon reversal of the (toroidal) magnetic field, $\mathbf{B}_T \propto \pm \hat{\mathbf{e}}_\zeta$.

The vertical drift (5.198) is ultimately responsible for the radial excursion of both passing and trapped orbits. The fact that this excursion is larger for trapped particles suggests that they exhibit a longer period of oscillation, also known as the *bounce time*, τ_b, compared to passing particles, which allows the vertical drift more time to pull the GC away from the flux surface. To confirm this prediction, we define the period of oscillation as the integral of $dt = dl/v_\parallel = d\theta/v^\theta$ over one poloidal turn of the orbit,[59]

$$\tau_b \equiv \oint \frac{dl}{v_\parallel} = \oint \frac{d\theta}{v_\parallel b^\theta} \approx \frac{B}{B_P} \int \frac{\sigma r d\theta}{v\sqrt{1 - \lambda(1 - \epsilon \cos\theta)}}. \tag{5.199}$$

Introducing the so-called trapping parameter, $k^2 \equiv [1 - \lambda(1 - \epsilon)]/2\epsilon\lambda$, this result may be expressed in terms of the complete elliptic integral of the first kind, $K(x = [0, 1])$, e.g. Arfken and Weber (2001),

$$\begin{aligned} \text{passing } (k > 1) \quad &\Rightarrow \quad \tau_\circ/\tau_\parallel \approx 2K(k^{-1})/\pi\sqrt{1 - \lambda(1 - \epsilon)}, \\ \text{trapped } (0 \le k < 1) \quad &\Rightarrow \quad \tau_\triangleright/\tau_\parallel \approx 4K(k)/\pi\sqrt{2\epsilon\lambda}, \end{aligned}$$

where $\tau_\parallel(v) \equiv 2\pi q R/v$ is the poloidal rotation period for particles with purely parallel velocities ($v_\parallel = v, \lambda = 0, k \to \infty$), $\tau_\circ(\lambda)$ is the rotation period for passing particles ($k > 1$) with finite perpendicular velocity and $\tau_\triangleright(\lambda)$ is the bounce time of trapped particles ($0 \le k < 1$). In the limits of strongly passing ($k \to \infty$), barely passing/barely trapped ($k \to \mp 1_\mp$) and deeply trapped ($k \to 0$) orbits, the rotation period/bounce time become[60]

$$\begin{aligned} \text{strongly passing } (k \to \infty) \quad &\Rightarrow \quad \tau_\circ/\tau_\parallel \approx 1, \\ \text{barely trapped } (k \to 1) \quad &\Rightarrow \quad (\tau_\circ, \tau_\triangleright)/\tau_\parallel \approx 2\ln[8/(1 - k)]/\pi\sqrt{\epsilon}, \\ \text{deeply trapped } (k \to 0) \quad &\Rightarrow \quad \tau_\triangleright/\tau_\parallel \approx 2k/\sqrt{\epsilon}. \end{aligned}$$

As expected, we find that $\tau_\circ$ approaches $\tau_\parallel$ for strongly passing orbits ($k \to \infty$) and that it takes an infinite time to complete a critical orbit ($k = 1$), although $\tau_\circ$ and $\tau_\triangleright$ approach infinity only logarithmically as $k \to 1$, i.e. the integral across

[59] This expression is analogous to that obtained for the period of a *simple pendulum*, e.g. Goldstein (1980). In fact, the poloidal dynamics of passing and trapped orbits in a small ϵ tokamak are almost identical to those of a simple pendulum, offering a very useful physical model.

[60] Here we use the asymptotic limits $K(k \to 0) \approx \frac{\pi}{2}(1 + \frac{1}{4}k^2)$ and $K(k \to 1) \to \ln(4/\sqrt{1 - k^2})$.

the singularity at $k = 1$ is finite. In general, it takes much longer to complete a typical trapped orbit ($\sqrt{\epsilon} \ll k < 1$) than a typical passing orbit ($k - 1 \gg \sqrt{\epsilon}$), i.e. $\tau_{\triangleright}/\tau_{\circ} \sim 1/\sqrt{\epsilon} \gg 1$, which explains the larger radial excursion of trapped orbits, $\Delta r_{\triangleright}/\Delta r_{\circ} \approx \tau_{\triangleright}/\tau_{\circ} \sim 1/\sqrt{\epsilon} \gg 1$.[61]

5.3.3.2 Radial flows in an axis-symmetric, toroidal plasma

In the above discussion, we neglected the effect of collisions between test particles, whose passing/trapped orbits we analyzed, and field particles representing the plasma background. In reality, such collisions tend to alter test particle orbits by changing v, λ and ψ. Of particular importance within this process, which, as always, is dominated by grazing collisions, is velocity space diffusion, which modifies λ (and k), and after a sufficient time can change the character of the orbit, i.e. scatter particles from passing onto trapped orbits (particle trapping) or vice versa (particle de-trapping).

The importance of such collisional deflection varies depending on the magnitude of the plasma collisionality, ν_s^* (2.13). It is conventional to replace this quantity with a so-called *neoclassical collisionality*,

$$\nu_{neo,s}^* \equiv \nu_{\triangleright s}/\omega_{\triangleright s} = (\nu_s qR/v_{Ts})\epsilon^{-3/2} \sim \nu_s^* \epsilon^{-3/2}, \tag{5.200}$$

where $\nu_{\triangleright s} = \nu_s/\epsilon$ is the effective de-trapping frequency, i.e. a collision frequency for scattering through a finite angle, $\Delta\theta = \sqrt{\epsilon}$, and $\omega_{\triangleright s} = \sqrt{\epsilon}\omega_{\| s}(v_{Ts})$ is the typical bounce frequency with $\omega_{\| s} = 2\pi/\tau_{\| s} = v_s/qR$ being the poloidal rotation frequency for strongly passing particles.

In a small ϵ tokamak, there are three distinct collisionality regimes,

$$\text{banana regime } (\circ, \triangleright \text{ completed}) \quad \Rightarrow \quad \nu_{neo}^* \ll 1, \tag{5.201}$$

$$\text{plateau regime } (\circ \text{ completed}, \triangleright \text{ interrupted}) \quad \Rightarrow 1 \ll \nu_{neo}^* \ll \epsilon^{-3/2}, \tag{5.202}$$

$$\text{P–S regime } (\circ, \triangleright \text{ interrupted}) \quad \Rightarrow \quad \epsilon^{-3/2} \ll \nu_{neo}^*. \tag{5.203}$$

The *banana* regime corresponds to completion of both passing and trapped orbits, the *plateau* regime to completion of passing, but interruption of trapped orbits, and the *Pfirsch–Schlüter* regime to interruption of both types of orbits. For inverse aspect ratios greater than ~0.2, the plateau regime plays a comparatively minor role, and one is justified in speaking of the *collisionless* regime ($\nu^* \ll 1$) and the *collisional* regime ($\nu^* \gg 1$). It should be stressed that the above divisions refer to thermal particles only. Since the collision frequency decreases with velocity as $\nu \propto v^{-3}$, the velocity-dependent collisionality scales as $\nu^*(v) \sim qR\nu/v \propto v^{-4}$. As a

[61] However, since there are fewer trapped particles, $f_{\triangleright}/f_{\circ} \approx f_{\triangleright} \sim \sqrt{\epsilon}$, the sum of radial excursions is comparable for passing and trapped orbits, i.e. $f_{\triangleright}\Delta r_{\triangleright} \sim f_{\circ}\Delta r_{\circ}$.

result, sufficiently fast particles ($v \gg v_t$) are always collisionless, while sufficiently slow particles ($v \ll v_t$) are always collisional!

The key insight which led to the development of neoclassical theory is the somewhat surprising fact that in the presence of toroidal flux surfaces, the lowest-order GC-distribution is Maxwellian, $\overline{f}_s(\mathbf{v}) = f_{Ms}(\mathbf{v}) + O(\delta_s)$, irrespective of the collisionality. To prove this result, it suffices to multiply the zeroth-order drift-kinetic equation, $v_{\parallel s}\nabla_{\parallel}\overline{f}_{s0} = C_s(\overline{f}_{s0})$, (2.118), by $\ln \overline{f}_{s0}$, integrate over velocity space using (5.193), and average over the flux surface using (3.17). The flux surface average on the left-hand side reduces to a form $\langle B\nabla_{\parallel}\mathcal{A}\rangle_B$ with $\mathcal{A} = \int \overline{f}_{s0}(\ln \overline{f}_{s0} - 1)\mathrm{d}\lambda\mathcal{E}\mathrm{d}\mathcal{E}$, which vanishes identically since $\mathcal{A}$ is single valued, recall (2.275). As a result the right-hand side, $\langle \int C_s(f_{s0}) \ln \overline{f}_{s0}\mathrm{d}\mathbf{v}\rangle_B$, must also vanish. Since $C_i \approx C_{ii}$, then by *Boltzmann's H-theorem*, e.g. Huang (1987), we conclude that $\overline{f}_{i0}$ must be Maxwellian.

The situation is more complicated for electrons, since $C_e = C_{ee} + C_{ei}$ consists of two parts, and the H-theorem alone does not imply $\overline{f}_{e0} = f_{Me}$. However, since for slowly moving ions, C_{ei} is given roughly by the self-adjoint *Lorentz* collision operator C^0_{ei} (5.78), for which $\int C_{ei}(f_{e0}) \ln \overline{f}_{e0}\mathrm{d}\mathbf{v} = 0$ if, and only if, $\overline{f}_{e0} = f_{Me}$, we again find that $\overline{f}_{e0}$ must be Maxwellian. In short,

$$\overline{f}_{s0}(\mathbf{v}) = f_{Ms}(\mathbf{v}, \psi) = (\sqrt{\pi} v_{Ts})^{-3} n_s \exp(-v^2/v^2_{Ts}), \tag{5.204}$$

$$n_s = N_s(\psi)\exp(-e_s\varphi/T_s), \qquad v^2_{Ts} = 2T_s(\psi)/m_s, \tag{5.205}$$

where $N_s(\psi)$ and $T_s(\psi)$ are flux surface labels.[62] The particle densities are related by the quasi-neutrality constraint, $n_e = Zn_i$, which by (5.205) implies $\varphi \approx \varphi_0(\psi)$, i.e. that the electric potential is an approximate flux surface label, with only a small deviation of order $e\varphi_1/T_e \sim \delta_i$.

The above result is clearly independent of ν^*, which represents the ratio of magnitudes of the collisional and advective terms in $v_{\parallel s}\nabla_{\parallel}\overline{f}_{s0} = C_s(\overline{f}_{s0})$. For $\nu^* \ll 1$, the left-hand side dominates, so that $\nabla_{\parallel}\overline{f}_{s0} \approx 0$, thus establishing a poloidally uniform $\overline{f}_{s0}(\psi)$ within a few bounce times; collisions then relax this distribution towards a Maxwellian, $\overline{f}_{s0}(\psi) \rightarrow f_{Ms}(\psi)$, on the collisional time scale, $1/\nu_{\triangleright} = \nu^*_{neo}/\omega_{\triangleright}$. The sequence of events is reversed for $\nu^* \gg 1$, when the right-hand side dominates, so that $C_s(\overline{f}_{s0}) \approx 0$. This first establishes a non-uniform LTE, $\overline{f}_{s0} \approx f_{Ms}(\psi, \theta)$, which parallel advection, or rather diffusion, then equilibrates poloidally, $f_{Ms}(\psi, \theta) \rightarrow f_{Ms}(\psi)$, on the diffusive time scale $\nu^{*2}/\nu_{\triangleright}$. In either case, relaxation towards $f_{Ms}(\psi)$ occurs on time scales much shorter than those assumed under the transport ordering (5.1), so that $\overline{f}_{s0} = f_{Ms}(\psi)$ may be assumed to lowest order in δ_s.

62 Although we wrote the result in symmetry co-ordinates, $r_f = r_0 = \psi$, it is valid more generally.

Consequently, neoclassical theory is concerned only with the relatively slow transport processes, i.e. radial collisional flows, which lead to the evolution of flux surface labels, $n_s(\psi)$, $T_s(\psi)$ and $\varphi(\psi)$. These processes (flows) are governed by the first-order drift-kinetic equation (2.118),

$$v_\parallel \nabla_\parallel \overline{f}_{s1} + \mathbf{U}_\perp \cdot \nabla \overline{f}_{s0} + e_s v_\parallel E_\parallel^A \partial_\mathcal{E} \overline{f}_{s0} = C_s^l(\overline{f}_{s1}), \qquad (5.206)$$

where $E_\parallel^A = -\mathbf{b} \cdot \partial_t \mathbf{A}$ is the inductive part of the parallel electric field and C_s^l is the linearized collision operator. The first term in (5.206) represents advection of $\overline{f}_{s1}$ due to parallel GC-motion, while the second term represents advection of $\overline{f}_{s0}$ due to the perpendicular GC-drift (5.198). Since $\nabla \overline{f}_{s0}$ is normal to the flux surface, this term may be written as $(\mathbf{U}_\perp \cdot \nabla\psi)\partial_\psi \overline{f}_{s0}$. The radial component of the GC-drift can be expressed in terms of a parallel gradient by starting from (2.86), exploiting the ambipolarity of radial flows, $J^r = \mathbf{J} \cdot \nabla\psi = 0$, the expression for the vector product $\mathbf{b} \times \nabla\psi$ (3.55), and the assumption of axis-symmetry, $\hat{\mathbf{e}}_\zeta \cdot \nabla = \partial_\zeta = 0$, with the result

$$U^r = \mathbf{U}_\perp \cdot \nabla\psi = \left[v_\parallel \nabla(\frac{v_\parallel}{\Omega}) \times \mathbf{b}\right] \cdot \nabla\psi = v_\parallel (\mathbf{b} \times \nabla\psi) \cdot \nabla\left(\frac{v_\parallel}{\Omega}\right) = v_\parallel I \nabla_\parallel \left(\frac{v_\parallel}{\Omega}\right).$$

Inserting into (5.206), and evaluating the third term using (5.204), we find a magnetic differential equation reminiscent of the Spitzer problem,

$$v_\parallel \nabla_\parallel \left(\overline{f}_{s1} + \frac{I v_\parallel}{\Omega_s} \partial_\psi \overline{f}_{Ms}\right) - \frac{e_s E_\parallel^A}{T_s} v_\parallel \overline{f}_{Ms} = C_s^l(\overline{f}_{s1}). \qquad (5.207)$$

The evolution of flux surface labels is governed by the surface integrals of radial fluxes, e.g. the radial density profile evolves in response to flux surface averaged radial particle flux, multiplied by the flux surface area,

$$\int n_s \mathbf{V}_s \cdot \mathrm{d}\mathbf{S} = \oint \frac{n_s \mathbf{V}_s \cdot \nabla\psi}{\mathbf{B} \cdot \nabla\theta} \mathrm{d}\theta \mathrm{d}\zeta = \oint \frac{n_s V_s^r}{B^\theta} \mathrm{d}\theta \mathrm{d}\zeta = \langle n_s V_s^r \rangle \oint \frac{\mathrm{d}\theta \mathrm{d}\zeta}{B^\theta}.$$

Using the curvilinear vector product (B.9) and recalling that $B^\theta = 1/\sqrt{g_0}$, (3.13), it is possible to express the *radial, contra-variant* component of $\mathbf{V}_s$ as a *toroidal, co-variant* component of $\mathbf{V}_s \times \mathbf{B}$,[63]

$$V_s^r = (\mathbf{V}_s \times \mathbf{B})_\zeta = R\hat{\mathbf{e}}_\zeta \cdot (\mathbf{V}_s \times \mathbf{B}) = R^2 (\mathbf{V}_s \times \mathbf{B})^\zeta, \qquad (5.208)$$

$$\mathcal{A}_\zeta = b_\zeta \mathcal{A}, \qquad b_\zeta = B_\zeta / B = I/B = RB_T/B = Rb_T. \qquad (5.209)$$

Inserting $\mathbf{V}_s \times \mathbf{B}$ from the momentum equation (2.164), and assuming axis-symmetry so that $\hat{\mathbf{e}}_\zeta \cdot \nabla p_s = \partial_\zeta p_s = 0$, we find the following relation,

$$e_s \langle n_s V_s^r \rangle = m_s \partial_t \langle n_s \mathbf{V}_{s\zeta} \rangle + \langle \nabla \cdot \boldsymbol{\pi}_s \rangle_\zeta - e_s \langle n_s E_\zeta^A \rangle - \langle F_{s\zeta} \rangle \approx -e_s \langle n_s E_\zeta^A \rangle - \langle F_{s\zeta} \rangle,$$

[63] It is straightforward to verify this expression using basic vector calculus, e.g. $(\mathbf{V}_s \times \mathbf{B}) \cdot R\hat{\mathbf{e}}_\zeta = (\mathbf{V}_s \times \hat{\mathbf{e}}_\zeta \times \nabla\psi) \cdot \hat{\mathbf{e}}_\zeta = (V_s^r \hat{\mathbf{e}}_\zeta - RV_s^\zeta \nabla\psi) \cdot \hat{\mathbf{e}}_\zeta = V_s^r$.

Table 5.2. *Comparison of classical,* $\mathbf{E} \times \mathbf{B}$ *and neoclassical radial flows.*

Property	Classical	$\mathbf{E} \times \mathbf{B}$	Neoclassical
kinetic origin	$\widetilde{f}_{s1}(\mathbf{x}) \approx -\mathbf{r} \cdot \nabla \overline{f}_{s0}(\mathbf{x})$	$E_T^A = -\partial_t A_T$	$\overline{f}_{s1}(\mathbf{R}) \approx \overline{f}_s - \overline{f}_{Ms}$
GC-drifts	$F_{s\wedge T}/e_s B_P$	$E_{\wedge T}^A/B \approx b_P^2 E_T^A/B_P$	$-(F_{s\parallel T} + e_s E_{\parallel T}^A)/e_s B_P$
fluid origin	friction of gyro-flows	electric drift due to diamagnetic EMF	friction of GC-flows and W–G pinch

where in the last expression we neglected the contribution of inertial and viscous terms, which are smaller by $O(\delta_s)$. In other words, the average radial flow of species s originates in the toroidal inductive electric field and in the toroidal projection of the collisional friction $\mathbf{F}_s$. Since $\nabla\psi \approx RB_P$, we find

$$\langle n_s V_{s\perp} \rangle = \frac{\langle n_s V_s^r \rangle}{|\nabla\psi|} \approx -\frac{\langle n_s E_T^A \rangle + \langle F_{sT} \rangle / e_s}{B_P} \approx -\frac{q}{\epsilon} \left(\frac{\langle n_s E_T^A \rangle + \langle F_{sT}/e_s \rangle}{B_T} \right).$$

It is instructive to break up the toroidal components into their parallel and diamagnetic contributions, i.e. $F_{s\zeta} = F_{s\parallel\zeta} - F_{s\wedge\zeta}$, etc. so that

$$\Gamma_s^r \equiv \langle n_s V_s^r \rangle \approx \langle F_{s\wedge\zeta} \rangle / e_s + \langle n_s E_{\wedge\zeta}^A \rangle - \langle F_{s\parallel\zeta}/e_s + n_s E_{\parallel\zeta}^A \rangle = \Gamma_s^{r,cl} + \Gamma_s^{r,E\times B} + \Gamma_s^{r,neo}.$$

The contra-variant radial particle flow Γ_s^r can thus be decomposed into three parts, see Table 5.2: (i) the classical flow, $\Gamma_s^{r,cl}$, due to collisional friction, $F_{s\wedge}$, between counter-rotating gyro-flows, which leads to an effective radial flow, $\langle F_{s\wedge T} \rangle / e_s B_P$;[64] (ii) the $E \times B$ flow, $\Gamma_s^{r,E\times B}$, due to a radial electric drift, $\langle E_{\wedge T}^A \rangle / B$, caused by the diamagnetic component of the toroidal (inductive) electric field;[65] and (iii) the neoclassical flow, $\Gamma_s^{r,neo}$,

$$\Gamma_s^{r,neo} = -\langle F_{s\parallel\zeta}/e_s + n_s E_{\parallel\zeta}^A \rangle = -I \langle (F_{s\parallel} + e_s n_s E_\parallel^A)/e_s B \rangle, \qquad (5.210)$$

due to two contributions: (iiia) the collisional friction $F_{s\parallel}$ between parallel return flows (of guiding centres), which leads to an effective radial flow $-\langle F_{s\parallel T} \rangle / e_s B_P$; and (iiib) the parallel component of E_T^A, which leads to toroidal advection of (trapped) particles with an electric drift velocity,[66]

$$V_\psi \equiv -\partial_t \psi / \nabla\psi \approx -\langle E_{\parallel T}^A \rangle / B_P \approx -E_T^A / B_P, \qquad V_\perp^{WG} \approx f_\triangleright V_\psi, \qquad (5.211)$$

where the sign of V_ψ reflects the fact that $|B_P|$ and $|\psi|$ increase with r. This so-called *Ware–Galeev pinch* represents an inward movement of banana GCs which

64 This is just the cylindrical result derived previously in (5.142).

65 This term, expressed more generally as $\langle n_s E_{\wedge\zeta} \rangle$, becomes dominant when the electrostatic electric field $E_\wedge = -\nabla\varphi \cdot \hat{\mathbf{e}}_\wedge$ and density n_s fluctuate out of phase due to plasma turbulence.

66 Note that the W–G pinch $\sim f_\triangleright V_\psi$ is larger than the $E \times B$ drift by a factor $f_\triangleright / b_P^2 \sim q^2/\epsilon^{3/2} \gg 1$.

are effectively tied to the flux surface labelled by ψ. To understand this effect, we first note that the (magnitude of) the poloidal flux increases due to a co-current toroidal EMF, $\partial_t\psi = -\partial_t A_\zeta = E_\zeta^A = RE_T^A$, so that the flux surface labelled by ψ moves inward with a velocity V_ψ (5.211); recall that in order to maintain the tokamak discharge in magnetic equilibrium, i.e. to keep $B_P \approx \nabla\psi/R$ at a constant value, poloidal flux must be constantly injected by the toroidal EMF, thus replenishing the flux destroyed by dissipative effects (Joule heating). Since the canonical toroidal momentum $p_\zeta = mv_\parallel b_\zeta - e_s\psi$ is conserved in GC-motion, the tips of the banana orbits, for which $v_\parallel = 0$, move radially so as to preserve $p_\zeta = -e_s\psi$, i.e. experience an inward pinch equal to V_ψ (5.211). The same is not true for passing orbits, since their parallel velocity increases according to $m\dot{v}_\parallel = e_s E_\parallel^A$, thus exactly compensating the increase of ψ to keep p_ζ constant. In other words, only trapped particles are drawn towards the centre of the plasma along with the flux surface labelled by ψ.

Since parallel return flows are driven by toroidal curvature, we may conclude that *neoclassical* radial flows are the toroidal plasma counterpart of *classical* radial flows in a cylindrical plasma, e.g. (5.142). On the kinetic level, neoclassical flows originate in the departure of the GC-distribution from a Maxwellian, $\overline{f}_{s1}(\mathbf{R}) \approx \overline{f}_s(\mathbf{R}) - \overline{f}_{Ms}(\mathbf{R})$, in contrast to classical flows which arise due to the difference between particle and GC-distributions, $\widetilde{f}_{s1}(\mathbf{x}) = f_s(\mathbf{x}) - f_s(\mathbf{R}) \approx -\mathbf{r}\cdot\nabla\overline{f}_{s0}(\mathbf{x})$. The neoclassical radial flows of particles $\langle n_s V_{s\perp}^{neo}\rangle = \langle n_s V_s^{r,neo}\rangle/|\nabla\psi|$ and heat $\langle q_{s\perp}^{neo}\rangle = \langle q_s^{r,neo}\rangle/|\nabla\psi|$ may then be calculated from the velocity space integrals of the advection of these quantities by the radial GC-drift, $U^r/|\nabla\psi|$, weighted by $\overline{f}_{s1}$ [67],

$$\left\langle \begin{matrix} n_s V_s^r \\ q_s^r \end{matrix} \right\rangle^{neo} = \left\langle \int \begin{pmatrix} U_s^r \\ (\frac{1}{2}m_s v^2 - \frac{5}{2}T_s)U_s^r \end{pmatrix} \overline{f}_{s1}\mathrm{d}\mathbf{v} \right\rangle. \tag{5.212}$$

The first of these reduces to the previously obtained result (5.210), while the second may be expressed in terms of the parallel component of the *energy weighted friction*, $\mathbf{G}_s \equiv \mathbf{C}_{s3}$, defined in (2.153) or, more compactly, in terms of the normalized *heat weighted friction*, defined as $\mathbf{H}_s \equiv \mathbf{G}_s/v_{ts}^2 - \frac{5}{2}\mathbf{F}_s$, [68]

$$\frac{\langle q_s^r\rangle^{neo}}{T_s} = \frac{\langle Q_s^r\rangle^{neo}}{T_s} - \frac{5}{2}\langle n_s V_s^r\rangle^{neo} = -\langle H_{s\parallel\zeta}/e_s\rangle = -I\langle H_{s\parallel}/e_s B\rangle. \tag{5.213}$$

Evidently, neoclassical radial flows of particles and heat, (5.210)–(5.213), are proportional to the parallel friction, $F_{s\parallel}$, and heat weighted friction, $H_{s\parallel}$, respectively, both of which vanish identically as $\nu_s^* \to 0$. [69]

67 The integrals may be evaluated by parts, using (5.193), (5.207) and the axis-symmetric U_s^r.

68 The same result follows from $Q_s^r = (\mathbf{Q}_s \times \mathbf{B})_\zeta$, cf. (5.208), evaluated with (2.166).

69 Momentum conservation in Coulomb collisions, $\sum \mathbf{F}_s = 0$, and quasi-neutrality, $\sum e_s n_s = 0$, thus imply that both classical and neoclassical flows are ambipolar, $\langle J^r\rangle = \sum\langle e_s\Gamma_s^r\rangle = 0$.

The above statements pertain to all regimes of collisionality. Nonetheless, different strategies are required to calculate $F_{s\|}$ and $H_{s\|}$ in the collisionless and the collisional regimes. For $\nu_s^* \ll 1$, the first-order drift-kinetic equation (5.207) must be solved for $\overline{f}_{s1}$, while for $\nu_s^* \gg 1$, the kinetic problem may be largely circumvented by solving the Braginskii equations, Section 5.3.2.1, instead. For some applications, e.g. for a multi-species plasma, it is useful to express $F_{s\|}$ and $H_{s\|}$ in terms of gyro-tropic tensors introduced in (2.189). Taking the parallel projection of (2.170) and neglecting the small temporal and inertial terms, the parallel friction and electric force appearing in (5.210) reduce to the parallel hydrodynamic and gyro-viscous forces,

$$\mathcal{F}_{s\|} \equiv F_{s\|} + e_s n_s E_\|^A \approx \mathbf{b} \cdot \nabla \cdot \mathsf{p}_s = (\nabla \cdot \mathsf{p}_s)_\| = \nabla_\| p_s + (\nabla \cdot \boldsymbol{\pi}_{*s})_\|, \qquad (5.214)$$

where $\boldsymbol{\pi}_{*s} = \mathsf{p}_{gts} - p\mathsf{I}$ is the gyro-viscous stress tensor (2.234),

$$\boldsymbol{\pi}_{*s} = p_{\|s}\mathbf{bb} + p_{\perp s}(\mathsf{I} - \mathbf{bb}) - p_s\mathsf{I} = p_{\Delta s}(\mathbf{bb} - \tfrac{1}{3}\mathsf{I}), \quad p_{\Delta s} = p_{\|s} - p_{\perp s}, \quad (5.215)$$

which originates in the anisotropy of $\overline{f}_s$. Since pressure anisotropy, $p_{\Delta s}/p_s$, decreases with increasing ν_s^*, we expect the hydrodynamic force in (5.214) to dominate in the collisional (P–S) regime and the gyro-viscous force to dominate in the collisionless (B–P) regime. It is therefore convenient to rewrite the radial neoclassical flow as a sum of these two contributions,

$$\Gamma_s^{r,neo} = \Gamma_s^{r,PS} + \Gamma_s^{r,BP} = -I\langle \mathcal{F}_{s\|}\mathcal{G}_B/e_s B\rangle - I\langle \mathcal{F}_{s\|}B\rangle/e_s\langle B^2\rangle, \qquad (5.216)$$

where $\mathcal{G}_B$ is defined below in (5.221). It is easy to show that the first term above depends mainly on $\nabla_\| p_s$ and the second only on

$$(\nabla \cdot \boldsymbol{\pi}_{*s})_\| = \mathbf{b} \cdot \nabla \cdot [p_{\Delta s}(\mathbf{bb} - \tfrac{1}{3}\mathsf{I})] = \tfrac{2}{3}\nabla_\| p_{\Delta s} - p_{\Delta s}\nabla_\| \ln B. \qquad (5.217)$$

In particular, evaluating $\langle \mathcal{F}_{s\|}B\rangle$ using (5.214)–(5.217), and noting that the hydrodynamic force vanishes on averaging because $\langle B\nabla_\| \mathcal{A}\rangle = 0$, we find

$$\langle \mathcal{F}_{s\|}B\rangle = \langle B\nabla_\| p_s + B(\nabla \cdot \boldsymbol{\pi}_{*s})_\|\rangle = \langle B(\nabla \cdot \boldsymbol{\pi}_{*s})_\|\rangle = -\langle p_{\Delta s}\nabla_\| B\rangle. \qquad (5.218)$$

The absence of $\nabla_\| p_s$ from this expression suggests that although both terms in (5.216) are driven by a combination of toroidal curvature and inter-particle collisions, they correspond to essentially different physics:

(i) In the collisional (P–S) regime, toroidal curvature gives rise to a vertical GC-drift, which drives a parallel return flow $n_s V_{s\|}$ (3.64); inter-particle collisions then impede (and dissipate) this flow, creating a parallel frictional force $F_{\|s} \propto \nu_s n_s \Delta V_{s\|}$ and setting up a parallel pressure gradient $\nabla_\| p_s$.

(ii) In the collisionless (B–P) regime, toroidal curvature and the related magnetic well produce an anisotropic GC-distribution (recall our discussion of passing and trapped orbits), and the resulting difference in parallel and perpendicular pressures produces the parallel gyro-viscous force (5.218).

As before, an analogous argument may be applied to the neoclassical radial heat flow, with $H_{\|s}$ and y_{*s}/v_{ts}^2 replacing $F_{\|s}$ and π_{*s}, respectively; e.g. the parallel projection of (2.166) in combination with (5.214), yields

$$v_{ts}^2 H_{s\|} \approx \mathbf{b} \cdot (\nabla \cdot \mathsf{y}_{*s}) = (\nabla \cdot \mathsf{y}_{*s})_{\|}, \tag{5.219}$$

which in the collisional regime, with the parallel heat flow, $q_{s\|}$ (3.65), driven by the parallel temperature gradient, $\nabla_{\|} T_s$, becomes $H_{s\|} \propto \nu_s \Delta q_{s\|}$.

It is possible to construct a rigorous theory of neoclassical transport in a multi-species plasma using the above, gyro-fluid approach, e.g. see Hirschman and Sigmar (1981a). While the resulting theory is impressive in its scope and accuracy, the underlying physics tends to become obscured by the complexities of tensorial algebra. Below, we prefer the underlying kinetic treatment, which provides direct insight into the role of passing and trapped particles in the collisionless (B–P) regime. However, we begin with a discussion of the collisional (P–S) regime, which is frequently applicable to the edge of the plasma and thus more relevant to power exhaust.

5.3.3.3 Radial flows in the collisional (Pfirsch–Schlüter) regime

Let us first calculate the ion heat flow, which to a good approximation depends only on the ion distribution. Starting from the ion energy conservation equation, (2.171) with $s = i$ and Braginskii heat flow (5.163)–(5.164), we find that $\nabla_{\|} q_{\|i0} = 0$ and hence $T_{i0} = T_{i0}(\psi)$ to zeroth order in δ_i, and

$$\nabla \cdot (\mathbf{q}_{\|i1} + \mathbf{q}_{*i0}) = \nabla \cdot (\kappa_{\|i} \mathbf{b} \nabla_{\|} T_{i1} + \kappa_{\wedge i} \mathbf{b} \times \nabla T_{i0}) = 0, \tag{5.220}$$

to first order in δ_i. In toroidal geometry, the diamagnetic heat flow has a finite divergence, $\nabla \cdot \mathbf{q}_{*i0} \neq 0$, which requires a parallel Pfirsch–Schlüter heat flow (3.65), and gives rise to a parallel ion temperature gradient,[70]

$$\nabla_{\|} T_{i1} = \left(\frac{I \kappa_{\wedge i}}{\kappa_{\|i}} \right) \mathcal{G}_B T_{i0}', \qquad \mathcal{G}_B \equiv 1 - \frac{B^2}{\langle B^2 \rangle}, \qquad ' = \mathrm{d}_{\psi}. \tag{5.221}$$

Here we evaluated the integration constant from the constraint that T_{i1} is single valued, and hence, must satisfy the identity $\langle B \nabla_{\|} T_{i1} \rangle = 0$.

Inserting this gradient into (5.163), we find the desired radial heat flow,[71]

$$q_i^r = (-\kappa_{\perp i} \nabla_{\perp} T_{i0} + \kappa_{\wedge i} \mathbf{b} \times \nabla T_{i1})^r = - \left[\kappa_{\perp i} |\nabla \psi|^2 + \left(I^2 \kappa_{\wedge i}^2 / \kappa_{\|i} \right) \mathcal{G}_B \right] T_{i0}',$$

[70] As a result, T_i varies poloidally by a factor $T_{i1}/T_{i0} \sim (\nu_i/\Omega_i)(L_{\|}/L_{\perp})(B/B_P) \sim \delta_i \nu_i^* q/\epsilon$.

[71] Recall that in axis-symmetric systems ($\nabla_{\zeta} = 0$) with a helical field, a parallel gradient reflects *poloidal* variation ($\nabla_{\|}/\nabla_{\theta} = b_P \approx |\nabla\psi|/I$), so that $\nabla T_{i1} \propto \hat{\mathbf{e}}_{\theta}$ and $\mathbf{b} \times \nabla T_{i1} \propto \hat{\mathbf{e}}_{\perp}$ is finite.

where the first term is the classical, and the second the neoclassical (P–S) contribution. Both are ultimately driven by the zeroth-order radial temperature gradient: the former directly via $\nabla_\perp T_{i0} \propto T'_{i0}$, and the latter indirectly via the generation of parallel return flows, $\mathbf{b} \times \nabla T_{i1} \propto \nabla_\| T_{i1} \propto T'_{i0}$. It is worth noting that since $\kappa_{\wedge i} \propto B^{-1}$, then $q_i^{r,PS}$ varies poloidally as $B^{-2} - \langle B^2\rangle^{-1} = \mathcal{G}_B/B^2$. Since the flux surface average of this quantity is positive definite, ion heat always flows down the ion temperature gradient.

In a small ϵ tokamak, $\mathcal{G}_B = 2\epsilon\cos\theta + O(\epsilon^2)$, cf. (3.118), and

$$q_{\perp i}^{PS} = q_i^{r,PS}/|\nabla\psi| = -1.6q^2[\epsilon^{-1}\cos\theta + O(1)]q_{\perp i}^{cl}, \tag{5.222}$$

where we evaluated $\kappa_{\wedge i}^2/\kappa_{\| i} \approx 0.8\kappa_{\perp i}$ for a simple plasma using (5.164) and introduced the classical ion heat flow, $q_{\perp i}^{cl} = \kappa_{\perp i}\nabla_\perp T_{i0}$. This indicates an *outflow* of heat on the outboard side, and an *inflow* on the inboard side, of the torus. On taking a flux surface average, these two large, $O(\epsilon^{-1})$, flows nearly cancel, leaving a small, net outflow. To the same accuracy, $\langle \mathcal{G}_B/B^2\rangle \approx 2\epsilon^2/B_0^2 + O(\epsilon^3)$ and the flux surface average of q_i^r becomes,

$$\langle q_{\perp i}\rangle = \left\langle q_{\perp i}^{cl}\right\rangle + \left\langle q_{\perp i}^{PS}\right\rangle = -(1+1.6q^2)\kappa_{\perp i}\nabla_\perp T_{i0}. \tag{5.223}$$

We thus find that the $O(1)$ constant in (5.222) is in fact equal to one, and the neoclassical (P–S) ion heat conductivity is larger than the classical value by a factor of $1.6q^2$. This neoclassical enhancement is most pronounced in the edge of the plasma, where q is largest, e.g. for typical values of $q_a \sim$ 3–5, one finds $1.6q_a^2 \sim$ 15–40 so that the P–S flow completely dominates.[72]

We next turn to consider the transport of species s which in general depends on the velocity distributions of all species s'. The kinetic problem may be tackled using the, by now familiar, Chapman–Enskog expansion in $\Delta_s = \lambda_s/L_\| \ll 1$; the ordering with respect to δ_s and Δ_s will be designated by sub- and superscripts, respectively. To lowest order, the collision term dominates, $C_s\left(f_{s1}^{(-1)}\right) = 0$, so that $f_{s1}^{(-1)} = f_{Ms}(\psi,\theta)$. The next order correction $f_{s1}^{(0)}$ is determined by a parallel Spitzer problem (5.119),

$$\left[A_{s1\|} + \left(x_s^2 - \tfrac{5}{2}\right)A_{2s\|}\right]v_\| f_{Ms}(\psi,\theta) = C_s^l\left(f_{s1}^{(0)}\right), \tag{5.224}$$

where $x_s = (v/v_{T_s})^2$ is the normalized speed and the thermodynamic forces, $\mathbf{A}_{s1}$ and $\mathbf{A}_{s2}$, are given by (5.122). The linear nature of the collision operator implies that $f_{s1}^{(0)}$ must be a linear function of $A_{s1\|}$ and $A_{s2\|}$. These two forces, in turn, can

[72] This enhancement can be even more pronounced in rapidly rotating plasmas, where centrifugal forces introduce a large $O(\epsilon^{-1})$ enhancement of the ion heat outflow on the outboard side.

be related to the friction $F_{s\parallel}$ and heat friction $H_{s\parallel}$ by multiplying (5.224) by 1 and $x_s^2 - \frac{5}{2}$, respectively, and integrating over velocity space,

$$F_{s\parallel} = p_s A_{s1\parallel}, \qquad H_{s\parallel} = \tfrac{5}{2} n_s A_{s2\parallel}. \tag{5.225}$$

The friction(s) can thus be written directly in terms of the parallel flows,

$$\begin{pmatrix} F_{s\parallel} \\ H_{s\parallel} \end{pmatrix} = \sum_{s'} \begin{pmatrix} \ell_{11} & \ell_{12} \\ \ell_{21} & \ell_{22} \end{pmatrix}_{ss'} \begin{pmatrix} V_{s'\parallel} \\ \frac{2}{5} q_{s'\parallel}/p_{s'} \end{pmatrix} \equiv \sum_{s'} \boldsymbol{\ell}_{ss'} \cdot \boldsymbol{\mathcal{V}}_{s'}, \tag{5.226}$$

where $\ell_{ij,ss'}$ are the (inverse) transport coefficients and the matrix $\boldsymbol{\ell}_{ss'}$ is the (inverse) transport matrix, i.e. it is proportional to the inverse of the transport matrix $\boldsymbol{\mathcal{L}}_{ss'}$ which relates flows to forces as $\boldsymbol{\Gamma}_s = -\sum \boldsymbol{\mathcal{L}}_{ss'} \cdot \mathbf{A}_{s'}$.

Inserting (5.226) into (5.210) and (5.213), we find an expression for the neoclassical (P–S) radial flows in terms of the parallel return (P–S) flows,

$$\boldsymbol{\Gamma}_s^{r,neo} \equiv \begin{pmatrix} \langle n_s V_s^r \rangle \\ \langle q_s^r \rangle / T_s \end{pmatrix}^{neo} = -\frac{I}{e_s} \sum_{s'} \left\langle \boldsymbol{\ell}_{ss'} \cdot \boldsymbol{\mathcal{V}}_{s'}^{PS} \frac{\mathcal{G}_B}{B} \right\rangle. \tag{5.227}$$

Combing this expression with the previously calculated P–S flows, (3.64) and (3.65), we find the desired relation between radial flows and gradients,

$$\boldsymbol{\Gamma}_s^{r,neo} = \frac{I^2}{e_s} \left\langle \frac{\mathcal{G}_B}{B^2} \right\rangle \sum_{s'} \frac{\boldsymbol{\ell}_{ss'}}{e_{s'}} \cdot \begin{pmatrix} p_{s'}'/n_{s'} \\ T_{s'}' \end{pmatrix} - I n_s \left\langle \frac{E_\parallel^A \mathcal{G}_B}{B} \right\rangle \begin{pmatrix} 1 \\ 0 \end{pmatrix}, \tag{5.228}$$

and the problem reduces to the determination of the matrix, $\boldsymbol{\ell}_{ss'}$.

To calculate the above matrices for electron transport ($s = e, s' = e, i$) in a simple plasma with ions of arbitrary charge Z, we exploit the Braginskii expressions for $F_{e\parallel}$ and $q_{e\parallel}$ in terms of $\Delta V_\parallel = V_{e\parallel} - V_{i\parallel}$ and $\nabla_\parallel T_e$, i.e. (5.152)–(5.153) and (5.163)–(5.166). Rearranging terms in accordance with (5.225)–(5.226), the electron matrices $\boldsymbol{\ell}_{ee}$ and $\boldsymbol{\ell}_{ei}$ may be inferred as[73]

$$\frac{\boldsymbol{\ell}_{ee}}{m_e n_e \nu_e} = \begin{pmatrix} -\left(r_1^{-1} + r_2^2/r_3\right) & \frac{5}{2}(r_2/r_3) \\ \frac{5}{2}(r_2/r_3) & -\left(\frac{5}{2}\right)^2 r_3^{-1} \end{pmatrix}, \quad \frac{\boldsymbol{\ell}_{ei}}{m_e n_e \nu_e} = \begin{pmatrix} r_1^{-1} + r_2^2/r_3 & 0 \\ -\frac{5}{2}(r_2/r_3) & 0 \end{pmatrix},$$

where r_1, r_2 and r_3 are given in Table 5.1 as a function of Z. Inserting these into (5.228), gives an explicit form of the electron neoclassical flows,

$$\boldsymbol{\Gamma}_e^{r,neo} = \frac{I^2 m_e \nu_e}{e^2} \left\langle \frac{\mathcal{G}_B}{B^2} \right\rangle \left(\frac{\boldsymbol{\ell}_{ee}}{m_e n_e \nu_e} \right) \cdot \begin{pmatrix} p_e' + p_i' \\ n_e T_e' \end{pmatrix} - I n_e \left\langle \frac{E_\parallel^A \mathcal{G}_B}{B} \right\rangle \begin{pmatrix} 1 \\ 0 \end{pmatrix}.$$

73 The zero elements in $\boldsymbol{\ell}_{ei}$ reflect the absence of $\nabla_\parallel T_i$ in the expressions for $F_{e\parallel}$ and $q_{e\parallel}$.

For a small ϵ tokamak, the last term may be neglected, to yield

$$\begin{pmatrix} \langle V_{\perp e}\rangle \\ \langle q_{\perp e}\rangle/p_e \end{pmatrix}^{neo} = 2q^2 D_{\perp e}\left(\frac{\boldsymbol{\ell}_{ee}}{m_e n_e \nu_e}\right)\cdot\begin{pmatrix} (p'_e + p'_i)/p_e \\ T'_e/T_e \end{pmatrix}, \tag{5.229}$$

where $D_{\perp e} = \rho_{te}^2 \nu_e$ was defined on page 198 and the prime now means $' = \mathrm{d}_r$. Comparing (5.229) with (5.184), we once again find that neoclassical flows are larger than the corresponding classical flows by a factor of $\sim 2q^2$. As a result, $q_{\perp i}^{neo}/q_{\perp e}^{neo} \sim q_{\perp i}^{cl}/q_{\perp e}^{cl} \sim (m_i/m_e)^{1/2} \gg 1$, so that power exhaust in the absence of turbulence is dominated by neoclassical ion heat conduction.

Finally, just as for classical flows, see page 199, we briefly discuss neoclassical flows of bulk (hydrogenic) ions in a multi-species plasma containing high-Z impurities with $Z_z \gg 1$. As before, the i–z and e–i Spitzer problems are entirely analogous and $\boldsymbol{\Gamma}_i^{r,neo}$ may be obtained from $\boldsymbol{\Gamma}_e^{r,neo}$, (5.228), by making the substitutions $i \to e$, $z \to i$. Thus, for a small ϵ tokamak,

$$\begin{pmatrix} \langle V_{\perp i}\rangle \\ \langle q_{\perp i}\rangle/p_i \end{pmatrix}^{neo} = 2q^2 D_{\perp i}\left(\frac{\boldsymbol{\ell}_{ii}}{m_i n_i \nu_i}\right)\cdot\begin{pmatrix} p'_i/p_i - (T_z/Z_z T_i)p'_z/p_z \\ T'_i/T_i \end{pmatrix},$$

where $D_{\perp i} = \rho_{ti}^2 \nu_{iz}$ was defined in (5.188) and the matrix coefficients are identical (in terms of r_1, r_2 and r_3) to those of $\boldsymbol{\ell}_{ee}/m_e n_e \nu_e$ in (5.229), but with the three constants now evaluated from Table 5.1 using the relative ion charge $\alpha_{zi} \equiv n_z e_z^2/n_i e_i^2$ instead of the bulk ion charge $Z = e_i/e$. This substitution is required to ensure an identical ratio of the dominant collision operators in the two Spitzer problems, i.e. $C_{iz}/C_{ii} = \nu_{iz}/\nu_{ii} \sim \alpha_{iz}$ on the one hand and $C_{ei}/C_{ee} = \nu_{ei}/\nu_{ee} \sim n_i e_i^2/n_e e^2 = e_i/e = Z$ on the other.[74]

As ever, the neoclassical (P–S) radial flows of ions $\langle n_i V_{\perp i}\rangle^{neo}$ and ion heat $\langle q_{\perp i}\rangle^{neo}$ are larger than the classical values by the familiar factor $\sim 2q^2$. The same enhancement factor applies to the inflow (pinch) of high-Z impurities, $\langle n_z V_{\perp z}\rangle^{neo} \approx -\langle n_i V_{\perp i}\rangle^{neo}/Z_z$, required by the smallness of the electron flow, $D_{\perp e}/D_{\perp i} \sim (m_e/m_i)^{1/2}$, and the quasi-neutrality constraint, $\langle J^r\rangle = 0$. Since both flows are augmented by the same factor, neoclassical peaking of impurity concentration, (5.189), is equal to the classical value.[75] Finally, comparing the ion heat diffusivities for pure and impure plasmas,

$$\chi_{\perp i}^{neo} = 1.6(q\rho_{ti})^2\nu_i, \qquad \chi_{\perp iz}^{neo} = 3.9(q\rho_{ti})^2\nu_{iz} = 3.9\alpha_{iz}(q\rho_{ti})^2\nu_{ii},$$

we note that i–z collisions enhance bulk ion heat transport by a factor of $\approx 3.5\alpha_{iz} \sim 3(Z_{eff} - 1)$, which is typically $\sim$ 3–6 in the edge of the plasma.

[74] For hydrogenic bulk ions, the relative ion charge reduces to $\alpha_{zi} = Z_{eff} - 1$, which is typically $\sim$ 1–2 in the edge of the plasma. It thus satisfies the requirement $\alpha_{iz} \gg (m_e/m_i)^{1/2}$, which is necessary for $C_{ie} \ll C_{ii}, C_{iz}$, so that i–e collisions can be neglected in the impurity problem.

[75] Since $Z_z \gg 1$, this impurity influx is driven predominantly by the bulk ion pressure gradient.

5.3.3.4 Radial flows in the collisionless (banana-plateau) regime

Neoclassical transport in the collisionless (banana-plateau) regime has an inherently kinetic character, involving collisional interactions between passing and trapped particles. Calculation of the parallel frictions, $F_{s\|}$ and $H_{s\|}$, and hence, of the radial flows, requires a solution of the first-order drift-kinetic equation, (5.207), for the perturbed distribution, $\overline{f}_{s1}$ (one for each species). Since the related analysis involves substantially more space than can be allocated here, we restrict ourselves to a presentation (without proof), and a brief discussion, of the final results for a small ϵ tokamak, referring the reader to the references in Section 5.4 for both the account of the derivation and for results pertaining to arbitrary aspect ratios, non-circular poloidal cross-sections and non-axis-symmetric geometries, e.g. stellarators.

In the plateau regime, neoclassical transport becomes dominated by a small subset of passing particles whose effective collision frequency, $\nu_{\triangleright s} = \nu_s/\epsilon$, is close to their poloidal rotation frequency $\omega_{\|s} = v_s/qR$. As a result of this resonance between $\nu_{\triangleright s}$ and $\omega_{\|s}$, radial flows become independent of the nature of the collisional process, i.e. of the collision operator $C_s(f_s)$.[76] The resonance can be exploited to calculate the perturbation $\overline{f}_{1s}$, and the associated neoclassical radial flows. It is convenient to express these directly in terms of the radial density and temperature gradients,[77]

$$\begin{pmatrix} \langle V_{\perp e}\rangle^{\circ} \\ \langle q_{\perp e}\rangle^{\circ}/p_e \\ \langle q_{\perp i}\rangle^{\circ}/p_i \end{pmatrix} = -D^{\circ}_{\perp e}\begin{pmatrix} 1 & 3/2 & 3\vartheta/2 \\ 1/2 & 15/4 & 3\vartheta/4 \\ 0 & 0 & 3\varsigma_{\circ} \end{pmatrix}\begin{pmatrix} (1+\vartheta)n_e'/n_e \\ T_e'/T_e \\ T_i'/T_i \end{pmatrix},$$

where $D^{\circ}_{\perp e} \equiv (q\rho_{te})^2\omega_{\|te}\sqrt{\pi/2}$ is the plateau electron diffusivity, $\omega_{\|ts} = \omega_{\|s}(v_{ts}) = v_{ts}/qR$ the thermal poloidal rotation frequency for species s, and ϑ and $\varsigma_{\circ}$ involve ratios of electron and ion temperatures, masses and charges: $\vartheta \equiv T_i/ZT_e$ and $\varsigma_{\circ} \equiv \vartheta^{3/2}\sqrt{m_i/Zm_e} \gg 1$.

The plateau-regime electron and ion heat diffusivities are now found as

$$\chi^{\circ}_{\perp e} = 3.75D^{\circ}_{\perp e}, \qquad \chi^{\circ}_{\perp i} = 3\varsigma_{\circ}D^{\circ}_{\perp e} = 3\sqrt{\pi/2}(q\rho_{ti})^2\omega_{\|ti}, \tag{5.230}$$

respectively. As expected, these transport coefficients do not dependent on ν_s, and correspond (aside from numerical factors of order unity) to the P–S coefficients with ν_s replaced by $\omega_{\|ts}$.

The situation becomes more complicated in the banana regime, where both passing and trapped particles must be considered. The drift-kinetic equation may be

[76] That is, radial flows form a plateau when plotted vs. ν^*_{neo}; hence, the label *plateau* regime.

[77] Note that the resulting matrix is not symmetric, since the chosen driving terms differ from the thermodynamic forces, $\mathbf{A}_s$. However, Onsager symmetry is easily demonstrated, by transforming the above transport relations into the $\mathbf{A}_s$-form, cf. (5.184).

solved by employing an expansion in the small parameter $1/\Delta_s \sim \nu_s^* \ll 1$, i.e. the inverse of the Chapman–Enskog expansion. For a simple, hydrogenic plasma ($Z = 1$) the resulting radial flows are found as

$$\begin{pmatrix} \langle V_{\perp e}\rangle^{\triangleright} \\ \langle q_{\perp e}\rangle^{\triangleright}/p_e \\ \langle q_{\perp i}\rangle^{\triangleright}/p_i \end{pmatrix}$$

$$\approx -D_{\perp e}^{\triangleright} \begin{pmatrix} 1.53 & -0.59 & -0.26\vartheta \\ -2.12 & 2.51 & 0.37\vartheta \\ 0 & 0 & 0.92\varsigma_{\triangleright} \end{pmatrix} \begin{pmatrix} (1+\vartheta)n_e'/n_e \\ T_e'/T_e \\ T_i'/T_i \end{pmatrix} + \boldsymbol{\Gamma}_{\perp}^{WG},$$

where $\varsigma_{\triangleright} \equiv v_{te}/v_{ti} \gg 1$ is the ratio of electron and ion thermal speeds and $\boldsymbol{\Gamma}_{\perp}^{WG}$ are the *Ware–Galeev* flows, already introduced on page 207,

$$\boldsymbol{\Gamma}_{\perp}^{WG} \equiv \begin{pmatrix} \langle V_{\perp e}\rangle^{WG} \\ \langle q_{\perp e}\rangle^{WG}/p_e \\ \langle q_{\perp i}\rangle^{WG}/p_i \end{pmatrix} \approx \frac{f_{\triangleright} E_{\parallel}^A}{B_P} \begin{pmatrix} -1.66 \\ 1.19 \\ 0 \end{pmatrix}. \tag{5.231}$$

Here $D_{\perp e}^{\triangleright} \equiv f_{\triangleright}\rho_{Pe}^2\nu_e = f_{\triangleright}(q\rho_{te}/\epsilon)^2\nu_e$ is the banana-regime electron diffusivity (in the limit $Z \to \infty$), $\rho_{Pe} = \rho_{te}/b_P = \rho_{te}q/\epsilon$ is the electron poloidal gyro-radius, and $f_{\triangleright}$ is the *effective* trapped particle fraction, defined as

$$f_{\triangleright} \equiv 1 - f_{\circ} \equiv 1 - \frac{3}{4}\int_0^{\lambda_c} \langle\sqrt{1-\lambda B/B_0}\rangle^{-1}\lambda d\lambda \approx 1.46\sqrt{\epsilon}, \tag{5.232}$$

with the last expression evaluated for a small ϵ tokamak. In the opposite limit ($Z \to \infty$) the above matrices become,

$$\begin{pmatrix} 1 & -1/2 & -0.173\vartheta \\ -3/2 & 7/4 & 0.26\vartheta \\ 0 & 0 & 0.92\varsigma_{\triangleright} \end{pmatrix}, \quad \begin{pmatrix} -1 \\ 0 \\ 0 \end{pmatrix}.$$

In this limit, only the particle W–G flow, also known as the *Ware pinch* remains. It represents an inward drift of *trapped* particles due to a temporal increase of the poloidal flux ψ caused by the toroidal inductive electric field E_T^A together with the conservation of canonical toroidal momentum p_{ζ}, see Table 5.2, and discussion on page 207.[78]

Not surprisingly, one finds that trapped particles, which make larger excursions away from the flux surface, dominate the radial flows, i.e.

$$D_{\perp}^{\triangleright}/D_{\perp}^{\circ} \sim f_{\triangleright}\Delta r_{\triangleright}^2\nu_{\triangleright}/f_{\circ}\Delta r_{\circ}^2\nu_{\circ} \sim \Delta r_{\triangleright}\nu_{\triangleright}/\Delta r_{\circ}\nu_{\circ} \sim \epsilon^{-3/2} \gg 1, \tag{5.233}$$

[78] Note that passing particles are not affected by this pinch, since their parallel velocity is allowed to increase in response to E_T^A.

where we used $f_\triangleright/f_\circ \sim \sqrt{\epsilon}$, $\nu_\triangleright/\nu_\circ \sim 1/\epsilon$, and (5.196)–(5.197). Although the sum of radial excursion is comparable for passing and trapped orbits ($f_\triangleright \Delta r_\triangleright \sim f_\circ \Delta r_\circ$), the latter dominate radial transport due to the square dependence of the diffusion coefficient on the radial step size. For instance, $D^\triangleright_{\perp e}$ represents a random walk of trapped electrons with a radial step equal to their banana width, $\Delta r_{\triangleright e} \sim \rho_{Pe}$, and a step time equal to the inverse of the effective de-trapping frequency, $\nu_{\triangleright e}$; as a result, it is greater than the P–S value by the large factor $\epsilon^{-3/2} \gg 1$. The same is true of the banana-regime electron and ion heat diffusivities, which may be inferred from the above as

$$\chi^\triangleright_{\perp e} = 0.54 r_4 D^\triangleright_{\perp e} = 0.54 r_4 f_\triangleright \rho^2_{Pe} \nu_e = 0.79 r_4 \sqrt{\epsilon}(q\rho_{te}/\epsilon)^2 \nu_e, \tag{5.234}$$

$$\chi^\triangleright_{\perp i} = 0.92 \varsigma_\triangleright D^\triangleright_{\perp e} = 0.92 f_\triangleright \rho^2_{Pi} \nu_i = 1.35\sqrt{\epsilon}(q\rho_{ti}/\epsilon)^2 \nu_i, \tag{5.235}$$

where r_4 is given in Table 5.1 and $\chi^\triangleright_{\perp i}/\chi^\triangleright_{\perp e} \sim \chi^{cl}_{\perp i}/\chi^{cl}_{\perp e} \sim \sqrt{m_i/m_e} \gg 1$.

Finally, the average radial ion heat flow for arbitrary aspect ratio and plasma shape may be calculated as (Taguchi, 1988),

$$\frac{\langle q^r_i \rangle^\triangleright}{p_i} = -2\left(\frac{I v_{ti}}{\Omega_{i0}}\right)^2 \nu_i \left(\left\langle \frac{B_0^2}{B^2}\right\rangle - \frac{f_\circ}{f_\circ + 0.462 f_\triangleright}\right)\frac{T_i'}{T_i}, \tag{5.236}$$

where $\Omega_{i0} = e_i B_0/m_i$ is the ion gyro-frequency at the nominal field B_0; this expression is valid in both the $\epsilon \to 0$ ($f_\triangleright \to 0$) and $\epsilon \to 1$ ($f_\triangleright \to 1$) limits.

5.3.3.5 *Flux surface flows in axis-symmetric geometry*

As our final topic, we examine the ion and electron flows within an axis-symmetric flux surface in the three collisionality regimes. The character of these flows was already discussed in Section 3.2.2, where the particle flow of species s was expressed as a sum of toroidal and parallel flows, (3.62),

$$n_s \mathbf{V}_s = n_s \omega_s(\psi) R\hat{\mathbf{e}}_\zeta + K_s(\psi) B\mathbf{b}, \qquad \omega_s(\psi) = -\varphi' - p_s'/n_s e_s, \tag{5.237}$$

with the former representing rigid-body *toroidal rotation* with a frequency ω_s, and the latter *poloidal rotation* with a velocity, $V_{Ps} = (K_s/n_s)B_P$. The related parallel flow (3.60), consists of a return (P–S) part with a vanishing flux surface average, $\langle V^{PS}_{\parallel s} B\rangle = 0$, and an average (Ohmic) part, $\langle V^{\Omega}_{\parallel s} B\rangle = (K_s/n_s)\langle B^2\rangle$, which represents average poloidal rotation, i.e.

$$\langle V_{Ps}\rangle = (K_s/n_s)\langle B_P\rangle, \qquad \langle V_{\parallel s} B\rangle = \omega_s R B_T + (K_s/n_s)\langle B^2\rangle. \tag{5.238}$$

It is worth stressing that ω_s reflects the first-order gyro-flow (5.142), which in the assumed drift ordering, consists of the electric and diamagnetic drifts, see (2.208).[79] As a result, toroidal rotation depends only indirectly on collisionality, while poloidal rotation is a strong function of ν_s^*.

In the collisional (P–S) regime, poloidal rotation is unspecified, or rather is determined by additional physics, such as external momentum injection or viscous transport from neighbouring flux surfaces. In the plateau regime, the free movement of passing particles facilitates the evolution of poloidal flows, which then influence the average radial flows via the parallel friction experienced by the moving guiding centres. Hence, the average poloidal velocity may be determined from the ambipolarity constraint, $\langle J^r \rangle = 0$.

The situation is similar in the banana regime, where again only passing particles contribute to poloidal rotation. Clearly, trapped particles cannot rotate poloidally, being confined by the magnetic well. However, they can, and do, contribute to toroidal rotation via a gradual precession of the tips of their banana orbits, which represent a toroidal drift of the guiding centre of the banana orbit, and by an effective diamagnetic flow of banana guiding centres, see below. For a simple plasma in a small ϵ tokamak, the poloidal velocity in the two regimes may be calculated as

$$K_i(\psi)/n_i \approx c_\circ T_i'/e_i B_0, \quad \Rightarrow \quad V_{Pi} \approx c_\circ b_P T_i'/e_i \approx c_\circ \rho_{ti} v_{ti} T_i'/T_i \sim c_\circ \delta_i v_{ti},$$

where $c_\circ = -1/2$ in the plateau regime, and $c_\circ = 1.17 f_\circ$ in the banana regime; note that the latter is proportional to the passing particle fraction, so that V_{Pi} vanishes (as it must) in the limit $\epsilon \to 1$ ($f_\triangleright \to 1$, $f_\circ \to 0$).

As expected, we find the poloidal flows to be drift-ordered and smaller than the corresponding toroidal and parallel flows by a factor $b_P = \epsilon/q$, i.e. $V_{Pi} \sim b_P V_{Ti} \sim b_P V_{\parallel i} \sim \delta_i v_{ti}$.[80] More importantly, we find that *poloidal* rotation is driven solely by the ion temperature gradient, in contrast to *toroidal* rotation, which is driven by the ion pressure and electric potential gradients which enter into the definition of $\omega_i(\psi)$, see (5.237).

The average flow of charge associated with $\langle V_{\parallel s} B \rangle$ is just the average current $\langle J_\parallel B \rangle$ appearing in (3.57). In the collisional (P–S) regime, $\langle J_\parallel B \rangle = \langle J_\parallel^\Omega B \rangle$ equals the average Ohmic current, i.e. the flow of electrons in response to the parallel component of the inductive electric field,

$$J_\parallel^\Omega = \sigma_e E_\parallel^A, \qquad \langle J_\parallel^\Omega B \rangle = \sigma_e \langle E_\parallel^A B \rangle, \qquad \sigma_e = \sigma_{\parallel e} = r_1 e^2 n_e / m_e \nu_e \qquad (5.239)$$

where σ_e is the parallel electrical conductivity (5.154). This result remains roughly valid in the plateau regime, but not in the banana regime, in which particle trapping modifies the average parallel current in two ways:

[79] This perpendicular flow can be written as $\mathbf{V}_{\wedge s} = \omega_s R \hat{\mathbf{e}}_\wedge = \omega_s R(\hat{\mathbf{e}}_\zeta - b_T \mathbf{b})$, (3.58).

[80] This is of course consistent with the assumed transport ordering (5.1), which requires all particle flows within the flux surface to be drift-ordered, irrespective of the collisionality regime.

(i) since only passing particles contribute to the parallel electrical conductivity, the Ohmic current is reduced by the passing fraction, $f_\circ$;

(ii) the presence of radial gradients of $n_s(\psi)$ or $T_s(\psi)$, i.e. of the density or average energy of banana GCs, implies more particles (for $n_s' < 0$) or more energetic particles (for $T_s' < 0$) on inbound, compared to outbound, banana orbits, e.g. $n_e', T_e' < 0$ results in more counter-moving than co-moving electrons, when viewed at a given location at the outer mid-plane.

The related perturbation of the GC-distribution can be estimated as

$$\overline{f}_{s1}^{\triangleright} \approx \overline{f}_s(\psi + \Delta\psi) - \overline{f}(\psi) \approx \Delta\psi\, \partial_\psi f_{Ms} = -\frac{I v_\parallel}{\Omega_s} \partial_\psi f_{Ms}, \tag{5.240}$$

which is the same as the perturbation due to a GC-drift, see (5.207). When integrated over **v** and summed over all (both) species, this perturbation gives rise to a toroidal 'diamagnetic' current in close analogy to the regular diamagnetic current, see Section 2.4.2.3, and in agreement with (5.237),

$$J_\parallel^{\triangleright} = \sum_s e_s \int v_\parallel \overline{f}_{s1}^{\triangleright} \mathrm{d}\mathbf{v} = -Rb_T \sum_s p_s'(\psi) = Rb_T \sum_s e_s n_s \omega_s(\psi). \tag{5.241}$$

Comparison with (3.51)–(3.52) confirms that this current is purely toroidal, i.e. $K(\psi) = 0$, $J_T^{\triangleright} = -p'R$ and $\mathbf{J}^{\triangleright} = J_T^{\triangleright}\hat{\mathbf{e}}_\zeta$. Indeed, in a tokamak with unit aspect ratio ($\epsilon = 1$), in which all the particles are trapped, (5.241) constitutes the entire plasma current. For $\epsilon < 1$, $J_T^{\triangleright}$ is collisionally coupled to the passing particles, driving an additional parallel (poloidal) current, which is larger than its 'diamagnetic' seed by a factor $1/\sqrt{\epsilon}$. The net parallel current produced in this way is generally known as the *bootstrap current* and is denoted by $J_\parallel^{bs}$. It is defined as the self-generated (non-inductive) toroidal current driven by neoclassical radial diffusion, i.e. due to collisional coupling of $J_T^{\triangleright}$ and J_P°, whose sum is the net $J_\parallel^{bs}$. The label *bootstrap* emphasizes the fact that it is generated by the confined plasma itself. However, this terms is a little misleading, since it ignores the even more basic 'bootstrap' effect of the plasma, namely the establishment of the ordinary (quasi-poloidal) diamagnetic current. It should be stressed that $J_\parallel^{bs}$ flows in the same direction as $J_\parallel^{\Omega}$, thus increasing the poloidal magnetic field and improving plasma confinement. More importantly, it reduces the amount of external current drive (inductive or auxiliary) needed to maintain the plasma in equilibrium – a major benefit for achieving steady tokamak operation!

A crude estimate of the bootstrap current in a small ϵ tokamak, in which $f_\triangleright \sim \sqrt{\epsilon} \ll 1$, was already given in (4.206), as the 'diamagnetic current' (5.241) multiplied by the trapped particle fraction $f_\triangleright$. A more accurate calculation yields the following average parallel current, which is made up of both bootstrap and Ohmic contributions, i.e. $\langle J_\parallel \rangle = \langle J_\parallel^{bs} \rangle + \langle J_\parallel^{\Omega} \rangle$, or

$$\langle J_\parallel \rangle = -\frac{f_\triangleright p_e}{B_P}\left[c_n(1+\vartheta)\frac{n_e'}{n_e} + c_{Te}\frac{T_e'}{T_e} + c_{Ti}\frac{T_i'}{T_i}\right] + (1 - f_\triangleright c_E)\sigma_e E_\parallel^A,$$

with $(c_n, c_{Te}, c_{Ti}, c_E) = (1, 1, -0.173, 1)$ in the Lorentz limit $(Z \to \infty)$ and $(1.67, 0.47, -0.29, 1.31)$ in the hydrogenic limit $(Z = 1)$. It is worth noting that the bootstrap current due to the density gradient is the Onsager symmetric counterpart of the W–G pinch due to a parallel inductive field.[81]

The ratio of the two terms above, i.e. of the bootstrap and Ohmic currents may be estimated as $J_\parallel^{bs}/J_\parallel^\Omega \sim (f_\triangleright/f_\circ)\beta_P \sim \sqrt{\epsilon}\beta_P$. Recalling that $J_\parallel^{PS}/J_\parallel^\Omega \sim \epsilon\beta_P$, see page 100, we find the following ordering of the magnitudes of Ohmic, bootstrap and Pfirsch–Schlüter currents:

$$J_\parallel^\Omega : J_\parallel^{bs} : J_\parallel^{PS} \sim 1 : \sqrt{\epsilon}\beta_P : \epsilon\beta_P. \tag{5.242}$$

Since $\beta_P \sim 1$ in the (Ohmic) tokamak ordering, the three currents are ordered by $\sqrt{\epsilon}$. In most of today's large tokamaks, this quantity is actually not small near the edge of the plasma, e.g. $\epsilon_a = a/R_0 \approx 0.3$, $\sqrt{\epsilon_a} \approx 0.6$ for both JET and ITER, see Table 8.1, so that the three currents in (5.242) become comparable. Of course, close to the plasma centre $\epsilon = r/R_0$ is always much less than one and the Ohmic current clearly dominates.

5.4 Further reading

There are many fine treatments of physical kinetics in gases and plasmas. The following are some of the best accounts. *Coulomb collisions*: Landau and Lifschitz (1960); Goldstein (1980). *Kinetic theory of gases*: Chapman and Cowling (1939); Lifschitz and Pitaevskii (1981); Gombosi (1994). *Collisional transport in gases/neutral fluids*: Bird *et al.* (1960); Streeter *et al.* (1962); Batchelor (1967). *Kinetic theory of plasmas*: Shkarofsky *et al.* (1966); Ichimaru (1973, 1992). *Classical and neoclassical transport in magnetized plasmas*: Braginskii (1965); Hinton and Hazeltine (1976); Galeev and Sagdeev (1979); Hirschman and Sigmar (1981a); Hinton (1983); Balescu (1988a,b); Hazeltine and Meiss (1992); Helander and Sigmar (2002).

81 This instance of Onsager symmetry is easily demonstrated by observing that the transport coefficients relating flows and forces in the two processes are identical for all values of Z.

6

Turbulent transport in magnetized plasmas

'... where water turbulence originates, where it persists, and where it dies away ...'

Leonardo Da Vinci (c. 1500)

'At infinite Reynolds numbers, all the small scale statistical properties (of a turbulent flow) are uniquely and universally determined by the (eddy) scale ℓ and the mean energy dissipation rate ε. At very high, but finite Reynolds numbers, they are determined by ℓ, ε and the kinematic viscosity ν.'

Kolmogorov (1941)

In the previous chapter, we investigated the relatively slow, transport-ordered, evolution of quasi-equilibrium plasma quantities due to radial collisional flows. In this chapter we consider similar evolution due to radial flows associated with plasma turbulence. As always, we will first try to shed some light on the issue by investigating the relatively simple case of hydrodynamic turbulence, Section 6.1, then proceed to turbulence in magneto-hydrodynamics, Section 6.2, and finally, turn to the dominant process in magnetized plasmas, namely, drift-wave turbulence, Section 6.3.

6.1 Hydrodynamic turbulence

What is turbulence?[1] We know it when we see it, yet it is not easy to define. A compact, yet accurate, definition has been formulated by Corrsin (1961),

'Incompressible hydrodynamic turbulence is a spatially complex distribution of vorticity which advects itself in a chaotic manner in accordance with (4.13). The vorticity field is

[1] The general concept of turbulence is of course not restricted to fluid mechanics, but applies to all dynamical systems with many degrees of freedom. Unless otherwise stated, however, we will henceforth use the term *turbulence* specifically in relation to gas, liquid or plasma dynamics.

random in both space and time, and exhibits a wide and continuous distribution of length and time scales.'

The appearance of vorticity, $\mathbf{\Omega}$, as opposed to velocity, $\mathbf{V}$, in the above can be traced to Kelvin's vorticity theorem (4.13)–(4.15), which states that $\mathbf{\Omega}$ can only be created/destroyed by vortex stretching/compression, $\mathbf{\Omega} \cdot \nabla \mathbf{V}$, and viscous damping, $\nu \nabla^2 \mathbf{\Omega}$, and is otherwise, merely advected by the flow. In other words, vorticity lines (tubes) are *frozen into* an HD flow, in exactly the same way as magnetic lines (tubes) are frozen into an MHD flow, making $\mathbf{\Omega} = \nabla \times \mathbf{V}$ on the one hand, and $\mathbf{B} = \nabla \times \mathbf{A}$ on the other, into local quantities of the flow. The same is evidently not true of the velocity $\mathbf{V}$ which is governed by the N–S equation (4.2) in which the pressure appears explicitly. As a result, the divergence of $\mathbf{V} \cdot \nabla \mathbf{V}$ sends out sound waves, which in an incompressible flow ($V \ll V_S$), propagate nearly instantaneously to set up pressure gradients, $\sim \nabla^2 (p/\rho)$, far away from the original flow. These pressure disturbances then drive flows in their vicinity, effectively coupling the entire velocity field and making $\mathbf{V}$ into a highly non-local quantity.

According to Corrsin's definition, turbulence involves self-advection of vorticity over a wide range of scales and hence the dynamical interaction of the many vortices composing a turbulent flow field. This implied presence of many degrees of freedom is what distinguishes *turbulence* from *chaos*. Recall that dynamical chaos refers only to extreme sensitivity to initial conditions, which can be achieved with as few as three degrees of freedom, e.g. three bodies coupled by gravity. This largely explains why recent progress in *chaos* theory has not produced a definitive theory of *turbulence*.

Although Corrsin's definition contains all the key features of hydrodynamic turbulence, its very terseness makes these somewhat difficult to unravel. A more substantial definition was put forward by Tennekes and Lumley (1972), who suggest that all turbulent flows are characterized by:

- *motion* – turbulence is a property of the flow and not of the fluid;
- *irregularity* or *randomness*, which calls for a statistical description,[2] i.e. a theory of turbulence aims to predict probabilities rather than outcomes;
- *rapid mixing* of local quantities, which increases their effective diffusivity.[3] From the practical point of view, this enhancement of transport is the most important difference between laminar and turbulent flows;
- *large Reynolds numbers*, $\mathrm{Re}_{\mathcal{K}} = \ell_{\mathcal{K}} V_{\mathcal{K}} / \nu$, which imply that the Navier–Stokes equation, (4.2), or (4.13), is dominated by the inertial (advective) term, $\mathbf{V} \cdot \nabla \mathbf{V} = \frac{1}{2} \nabla V^2 -$

[2] There is an interesting historical analogy between turbulence theory and both statistical and quantum mechanics: in the early days of their development, each field was treated with suspicion, and even discounted as pseudo-science, on account of the perceived lack of determinism.

[3] An experience easily verified by stirring a cup of tea to more quickly dissolve a drop of cream.

$\mathbf{V}\times\mathbf{\Omega}$, over a large range of spatial scales, aptly known as the *inertial sub-range*, defined as $\ell_{\mathcal{K}} \ll \ell \ll \ell_\varepsilon$, where $\ell_{\mathcal{K}}$ and ℓ_ε are the integral and dissipative scales, respectively, see page 231;

- *non-linearity* of this inertial term, which both creates vorticity, via vortex (vorticity tube, turbulent *eddy*) stretching, $\mathbf{\Omega}\cdot\nabla\mathbf{V}$, and advects vorticity, via vortex–vortex interaction, $\mathbf{V}\cdot\nabla\mathbf{\Omega}$, and thus produces highly rotational, fluctuating flows composed of vortices of sizes ranging from $\ell_{\mathcal{K}}$ to ℓ_ε;
- *collisional dissipation*, e.g. by viscous heating in incompressible flows,

$$\tfrac{1}{2}\partial_t V^2 + \nabla\cdot\left[\left(\tfrac{1}{2}V^2 + p/\rho\right)\mathbf{V} + \nu\mathbf{\Omega}\times\mathbf{V}\right] = -\varepsilon, \tag{6.1}$$
$$\varepsilon \equiv \pi : \{\nabla\mathbf{V}\}^*/\rho = 2\nu\{\nabla\mathbf{V}\}^* : \{\nabla\mathbf{V}\}^* = \tfrac{1}{2}\nu\mathsf{W} : \mathsf{W} = \nu\mathbf{\Omega}\cdot\mathbf{\Omega} = \nu\Omega^2,$$

which transfers kinetic into thermal energy at sufficiently small scales. The scale at which dissipative and inertial terms become comparable, $\mathrm{Re}_\varepsilon = \ell_\varepsilon V_\varepsilon/\nu \sim 1$, is known as the *dissipative scale*, ℓ_ε.

It can be added that *fluid turbulence* assumes the applicability of *continuum* mechanics, i.e. of the fluid approximation, at all spatial scales. Finally, we note that $V_{\mathcal{K}}$ and V_ε, entering the definition of $\mathrm{Re}_{\mathcal{K}}$ and Re_ε, represent the root-mean-square (RMS) fluctuating velocities at scales $\ell_{\mathcal{K}}$ and ℓ_ε,

$$V_{\mathcal{K}} = \sqrt{\langle \widetilde{V}^2[\ell\sim\ell_{\mathcal{K}}]\rangle}, \qquad V_\varepsilon = \sqrt{\langle \widetilde{V}^2[\ell\sim\ell_\varepsilon]\rangle}, \tag{6.2}$$

typically defined by low (high) pass filtering of $\widetilde{V}$ for $\ell > \ell_{\mathcal{K}}$ ($\ell < \ell_\varepsilon$),[4] or as the Fourier components of $\widetilde{V}$ at wave numbers $k \sim 1/\ell_{\mathcal{K}}$ ($k \sim 1/\ell_\varepsilon$).

6.1.1 Transition to turbulence in hydrodynamics

There are many routes to hydrodynamic turbulence, depending on the character of the laminar flow, i.e. its geometry, topology, boundary conditions, etc. Yet all these paths have one thing in common, namely, they can be traced to the onset and non-linear evolution of R–T and K–H instabilities above some critical value of the Reynolds number, Re_{cr}.[5] In some instances, these instabilities saturate due to higher-order non-linearities, in others they grow exponentially, and in still others quicker than exponentially, attaining infinite amplitude in a finite time, i.e. they exhibit what is known as a *finite time singularity* (FTS). In the case of saturated instabilities, the route to turbulence occurs via a sequence of ever more complicated laminar flows, each of which becoming unstable above its own critical value,

[4] We will find that $V_{\mathcal{K}} \approx \sqrt{2\mathcal{K}}$, where $\mathcal{K} \equiv \frac{1}{2}\langle\widetilde{V}^2\rangle$ is the turbulent kinetic energy per unit mass

[5] In some flows, the Reynolds number Re is replaced by other dimensionless parameters, e.g. the Rayleigh number Ra in thermal convection, the Taylor number Ta in sheared rotation, etc.

Re_j, where $\mathrm{Re}_j > \mathrm{Re}_{j-1}$. As j increases, the gap between these critical values, i.e. $\Delta\mathrm{Re}_j = \mathrm{Re}_{j+1} - \mathrm{Re}_j$, decreases, producing an ever quicker succession of bifurcations. These can be viewed as an overlap of unstable modes, or in the language of classical mechanics, of the resonant islands on the n-tori, which develop chaotic regions near their X-points, overlap with neighbouring islands, and eventually destroy the n-tori. In this way, when $\mathrm{Re}_{\mathcal{K}}$ exceeds some threshold value Re_*, fully developed turbulence ensues.[6]

The above picture of the route to turbulence was first formulated by Landau (1944) who argued that close to the marginal stability threshold, i.e. for values of the Reynolds number slightly larger than some critical value Re_1, the square of the amplitude $\mathcal{A}(t)$ of an unstable mode,

$$\mathcal{A}(t) = \mathcal{A}_0 \mathrm{e}^{\mathrm{i}\omega t} = \mathcal{A}_0 \mathrm{e}^{\mathrm{i}(\omega_r + \mathrm{i}\omega_i)t} = \mathcal{A}_0 \mathrm{e}^{\gamma t} \mathrm{e}^{\mathrm{i}\omega_r t}, \qquad \gamma = -\omega_i, \tag{6.3}$$

would increase as $\mathcal{A}^2$ and increase/decrease as $\mathcal{A}^4, \mathcal{A}^6, \ldots$, i.e.

$$\mathrm{d}_t \mathcal{A}^2 = 2\gamma \mathcal{A}^2 - \alpha_2 \mathcal{A}^4 - \alpha_3 \mathcal{A}^6 + \cdots \approx 2\gamma \mathcal{A}^2 - \alpha_2 \mathcal{A}^4. \tag{6.4}$$

Here $\gamma = -\omega_i \propto \mathrm{Re}_{\mathcal{K}} - \mathrm{Re}_1$ is the growth rate of the mode, which becomes unstable whenever $\gamma > 0$, while α_j are the so-called *Landau constants*, which contain the non-linear properties of the instability and, depending on their sign, can either accelerate ($\alpha_j < 0$) or decelerate ($\alpha_j > 0$) its growth.

The lowest-order form of (6.4), which can be solved exactly as,

$$\mathcal{A}^2/\mathcal{A}_0^2 = \mathrm{e}^{2\gamma t}/[1 + \lambda(\mathrm{e}^{2\gamma t} - 1)], \qquad \lambda = \alpha_2 \mathcal{A}_0^2 / 2\gamma, \tag{6.5}$$

behaves very differently depending on the sign of γ and α_2. Its solutions may be divided into *super-critical* ($\mathrm{Re}_{\mathcal{K}} > \mathrm{Re}_1, \gamma > 0$) and *sub-critical* ($\mathrm{Re}_{\mathcal{K}} < \mathrm{Re}_1, \gamma < 0$). Let us first consider the former. When $\alpha_2 > 0$, so that the mode is *non-linearly* stable, $\mathcal{A}(t)$ saturates at a value of $\mathcal{A}_\infty \equiv \sqrt{2\gamma/\alpha_2}$, which increases as $\mathcal{A}_\infty \propto \sqrt{\mathrm{Re}_{\mathcal{K}} - \mathrm{Re}_1}$, i.e. the mode suffers a *super-critical bifurcation* between $\mathcal{A}_0$ and $\mathcal{A}_\infty$. When $\alpha_2 < 0$, and the non-linearity accentuates the linear instability, $\mathcal{A}(t)$ grows faster than exponentially, becoming infinite in a finite time, $t_* = \ln[1 + 1/\lambda]/2\gamma$. Of course, as $\mathcal{A}(t) \to \infty$, the higher-order terms in (6.4) can saturate the instability, provided that one of $\alpha_3, \alpha_4, \ldots$ is positive. As $\mathrm{Re}_{\mathcal{K}}$ increases, one observes a sequence of super-critical bifurcations, which lead to ever more complicated laminar flows, terminating in fully developed turbulence above Re_*.

Let us next examine the sub-critical behaviour of (6.5), when the mode is *linearly* stable. In this case, the mode amplitude decays exponentially to zero whenever

[6] Such behaviour is well documented in chaos theory. For instance, the examination of the so-called *logistic equation*, $x_{l+1} = a x_l (1 - x_l)$, $1 < a \le 4$, reveals an *infinite* series of *period-doubling* bifurcations for $3 < a < 3.57$, which lead to dynamical chaos for $3.57 < a \le 4$.

Table 6.1. *Summary of Landau's theory of transition to turbulence.*

	$\alpha_2 > 0$ (stabilizing) non-linearly stable	$\alpha_2 < 0$ (destabilizing) non-linearly unstable
$\gamma < 0$ (sub-critical) linearly stable	$\mathcal{A}(\infty) \to 0$	$\mathcal{A}_0 < \mathcal{A}_\infty \Rightarrow \mathcal{A}(\infty) \to 0$ $\mathcal{A}_0 > \mathcal{A}_\infty \Rightarrow \mathcal{A}(t_*) \to \infty$
$\gamma > 0$ (super-critical) linearly unstable	$\mathcal{A}(\infty) \to \mathcal{A}_\infty \equiv \sqrt{2\gamma/\alpha_2}$	$\mathcal{A}(t_*) \to \infty$

$\alpha_2 > 0$, or $\alpha_2 < 0$ and $\mathcal{A}_0 < \mathcal{A}_\infty$. On the other hand, when $\alpha_2 < 0$ and $\mathcal{A}_0 > \mathcal{A}_\infty$, i.e. when the non-linearity is destabilizing and the perturbation is sufficiently strong, the mode exhibits a *sub-critical bifurcation* towards an FTS. This corresponds to a sudden bursting forth of turbulence, *ex nihilo* so to speak, without any obvious pre-cursors in the form of a sequence of period-doubling bifurcations and laminar flows of ever increasing complexity. Instead, the laminar flow is destroyed, and replaced by a turbulent flow, all at once! The main lesson to be learned here is that *linear* stability does not guarantee *non-linear* stability, and that the latter, even if present for a given $\mathrm{Re}_\mathcal{K}$ and $\mathcal{A}_0$, is fragile, at best, for larger values.

The overall behaviour of the various *bifurcations*, or *phase transitions*, resulting from Landau's simple theory, (6.4)–(6.5), are summarized in Table 6.1. It is remarkable that despite its simplicity, this theory captures many features of transition to turbulence in actual flows: e.g. (i) a flow in a pipe or in a boundary layer achieves turbulence via intermittent 'bursting forth' of turbulent 'slugs', i.e. via the sub-critical route; (ii) shear rotational or thermal convective flows, arrive at turbulence through a sequence of complex laminar flows, i.e. via the super-critical route; and (iii) external flow over a wing exhibits features of both types of transitions. We may conclude that the onset of turbulence can be either gradual or sudden, or any mixture of the two, depending on the details of the flow.

6.1.2 HD turbulence in 3D

The N–S equation, (4.2), for an incompressible ($\nabla \cdot \mathbf{V} = 0$) flow becomes

$$\rho \mathrm{d}_t \mathbf{V} + \nabla \cdot (p\mathsf{I} + \boldsymbol{\pi}) = \rho \mathbf{g}, \qquad \mathrm{d}_t = \partial_t + \mathbf{V} \cdot \nabla, \tag{6.6}$$

$$\boldsymbol{\pi} = -\mu \mathsf{W} = -2\mu\{\nabla\mathbf{V}\}^* = -\mu(\nabla\mathbf{V} + \nabla\mathbf{V}^\dagger) = -\mu(\partial_i V_j + \partial_j V_i). \tag{6.7}$$

In a turbulent flow, we expect the flow velocity to fluctuate randomly about some average value, motivating a mean-field description,

$$\mathbf{V} = \overline{\mathbf{V}} + \widetilde{\mathbf{V}}, \qquad \langle \mathbf{V} \rangle = \langle \overline{\mathbf{V}} \rangle = \overline{\mathbf{V}}, \qquad \langle \widetilde{\mathbf{V}} \rangle = 0. \tag{6.8}$$

Here, as always, the bar denotes the *average* and tilde the *fluctuating* quantity.[7] Inserting (6.8) into (6.6), assuming $\overline{\mu} = \mu$, and averaging, we find

$$\rho \overline{\mathrm{d}}_t \overline{\mathbf{V}} + \nabla \cdot (\overline{p}\mathsf{I} + \overline{\boldsymbol{\pi}} + \boldsymbol{\pi}_{tb}) = \rho \overline{\mathbf{g}}, \qquad \overline{\mathrm{d}}_t = \partial_t + \overline{\mathbf{V}} \cdot \nabla, \tag{6.9}$$

$$\overline{\boldsymbol{\pi}} = -\langle \mu \mathsf{W} \rangle = -\mu \overline{\mathsf{W}} = -2\mu \{\nabla \overline{\mathbf{V}}\}^*, \qquad \boldsymbol{\pi}_{tb} \equiv \rho \langle \widetilde{\mathbf{V}}\widetilde{\mathbf{V}} \rangle, \tag{6.10}$$

which is just the N–S equation evaluated for the average quantities,[8] plus an additional term, $\boldsymbol{\pi}_{tb}$, known as the *Reynolds stress*, which represents the mean momentum flux due to turbulent fluctuations.

The central message of (6.9) is the following: whereas the R–T and K–H instabilities of the *mean flow*, $\overline{\mathbf{V}}$, provide a continual drive for the *turbulent flow*, $\widetilde{\mathbf{V}}$, the latter can, and does, alter the mean flow via the Reynolds stress. In other words, there is a strong two-way interaction between the two flows, $\overline{\mathbf{V}} \leftrightarrow \widetilde{\mathbf{V}}$, which is the very essence of turbulent *shear flow* dynamics.

The dynamical evolution of $\boldsymbol{\pi}_{tb}$ follows from the second moment (stress tensor) conservation equation, which after some algebra is obtained as

$$\begin{aligned} \overline{\mathrm{d}}_t \boldsymbol{\pi}_{tb} = {} & -\{\boldsymbol{\pi}_{tb} \cdot \nabla \overline{\mathbf{V}}\}^* - \nabla \cdot \rho \langle \widetilde{\mathbf{V}}\widetilde{\mathbf{V}}\widetilde{\mathbf{V}} \rangle - \{\nabla \langle \widetilde{p}\widetilde{\mathbf{V}} \rangle\}^* + 2\langle \widetilde{p}\{\nabla \widetilde{\mathbf{V}}\}^* \rangle \\ & + \nu \nabla^2 \boldsymbol{\pi}_{tb} - 2\mu \langle \nabla \widetilde{\mathbf{V}} \nabla \widetilde{\mathbf{V}} \rangle, \end{aligned} \tag{6.11}$$

$$\begin{aligned} \rho \overline{\mathrm{d}}_t \mathcal{K} = {} & -\boldsymbol{\pi}_{tb} : \{\nabla \overline{\mathbf{V}}\}^* - \nabla \cdot \langle \widetilde{\mathbf{V}} \cdot [(\mathcal{K} + \widetilde{p})\mathsf{I} + \widetilde{\boldsymbol{\pi}}] \rangle \\ & -2\mu \langle \{\nabla \widetilde{\mathbf{V}}\}^* : \{\nabla \widetilde{\mathbf{V}}\}^* \rangle, \end{aligned} \tag{6.12}$$

where $\widetilde{\boldsymbol{\pi}} = -\mu \widetilde{\mathsf{W}} = -2\mu \{\nabla \widetilde{\mathbf{V}}\}^*$ is the fluctuating viscous stress and $\mathcal{K} = \frac{1}{2}\langle \widetilde{V}^2 \rangle = \frac{1}{2}\mathrm{Tr}\, \boldsymbol{\pi}_{tb}/\rho$ is the turbulent kinetic energy per unit mass. In deriving (6.12), we simplified the viscous terms to obtain the turbulent viscous dissipation in the form derived in (6.1). Note that (6.11)–(6.12) introduce the third moments, $\langle \widetilde{\mathbf{V}}\widetilde{\mathbf{V}}\widetilde{\mathbf{V}} \rangle$ and $\langle \widetilde{\mathbf{V}} \cdot \widetilde{\boldsymbol{\pi}} \rangle$, which in turn are related to $\langle \widetilde{\mathbf{V}}\widetilde{\mathbf{V}}\widetilde{\mathbf{V}}\widetilde{\mathbf{V}} \rangle$ via the third moment conservation equation, and so on.

The resulting infinite chain of *partial differential* equations may be written symbolically, neglecting differential operators and numerical constants, as

$$\partial_t \langle \widetilde{\mathbf{V}}\widetilde{\mathbf{V}} \rangle \sim \langle \widetilde{\mathbf{V}}\widetilde{\mathbf{V}}\widetilde{\mathbf{V}} \rangle + \nu \langle \widetilde{\mathbf{V}}\widetilde{\mathbf{V}} \rangle + \rho^{-1} \langle \widetilde{p}\widetilde{\mathbf{V}} \rangle, \tag{6.13}$$

$$\partial_t \langle \widetilde{\mathbf{V}}\widetilde{\mathbf{V}}\widetilde{\mathbf{V}} \rangle \sim \langle \widetilde{\mathbf{V}}\widetilde{\mathbf{V}}\widetilde{\mathbf{V}}\widetilde{\mathbf{V}} \rangle + \nu \langle \widetilde{\mathbf{V}}\widetilde{\mathbf{V}}\widetilde{\mathbf{V}} \rangle + \rho^{-1} \langle \widetilde{p}\widetilde{\mathbf{V}}\widetilde{\mathbf{V}} \rangle, \quad \text{etc.} \tag{6.14}$$

The ratio $\widetilde{p}/\rho$ can be obtained from the divergence of the N–S equation, $\nabla^2 \widetilde{p}/\rho = -\nabla \cdot [\widetilde{\mathbf{V}} \cdot \nabla \widetilde{\mathbf{V}}]$, and integrated using the *Biot–Savart's law*,

$$\frac{\widetilde{p}(\mathbf{x})}{\rho} \approx \frac{1}{4\pi} \int \nabla \cdot (\widetilde{\mathbf{V}} \cdot \nabla \widetilde{\mathbf{V}}) \frac{\mathrm{d}\mathbf{x}'}{|\mathbf{x} - \mathbf{x}'|}. \tag{6.15}$$

[7] Here $\langle \cdot \rangle$ represents the ensemble average, $\langle \cdot \rangle_{ens}$, to emphasize similarities with statistical mechanics. In practice, the time average, in either frame of reference, is usually sufficient.

[8] Since cross terms involving fluctuating and average quantities must vanish.

Hence, terms involving $\widetilde{p}/\rho$ introduce spatial integrals of $\nabla^2\widetilde{\mathbf{V}}\widetilde{\mathbf{V}}$ so that the nth order equation in the above chain may be written symbolically as

$$\partial_t\langle\widetilde{\mathbf{V}}^n\rangle \sim \langle\widetilde{\mathbf{V}}^{n+1}\rangle + \nu\langle\widetilde{\mathbf{V}}^n\rangle + \int\langle\widetilde{\mathbf{V}}^{n+1}\rangle d\mathbf{x}'. \tag{6.16}$$

One is thus faced with an infinite chain of *integro-differential* equations, the closure of which is the central problem of turbulence theory.[9]

Typically, the above chain is simply truncated by a number of ad hoc assumptions. Below we consider several of the most widely used truncation closure schemes, or *single-point models*, in order of decreasing complexity (Davidson, 2004). One common scheme is to adopt the *quasi-normal approximation*, $\langle\widetilde{\mathbf{V}}\widetilde{\mathbf{V}}\widetilde{\mathbf{V}}\widetilde{\mathbf{V}}\rangle \propto \langle\widetilde{\mathbf{V}}\widetilde{\mathbf{V}}\rangle^2$, and thus truncate the chain at (6.14). Unfortunately, this implies a *symmetrical* distribution of the velocity fluctuations, which as we will see shortly, are necessarily *skewed*.

Perhaps the best single-point closure is the so-called *Reynolds stress* model,

$$\overline{\mathrm{d}}_t\boldsymbol{\pi}_{tb} = \langle\widetilde{p}\widetilde{\mathsf{W}}\rangle + \rho\boldsymbol{\Pi} - \tfrac{2}{3}\rho\varepsilon\mathsf{I} + \nabla\cdot(c_1\boldsymbol{\alpha}\cdot\nabla\boldsymbol{\pi}_{tb}), \tag{6.17}$$

$$\overline{\mathrm{d}}_t\varepsilon = \nabla\cdot(c_2\boldsymbol{\alpha}\cdot\nabla\varepsilon) + c_3\mathcal{G}\varepsilon/\mathcal{K} - c_4\varepsilon^2/\mathcal{K}, \tag{6.18}$$

$$\langle\widetilde{p}\widetilde{\mathsf{W}}\rangle = 2\langle\widetilde{p}\{\nabla\widetilde{\mathbf{V}}\}^*\rangle = -\rho c_5\left(\langle\mathbf{V}\mathbf{V}\rangle - \tfrac{2}{3}\mathcal{K}\mathsf{I}\right)\varepsilon/\mathcal{K} - \rho c_6\left(\boldsymbol{\Pi} - \tfrac{2}{3}\mathcal{G}\mathsf{I}\right), \tag{6.19}$$

where $\boldsymbol{\alpha} \equiv \langle\mathbf{V}\mathbf{V}\rangle\mathcal{K}/\varepsilon = [\overline{\mathbf{V}}\,\overline{\mathbf{V}} + \langle\widetilde{\mathbf{V}}\widetilde{\mathbf{V}}\rangle]\mathcal{K}/\varepsilon$ is a combined stress tensor, $\rho\boldsymbol{\Pi} \equiv -\{\boldsymbol{\pi}_{tb}\cdot\nabla\overline{\mathbf{V}}\}^*$ is a symmetric scalar product of the Reynolds stress and the mean flow velocity gradient, $\mathcal{G} \equiv -\boldsymbol{\pi}_{tb} : \{\nabla\overline{\mathbf{V}}\}^*/\rho = \frac{1}{2}\boldsymbol{\pi}_{tb} : \overline{\mathsf{W}}/\rho = \frac{1}{2}\mathrm{Tr}\,\boldsymbol{\Pi}$ is the rate at which kinetic energy per unit mass $\mathcal{K}$ is generated and $\varepsilon = \frac{1}{2}\nu\langle\widetilde{\mathsf{W}} : \widetilde{\mathsf{W}}\rangle = 2\nu\langle\{\nabla\widetilde{\mathbf{V}}\}^* : \{\nabla\widetilde{\mathbf{V}}\}^*\rangle$ is the rate at which it is converted into heat. The dimensionless constants $c_j \sim 1$ are treated as adjustable parameters[10] to be determined from comparison with experimental measurements and/or direct numerical simulations.

The key shortcoming of the Reynolds stress model (6.17)–(6.18) is the ad hoc nature of the expression (6.19), which estimates the average correlation between the fluctuating pressure $\widetilde{p}$ and rate-of-strain $\widetilde{\mathsf{W}} = 2\{\nabla\widetilde{\mathbf{V}}\}^*$; indeed, the many variants of this model employ different expressions for the 'slow' and 'fast' parts of $\langle\widetilde{p}\widetilde{\mathsf{W}}\rangle$, see Hanjalic (2002). Nonetheless, when combined with *computational fluid dynamics* (CFD), the model has been remarkably successful in reproducing the experimental observations, although the success of each model variant has been limited to a specific type of flow.

In a simpler version of (6.17)–(6.19), known as the $\mathcal{K}-\varepsilon$ model, the evolution equation for $\boldsymbol{\pi}_{tb}$ is replaced with that for the kinetic energy $\mathcal{K}$, which is then used to construct the turbulent viscosity ν_{tb}, see (6.35). This in turn is related

[9] The problem bears close resemblance to the fundamental closure problem of gas dynamics.
[10] Although c_1 and c_2 are usually kept fixed at 0.22 and 0.15, respectively.

to π_{tb} via a so-called *Prandtl–Boussinesq* approximation,[11] which assumes the Reynolds stress, π_{tb}, to be proportional to the *average* rate-of-strain tensor, $\overline{\mathsf{W}}$, (4.5), and the *turbulent* kinetic energy, $\mathcal{K} = \frac{1}{2}\langle\widetilde{V}^2\rangle$. The resulting model equations, corresponding to (6.17)–(6.19), become

$$\pi_{tb}/\rho = \langle\widetilde{\mathbf{V}}\widetilde{\mathbf{V}}\rangle = \tfrac{2}{3}\mathcal{K}\mathsf{I} - \nu_{tb}\overline{\mathsf{W}}, \qquad \overline{\mathsf{W}} = 2\{\nabla\overline{\mathbf{V}}\}^* = \nabla\overline{\mathbf{V}} + \nabla\overline{\mathbf{V}}^\dagger, \quad (6.20)$$

$$\nu_{tb} = c_0\mathcal{K}^2/\varepsilon, \qquad \nu_{\mathcal{K}} = c_1\mathcal{K}^2/\varepsilon, \qquad \nu_\varepsilon = c_2\mathcal{K}^2/\varepsilon, \quad (6.21)$$

$$\overline{\mathrm{d}_t}\mathcal{K} = \nabla\cdot[(\nu+\nu_{\mathcal{K}})\nabla\mathcal{K}] + \mathcal{G} - \varepsilon, \quad (6.22)$$

$$\overline{\mathrm{d}_t}\varepsilon = \nabla\cdot[(\nu+\nu_\varepsilon)\nabla\varepsilon] + c_3\mathcal{G}\varepsilon/\mathcal{K} - c_4\varepsilon^2/\mathcal{K}, \quad (6.23)$$

where c_j are dimensionless (free) parameters of order unity and $\nu_{tb} = \mu_{tb}/\rho$, $\nu_{\mathcal{K}}$ and ν_ε are *turbulent*, or *eddy, diffusivities* of momentum, kinetic energy and the dissipation rate. Although somewhat less accurate than (6.17)–(6.19), the $\mathcal{K}$–ε model has scored many successes and remains the most popular method of handling turbulent flows in engineering applications.[12]

Taken by itself, with the eddy viscosity ν_{tb} treated as a free parameter, (6.20) represents what is perhaps the simplest turbulent closure scheme of all, generally known as the *Prandtl–Boussinesq model*. The $\mathcal{K}$-term, together with the traceless nature of $\overline{\mathsf{W}}$, ensure that $\mathrm{Tr}\,\pi_{tb}/\rho = 2\mathcal{K} = \langle\widetilde{V}^2\rangle$, as required by the definitions of π_{tb} and $\mathcal{K}$. Inserting (6.20) into (6.9) we find,

$$\overline{\mathrm{d}_t}\,\overline{\mathbf{V}} + \nabla\cdot\left[\left(\overline{p}/\rho + \tfrac{2}{3}\mathcal{K}\right)\mathsf{I} - (\nu+\nu_{tb})\overline{\mathsf{W}}\right] = \overline{\mathbf{g}}, \quad (6.24)$$

or setting $\overline{\mathbf{g}} = 0$ and evaluating the divergence,

$$\overline{\mathrm{d}_t}\,\overline{\mathbf{V}} + \nabla\left(\overline{p}/\rho + \tfrac{2}{3}\mathcal{K}\right) = (\nu+\nu_{tb})\nabla^2\overline{\mathbf{V}}, \quad (6.25)$$

with the effect of enhancing the collisional viscosity, $\nu = \mu/\rho$, by the typically much larger turbulent viscosity, $\nu_{tb} = \mu_{tb}/\rho$, and increasing the mean fluid temperature, $\overline{p}/\rho = \overline{T}/m$, by the turbulent 'temperature', $\frac{2}{3}\mathcal{K} = \frac{1}{3}\langle\widetilde{V}^2\rangle$.

6.1.2.1 *Fluid frame, temporal correlations: passive scalar diffusion*

To estimate ν_{tb}, it is useful to consider the trajectory of a fluid element moving in a turbulent flow field. We denote its displacement from its initial ($t = 0$) location by $\mathbf{r}(t)$ and its velocity by $\mathbf{v}(t) = \mathrm{d}_t\mathbf{r}(t)$, using lower case to distinguish from the flow velocity $\mathbf{V}$.[13] Assuming the average flow to be stagnant, $\overline{\mathbf{V}} = 0$, and the turbulent flow field to be homogeneous and isotropic, $\langle\mathbf{v}(t)\rangle = 0$, we can calculate the

[11] It was first proposed by Boussinesq (circa 1870), and developed by Prandtl in the 1920s.

[12] The success of both the single-point models suggests that they capture much of the physics governing hydrodynamic turbulence. We will return to this theme in the following section.

[13] Nonetheless, $\mathbf{v}(t)$ is clearly the Lagrangian analogue of the Eulerian fluctuating velocity, $\widetilde{\mathbf{V}}(\mathbf{x})$, in the sense that the ensemble average of both quantities is zero.

average (e.g. over an ensemble of similar flows) displacement of the fluid element after time t, as

$$\langle \mathbf{r}(t) \rangle = \int_0^t \langle \mathbf{v}(t_1) \rangle \mathrm{d}t_1 = 0. \tag{6.26}$$

In other words, there is no average (mean) displacement of the turbulent fluid element, which executes a 3D random walk around its initial location.[14]

The same is not true for the *average square* displacement,

$$\langle r^2(t) \rangle = \langle \mathbf{r}(t) \cdot \mathbf{r}(t) \rangle = \int_0^t \int_0^t \langle \mathbf{v}(t_1) \cdot \mathbf{v}(t_2) \rangle \mathrm{d}t_2 \mathrm{d}t_1, \tag{6.27}$$

which we expect to increase with time, provided the velocity of the fluid element has a finite *memory* of its past value, i.e. provided $\mathbf{v}(t_1)$ and $\mathbf{v}(t_2)$ are statistically correlated over some time interval $\Delta t = t_2 - t_1$. We expect this correlation, expressed in terms of the temporal *correlation function*,

$$C(\Delta t) \equiv \langle \mathbf{v}(t_1) \cdot \mathbf{v}(t_2) \rangle / \langle v^2 \rangle, \qquad \Delta t = t_2 - t_1, \tag{6.28}$$

to be equal to unity for $\Delta t = 0$ and decrease monotonically with $|\Delta t|$, vanishing in the limits $\Delta t \to \pm\infty$. Defining the *correlation time* as $\tau_{cor} \equiv \int_0^\infty C(\tau)\mathrm{d}\tau$, we can summarize the asymptotic behaviour of $C(\Delta t)$ as

$$C(\Delta t \ll \tau_{cor}) \to 1, \qquad C(\Delta t \sim \tau_{cor}) \sim 1/2, \qquad C(\Delta t \gg \tau_{cor}) \to 0.$$

Combining (6.27) and (6.28) we find an expression for the mean square displacement as a function of the temporal correlation $C(\Delta t)$,

$$\langle r^2(t) \rangle = \int_0^t \langle v^2 \rangle \int_0^t C(\Delta t) \mathrm{d}t_2 \mathrm{d}t_1. \tag{6.29}$$

For times much shorter than τ_{cor}, the above reduces to $\langle r^2 \rangle = \langle v^2 \rangle t^2$, so that the RMS distance increases linearly with time. For times much longer than τ_{cor}, the inner integral in (6.29) yields the correlation time, so that

$$\langle r^2(t) \rangle = \int_0^t \langle v^2 \rangle \int_{-\infty}^{\infty} C(\Delta t) \mathrm{d}t_2 \mathrm{d}t_1 \approx \int_0^t 2\langle v^2 \rangle \tau_{cor} \mathrm{d}t_1, \tag{6.30}$$

and the RMS distance increases as the square root of the time. Such increase is a hallmark of a diffusive process, with an effective diffusivity of[15]

[14] Note the analogy between this random walk of the fluid element and that of a test particle due to multiple collisions with background (field) particles, i.e. so-called *Brownian motion*.

[15] This definition of D_{tb} corresponds to a broadening of an isotropic density profile of some passive scalar density $n_p(r)$ satisfying $\partial_t n_p = D_{tb} \nabla^2 n_p$, i.e. $\langle r^2(t) \rangle = \int_0^\infty r^2 n_p \mathrm{d}\mathbf{x} / \int_0^\infty n_p \mathrm{d}\mathbf{x} = 6 D_{tb} t$. Note that D_{tb} scales linearly with τ_{cor}, and hence vanishes as the 'memory' of the fluid element and its average radial step size, or correlation length, $\lambda_{cor} \approx \tau_{cor} v_{rms}/\sqrt{3}$, tend towards zero.

Table 6.2. *Similarities between turbulent and collisional diffusion.*

Transport	Time	Speed	Distance	Diffusivity
collisional	τ_s	v_{ts}	$\lambda_s \sim \tau_s v_{ts}$	$D_s \sim v_{ts}\lambda_s \sim v_{ts}^2 \tau_s \sim \lambda_s^2/\tau_s$
turbulent	τ_{cor}	$v_{rms} \equiv \sqrt{\langle v^2 \rangle}$	$\lambda_{cor} \sim \tau_{cor} v_{rms}$	$D_{tb} \sim v_{rms}\lambda_{cor} \sim \langle v^2 \rangle \tau_{cor} \sim \lambda_{cor}^2/\tau_{cor}$

$$D_{tb} \equiv \langle r^2(t) \rangle /6t \approx \tfrac{1}{3}\langle v^2 \rangle \tau_{cor} \approx \tfrac{1}{3} V_{\mathcal{K}}^2 \tau_{cor} \approx \lambda_{cor}^2/\tau_{cor}, \tag{6.31}$$

where we replaced $\langle v^2 \rangle = v_{rms}^2$ by the turbulent kinetic energy, $V_{\mathcal{K}}^2 \approx 2\mathcal{K}$.

The same diffusivity applies for any quantity *passively* advected by the turbulence, i.e. carried along with the fluid element velocity $\mathbf{v}$, without altering this velocity. To a fair approximation this applies to both momentum and heat, the diffusivities of which are therefore roughly equal to D_{tb}, i.e. $\nu_{tb} \approx \chi_{tb} \approx D_{tb}$. It is worth observing the kinematic correspondence between turbulent and collisional diffusion, see Section 5.1.1 and Table 6.2. In particular, the correspondence between the turbulent *correlation length*, $\lambda_{cor} \sim \tau_{cor} V_{\mathcal{K}}$, and the collisional *mean free path*, $\lambda_c \sim \tau_s v_{ts}$, was used by Prandtl (1925) to develop his famous *mixing length* model,

$$D_{tb} \sim V_{\mathcal{K}} \ell_m \sim V_{\mathcal{K}}^2 \tau_m \sim \ell_m^2/\tau_m, \qquad \ell_m \sim V_{\mathcal{K}} \tau_m, \tag{6.32}$$

where the mixing length, $\ell_m \sim \ell_{\mathcal{K}}$, and time, $\tau_m \sim \tau_{\mathcal{K}} \approx \ell_{\mathcal{K}}/V_{\mathcal{K}}$, represent the size and *turn-over* (distortion) time of the large eddies. Depending on the degree of shear in the mean flow, these values may be estimated as,

$$|\nabla \overline{V}/\overline{V}| \sim |\nabla V_{\mathcal{K}}/V_{\mathcal{K}}| \Rightarrow \tau_m \sim 1/|\nabla \overline{V}|, \tag{6.33}$$

$$|\nabla \overline{V}/\overline{V}| \ll |\nabla V_{\mathcal{K}}/V_{\mathcal{K}}| \Rightarrow \tau_m \sim 1/|\nabla V_{\mathcal{K}}| \sim \ell_{\mathcal{K}}/V_{\mathcal{K}} \sim \tau_{\mathcal{K}}, \tag{6.34}$$

which correspond to *shear* and *homogeneous* flows, respectively. Since $\tau_m \sim \ell_m/\mathcal{K}^{1/2}$ and $\varepsilon \sim \mathcal{K}/\tau_m$, the above estimates lead to the scaling law, $\nu_{tb} \propto \mathcal{K}^2/\varepsilon$, (6.23), which forms an essential part of the $\mathcal{K}-\varepsilon$ model, (6.20)–(6.23),

$$\nu_{tb} \sim D_{tb} \sim \ell_m^2/\tau_m \sim \mathcal{K}\tau_m \sim \mathcal{K}^{1/2}\ell_m \sim \mathcal{K}^2/\varepsilon. \tag{6.35}$$

6.1.2.2 Laboratory frame, spatial correlations: direct energy cascade

The *single-point* turbulence models considered above, which referred to a single point in co-ordinate space, may be extended into *two-point* models by considering the spatial variation of the turbulence. To this end, let us examine the statistical properties of the turbulence in the laboratory frame of reference, specifically, the probability of velocity fluctuations at some location $\mathbf{x}$ conditional upon the fluctuations at a different location $\mathbf{x} + \boldsymbol{\ell}$. This property is conveniently expressed in terms of the *correlation tensor*,[16]

$$\mathsf{C}(\mathbf{x}, \boldsymbol{\ell}) = \langle \widetilde{\mathbf{V}}(\mathbf{x})\widetilde{\mathbf{V}}(\mathbf{x} + \boldsymbol{\ell})\rangle, \tag{6.36}$$

which is clearly symmetric with respect to the separation vector $\boldsymbol{\ell}$ and satisfies $\nabla_{\boldsymbol{\ell}} \cdot \mathsf{C}(\mathbf{x}, \boldsymbol{\ell}) = 0$ for incompressible flows. It is worth observing that if $\mathsf{C}(\mathbf{x}, \boldsymbol{\ell})$ could be determined, either by direct measurement or by numerical simulations, then $\boldsymbol{\pi}_{tb}$ and $\mathcal{K}$ follow immediately from

$$\boldsymbol{\pi}_{tb}(\mathbf{x})/\rho = \langle \widetilde{\mathbf{V}}(\mathbf{x})\widetilde{\mathbf{V}}(\mathbf{x})\rangle = \mathsf{C}(\mathbf{x}, 0), \qquad 2\mathcal{K} = \langle \widetilde{V}^2(\mathbf{x})\rangle = \mathrm{Tr}\,\mathsf{C}(\mathbf{x}, 0). \tag{6.37}$$

To pursue this approach further, we again consider the case of homogeneous, isotropic turbulence, embedded in an otherwise stagnant fluid ($\overline{\mathbf{V}} = 0, \widetilde{\mathbf{V}} = \mathbf{V}$). In this case, the correlation tensor depends only on $\ell = |\boldsymbol{\ell}|$,

$$\mathsf{C}(\ell) = C_\Delta(\ell)\frac{\boldsymbol{\ell\ell}}{\ell^2} + C_\perp(\ell)\mathsf{I} = \frac{1}{3}\langle V^2\rangle\left([f(\ell) - g(\ell)]\frac{\boldsymbol{\ell\ell}}{\ell^2} + g(\ell)\mathsf{I}\right), \tag{6.38}$$

$$C_\parallel(\ell) = \frac{1}{3}\langle V^2\rangle f(\ell), \quad C_\perp(\ell) = \frac{1}{3}\langle V^2\rangle g(\ell), \quad C_\Delta = C_\parallel - C_\perp, \tag{6.39}$$

where $\parallel$ and $\perp$ refer to the separation $\boldsymbol{\ell}$, so that $C_\parallel(\ell)$ and $C_\perp(\ell)$, and their normalized values $f(\ell)$ and $g(\ell)$,[17] represent *longitudinal* and *transverse* components of the correlation tensor. Since $\mathsf{C}(\ell)$ is isotropic and incompressible, it follows that $C_\perp(\ell)$ can be expressed in terms of $C_\parallel(\ell)$ as

$$g(\ell) = [\ell^2 f(\ell)]'/2\ell = f(\ell) + \tfrac{1}{2}\ell f'(\ell), \qquad C_\Delta = -\tfrac{1}{2}\ell C_\parallel', \qquad ' = \mathrm{d}_\ell.$$

The Fourier transform of $\mathsf{C}(\ell)$, which we denote by $\mathsf{C}_k(k)$,[18] is now found as

$$\mathsf{C}_k(k) = (2\pi)^{-3}\int_0^\infty \mathsf{C}(\ell)\mathrm{e}^{\mathrm{i}\mathbf{k}\cdot\boldsymbol{\ell}}\mathrm{d}\mathbf{k} = \frac{E_k(k)}{4\pi k^2}\left(\mathsf{I} - \frac{\mathbf{kk}}{k^2}\right), \tag{6.40}$$

where the final expression follows on account of the assumed isotropy and incompressibility of $\mathsf{C}(\ell)$ and $\mathsf{C}_k(k)$, i.e. $\mathbf{k}\cdot\mathsf{C}_k(k) = 0$. The new function $E_k(k)$ represents the *turbulent energy spectrum*, as is easily verified by calculating the ensemble average of the turbulent kinetic energy per unit mass,

$$2\mathcal{K} = \langle V^2\rangle = \langle \widetilde{V}^2\rangle = \mathrm{Tr}\,\mathsf{C}(0) = \int_0^\infty \mathrm{Tr}\,\mathsf{C}_k(k)\mathrm{d}\mathbf{k} = 2\int_0^\infty E_k(k)\mathrm{d}k. \tag{6.41}$$

[16] Once again, we denote the correlations between turbulent *fluid motions* by the symbol $\mathsf{C}(\boldsymbol{\ell})$ in order to highlight their similarity to the correlations between individual *particle motions*, measured by the collision operator $\mathcal{C}_{ss'}$. Since the latter originate in inter-particle collisions, we may view the former as arising due to collisions between turbulent *vortices (eddies)*.

[17] The normalization factor is chosen such that $f(0) = g(0) = 1$, while clearly $f(\infty) = g(\infty) = 0$.

[18] The reverse transform is given by $\mathsf{C}(\ell) = \int \mathsf{C}_k(k)\mathrm{e}^{-\mathrm{i}\mathbf{k}\cdot\boldsymbol{\ell}}\mathrm{d}\mathbf{k}$.

Here the last term is evaluated by substituting (6.40), replacing $\mathbf{dk}$ by the isotropic volume element $4\pi k^2 \mathrm{d}k$ and noting that $\mathrm{Tr}(\mathsf{I} - \mathbf{kk}/k^2) = 3 - 1 = 2$.

Similarly, $k^2 E_k(k)$ represents the *turbulent enstrophy spectrum*, where the enstrophy, $\Omega^2/2$, measures the intensity of the vorticity, $\mathbf{\Omega}$,

$$2\mathcal{O} = \langle \mathbf{\Omega} \cdot \mathbf{\Omega} \rangle = \langle \Omega^2 \rangle = \langle \widetilde{\Omega}^2 \rangle = 2 \int_0^\infty E_k(k) k^2 \mathrm{d}k. \qquad (6.42)$$

There is a dimensional argument, first put forward by Kolmogorov (1941), for the functional dependence of $E_k(k)$ on the wave number k, which roughly corresponds to eddies of size $\ell \sim 1/k$, and the local energy generation–dissipation rate per unit mass, ε, in fully developed, steady, 3D turbulence, assumed to be both homogeneous and isotropic.[19] The argument rests on a simple picture of energy transfer in such an idealized flow:

(i) Kinetic energy is uniformly injected into the flow at the large (integral) scale, $\ell_{\mathcal{K}} \sim 1/k_{\mathcal{K}}$, e.g. by a random body force, at a constant rate per unit mass of $\mathcal{G}$. The *integral scale*, defined as $\ell_{\mathcal{K}} \equiv \int_0^\infty f(\ell)\mathrm{d}\ell \approx \lambda_{cor}$, is assumed to be large enough to ensure that $\mathrm{Re}_{\mathcal{K}} = \ell_{\mathcal{K}} V_{\mathcal{K}}/\nu \gg 1$.

(ii) Kinetic energy is then transmitted from larger to smaller scales by a combination of the non-linear terms, namely vortex stretching, $\mathbf{\Omega} \cdot \nabla \mathbf{V}$, and advection, $\mathbf{V} \cdot \nabla \mathbf{\Omega}$, representing an effective 'collision' of two, or more, vortices. This non-linear *vortex 'collision'* is most effective when the two, or more, vortices (eddies) have comparable *size*, since according to detailed balance relations, the energy exchange between the triplet of wave vectors $\mathbf{k}_1, \mathbf{k}_2, \mathbf{k}_3$ is largest when k_1, k_2 and k_3 have comparable magnitude.[20] This transfer of energy from small to large k, i.e. from large, $\ell_{\mathcal{K}}$, to small, ℓ_ε, eddies via a sequence of intermediate-sized eddies, ℓ, which occurs at a constant rate per unit mass, $\Pi_k = \varepsilon$, and is known as a *direct energy cascade*. The concept of an eddy cascade was first formulated by Richardson (1922), although the notion dates back to Leonardo Da Vinci. The term *cascade* refers to transfer of some conserved (invariant) quantity of the turbulent flow, e.g. the kinetic energy $\mathcal{K}$, either from large to small scales (direct cascade) or from small to large scales (inverse cascade), via interactions of comparably sized eddies, i.e. a spectral flow Π_k via interactions which are local in $\mathbf{k}$-space. It is a direct consequence of turbulent diffusion of fluid elements, whose RMS displacement $\sqrt{\langle r^2(t) \rangle}$ increases as (6.27), thereby stretching the vortex tubes embedded in the flow and so reducing their cross-sectional area.

(iii) Finally, at some small, *dissipative scale*, ℓ_ε, defined by $\mathrm{Re}_\varepsilon = \ell_\varepsilon V_\varepsilon/\nu \sim 1$, the kinetic energy of the turbulent flow is dissipated by viscous heating, at a constant rate per unit mass, $\varepsilon \approx \nu \Omega_\varepsilon^2$, see (6.1).

[19] In this idealized case, the energy must be locally generated ($\mathcal{G}$), transported between the scales (Π_k) and dissipated (ε) at the same rate, $\mathcal{G} = \Pi_k = \varepsilon$.

[20] It is worth noting the analogy with *inter-particle collision*, in which energy is transferred most effectively when the particles have comparable *masses*!

In his seminal papers, Kolmogorov argued that, at any given scale, the turbulent eddies are characterized by only two quantities: their transverse size, ℓ (wave number, k) and their characteristic velocity, $V_\ell \equiv V$, so that the energy transfer rate, ε, must also depend on only ℓ and V_ℓ. On dimensional grounds, this requires a unique combination of these three quantities,

$$\mathcal{G} = \Pi_k = \varepsilon \sim V_\mathcal{K}^3/\ell_\mathcal{K} \sim V_\ell^3/\ell \sim V_\varepsilon^3/\ell_\varepsilon. \tag{6.43}$$

This so-called *Kolmogorov's (K-41) scaling law* or *law of similarity* is stated explicitly in the quote on page 220. Since $\varepsilon \sim$ const, it implies a cube root dependence of the velocity on the size of turbulent eddies, $V_\ell \propto \ell^{1/3}$, across the entire inertial range, $\ell_\mathcal{K} \gg \ell \gg \ell_\varepsilon$. Alternatively, it can be recast as

$$E_k(k)k \sim V_\ell^2 \sim (\varepsilon\ell)^{2/3} \sim (\varepsilon/k)^{2/3} \qquad \Rightarrow \qquad E_k(k) \approx \varepsilon^{2/3}k^{-5/3}, \tag{6.44}$$

which predicts that the turbulent energy spectrum, $E_k(k)$, increases as the $5/3$ power of the eddy size, $\ell \sim 1/k$, i.e. the kinetic energy of a turbulent flow is concentrated preferentially in the largest eddies, $2\mathcal{K} = \langle V^2\rangle \approx V_\mathcal{K}^2$.[21] By the same argument, one finds $k^2E_k(k) \sim \varepsilon^{2/3}k^{1/3}$, so that the vorticity (enstrophy) is concentrated preferentially in the smallest eddies, $\langle\Omega^2\rangle \approx \Omega_\varepsilon^2$, where it is most effectively dissipated, i.e. $\varepsilon = \nu\langle\Omega^2\rangle \approx \nu\Omega_\varepsilon^2$.

It is perhaps surprising to find that the energy dissipation rate does not depend on the viscosity, but is entirely prescribed by the integral scale quantities, $V_\mathcal{K}$ and $\ell_\mathcal{K}$ according with (6.43). The resolution of this apparent paradox is simple: the dissipative scale, ℓ_ε, must increase with viscosity ($\mathrm{d}_\nu\ell_\varepsilon > 0$), so that as ν is reduced, ever smaller tubes (thinner sheets) of vorticity are generated, until the required dissipation rate, $\mathcal{G} \sim \varepsilon$, is achieved. Hence the distribution of $\mathbf{\Omega}$ and ε becomes ever more spotty (in space) and bursty (in time) as either $\mathrm{Re}_\mathcal{K}$ or k are increased, i.e. as one reduces the viscosity and/or samples ever smaller scales, see Section 6.1.2.3 below.

The second apparent contradiction concerns the source of vorticity in an initially irrotational ($\mathbf{\Omega} = 0$), turbulent flow, e.g. what causes the shedding of vortices in an external flow over a cylinder? Here the answer lies at the fluid–solid boundary, where vorticity is created (by velocity shear) and then diffuses into the flow across the viscous boundary layer, according with

$$\tfrac{1}{2}\mathrm{d}_t\Omega^2 = \mathbf{\Omega}^2 : \{\nabla\mathbf{V}\}^* - \nu|\nabla\times\mathbf{\Omega}|^2 + \nu\nabla\cdot[\mathbf{\Omega}\times(\nabla\times\mathbf{\Omega})], \tag{6.45}$$

$$\mathrm{d}_t\mathcal{O} = \tfrac{1}{2}\mathrm{d}_t\langle\Omega^2\rangle = \langle\mathbf{\Omega}^2 : \{\nabla\mathbf{V}\}^*\rangle - \nu\langle|\nabla\times\mathbf{\Omega}|^2\rangle. \tag{6.46}$$

[21] The K-41 scaling law has been verified experimentally in the inertial sub-range. It should be added that for $\ell_\varepsilon \gg \ell$, $E_k(k)$ changes from a power law to an exponential decay.

The first two terms on the right-hand side represent vortex stretching and viscous damping. The former originates in a general decomposition of the distortion of a fluid element, measured by $\nabla\mathbf{V}$, into local rotation (spin), $\mathbf{\Omega}$, and deformation (strain), $\{\nabla\mathbf{V}\}^*$, whose double product gives the generation/destruction of Ω^2. The latter indicates that $\mathcal{O}$ is dissipated by viscosity (friction) in regions of intense *palinstrophy*, $\langle|\nabla \times \mathbf{\Omega}|^2\rangle \equiv 2\mathcal{P}$, similar to $\mathcal{K}$, which is dissipated in regions of intense *enstrophy*, $\langle|\nabla \times \mathbf{V}|^2\rangle = \langle\Omega^2\rangle = 2\mathcal{O}$,

$$d_t\mathcal{K} = \tfrac{1}{2}d_t\langle V^2\rangle = -\nu\langle\Omega^2\rangle = -2\nu\mathcal{O} = -\varepsilon. \tag{6.47}$$

Kolmogorov's scaling law can be used to infer the dependance of the eddy size and velocity on the Reynolds number across the inertial sub-range,

$$\mathrm{Re}_\ell \sim \ell V_\ell/\nu \sim V_\ell^4/\varepsilon\nu \sim \ell^{4/3}\varepsilon^{1/3}/\nu, \qquad \ell_\mathcal{K} \ll \ell \ll \ell_\varepsilon, \tag{6.48}$$

$$\ell/\ell_\varepsilon \sim \mathrm{Re}_\ell^{3/4}, \qquad \ell_\varepsilon \sim (\nu^3/\varepsilon)^{1/4}, \tag{6.49}$$

$$V_\ell/V_\varepsilon \sim \mathrm{Re}_\ell^{1/4}, \qquad V_\varepsilon \sim (\nu\varepsilon)^{1/4}, \tag{6.50}$$

$$\tau_\ell/\tau_\varepsilon \sim \mathrm{Re}_\ell^{1/2}, \qquad \tau_\varepsilon \sim (\nu/\varepsilon)^{1/2}, \tag{6.51}$$

where we used the fact that $\mathrm{Re}_\varepsilon = \ell_\varepsilon V_\varepsilon/\nu \sim 1$ to estimate the dissipative scale, ℓ_ε, velocity, V_ε and eddy turn-over time, $\tau_\varepsilon = \ell_\varepsilon/V_\varepsilon$. The large variation of scales comprising the inertial sub-range, implied by (6.49), e.g. for typical turbulent flows, one has $\log_{10}\mathrm{Re}_\mathcal{K} \sim 6$–$8$, $\log_{10}(\ell_\mathcal{K}/\ell_\varepsilon) \sim 4.5$–$6$ and $\log_{10}(\tau_\mathcal{K}/\tau_\varepsilon) \sim 3$–$4$, is precisely the reason why a comprehensive theory of turbulence is still beyond our reach in all but a few simple cases and/or smallish Reynolds numbers. In other words, the progress in numerical simulations is limited by the fact that CPU time scales roughly as $(\ell_\mathcal{K}/\ell_\varepsilon)^3(\tau_\mathcal{K}/\tau_\varepsilon) \sim \mathrm{Re}_\mathcal{K}^3$, which requires a CPU speed of $\sim 10^{18-24}$ flops! Moreover, since one needs $L_{grid} \gg \ell_\mathcal{K}$ to prevent distortion of the mean flow, this increases the demands on the CPU speed still further.

The above dimensionless scalings explain the phrase *law of similarity* when applied to Kolmogorov's spectrum, (6.43). Indeed, the physical assumption underlying both (6.43)–(6.44) and (6.48)–(6.51) is the notion of *scale invariance* of the turbulent flow within the inertial sub-range, $\ell_\mathcal{K} \ll \ell \ll \ell_\varepsilon$, i.e. of a fractal-like *self-similarity* between the various sized eddies. The idea can be ultimately traced to the fact that the Euler equation is invariant under the following transformations of distance, velocity and time,

$$\mathbf{x} \to \lambda\mathbf{x}, \qquad \mathbf{V} \to \lambda^n\mathbf{V}, \qquad t \to \lambda^{1-n}t, \tag{6.52}$$

where λ is an arbitrary scaling factor and n is some scaling exponent.

Let us define the *pth order, longitudinal structure function* as $\langle(\Delta V_\parallel)^p\rangle$, where $\Delta V_\parallel(\ell) \equiv V_\parallel(\ell) - V_\parallel(0) = [\mathbf{V}(\mathbf{x}+\boldsymbol{\ell}) - \mathbf{V}(\mathbf{x})]\cdot\boldsymbol{\ell}$ is the difference in the longitudinal

Table 6.3. *Comparison of spatial scales/ranges in HD turbulence.*

	Range of scales	$\langle(\Delta V_\parallel)^2\rangle$	Comments
integral scale	$\ell_\mathcal{K} \sim \ell$	$V_\mathcal{K}^2 f(\ell/\ell_\mathcal{K})$	$2\mathcal{K} = \langle V^2\rangle \approx V_\mathcal{K}^2$
inertial sub-range	$\ell_\mathcal{K} \gg \ell \gg \ell_\varepsilon$	$\beta_2(\varepsilon\ell)^{2/3}$	$E_k(k) \sim \varepsilon^{2/3}k^{-5/3}$
universal equilibrium	$\ell_\mathcal{K} \gg \ell$	$V_\varepsilon^2 f(\ell/\ell_\varepsilon)$	$\tau_\ell = \ell/V \ll \tau_\mathcal{K}$
dissipative scale	$\ell_\varepsilon \sim \ell$	$V_\varepsilon^2 f(1)$	$2\mathcal{O} = \langle\Omega^2\rangle \approx \Omega_\varepsilon^2$
dissipative sub-range	$\ell_\varepsilon \gg \ell$	V_ε^2	$E_k(k) \propto \exp(-\alpha k)$

velocity.[22] Then, making the assumption of scale invariance in the inertial sub-range, (6.48)–(6.52), it is possible to show that

$$\langle(\Delta V_\parallel)^p(\ell\lambda)\rangle = \lambda^{pn}\langle(\Delta V_\parallel)^p(\ell)\rangle, \qquad \langle(\Delta V_\parallel)^p\rangle(\ell) = \beta_p(\varepsilon\ell)^{\zeta_p}, \quad (6.53)$$

where β_p are dimensionless constants of order unity. Combining the above with $V_\ell \sim \ell^{1/3}$, (6.43), indicates that $n = \zeta_p/p = 1/3$ provided $\ell_\mathcal{K} \ll \ell \ll \ell_\varepsilon$.

The lowest instances of (6.53), which include the exact (and non-trivial) result, $\beta_3 = -4/5$, are known as the 2/3 *rd* and 4/5 *th laws*, respectively,[23]

$$\langle(\Delta V_\parallel)^2(\ell)\rangle = \beta_2(\varepsilon\ell)^{2/3}, \qquad \langle(\Delta V_\parallel)^3(\ell)\rangle = -\frac{4}{5}\varepsilon\ell, \quad (6.54)$$

with the former being equivalent to Kolmogorov's law of similarity, (6.44); the form of $\langle(\Delta V_\parallel)^2\rangle$ at various scales is summarized in Table 6.3.

Scale invariance is ultimately violated because ε, (6.1), is not uniform, as assumed in (6.43), but is rather concentrated in regions of intense vorticity (sharp velocity gradients).[24] Since vorticity tubes become longer and thinner, asymptotically reducing to lines or sheets, as we approach the dissipative scale, ℓ_ε, we may expect the flow to become more spotty and bursty at smaller ℓ and t, as is indeed observed in experiments.[25] To explore this effect, we must first develop a statistical description of turbulence.

22 Note that for small ℓ, the statistical properties of $\Delta V_\parallel$ are identical to those of $\nabla_\parallel V_\parallel = \partial_\ell V_\parallel$. Alternatively, we could define the pth order structure function as $\langle(\Delta\mathbf{V})^p\rangle$, where $\Delta\mathbf{V}(\boldsymbol{\ell}) \equiv \mathbf{V}(\mathbf{x}+\boldsymbol{\ell}) - \mathbf{V}(\mathbf{x})$, e.g. second–order functions are related as $\langle(\Delta\mathbf{V})^2\rangle = \ell^{-2}d_\ell[\ell^3\langle(\Delta V_\parallel)^2\rangle]$.

23 Similarly, Kolmogorov's law of similarity, (6.44), is frequently referred to as the *5/3rd law*.

24 A fact first recognized by Landau shortly after the publication of Kolmogorov's 1941 papers. Put another way, β_p and ε are not constant, as assumed by Kolmogorov, but rather depend on ℓ.

25 Indeed, both spatial and temporal intermittency of turbulent fluctuations can be traced to vortex stretching as the mechanism of the direct energy cascade.

6.1.2.3 Probability distribution functions (PDFs): intermittency

One of the aims of all turbulence theories is to determine the *probability distribution functions* (PDFs) of fluctuating quantities, $\widetilde{\mathcal{A}}$, denoted by $P(\widetilde{\mathcal{A}})$, from which all statistical moments $\langle \widetilde{\mathcal{A}}^n \rangle$ can be determined, e.g. their variance, $\sigma^2 \equiv \langle \widetilde{\mathcal{A}}^2 \rangle$, skewness, $S \equiv \langle \widetilde{\mathcal{A}}^3 \rangle / \sigma^3$, and flatness, $F \equiv \langle \widetilde{\mathcal{A}}^4 \rangle / \sigma^4$.[26]

We begin with three general remarks: (i) for a *Gaussian* (*normal*) PDF,

$$P_G(\widetilde{\mathcal{A}}; \sigma, a = 0) = (2\pi\sigma^2)^{-1/2} \exp[-\widetilde{\mathcal{A}}^2/2\sigma^2], \qquad a = \langle \widetilde{\mathcal{A}} \rangle = 0, \tag{6.55}$$

the skewness and flatness are $S = 0$ and $F = 3$; (ii) $S < 0$ indicates that $P(\widetilde{\mathcal{A}})$ has a maximum value for $\widetilde{\mathcal{A}} > 0$, so that the median of $\widetilde{\mathcal{A}}$ is larger than its mean, which remains equal to zero, and the distribution is skewed towards positive values of $\widetilde{\mathcal{A}}$; and (iii) $F > 3$ implies a higher probability of large amplitude excursions of $\widetilde{\mathcal{A}}$ than would be expected from the exponential fall-off implied by $P_G(\widetilde{\mathcal{A}})$, i.e. a more bursty, or *intermittent*, time series, $\widetilde{\mathcal{A}}(t)$. In other words, an intermittent PDF has a higher peak, thinner waist and longer (broader) tails than a Gaussian PDF of equal variance.

Let us again return to fully developed, homogeneous, isotropic and steady turbulence. Since the fluctuating velocity at *any one point* represents a random superposition of a large number of statistically independent eddies (of all sizes), then from the *central limit theorem* we would expect $P(\mathbf{V}) = P_G(V)$, with a zero mean and a variance equal to $\langle V^2 \rangle$. Such Gaussian PDFs of single-point fluctuating velocities are indeed widely observed experimentally. They are clearly consistent with the assumed isotropy and uniformity of the turbulent flow, which requires all odd moments of $\mathbf{V}$ to vanish, i.e. $\langle \mathbf{V}^{2p+1} \rangle = 0$ for all integers $p \geq 0$, and all even moments $\langle \mathbf{V}^{2p} \rangle$ to depend only on the mean value, $2\mathcal{K} = \langle V^2 \rangle \approx V_{\mathcal{K}}^2$.

In contrast to single-point recordings, the correlation between fluctuating velocities at *any two points* separated by ℓ, reflects the dynamics of the turbulent eddies of this size. Hence, we may expect $P(\Delta V_\parallel)$ not only to depend on ℓ, but also to differ from $P_G(\Delta V_\parallel)$, (6.55). To lowest order this may be characterized by departures of the skewness S and flatness F,

$$S \equiv \langle (\Delta V_\parallel)^3 \rangle / \langle (\Delta V_\parallel)^2 \rangle^{3/2}, \qquad F \equiv \langle (\Delta V_\parallel)^4 \rangle / \langle (\Delta V_\parallel)^2 \rangle^2, \tag{6.56}$$

from their Gaussian values of 0 and 3, respectively.[27] Indeed, the 2/3rd and 4/5th laws, (6.54), imply negative skewness of $P(\Delta V_\parallel)$ with $S = \beta_3/\beta_2^{3/2} \approx -0.3$,[28] roughly in line with experimental observations.

[26] Obviously the mean of $\widetilde{\mathcal{A}}$ is equal to zero by definition, i.e. $\langle \widetilde{\mathcal{A}} \rangle \equiv 0$.
[27] Clearly, for small ℓ, we expect $P[(\Delta V_\parallel)^p] \approx P[(\nabla_\parallel V_\parallel)^p]$ and $\langle (\Delta V_\parallel)^p \rangle \approx \langle (\nabla_\parallel V_\parallel)^p \rangle$.
[28] Since $\beta_3 = -4/5$ exactly and $\beta_2 \approx 2$ inferred from experimental measurements.

Of course, it is flatness ($F > 3$), rather than skewness ($S \neq 0$), which is an indicator of intermittency. More accurately, the degree of intermittency can be quantified by the departure of the pth order spectral exponents, ζ_p, (6.53), from their Gaussian values of $\zeta_p = p/3$, with $\zeta_p < p/3$ indicating more intermittent behaviour than predicted by $P_G(\Delta V_\parallel)$.[29]

Recall that the law of similarity (6.53) assumes the constants β_p and the local energy dissipation rate, ε, to be independent of ℓ, which implicitly assumes the vortices to be entirely space filling at all scales. All intermittency theories recognize the latter not to be the case, i.e. that vorticity does not fill space uniformly, but tends to be concentrated in sparse, intense patches due to the stretching of vortices into ever thinner and longer structures, with this tendency increasing towards the smaller scales. This key insight of *scale-dependence* is translated into different modifications to (6.53).

The earliest attempt at capturing intermittency was the *refined similarity hypothesis* formulated by Kolmogorov (1962) and Obukhov (1962). It states that since ε is a function of ℓ, then $\varepsilon^{p/3}$ in (6.53) must be replaced by $\langle \varepsilon_\ell^{p/3} \rangle$ where $\varepsilon_\ell(\ell)$ is a spatial average of ε over a sphere of radius ℓ,

$$\langle (\Delta V_\parallel)^p(\ell) \rangle = \beta_p \left\langle \varepsilon_\ell^{p/3} \right\rangle \ell^{p/3}, \qquad \ell_K \ll \ell \ll \ell_\varepsilon. \tag{6.57}$$

The new expression is evaluated as follows: (i) for $p = 3$, it reduces to $\langle \varepsilon_\ell \rangle = \varepsilon$, consistent with the 4/5th law, (6.54); (ii) for $p = 6$, a freely adjustable *intermittency exponent*, $\mu \equiv \mu_2$, is introduced, defined by[30]

$$\langle \varepsilon_\ell^m \rangle \sim \varepsilon^m (\ell_K/\ell)^{\mu_m} \propto \ell^{-\mu_m}, \qquad \mu_m = m - \zeta_{3m}, \qquad \zeta_p = p/3 - \mu_{p/3}; \tag{6.58}$$

and (iii) for other values of $p \geq 4$, $\left\langle \varepsilon_\ell^{p/3} \right\rangle$ is calculated from a specified PDF of ε_ℓ, which the authors assumed to follow a *log-normal* distribution,

$$P(\varepsilon_\ell) = P_G(\ln \varepsilon_\ell)/\varepsilon_\ell = \varepsilon_\ell^{-1} \left(2\pi\sigma_\ell^2\right)^{-1/2} \exp\left[-(\ln \varepsilon_\ell - a_\ell)^2/2\sigma_\ell^2\right], \tag{6.59}$$

where $a_\ell = -\sigma_\ell^2/2$ is the mean value of $\ln \varepsilon_\ell$. This result, which has the virtue of preventing negative excursions of ε_ℓ, corresponds to a process of self-similar eddy fragmentation (Yaglom, 1966), in which the normalized energy dissipation rate, $\varepsilon_j/\varepsilon_0$, at scale $\ell_j = \ell_0/2^j$ can be written as the product of independent random variables, $\chi_j = \varepsilon_j/\varepsilon_{j-1}$,

$$\varepsilon_j/\varepsilon_0 = \prod \chi_j \qquad \Rightarrow \qquad \ln(\varepsilon_j/\varepsilon_0) = \sum \chi_j, \tag{6.60}$$

[29] In practice, the departure from $\zeta_p = p/3$ can only be detected for $p \geq 5-6$, e.g. $\zeta_{12} \approx 2.8$.
[30] Note that $\mu_m = 0$ in self-similar (non-intermittent) analysis in which $\zeta_p = p/3$ and/or $\zeta_{3m} = m$.

and hence $\ln(\varepsilon_j/\varepsilon_0)$ as their sum. By the central limit theorem this implies that $P(\varepsilon_j)\mathrm{d}\varepsilon_j = P_G(\ln\varepsilon_j)\mathrm{d}\ln\varepsilon_j$, so that $P(\varepsilon_j) = P_G(\ln\varepsilon_j)/\varepsilon_j$, as in (6.59).

Using (6.59), it is possible to evaluate the averages in (6.57) to find

$$\langle\varepsilon_\ell^m\rangle/\varepsilon^m = \left[\langle\varepsilon_\ell^2\rangle/\varepsilon^2\right]^{m(m-1)/2} = (\ell_K/\ell)^{\mu_m}, \qquad \mu_m = \tfrac{1}{2}\mu m(m-1),$$
$$\langle(\Delta V_\parallel)^p\rangle(\ell) \propto (\varepsilon\ell)^{p/3}(\ell_K/\ell)^{\mu p(p-3)/18} \propto \ell^{\zeta_p}. \tag{6.61}$$

The resulting spectral exponents, $\zeta_p = p/3 - \mu(p-3)p/18$, with the choice of $\mu \approx 0.2$, are in fair agreement with experimental data up to $p \approx 12$,[31] above which the agreement becomes noticeably degraded. Indeed, for $p > (6/\mu + 3)/2 \approx 16$ one finds $\mathrm{d}_p\zeta_p < 0$ which is physically inadmissable.

The second notable theory of intermittency (She and Leveque, 1994), attempts to correct this error in the prediction of the tail of $P(\varepsilon_\ell)$ by suggesting instead a relation for the relative moments of ε_ℓ,

$$\varepsilon_\ell^{(m+1)} \propto \left(\varepsilon_\ell^{(m)}\right)^\beta \left(\varepsilon_\ell^{(\infty)}\right)^{1-\beta}, \qquad \varepsilon_\ell^{(m)} \equiv \langle\varepsilon_\ell^{m+1}\rangle/\langle\varepsilon_\ell^m\rangle, \tag{6.62}$$

where $0 < \beta < 1$ is an intermittency parameter, with $\beta = 1$ corresponding to self-similarity. It can be shown (Dubrulle, 1994) that this relation corresponds to a generalized *Poisson* distribution for $\ln\pi_\ell$, and hence a *log-Poisson* distribution for $P(\pi_\ell)$, where $\pi_\ell \equiv \varepsilon_\ell/\varepsilon_\ell^{(\infty)}$ is a dimensionless dissipation rate and $Y \equiv \ln\pi_\ell/\ln\beta$ is its normalized logarithm,

$$P(\pi_\ell)\mathrm{d}\pi_\ell = P_X(Y)\mathrm{d}Y, \qquad P_X(Y) = \int \frac{a^Z \mathrm{e}^{-a}}{Z!} X(Y-Z)\mathrm{d}Z. \tag{6.63}$$

Here $a = \langle Y\rangle$ is the expectation value, i.e. the mean, of Y,

$$a = \langle Y\rangle = \int Y P_X(Y)\mathrm{d}Y = -\ln\left(\langle\pi_\ell\rangle \Big/ \int \beta^Z X(Z)\mathrm{d}Z\right)/(1-\beta), \tag{6.64}$$

and $X(Y) > 0$ is an arbitrary PDF, normalized to unity, $\int X(Y)\mathrm{d}Y = 1$.[32] Recall that a Poisson distribution, $P_X[N(t) = n] = (\lambda t)^n \mathrm{e}^{-\lambda t}/n!$, gives the probability of n occurrences after time t, where λ is the probability of an occurrence per unit time, i.e. the occurrence rate. It can be shown that the waiting times between successive events are then distributed exponentially as $P[t] = \lambda\mathrm{e}^{-\lambda t}$. Such a *Poisson process* has *no memory* of past occurrences, e.g. $N(t_1) - N(t_0)$ is independent of $N(t_0)$ where $t_1 > t_0$, and is thus frequently found to describe a sequence of unrelated events.

[31] Indeed, the *log-normal* model is one of the most successful intermittency theories, which is remarkable considering it was virtually the earliest attempt at tackling the problem.

[32] Since the Poisson distribution is actually defined in terms of natural numbers $n > 0$ as $P_X(n) = a^n \mathrm{e}^{-a}/n!$, where $a = \langle n\rangle$ is again the mean of $P_X(n)$, it follows that (6.63) assumes that $a \gg 1$ so that the sum over n may be approximated as an integral over Z.

The resulting intermittency exponents, μ_m, (6.58), may be calculated as

$$\mu_m = x\left[m - (1-\beta^m)/(1-\beta)\right] = C_0\left[(1-\beta)m - (1-\beta^m)\right], \qquad (6.65)$$

where x is the scaling exponent of $\varepsilon_\ell^{(\infty)} \propto \ell^{-x}$ and $C_0 = D-d$ is the co-dimension of the most intermittent structures, d being their dimensions and D that of the flow itself. Since β, x and $C_0 = x/(1-\beta)$ are properties of the turbulence, they are not free parameters in the model, but must be obtained from additional geometrical and physical arguments. For instance, for 3D HD flows the small scale structures are vorticity lines, so that $D = 3$, $d = 1$ and $C_0 = 3 - 1 = 2$. Moreover, assuming they follow the law of similarity, (6.44), one has $x = 2/3 = \beta$. Substituting into (6.65) yields the corresponding intermittency, μ_m, and spectral, $\zeta_p = p/3 - \mu_{p/3}$, exponents,

$$\mu_m = \tfrac{2}{3}m - 2[1-(2/3)^m], \qquad \zeta_p = p/9 + 2[1-(2/3)^{p/3}], \qquad (6.66)$$

which are in good agreement with experimental and numerical data for all values p. Moreover, unlike the results of the log-normal model, the log-Poisson ζ_p satisfy $d_p\zeta_p > 0$ for all p and are thus physically admissible.

The final theory, or rather class of theories, which deserve our attention, are the *multi-fractal* models, the simplest of which is the $\hat{\beta}$ model. Multi-fractal models assume a spatial dependence of β_p in (6.57), i.e. $\beta = \beta_p(\ell)$, and try to estimate this dependence based on the assumed break-up of the turbulent eddies, and hence the degree to which they are space filling at small scales, by employing *fractal* techniques. In the $\hat{\beta}$ model, this fragmentation is characterized by a non-integer *fractal dimension*, $\mathcal{D} \leq D$,[33] so that $\hat{\beta} = 2^{\mathcal{D}-D} \leq 1$ is the fractional reduction in the volume of an eddy as it breaks up into two smaller eddies, with $D = 3$ for 3D flows. By equating the spectral energy flux for a jth generation eddy, $\Pi_j \sim \hat{\beta}^j V_j^3/\ell_j$, to the corresponding energy dissipation rate, $\varepsilon \sim (\varepsilon\ell_j)^{2/3}\hat{\beta}^{j/3}$, it can be shown that,[34]

$$\langle(\Delta V_\parallel)^p(\ell)\rangle \propto (\varepsilon\ell)^{p/3}(\ell/\ell_K)^{(3-p)(3-\mathcal{D})/3}. \qquad (6.67)$$

The resulting spectral exponents, $\zeta_p = (3-\mathcal{D}) + (\mathcal{D}-2)p/3$ are in fair agreement with experimental data up to $p \approx 8$ for $\mathcal{D} \approx 2.8$; this agreement can be extended to higher p by employing additional fractal dimensions.

For a review of intermittency theories, the most important of which are summarized in Table 6.4 and Fig. 6.1, the reader is referred to Frisch (1995).

33 $\mathcal{D}$ is treated as a free parameter which is equal to D for space filling eddies, when $P = P_G$ and $F = 3$, and is smaller than D for spotty (intermittent) eddies, $F < 3$.

34 Note that the correction to $\langle(\Delta V_\parallel)^p\rangle$ vanishes when $\mathcal{D} = 3$ and/or $p = 3$, the former due to $P(\Delta V_\parallel) = P_G(\Delta V_\parallel)$ for space filling eddies, the latter due to the 4/5th law.

Table 6.4. *Comparison of intermittency theories of 3D HD turbulence.*

Theory	Spectral exponents, ζ_p	$d_p\zeta_p > 0$	Comments
scale invariant	$p/3$	all $p > 0$	Gaussian PDF (non-intermittent)
log-normal	$p/3 + \mu(3-p)p/18$	$p < 16$	$\mu \approx 0.2$, good for $p \leq 10$
log-Poisson	$p/9 + 2[1-(2/3)^{p/3}]$	all $p > 0$	good agreement for all p
$\hat{\beta}$, bi-fractal	$(3-\mathcal{D}) + (\mathcal{D}-2)p/3$	all $p > 0$	$\mathcal{D} \approx 2.8 \leq 3$, good for $p \leq 8$

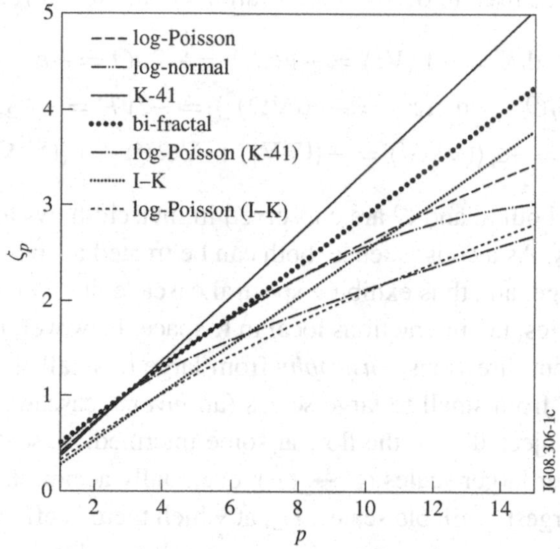

Fig. 6.1. Variation of ζ_p with p in various intermittency theories. For a discussion of K-41 and I–K theories, see pages 231 and 248, respectively.

6.1.3 HD turbulence in 2D

In the previous section, we implicitly assumed that the flow evolves in three-dimensional space. However, there are many circumstances when the flow velocity in one of the spatial dimensions is much weaker than in the other two, or even vanishes identically. Examples of the former, *quasi-2D*, *reduced HD*, or *Boussinesq* case ($\partial_z \ll \partial_x, \partial_y$), include large-scale planetary (atmospheric, oceanic) flows and a variety of boundary or shear layer flows; the latter, *perfect 2D* case ($\partial_z = 0$), is rare, but can be realized under special circumstances, e.g. in soap bubble films. Most importantly for us, turbulence in strongly magnetized plasmas can be approximated as a quasi-2D flow perpendicular to the magnetic field. Therefore we close our discussion of HD turbulence with a look at its general properties in 2D.

To begin with, 2D flows can indeed become turbulent, under the strict definition(s) of the term put forward in Section 6.1, e.g. they exhibit both spatial and temporal randomness, involve self-advection of vorticity over a wide range of scales, etc. The main difference from their 3D counterparts is the absence of vortex stretching, since the term $\mathbf{\Omega}\cdot\nabla\mathbf{V}$ in (4.13) is either small (quasi-2D) or zero (perfect 2D).[35] As a result, vorticity evolves according with a simple advective-diffusive equation,

$$\mathrm{d}_t\Omega = \nu\nabla^2\Omega, \qquad \mathrm{d}_t\mathbf{V} + \nabla(p/\rho) = -\nu\nabla\times\mathbf{\Omega}, \tag{6.68}$$

and so is passively carried by the flow and only destroyed by viscous heating.

The above can be used to derive the evolution equations for $\mathcal{K}$, $\mathcal{O}$ and $\mathcal{P}$,

$$\mathrm{d}_t\mathcal{K} = \tfrac{1}{2}\mathrm{d}_t\langle V^2\rangle = -\nu\langle\Omega^2\rangle = -2\nu\mathcal{O} = -\varepsilon, \tag{6.69}$$

$$\mathrm{d}_t\mathcal{O} = \tfrac{1}{2}\mathrm{d}_t\langle\Omega^2\rangle = -\nu\langle(\nabla\Omega)^2\rangle = -2\nu\mathcal{P} = -\varepsilon_\Omega, \tag{6.70}$$

$$\mathrm{d}_t\mathcal{P} = \tfrac{1}{2}\mathrm{d}_t\langle(\nabla\Omega)^2\rangle = -\langle\{\nabla\mathbf{V}\}^* : (\nabla\Omega)^2\rangle - \nu\langle(\nabla^2\Omega)^2\rangle, \tag{6.71}$$

which imply that both $\mathcal{K}$ and $\mathcal{O}$ are conserved in inviscid flows and are only dissipated by viscosity. As a consequence, both can be treated as dynamically invariant in the inertial range, and thus exhibit a spectral cascade due to interaction of comparably sized eddies, i.e. interactions local in **k**-space. However, the two quantities cascade in different directions: *enstrophy* from large to small scales (a *direct* cascade) and *energy* from small to large scales (an *inverse* cascade). In other words, when energy is 'injected' into the flow at some intermediate scale, $\ell_\Omega < \ell < \ell_\mathcal{K}$, it is carried to ever larger scales ($\ell \to \ell_\mathcal{K}$), eventually accumulating, i.e. forming vortices, at the largest available scales, $\ell_\mathcal{K}$, at which there is effectively no dissipation.[36] In contrast, enstrophy is carried to ever smaller scales ($\ell \to \ell_\Omega$), where it is dissipated by viscosity. The above process is known as a *dual cascade*.

There are several common explanations for the inverse energy cascade, all of which equally valid: (i) the observed tendency of two like-signed vortices to merge, or coalesce, into a single, larger vortex, suggests the formation of ever larger structures; (ii) as a patch of vorticity is deformed and sheared in the 2D plane, its planar length increases, due to the random walk of its fluid elements, while its planar thickness decreases, due to conservation of Ω, which is finally redistributed by viscous diffusion over the entire region. As a result, the effective size (extent) of the vorticity patch increases;[37] (iii) writing (6.69)–(6.70) in terms of the spectral energy density,

[35] To simplify the discussion, we assume the latter (perfect 2D) case below, for which $\mathbf{V}(x, y) = V_x(x, y)\hat{\mathbf{e}}_x + V_y(x, y)\hat{\mathbf{e}}_y$, $\mathbf{\Omega}(x, y) = \nabla\times\mathbf{V} = \Omega(x, y)\hat{\mathbf{e}}_z$ and $\mathbf{\Omega}\cdot\nabla\mathbf{V} = \Omega\partial_z\mathbf{V}(x, y) = 0$. Defining the stream function ϕ as $\mathbf{V} = -\nabla\times(\phi\hat{\mathbf{e}}_z) = \hat{\mathbf{e}}_z\times\nabla\phi$, allows the vorticity to be expressed as a Laplacian of ϕ, i.e. $\Omega = \nabla^2\phi$. Note that V_z = const does not alter the above results.

[36] Classic instances of this process are the large, coherent structures generated in 2D shear flows, such as tropical cyclones in the Earth's atmosphere or the giant 'eye' storm of Jupiter.

[37] The process can be described as the *filamentation* of vorticity.

$$\mathrm{d}_t \int E_k(k)\mathrm{d}k = -2\nu \int E_k(k)k^2\mathrm{d}k, \quad \mathrm{d}_t \int E_k(k)k^2\mathrm{d}k = -2\nu \int E_k(k)k^4\mathrm{d}k,$$

with all integrals evaluated between 0 and ∞, and noting that although both decrease monotonically with time in freely decaying turbulence, the latter must do so much more rapidly (due to sharper gradients involved in $\mathcal{P}$),

$$\mathrm{d}_t\mathcal{K} = O(\nu^a), \qquad \mathrm{d}_t\mathcal{O} = O(\nu^b), \qquad a \approx 1, \quad b \ll 1, \tag{6.72}$$

it becomes clear that the above relation requires the median of $E_k(k)$, denoted by $k_\mathcal{K} \sim 1/\ell_\mathcal{K}$, to decrease with time and hence for the integral (energy containing) scale to increase with time;[38] (iv) finally, perhaps the most elegant, and satisfying, argument begins by noting that the *equilibrium* phase space distribution of the system is given by the *Gibbs distribution*,

$$\mathcal{F}(k) = Z^{-1}\exp(-\alpha\mathcal{K} - \beta\mathcal{O}), \tag{6.73}$$

$$\mathcal{K} = \tfrac{1}{2}\langle V^2\rangle \propto \tfrac{1}{2}\textstyle\int V_\mathbf{k}^2\mathrm{d}\mathbf{k}, \qquad \mathcal{O} = \tfrac{1}{2}\langle\Omega^2\rangle \propto \tfrac{1}{2}\textstyle\int \Omega_\mathbf{k}^2\mathrm{d}\mathbf{k}, \tag{6.74}$$

where $\mathcal{K}$ and $\mathcal{O}$ are the dynamical invariants, the parameters α and β are functions of these invariants and Z is the partition function, i.e. a normalization constant. The equilibrium energy spectrum can then be solved as $E_k(k) = (\alpha + \beta' k^2)^{-1}$, which is a *decreasing* function of k. Since the direction of the energy cascade must be such as to relax the system towards equilibrium, energy must flow from large to small k.

It is interesting to repeat the above argument for 3D turbulence, in which case the *enstrophy*, $\mathcal{O}$, must be replaced by the *kinetic helicity*, $\mathcal{H}^V = \frac{1}{2}\langle\mathbf{V}\cdot\mathbf{\Omega}\rangle \propto \frac{1}{2}\int \mathbf{V}_\mathbf{k}\cdot\mathbf{\Omega}_{-\mathbf{k}}\mathrm{d}\mathbf{k}$, which is the second dynamical invariant. Solving the corresponding Gibbs distribution yields $E_k(k) = \alpha(\alpha^2 - \beta^2k^2)^{-1}$, which is an *increasing* function of k, indicating a *direct* energy cascade.[39]

It is possible to repeat the dimensional arguments à *la Kolmogorov* to obtain the energy spectrum, $E_k(k)$, and the structure functions, $\langle(\Delta V_\parallel)^p\rangle(\ell)$, in the inertial range of 2D turbulence. This was first done by Kraichnan (1967) and Batchelor (1969), who argued that if the break-up of vorticity is indeed a cascade process,[40] i.e. local in **k**-space, then $E_k(k)$ must depend only on the wave number, k, and the *enstrophy* dissipation rate, ε_Ω, (6.70), which replaces the energy dissipation rate, ε, in the 3D analysis. With these assumptions, it is easy to obtain the results shown in Table 6.5. Although these are roughly in line with experimental data, e.g.

[38] In fact, this result can be obtained directly as $\mathrm{d}_t \int E_k(k)(k_\mathcal{K} - k)^2\mathrm{d}k = -V_\mathcal{K}^2\mathrm{d}_t k_\mathcal{K}^2 > 0$.

[39] Similar arguments may be constructed to show that $\mathcal{O}$ in 2D, and $\mathcal{H}^V$ in 3D, exhibit *direct* cascades from large to small scales, the former associated with shearing of vorticity into thin sheets, the latter with stretching of vorticity into thin tubes, see Table 6.5.

[40] This point is indeed debatable as it can be shown that the planar shearing of a 2D vortex introduces a coupling of various scales, and thus a highly non-local interaction in **k**-space. Nonetheless, the idea of an enstrophy cascade is a useful construct, and the inertial range results, e.g. $E_k(k) \sim k^{-n}$, $n = 3$, are indeed close to those found in the experiments, $n \sim 3$–4.

Table 6.5. *Comparison of 2D and 3D HD turbulence. Lower half shows self-similar (non-intermittent), inertial range predictions.*

	3D HD	2D HD
vortex	stretching into thin tubes	shearing into thin sheets
first cascade	energy, $\langle V^2 \rangle$; direct	energy, $\langle V^2 \rangle$; inverse, $k < k_{in}$
second cascade	kinetic helicity, $\langle \mathbf{V} \cdot \mathbf{\Omega} \rangle$; direct	enstrophy, $\langle \Omega^2 \rangle$; direct, $k > k_{in}$
dissipation of	energy, $\ell_\varepsilon \sim (\nu^3/\varepsilon)^{1/4}$	enstrophy, $\ell_\Omega \sim (\nu^3/\varepsilon_\Omega)^{1/6}$
self-organization	no	yes
$E_k(k)$	$\sim \varepsilon^{2/3} k^{-5/3}$	$\sim \varepsilon^{2/3} k^{-5/3}, k < k_{in}$ $\sim \varepsilon_\Omega^{2/3} k^{-3} [\ln(k/k_{in})]^{-1/3}, k > k_{in}$
$\langle (\Delta V_\parallel)^p \rangle$	$\sim (\varepsilon \ell)^{p/3}$	$\sim (\varepsilon \ell)^{p/3}, k < k_{in}$ $\sim (\varepsilon_\Omega \ell^3)^{p/3}, k > k_{in}$
$\langle (\Delta V_\parallel)^3 \rangle$	$-\frac{4}{5}\varepsilon\ell$	$\frac{3}{2}\varepsilon\ell, k < k_{in}$
eddy size	$\ell/\ell_\varepsilon \sim \mathrm{Re}_\ell^{3/4}$	$\ell/\ell_\Omega \sim \mathrm{Re}_\ell^{1/2}$
eddy velocity	$V_\ell/V_\varepsilon \sim \mathrm{Re}_\ell^{1/4}$	$V_\ell/V_\Omega \sim \mathrm{Re}_\ell^{1/2}$
turn-over time	$\tau_\ell/\tau_\varepsilon \sim \mathrm{Re}_\ell^{1/2}$	$\tau_\ell/\tau_\Omega \sim \mathrm{Re}_\ell^0 = 1$

$E_k(k) \sim k^{-n}$, with $n = 3$ predicted vs. $n \sim 3$–4 measured, the agreement is worse than in the 3D case, where $n \approx 5/3$ is found to high accuracy. It is therefore encouraging that relaxing the cascade assumption introduces a small logarithmic correction to $E_k(k)$, improving the agreement with experiment.

Perhaps the most startling result of the 2D inertial range is the fact that the eddy turn-over time is independent of the eddy size, $\tau_\ell/\tau_\Omega \sim \mathrm{Re}_\ell^0 = 1$, in contrast, to its 3D counterpart, where small eddies evolve much faster than large ones, $\tau_\ell/\tau_\varepsilon \sim \mathrm{Re}_\ell^{1/2} \sim (\ell/\ell_\varepsilon)^{2/3}$, and thus effectively equilibrate on the evolution time scale of the large eddies. Hence, somewhat paradoxically, the evolution of HD turbulence is more complicated in 2D than in 3D, despite, and because of, the reduced dimensionality of the system.

The primary example of this complexity of 2D turbulence is the appearance of the already mentioned, long-lived, *coherent structures*, i.e. of large eddies which persist for tens, or even hundreds, of turn-over times.[41] These structures appear both in freely decaying 2D turbulence, in which case they survive the enstrophy cascade and represent the remnants of the original large eddies, and in forced 2D turbulence, in which they are generated as part of the inverse energy cascade. In either case, the merging of like-signed vortices increases their size and decreases their number.

[41] The tendency to produce large scale, coherent structures is generally known as *self-organization* of the flow. Such structures are also observed in quasi-2D turbulence, a good example being the so-called *hairpin vortices* in either boundary layers or shear layers in quasi-2D flows.

Coherent structures, which are large, intense and relatively sparse,[42] represent quiescent islands within a sea of turbulence, from which they are largely decoupled, i.e. dynamically isolated. Since they persist for many turn-over times, $\Delta t \gg \tau_\ell = \ell/V$, coherent structures can travel over distances many times their size, $\Delta r \gg \ell$, thus significantly increasing the net turbulent diffusion (random advective transport) within the 2D plane. Their ubiquitous appearance in 2D flows largely explains why these are generally more intermittent than comparable 3D flows, i.e. why $\zeta_p^{2D}/p < \zeta_p^{3D}/(p/3)$ in $\langle(\Delta V_\parallel)^p\rangle \propto \ell^{\zeta_p}$. As in the 3D case, the degree of intermittency increases towards the small scales, especially in the enstrophy cascade ($k > k_{in}$), but is also noticeable in the energy cascade ($k < k_{in}$) provided the coherent structures have had sufficient time to form at $k \sim k_K$.[43] For these reasons, theories of 2D intermittency are typically more complicated than their 3D counterparts, employing such concepts as negative temperature, maximum entropy and minimum enstrophy, e.g. see Frisch (1995) or Davidson (2004).

6.2 MHD turbulence

The reader may well wonder why we have dwelt so long on the discussion of turbulence in a neutral fluid. Surely in going from HD to MHD, the appearance of freely moving charges and a freely evolving magnetic field, must introduce a whole new range of turbulent phenomena which will render much of the foregoing discussion irrelevant? The answer is both yes and no. New phenomena are certainly introduced, and they do alter the picture of turbulence described in the previous section, in some respects fairly dramatically. However, it turns out that MHD turbulence is quite rare in fusion plasmas, or at least its impact on power exhaust is at best modest, due to the presence of strong external toroidal fields and low-β imposed by MHD stability. Indeed, the dominant transport mechanism, that of drift-wave turbulence, is in many respects more akin to HD than to MHD turbulence! For these reasons, we cast only a passing glance at the latter below, referring the interested reader to an excellent review of this topic by Biskamp (2003).

The transition from laminar to turbulent MHD flows results due to the non-linear evolution of R–T (∇p), K–H ($\nabla \mathbf{V}$) and *tearing* ($\nabla \mathbf{J}$) instabilities, see quote on page 107, which also provide the continual drive for sustaining fully developed, steady turbulence. As in HD, dissipation of dynamical invariants (kinetic, magnetic and mixed) occurs due to inter-particle collisions, which underlie the electrical resistivity (magnetic diffusivity), $D_\eta = \eta/\mu_0$, and the kinematic viscosity (kinetic diffusivity), $D_V = \nu$. The degree of dissipation must hence be measured by the

42 In this respect, they resemble a gas of freely evolving molecules constrained to evolve in 2D.
43 Interestingly enough, before their formation, the energy cascade is very nearly self-similar!

magnetic Reynolds number, $\mathrm{Re}_B = VL/D_\eta$, as well as the *kinetic* Reynolds number, $\mathrm{Re}_V = VL/D_V$.[44] In non-uniform flows/fields, the length L appearing above can be taken as the gradient length of the mean flow, $\overline{\mathbf{V}}$, and/or magnetic field, $\overline{\mathbf{B}}$. Below we assume Re_V and Re_B to be comparable, i.e. for the *magnetic Prandtl number*, $\mathrm{Pr}_B = \mathrm{Re}_B/\mathrm{Re}_V = D_V/D_\eta$, to be close to unity.

Repeating the mean field analysis of Section 6.1.2 for the resistive MHD equations, and setting $\rho = 1$ and $\mu_0 = 1$ for simplicity, one finds

$$\overline{\mathrm{d}}_t \overline{\mathbf{V}} + \nabla \overline{p} - \overline{\mathbf{B}} \cdot \nabla \overline{\mathbf{B}} + \nabla \cdot \left(\boldsymbol{\pi}^V_{tb} + \boldsymbol{\pi}^B_{tb} \right) = \nu \nabla^2 \overline{\mathbf{V}}, \tag{6.75}$$

$$\partial_t \overline{\mathbf{B}} - \nabla \times \left(\overline{\mathbf{V}} \times \overline{\mathbf{B}} + \boldsymbol{\epsilon}_{tb} \right) = \eta \nabla^2 \overline{\mathbf{B}}, \tag{6.76}$$

where as before $\overline{\mathrm{d}}_t = \partial_t + \overline{\mathbf{V}} \cdot \nabla$ is the mean flow advective derivative. The turbulent stress now consists of the Reynolds stress, $\boldsymbol{\pi}^V_{tb} = \langle \widetilde{\mathbf{V}}\widetilde{\mathbf{V}} \rangle$, and the Maxwell's stress, $\boldsymbol{\pi}^B_{tb} = -\langle \widetilde{\mathbf{B}}\widetilde{\mathbf{B}} \rangle$, representing the correlations of fluctuating velocity and magnetic field, respectively. Similarly, the mean electromotive force, $\overline{\mathbf{V}} \times \overline{\mathbf{B}}$, in the induction equation is now supplemented by the turbulent electromotive force, $\boldsymbol{\epsilon}_{tb} = \langle \widetilde{\mathbf{V}} \times \widetilde{\mathbf{B}} \rangle$, which arises due to perpendicular correlations of $\widetilde{\mathbf{V}}$ and $\widetilde{\mathbf{B}}$. These three terms represent the back-reaction of the MHD turbulence on the mean flow and field.

The various turbulent closure schemes discussed in Section 6.1.2, such as Prandtl's mixing length approximation or the $\mathcal{K}$–ε model, are easily extended to the case of MHD. Thus, $\boldsymbol{\pi}^V_{tb} + \boldsymbol{\pi}^B_{tb}$ give rise to a turbulent viscosity, ν^V_{tb}, a turbulent cross-viscosity, ν^B_{tb} and a turbulent energy density, $\mathcal{K} = \frac{1}{2}\langle \widetilde{\mathbf{V}}^2 - \widetilde{\mathbf{B}}^2 \rangle$ representing the difference between the kinetic and magnetic turbulent energies. Similarly, $\boldsymbol{\epsilon}_{tb}$ gives rise to a turbulent resistivity, η^B_{tb}, turbulent cross-resistivity, η^V_{tb}, and an additional term, α_{tb}. With these changes the mean field MHD equations (6.75)–(6.76) become,

$$\overline{\mathrm{d}}_t \overline{\mathbf{V}} + \nabla \left(\overline{p} + \tfrac{2}{3}\mathcal{K} \right) - \overline{\mathbf{B}} \cdot \nabla \overline{\mathbf{B}} = \left(\nu + \nu^V_{tb} \right) \nabla^2 \overline{\mathbf{V}} + \nu^B_{tb} \nabla^2 \overline{\mathbf{B}}, \tag{6.77}$$

$$\partial_t \overline{\mathbf{B}} - \nabla \times \left(\overline{\mathbf{V}} \times \overline{\mathbf{B}} \right) = \alpha_{tb} \overline{\mathbf{J}} + \eta^V_{tb} \nabla^2 \overline{\mathbf{V}} + \left(\eta + \eta^B_{tb} \right) \nabla^2 \overline{\mathbf{B}}. \tag{6.78}$$

These can be simplified by neglecting the off-diagonal coefficients, ν^B_{tb}, η^V_{tb} and α_{tb}, which are found to be significantly smaller than both ν^V_{tb} and η^V_{tb},

$$\overline{\mathrm{d}}_t \overline{\mathbf{V}} + \nabla \left(\overline{p} + \tfrac{2}{3}\mathcal{K} \right) - \overline{\mathbf{B}} \cdot \nabla \overline{\mathbf{B}} = \left(\nu + \nu^V_{tb} \right) \nabla^2 \overline{\mathbf{V}}, \tag{6.79}$$

$$\partial_t \overline{\mathbf{B}} = \nabla \times \left(\overline{\mathbf{V}} \times \overline{\mathbf{B}} \right) + \left(\eta + \eta^B_{tb} \right) \nabla^2 \overline{\mathbf{B}}. \tag{6.80}$$

[44] The magnetic Reynolds number is sometimes incorrectly defined with the flow speed, V, replaced by the Alfvén speed, V_A, which forms the *Lundquist number*, $\mathrm{Lu} \sim V_A L/D_\eta$, (2.258).

Note that (6.79) is identical to the HD result (6.25), except for the mean field magnetic tension, $\overline{\mathbf{B}} \cdot \nabla \overline{\mathbf{B}}$, which also modifies the turbulent 'temperature'. Similarly, (6.80) is identical to (2.257), but for the turbulent resistivity.

One can distinguish three cases of MHD turbulent flows on the basis of their geometry (dimensionality) with respect to the mean field: (i) vanishing mean field, $\overline{\mathbf{B}} = 0$, representing isotropic, 3D MHD; (ii) dominant mean field, $\widetilde{\mathbf{B}} \ll \overline{\mathbf{B}}$, $\nabla_{\parallel} \ll \nabla_{\perp}$, giving rise to flute-reduced MHD, see Section 4.2.5; and (iii) infinite mean field, $\overline{\mathbf{B}} \to \infty$, $\nabla_{\parallel} = 0$, $\nabla = \nabla_{\perp}$, representing the perfect 2D MHD case. We will consider each of these briefly below.

6.2.1 MHD turbulence in 3D

There are three dynamical invariants in 3D MHD flows:[45] the total energy, $\mathcal{E}$, the *magnetic helicity*, $\mathcal{H}^B$, and the *cross-helicity*, $\mathcal{H}^X$,

$$\mathcal{E} = \mathcal{K} + \mathcal{B} = \tfrac{1}{2}\langle V^2 \rangle + \tfrac{1}{2}\langle B^2 \rangle \propto \tfrac{1}{2} \int V_{\mathbf{k}}^2 \mathbf{dk} + \tfrac{1}{2} \int B_{\mathbf{k}}^2 \mathbf{dk}, \tag{6.81}$$

$$\mathcal{H}^B = \tfrac{1}{2}\langle \mathbf{A} \cdot \mathbf{B} \rangle \propto \tfrac{1}{2} \int \mathbf{A}_{\mathbf{k}} \cdot \mathbf{B}_{-\mathbf{k}} \mathbf{dk}, \tag{6.82}$$

$$\mathcal{H}^X = \tfrac{1}{2}\langle \mathbf{V} \cdot \mathbf{B} \rangle \propto \tfrac{1}{2} \int \mathbf{V}_{\mathbf{k}} \cdot \mathbf{B}_{-\mathbf{k}} \mathbf{dk}, \tag{6.83}$$

as can be seen from the evolution equations of these quantities,

$$\mathrm{d}_t \mathcal{E} = -D_\eta \langle J^2 \rangle - D_V \langle \Omega^2 \rangle = -\varepsilon_B - \varepsilon_V, \tag{6.84}$$

$$\mathrm{d}_t \mathcal{H}_B = -D_\eta \langle \mathbf{J} \cdot \mathbf{B} \rangle = -\varepsilon_H, \tag{6.85}$$

$$\mathrm{d}_t \mathcal{H}_X = -\tfrac{1}{2}(D_\eta + D_V)\langle \mathbf{J} \cdot \boldsymbol{\Omega} \rangle = -\varepsilon_X. \tag{6.86}$$

The latter imply that $\mathcal{E}$, $\mathcal{H}_B$ and $\mathcal{H}_X$ are conserved in ideal MHD flows and are only dissipated due to finite viscosity, ν, and/or resistivity, η.

Assuming global isotropy, the equilibrium spectra of the three dynamical invariants may again be calculated from the Gibbs phase space distribution,

$$\mathcal{F}(k) = Z^{-1} \exp(-\alpha \mathcal{E} - \beta \mathcal{H}^B - \gamma \mathcal{H}^X), \tag{6.87}$$

with the following result (Frisch *et al.*, 1975),

$$E_k^V(k) = \alpha^{-1} \left[1 + \tan^2 \varphi (1 - k_c^2/k^2)^{-1}\right], \tag{6.88}$$

$$E_k^B(k) = \left[\alpha \cos^2 \varphi (1 - k_c^2/k^2)\right]^{-1}, \tag{6.89}$$

$$H_k^B(k) = -(k_c/k^2) E_k^B(k), \quad H_k^X(k) = -(\gamma/\alpha) E_k^B(k), \tag{6.90}$$

[45] To simplify the resulting expressions, let us assume an incompressible flow ($\nabla \cdot \mathbf{V} = 0$) with unit density ($\rho = 1$), and set the permeability of vacuum to one ($\mu_0 = 1$), which corresponds to normalization of the time and velocity by the *Alfvén* time and speed, (2.253).

Table 6.6. *Invariants and cascade directions (direct ↘ and inverse ↖) in HD and MHD turbulence; after Biskamp (1993).*

	3D	2D
HD	$\langle V^2\rangle \searrow, \langle \mathbf{V}\cdot\mathbf{\Omega}\rangle \searrow$	$\langle V^2\rangle \nwarrow, \langle \Omega^2\rangle \searrow$
MHD	$\langle V^2+B^2\rangle \searrow, \langle \mathbf{V}\cdot\mathbf{B}\rangle \searrow, \langle \mathbf{A}\cdot\mathbf{B}\rangle \nwarrow$	$\langle V^2+B^2\rangle \searrow, \langle \mathbf{V}\cdot\mathbf{B}\rangle \searrow, \langle \psi^2\rangle \nwarrow$

where $k_c = \beta/\alpha\cos^2\varphi$, $\sin\varphi = \gamma/2\alpha$ are functions of α, β, γ and are independent of k. The dynamical invariants in 2D MHD are identical to those in 3D MHD, with the exception that the magnetic helicity is replaced by the mean square (parallel) magnetic potential, $\mathcal{A}_\parallel = \frac{1}{2}\langle\psi^2\rangle$, which evolves according with $\mathrm{d}_t\mathcal{A}_\parallel = -D_\eta\langle B^2\rangle = -\varepsilon_\mathcal{A}$, while $\mathcal{E}$ and $\mathcal{H}^X$ become[46]

$$\mathcal{E} = \mathcal{K} + \mathcal{B} = \tfrac{1}{2}\langle(\nabla\phi)^2\rangle + \tfrac{1}{2}\langle(\nabla\psi)^2\rangle, \quad \mathcal{H}^X = -\tfrac{1}{2}\langle\Omega\psi\rangle. \tag{6.91}$$

The resulting Gibbs equilibrium spectra for 2D MHD are found from (6.87)–(6.90) with $\mathcal{H}^B \to \mathcal{A}_\parallel$, $k_c \to -k_c$ and $H_k^B(k) \to A_k(k) = k^{-2}E_k^B(k)$.

As in the HD case, the Gibbs equilibrium spectra (6.87)–(6.90) yield the cascade directions of the dynamical invariants, see Table 6.6. Evidently, MHD turbulence differs substantially from its HD counterpart, and unlike the latter, is quite similar in 2D and 3D flows. In both cases, it exhibits a *direct* energy cascade of total (kinetic + magnetic) energy and cross-helicity, and an *inverse* cascade of magnetic helicity, that is, of the purely magnetic invariant. Thus, irrespective of the dimensionality of the flow, MHD turbulence leads to self-organization of the magnetic field, i.e. to the formation of coherent, large scale *magnetic structures*.[47] In contrast, HD turbulence is essentially different in 3D and 2D, with an inverse energy cascade, self-organization and coherent vortices appearing only in 2D flows. It is interesting to note that the relation between (magnetic) helicity and (total) energy in MHD is identical to that between energy and enstrophy in 2D HD flows. Thus based on (6.72) with $\mathcal{K} \to \mathcal{A}_\parallel$, $\mathcal{O} \to \mathcal{E}$ and $\nu = D_V \to D_\eta$ we would expect $\mathcal{E}$ to decay much more rapidly than $\mathcal{A}_\parallel$ in freely evolving MHD turbulence, as is indeed observed in numerical simulations.

The Gibbs equilibrium spectra (6.88)–(6.89), whose difference yields

$$E_k^B - E_k^V = \alpha^{-1}\left(k^2/k_c^2 - 1\right)^{-1} \geq 0, \qquad k_c < k_{min} \leq k \leq k_{max}, \tag{6.92}$$

reveal that while magnetic, $E_k^B(k)$, and kinetic, $E_k^V(k)$, energies of MHD turbulence are strongly coupled and effectively equipartitioned $\left(E_k^B \approx E_k^V\right)$ at small

[46] Note that $\langle(\nabla\phi)^2 + (\nabla\psi)^2\rangle$, $\langle\Omega\psi\rangle$ and $\langle\psi^2\rangle$ are not conserved in quasi-2D or reduced MHD.

[47] In forced MHD turbulence, this process is known as a *magnetic dynamo*. It is the underlying explanation of both planetary (terrestrial) and stellar (solar) magnetism.

scales ($k \sim k_{min}$), they become decoupled, with the magnetic energy becoming dominant $\left(E_k^B \gg E_k^V\right)$ at large scales ($k \sim k_{max}$). As a result, the energy of steady MHD turbulence is preferentially concentrated in large scale magnetic structures. Since these structures are quasi-static (long lived), they must be nearly force free ($\mathbf{J}\times\mathbf{B} \to 0$) and thus require near-perfect alignment of **J** and **B**.[48] This process can be ultimately traced to the conservation of $\mathcal{E}$ and $\mathcal{H}_B$. Similarly, the conservation of $\mathcal{E}$ and $\mathcal{H}_X$ results in the dynamical alignment of **V** and **B** at small scales, since $\mathcal{H}_X$ exhibits a direct cascade,

$$\mathcal{H}_B = \tfrac{1}{2}\langle \mathbf{A}\cdot\mathbf{B}\rangle \nwarrow \quad \Rightarrow \quad \mathbf{J}\times\mathbf{B} \approx 0, \quad \mathbf{J} \propto \mathbf{B} \quad \text{at} \quad k \sim k_{min}, \tag{6.93}$$

$$\mathcal{H}_X = \tfrac{1}{2}\langle \mathbf{V}\cdot\mathbf{B}\rangle \searrow \quad \Rightarrow \quad \mathbf{V}\times\mathbf{B} \approx 0, \quad \mathbf{V} \propto \mathbf{B} \quad \text{at} \quad k \sim k_{max}. \tag{6.94}$$

The latter process, which represents the dynamical alignment of $\widetilde{\mathbf{V}}$ and $\widetilde{\mathbf{B}}$ (and hence the equipartition of turbulent kinetic and magnetic energies) at small scales, is known as the *Alfvén effect*.[49] Not surprisingly, the label refers to the corresponding quasi-equilibrium, i.e. the lowest energy configuration for a given value of $\mathcal{H}_X$, known as a *pure Alfvénic* state, which consists of linearly polarized *Alfvén waves* ($\widetilde{\mathbf{V}} = \pm\widetilde{\mathbf{B}}$).[50] For this reason, MHD turbulence is frequently studied in the so-called *Elsässer variables*,

$$\mathbf{z}^{\pm} = \widetilde{\mathbf{V}} \pm \widetilde{\mathbf{B}}, \qquad \mathbf{V}_A = \overline{\mathbf{B}}, \qquad (\rho = 1,\ \mu_0 = 1), \tag{6.95}$$

in which the (resistive) MHD equations, recall (2.246)–(2.247), become[51]

$$\nabla\cdot\mathbf{z}^{\pm} = 0, \quad \widetilde{\mathbf{V}} = \tfrac{1}{2}(\mathbf{z}^+ + \mathbf{z}^-), \quad \widetilde{\mathbf{B}} = \tfrac{1}{2}(\mathbf{z}^+ - \mathbf{z}^-) \tag{6.96}$$

$$\left[\partial_t \mp (\mathbf{V}_A + \mathbf{z}^{\mp})\cdot\nabla\right]\mathbf{z}^{\pm} = -\nabla P + D_+\nabla^2\mathbf{z}^{\pm} + D_-\nabla^2\mathbf{z}^{\mp}, \tag{6.97}$$

$$P = p + \tfrac{1}{2}B^2, \quad D_+ = \tfrac{1}{2}(\nu+\eta), \quad D_- = \tfrac{1}{2}(\nu-\eta). \tag{6.98}$$

The mean field MHD equations (6.75)–(6.76) now simplify to

$$(\partial_t + \overline{\mathbf{z}^{\pm}}\cdot\nabla)\overline{\mathbf{z}^{\pm}} + \nabla\overline{p} + \nabla\cdot\langle\widetilde{\mathbf{z}^{\pm}}\widetilde{\mathbf{z}^{\mp}}\rangle = D_+\nabla^2\overline{\mathbf{z}^{\pm}} + D_-\nabla^2\overline{\mathbf{z}^{\mp}}, \tag{6.99}$$

while the invariants $\mathcal{E}$ and $\mathcal{H}_X$ may be expressed symmetrically as

$$\mathcal{E} = \mathcal{E}^+ + \mathcal{E}^-, \quad \mathcal{H}_X = \tfrac{1}{2}(\mathcal{E}^+ - \mathcal{E}^-), \quad \mathcal{E}^{\pm} = \tfrac{1}{4}\langle(z^{\pm})^2\rangle, \quad \mathrm{d}_t\mathcal{E}^{\pm} \equiv -\varepsilon_{\pm}.$$

Aside from highlighting a degree of symmetry between $\widetilde{\mathbf{V}}$ and $\widetilde{\mathbf{B}}$, which is implicit in the MHD model, the Elsässer formulation elucidates the nature of MHD turbulence at small scales by identifying its dynamical modes as Alfvén waves, $\mathbf{z}^{\pm}$,

[48] It turns out that the corresponding quasi-equilibrium, which is the lowest energy configuration for a given value of $\mathcal{H}_B$, is both stagnant ($\mathbf{V} \approx 0$) and linear ($\mathbf{J} \propto \mathbf{B}$), so that $\nabla\times\mathbf{B} \propto \mathbf{B}$.

[49] Essentially, it implies the coupling of small-scale kinetic and magnetic fluctuations, $\widetilde{\mathbf{V}}$ and $\widetilde{\mathbf{B}}$, by a large-scale field, $\overline{\mathbf{B}}_{\ell}$, which represents the mean field averaged over the volume of the eddy.

[50] Once again, we assume $\rho = 1$ and $\mu_0 = 1$ to simplify the notation. Such pure Alfvénic turbulence with $\widetilde{\mathbf{V}} \approx \pm\widetilde{\mathbf{B}}$ is indeed observed in satelite measurements of the solar wind.

[51] To simplify notation, we denote the diffusivities D_V and D_η in (6.84) by ν and η, respectively.

travelling parallel, $\mathbf{z}^-$, and anti-parallel, $\mathbf{z}^+$, to the background field, $\mathbf{V}_A = \overline{\mathbf{B}}$, and interacting only with counter-propagating waves via the (non-linear) advective terms, $\mathbf{z}^\pm \cdot \nabla \mathbf{z}^\mp$. It is this non-linear coupling of oppositely moving waves which is responsible for the Alfvén effect.

Due to the appearance of the mean magnetic field in (6.97) one cannot repeat the Kolmogorov-like dimensional analysis to determine the inertial-range energy spectrum of MHD turbulence. However, this spectrum can be calculated using a simple dynamical model based on the Elsässer formulation. Introducing the Alfvén 'collision' and turn-over times for a small scale MHD eddy of size $\ell \sim 1/k$, and noting that $\tau_A \ll \tau_\ell^\pm$ in weak turbulence,

$$\tau_A = \ell/V_A, \qquad \tau_\ell^\pm = \ell/z_\ell^\mp, \qquad \widetilde{B}/\overline{B} \ll 1 \quad \rightarrow \quad \tau_A \ll \tau_\ell^\pm, \tag{6.100}$$

the energy transfer times, $\tau_\ell^{\varepsilon\pm}$, and rates, $\Pi_k^\pm$, may be estimated as,

$$\tau_\ell^{\varepsilon\pm} \approx N^\pm \tau_A \approx \left(\tau_\ell^\pm\right)^2/\tau_A, \qquad \Pi_k^\pm \approx \left(z_\ell^\pm\right)^2/\tau_\ell^{\varepsilon\pm} \approx (z^+)^2(z^-)^2\tau_A/\ell^2 \approx \Pi_k^\mp,$$

where $N^\pm = \left(z_\ell^\pm/\Delta z_\ell^\pm\right)^2 \approx \left(\tau_\ell^\pm/\tau_A\right)^2$ is the number of Alfvén 'collisions' needed to change $z_\ell^\pm$ by a factor of two.[52] One thus finds that $\mathcal{E}_+$ and $\mathcal{E}_-$ are transferred in $\mathbf{k}$-space at approximately the same rate, $\Pi_k^+ \approx \Pi_k^- \approx \Pi_k/2$, and must hence be injected and dissipated at an equal rate, $\varepsilon_\ell^+ \approx \varepsilon_\ell^- \approx \varepsilon/2$, both of which are constant across the inertial range, i.e. $\Pi_k = \varepsilon$. If we now consider the case of weak correlation between $\mathbf{V}$ and $\mathbf{B}$, so that $z_\ell^+ \approx z_\ell^- \approx z_\ell$, then the corresponding energy spectrum is easily determined as,

$$\varepsilon \sim z_\ell^4\tau_A/\ell^2 \sim z_\ell^4/V_A\ell, \qquad z_\ell \sim (\varepsilon V_A \ell)^{1/4} \sim (\varepsilon V_A/k)^{1/4}, \tag{6.101}$$

$$E_k^+(k) \sim E_k^-(k) \sim E_k(k)/2 \sim z_\ell^2/k \sim (\varepsilon V_A)^{1/2}k^{-3/2}. \tag{6.102}$$

This result, also known as the *I–K spectrum*, was first derived, by somewhat different routes, by Iroshnikov (1964) and Kraichnan (1965). Note that the k-dependence in (6.102) is somewhat weaker than in the K-41 spectrum ($3/2 < 5/3$), due to the $\tau_\ell^\pm/\tau_A \propto k^{1/4}$ correction to the energy transfer time, $\tau_\ell^{\varepsilon\pm}$, which becomes more pronounced at smaller scales.

When satellite measurements of the energy spectrum in the solar wind first became available,[53] they revealed, somewhat surprisingly, that $E_k(k)$ in 3D MHD turbulence appears to follow the K-41, rather than the I–K, spectrum. More recently, this finding was confirmed using direct numerical simulations. The explanation to this puzzling result lies in the inherent anisotropy of all MHD flows. Thus, even though 3D MHD turbulence may be globally isotropic, with no preferred direction to the average field over the volume of an eddy, $\overline{\mathbf{B}}_\ell \equiv \langle\widetilde{\mathbf{B}}\rangle_\ell$,

[52] The quadratic dependence reflects the Fokker–Planck nature of this gentle 'collision' process.
[53] Solar wind is still the best laboratory for fully developed 3D MHD turbulence.

nonetheless, its mere presence, combined with the fact that parallel deformations are counteracted by magnetic tension, introduces anisotropy in the fluctuating quantities, $\mathbf{z}^{\pm} = \widetilde{\mathbf{V}} \pm \widetilde{\mathbf{B}}$, which become largely constrained to the local $\perp$-plane, i.e. take on a familiar flute-like form with $k_\perp \gg k_\parallel$.[54] We may expect the energy cascade to evolve primarily in this plane, i.e. with a K-41 spectrum with respect to $k_\perp$, and for the rates of eddy deformation in the $\parallel$ and $\perp$ directions to be *critically balanced*,

$$\varepsilon \sim z_{\ell_\perp}^2/\tau_{\ell_\perp} \sim z_{\ell_\perp}^3/\ell_\perp, \qquad \tau_{\ell_\perp} \sim \ell_\perp/z_{\ell_\perp} \sim \ell_\parallel/V_A \sim \tau_{\ell_\parallel}, \tag{6.103}$$

so that the Alfvén effect does not dominate the spectral transfer of energy. Using (6.103), we obtain the anisotropy of MHD turbulent eddies,

$$k_\perp/k_\parallel \sim \ell_\parallel/\ell_\perp \sim (\ell_{\mathcal{K}}/\ell_\perp)^{1/3}, \qquad \ell_{\mathcal{K}} = \mathcal{E}^{3/2}/\varepsilon \approx V_A^3/\varepsilon, \tag{6.104}$$

which are roughly isotropic on the integral scale, $\ell_{\mathcal{K}}$, but become strongly anisotropic on the dissipative scale, ℓ_ε. The corresponding parallel energy spectrum is just the K-41 spectrum with $k_\perp$ eliminated using (6.104),

$$E_k(k_\perp) \sim \varepsilon^{2/3} k_\perp^{-5/3} \quad \Rightarrow \quad E_k(k_\parallel) \sim \varepsilon^{3/2}(V_A k_\parallel)^{-5/2}, \tag{6.105}$$

so that E_k decays faster with $k_\parallel$ than with $k_\perp$, consistent with (6.104). As in hydrodynamics, the decay is exponential for $k > k_\varepsilon \sim 1/\ell_\varepsilon$.

The dissipative scale, ℓ_ε, corresponding to the I–K spectrum may be obtained by assuming $\varepsilon_B \approx \varepsilon_V \approx \varepsilon/2$, $\mathrm{Pr} = \nu/\eta \approx 1$ and equating τ_ℓ^ε with ℓ_ε^2/ν, with the result $\ell_\varepsilon \sim (\nu^2 V_A/\varepsilon)^{1/3}$, which is somewhat larger than the K-41 dissipative scale, (6.49). Using $V_\ell \sim (\varepsilon V_A \ell)^{1/4}$, (6.101), the Reynolds number corresponding to the I–K cascade can be expressed as,

$$\mathrm{Re}_\ell = \ell V_\ell/\nu \sim \ell^{5/4}(\varepsilon V_A)^{1/4}/\nu \sim V_\ell^5/\varepsilon V_A \nu, \tag{6.106}$$

so that $\ell/\ell_\varepsilon \sim \mathrm{Re}_\ell^{4/5}$ compared to the K-41 result, $\ell/\ell_\varepsilon \sim \mathrm{Re}_\ell^{3/4}$, see Table 6.5.[55] However, since V_A is related to the integral scale $\ell_{\mathcal{K}}$ and $\mathrm{Re}_{\mathcal{K}}$ by

$$\ell_{\mathcal{K}} \sim \mathcal{E}^{3/2}/\varepsilon \sim V_A^3/\varepsilon, \qquad \mathrm{Re}_{\mathcal{K}} \sim \mathcal{E}^2/\nu\varepsilon \sim V_A^4/\nu\varepsilon, \tag{6.107}$$

the ratio of integral and dissipative scales, $\ell_{\mathcal{K}}/\ell_\varepsilon \sim \mathrm{Re}_{\mathcal{K}}^{3/4-1/12} = \mathrm{Re}_{\mathcal{K}}^{2/3}$, exhibits a somewhat weaker scaling with $\mathrm{Re}_{\mathcal{K}}$ than with Re_ℓ. Obviously, the above analysis must be modified for non-unit Prandtl numbers, e.g. when $\eta \gg \nu$, $\mathcal{B}$ is dissipated at a much larger scale than $\mathcal{K}$.

54 In particular, perpendicular motions become dominant on the smallest scales.
55 Also $V_\ell/V_\varepsilon \sim \mathrm{Re}_\ell^{1/5}$ and $\tau_\ell/\tau_\varepsilon \sim \mathrm{Re}_\ell^{3/5}$, so that smaller eddies evolve faster than larger ones.

In close analogy with hydrodynamics, 3D MHD turbulence is not self-similar, but becomes increasingly intermittent as one approaches the dissipative scale. Indeed, both numerical simulations and satellite measurements of the solar wind indicate it to be more intermittent than HD turbulence of comparable Reynolds number, which is due in part to the Alfvén effect and in part to the shape of the smallest eddies, which are more tube-like in 3D HD and more sheet-like in 3D MHD.[56] As before, intermittency may be quantified as the departure of the actual spectral exponents, ζ_p, from their scale-invariant values, for which we have two reference points,[57]

$$\text{K-41}: \quad \langle z_\ell^p \rangle \sim (\varepsilon \ell)^{\zeta_p}, \qquad \zeta_p = p/3, \tag{6.108}$$

$$\text{I–K}: \quad \langle z_\ell^p \rangle \sim (\varepsilon V_A \ell)^{\zeta_p}, \qquad \zeta_p = p/4, \tag{6.109}$$

corresponding to the two energy spectra relevant to (weakly correlated) MHD turbulence. The degree of intermittency is typically modelled using the same techniques used for hydrodynamics, recall Section 6.1.2.3, one of the most successful being the *log-Poisson* model.

Since only the K-41 energy spectrum is in agreement with observations, and indeed with an exact relation, analogous to the 4/5th law in hydrodynamics, (6.54), which requires that $\zeta_3 = 1$, we are justified in repeating the log-Poisson analysis, as outlined on page 237, only for this case.[58] Assuming the smallest structures to be current sheets ($d = 2$, $C_0 = 3 - 2 = 1$), which follow the K-41 spectrum ($x = 2/3$), we now have $\beta = 2/3$ and hence

$$\mu_m = \tfrac{2}{3}m - 1 + (1/3)^m, \qquad \zeta_p = p/9 + 1 - (1/3)^{p/3}. \tag{6.110}$$

The resulting ζ_p increase monotonically with p,[59] but less rapidly than (6.66), indicating a greater level of intermittency in MHD compared to HD.

Numerical simulations of globally isotropic, 3D MHD turbulence (Müller *et al.*, 2003) reveal that (6.110) is in excellent agreement with the perpendicular spectral exponents, $\zeta_{p\perp}$, consistent with the assumed K-41 spectrum in the plane perpendicular to the local, eddy-averaged field, $\overline{\mathbf{B}}_\ell$.[60] When a constant ambient field, $\mathbf{B}_0$, was introduced in the simulations, $\zeta_{p\perp}$ ($\zeta_{p\parallel}$) decreased (increased) moderately, with only a $\sim$ 20% variation between the globally isotropic ($\mathbf{B}_0 = 0$) and quasi-2D

56 The latter result is perhaps surprising in view of the tearing instability, which might be expected to rip these structures into tube-like filaments. Nonetheless, the stability of micro-current sheets against the tearing mode was clearly demonstrated in 3D numerical simulations.

57 The I–K result follows directly from (6.101), which implies $z_\ell \propto \ell^{1/4}$.

58 It is left as an exercise to the reader to confirm that application to the I–K spectrum leads to $\mu_m = m/2 - 1 + (1/2)^m$ and $\zeta_p = p/8 + 1 - (1/2)^{p/4}$, which violate the exact relation, $\zeta_3 = 1$.

59 This, and the fact they satisfy the exact relation, $\zeta_3 = 1$, makes them physically admissible.

60 The parallel exponents, $\zeta_{p\parallel}$, are $\sim$ 15% larger. Both exponents are defined as $\langle z^p_{\perp\parallel} \rangle \propto \ell^{\zeta_{p\perp\parallel}}$.

($\mathbf{B}_0/\overline{\mathbf{B}}_\ell = 10$) MHD turbulence. The perfect 2D case ($\mathbf{B}_0/\overline{\mathbf{B}}_\ell \to \infty$) will be treated below.

6.2.2 MHD turbulence in 2D

Since many of the derivations of the previous section did not depend on the geometry of the flow, they should remain valid for both 2D and 3D MHD flows, e.g. the dynamical invariants and spectral cascades are quite similar in both cases, see Table 6.6, as is the tendency to form large-scale, coherent magnetic structures and to align $\widetilde{\mathbf{V}}$ and $\widetilde{\mathbf{B}}$ at small scales. Nonetheless, there are some important differences between the two types of flows.

Let us begin with the quasi-2D MHD equations for a strongly magnetized plasma, with the simplifying assumptions of a dominant, constant mean field, $\overline{\mathbf{B}} = B_0\hat{\mathbf{e}}_z$, uniform pressure, $p_0 = \text{const}$, and zero mean flow, $\overline{\mathbf{V}} = 0$.[61] Subjecting (2.246)–(2.248), with additional viscous and resistive terms, to flute-like perturbations, $\nabla_\parallel = \hat{\mathbf{e}}_z\partial_z \ll \nabla_\perp = \hat{\mathbf{e}}_x\partial_x + \hat{\mathbf{e}}_y\partial_y$, it can be shown that to lowest order in $\epsilon \sim \ell_\perp/\ell_\parallel \sim B_\perp/B_0$, the 3D MHD equations in a uniform magnetic field take on the *reduced MHD* form (Strauss, 1976),[62]

$$\widetilde{\mathbf{V}} = \widetilde{\mathbf{V}}_\perp = \hat{\mathbf{e}}_z \times \nabla_\perp\phi, \qquad \widetilde{\mathbf{B}} = \widetilde{\mathbf{B}}_\perp = \hat{\mathbf{e}}_z \times \nabla_\perp\psi, \tag{6.111}$$

$$\mathrm{d}_t\psi = B_0\partial_z\phi + \eta\nabla_\perp^2\psi, \qquad \Omega = \Omega_z = \nabla_\perp^2\phi, \qquad J = J_z = \nabla_\perp^2\psi, \tag{6.112}$$

$$\mathrm{d}_t\Omega = \mathbf{B}\cdot\nabla J + \nu\nabla_\perp^2\Omega = B_0\partial_z J + \widetilde{\mathbf{B}}\cdot\nabla_\perp J + \nu\nabla_\perp^2\Omega, \tag{6.113}$$

where $\mathrm{d}_t = \partial_t + \widetilde{\mathbf{V}}\cdot\nabla_\perp$ is the advective derivative. Note that the perturbed velocity and magnetic field are expressed as the perpendicular gradients of the stream and flux functions, $\phi = \varphi/B_0$ and $\psi = -A_z$, respectively, while the parallel vorticity and current are given by the Laplacian of ϕ and ψ. For $\psi = \text{const}$, (6.113) becomes the vorticity equation of 2D hydrodynamics, while for $\partial_z = 0$ (6.112)–(6.113) take on the (perfect) 2D MHD form,

$$\mathrm{d}_t\psi = \eta\nabla^2\psi, \qquad \mathrm{d}_t\Omega = \widetilde{\mathbf{B}}\cdot\nabla J + \nu\nabla^2\Omega, \qquad \nabla = \nabla_\perp. \tag{6.114}$$

Unlike 2D HD, in which vorticity is merely advected by the flow and destroyed by diffusion, 2D MHD allows for the creation of vorticity by the Lorentz force, $\widetilde{\mathbf{J}}\times\widetilde{\mathbf{B}}$, represented in (6.114) by $\widetilde{\mathbf{B}}\cdot\nabla J$. In contrast, the induction equation reduces to a purely advective-diffusive form, with ψ playing the role of an *active* scalar, coupled to the flow by the Lorentz force.

[61] In reality, the gradients of the mean quantities provide the free energy which drives the turbulence, see page 243, and are neglected here only to better focus on fluctuating quantities.

[62] These equations can also be obtained from those of flute-reduced MHD for toroidal geometry, Section 4.2.5, with $\mathbf{B}_0 = \text{const}$, $R_0 \to \infty$, $\kappa = 0$, $\partial_\zeta = 0$, $\rho_0 = 1 = \mu_0$, $\chi_\perp = 0$ and $\phi = \breve{\varphi}_2$.

Since the derivation of the I–K energy spectrum relies on a dominant magnetic field, so that $\tau_A \ll \tau_\ell$, (6.100), we may expect this result to be more applicable to globally anisotropic (quasi-2D or 2D), than to globally isotropic (3D) flows. On the other hand, one might expect the increased anisotropy to further constrain the energy cascade to the perpendicular plane, thus favouring the K-41 spectrum. In reality, the competition between these two processes determines the actual cascade dynamics, which are hence difficult to predict based on such simple arguments. To resolve the issue one must turn to numerical simulations, which indicate good, if not quite perfect, agreement with the I–K energy spectrum (6.102), suggesting that the Alfvén effect is indeed the dominant process in the energy cascade.

In the simulations discussed at the end of the previous section, we already saw that MHD turbulence became more intermittent following the introduction of a strong ambient field, $\mathbf{B}_0$, with $\zeta_{p\perp}$ reduced moderately from the 3D (globally isotropic) values. As $\mathbf{B}_0 \to \infty$, $\zeta_{p\perp}$ are reduced further, asymptotically approaching their 2D MHD limits. These are found to be ~35% and ~15% lower than the 3D and quasi-2D values, respectively, indicating a substantial increase in intermittency levels compared to 3D MHD.

To conclude our discussion of MHD turbulence, let us remark on its relevance to magnetically confined plasmas. Since the existence of magnetic flux surfaces presupposes relatively slow, drift-ordered flows, recall Chapter 3, we would not expect MHD flows to dominate radial transport in the presence of MHD equilibrium. Indeed, even MHD instabilities, recall Chapter 4, rarely produce a coupling of highly disparate scales and the associated spectral cascades, which characterize turbulent flows.[63]

6.3 DHD turbulence

In the preceding sections we introduced the basic concepts of fluid turbulence and explored its properties in the context of both hydrodynamics and MHD. We are now ready to investigate the turbulent flows in magnetically confined plasmas, specifically tokamaks, which dominate the radial *transport* of mass, momentum and energy, and thus determine the degree of *confinement* and *exhaust* of these quantities in/from the plasma column. The dominant dynamical modes of these flows, which are drift- rather than MHD-ordered, are the various drift-waves and

[63] One of the rare instances of MHD turbulence in low-β_T tokamaks are the already mentioned Alfvén cascades, Section 4.2.6, which occur in the central part of the plasma column when the magnetic shear is reversed, i.e. when $q' < 0$. On the other hand, MHD turbulence is quite prevalent in reversed field pinches and to a lesser extent in $\epsilon_a \sim 1$ (spherical) tokamaks, both of which allow for much higher β_T than conventional, $\epsilon_a \sim 0.3$, designs.

instabilities introduced in Section 4.3. Their spatial and temporal scales are given roughly by $\rho_S = c_S/\Omega_i$ and $\omega_{tS} = c_S/L_\perp$ with $L_\perp \approx |\nabla \ln p|^{-1}$, respectively, see (4.225),[64]

$$k_\wedge \rho_S \sim 1, \qquad \omega \sim \omega_{*n} \sim \omega_{tS}, \qquad k_\parallel c_S \ll \omega \ll k_\parallel v_{te}. \tag{6.115}$$

and are thus short compared to those of the background gradient, $k_\wedge L_\perp \sim \Omega_i/\omega_{tS} \sim \delta_i^{-1} \gg 1$. In other words, they involve small-scale ($\sim\rho_S$), low-frequency ($\sim\omega_{tS}$) motions of turbulent eddies of the electric drift velocity, which in the drift ordering is primarily electrostatic, $\mathbf{V}_E = \mathbf{b}\times\nabla\varphi/B$. The relative fluctuation amplitudes are comparable to δ_i and are thus small in a magnetized plasma, with typical values of ~1–10%,

$$e\widetilde{\varphi}/T_e \sim \widetilde{p}_e/p_e \sim \widetilde{n}_e/n_e \sim \rho_S/L_\perp \sim \delta_i \sim 0.01\text{–}0.1. \tag{6.116}$$

This relation is a consequence of the drift ordering, $\widetilde{V}_E \sim \widetilde{V}_{*i} \sim \delta_i V_S \sim \overline{T}_e/eB_0L_\perp$, where $\widetilde{V}_E \sim k_\wedge\widetilde{\varphi}/B_0$ and $\widetilde{V}_{*i} \sim k_\wedge\widetilde{p}_i/e\overline{n}_e B_0$, so that $e\widetilde{\varphi}/\overline{T}_e \sim \widetilde{p}_e/\overline{p}_e \sim 1/k_\wedge L_\perp \sim \rho_{ti}/L_\perp \sim \delta_i$; the last result makes use of the characteristic spatial scale of drift-waves, $k_\wedge\rho_{ti} \sim 1$. Similarly, under the MHD ordering, $\widetilde{V}_E \sim V_S \gg \widetilde{V}_{*i} \sim \delta_i V_S$, one finds $e\widetilde{\varphi}/\overline{T}_e \sim 1 \gg \widetilde{p}_e/\overline{p}_e \sim \delta_i$.[65]

The non-linear growth, interaction and saturation of these modes is responsible for the complicated radial flows known as *drift-wave* turbulence or, more generally, *drift-fluid* or *DHD turbulence*. Below we introduce this subject with an emphasis on the tokamak plasma edge, closely following the account given in a series of excellent papers by Scott and co-workers.

6.3.1 Drift-fluid turbulence

The bulk of our discussion will centre on the simplest, drift-fluid approximation. First, we'll derive a global drift-fluid model, (Section 6.3.1.1), then we'll reduce it into a local (fixed background) version (Section 6.3.1.3) and, finally, we'll investigate the latter using numerical simulations (Section 6.3.1.4). In the latter part of the section we'll briefly discuss how the obtained results are modified by the inclusion of gyro-fluid and kinetic effects.

64 Here we use the warm ion version of (4.225), replacing the cold ion sound speed, $c_{Se}^2 = p_e/\rho = ZT_e/m_i$, by the warm ion sound speed, $c_S^2 = \left(\frac{5}{3}p_i + p_e\right)/\rho$, in the definition of ρ_S and ω_{tS}.

65 Since the drift-wave phase shift, Δ_1, (4.231), decreases with $L_\perp \sim a$ (the gradient provides the free energy for the turbulence) and with $L_\wedge^2$ (shorter scales being most unstable), while increasing with $L_\parallel^2 \sim (qR)^2$ (parallel response provides the dissipation), and hence with the ratio, $L_\parallel^2/L_\perp$, the relative fluctuations, (6.116), increase towards the edge of the plasma.

6.3.1.1 Global drift-fluid model

Before considering plasma turbulence in the drift-fluid approximation, let us update the six-field (φ, n_e, $J_\parallel$, $V_{\parallel i}$, p_e, p_i) drift-fluid model derived in Section 2.4.2.7 with three physical effects likely to play a role in edge plasma dynamics: (i) the ion polarization drift; (ii) electromagnetic effects (Alfvén waves, magnetic flutter); and (iii) collisional dissipation and transport, recall Chapter 5. The derivation, as outlined in Scott (2003a), follows the same iterative procedure as that in Section 2.4.2.7, but leads to a more generally valid set of drift-fluid equations, i.e. encompassing all fluid-like micro-instabilities, including drift-waves and interchange modes. Specifically, it contains the reduced MHD model as a complete subset, thus allowing the study of both MHD and drift-fluid phenomena, including any transition between the two, using the same set of equations; by employing such a superset of reduced MHD and drift models, we guard against preselecting any dynamical mode, which could prejudge the outcome of the analysis. Finally, the resulting model is *global*, rather than *local*, in that it follows the evolution of complete quantities, $\mathcal{A} = \overline{\mathcal{A}} + \widetilde{\mathcal{A}}$, rather than fluctuations, $\widetilde{\mathcal{A}}$, evolving on prescribed profiles, $\overline{\mathcal{A}}$, i.e. it does not invoke the thin-layer (Boussinesq) approximation, frequently used as the starting point for (local) turbulence analysis. As such, it does not presuppose a local relation between turbulent flows and mean profile gradients, $\mathbf{\Gamma}_{tb} \propto \nabla\overline{\mathcal{A}}$, evaluated at a given location.[66]

We begin by assuming a quasi-static perpendicular force balance,

$$\mathbf{E}_\perp \approx -\nabla_\perp\varphi, \qquad E_\parallel = -\nabla_\parallel\varphi - \partial_t A_\parallel, \tag{6.117}$$

and write the magnetic field in terms of equilibrium and perturbed parts,[67]

$$\mathbf{B} \approx \mathbf{B}_0 + \mathbf{B}_\perp + \widetilde{B}\mathbf{b}_0, \qquad \mathbf{B}_\perp = -\mathbf{b}_0 \times \nabla A_\parallel, \qquad \mathbf{b}_0 = \mathbf{B}_0/B_0,$$
$$\mathbf{b} \approx \mathbf{b}_0 + \mathbf{B}_\perp/B_0, \qquad \nabla_\parallel \approx \mathbf{b}\cdot\nabla, \qquad \nabla_\perp = -\mathbf{b}_0 \times \mathbf{b}_0 \times \nabla.$$

The first-order perpendicular flows are purely diamagnetic, (2.206)–(2.232),

$$\mathbf{V}_{*s} = \frac{\mathbf{b}_0}{n_s e_s B_0} \times \nabla p_s, \qquad \mathbf{J}_* = \frac{\mathbf{b}_0}{B_0} \times \nabla p, \qquad \mathbf{Q}_{*s} = \frac{5}{2}\frac{\mathbf{b}_0}{e_s B_0} \times \nabla(p_s T_s),$$

where $p = p_i + p_e$ is the total pressure. To ensure charge and energy conservation, we also retain the (second-order) polarization drift in the ion dynamics, so that the net ion and electron flow velocities become

[66] Nonetheless, this *local* flux-force relation, which leads to a diffusive, mixing-length transport process is generally recovered in the thin layer limit, provided the size and turn-over time of the eddies are small compared to the length scale and evolution time of the background profiles. Recall that this relation is also the starting ansatz of most collisional transport theories.

[67] Note that $\widetilde{\varphi}/B_0$ and $-\widetilde{A}_\parallel/B_0$ act as stream functions for $\mathbf{V}_E$ and $\mathbf{b}_\perp$.

$$\mathbf{V}_i \approx V_{\|i}\mathbf{b} + \mathbf{V}_{\perp i} + \mathbf{V}_{pi}, \qquad \mathbf{V}_e \approx V_{\|e}\mathbf{b} + \mathbf{V}_{\perp e}, \qquad V_{\|e} = V_{\|i} - J_\|/en_e,$$
$$\mathbf{V}_E \approx \frac{\mathbf{b}_0}{B_0} \times \nabla\varphi, \qquad \mathbf{V}_{\perp s} \approx \mathbf{V}_E + \mathbf{V}_{*s}. \tag{6.118}$$

Specifically, we retain the full ion velocity, $\mathbf{V}_i$, in the advective derivative, $\mathrm{d}_t = \partial_t + \mathbf{V}_i \cdot \nabla$, although it can be omitted when combined with other small quantities, e.g. the drift energy is given to a high accuracy by the square of the lowest-order ion drift velocity, $\mathbf{V}_{\perp i}$.[68] The polarization drift, and the related current, are defined by the implicit relation,

$$\mathbf{J}_p \approx en_e\mathbf{V}_{pi} = \rho\frac{\mathbf{b}_0}{B_0} \times \mathrm{d}_t\mathbf{V}_{\perp i} + \mathbf{J}_\pi, \qquad \mathbf{J}_\pi = \frac{\mathbf{b}_0}{B_0} \times (\nabla \cdot \boldsymbol{\pi}_*), \tag{6.119}$$

where $\rho = n_i m_i$ is the mass density and $\boldsymbol{\pi}_*$ is the gyro-viscous stress with respect to the first-order velocity, $V_{\|i}\mathbf{b} + \mathbf{V}_{\perp i}$. Most importantly, the polarization current enters the charge conservation (quasi-neutrality) equation,[69]

$$\nabla \cdot \mathbf{J} = 0, \qquad \mathbf{J} = J_\|\mathbf{b} + \mathbf{J}_* + \mathbf{J}_p. \tag{6.120}$$

Since we derive additional evolution equations for $J_\|$, $V_{\|i}$, p_e and p_i below, (6.120) effectively specifies the evolution of φ, $\mathbf{V}_E$ and $\mathbf{V}_{\perp i}$, and thus represents the conservation of perpendicular ion momentum.

Evolution equations for the remaining state variables are obtained by the same procedure, i.e. by substituting (6.118)–(6.120) into the fluid conservation equations, (2.163)–(2.165). Neglecting volumetric sources, one thus obtains the following conservation/evolution equations:

- conservation of mass, evolution of $n_e = Zn_i$:

$$\partial_t n_e + \nabla \cdot (n_e\mathbf{V}_e) = 0; \tag{6.121}$$

- conservation of parallel electron momentum, evolution of $J_\|$:

$$E_\| = -\partial_t A_\| - \nabla_\|\varphi = (F_{\|e} - \nabla_\| p_e)/en_e, \tag{6.122}$$
$$\mu_0 J_\| = -\nabla_\perp^2 A_\| = -\nabla \cdot \nabla_\perp A_\|; \tag{6.123}$$

- conservation of parallel ion momentum, evolution of $V_{\|i}$:

$$\rho \mathrm{d}_t V_{\|i} + \mathbf{b} \cdot (\nabla \cdot \boldsymbol{\pi}_*) = -\nabla_\|(p_e + p_i) + \nabla \cdot (\mu_\|\mathbf{b}\nabla_\| V_{\|i}); \tag{6.124}$$

- conservation of electron energy, evolution of p_e:

$$\tfrac{3}{2}\partial_t p_e + \nabla \cdot \left[\tfrac{3}{2}p_e\mathbf{V}_E + \left(\tfrac{5}{2}p_e V_{\|e} + q_{\|e}\right)\mathbf{b} - \tfrac{5}{2}(\mathbf{b}_0/eB_0) \times \nabla(p_e T_e)\right]$$
$$= V_{\|e}\nabla_\| p_e - p_e\nabla \cdot \mathbf{V}_E + F_{\|e}J_\|/en_e - Q_i; \tag{6.125}$$

[68] Henceforth, we reserve the symbol $\mathbf{V}_{\perp s}$ for the first-order perpendicular flow (drift) velocity.
[69] Indeed, plasma remains quasi-neutral, $(n_e - Zn_i)/n_e \ll 1$, and the charge density vanishingly small, to high accuracy, specifically to order $(V_A/c)^2 \ll 1$, at all scales down to ρ_S and ω_{tS}.

• conservation of ion energy, evolution of p_i:

$$\tfrac{3}{2}\partial_t p_i + \nabla\cdot\left[\tfrac{3}{2}p_i\mathbf{V}_E + \tfrac{5}{2}p_i(V_{\|i}\mathbf{b}+\mathbf{V}_{pi}) + q_{\|i}\mathbf{b} + \tfrac{5}{2}(\mathbf{b}_0/eB_0)\times\nabla(p_iT_i) + \mathbf{q}_i^{neo}\right]$$
$$= (V_{\|i}\mathbf{b}+\mathbf{V}_{pi})\cdot\nabla p_i - p_i\nabla\cdot\mathbf{V}_E + \mu_\|\,|\nabla_\| V_{\|i}|^2 + \mathcal{Q}_i. \quad (6.126)$$

Assuming at least a moderate level of collisionality (i.e. ruling out $\nu^* \ll 1$), the parallel friction, $F_{\|e}$, parallel heat flows, $q_{\|s}$, and perpendicular ion heat flow, $\mathbf{q}_i^{neo} = q_{\perp i}$, can be estimated by the (flux-limited) Braginskii expressions,[70] see Section 5.3.2.1, including a neoclassical, Pfirsch–Schlüter correction for the perpendicular ion heat flow,[71] (5.223),

$$F_{\|e} = en_e\eta_\|\left[J_\| + \tfrac{0.71}{3.2}\left(0.71J_\| + \tfrac{e}{T_e}q_{\|e}\right)\right], \quad (6.127)$$

$$q_{\|e} = -\kappa_{\|e}\nabla_\| T_e - 0.71T_eJ_\|/e, \qquad q_{\|i} = -\kappa_{\|i}\nabla_\| T_i, \quad (6.128)$$

$$q_i^{neo} \approx q_{\perp i}^{cl} + q_{\perp i}^{PS} \approx -\kappa_{\perp i}(1+1.6q^2)\nabla_\perp T_i. \quad (6.129)$$

We would next like to confirm the conservation of the total energy density,

$$\mathcal{E} = \tfrac{1}{2}\rho V_{\perp i}^2 + \tfrac{1}{2}\rho V_{\|i}^2 + \tfrac{3}{2}(p_i+p_e) + \tfrac{1}{2}B_\perp^2/\mu_0, \quad (6.130)$$

which, term by term, corresponds to the drift energy, the acoustic energy, the (ion and electron) thermal energy and the magnetic energy. The evolution of the thermal energies is given by (6.125)–(6.126), while selective elimination yields conservation equations for the perpendicular kinetic (drift) energy,

$$\tfrac{1}{2}\partial_t\left(\rho V_\perp^2\right) + \nabla\cdot\left[\tfrac{1}{2}\rho V_{\perp i}^2\mathbf{V}_i + \mathbf{V}_{\perp i}\cdot\boldsymbol{\pi}_* + \varphi\mathbf{J}_\perp + p\mathbf{V}_E\right]$$
$$= \boldsymbol{\pi}_* : \nabla\mathbf{V}_{\perp i} + J_\|\nabla_\|\varphi + p\nabla\cdot\mathbf{V}_E - \mathbf{V}_{pi}\cdot\nabla p_i, \quad (6.131)$$

the parallel kinetic (acoustic or sound wave) energy,

$$\tfrac{1}{2}\partial_t\left(\rho V_{\|i}^2\right) + \nabla\cdot\left[\tfrac{1}{2}\rho V_{\|i}^2\mathbf{V}_i + V_{\|i}\mathbf{b}\cdot\boldsymbol{\pi}_* + V_{\|i}\mathbf{b}\mu_\|\nabla_\| V_{\|i}\right]$$
$$= \boldsymbol{\pi}_* : \nabla(V_{\|i}\mathbf{b}) - V_{\|i}\nabla_\| p - \mu_\|\,|\nabla_\| V_{\|i}|^2, \quad (6.132)$$

and the perpendicular magnetic (Alfvén wave) energy,

$$\tfrac{1}{2}\partial_t B_\perp^2 - \nabla\cdot(\mathbf{b}\partial_t A_\| \times \mathbf{B}_\perp) = \mu_0 J_\|\partial_t A_\|$$
$$= \mu_0 J_\|[(\nabla_\| p_e - F_{\|e})/en_e - \nabla_\|\varphi]. \quad (6.133)$$

All five energy equations are written in conservative form, i.e. consist of a partial time derivative of the energy in question, a divergence of the corresponding energy

[70] The resistivity, $\eta_\| = 1/\sigma_{\|e}$, parallel and perpendicular heat conductivities, $\kappa_{\|s}$ and $\kappa_{\perp s}$, and the parallel ion viscosity, $\mu_{\|i}$, are given by (5.154), (5.164), (5.165) and (5.161), respectively. The numerical constants in (6.127) are evaluated for hydrogenic ions; for $Z > 1$, see Table 5.1. Note that we retain $\mathbf{q}_i^{neo}$ as the largest neoclassical term, despite being second order in δ_i.

[71] For $\nu^* < 1$, this should be replaced by a banana-plateau expression, see Section 5.3.3.4.

Table 6.7. *Drift-fluid energy transfer channels.*

Transfer term	Energy coupling	Fluctuation coupling	Mechanism
$J_\parallel \nabla_\parallel \varphi$	$V_\perp \leftrightarrow B_\perp$	$\widetilde{\varphi} \leftrightarrow \widetilde{J}_\parallel$	Alfvén wave
$(J_\parallel/en_e)\nabla_\parallel p_e$	$B_\perp \leftrightarrow p_e$	$\widetilde{J}_\parallel \leftrightarrow \widetilde{p}_e$	adiabatic electrons
$(J_\parallel/en_e)F_{\parallel e}$	$B_\perp \leftrightarrow p_e$	$\widetilde{J}_\parallel \leftrightarrow \widetilde{p}_e$	resistive dissipation
$V_{\parallel i}\nabla_\parallel p_e$	$V_{\parallel i} \leftrightarrow p_e$	$\widetilde{V}_{\parallel i} \leftrightarrow \widetilde{p}_e$	[sound wave]
$V_{\parallel i}\nabla_\parallel p_i$	$V_{\parallel i} \leftrightarrow p_i$	$\widetilde{V}_{\parallel i} \leftrightarrow \widetilde{p}_i$	
$p_e\nabla\cdot\mathbf{V}_E$	$V_{\perp i} \leftrightarrow p_e$	$\widetilde{\varphi} \leftrightarrow \widetilde{p}_e$	[$\mathbf{V}_E$ compression due to
$p_i\nabla\cdot\mathbf{V}_E$	$V_{\perp i} \leftrightarrow p_i$	$\widetilde{\varphi} \leftrightarrow \widetilde{p}_i$	curvature: interchange]
$\mathbf{V}_{pi}\cdot\nabla p_i$	$V_{\perp i} \leftrightarrow p_i$	$\widetilde{\varphi} \leftrightarrow \widetilde{p}_i$	polarization advection
$\mu_\parallel\lvert\nabla_\parallel V_{\parallel i}\rvert^2$	$V_{\parallel i} \leftrightarrow p_i$	$\widetilde{V}_{\parallel i} \leftrightarrow \widetilde{p}_i$	viscous dissipation
$\mathcal{Q}_i$	$p_e \leftrightarrow p_i$	$\widetilde{p}_e \leftrightarrow \widetilde{p}_i$	collisional relaxation

flow (sum of *transport* terms) and a volumetric source (sum of *transfer* terms). The latter are summarized in Table 6.7; all the non-dissipative transfer terms in this table have their origin in the Joule heating term, $e_s n_s \mathbf{V}_s \cdot \mathbf{E}$ for species s, or $\mathbf{J}\cdot\mathbf{E}$ for their sum. They represent transfer terms between the various energies in (6.130), i.e. appear with opposite signs in different energy equations and thus cancel upon summation. As a result, the total energy is conserved to high accuracy,

$$\partial_t \mathcal{E} + \nabla\cdot\Big[\frac{3}{2}p\mathbf{V}_E + q_\parallel \mathbf{b} + \frac{5}{2}(\varphi J_\parallel + p_e V_{\parallel e} + p_i V_{\parallel i})\mathbf{b} + \frac{5}{2}p_i\mathbf{V}_{pi}$$
$$+\frac{5}{2}(\mathbf{b}_0/eB_0)\times\nabla(p_iT_i - p_eT_e) + \frac{1}{2}\rho\left(V_{\parallel i}^2 + V_{\perp i}^2\right)\mathbf{V}_i + \boldsymbol{\pi}_*\cdot(V_{\parallel i}\mathbf{b} + \mathbf{V}_{\perp i})$$
$$+(p\mathbf{V}_E + \varphi\mathbf{J}_\perp) - \mathbf{b}\partial_t A_\parallel \times \mathbf{B}_\perp/\mu_0 + \mathbf{q}_i^{neo}\Big] = 0, \tag{6.134}$$

and evolves purely due to *transport* effects, which appear in the divergence in order of descending magnitude. The first two terms, representing $\mathbf{E}\times\mathbf{B}$ advection and parallel conduction, usually dominate the net energy flow, while the next two, namely the parallel (Alfvénic and acoustic) waves and polarization advection, are generally much smaller.[72] The remaining terms, i.e. the second and third lines in (6.134), are nearly divergence free, unless one is considering transonic or quasi-laminar flows.

It is important to note that the third but last term in (6.134) includes a factor $p\mathbf{V}_E$, whose divergence nearly cancels with that of $\varphi\mathbf{J}_\perp$, to leave a pre-factor of $3/2$, rather than $5/2$, in the leading $\mathbf{V}_E$ advective term. This so-called *Poynting cancellation* is a characteristic feature of $\mathbf{V}_E$ advection. To demonstrate this cancellation, we begin by taking the divergence of

[72] The parallel energy flow in (6.134) includes the perturbed magnetic field, $\mathbf{b}_\perp \approx \mathbf{B}_\perp/B_0$, which leads to a *perpendicular* deflection of the *parallel* energy flow, $q_\parallel + \frac{5}{2}(\varphi J_\parallel + p_e V_{\parallel e} + p_i V_{\parallel i})$.

$$\nabla \times (\varphi \mathbf{B}) = \nabla\varphi \times \mathbf{B} + \varphi \nabla \times \mathbf{B}, \tag{6.135}$$

recalling Ampere's law and noting that $\nabla \cdot \nabla \times = 0$, to find

$$\nabla \cdot (\varphi \mathbf{J}) = \nabla \cdot (\varphi \nabla \times \mathbf{B}/\mu_0) = -\nabla \cdot (\nabla\varphi \times \mathbf{B}/\mu_0) = \nabla \cdot (\mathbf{E} \times \mathbf{B}/\mu_0) = \nabla \cdot \mathbf{S},$$

where $\mathbf{S} = \mathbf{E} \times \mathbf{B}/\mu_0$ is the electromagnetic, or *Poynting*, energy flow. Since we are mainly interested in fluctuating $\mathbf{E} \times \mathbf{B}$ flows, we focus on first-order fluctuations of the electric potential, $\widetilde{\varphi}$, parallel field, $\widetilde{\mathbf{B}}_{\|} = \widetilde{B}_{\|}\mathbf{b}_0$, and pressure, $\widetilde{p}$, see page 258, assuming the field direction to be constant, $\mathbf{b}_0$. Hence,

$$\nabla \cdot (\widetilde{\varphi}\widetilde{\mathbf{J}}_\perp + \widetilde{p}\widetilde{\mathbf{V}}_E) = \nabla \cdot \left[\widetilde{\mathbf{E}} \times \mathbf{b}_0 \left(\widetilde{B}_{\|}/\mu_0 + \widetilde{p}/B_0\right)\right] \approx 0, \tag{6.136}$$

which vanishes because of a perpendicular force balance between fluctuating thermal and magnetic pressures,[73] recall (2.47) and (4.87),

$$\nabla_\perp (p + B^2/2\mu_0) \approx 0 \quad \Rightarrow \quad \widetilde{p} + B_0 \widetilde{B}_{\|}/\mu_0 \approx 0; \tag{6.137}$$

In other words, the work, $\widetilde{p}\widetilde{\mathbf{V}}_E$, is balanced by the perpendicular Poynting flux, $\widetilde{\mathbf{S}}_\perp = \widetilde{\varphi}\widetilde{\mathbf{J}}_\perp$, while the parallel Poynting flux, $\widetilde{\mathbf{S}}_{\|} = \widetilde{\mathbf{E}} \times \widetilde{\mathbf{B}}_\perp = \widetilde{\varphi}\widetilde{\mathbf{J}}_{\|}$ represents the parallel flow of energy associated with a shear Alfvén wave.

It is also worth noting that expanding $\nabla \cdot (\varphi \mathbf{J})$, using (6.120) to eliminate $\nabla \cdot \mathbf{J} = 0$, and rewriting the diamagnetic term $\mathbf{J}_* \cdot \nabla\varphi$ as $-\mathbf{V}_E \cdot \nabla p$, we find,

$$\nabla \cdot (p\mathbf{V}_E + \varphi \mathbf{J}) = J_{\|}\nabla_{\|}\varphi + \mathbf{J}_p \cdot \nabla\varphi + p\nabla \cdot \mathbf{V}_E, \tag{6.138}$$

$$\nabla \cdot (p\mathbf{V}_E + \varphi \mathbf{J}_\perp) = \varphi \nabla_{\|} J_{\|} + \mathbf{J}_p \cdot \nabla\varphi + p\nabla \cdot \mathbf{V}_E, \tag{6.139}$$

implying that the transfer terms on the right-hand side are marginally balanced.

Equations (6.119)–(6.129) for $\{\varphi, n_e, J_{\|}, V_{\|i}, p_e, p_i\}$ are a more general version of the six-field drift-fluid model derived in Section 2.4.2.7. They will form the basis of our discussion of DHD turbulence below.

6.3.1.2 *Mean field approximation*

We would next like to decompose the state variables $\mathcal{A} = \{\varphi, n_e, J_{\|}, V_{\|i}, p_e, p_i\}$ into the mean $\overline{\mathcal{A}} = \langle \mathcal{A} \rangle$ and fluctuating $\widetilde{\mathcal{A}} = \mathcal{A} - \overline{\mathcal{A}}$ parts, $\langle \widetilde{\mathcal{A}} \rangle = 0$, where $\langle \mathcal{A} \rangle$ denotes an ensemble average, and derive evolution equations for each part, i.e. to extend the mean field analysis developed in the context of HD and MHD, recall Sections 6.1.2 and 6.2, to DHD turbulence. To this end, we assume the existence of a low-beta tokamak equilibrium, e.g. (3.29) or (4.79) with $\overline{\beta} = 2\mu_0 \overline{p}/B_0^2 \sim O\left(\epsilon_a^2\right)$, and force balance satisfied by the *mean* quantities, $\overline{\mathcal{A}}$. These take the form

[73] This quasi-equilibrium is maintained by compressional Alfvén waves, provided that the oscillations are slow enough, $\omega^2 \ll k_\perp^2 V_A^2$, as is usually the case under the drift ordering.

of flux surface labels, $\overline{\mathcal{A}}(r_f)$, where r_f is a general radial co-ordinate.[74] Finally, we restrict the discussion to the edge plasma of a small ϵ_a tokamak, in which the perpendicular gradient length $L_\perp \approx |\nabla_\perp \ln \overline{p}|^{-1}$ is typically much shorter than the minor radius, $L_\perp \sim 0.1a$. Since $\nabla \cdot \mathbf{V}_E \sim \mathbf{V}_E \cdot \nabla \ln B_0 \sim \mathbf{V}_E / R_0$, so that

$$p\nabla \cdot \mathbf{V}_E / \mathbf{V}_E \cdot \nabla p \sim L_\perp / R_0 \sim 0.1\epsilon_a \ll 1, \tag{6.140}$$

$\mathbf{V}_E$ becomes nearly incompressible, justifying the omission of $\mathbf{V}_E$-compressional terms in the mean field equations, aside from the mean average of $\widetilde{p}\,\nabla \cdot \widetilde{\mathbf{V}}_E$, which is at the heart of the interchange/curvature energy coupling between $\widetilde{p}$ and $\widetilde{\varphi}$.[75] This result is of course consistent with flute-like displacements, $k_\parallel \ll k_\perp$, implied by the small ϵ_a tokamak equilibrium, see Section 4.2.5, which constrain the turbulent flows primarily to the 2D ($\perp$) drift plane.

It is worth observing that this approximate incompressibility of $\mathbf{V}_E$ is in stark contrast to the generally highly compressible diffusive velocity, $\mathbf{V}_D = -D(n, T)\nabla \ln n$, which is frequently used as the ansatz for anomalous transport; the divergence of the latter vanishes only for specific $D(n, T)$, while for $D = \text{const}$, it reduces to $\nabla \cdot \mathbf{V}_D = -D\nabla^2 \ln n \sim \mathbf{V}_D / L_\perp$.

The conservation/evolution equations for the mean quantities are obtained by substituting $\mathcal{A}(\psi, \zeta, \theta) = \overline{\mathcal{A}}(\psi) + \widetilde{\mathcal{A}}(\psi, \zeta, \theta)$ into (6.120)–(6.126), neglecting $\nabla \cdot \overline{\mathcal{A}\widetilde{\mathbf{V}}}_E$ and ensemble averaging. In the same order as presented earlier, we find the mean charge conservation, $\nabla \cdot \overline{\mathbf{J}}_p \approx 0$, where

$$\overline{\mathbf{J}}_p = e\overline{n}_e \overline{\mathbf{V}}_p = \frac{\mathbf{b}_0}{B_0} \times \left(\overline{\rho}\, \overline{\mathrm{d}}_t \overline{\mathbf{V}}_{\wedge i} + \nabla \cdot \boldsymbol{\pi}_{tb\perp} + \nu_{neo} \overline{\rho}\, \overline{V}_i^\theta \nabla\theta \right) \tag{6.141}$$

is the mean polarization current, in which $\overline{\mathrm{d}}_t = \partial_t + \mathbf{V}_i^{tb} \cdot \nabla$ is the flux surface average advective derivative with $\mathbf{V}_i^{tb}$ given by (6.148),

$$\overline{\mathbf{V}}_{\perp i} = \overline{\mathbf{V}}_{\wedge i} = \overline{\mathbf{V}}_E + \overline{\mathbf{V}}_{*i} = \frac{\mathbf{b}_0}{B_0} \times \left(\nabla_\perp \overline{\varphi} + \frac{\nabla_\perp \overline{p}_i}{e\overline{n}_e} \right), \tag{6.142}$$

is the mean perpendicular ion flow velocity, which lies within the flux surface in the diamagnetic direction, $\hat{\mathbf{e}}_\wedge$,

$$\overline{V}_i^\theta = \left(\overline{V}_{\parallel i}\mathbf{b} + \overline{\mathbf{V}}_{\wedge i} \right) \cdot \nabla\theta = \overline{V}_{\parallel i}(\mathbf{b} \cdot \nabla\theta) + \overline{V}_{\wedge i}(\widehat{\mathbf{e}}_\wedge \cdot \nabla\theta) = \overline{V}_{\parallel i} b^\theta + \overline{V}_{\wedge i} b^\zeta,$$

74 For example in symmetry co-ordinates, see Section 3.2.2, r_f becomes the poloidal flux, $\psi = \psi_P$.

75 This compressibility can be expressed as $\mathcal{K}\varphi$, where $\mathcal{K} = \nabla \cdot [(\mathbf{b}_0/B_0) \times \nabla]$ is the the *curvature operator*. Hence, its poloidal variation is the same as that of the diamagnetic current, $\nabla \cdot \mathbf{J}_* = \mathcal{K}p$, i.e. it nearly vanishes near the outer and inner mid-planes and is largest at the top and bottom of the torus. This is easily shown from $\overline{\mathbf{V}}_E = (\varphi'/B_0)\hat{\mathbf{e}}_\wedge$ so that $\nabla \cdot \overline{\mathbf{V}}_E = \varphi' \hat{\mathbf{e}}_\perp \cdot [\nabla B^{-1} \times \mathbf{b}] \propto \varphi' \hat{\mathbf{e}}_\perp \cdot [\hat{\mathbf{e}}_R \times \mathbf{b}] \propto \sin\theta$, where θ is the poloidal angle.

is the poloidal projection of the mean ion flow velocity and ν_{neo} is the neoclassical kinematic viscosity, i.e. a coefficient of poloidal friction. Written explicitly, this mean quasi-neutrality relation, $\nabla \cdot \overline{\mathbf{J}}_p \approx 0$, becomes,

$$\nabla \cdot \frac{\rho}{B_0^2}\partial_t\left(\nabla_\perp\overline{\varphi} + \frac{\nabla_\perp \overline{p}_i}{e\overline{n}_e}\right) = \nabla \cdot \frac{\mathbf{b}_0}{B_0} \times \left(\overline{\rho}\mathbf{V}_i^{tb} \cdot \nabla\overline{\mathbf{V}}_{\wedge i} + \nabla \cdot \boldsymbol{\pi}_{tb\perp} + \nu_{neo}\overline{\rho}\,\overline{V}_i^{\theta}\nabla\theta\right),$$

which, as before, specifies the evolution of $\overline{\varphi}$ and the mean ion flow, $\overline{\mathbf{V}}_{\wedge i}$.

The remaining conservation/evolution equations are similarly obtained as:

- conservation of mean mass, evolution of $\overline{n}_e = Z\overline{n}_i$:

$$\partial_t\overline{n}_e + \nabla \cdot \boldsymbol{\Gamma}_{tb} = 0; \tag{6.143}$$

- conservation of mean parallel electron momentum, evolution of $\overline{J}_\|$:[76]

$$\partial_t\overline{A}_\| \approx E_\|^A - \eta_\|\overline{J}_\| + E_\|^{tb}, \qquad \mu_0\overline{J}_\| = -\nabla_\perp^2\overline{A}_\|; \tag{6.144}$$

- conservation of mean parallel ion momentum, evolution of $\overline{V}_{\|i}$:

$$\overline{\rho}\,\mathrm{d}_t\overline{V}_{\|i} + \nabla \cdot \boldsymbol{\pi}_{tb\|} = -\nu_{neo}\overline{\rho}\,\overline{V}_i^{\theta}b^{\theta}; \tag{6.145}$$

- conservation of mean electron energy, evolution of $\overline{p}_e$:

$$\tfrac{3}{2}\partial_t\overline{p}_e + \nabla \cdot \mathbf{Q}_e^{tb} = \mathcal{T}_{ei}^{tb} + \eta_\|\overline{J}_\|^2 - \overline{\mathcal{Q}}_i; \tag{6.146}$$

- conservation of mean ion energy, evolution of $\overline{p}_i$:

$$\frac{3}{2}\partial_t\overline{p}_i + \nabla \cdot (\mathbf{Q}_i^{tb} + \overline{\mathbf{q}}_i^{neo}) = -\mathcal{T}_{ei}^{tb} - \boldsymbol{\pi}_{tb\perp} : \nabla\overline{\mathbf{V}}_{\wedge i} - \boldsymbol{\pi}_{tb\|} \cdot \nabla\overline{\mathbf{V}}_{\|i}$$
$$+\nu_{neo}\overline{\rho}\,\overline{V}_i^{\theta}\overline{V}_i^{\theta} + \overline{\mathbf{V}}_p \cdot \nabla\overline{p}_i + \overline{\mathcal{Q}}_i. \tag{6.147}$$

The turbulence-related terms appearing in (6.143)–(6.147) include

$$\boldsymbol{\Gamma}_{tb}(\psi) = \overline{n}_i(\psi)\mathbf{V}_i^{tb}(\psi) = \langle\widetilde{n}_e\widetilde{\mathbf{V}}_E\rangle + \overline{n}_i\langle\widetilde{V}_{\|i}\widetilde{\mathbf{b}}_\perp\rangle + \overline{V}_{\|i}\langle\widetilde{n}_i\widetilde{\mathbf{b}}_\perp\rangle, \tag{6.148}$$
$$\mathbf{Q}_s^{tb}(\psi) = \tfrac{3}{2}\langle\widetilde{p}_s\widetilde{\mathbf{V}}_E\rangle + \langle\widetilde{q}_{\|s}\widetilde{\mathbf{b}}_\perp\rangle + \tfrac{5}{2}\overline{p}_s\langle\widetilde{V}_{\|i}\widetilde{\mathbf{b}}_\perp\rangle, \tag{6.149}$$
$$\boldsymbol{\pi}_{tb\perp}(\psi) = \overline{\rho}\,\langle\widetilde{\mathbf{V}}_E\widetilde{\mathbf{V}}_E\rangle, \qquad \boldsymbol{\pi}_{tb\|}(\psi) = \overline{\rho}\,\langle\widetilde{V}_{\|i}\widetilde{\mathbf{V}}_E\rangle, \tag{6.150}$$

which are the flux-surface-averaged mean radial turbulent flows of particles, energy, and perpendicular and parallel momentum,[77]

$$\mathcal{T}_{ei}^{tb} = \langle\widetilde{V}_{\|i}\nabla_\|\widetilde{p}_e\rangle - \langle\widetilde{p}_e\nabla \cdot \widetilde{\mathbf{V}}_E\rangle - \langle\widetilde{J}_\|\nabla_\|\widetilde{\varphi}\rangle, \tag{6.151}$$
$$= -\langle\widetilde{V}_{\|e}\nabla_\|\widetilde{p}_e\rangle - \langle\widetilde{p}_e\nabla \cdot \widetilde{\mathbf{V}}_E\rangle - \langle\widetilde{J}_\|\nabla_\|\widetilde{\varphi}\rangle + \langle\widetilde{J}_\|\nabla_\|\widetilde{p}_e\rangle/e\overline{n}_e,$$

76 Here $E_\|^A$ represents the parallel component of the toroidal inductive electric field in a tokamak.

77 The turbulent momentum flows are of course just the $\perp\perp$ and $\| \perp$ Reynolds stresses.

which is the mean energy transfer term from the electrons to the ions,[78] and

$$E_{\parallel}^{tb} = (en_e)^{-1}\nabla \cdot \langle \widetilde{p}_e \widetilde{\mathbf{b}}_{\perp} \rangle - \nabla \cdot \langle \widetilde{\varphi} \widetilde{\mathbf{b}}_{\perp} \rangle \tag{6.152}$$

which is the parallel electric field induced by magnetic fluctuations.

Although, for completeness, we retain magnetic flutter effects in (6.148)–(6.152), it should be stressed that these are found to be small,

$$|\widetilde{\mathbf{b}}_{\perp}| \sim 10^{-5}\text{–}10^{-4} \ll 10^{-2}\text{–}10^{-1} \sim \delta_i \sim \{e\widetilde{\varphi}/\overline{T}_e,\ \widetilde{p}_e/\overline{p}_e, \ldots\}, \tag{6.153}$$

in low-beta tokamak equilibrium, and are only weakly correlated with plasma fluctuations, e.g. $\widetilde{n}$, $\widetilde{\varphi}$, $\widetilde{V}_{\parallel i}$, $\widetilde{p}_e$, $\widetilde{q}_{\parallel e}$, etc. and thus nearly vanish on flux surface averaging. Consequently, both transport and transfer effects associated with magnetic flutter, i.e. all terms involving $\widetilde{\mathbf{b}}_{\perp}$, may be typically neglected. There are, of course, situations when these effects are not small and must be retained. For instance, when the magnetic flux surfaces are strongly perturbed (by either MHD instabilities or external resonant magnetic perturbations), or even destroyed (by magnetic island growth and overlap), radial transport due to magnetic flutter can become comparable to, or even exceed, that due to $\mathbf{E} \times \mathbf{B}$ turbulent advection.

The conservation/evolution equations for the fluctuating quantities, $\widetilde{A}$, are obtained by subtracting equations for $\mathcal{A}$ and $\overline{A}$. For instance, the perpendicular and parallel fluctuating velocities evolve according with: conservation of fluctuating charge/current, $\nabla \cdot \widetilde{\mathbf{J}} = \nabla \cdot (\widetilde{J}_{\parallel}\mathbf{b} + \widetilde{\mathbf{J}}_* + \widetilde{\mathbf{J}}_p) = 0$, and hence the evolution of $\widetilde{\mathbf{V}}_E$ and $\widetilde{\varphi}$, which can be written explicitly as

$$\nabla \cdot \frac{\overline{\rho}}{B_0^2}\partial_t \nabla_{\perp}\widetilde{\varphi} \approx \nabla \cdot \left\{ \frac{\mathbf{b}_0}{B_0} \times \left[\overline{\rho}\,\overline{\mathbf{V}}_E \cdot \nabla \widetilde{\mathbf{V}}_E + \overline{\rho}\,\widetilde{\mathbf{V}}_E \cdot \nabla(\widetilde{\mathbf{V}}_E + \overline{\mathbf{V}}_{\wedge i}) + \nabla \widetilde{p} \right] + \widetilde{J}_{\parallel}\mathbf{b} \right\};$$

and conservation of fluctuating parallel ion momentum, evolution of $\widetilde{V}_{\parallel i}$:

$$\overline{\rho}\,\overline{\mathrm{d}}_t \widetilde{V}_{\parallel i} + \overline{\rho}\,\widetilde{\mathbf{V}}_E \cdot \nabla(\widetilde{V}_{\parallel i} + \overline{V}_{\parallel i}) = -\nabla_{\parallel}\widetilde{p} + \mu_{\parallel}\nabla_{\parallel}^2 \widetilde{V}_{\parallel i}. \tag{6.154}$$

The conservation/evolution equations for the corresponding mean and fluctuation energies may be obtained from (6.131)–(6.132):

- conservation of mean perpendicular (drift) ion kinetic energy:

$$\begin{aligned} &\tfrac{1}{2}\partial_t \left(\overline{\rho}\,\overline{V}_{\wedge i}^2\right) + \nabla \cdot \left[\tfrac{1}{2}\overline{\rho}\,\mathbf{V}_i^{tb}\overline{V}_{\wedge i}^2 + \boldsymbol{\pi}_{tb\perp} \cdot \overline{\mathbf{V}}_{\wedge i} + \overline{\varphi}\,\overline{\mathbf{J}}_p \right] \\ &\quad = \boldsymbol{\pi}_{tb\perp} : \nabla \overline{\mathbf{V}}_{\wedge i} - \nu_{neo}\overline{\rho}\,\overline{V}_i^{\theta}\,\overline{V}_{\wedge i}^{\theta} - \overline{\mathbf{V}}_p \cdot \nabla \overline{p}_i; \end{aligned} \tag{6.155}$$

[78] This form of $\mathcal{T}_{ei}^{tb}$ makes use of the fluctuating energy transfer relations, see (6.157)–(6.159).

- conservation of mean parallel (acoustic) ion kinetic energy:

$$\tfrac{1}{2}\partial_t\left(\overline{\rho}\overline{V}_{\|i}^2\right)+\nabla\cdot\left[\tfrac{1}{2}\overline{\rho}\,\mathbf{V}_i^{tb}\overline{V}_{\|i}^2+\boldsymbol{\pi}_{tb\|}\cdot\overline{\mathbf{V}}_{\|i}\right]$$
$$=\boldsymbol{\pi}_{tb\|}\cdot\nabla\overline{V}_{\|i}-\nu_{neo}\overline{\rho}\overline{V}_i^{\theta}\overline{V}_{\|i}^{\theta}; \qquad (6.156)$$

- conservation of fluctuation perpendicular (drift) ion kinetic energy:

$$\tfrac{1}{2}\partial_t\left(\overline{\rho}\,\langle\widetilde{V}_{\perp i}^2\rangle\right)+\nabla\cdot\left[\tfrac{1}{2}\overline{\rho}\,\mathbf{V}_i^{tb}\langle\widetilde{V}_{\perp i}^2\rangle+\tfrac{1}{2}\overline{\rho}\,\langle\widetilde{\mathbf{V}}_E\widetilde{V}_{\perp i}^2\rangle+\langle\widetilde{\varphi}\widetilde{\mathbf{J}}\rangle+\langle\widetilde{p}\widetilde{\mathbf{V}}_E\rangle\right]$$
$$=-\boldsymbol{\pi}_{tb\perp}:\nabla\overline{\mathbf{V}}_{\wedge i}+\langle\widetilde{J}_{\|}\nabla_{\|}\widetilde{\varphi}\rangle-\langle\widetilde{p}\,\nabla\cdot\widetilde{\mathbf{V}}_E\rangle; \qquad (6.157)$$

- conservation of fluctuation parallel (acoustic) ion kinetic energy:

$$\tfrac{1}{2}\partial_t\left(\overline{\rho}\,\langle\widetilde{V}_{\|i}^2\rangle\right)+\nabla\cdot\left[\tfrac{1}{2}\overline{\rho}\,\mathbf{V}_i^{tb}\langle\widetilde{V}_{\|i}^2\rangle+\tfrac{1}{2}\overline{\rho}\,\langle\widetilde{\mathbf{V}}_E\widetilde{V}_{\|i}^2\rangle-\mu_{\|}\langle\widetilde{V}_{\|i}\nabla_{\|}\widetilde{V}_{\|i}\rangle\mathbf{b}\right]$$
$$=-\boldsymbol{\pi}_{tb\|}\cdot\nabla\overline{V}_{\|i}-\langle\widetilde{V}_{\|i}\nabla_{\|}\widetilde{p}\rangle-\mu_{\|}\langle|\nabla_{\|}\widetilde{V}_{\|i}|^2\rangle; \qquad (6.158)$$

- conservation of fluctuation perpendicular magnetic (Alfvén wave) energy:

$$\tfrac{1}{2}\partial_t\langle\widetilde{B}_{\perp}^2\rangle-\nabla\cdot(\partial_t\widetilde{A}_{\|}\nabla_{\perp}\widetilde{A}_{\|})=\mu_0\langle\widetilde{J}_{\|}\partial_t\widetilde{A}_{\|}\rangle$$
$$=\mu_0\langle\widetilde{J}_{\|}(\nabla_{\|}\widetilde{p}_e-\widetilde{F}_{\|e})\rangle/e\overline{n}_e-\mu_0\langle\widetilde{J}_{\|}\nabla_{\|}\widetilde{\varphi}\rangle. \qquad (6.159)$$

Provided the fluctuations are small, $\widetilde{A}/\overline{A}<0.1$, their energy content is negligible compared to that of the mean quantities, $\langle\widetilde{A}^2\rangle/\overline{A}^2\ll 1$. As a result, the expressions on the left-hand side in (6.157)–(6.159) can be neglected, leaving a series of quasi-static balance relations between the energy transfer terms appearing on the right-hand side, i.e. fluctuations have a high energy 'conductance' but a small energy 'capacitance' and merely act as a conduit for energy transfer between the mean quantities. This purely 'transferring' character of the fluctuations, i.e. the fact that the sum of all transfer terms vanishes is best illustrated by writing down the total mean energy balance, obtained from the ensemble average of (6.134),

$$\partial_t\overline{\mathcal{E}}+\nabla\cdot\left(\mathbf{Q}_e^{tb}+\mathbf{Q}_i^{tb}+\mathbf{q}_i^{neo}\right)\approx\eta_{\|}\overline{J}_{\|}^2,\qquad \mathcal{E}\approx\overline{\mathcal{E}}, \qquad (6.160)$$

so that no energy is lost to the fluctuations.[79] Instead, mean energy is only injected, mainly by the mean Joule heating (with $\overline{J}_{\|}$ driven by $E_{\|}^A$), and transported, mainly by turbulent advection, $\mathbf{Q}^{tb}\approx\frac{3}{2}\langle\widetilde{p}\widetilde{\mathbf{V}}_E\rangle$.

The energy transfer channels appearing in (6.146)–(6.159) include the ensemble averages of those listed in Table 6.7, plus additional terms composing (i) the Reynolds stress, $\boldsymbol{\pi}_{tb}:\nabla\overline{\mathbf{V}}$, and (ii) neoclassical dissipation, $\nu_{neo}\overline{\rho}\,\overline{V}_i^{\theta}\overline{V}_i^{\theta}$, which transfer energy between the mean and fluctuating ion flows, $\overline{V}_i\leftrightarrow\widetilde{V}_i$, and between the mean ion flow and pressure, $\overline{V}_i\leftrightarrow\overline{p}_i$. While (ii) represents viscous heating

[79] Indeed, the total fluctuating energy balance could be written as $0\approx 0$, with fluctuating energy content, transport and transfer all vanishing to the same order of accuracy.

and thus tends to dampen the mean ion flow, $\overline{V}_i \rightarrow \overline{p}_i$, (i) allows the free energy associated with the radial pressure gradient to drive a mean ion flow, $\overline{p}_i \rightarrow \overline{V}_i$, in both the diamagnetic and parallel directions.[80] The importance of this diamagnetic, also known as *zonal*, flow $\overline{V}_{\wedge i}$, to the overall turbulence dynamics has only recently been fully appreciated. The label *zonal* flow refers to latitudinal flows observed in the atmospheres of the Jovian planet; due to the close correspondence between, on the one hand, the *Coriolis* force and *Rossby* waves in a rotating fluid, and on the other, the *Lorentz* force and *drift*-waves in a magnetized plasma, such geostrophic zonal flows are very similar to the diamagnetic mean flows in a tokamak. The associated flow shear, $\overline{V}'_{\wedge i}$, tends to de-correlate (break up) the turbulent eddies, thereby reducing the radial mixing length of the turbulence and the associated radial transport. Hence, the interaction of the fluctuations and zonal flows offers a self-governing (negative feedback) mechanism in the turbulent flow evolution, which tends to be governed by the familiar *predator–prey* dynamics (Diamond *et al.*, 2005). This mechanism is thought to be a necessary, if not a sufficient, condition for the formation of internal (ITB) and edge (ETB) transport barriers in tokamaks, in which turbulent transport can be reduced down to neoclassical levels.

The $\mathbf{E} \times \mathbf{B}$ zonal flow, $\overline{V}_{E\wedge}$, corresponds to a mean radial electric field, $\overline{E}_\perp = -\nabla_\perp \overline{\varphi}$, whose divergence, $\nabla \cdot \overline{\mathbf{E}}_\perp = -\nabla^2_\perp \overline{\varphi} = \overline{\varrho}/\epsilon_0$, requires a small, but finite, charge density, $\overline{\varrho}$. This space charge, being a factor $(V_A/c)^2$ smaller than the polarization current, can be neglected from the quasi-neutrality relation, which remains $\nabla \cdot \mathbf{J} \approx 0$ to high accuracy. However, it does lead to a significant mean parallel vorticity, $\overline{\Omega}_E = \mathbf{b}_0 \cdot \nabla_\perp \times \overline{\mathbf{V}}_{E\wedge} = \nabla^2_\perp \overline{\varphi}/B_0 = -\overline{\varrho}/\epsilon_0 B_0$, (4.92), which is just the zonal flow in question.

A similar situation applies to all other flow generating mechanisms, which may be represented by arbitrary sources or sinks, ϱ_{ext}, in the charge conservation equation, $\nabla \cdot \mathbf{J} = \varrho_{ext}$. This space charge, which generates a torque in the diamagnetic direction, giving rise to a radial electric field and a parallel vorticity, $\overline{\Omega}_E = -\varrho_{ext}/\epsilon_0 B_0$, is nonetheless quickly counteracted by additional currents, which re-establish $\nabla \cdot \mathbf{J} \approx 0$.[81] In other words, the removal of charge from a plasma filament causes it to rotate in the $\wedge$ direction, gaining angular momentum via the Lorentz force, $\mathbf{J} \times \mathbf{B}_0$, from the background magnetic field, $\mathbf{B}_0$, to which the drift fluid is always anchored.

The radial transport of ion momentum occurs due to a net *viscosity*, representing the *turbulent* (Reynolds) stresses, $\boldsymbol{\pi}_{tb\perp}(\psi) = \overline{\rho}\,\langle \widetilde{\mathbf{V}}_E \widetilde{\mathbf{V}}_E \rangle$ and $\boldsymbol{\pi}_{tb\parallel}(\psi) = \overline{\rho}\,\langle \widetilde{V}_{\parallel i} \widetilde{\mathbf{V}}_E \rangle$, and the gyro-viscous stress, $\boldsymbol{\pi}_*$, linked to pressure anisotropy. It does

80 Recall that $\wedge$ and $\parallel$ are close to, but not identical with, the poloidal and toroidal directions.

81 As with the fluctuation energy content, we may describe the plasma as having a low capacitance and a large conductance (high throughput) for electric charge.

not involve the net *resistivity*, which represents a momentum conserving friction between the ion and electron fluids. Thus, as with collisional transport, we find that turbulent resistivity determines particle diffusion, while turbulent viscosity determines momentum diffusion.

In this context it is also worth examining the radial transport of magnetic flux, which is governed by (6.144) with $\eta_\parallel$ representing some neoclassical resistivity and $E_\parallel^{tb}$ the turbulent induced parallel electric field, given by (6.152). Due to the smallness of magnetic flutter effects, (6.153), and the comparatively strong collisional magnetic diffusivity, $\eta_\parallel/\mu_0 \approx D^{cl}_{\perp e}/\beta_e$, magnetic diffusion is comparable to, and frequently greater than, the turbulent drive, $(\eta_\parallel/\mu_0)\nabla_\perp^2 A_\parallel \sim E_\parallel^{tb}$.[82] Therefore, in contrast to collisional transport, which causes magnetic flux to diffuse more quickly than plasma mass, momentum or energy, DHD turbulence drives a strong radial diffusion of plasma quantities, but leaves the magnetic flux nearly unaffected; in between these two limits, lies MHD turbulence, in which the magnetic flux is 'frozen' into the plasma, so that the two diffuse at the same rate.[83] This explains the otherwise curious tokamak observation that current relaxes on a time scale consistent with *neoclassical* diffusion, while particles, momentum and energy relax on time scales much shorter than predicted by neoclassical theory; it is now widely accepted that this *anomalous* plasma transport (much faster than neoclassical) is indeed a consequence of DHD turbulence.

Let us next consider the pathways for energy flow between the mean ion and electron pressures, and more importantly, from the pressures into the turbulent flow. Roughly speaking, these correspond to the three terms in $\mathcal{T}_{ei}^{tb}$: (i) the Alfvén wave, $\langle \widetilde{J}_\parallel \nabla_\parallel \widetilde{\varphi}\rangle$, which together with adiabatic electron response, $\langle \widetilde{J}_\parallel \nabla_\parallel \widetilde{p}_e\rangle / e\overline{n}_e$, constitute the *drift-wave* energy coupling; effectively the pressure disturbances launch shear-Alfvén waves along the guide field, $\mathbf{B}_0$, ensuring that $\widetilde{p}_e/\overline{p}_e \sim e\widetilde{\varphi}/\overline{T}_e$ are comparable and correlated; the energy transfer represents the work done by the polarization drift on the ion fluid; (ii) the compressional work due to magnetic curvature, $\langle \widetilde{p}_e \nabla \cdot \widetilde{\mathbf{V}}_E\rangle$ and $\langle \widetilde{p}_i \nabla \cdot \widetilde{\mathbf{V}}_E\rangle$, which underlie the *interchange* energy coupling;[84] and (iii) the sound wave, $\langle \widetilde{V}_{\parallel i} \nabla_\parallel \widetilde{p}_e\rangle$ and $\langle \widetilde{V}_{\parallel i} \nabla_\parallel \widetilde{p}_i\rangle$, which is normally weaker than (i) and (ii), but can become dominant for ion temperature gradient (ITG) modes. These three channels can be summarized symbolically as

$$\text{drift-wave (DW)}: \quad \widetilde{p}_e \leftrightarrow \widetilde{J}_\parallel \leftrightarrow \widetilde{\varphi}, \qquad \overline{p}_e \leftrightarrow \overline{B}_\perp \leftrightarrow \overline{V}_{\wedge i}, \tag{6.161}$$

$$\text{interchange } (\kappa): \quad \widetilde{p}_e \leftrightarrow \widetilde{\varphi} \leftrightarrow \widetilde{p}_i, \qquad \overline{p}_e \leftrightarrow \overline{V}_{\wedge i} \leftrightarrow \overline{p}_i, \tag{6.162}$$

$$\text{sound wave (SW)}: \quad \widetilde{p}_e \leftrightarrow \widetilde{V}_{\parallel i} \leftrightarrow \widetilde{p}_i, \qquad \overline{p}_e \leftrightarrow \overline{V}_{\parallel i} \leftrightarrow \overline{p}_i. \tag{6.163}$$

[82] Especially in the relatively cool (and hence resistive) edge plasma; recall that $\eta_\parallel \propto ZT_e^{-3/2}$.

[83] In the language of Chapter 2, the collisional transport increases the mobility of magnetic flux tubes, DHD turbulence that of plasma filaments, and MHD turbulence that of both.

[84] These terms can be rewritten as $\langle \widetilde{p}_e \mathcal{K} \widetilde{\varphi}\rangle$ and $\langle \widetilde{p}_i \mathcal{K} \widetilde{\varphi}\rangle$ and thus vanish as $\kappa \to 0$.

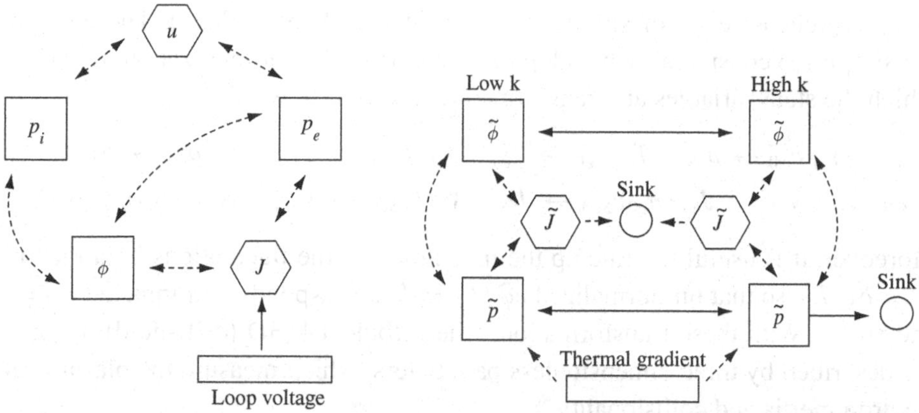

Fig. 6.2. Schematic representation of energy transfer between mean (left) and fluctuating (right) quantities in DHD turbulence. In the latter case, the generally smaller, acoustic coupling is omitted, and the ion and electron pressures are combined. Loop voltage and thermal gradients provide the energy source of mean and fluctuating quantities, respectively. Collisional dissipation (resistive and viscous) provides the energy sink for the latter, while mean energies are only reduced due to transport effects, (6.160). On the right, the spectral cascades between low and high k, due to $\widetilde{\mathbf{V}}_E$ and $\widetilde{\mathbf{V}}_p$ advection, are also indicated. Reused with permission from Scott (2003a), ©AIP.

They are represented schematically in Fig. 6.2 for both mean and fluctuating quantities. Since the adiabatic electron response, $\widetilde{p}_e \leftrightarrow \widetilde{J}_\parallel$, is a direct consequence of the drift ordering, the DW coupling, $\widetilde{p}_e \leftrightarrow \widetilde{J}_\parallel \leftrightarrow \widetilde{\varphi}$, is absent from the MHD model, which thus lacks an important channel for energy flow from the background gradient, $\nabla_\perp \overline{p}$, into turbulent motion, $\widetilde{\mathbf{V}}_E$. Nonetheless, MHD contains the Alfvénic, acoustic and interchange couplings. Indeed, the last of these, representing the R–T instability due to radial pressure gradients and magnetic curvature, is a perennial feature of both MHD and DHD turbulence in magnetically confined plasmas.

6.3.1.3 Local drift-fluid models

We would next like to make use of the above equations to compute the turbulent transport in the edge plasma of a small ϵ tokamak. Ideally, this should include a simultaneous solution of both the mean and fluctuating quantities. Such *global* computations are only now becoming possible and are therefore a topic of active research. However, it is also instructive to consider *local* computations in which the prescribed background profiles are held fixed and only the fluctuations, which are assumed to obey the flute-ordering, $\nabla_\parallel/\nabla_\perp \sim L_\perp/L_\parallel \ll 1$, (4.57), are evolved in time. For this purpose, one can use either the six- or four-field models derived above, both of which capture the essential drift-wave and interchange dynamics; we will base our discussion largely on the four-field model, closely following the comprehensive set of simulations presented in Scott (2005a). The equations are simplified considerably by adopting the gyro-Bohm normalization, (6.217), in which the state variables are transformed in accordance with[85]

$$n_e/n_{e0} \to n_e, \quad T_e/T_{e0} \to T_e, \quad \tau T_i/T_{i0} \to T_i, \quad p_e/p_{e0} \to p_e,$$
$$e\varphi/T_{e0} \to \varphi, \quad J_\parallel/en_{e0}c_{Se0} \to J_\parallel, \quad V_{\parallel i}/c_{Se0} \to V_{\parallel i}, \quad A_\parallel/\hat{\beta}B_0\rho_{S0} \to A_\parallel.$$

Moreover, it is useful to scale up the magnitude of the fluctuations by a factor of $\delta_i = \overline{\rho}_S/L_\perp$ so that un-normalized $e\widetilde{\varphi}/\overline{T}_e = \delta_i$ corresponds to normalized $\widetilde{\varphi} = 1$, and so on. With these transformations, the turbulent DHD (drift-fluid) dynamics are described by three dimensionless parameters, which measure the plasma beta, electron inertia and collisionality,[86]

$$\hat{\beta} = \overline{\beta}_e\hat{\epsilon}, \qquad \hat{\mu} = (m_e/m_i)\hat{\epsilon}, \qquad \hat{\nu} = \hat{\nu}_e = \hat{C}/\hat{\eta}\hat{\mu} = \nu_{ei}/\omega_{tS}, \tag{6.164}$$

with $\overline{\beta}_e = \mu_0\overline{p}_e/B_0^2 \ll 1$ being the electron beta, $\hat{\epsilon} = (qR_0/L_\perp)^2 \gg 1$ the square of the ratio of parallel and perpendicular gradient lengths, ν_{ei} the electron–ion collision frequency and $\omega_{ts} = \overline{c}_{Se}/L_\perp \sim \omega_*$ the characteristic gyro-Bohm frequency. The background profiles are parameterized by inverse gradient lengths and normalized by a nominal perpendicular length, $L_\perp$,

$$-\nabla_\perp\overline{n}_e = \omega_n, \qquad -\nabla_\perp\overline{T}_e = \omega_e, \qquad -\nabla_\perp\overline{T}_i = \tau\omega_i, \tag{6.165}$$
$$-\nabla_\perp\overline{p}_e = \omega_{pe} = \omega_n + \omega_e, \qquad -\nabla_\perp\overline{p}_i = \omega_{pi} = \tau(\omega_n + \omega_i), \tag{6.166}$$

which for $T_i = 0$ and $T_e = \text{const}$ reduce to $\tau = \omega_e = \omega_{pi} = 0$ and $\omega_{pe} = \omega_n$; similarly, parallel gradients are normalized by $L_\parallel/2\pi = qR_0$.

Magnetic curvature enters the model equations via the *curvature operator*, $\mathcal{K} = \nabla\cdot[(\mathbf{b}_0/B_0)\times\nabla]$, which in a small ϵ tokamak, can be approximated as

$$\mathcal{K} \approx 2(\mathbf{b}_0/B_0)\cdot(\boldsymbol{\kappa}\times\nabla) \to \omega_B(\sin\theta\nabla_\perp + \cos\theta\nabla_\wedge), \tag{6.167}$$

where the final expression is given in normalized units, in which the curvature is measured by $\omega_B = 2L_\perp/R_0$;[87] note that at the outer ($\theta = 0$) and inner ($\theta = \pi$) mid-planes, $\mathcal{K} = \pm\omega_B\nabla_\wedge$ involves only diamagnetic gradients. All other finite toroidicity effects can be neglected in the small ϵ limit.

The above parameters can be combined to form the *ideal ballooning* parameter, α_B, and the *resistive ballooning* parameter, ν_B, which determine the transition from DHD to MHD dynamics, see Section 6.3.1.4,

[85] Note that $\widetilde{p}_e = \widetilde{n}_e + \widetilde{T}_e$ and $\widetilde{p}_i = \tau\widetilde{n}_e + \widetilde{T}_i$ in normalized units, where $\tau = \overline{T}_i/\overline{T}_e$.
[86] We also introduce the Braginskii coefficients: $\hat{\eta} = 0.51$, $\hat{\alpha}_e = 0.71$, $\hat{\kappa}_e = 3.2$, $\hat{\kappa}_i = 3.9$.
[87] This expression is easily derived for small ϵ tokamak with circular flux surfaces, in which the curvature is primarily toroidal. Recalling the results of Section 3.3.3, we find that $\mathcal{K} \approx [\nabla\times\mathbf{b}_0 - (\nabla B_0/B_0)\times\mathbf{b}_0]/B_0\cdot\nabla \approx -(\mathbf{b}_0/B_0)\times(\boldsymbol{\kappa} + \nabla B_0/B_0)\cdot\nabla \approx -(\mathbf{b}_0/B_0)\cdot 2\boldsymbol{\kappa}\times\nabla$, where $\boldsymbol{\kappa} \approx -\nabla R/R_0$. Inserting $R = R_0 + r\cos\theta$ and $\nabla = \partial_r\hat{\mathbf{e}}_r + \partial_\theta\hat{\mathbf{e}}_\theta/r$, we find (6.167).

$$\alpha_B = -q^2 R_0 \nabla\beta = \hat{\beta}\omega_B[(\omega_n + \omega_e) + \tau(\omega_n + \omega_i)], \qquad \nu_B = \hat{C}\omega_B. \quad (6.168)$$

With the above transformations, the six-field model becomes (Scott, 1998):

$$B_0^{-2}\left[d_t \nabla_\perp^2(\widetilde{\varphi} + \widetilde{p}_i) + \nabla\mathbf{V}_E : \nabla^2\widetilde{p}_i\right] \approx \nabla_\parallel \widetilde{J}_\parallel - \mathcal{K}(\widetilde{p}_e + \widetilde{p}_i), \quad (6.169)$$

$$d_t(\overline{n}_e + \widetilde{n}_e) \approx \nabla_\parallel(\widetilde{J}_\parallel - \widetilde{V}_{\parallel i}) - \mathcal{K}(\widetilde{p}_e - \widetilde{\varphi}), \quad (6.170)$$

$$\tfrac{3}{2}d_t(\overline{T}_e + \widetilde{T}_e) \approx \nabla_\parallel(\widetilde{J}_\parallel - \widetilde{V}_{\parallel i} - \widetilde{q}_{\parallel e}) - \mathcal{K}(\widetilde{p}_e - \widetilde{\varphi}) - \tfrac{5}{2}\mathcal{K}\widetilde{T}_e, \quad (6.171)$$

$$\tfrac{3}{2}d_t(\overline{T}_i + \widetilde{T}_i) \approx \nabla_\parallel[\tau(\widetilde{J}_\parallel - \widetilde{V}_{\parallel i}) - \widetilde{q}_{\parallel e}] - \tau\mathcal{K}(\widetilde{p}_e - \widetilde{\varphi}) + \tfrac{5}{2}\mathcal{K}\widetilde{T}_i, \quad (6.172)$$

$$\hat{\beta}\partial_t\widetilde{A}_\parallel + \hat{\mu}d_t\widetilde{J}_\parallel \approx \nabla_\parallel(\overline{p}_e + \widetilde{p}_e - \widetilde{\varphi}) - \widetilde{F}_{\parallel e}, \quad (6.173)$$

$$\hat{\epsilon}d_t\widetilde{V}_{\parallel i} \approx -\nabla_\parallel(\overline{p}_e + \overline{p}_i + \widetilde{p}_e + \widetilde{p}_i) + \mu_\parallel\nabla_\parallel^2\widetilde{V}_{\parallel i}, \quad (6.174)$$

where $d_t = \partial_t + \mathbf{V}_E \cdot \nabla$ is the advective derivative in the drift plane and $\widetilde{J}_\parallel = -\nabla_\perp^2\widetilde{A}_\parallel$ is the parallel current.[88] Closure is achieved by collisional estimates for parallel friction and ion and electron heat fluxes,

$$\widetilde{F}_{\parallel e} = \hat{\mu}\hat{\nu}_e[\hat{\eta}\widetilde{J}_\parallel + (\hat{\alpha}_e/\hat{\kappa}_e)(\widetilde{q}_{\parallel e} + \hat{\alpha}_e\widetilde{J}_\parallel)], \quad (6.175)$$

$$(\hat{\mu}d_t + \mathcal{A}_{Le})\widetilde{q}_{\parallel e} = -\tfrac{5}{2}[\nabla_\parallel(\overline{T}_e + \widetilde{T}_e) + (\hat{\mu}\hat{\nu}_e/\hat{\kappa}_e)(\widetilde{q}_{\parallel e} + \hat{\alpha}_e\widetilde{J}_\parallel)], \quad (6.176)$$

$$(\hat{\epsilon}d_t + \mathcal{A}_{Li})\widetilde{q}_{\parallel i} = -\tfrac{5}{2}[\tau\nabla_\parallel(\overline{T}_i + \widetilde{T}_i) + (\hat{\epsilon}\hat{\nu}_i/\hat{\kappa}_i)\widetilde{q}_{\parallel i}], \quad (6.177)$$

including the effect of Landau damping, expressed by the operators,

$$\mathcal{A}_{Le} = \hat{\mu}^{1/2}\left(1 - 0.2\nabla_\parallel^2\right), \qquad \mathcal{A}_{Li} = (\tau\hat{\epsilon})^{1/2}\left(1 - 0.2\nabla_\parallel^2\right). \quad (6.178)$$

Neglecting $d_t q_{\parallel s}$ and $\mathcal{A}_{Ls}$, yields the familiar Braginskii estimates, with

$$\hat{\nu}_e = \nu_e L_p/c_{Se} = \nu_e/\omega_{tS}\omega_p, \qquad \hat{\nu}_i = \nu_i L_p/c_{Se} = \nu_i/\omega_{tS}\omega_p. \quad (6.179)$$

In the case of cold ions and isothermal electrons, (6.169)–(6.174) reduce to the four-field model for $\widetilde{n}_e = \widetilde{p}_e$, $\widetilde{\Omega}_E(\widetilde{\varphi})$, $\widetilde{J}_\parallel(\widetilde{A}_\parallel)$ and $V_{\parallel i}$, (Scott, 1997),

$$d_t\widetilde{\Omega}_E \approx \nabla_\parallel\widetilde{J}_\parallel - \mathcal{K}\widetilde{p}_e, \quad (6.180)$$

$$d_t(\overline{p}_e + \widetilde{p}_e) \approx \nabla_\parallel(\boldsymbol{\widetilde{J}_\parallel} - \widetilde{V}_{\parallel i}) - \mathcal{K}(\boldsymbol{\widetilde{p}_e} - \widetilde{\varphi}), \quad (6.181)$$

$$\hat{\beta}\partial_t\widetilde{A}_\parallel + \hat{\mu}d_t\widetilde{J}_\parallel \approx \nabla_\parallel(\boldsymbol{\overline{p}_e} + \boldsymbol{\widetilde{p}_e} - \widetilde{\varphi}) - \hat{C}\widetilde{J}_\parallel, \quad (6.182)$$

$$\hat{\epsilon}d_t\widetilde{V}_{\parallel i} \approx -\nabla_\parallel(\overline{p}_e + \widetilde{p}_e) + \mu_\parallel\nabla_\parallel^2\widetilde{V}_{\parallel i}, \quad (6.183)$$

where $\widetilde{\Omega}_E = \nabla_\perp^2\widetilde{\varphi}/B_0^2$ is the $\mathbf{E}\times\mathbf{B}$ vorticity. Four of the terms above are highlighted in bold to emphasize their importance for drift-fluid dynamics; indeed, omission of these terms, which represent the adiabatic electron, or drift-wave coupling, $\widetilde{p}_e \leftrightarrow \widetilde{J}_\parallel \leftrightarrow \widetilde{\varphi}$, yields the (reduced) resistive MHD model, derived in Section 4.2.5, which is thus an entirely included subset of the four-field DHD model. Since the resistive

[88] Recall that $\widetilde{\varphi}/B_0$ and $-\widetilde{A}_\parallel/B_0$ act as stream functions for $\mathbf{V}_E$ and $\mathbf{b}_\perp$.

MHD model captures the interchange, but not drift-wave, dynamics, the four bold terms in (6.180)–(6.183) are evidently the distinguishing features of DHD.

Finally, neglecting the parallel direction, i.e. replacing the operator $-\hat{C}^{-1}\nabla_{\|}^{2}$ by the constant $C_{\|}$, as well as omitting ion and electron parallel momentum and electron beta ($V_{\|i} = \hat{\beta} = \hat{\mu} = 0$), yields the rudimentary two-field model,

$$\mathrm{d}_t\widetilde{\Omega}_E \approx \boldsymbol{C_{\|}(\widetilde{\varphi} - \widetilde{p}_e)} - \mathcal{K}\widetilde{p}_e + \nu_{\varphi}\widetilde{\varphi}, \tag{6.184}$$

$$\mathrm{d}_t(\overline{p}_e + \widetilde{p}_e) \approx \boldsymbol{C_{\|}(\widetilde{\varphi} - \widetilde{p}_e)} - \mathcal{K}(\widetilde{p}_e - \widetilde{\varphi}) - \omega_p\nabla_{\wedge}\widetilde{\varphi}, \tag{6.185}$$

with the curvature operator reducing to $\mathcal{K} = \omega_B\nabla_{\wedge}$. The pure drift-wave (Hasegawa–Wakatani) model is obtained by setting $\omega_B = \nu_{\varphi} = 0$, while the pure interchange (MHD) model follows from $C_{\|} = 0$. As we will see presently, these two limiting cases provide a useful benchmark with which to compare the mixed mode (DHD = DW + MHD) drift-fluid dynamics.

Energy flows between the fluctuating quantities are the same as those outlined in the previous section and illustrated schematically in Fig. 6.2. For instance, the four-field model fluctuating energies become,[89]

$$\tfrac{1}{2}\partial_t\langle\widetilde{V}_E^2\rangle \approx \boldsymbol{\langle\widetilde{J}_{\|}\nabla_{\|}\widetilde{\varphi}\rangle} - \langle\widetilde{p}_e\mathcal{K}\widetilde{\varphi}\rangle, \tag{6.186}$$

$$\tfrac{1}{2}\partial_t\langle\widetilde{p}_e^2\rangle \approx -\boldsymbol{\langle\widetilde{J}_{\|}\nabla_{\|}\widetilde{p}_e\rangle} + \langle\widetilde{V}_{\|i}\nabla_{\|}\widetilde{p}_e\rangle + \langle\widetilde{p}_e\mathcal{K}\widetilde{\varphi}\rangle - \omega_p\langle\widetilde{p}_e\nabla_{\wedge}\widetilde{\varphi}\rangle, \tag{6.187}$$

$$\tfrac{1}{2}\partial_t\langle\widetilde{J}_{\|}(\hat{\beta}\widetilde{A}_{\|} + \hat{\mu}\widetilde{J}_{\|})\rangle \approx \boldsymbol{\langle\widetilde{J}_{\|}\nabla_{\|}\widetilde{p}_e\rangle} - \boldsymbol{\langle\widetilde{J}_{\|}\nabla_{\|}\widetilde{\varphi}\rangle} - \hat{C}\langle\widetilde{J}_{\|}^2\rangle, \tag{6.188}$$

$$\tfrac{1}{2}\partial_t\langle\hat{\epsilon}\widetilde{V}_{\|i}^2\rangle \approx -\langle\widetilde{V}_{\|i}\nabla_{\|}\widetilde{p}_e\rangle + \omega_p\hat{\beta}\langle\widetilde{V}_{\|i}\nabla_{\wedge}\widetilde{A}_{\|}\rangle - \mu_{\|}\langle|\nabla_{\|}\widetilde{V}_{\|i}|^2\rangle, \tag{6.189}$$

where once again we highlight the drift-wave energy coupling in bold type; removal of these terms yields the energy relations for (reduced) resistive MHD. The meaning of the various terms is the same as in Table 6.7, with the $\mathbf{E}\times\mathbf{B}$ compression, $\widetilde{p}_e\nabla\cdot\widetilde{\mathbf{V}}_E$, represented by the curvature term, $\widetilde{p}_e\mathcal{K}\widetilde{\varphi}$. The two ω_p terms, of which $\omega_p\langle\widetilde{p}_e\nabla_{\wedge}\widetilde{\varphi}\rangle$ is by far the larger, represent the source of free energy, while the $\hat{C}$ and $\mu_{\|}$ terms, represent the energy dissipation.

6.3.1.4 Numerical simulations of drift-fluid edge plasma turbulence

The local turbulence models[90] derived in the previous section form the bases of a series of numerical simulations (Kendl and Scott, 2005; Reiser and Scott, 2005; Scott, 2005a), which taken together represent perhaps the most comprehensive computational study of edge plasma turbulence to date. The equations were solved in flux tube co-ordinates, with a finite magnetic shear, $\hat{s} = \mathrm{d}\ln q/\mathrm{d}\ln r$, typically set at unity, on a 3D ($\perp$, $\wedge$, $\|$) computational mesh consisting of $64\times256\times16$ cells,

[89] Here $\langle\cdot\rangle$ represent integrals over the whole domain; hence the absence of transport terms.

[90] We denote the two-, four- and six-field drift-fluid (DHD) models as DF2, DF4 and DF6, respectively, and the corresponding magnetic-fluid (resistive MHD) models as MF2, MF4, MF6.

with typical cell size in the drift plane having linear dimensions of $\sim \rho_S \times \rho_S$, or $\sim 1 \times 1$ in normalized units – clearly, the resolution of small scales, down to ρ_S, is necessary for capturing drift-wave dynamics. The relevant dimensionless parameters were typically chosen at nominal values for an (L-mode) edge plasma of a medium sized tokamak:[91]

$$\overline{\beta}_e \approx 5 \times 10^{-5}, \qquad \hat{\epsilon} \approx 2 \times 10^4, \qquad \omega_p = 1, \qquad \omega_B = 0.05, \tag{6.190}$$

$$\hat{\beta} \approx 1, \qquad \hat{\mu} \approx 5, \qquad \hat{\nu} \approx 3, \qquad \hat{C} = \hat{\eta}\hat{\mu}\hat{\nu} \approx 7.5. \tag{6.191}$$

As was mentioned in the previous section, the transition from DHD to MHD physics, i.e. from drift-wave (DW) to interchange (κ) turbulence and hence from DF(2,4,6) to MF(2,4,6) equations, is governed by the *ideal ballooning* parameter, α_B, and the *resistive ballooning* parameter, ν_B, defined by (6.168). It is therefore hardly surprising that, in rough agreement with the first stability boundary, (4.162), one finds rapid growth of ideal ballooning modes for $\alpha_B/\hat{s} > 0.5$, leading to an equally rapid increase in turbulent transport above this threshold, e.g. in DF6 simulations, the turbulent diffusivity, $D_\perp^{tb} = V_\perp^{tb} L_\perp$, increases from $\sim 0.2\,\mathrm{m^2/s}$ for $\alpha_B = 0.3$ to $\sim 10\,\mathrm{m^2/s}$ for $\alpha_B = 0.6$, with similar increase in the ion and electron heat diffusivities. In contrast, resistive ballooning modes give rise to a gradual transition from drift-wave to interchange dominated turbulence with increasing $\hat{\nu}$. This is illustrated in Fig. 6.3, which shows the variation with $\hat{\nu}$ of the effective radial velocity, expressed in units of the gyro-Bohm velocity,

$$V_\perp^{gB} = D_\perp^{gB}/L_\perp = \rho_S^2 c_{Se}/L_\perp^2 = \delta_i^2 c_{Se}, \tag{6.192}$$

and the ratio of potential and pressure fluctuation amplitudes, $A_\varphi/A_p = |\widetilde{\varphi}|^2/|\widetilde{p}_e|^2$, in DF(4,6) and MF(4,6) simulations. In the latter case, $V_\perp$ remains largely unaffected by $\hat{\nu}$, with a maximum at $\nu_B = \hat{\nu}/8 \sim 1$–$3$,[92] while in the former, $V_\perp$ increases with $\hat{\nu}$ in the DW regime ($\nu_B < 1$ for DF4 and $\nu_B < 0.5$ for DF6), gradually saturating at several $V_\perp^{gB}$ in the κ regime. The fluctuating amplitudes tell a similar story: in DF(4,6) simulations we observe a transition from $A_\varphi \sim A_p$ in the DW regime, where $\widetilde{\varphi} \sim \widetilde{p}_e$ is imposed by quasi-adiabatic electrons, to $A_\varphi > A_p$ in the κ regime, converging to the MF(4,6) results. It is significant that the MHD ordering at the heart of the MF(4,6) models assumes $\widetilde{\varphi} \gg \widetilde{p}_e$, and hence $A_\varphi \gg A_p$; since the MF(4,6) simulations obtain the opposite result at low $\hat{\nu}$, they contradict the model assumptions and are thus physically inadmissable in the DW regime.

91 For example, the typical Asdex Upgrade (L-mode) edge plasma with $n_e = 3 \times 10^{19}\,\mathrm{m^{-3}}$, $T_e = 70\,\mathrm{eV}$, $B_0 = 2.5\,\mathrm{T}$, $R_0 = 1.65\,\mathrm{m}$, $R_0/a = 3$, $q = 3.5\,\mathrm{m}$, $L_\perp = L_p = 4\,\mathrm{cm}$, see Scott (2005a). Ion viscosity, $\mu_\parallel$, was set equal to zero, relying on the numerical (grid) dissipation at ρ_S scales.

92 Note that $V_\perp$ is larger in the six-field, compared to the four-field, model due to additional free energy supplied by the ion pressure gradient. This applies to both DF and MF simulations.

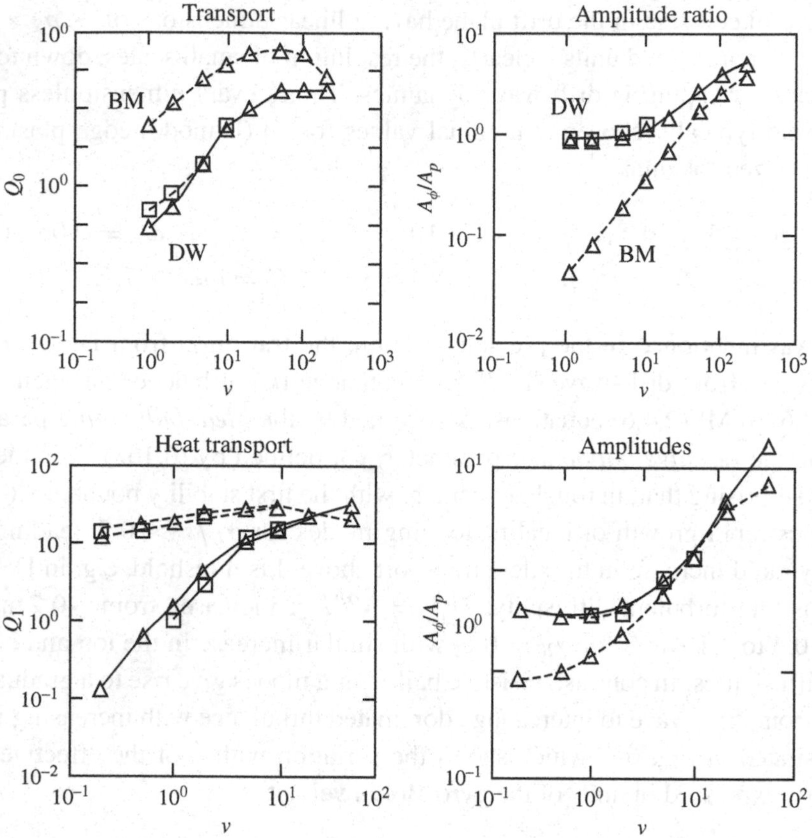

Fig. 6.3. Radial energy transport and ratio of potential and pressure fluctuation amplitudes vs. the collisionality $\hat{\nu}$ in four-field (top) and six-field (bottom) simulations. The full DHD (DF4 and DF6) results are labelled as DW, while the reduced MHD (MF4 and MF6) results are labelled as BM. Reused with permission from Scott (2005a), ©AIP.

Since (6.190)–(6.191) correspond to $\alpha_B \sim 0.05$ in DF4 ($\alpha_B \sim 0.2$ in DF6) and $\nu_B \sim 0.37$ in both models, we may conclude that typical (L-mode) edge plasma profiles are well below the ideal ballooning threshold ($\alpha_B \sim 0.5$), but are close to, yet somewhat below, the resistive ballooning threshold ($\nu_B \sim 0.5$–1). In other words, they are governed (and largely dominated) by DW dynamics, and as such, require the full DHD description, i.e. DF rather than MF simulations. It also means that the primary driving force for edge plasma turbulence is the electron non-adiabaticity, $\widetilde{h}_e = \widetilde{p}_e - \widetilde{\varphi}$, which is the sole force potential in DW dynamics, appearing as $\nabla_\parallel \widetilde{h}_e$ and $C_\parallel \widetilde{h}_e$ in DF4 and DF2 models, respectively, e.g. recall (6.181) and (6.185).

With this in mind, let us examine the transition from a laminar (and radially stagnant) to a fully turbulent edge plasma using the DF4 model. We choose the

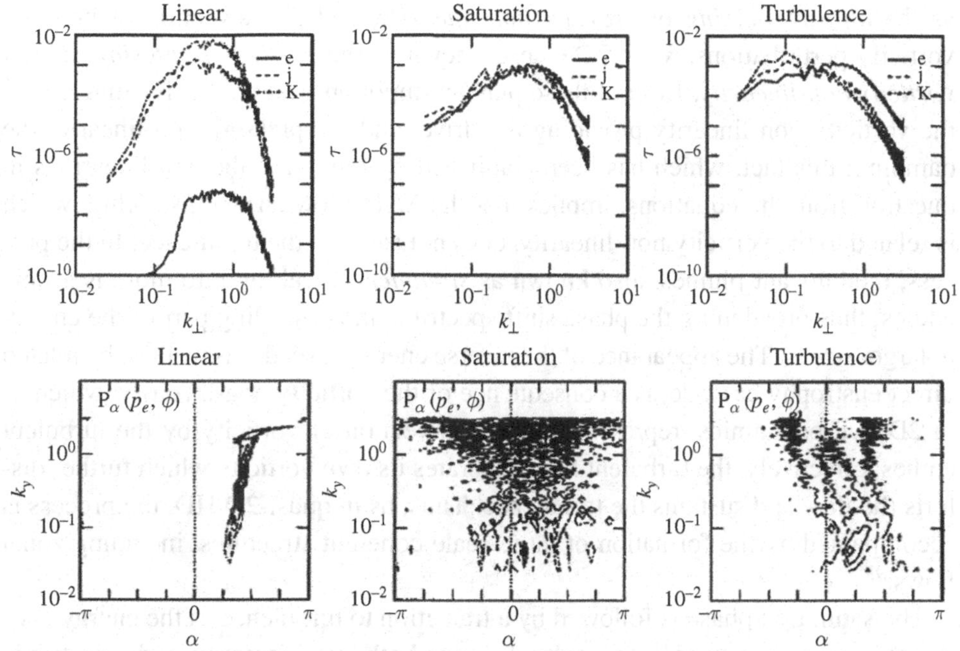

Fig. 6.4. Transition to turbulence in DF4 simulations with $\nu_B = \hat{C}\omega_B = 1.25$: linear instability (left), its saturation (middle) and fully developed turbulence (right). Upper frames show $k_\perp$ spectra of the three terms in the evolution equation for $\widetilde{\Omega}_E$: non-linear polarization (self-advection), $\langle \widetilde{\varphi} \mathbf{V}_E \cdot \nabla \widetilde{\Omega}_E \rangle$ (e), net non-adiabaticity, $\langle \widetilde{\varphi} \nabla_\parallel \widetilde{J}_\parallel \rangle$ (j), and the interchange drive, $\langle \widetilde{\varphi} \mathcal{K} \widetilde{p}_e \rangle$ (K). Lower frames show $k_\wedge$ spectra of the phase shift between $\widetilde{p}_e$ and $\widetilde{\varphi}$. Reused with permission from Scott (2005a), ©AIP.

background parameters to be well below the ideal ballooning threshold and roughly at the resistive ballooning threshold, i.e. with (6.190)–(6.191) but $\hat{\nu} = 10$ ($\nu_B = 1.25$). The three stages in the transition are illustrated in Fig. 6.4, which shows: (i) the growth of the linear instability; (ii) its non-linear saturation; and (iii) fully developed turbulence. The linear growth phase is dominated by the interchange instability, which is largest in the range $0.3 < (k_\wedge, k_\perp) < 1$, with k in units of $1/\rho_S$. During this phase, the divergence of the diamagnetic current, i.e. the curvature drive, $\mathcal{K} p_e$, is roughly balanced by the divergence of the parallel current, $\nabla_\parallel \widetilde{J}_\parallel$, which leads to a large, and strongly ballooned, electron non-adiabaticity, $\widetilde{h}_e \sim \widetilde{p}_e \gg \widetilde{\varphi}$.

As the R–T instability evolves, which proceeds by the formation of characteristic radially aligned structures, effectively *buoyant plumes*, the polarization current increases, until its divergence competes with, and eventually exceeds, that of the diamagnetic current. At this point, the linear instability saturates due to two main non-linearities: the advection of pressure perturbations, $\widetilde{\mathbf{V}}_E \cdot \nabla \widetilde{p}_e$, also known as the *diffusive mixing* or *pressure non-linearity*, and the advection of pressure vorticity perturbations, $\widetilde{\mathbf{V}}_E \cdot \nabla\widetilde{\Omega}_E$, also known as the *vorticity scattering*, or just *vorticity non-linearity*. Both of these play an important role in the dynamics, with the vorticity non-linearity providing the drive, and the pressure non-linearity the damping; this fact, which has been confirmed by removing the non-linearities in question from the equations, implies that the K–H (flow shear) instability, which is related to the vorticity non-linearity, does not saturate the turbulence. In the process, the buoyant plumes, also known as *streamers*, break up into more isotropic eddies, thus broadening the phase shift spectrum and cascading part of the energy to larger scales. The appearance of the inverse energy cascade, as well as the related direct enstrophy cascade, is a consequence of the vorticity non-linearity, which as in 2D hydrodynamics, represents the self-advection of vorticity by the turbulent eddies; effectively, the turbulent flow generates its own vorticity which further distorts the flow and sustains the turbulence. Thus, as in quasi-2D HD, the process is accompanied by the formation of large-scale coherent structures, including zonal flows.[93]

The saturation phase is followed by a transition to turbulence as the energy continues to cascade to ever larger scales, forming both quasi-isotropic and anisotropic coherent structures, i.e. both large-scale eddies and zonal flows. When the cascade reaches the system size, in this case determined by the gradient length, $L_\perp$, the turbulent flow comes to a statistical equilibrium and the turbulence can be said to be fully developed. The final state, which includes the predator–prey dynamics between the zonal and radial flows, bears little, if any resemblance, to the linear instability, Fig. 6.4. Hence, knowing whether a plasma is linearly stable is of no help in determining whether it is non-linearly stable; recall our discussion in Section 6.1.1; indeed, it has been shown (Scott, 2002) that within the DF(4,6) system, the non-linear DW instabilities can develop and persist even in what was a linearly stable plasma, provided the initial perturbation is sufficiently large. Specifically, the most linearly unstable region, $0.3 < k_\perp < 1$, is now dominated by the non-linear polarization drive, with the interchange drive relegated to the longer wavelengths, $k_\perp < 0.3$; when plotted vs. $k_\wedge$ the polarization non-linearity is dominant across the the entire spectrum, see Fig. 6.5 (left).[94] As a result, the pressure and potential fluctuations are comparable, $\widetilde{p}_e \sim \widetilde{\varphi}$, and although both are still considerably ballooned, their difference, i.e. the electron non-adiabaticity, $\widetilde{h}_e \ll \widetilde{p}_e \sim \widetilde{\varphi}$, is only moderately ballooned, e.g. less than twice larger at the outboard, than on the inboard, of the torus.

Another consequence of this dominance of the polarization non-linearity at ρ_S scales is the characteristic spectrum of the phase shift between $\widetilde{p}_e$ and $\widetilde{\varphi}$: at small

[93] This also explains the large levels of intermittency typically found in edge plasma turbulence.
[94] It should be stressed that interchange forcing, and the related ideal/resistive MHD dynamics (ballooning modes), are always relevant at sufficiently large scales, typically below $k_\perp < 0.1$.

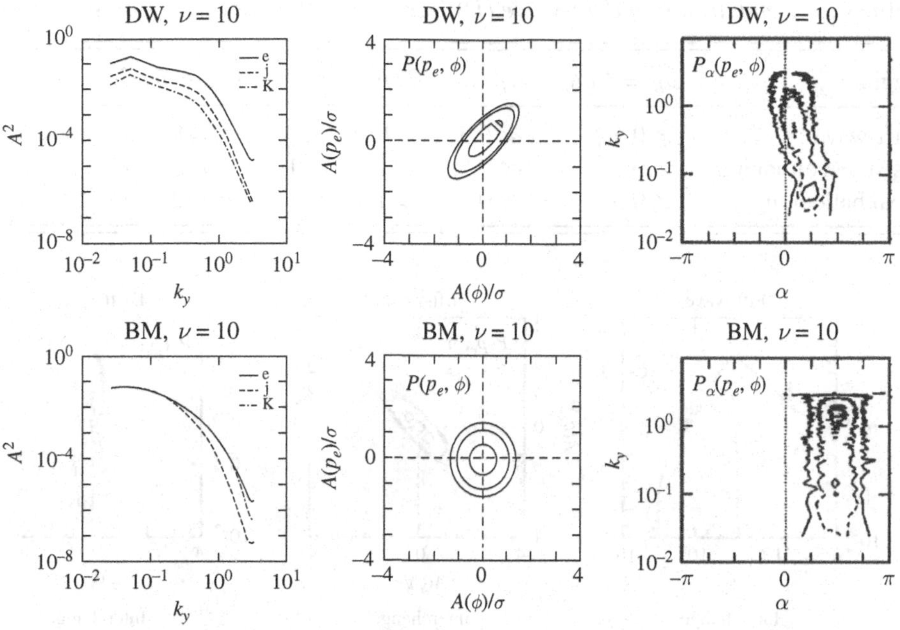

Fig. 6.5. Fully developed turbulence in the DF4 (top) and MF4 (bottom) simulations with $\hat{\nu} = 10$, $\nu_B = 1.25$: left frames show $k_\wedge$ spectra of the three terms in the evolution equation for $\widetilde{\Omega}_E$: polarization advection, $\langle \widetilde{\varphi} \mathbf{V}_E \cdot \nabla \widetilde{\Omega}_E \rangle$ (e), non-adiabaticity, $\langle \widetilde{\varphi} \nabla_\parallel \widetilde{J}_\parallel \rangle$ (j), and interchange drive, $\langle \widetilde{\varphi} \mathcal{K} \widetilde{p}_e \rangle$ (K); middle frames show cross-coherence between $\widetilde{p}_e$ and $\widetilde{\varphi}$, normalized by their standard deviations, and the right frames show $k_\wedge$ spectra of the phase shift between $\widetilde{p}_e$ and $\widetilde{\varphi}$. Reused with permission from Scott (2005a), ©AIP.

scales ($k_\wedge \sim 1$), this shift is typically less than $\pi/8$, whereas at larger scales ($k_\wedge < 0.1$) it tends to towards $\pi/2$. In fact, these two phase shifts are typical of the DW and κ regimes, respectively, see Table 6.8.

To better understand this phase shift, it is instructive to compare the steady turbulence obtained by DF4 and MF4 simulations for the same set of parameters. This is shown in Fig. 6.5 which contains two new diagnostics: the dynamical transfer spectra plotted vs. $k_\wedge$, rather than $k_\perp$, and the cross-coherence between $\widetilde{p}_e$ and $\widetilde{\varphi}$, normalized by their standard deviations. We see that in the absence of drift-wave dynamics $\widetilde{p}_e$ and $\widetilde{\varphi}$ become de-correlated (due to weak coupling via $\widetilde{J}_\parallel$), with $\widetilde{p}_e$ typically leading $\widetilde{\varphi}$ by $\pi/2$ at all scales. In contrast, adiabatic electrons establish a close correlation between the two (due to strong parallel coupling), with a small positive phase shift of $\sim \pi/8$ at ρ_S scales.[95] This correlation is progressively modified, and the phase shift increased, by interchange forcing at larger scales, at

[95] Note that the $\pi/8$ phase shift is characteristic of driven systems with strong dissipation, e.g. the temperature response in a 1D semi-infinite solid subject to a sinusoidal heat load; this explains why it is hottest at 1:30 p.m. during the day and in mid-August during the year.

Table 6.8. *Comparison of drift-wave (DW) and interchange (κ) regimes.*

Regime	$\nu_B = \hat{C}\omega_B$	$\alpha_B/\hat{s}$	$\widetilde{\varphi}/\widetilde{p}_e$	$\widetilde{h}_e$	Phase$(\widetilde{p}_e, \widetilde{\varphi})$	$\nabla_\parallel \widetilde{J}_\parallel$
drift-wave	$\ll 0.5$	$\ll 0.5$	~ 1	$\ll 1$	$\sim \pi/8$	$\sim \nabla \cdot \widetilde{\mathbf{J}}_p$
resistive ballooning	$\gg 0.5$	$\ll 0.5$	$\gg 1$	~ 1	$\sim \pi/2$	$\sim \nabla \cdot \widetilde{\mathbf{J}}_*$
ideal ballooning	$\ll 0.5$	$\gg 0.5$	$\gg 1$	~ 1	$\sim \pi/2$	$\sim \nabla \cdot \widetilde{\mathbf{J}}_*$

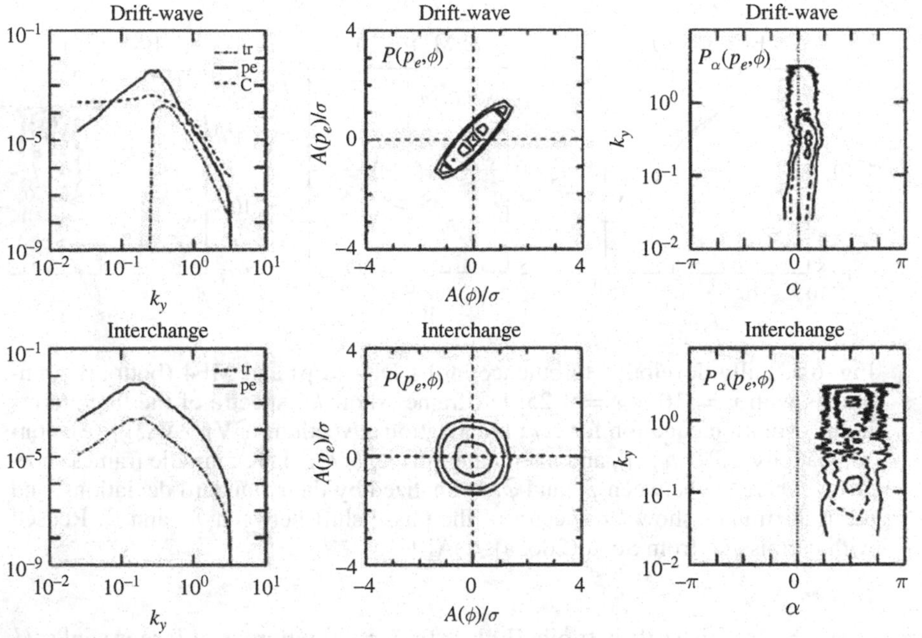

Fig. 6.6. Fully developed turbulence in the DF2 simulations ($C_\parallel = 0.05$, top) and MF2 simulations ($\omega_B = 5\gamma_\varphi = 0.05$, bottom): left frames show $k_\wedge$ spectra of the energy source (pe), transfer (tr) and dissipation (C), middle frames show cross-correlation between $\widetilde{p}_e$ and $\widetilde{\varphi}$, normalized by their standard deviations, and the right frames show $k_\wedge$ spectra of the phase shift between $\widetilde{p}_e$ and $\widetilde{\varphi}$. Reused with permission from Scott (2005a), ©AIP.

which the polarization non-linearity becomes ineffective. Indeed, the fundamental difference between drift-wave and interchange turbulence can be captured with the generic two-field model, i.e. by comparing the results of DF2 and MF2 simulations, Fig. 6.6.[96] These two case, which represent pure DW and pure κ turbulence, may be viewed as the asymptotes, or dynamical attractors, of DF4 turbulence in the limits of small and large ν_B, respectively.

[96] Note that the two models cannot be distinguished by the drive spectra alone, see Fig. 6.6.

Finally, let us examine the differences between the results of six-field (DF6, MF6) and four-field (DF4, MF4) simulations; recall that the latter assume cold ions and isothermal electrons ($\omega_e = \tau = 0$), while the former include warm ions and non-uniform temperature.[97] In the default cases, the ion and electron temperatures were set to be equal, $\tau = 1$, and their radial gradient was assumed to equal the density gradient, $\omega_n = \omega_e = \omega_i = 1$, so that the ballooning parameter, $\alpha_B = 4\hat{\beta}\omega_B\omega_n$, increased by a factor of 4 with respect to the DF4 value. The results of the six- and four-field simulations are broadly similar in the (parallel and spectral) structure of the dominant (DW and κ) modes, transition between the corresponding regimes with ν_B and α_B, and in the resulting levels of turbulent transport (Scott, 2005a). Nonetheless, there are a number of notable differences: (i) the presence of ion pressure gradient free-energy allows for direct interchange coupling between $\widetilde{p}_i$ and $\widetilde{\varphi}$, which for $\eta_i > 1$ becomes dominant, resulting in ion temperature gradient (ITG) dominated turbulence; (ii) on sub-ρ_S scales, $k_\wedge > 1$, the phase shifts between φ and n_e tend to peak at $\pm\pi$, due to finite gyro-radius effects, specifically the non-linearity in the ion gyro-viscosity. The tendency towards the $+\pi$ phasing is also evident in MF6 simulations at super-ρ_S scales, $0.3 < k_\wedge < 1$; (iii) resistive ballooning threshold is reduced from $\nu_B \sim 1$ in DF4 to $\nu_B \sim 0.5$ in DF6, while the ideal ballooning threshold is increased from $\alpha_B \sim 0.15$ in DF4 to $\alpha_B \sim 0.3$ in DF6. Moreover, the transition becomes more rapid in the latter case due to two effects: an additional source of free energy offered by the ion temperature gradient, and the finite gyro-radius effects in the vorticity, which now becomes $\widetilde{\Omega}_\parallel = \nabla_\perp^2(\widetilde{\varphi} + \widetilde{p}_i)$, and enhances the influence of the non-linear polarization drift/current; and finally (iv) the role of the non-adiabaticity, $\widetilde{h}_e$, in DF4, which governs the release of free energy associated with the electron temperature gradient, is now played by $\widetilde{T}_e$ itself in DF6, at least in the DW regime. Both of these quantities are much less ballooned than either $\widetilde{n}_i$ and, in the case of DF6, $\widetilde{T}_i$.

6.3.2 Gyro-fluid turbulence

Let us now turn to the next level of sophistication, retaining the fluid ansatz, but taking the moments of the non-linear gyro-kinetic equation (Hahm, 1988), obtained along the same lines as (2.129) derived in Section 2.3.2.2. This gyro-tropic moment approach is comparable to the one we considered for the parallel transport equations, recall Section 5.3.2.2, except we now have to deal with the added complication of perpendicular motion and finite ρ_{ti} effects, recall (2.260). It yields the so-called *gyro-fluid* model, whose electrostatic version is derived in Dorland

[97] The MF6 equations are obtained from the DF6 equations by a route analogous to the reduction of DF4 to MF4, including the transformations $\widetilde{\varphi} + \widetilde{p}_i \to \widetilde{\varphi}$, $\widetilde{\varphi} - \widetilde{p}_e \to \widetilde{\varphi}$, $\widetilde{p}_e + \widetilde{p}_i \to \widetilde{p}$, etc.

and Hammett (1993); Hammett *et al.* (1993) and Beer and Hammett (1996) and the electromagnetic one in Scott (2000); Snyder and Hammett (2001). The gyro-fluid approach can be used to obtain global equations for total (mean + fluctuating) quantities, which are typically reduced to local equations for the fluctuations, similar in form to the drift-fluid models DF4 and DF6 discussed in the previous section. Below we present one of the simplest variants, namely a six-field gyro-fluid model (GF6) containing electromagnetic effects, referring the reader to original publications for higher field variants, e.g. extension to multi-species, inclusion of finite temperatures, parallel heat flows, etc.

The GF6 (GEM3) equations are derived in Scott (2003b):

$$\mathrm{d}_{t,e}(\overline{n}_e + \widetilde{n}_e) \approx -B\nabla_\parallel(\widetilde{V}_{\parallel e}/B) - \mathcal{K}(\widetilde{n}_e - \widetilde{\varphi}), \tag{6.193}$$

$$\mathrm{d}_{t,i}(\overline{n}_i + \widetilde{n}_i) \approx -B\nabla_\parallel(\widetilde{V}_{\parallel i}/B) + \mathcal{K}(\tau_i\widetilde{n}_i + \widetilde{\varphi}_G), \tag{6.194}$$

$$\hat{\beta}\partial_t\widetilde{A}_\parallel - \hat{\mu}\mathrm{d}_{t,e}\widetilde{V}_{\parallel e} \approx \nabla_\parallel(\overline{n}_e + \widetilde{n}_e - \widetilde{\varphi}) + \mathcal{K}(2\hat{\mu}\widetilde{V}_{\parallel e}) - \hat{C}\widetilde{J}_\parallel, \tag{6.195}$$

$$\hat{\beta}\partial_t\widetilde{A}_\parallel + \hat{\epsilon}\mathrm{d}_{t,i}\widetilde{V}_{\parallel i} \approx -\nabla_\parallel(\tau_i\overline{n}_i + \tau_i\widetilde{n}_i + \widetilde{\varphi}_G) + \mathcal{K}(2\tau_i\hat{\epsilon}\widetilde{V}_{\parallel i}) - \hat{C}\widetilde{J}_\parallel, \tag{6.196}$$

where $\mathrm{d}_{t,e} = \partial_t + \mathbf{V}_E \cdot \nabla$ and $\mathrm{d}_{t,i} = \partial_t + \mathbf{V}_{E,i} \cdot \nabla$ are the usual $\mathbf{E} \times \mathbf{B}$ advective derivatives, which are now different for electrons and ions, however, because ions feel/sample a gyro-averaged, and hence diminished, electric potential,

$$\widetilde{\varphi}_G = \Gamma_0^{1/2}\left(k_\perp^2\rho_S^2\right)\widetilde{\varphi} = \Gamma_1\left(k_\perp^2\rho_S^2\right)\widetilde{\varphi}, \tag{6.197}$$

$$\Gamma_0\left(\rho_{ti}^2 k_\perp^2\right) \approx \left(1 - \rho_{ti}^2\nabla_\perp^2\right)^{-1}, \quad \Gamma_1 = \Gamma_0^{1/2} \approx \left(1 - \tfrac{1}{2}\rho_{ti}^2\nabla_\perp^2\right)^{-1}, \tag{6.198}$$

which is here written in terms of the gyro-average operator derived in Dorland and Hammett (1993) and Beer and Hammett (1996), whereas electrons feel/sample the raw potential, $\widetilde{\varphi}$;[98] clearly $\mathbf{V}_E = (\mathbf{b}/B) \times \nabla\widetilde{\varphi}$ and $\mathbf{V}_{E,i} = (\mathbf{b}/B) \times \nabla\widetilde{\varphi}_G$ are the electric drifts caused by the gradients of $\widetilde{\varphi}$ and $\widetilde{\varphi}_G$, respectively. Since the gyro-averaging operators Γ_0 and Γ_1 take as their argument the square of the fluctuating magnetization parameter,

$$b_i \equiv \widetilde{\delta}_i^2 = (\rho_{ti}/\widetilde{L}_\perp)^2 = \rho_{ti}^2 k_\perp^2, \tag{6.199}$$

the effect of gyro-averaging becomes negligible for disturbances with wavelengths larger than several ion gyro-radii, e.g. $\widetilde{\delta}_i \sim 0.1$ yields $\Gamma_0 \sim 1$.

The form of the *gyro-fluid* averaging operators is best understood by revisiting our earlier discussion of *gyro-kinetic* averaging, as defined in (2.121). There we encountered the Bessel functions, J_0 and J_1, taking $\rho_i k_\perp = v_\perp k_\perp/\Omega_i$ as their argument, recall (2.122)–(2.123),

$$\widetilde{\varphi}_G = J_0\widetilde{\varphi} = \sum J_0(\rho_i k_\perp)\widetilde{\varphi}_\mathbf{k} e^{i\mathbf{k}\cdot\mathbf{x}} \to \sum \Gamma_1(b_i)\widetilde{\varphi}_\mathbf{k} e^{i\mathbf{k}\cdot\mathbf{x}}, \tag{6.200}$$

98 This follows from the smallness of the electron gyro-radius, $\rho_{te}/\rho_{ti} \sim \sqrt{m_e/m_i} \ll 1$.

where the arrow indicates the transition from gyro-kinetic to gyro-fluid description, i.e. the average of $J_0(\rho_i k_\perp)$ over velocity space, which introduces the thermal gyro-radius, ρ_{ti}, in the argument b_i of $\Gamma_1(b_i)$. Higher-order gyro-averaging operators, denoted as $\Gamma_j(b_i)$, which appear along with higher gyro-fluid moments, are defined in terms of derivatives of $\Gamma_1(b)$,

$$\Gamma_2 = b\partial_b\Gamma_1, \qquad \Gamma_3 = \tfrac{1}{2}b\partial_b^2(b\Gamma_1). \tag{6.201}$$

The hyperbolic (advective) evolution equations for the densities, (6.193)–(6.194), and parallel velocities, (6.195)–(6.196), are closed by corresponding elliptic (diffusive) equations for the electric and magnetic potentials, i.e. by appropriate Maxwell's equations. Hence, the bare potential evolves according with the 'gyro-kinetic' Poisson's equation (Lee, 1983),

$$(\Gamma_0 - 1)\widetilde{\varphi}/\tau_i = \widetilde{n}_e - \Gamma_1\widetilde{n}_i \tag{6.202}$$

i.e. Poisson's equation, $\nabla^2\varphi = -\varrho$, for zero gyro-radius electrons and finite gyro-radius ions, which describes the effect of charge polarization. Finally, the magnetic vector potential evolves according with Ampere's law,

$$\widetilde{J}_\| = \widetilde{V}_{\|i} - \widetilde{V}_{\|e} = -\nabla_\perp^2\widetilde{A}_\|, \tag{6.203}$$

which represents the law of induction. The other terms appearing in (6.193)–(6.196) have the same meaning and definitions as in Section 6.3.1.3.

The GF6 model is easily extended to multiple ion species, e.g. plasma impurities, by defining species dependent quantities,

$$\mu_s = m_s/Z_s m_D = A/2Z_s, \qquad a_s = Z_s n_s/n_e, \qquad \tau_s = T_s/Z_s T_e, \tag{6.204}$$

and rewriting (6.193)–(6.196) as Scott (2005b)

$$\mathrm{d}_{t,s}(\overline{n}_s + \widetilde{n}_s) \approx -B\nabla_\|(\widetilde{V}_{\|s}/B) + \mathcal{K}(\tau_s\widetilde{n}_s + \widetilde{\varphi}_{G,s}), \tag{6.205}$$

$$\hat{\beta}\partial_t\widetilde{A}_\| + \hat{\epsilon}_s\mathrm{d}_{t,s}\widetilde{V}_{\|s} \approx -\nabla_\|(\tau_s\overline{n}_s + \tau_s\widetilde{n}_s + \widetilde{\varphi}_{G,s}) + \mathcal{K}(\tau_s\hat{\epsilon}_s\widetilde{V}_{\|s}) - \hat{C}\widetilde{J}_\|. \tag{6.206}$$

The advective derivatives are now different for each ion species,

$$\mathrm{d}_{t,s} = \partial_t + \mathbf{V}_{E,s}\cdot\nabla = \partial_t + \{\varphi_{G,s},\,\}, \qquad \nabla_\| = B^{-1}\partial_s - \{\hat{\beta}\widetilde{A}_\|,\,\}, \tag{6.207}$$

$$\widetilde{\varphi}_G = \Gamma_1(b_s)\widetilde{\varphi}, \qquad b_s = k_\perp^2\rho_s^2, \qquad \rho_s^2 = \mu_s\tau_s/B^2, \tag{6.208}$$

due to the different gyro-radius involved in the gyro-averaging of $\widetilde{\varphi}$, and hence in the definition of $\widetilde{\varphi}_{G,s}$; above we wrote both gradients in terms of the Poisson bracket with respect to drift plane co-ordinates,

$$\{\mathcal{A}, \mathcal{B}\} \equiv \delta_S(\partial_x\mathcal{A}\,\partial_y\mathcal{B} - \partial_y\mathcal{A}\,\partial_x\mathcal{B}). \tag{6.209}$$

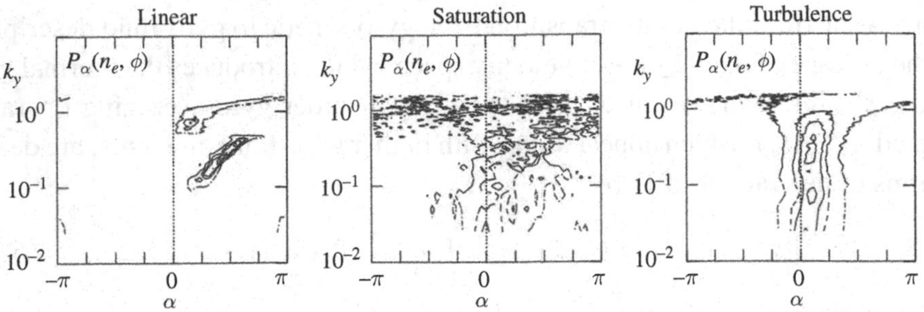

Fig. 6.7. Similar to Fig. 6.4, only in GF6 simulations ($\omega_B = 0.05$, $\nu_B = \hat{C}\omega_B = 0.125$, $\hat{\beta} = 1$, $\hat{\mu} = 5$, $\tau_i = 1$) showing $k_\wedge$ spectra of the phase shift between $\widetilde{n}_e$ and $\widetilde{\varphi}$ in linear instability (left), its saturation (middle) and fully developed turbulence (right). Note the transition from the linear dispersion relation (left) to a drift-wave mode structure, peaking at $\alpha \approx \pi/8$ (right). Reused with permission from Scott (2003b), ©AIP.

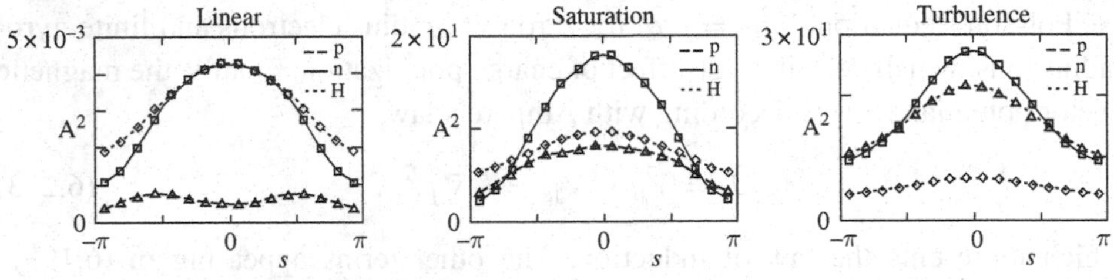

Fig. 6.8. Poloidal dependence ($\theta = 0$ denotes the outer mid-plane) of the squared amplitude of $\widetilde{n}_e(\square)$, $\widetilde{\varphi}(\Delta)$ and $\widetilde{h}_e = \widetilde{n}_e - \widetilde{\varphi}_e(\diamond)$ during the transition shown in Fig. 6.7. While $\widetilde{n}_e$ is strongly ballooned at all times, the ballooning character of $\widetilde{h}_e$ is pronounced in the linear phase and decreases in the turbulent phase, while that of $\widetilde{\varphi}$ has the opposite tendency. This is consistent with the adiabatic electron response in drift-wave dynamics, which imposes $\widetilde{h}_e \ll \widetilde{n}_e \sim \widetilde{\varphi}$ (Scott, 2003b).

Finally, the polarization and induction equations, (6.202)–(6.203), become

$$\widetilde{J}_\parallel = \sum a_s \widetilde{V}_{\parallel s} = -\nabla_\perp^2 \widetilde{A}_\parallel, \qquad \sum a_s \left[(\Gamma_0 - 1)\widetilde{\varphi}/\tau_s + \Gamma_1 \widetilde{n}_s\right] = 0. \qquad (6.210)$$

Turbulence simulations using the GF6 model are broadly consistent with those performed using DF4, see Section 6.3.1.4. For instance, the transition from linear instability to fully developed turbulence, investigated using the DF6 model in Fig. 6.4, is found to be similar in GF6 simulations, Fig. 6.7. The same is true of the poloidal structure of the fluctuations in both the linear and turbulent phases, see Fig. 6.8. There are of course quantitative differences, but the overall behaviour is generally preserved.

The GF6 model can be extended to include higher-order gyro-tropic moments, specifically the parallel and perpendicular temperatures and heat flux densities, to yield evolution equations for $\widetilde{n}_s$, $\widetilde{V}_{\|s}$, $\widetilde{T}_{\|s}$, $\widetilde{T}_{\perp s}$, $\widetilde{q}_{\|s}$, $\widetilde{q}_{\perp s}$, $\widetilde{\varphi}$ and $\widetilde{A}_{\|}$; for single ion species, this leads to 14 individual fields and hence a label GF14 by which we will designate this model.[99] The full account of the derivation may be found in Scott (2005c,d), with a concise summary of the model equations given in the appendix of Ribeiro and Scott (2008). To give a flavour of the modifications, we write down only the continuity and polarization equations, noting the appearance of the second-order gyro-averaging operator, Γ_2, (6.201), and additional $\mathbf{E} \times \mathbf{B}$ advection due to $\Gamma_2\widetilde{\varphi}$,

$$d_{t,s}(\overline{n}_s + \widetilde{n}_s) + \{\Gamma_2\widetilde{\varphi}, T_s + \widetilde{T}_{\perp s}\} \approx \tag{6.211}$$

$$-B\nabla_{\|}(\widetilde{V}_{\|s}/B) + \mathcal{K}\left(\left[\Gamma_1 + \tfrac{1}{2}\Gamma_2\right]\widetilde{\varphi} + \tfrac{1}{2}\tau_s[\widetilde{p}_{\|s} + \widetilde{p}_{\perp s}]\right),$$

$$\sum a_s\left[(\Gamma_0 - 1)\widetilde{\varphi}/\tau_s + \Gamma_1\widetilde{n}_s + \Gamma_2\widetilde{T}_{\perp s}\right] = 0. \tag{6.212}$$

The relationship between GF14 and GF6 is comparable to that between DF6 and DF4 in the sense that in both cases the simpler of the two models assumes cold ions and isothermal electrons, while the more complex model includes ion and electron temperature non-uniformity, and hence an important new source of free energy. This is particularly important for the ion temperature gradient which gives rise to a robust ITG instability for $\eta_i = d\ln T_i/d\ln n_i \gtrsim 2$, roughly in line with linear analysis, recall Section 4.3. The resulting increase in radial ion heat flow is shown in Fig. 6.9. For the transitional $\eta_i = 2$ case, ion heat transport increases with inductivity, $\hat{\beta}$, above some critical value of order unity, and with collisionality, $\hat{\nu}$, more gradually across more than one decade, see Fig. 6.10.

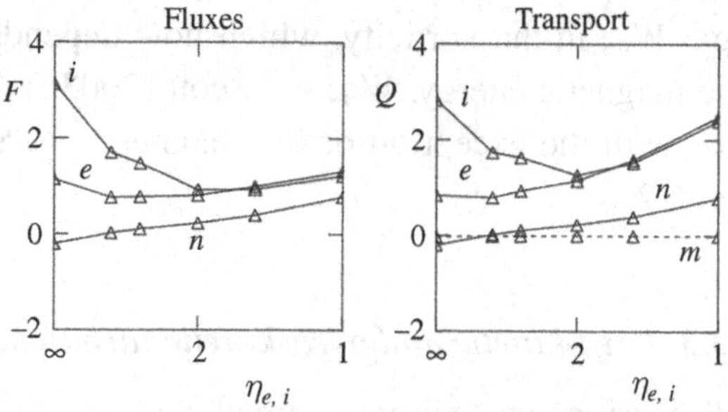

Fig. 6.9. Radial $\mathbf{E} \times \mathbf{B}$ transport of particles (n) as well as electron (e) and ion (i) heat as a function of $\eta_s = d\ln T_s/d\ln n_s$, ($\eta_i = \eta_e$), in GF14 simulations; left frame shows the conductive and right frame shows the total energy flows, including that associated with magnetic flutter (m). Drift-wave dynamics (DG) dominates for $\eta_i \lesssim 2$, while ITG dynamics takes over for $\eta_i \gtrsim 2$. Reused with permission from Scott (2003b), ©AIP.

[99] The GF6 and GF14 models are also known as GEM3 and GEM, respectively.

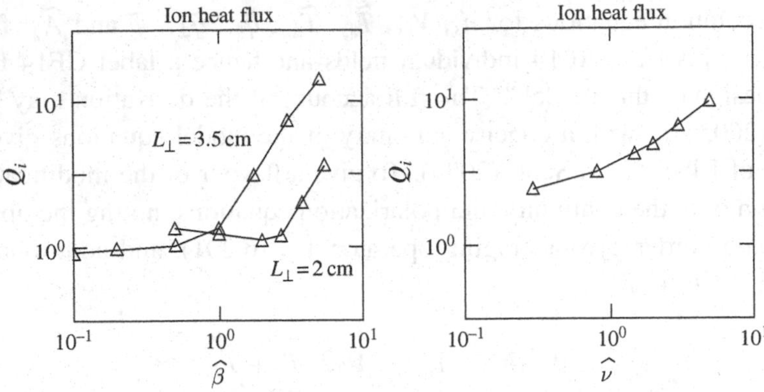

Fig. 6.10. Radial $\mathbf{E} \times \mathbf{B}$ transport of ion (i) energy as a function of inductivity, $\hat{\beta}$, and collisionality, $\hat{\nu}$, in GF14 simulations with $\eta_i = \eta_e = 2$. Reused with permission from Scott (2007), ©AIP.

In summary, gyro-fluid models are distinguished from their drift-fluid counterparts by the inclusion of finite-ρ_{ti} effects, such as gyro-averaging, and the appearance of gyro-tropic ion pressures and heat flux densities. Aside from these modifications, their energetics are broadly similar:

$$W_t = \sum \tfrac{1}{2} a_s \tau_s \left[\widetilde{n}_s^2 + \tfrac{1}{2}\widetilde{T}_{\parallel s}^2 + \widetilde{T}_{\perp s}^2\right], \quad W_v = \sum \tfrac{1}{2} a_s \mu_s \left[\widetilde{V}_{\parallel s}^2 + \tfrac{2}{3}\widetilde{q}_{\parallel s}^2 + \widetilde{q}_{\perp s}^2\right],$$
$$W_E = \sum \tfrac{1}{2}(a_s/\tau_s)(\Gamma_0 - 1)\widetilde{\varphi}^2 = \sum \tfrac{1}{2} a_s (\widetilde{n}_s \Gamma_1 \widetilde{\varphi} + \widetilde{T}_{\perp s}\Gamma_2 \widetilde{\varphi}), \quad (6.213)$$
$$W_B = \tfrac{1}{2}\beta_e |\nabla_\perp \widetilde{A}_\parallel|^2 = \tfrac{1}{2}\beta_e \widetilde{A}_\parallel \widetilde{J}_\parallel;$$

the thermal free energy, W_t, is contained in the densities and temperatures, the parallel flow energy, W_v, in the parallel flows of particles and heat, the perpendicular (drift) energy, W_E, in the vorticity, which now depends on the Laplacian of $\widetilde{\varphi}$ and $\widetilde{\varphi}_G$, and the magnetic energy, W_B, see Scott (2003b). The flows between these four reservoirs, with the exception of the magnetic energy, are represented schematically in Fig. 6.2.

6.3.3 Drift-kinetic and gyro-kinetic turbulence

In the preceding discussion we tacitly assumed that the distribution function, $f_s(\mathbf{x}, \mathbf{v})$, can be described by a small number of its lowest moments, e.g. $n_s(\mathbf{x})$, $\mathbf{V}_s(\mathbf{x})$, $T_s(\mathbf{x})$, $\mathbf{q}_s(\mathbf{x})$, either in isotropic or gyro-tropic form. In reality, there are situations in which the shape of $f_s(\mathbf{v})$ is not easily captured by its lowest moments and/or when the tails of the distribution play an important role in the dynamics. In such circumstances, a kinetic description, whether via the drift-kinetic (DK), (2.115), or the more accurate, gyro-kinetic (GK), (2.129), form becomes indispensable. It is particularly important to the study of transport processes in the edge region of a magnetized plasma, which is our main focus in this text. The general motivation for edge plasma kinetic simulation has been nicely expressed in Cohen and Xu (2008),

'In spite of the conventional wisdom that edge plasmas are fluid-like because they are relatively cold and collisional, kinetic simulation is essential for quantitative prediction of edge plasma phenomena. This is because the steep radial gradients in the edge, with scale lengths comparable to drift orbit widths, and because of the significant equilibrium temperature, density and potential variations along magnetic field lines on length scales comparable to collisional mean free paths. Because of these circumstances, kinetic simulation is needed for both plasma equilibrium (transport time-scale) and micro-turbulence (fluctuation time-scale) simulations, whereas for core plasmas, with characteristically much larger equilibrium radial scale lengths and confinement times, the transport processes can be reasonably well captured by fluid simulation, while kinetic simulation is important only for micro-turbulence. The use of kinetic simulations also removes the necessity of introducing approximate boundary conditions at material surfaces.

Properties of edge plasmas and the magnetic geometry of the edge create unique challenges for kinetic simulation that are not present in core plasmas. First, because of the overlap of spatial scales noted above, the distribution function in the edge is not *a priori* known. Second, the fluctuations in the edge are often not low amplitude, e.g. edge plasmas often contain large amplitude intermittent structures (blobs, ELMs)... Hence, δf techniques, in which the distribution function is expanded around a fixed, known background, f_0, are not justified for the edge.'

The above quote serves the double role of (i) outlining the key differences between core and edge plasma dynamics, and hence introducing the theme of the next chapter, and (ii) motivating the significant effort currently devoted to development and running of full-f gyro-kinetic codes, dedicated to simulating edge plasma phenomena.[100] Since the effort is very much ongoing, we will refrain from reviewing the preliminary results, and offer instead a brief glance at the problems to be overcome and the proposed solutions.

Let us begin with the observation that since the plasma kinetic equations are known: DK (2.115) and GK (2.129), the bulk of the work shifts from mathematical physics to scientific computing, i.e. to the formulation of efficient numerical algorithms for the solution of kinetic equations and their implementation on multi-processor architectures. The two competing, and complementary, solution approaches are the continuum and particle-in-cell (PiC) methods: the former relying on finite difference/volume discretization, the other on a variant of the

[100] The distinction between full-f and δf is basically the same as that between global (total quantities) and local (fluctuations only) fluid codes made in the previous section.

Monte-Carlo technique (Tajima, 1989).[101] The main advantage of continuum techniques is the fully deterministic (non-statistical) nature of the numerical scheme, whereas the main challenge is the computational cost involved in evaluating the collision operator, involving velocity space integrals, and more generally, in sufficiently resolving $f_s(\mathbf{v})$ in velocity space. Similarly, the main advantage of the PiC technique is its high flexibility and ability to be easily parallelized for high power computing, whereas its main disadvantage is the statistical noise, which tends to build up with time and can limit the overall length of the simulation.

To conclude the section, we note that kinetic codes tend to be broadly consistent with their fluid counterparts (typically within a factor of two) and to be broadly similar in the drift-fluid and gyro-fluid approximations. For instance, a recent comparison of several (EU) GF/GK codes describing gradient-driven plasma turbulence in the core and edge regions of tokamaks (Falchetto *et al.*, 2008) revealed good agreement in the prediction of core ITG driven turbulence (with adiabatic electrons, but without trapped particles) and fair agreement in the edge turbulence collisionality scaling, beta scaling (with cold ions and below the ideal MHD limit) and the non-linear mode structure; the edge turbulence comparison is shown in Fig. 6.11 and includes the DF6 and GF14 models described in the previous sections. These codes have been successful in reproducing the radial flows of particles and energy in the edge plasma of the Asdex Upgrade tokamak for experimentally

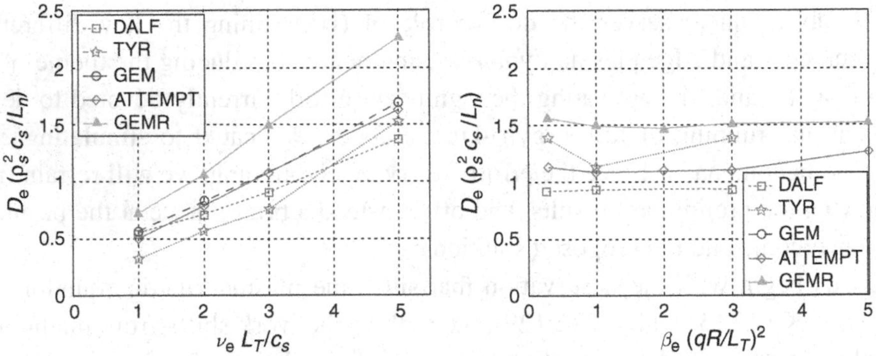

Fig. 6.11. Comparison of selected (EU) turbulence codes in the edge plasma region, showing a strong increase of the turbulent electron diffusivity with the collisionality, $\hat{\nu}$, and little change with the inductivity, $\hat{\beta}$, see (6.191); here DALF and GEM correspond to DF4/6 and GF14 of Section 6.3.2 (Falchetto *et al.*, 2008).

[101] For example, the former is used in TEMPEST (Xu *et al.*, 2007; Cohen and Xu, 2008) and FEFI (Scott, 2006) codes, while the latter is used in PARASOL (Takizuka, 2003), XGC (Chang *et al.* 2004, 2008; Chang and Ku 2006, 2008) and ELMFIRE (Heikkinen *et al.*, 2006, 2008; Henriksson *et al.*, 2006; Janhunen *et al.*, 2007; Leerink *et al.*, 2008).

measured profiles under L-mode conditions (Scott, 2006). Finally, as a first step on their validation, full-f GK codes have been found to capture the main neoclassical results, see Section 5.3.3, when fluctuations/turbulence were artificially suppressed (Janhunen *et al.*, 2007; Chang *et al.*, 2008).

6.4 Comparison of collisional and turbulent diffusivities

The most pessimistic estimate of the anomalous (turbulent) plasma diffusivity may be obtained from the so-called *Bohm normalization*,

$$\mathbf{x}/\rho_S \to \mathbf{x}, \qquad t\Omega_i \to t, \qquad e\varphi/T_e \to \varphi, \tag{6.214}$$

$$D_\perp^B = \rho_S^2 \Omega_i = \rho_S c_{Se} = c_{Se}^2/\Omega_i, \tag{6.215}$$

obtained by taking the radial step as the ion gyro-radius, $\rho_S = c_{Se}/\Omega_i$, evaluated at the cold ion sound speed, $c_{Se}^2 = T_e/m_i$, and the time step as the ion gyro-frequency, Ω_i.[102] The resulting *Bohm* diffusivity, $D_\perp^B = \rho_S c_{Se}$, is independent of $L_\perp$, and the corresponding energy confinement time, $\tau_E^B \approx L_\perp^2/\chi_\perp^B$ is independent of both plasma beta, β, and collisionality, ν^*,

$$\chi_\perp^B \sim D_\perp^B, \qquad \Omega_i \tau_E^B \approx (L_\perp/\rho_S)^2 \approx \delta_i^{-2} \approx \rho_*^{-2}, \tag{6.216}$$

and increases only *linearly* with the magnetic field.

Fortunately for the prospects of fusion energy, tokamak plasma turbulence does not follow the Bohm diffusivity. Rather it appears to scale with the more optimistic *gyro-Bohm* diffusivity, $D_\perp^{gB}$, obtained from the corresponding *gyro-Bohm* normalization, appropriate to drift-wave (DHD) turbulence,

$$\mathbf{x}/\rho_S \to \mathbf{x}, \qquad t\omega_{tS} \to t, \qquad e\varphi/T_e \to \varphi, \tag{6.217}$$

$$D_\perp^{gB} = \rho_S^2 \omega_{tS} = \delta_i D_\perp^B = c_{Se}^3/\Omega_i^2 L_\perp = \rho_S^3(\Omega_i/L_\perp), \tag{6.218}$$

where $\omega_{tS} = c_{Se}/L_\perp$ is sound wave transit frequency. Note that $D_\perp^{gB}$ is a factor of δ_i smaller than $D_\perp^B$,[103] and the corresponding energy confinement time $\tau_E^{gB} \approx L_\perp^2/\chi_\perp^{gB}$ is a factor of δ_i larger than τ_E^B,

$$\chi_\perp^{gB} \sim D_\perp^{gB}, \qquad \Omega_i \tau_E^{gB} \approx (L_\perp/\rho_S)^3 \approx \delta_i^{-3} \approx \rho_*^{-3}, \tag{6.219}$$

and increases *quadratically* with the magnetic field. It is worth stressing that both times depend only on the magnetization parameter, $\delta_i \sim \rho_*$.

[102] Alternatively, $D_\perp^B$ may be obtained by taking the radial step as the logarithmic average of ρ_S and $L_\perp$, and the time step as the sound wave transit frequency, $\omega_{tS} = c_{Se}/L_\perp$.

[103] For a simple plasma, the Bohm and gyro-Bohm diffusivities may be evaluated as $D_\perp^B \approx (ZT_e + T_i)/ZeB$ and $D_\perp^{gB} \approx (ZT_e + T_i)^{3/2} m_i^{1/2} (ZeB)^{-2} L_\perp^{-1}$, respectively.

Table 6.9. *Comparison of classical, neoclassical and turbulent diffusivities.*

Transport	Class	P–S	Plateau	Banana	Gyro-Bohm	Bohm
$D_{\perp e}, \chi_{\perp e} \sim$	$\rho_{te}^2 \nu_e$	$(q\rho_{te})^2 \nu_e$	$(q\rho_{te})^2 \omega_{\parallel e}$	$\sqrt{\epsilon}(q\rho_{te}/\epsilon)^2 \nu_e$	$\rho_S^2 \delta_i \Omega_i$	$\rho_S^2 \Omega_i$
$\chi_{\perp i} \sim$	$\rho_{ti}^2 \nu_i$	$(q\rho_{ti})^2 \nu_i$	$(q\rho_{ti})^2 \omega_{\parallel i}$	$\sqrt{\epsilon}(q\rho_{ti}/\epsilon)^2 \nu_i$		

Due to toroidicity effects, ρ_S is frequently replaced by $q_a \rho_S$ (recall the neoclassical correction), and the two energy confinement time scalings can be combined into the generic expression previously given in (1.19),

$$\Omega_i \tau_E \approx (q_a \rho_*)^{-x} \propto (q_a \sqrt{T}/aB_0)^{-x}, \tag{6.220}$$

where $x = 3$ for *gyro-Bohm* and $x = 2$ for *Bohm* scalings. These two values represent the most common mixing-length estimates of $D_\perp \sim \chi_\perp$ given by (6.216) and (6.219), in which $\ell_m \sim \rho_S$, $\tau_m^B \sim \Omega_i$ and $\tau_m^{gB} \sim \omega_{ts} \approx \delta_i \Omega_i$.

We conclude the chapter with a brief comparison of laminar and turbulent radial particle and heat diffusivities, see Table 6.9. There are however two caveats in such a comparison: (i) the Bohm and gyro-Bohm expressions do not reflect the actual turbulent diffusivities, but only serve as normalization factors (units) indicative of their overall scaling with size, field and plasma parameters; (ii) by introducing the tokamak magnetization parameter, $\rho_i^* = \rho_{ti}/a$, the radial gradient length will be effectively assumed as $L_\perp \sim a$, a significant over estimate in the edge of the plasma where more typically, $L_\perp/a \sim 0.1$. Nonetheless, to illustrate the order of magnitude of the various expressions, let us consider the ratio of selected turbulent and laminar diffusivities tabulated above, i.e. of $\chi_\perp^B$ and $\chi_{\perp i}^{cl}$,

$$\frac{\chi_{\perp i}^B}{\chi_{\perp i}^{cl}} \sim \frac{\rho_S^2 \Omega_i}{2\rho_{ti}^2 \nu_i} \sim \frac{\Omega_i}{2\nu_i} \sim \frac{\pi}{2} \frac{q/\varepsilon}{\rho_i^* \nu_i^*} \sim \frac{15}{\rho_i^* \nu_i^*}, \qquad \frac{\chi_{\perp i}^{gB}}{\chi_{\perp i}^{cl}} \sim \frac{15}{\nu_i^*}, \tag{6.221}$$

where $\rho_i^* = \rho_{ti}/a$ and $\nu_i^* = L_\parallel/\lambda_i \approx \pi q R(\nu_i/v_{ti})$ are the usual measures of magnetization and collisionality, and we estimated $q/\varepsilon \sim 3/0.3 \sim 10$ for conventional tokamaks. Evaluating this expression for typical JET parameters, see Table 8.1, under which $\rho_i^* \sim 10^{-3}$ and ν_i^* ranges from $\sim 10^{-3}$ in the core, through ~ 0.1 in the edge, to ~ 10 in the SOL, we find $\chi_{\perp i}^B/\chi_{\perp i}^{cl} \sim (10^7, 10^5, 10^3)$ and $\chi_{\perp i}^{gB}/\chi_{\perp i}^{cl} \sim (10^4, 10^2, 1)$ in the three regions.[104]

Similar ratios involving P–S, plateau and banana diffusivities are easily obtained from (6.221) by noting that

$$\chi_{\perp i}^{cl} : \chi_{\perp i}^{PS} : \chi_{\perp i}^{\circ} : \chi_{\perp i}^{\triangleright} \approx 1 : q^2 : q^2 : q^2 \varepsilon^{-3/2} \sim 1 : 10 : 10 : 50. \tag{6.222}$$

[104] It should be stressed, however, that the neoclassical expressions are not valid in the SOL.

Hence, in the edge of the plasma, neoclassical ion heat diffusivity becomes comparable to the gyro-Bohm diffusivity, and is likely to influence the ion power exhaust. The same is not true for electron power or particle exhaust, which involve the much smaller collisional diffusivities, $D_{\perp e}, \chi_{\perp e} \ll \chi_{\perp i}$.

Finally, there are three regions of the tokamak where the magnetized plasma criterion, $\delta_i \ll 1$, is only weakly satisfied: (i) the *internal transport barrier* (ITB) associated with the resonant flux surfaces in a reversed magnetic shear region; (ii) the *edge transport barrier* (ETB) associated with the high confinement regime; and (iii) the boundary plasma or the *scrape-off layer* (SOL), see Chapter 7. In all three regions, the radial gradient length is equal to several poloidal gyro-radii, so that $\delta_i = \rho_{ti}/L_\perp \sim \rho_{ti}/5\rho_P \sim \epsilon/5q \sim 0.02$; indeed, in the near-SOL region, δ_i can frequently exceed 0.1, introducing non-negligible corrections to many of the results, including the collision and turbulent flows, derived assuming a strongly magnetized plasma.

6.5 Further reading

There are many excellent texts dealing with turbulence in neutral fluids. The coverage of turbulence in plasmas is less extensive, especially for the case of most interest to power exhaust, namely strong turbulence in edge plasmas. A selection of the most relevant works in given below. *HD turbulence*: Batchelor (1953); Monin and Yaglom (1971); Tennekes and Lumley (1972); McComb (1992); Frisch (1995); Pope (2000); Davidson (2004). *MHD turbulence*: Batemann (1978); Biskamp (1993, 2003). *DHD turbulence*: Kadomtsev (1965); Sagdeev and Galeev (1969); Hasegawa (1975); Garbet and Waltz (1998); Mikhailovskii (1998); Garbet *et al.* (1999); Itoh *et al.* (2000); Weiland (2000); Scott (2001); Yoshizawa *et al.* (2003); Garbet (2006); Balescu (2008). *Computational fluid dynamics*: Anderson *et al.* (1984); Hirsch (1990a,b). *Computational plasma dynamics*: Tajima (1989); Birdsall and Langdon (1991).

7

Tokamak plasma boundary and power exhaust

'By doubting we come to questioning, and by questioning we perceive the truth.'
P. Abelard (c. 1120)

Having examined the equilibrium, stability and transport properties of magnetized plasmas, we are finally ready to tackle our ultimate aim, i.e. the exhaust of particles and power from the plasma itself and the limits imposed by the resulting plasma wall loads and thermo-mechanical material constraints on the reactor performance. Geographically, this question naturally leads us towards the plasma boundary, commonly known as the *scrape-off layer* (SOL), defined as the region of open field lines between the *last closed flux surface* (LCFS) and the vessel wall, which is the only place in a fusion reactor where the stellar world of hot (keV) plasmas meets the earthly world of cold (sub-eV) solids. As the name indicates the LCFS represents the transition between the regions of closed and open magnetic field lines, i.e. between the edge and SOL regions. This surface can be created either by inserting a solid object, known as the *limiter*, into the plasma, or by shaping the poloidal magnetic field with external current carrying coils to create a poloidal field null, or X-point, and a magnetic *separatrix*, and thus to divert the SOL plasma into a specifically designed structure, known as the *divertor*, see Section 7.1.3. Since the location and degree of plasma contact with the wall is determined by transport processes in the SOL, the understanding of these processes is critical for translating a plasma *load* limit (per component surface area) into a plasma *exhaust* limit (per plasma surface area at the LCFS), recall (1.16).

In this penultimate chapter, we address the power exhaust problem in three parts: (i) by introducing the physical process specific to, and the overall system dynamics of, the boundary plasma, Section 7.1; (ii) by investigating power exhaust in the *low confinement* (L-mode) regime, which is dominated by the turbulent transport in the edge-SOL region, Section 7.2; and (iii) by examining power exhaust in the *high confinement* (H-mode) regime, which is characterized by the reduction of turbulent transport, i.e. formation of the edge transport barrier, and its quasi-periodic destruction by *edge localized modes* (ELMs), which in turn lead to intense, transient plasma loads on the vessel wall, Section 7.3.

7.1 The scrape-off layer (SOL)

In the previous chapters, we have tacitly assumed the existence of a fully ionized plasma, well separated from material boundaries. In contrast, the SOL region involves both plasma–surface and plasma–neutral interactions which play a vital role in its overall dynamics.[1] Therefore, before discussing the latter, Section 7.1.4, let us first consider the interaction of ions and electrons with solids, Section 7.1.1, and atoms/molecules, Section 7.1.2. A book length treatment of all three topics may be found in Stangeby (2000).

7.1.1 Plasma–surface interactions

Unlike the core plasma, in which atomic and molecular physics plays only a minor role, SOL plasma behaviour is frequently dominated by the interaction of ions and electrons with neutrals and solid surfaces. A solid surface has a very different effect on a plasma than on a neutral gas; for the latter, it is largely a reflecting surface – although atoms and molecules can transiently stick to the wall and penetrate/diffuse into the bulk of the solid – for the former, it is always an absorbing surface – plasma ions and electrons recombine on the surface, and re-enter the plasma volume as neutrals. In both cases, particles moving towards it have an average velocity comparable to their respective sound speeds. Since there are no ions moving away from the absorbing surface, $f_i(v_\parallel > 0) = 0$, the plasma flow velocity in front of the surface is equal to the average ion velocity moving towards this surface. Plasma transport equations based on a weakly shifted Maxwellian $f_i(v_\parallel)$ are thus invalid in the near surface region; their inaccuracy may be estimated by a comparison with a fully kinetic solution.

The problem of plasma–solid contact is one of the oldest in plasma physics and remains one of the most elusive. Let us take $l = 0$ as the location of the absorbing surface, which we assume to be perpendicular to an otherwise uniform and constant magnetic field;[2] the parallel domain is closed by requiring vanishing gradients at infinity, $L_\parallel \gg \lambda_D$, $\nabla_\parallel(l = L_\parallel) \to 0$. The problem is described

[1] As a result, SOL phenomena exhibit a complexity not found anywhere else in the tokamak.
[2] Here l measures the distance along the field away from the solid surface.

by a kinetic equation (2.99) for ions, electrons and neutrals integrated over $\mathrm{d}\mathbf{v}_\perp = 2\pi v_\perp \mathrm{d}v_\perp$,[3]

$$[\partial_t + v_{\parallel i}\nabla_\parallel - (Ze/m_i)(\nabla_\parallel\varphi)\partial_{v_{\parallel i}}]f_i(l, v_{\parallel i}) = C_i(f_i) + I_i, \tag{7.1}$$

$$[\partial_t + v_{\parallel e}\nabla_\parallel + (e/m_e)(\nabla_\parallel\varphi)\partial_{v_{\parallel e}}]f_e(l, v_{\parallel e}) = C_e(f_e) + I_e, \tag{7.2}$$

$$[\partial_t + v_{\parallel n}\nabla_\parallel]f_n(l, v_{\parallel n}) = C_n(f_n) + I_n, \tag{7.3}$$

and by Poisson's equation for the electrostatic potential φ,

$$\nabla_\parallel^2\varphi = \frac{e}{\epsilon_0}(n_e - Zn_i) = \frac{e}{\epsilon_0}\left(\int f_e \mathrm{d}v_{\parallel e} - Z\int f_i \mathrm{d}v_{\parallel i}\right). \tag{7.4}$$

We next assume an ambipolar flow of charge (zero current) at the solid surface $l = 0$ and no reflection of charged particles, and, hence, no backward moving ions and electrons.[4] These boundary conditions can be expressed as

$$Z\int f_i v_{\parallel i}\mathrm{d}v_{\parallel i} = \int f_e v_{\parallel e}\mathrm{d}v_{\parallel e} = -\int f_n v_{\parallel n}\mathrm{d}v_{\parallel n}, \tag{7.5}$$

$$f_i(l = 0, v_{\parallel i} > 0) = 0 = f_e(l = 0, v_{\parallel e} > 0). \tag{7.6}$$

The neutrals have been included explicitly in the problem, since they enter both the source and collision terms in the ion and electron equations; we will discuss plasma–neutral interactions in Section 7.1.2.

Since $v_{te} \gg v_{ti}$, an initially neutral surface would quickly become negatively charged, $\varphi(l = 0) \equiv \varphi_0 < 0$. The negative floating potential opposes the electron outflow, until a steady-state is reached in which the electron and ion currents to the surface are equal, $Z\langle v_{\parallel i}\rangle = \langle v_{\parallel e}\rangle$. The profile and magnitude of this potential may be estimated by considering the Poisson equation in the near surface region. The smallness of the electron mass allows us to neglect the first two terms in (7.2) and the solution $f_e(v_{\parallel e})$ is easily obtained as a shifted Maxwellian reduced by the Boltzmann factor,

$$f_e(l, v_{\parallel e})/f_e^M(L_\parallel, v_{\parallel e}) = n_e(l)/n_e(L_\parallel) = \exp[e\varphi(l)/T_e]. \tag{7.7}$$

The ion distribution becomes entirely forward shifted at some location $l_\prec$, $f_i(0 < l < l_\prec, v_{\parallel i} > 0) = 0$, with the region $0 < l < l_\prec$ known as the electrostatic sheath, Fig. 7.1. Since we expect $l_\prec$ to be of the order of the Debye length, λ_D, (2.5), which is typically much shorter than any collisional mean free path, collisions and sources

[3] For simplicity, we assume a Maxwellian distribution in $v_\perp$, normalized to unity, and symmetry in the drift plane so that $\nabla_\perp = 0$, $\langle v_\perp\rangle = 0$ and $\langle \mathbf{v} \times \mathbf{B}\rangle = 0$.

[4] In reality, a small fraction of ions and electrons are reflected from the surface, although the great majority recombine into atoms and/or molecules and return into the plasma as neutrals. The solid surface thus acts as a *local sink* for the plasma density, while the returning neutrals provide a corresponding *volumetric source* of equal magnitude. In steady-state, the approximate balance between plasma outflow and neutral inflow is known as *plasma recycling*.

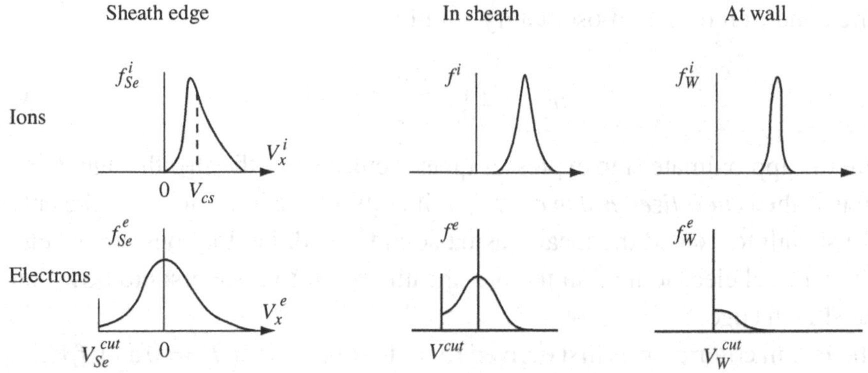

Fig. 7.1. Schematic evolution of ion and electron velocity distributions, $f_i(v_\|)$ and $f_e(v_\|)$, across the electrostatic sheath. *Source*: Stangeby (2000).

may be neglected in the sheath itself, $C_s = I_s = 0$. There are various methods of analyzing the resulting kinetic equations, the simplest of which being the method of characteristics (Riemann and Meyer, 1996). One begins by noting that any ion trajectory satisfies the collisionless kinetic equation and that conservation of energy gives one such trajectory, $\frac{1}{2}m_i v_{\|i}(l)^2 + Ze\varphi(l) = \frac{1}{2}m_i v_{\|i\prec}^2 + Ze\varphi_\prec$ where $\mathcal{A}_\prec \equiv \mathcal{A}(l_\prec)$; based on this result one may construct a solution to $f_i(l, v_{\|i})$ from $v_{\|i}(l)$. In the absence of sources, particle continuity requires that $f_1(v_1)v_1 = f_2(v_2)v_2$, where $v_1 = v_{\|i}(l_1)$ and $v_2 = v_{\|i}(l_2)$ are related by the trajectory $v_{\|i}(l)$,

$$\frac{f_i(l, v_{\|i})}{f_{i\prec}(v_{\|i})} = \frac{v_{\|i\prec}}{v_{\|i}(l)} = \left[1 + \frac{Ze\varphi_{SOL}}{\frac{1}{2}m_i v_{\|i}^2}\right]^{1/2}, \tag{7.8}$$

and $\varphi_{SOL}(l) = \varphi(l) - \varphi_\prec > 0$. Inserting $n_e(l)$ and $n_i(l)$ into Poisson's equation and expanding to first order in φ_{SOL}, one finds

$$\begin{aligned}
\nabla_\|^2 \varphi_{SOL} &= \frac{e}{\epsilon_0}\left[n_{e\prec}\left(1 + \frac{e\varphi_{SOL}}{T_e}\right) - Z\int f_{i\prec}(v_{\|i})\left(1 + \frac{Ze\varphi_{SOL}}{m_i v_{\|i}^2}\right) dv_{\|i}\right] \\
&= \frac{e^2}{\epsilon_0}\varphi_{SOL}\left[\frac{n_{e\prec}}{T_e} - \frac{Z^2 n_{i\prec}}{m_i}\int \frac{f_{i\prec}(v_{\|i})}{v_{\|i}^2} dv_{\|i}\right] \\
&= \frac{\varphi_{SOL}}{\lambda_D^2}\left[1 - \frac{Z n_{i\prec}}{n_{e\prec}} c_{Se}^2 \langle v_{\|i}^{-2}\rangle\right],
\end{aligned} \tag{7.9}$$

where $c_{Se}^2 = ZT_e/m_i$ is the cold ion plasma sound speed.[5]

[5] We thus find that $l_\prec \sim O(\lambda_D) \ll \lambda_{ei}$, confirming our collisionless ansatz.

The condition for a non-oscillatory solution,

$$\left\langle v_{\parallel i}^{-2}\right\rangle_{\prec}^{-1} \geq c_{Se}^2 \frac{Zn_{i\prec}}{n_{e\prec}} \approx c_{Se}, \tag{7.10}$$

where the approximate sign expresses quasi-neutrality at the sheath edge, $l = l_{\prec}$, is known as the *generalized Bohm criterion*. It requires the ion velocity at the entrance of the sheath to exceed the local plasma sound speed, i.e. the ions are accelerated by the parallel electric field in the pre-sheath region to a supersonic flow velocity at the sheath edge.

The Bohm criterion was first derived for cold ions, i.e. for $T_i = 0$ and $f_i(l, v_{\parallel i}) = n_i(l)\delta[v_{\parallel i} - V_{\parallel i}(l)]$, for which it becomes $V_{\parallel i\prec} \geq c_{Se}$. For cool ions, described by a shifted Maxwellian, $f_i(l, v_{\parallel i}) = n_i(l)\sqrt{\beta/\pi}\exp[-\beta(v_{\parallel} - V_{\parallel i}(l))^2]$, where $\beta(l) = m_i/2T_{\parallel i}(l) = 1/v_{\parallel Ti}^2$, such that $f_i(v_{\parallel} > 0) = 0$, and $f_i(v_{\parallel} < 0)$ is a broadened delta function of width $v_{\parallel ti}$, the Bohm criterion becomes $V_{\parallel i\prec}^2 - v_{\parallel ti}^2 > c_{Se}^2$; this may be written as $V_{\parallel i\prec} > [(ZT_e + T_{\parallel i})/m_i]^{1/2} \equiv c_{S\parallel}$, where $c_{S\parallel}$ is the parallel plasma sound speed. Hence, the flow velocity at the entrance into the sheath is increased by roughly the thermal spread of the parallel ion velocities. Defining a local parallel Mach number as $M_{\parallel} = V_{\parallel i}/c_S$, the Bohm criterion is often stated (with some inaccuracy, since $T_{i\parallel}$ need not equal T_i) as

$$M_{\parallel\prec} = (V_{\parallel i}/c_S)_{\prec} \geq 1, \qquad c_S^2 = (ZT_e + \gamma T_i)/m_i, \tag{7.11}$$

where γ is the polytropic exponent of the parallel ion flow.

The parallel current into the surface is just the difference between the ion and electron currents, $J_{\parallel\prec} = Ze(n_i V_{\parallel i})_{\prec} - e(n_e V_{\parallel e})_{\prec}$. Assuming sonic ion flow at the sheath edge, $M_{\parallel\prec} = 1$, and Maxwellian electrons, (7.7), we find

$$J_{\parallel\prec} = Zen_{i\prec}c_S - \tfrac{1}{4}en_{e\prec}\hat{c}_e\exp(-e\varphi_{sh}/T_e), \tag{7.12}$$

where $\hat{c}_e = (8T_e/\pi m_e)^{1/2} \approx 1.6v_{te}$ and $\varphi_{sh} = \varphi_{\prec} - \varphi_0 > 0$ is the potential drop across the sheath. Setting $J_{\parallel\prec} = 0$ in (7.12) and solving for φ_{sh} yields the sheath potential drop at a *floating* surface, $\varphi_{sh,fl} \equiv \varphi_{sh}(J_{\parallel\prec} = 0)$, and hence the corresponding floating surface potential, $\varphi_{0,fl} \equiv \varphi_0(J_{\parallel\prec} = 0)$, which can be expressed purely in terms of the mass and temperature ratios,

$$e\varphi_{sh,fl}/T_e \approx \tfrac{1}{2}\ln\left[2\pi(1 + T_i/ZT_e)(m_e/m_i)\right]. \tag{7.13}$$

In contrast, the potential drop across the SOL depends on the flow in the pre-sheath region ($l > l_{\prec}$), which is determined by the parallel Ohm's law. Typical values for a hydrogenic plasma with $T_i \sim T_e$ are $\varphi_{SOL}/T_e \sim 1/2$ and $\varphi_{sh,fl}/T_e \sim 2$–3, so that the potential drop in the upstream (pre-sheath) region is generally much smaller than the drop across the sheath itself.

When φ_{sh} deviates from $\varphi_{sh,fl}$, it leads to a net parallel current,

$$J_{\parallel\prec} \approx en_{e\prec}c_S[1 - \exp(-e\widehat{\varphi}_{sh}/T_e)], \tag{7.14}$$

where $\widehat{\varphi}_{sh} = \varphi_{sh} - \varphi_{sh,fl}$ is the departure of the sheath potential drop away from its floating value and we assumed $Zn_i \approx n_e$ at the sheath edge; note that $J_{\parallel\prec}(\widehat{\varphi}_{sh} = 0) = 0$ as required. When these departures are small, the exponential can be linearized around zero to yield a linear relation,

$$J_{\parallel\prec} \approx en_{e\prec}c_S(e\widehat{\varphi}_{sh}/T_e). \tag{7.15}$$

Since the deviation of the electric field at the sheath edge from its floating value can be approximated as $\widehat{E}_{\parallel sh} = -\nabla_\parallel\widehat{\varphi}_{sh} \approx \widetilde{\varphi}_{sh}/\lambda_{\parallel sh}$, where $\lambda_{\parallel sh} \sim l_\prec \sim 3\lambda_D$ is the parallel extent of the sheath,[6] we find that, for small $e\widehat{\varphi}_{sh}/T_e$, the sheath offers a simple resistance to the plasma current,

$$J_{\parallel\prec} \approx \widehat{E}_{\parallel sh}/\eta_{\parallel sh}, \qquad \eta_{\parallel sh} = \lambda_{\parallel sh}T_e/e^2 n_{e\prec}c_S. \tag{7.16}$$

Note that the sheath resistivity, $\eta_{\parallel sh}$, is a function of plasma quantities at the sheath edge, so that the associated Joule heating, also known as *sheath dissipation*, $\eta_{\parallel sh}J_{\parallel\prec}^2$, is also a purely local phenomenon.

We would next like to estimate the ion and electron power flowing into the sheath and the total power deposited onto the solid surface, which may include chemical energies. The electron energy flow at $l = 0$ is easily calculated with a cut-off Maxwellian distribution, $f_e(v_\parallel > 0) = 0$, while the energy flow at the sheath edge $l = l_\prec$ must be larger by the value of the sheath potential drop (which is repulsive for the electrons),

$$Q_{\parallel e0} = \frac{1}{2}m_e\int v^2 v_\parallel f_e(\mathbf{v})d\mathbf{v} = 2T_e\int v_\parallel f_e(\mathbf{v})dv_\parallel = 2T_e\Gamma_{\parallel e}, \tag{7.17}$$

and $Q_{\parallel e\prec} = Q_{\parallel e0} + e\varphi_{sh}\Gamma_{\parallel e} = (2T_{e\prec} + e\varphi_{sh})\Gamma_{\parallel e}$. Assuming a shifted Maxwellian distribution for $f_i(v_\parallel)$ with $M_\parallel(l_\prec) = 1$, one obtains an identical result, except the effect of the potential drop on the energy flow is reversed, since the sheath transfers part of the electron energy to the ions,

$$Q_{\parallel i0} = Q_{\parallel i\prec} + Ze\varphi_{sh}\Gamma_{\parallel i} = (2T_{i\prec} + Ze\varphi_{sh})\Gamma_{\parallel i}. \tag{7.18}$$

As a result, the ions strike the solid surface with an average energy of $\sim 2T_{i\prec} + 3ZT_{e\prec}$, or $\sim 5T_{e\prec}$ for hydrogenic plasmas. The total energy deposited on the surface per ion–electron pair is thus $\sim 2T_{i\prec} + (3Z+2)T_{e\prec} + E_{iz}$, where E_{iz} is the ionization potential of a neutral atom, and hence the energy released in recombination of the corresponding ion–electron pair.[7]

[6] For an oblique magnetic field, $l_\prec$ should be replaced by $l_\subset \sim 3\rho_S/\mathbf{b}\cdot\mathbf{n}$, which is the parallel extent of the magnetic pre-sheath, see below.

[7] For example, $E_{iz} \sim 13.6\,\text{eV}$ for a hydrogen atom.

In reality the presence of the parallel electric field in the pre-sheath, which leads to the supersonic ion flow, distorts $f_i(l_{\prec}, v_\parallel)$ away from the Maxwellian. To estimate the resulting correction, (7.1)–(7.3) must be solved on the entire domain, $0 < l < L_\parallel$, simultaneously. Although a general solution can only be obtained numerically, a variety of simplified cases can be solved analytically. Somewhat surprisingly, the resulting normalized potential drop, energy fluxes and the ion distribution function at the sheath edge are rather insensitive to a wide range of assumptions (Stangeby, 2000), so that (7.17)–(7.18) offer a fair approximation to the exact result.

So far, we assumed the magnetic field to be perpendicular to the solid surface, i.e. $\mathbf{b}\cdot\mathbf{n} = 1$, where $\mathbf{n}$ is a unit vector normal to the surface, so that the $\mathbf{v}\times\mathbf{B}$ forces could be neglected in the kinetic equations (7.1)–(7.2). In practice, we are more interested in the case of an oblique magnetic field, which allows the plasma wetted area to be maximized, thus reducing the plasma loads on the solid surface. Due to its practical importance, this problem has been studied in detail (Chodura, 1982). The trajectories of the ions and electrons in the near surface region were followed numerically, with φ determined self-consistently from the Poisson equation using the particle-in-cell method, and the magnetic field taken as constant at an angle ϑ to the surface normal. The resulting flow is characterized by two regions: (i) within a few Debye lengths of the surface $0 < l\mathbf{b}\cdot\mathbf{n} < l_{\prec}\mathbf{b}\cdot\mathbf{n} \sim 3\lambda_{\mathrm{D}}$, there appears the usual electrostatic sheath, with strong potential gradients and plasma flow normal to the solid surface; (ii) further out, up to a few ion gyro-radii at the local sound speed, $l_{\prec}\mathbf{b}\cdot\mathbf{n} < l\mathbf{b}\cdot\mathbf{n} < l_{\subset}\mathbf{b}\cdot\mathbf{n} \sim 3\rho_S$, where $\rho_S = c_S/\Omega_i$, a quasi-neutral magnetic pre-sheath region appears in which the net flow becomes aligned with the magnetic field, Fig. 7.2. The Bohm criterion with an oblique magnetic field, also known as the *Chodura criterion*, is identical to the normal field result, expect it applies at the entrance into the magnetic pre-sheath, $l \sim l_{\subset}$, rather than at the electrostatic sheath $l \sim l_{\prec}$,

$$\left\langle v_{\parallel i}^{-2}\right\rangle_{\subset} = \int f_{i\subset}(v_\parallel)v_\parallel^{-2}\mathrm{d}v_\parallel \le c_{Se}^{-2}. \tag{7.19}$$

To obtain the boundary conditions for the plasma fluid equations, we again consider the various length scales in the near surface region: the Debye length, $\lambda_D \sim l_{\prec}\mathbf{b}\cdot\mathbf{n}/3$, the ion gyro-radius, $\rho_S \sim l_{\subset}\mathbf{b}\cdot\mathbf{n}/3$, and the normal projection of the collisional mean free path, $\lambda_{ei}\mathbf{b}\cdot\mathbf{n}$ and the ionization mean free path, λ_{iz}.[8] Simple kinetic solutions, neglecting both source and collision terms, are clearly not valid for $l\mathbf{b}\cdot\mathbf{n} \gg \min(\lambda_{ei}\mathbf{b}\cdot\mathbf{n}, \lambda_{iz})$. On the other hand, the plasma fluid equations, which are a more efficient modelling tool, are fairly accurate in the pre-sheath region, $l \gg \max(l_{\prec}, l_{\subset})$. They must, however, be constrained by the kinetic

[8] A comparison of the kinetic and moment sheath expressions may be found in Stangeby (2000).

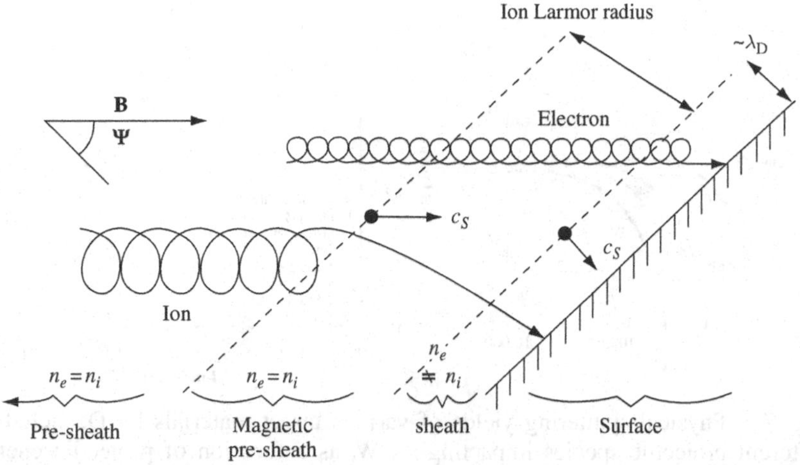

Fig. 7.2. Schematic representation of the pre-sheath, the magnetic (Chodura) pre-sheath and the electrostatic (Debye) sheath. *Source*: Stangeby (2000).

solution at some location in the region where both solutions are marginally valid, i.e. $l_{\prec} < l < l_{\mathrm{C}}$. The sheath edge, taken roughly as the location where the Bohm–Chodura criterion is satisfied, is a convenient location to effect this matching of solutions; it is both close enough to the surface that any sources or collisions within the sheath may be neglected, and far enough for quasi-neutrality and moment equations to be valid. In reality, some $f_i(v)$ distribution exists at this location which changes continuously into the forward shifted sheath $f_i(v)$ for $l < l_{\prec}$, and into gradually less distorted Maxwellian, $f_i(v)$, for $l > l_{\prec}$; this distribution is unknown to us. Instead, we have an approximation to $f_i(v)$ in the sheath region and again an approximation to its moments in the pre-sheath region. If we denote the latter by $f_s^{>}(v)$ and the former by $f_s^{<}(v)$, we need only ensure that the two agree in all their lowest moments, which can be expressed as

$$\langle \mathcal{A} \rangle_{f_s^>} = \langle \mathcal{A} \rangle_{f_s^<}, \qquad \langle \mathcal{A} \rangle_f = \int \mathcal{A} f(\mathbf{v}) \mathrm{d}\mathbf{v}, \qquad \mathcal{A} = \left\{1, v_\parallel, v_\parallel^2, v_\parallel v^2\right\}, \quad (7.20)$$

Since the pre-sheath moments express the density and parallel flows of mass, momentum and energy in the plasma fluid equations, while the sheath moments require the knowledge of $f_i^{<}(v)$ (to be obtained from a kinetic solution), the above matching conditions impose kinetic requirements onto the plasma fluid quantities. Thus, the pre-sheath momentum flow may constrained by imposing the Bohm–Chodura criterion on the plasma velocity, which requires a supersonic flow at the entrance into the sheath. Similarly, the pre-sheath energy fluxes are constrained by

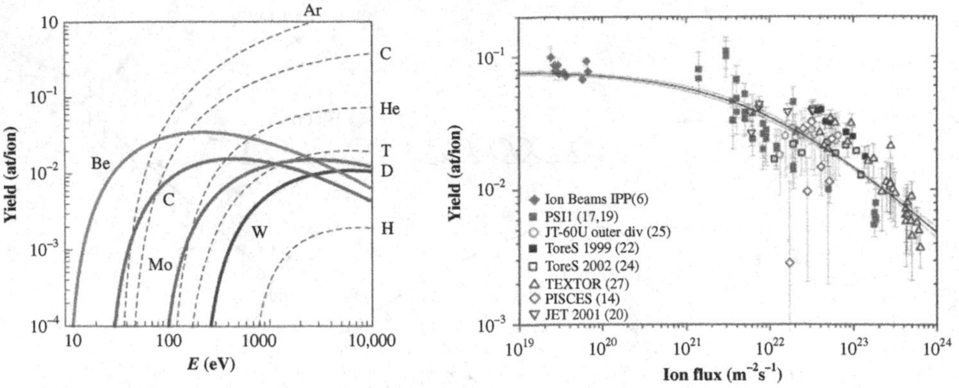

Fig. 7.3. Physical sputtering yields of various target materials by D, including different projectile species impacting on W, as a function of projectile energy (left), and chemical sputtering yields of C by D as a function of D flux density (right). *Sources*: (left) based on Eckstein (1997); (right) Roth *et al.* (2004).

introducing *energy sheath transmission coefficients* for ions γ_i and electrons γ_e, defined as

$$Q_{\parallel s} = \frac{1}{2} m_s \left\langle v_\parallel v^2 \right\rangle_{f_s} = \gamma_s p_s V_{\parallel s} = \gamma_s T_s \Gamma_{\parallel s}, \tag{7.21}$$

with all quantities evaluated at the sheath edge. These factors must be derived from kinetic solutions and are typically found as $\gamma_e \sim 5$ and $\gamma_i \sim 2-3$, in fair agreement to our simple analysis in (7.17)–(7.18).

There are several important consequences of plasma–surface interactions for a fusion reactor: (i) *physical sputtering* of the solid surfaces by the plasma, whose yield increases with projectile energy and mass, while decreasing with the target material atomic mass, Fig. 7.3; (ii) *chemical sputtering*, or erosion, of the solid surfaces by the chemically reactive plasma species, whose yield for carbon increases with D ion flux and is sensitive to C target temperature, Fig. 7.3; (iii) the resulting material erosion and migration, which leads to tritium retention via co-deposition with solid material, e.g. Be or C; this has important operational consequences as the in-vessel safety limit on mobilizable tritium is set at ~1 kg on ITER; and finally, (iv) the thermo-mechanical properties of the PFCs impose steady-state heat load limits of ~10 MW/m^2 and transient energy load limits of ~0.5 MJ/m^2 in 250 μs, recall Section 1.1. This value was chosen based on cyclical plasma gun tests of actual PFC components for ITER.[9]

[9] Interestingly, surface failure was observed at similar values of energy heat loads for both W and CFC clad components (micro-cracking in W and pan-fibre erosion for CFC), although the mode of failure was grossly different (Linke *et al.*, 2007); this is possibly linked to the similar binding energy of surface atoms, ~8 eV, and similar heat diffusivities, in both armour materials.

7.1.2 Plasma–neutral interactions

Due to the proximity of solid surfaces, the edge plasma always contains a small concentration of neutral species with which it interacts. These interactions provide sources (and sinks) of mass, momentum and energy for the plasma fluid. Below, we briefly examine the relevant atomic/molecular physics, determine the dominant processes by examining their cross-sections, and construct interaction source terms for the plasma fluid equations.

Plasma–neutral interactions can be classified according to the reacting species (neutrals, ions, electrons, photons) and the reaction type (elastic scattering, charge exchange, excitation/emission, dissociation/molecular formation, ionization/recombination, etc.). More generally, we can define $\mathbf{Y}(n)$ to be a set of all possible neutral, n-atom molecules, $X_{Z(1)}X_{Z(2)}\cdots X_{Z(n)}$ where X_Z is an atom of nuclear charge, Z, $\mathbf{Y}_0$ and $\mathbf{Y}_+$ as the sets of all neutral and ionized molecules, respectively, and $\mathcal{Y}$ is the set of all reactants,

$$\mathbf{Y}(n) = \{X_{Z(1)}X_{Z(2)}\cdots X_{Z(n)}|Z(i) \in (1,\ldots,N);\, i \in (1,\ldots,N)\}$$
$$\mathbf{Y}_0 = \bigcup_{n=1}^{n_{max}} \mathbf{Y}(n), \quad \mathbf{Y}_+ = \bigcup_{j=0}^{Nn_{max}} \mathbf{Y}_1^{j+}, \quad \mathcal{Y} = \{h\nu, e, \mathbf{Y}_0, \mathbf{Y}_+\}$$

Using this formulation, we can write down all possible k-body *encounters* by taking $k-1$ successive set products of $\mathcal{Y}$ with itself, $\mathcal{R} = \mathcal{Y}\otimes\mathcal{Y}\otimes\cdots\otimes\mathcal{Y}$, where each encounter can lead to several distinct *interactions* with a probability dependent on the relative energy of the reactants.

In the context of fusion research, we are interested primarily in hydrogenic ($Z = 1$) plasmas containing small quantities (a few %) of low-Z impurities and even smaller quantities ($\ll 1\%$) of medium- to high-Z impurities.[10] We can thus divide our discussion into hydrogen and impurity reactions. A comprehensive introduction to both topics may be found in Janev *et al.* (1988) and Janev (1995a).

7.1.2.1 Plasma–hydrogen interactions

The dominant atomic and molecular hydrogen reactions are summarized in Table 7.1. The corresponding cross-sections, $\sigma(v)$, a selection of which is shown in Fig. 7.4, have been measured experimentally over a wide range of energies and are generally in good agreement with quantum mechanical scattering predictions. In practice, semi-empirical formulas are used to fit the experimental data and thus to extrapolate the cross-sections beyond the experimentally available energy range, e.g. to sub-eV energies.

[10] Impurities are said to be either *intrinsic*, i.e. originating due to plasma–surface contact at the vessel wall or *extrinsic*, i.e. injected with the aim of increasing radiation from the plasma. The former commonly include Be, C, O and W, and the latter N, Ne and Ar.

Table 7.1. *Selected atomic/molecular hydrogen reactions (Janev, 1995b).*

Atomic: $\{h\nu, e, H^0, H^+\}$	Molecular: $\{h\nu, e, H_2, H_2^+\}$	Interaction
$H^0 + H^0 \rightarrow H^0 + H^0$	$H_2 + H_2 \rightarrow H_2 + H_2$	elastic scattering (ES)
$H^0 + e \rightarrow H^* + e$	$H_2 + e \rightarrow H_2^* + e$	excitation
	$H_2^+ + e \rightarrow H_2^{+*} + e$	molec. ion excitation
$H^* \rightarrow H^0 + h\nu$	$H_2^* \rightarrow H_2 + h\nu$	radiative de-excitation
	$H_2^* \rightarrow H^0 + H^0 + h\nu$	molec. radiative dissoc.
	$H_2^{+*} \rightarrow H^+ + H^0 + h\nu$	molec. ion radiative dissoc.
$H^0 + e \rightarrow H^+ + 2e$	$H_2 + e \rightarrow H_2^+ + 2e$	ionization (IZ)
	$H_2 + e \rightarrow H^+ + H^0 + 2e$	molec. dissoc. ioniz.
	$H_2 + e \rightarrow H^+ + H^+ + 3e$	molec. dissoc. double ioniz.
	$H_2 + e \rightarrow H^0 + H^-$	molec. dissoc. attach.
	$H_2 + e \rightarrow H^0 + H^0 + e$	molec. dissociation
	$H_2^+ + e \rightarrow H^+ + H^0 + e$	molec. ion dissoc.
	$H_2^+ + e \rightarrow H^+ + H^0 + 2e$	molec. ion dissoc. ioniz.
$H^0 + H^0 \rightarrow H_2 + h\nu$		molecular formation
$H^+ + e \rightarrow H^* + h\nu$	$H_2^+ + e \rightarrow H^0 + H^0$	2-body recombination
$H^+ + 2e \rightarrow H^* + e$		3-body recombination
$H^0 + H^+ \rightarrow H^+ + H^0$	$H_2 + H_2^+ \rightarrow H_2^+ + H_2$	charge exchange (CX)

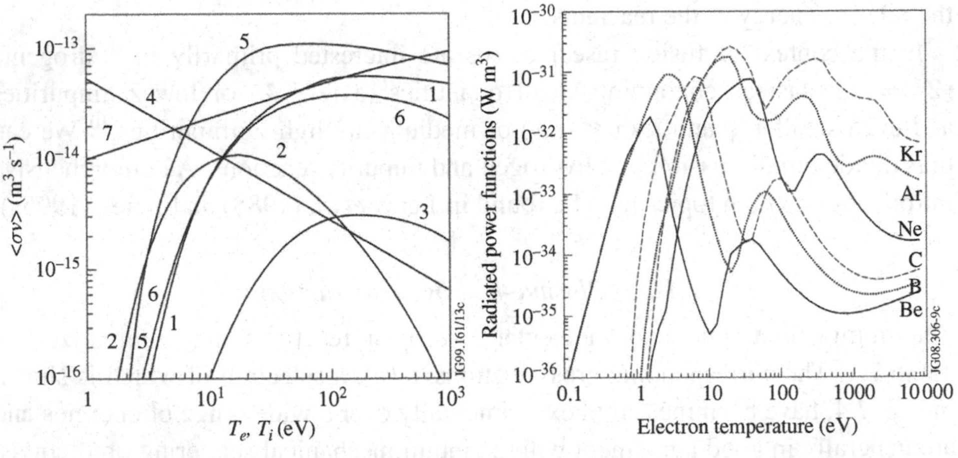

Fig. 7.4. The left frame shows selected atomic and molecular hydrogen thermal reaction rate coefficients, $\langle\sigma v\rangle$, for (1) molec. ioniz., (2) molec. dissoc., (3) molec. dissoc. ioniz., (4) molec. form., (5) molec. ion dissoc., (6) atomic ioniz. and (7) charge exchange; see Table 7.1. The right frame shows radiative loss functions, $f_Z(T_e) = L_Z(T_e)n_e = \sum P_{rad}\left(X_Z^{j+}\right)/n_Z$, of selected impurities (Be, B, C, Ne, Ar, Kr) assuming coronal equilibrium. *Source*: Stangeby (2000) and M. O'Mullane.

Let us examine the effect of these reactions on the plasma fluid. In a uniform plasma, the ion–neutral balance is governed by competition between ionization, dissociation and recombination processes. In its simplest form, this leads to the well-known *Saha equation* for the degree of ionization of a gas/plasma mixture, which can be obtained by estimating the ionization and recombination cross-sections for hydrogen atoms (Saha, 1921),

$$n_i n_e / n_n = 2(g_1/g_0)(m_e T_e / 2\pi)^{3/2} \hbar^{-3} \exp(-E_{iz}/T_e), \tag{7.22}$$

In the above n_e, n_i, n_n are the ion, electron and neutral particle densities, E_{iz} is the ionization potential, and g_0, g_1 are statistical factors. For $T_e \gg E_{iz}$ a hydrogenic plasma becomes fully ionized. The ion–neutral balance in a non-uniform plasma includes the transport of neutral species, which is best calculated using Monte-Carlo simulation of neutral trajectories.[11]

The energy spectrum of the hydrogen atom in a field-free environment is described by the quantum numbers (n, l, s), which determine the orbital electron energies and photon energies for allowed transitions. The hydrogen atom is most readily ionized by electron impact, resulting in a volumetric plasma source. Similarly, atomic recombination, which can occur as either a two-body (radiative) or a three-body (dielectronic) process, with the latter becoming significant at higher electron densities, provides a volumetric plasma sink. In contrast, elastic scattering (ES) and charge exchange (CX) between hydrogen atoms and ions does not affect the particle balance, but transfers both momentum and energy; since σ_{ES} and σ_{CX} are only weakly dependent on energy, while σ_{iz} follows an Arrhenius relation, $\exp(-E_{iz}/T_e)$, charge exchange becomes dominant at electron temperatures below ~5 eV, allowing a hydrogen atom to suffer many CX collisions before being ionized.

The removal of momentum and energy by the recycling neutrals is a key mechanism in the *detachment* of the divertor plasma from the solid surface, see Section 7.1.4. At low to moderate densities, the SOL plasma is nearly transparent to the fast neutrals created in the CX collisions between cold neutrals and fast ions, which deposit virtually all their newly acquired momentum and energy directly on the vessel wall. At high densities the plasma becomes opaque to even these fast neutrals, so that their momentum and energy is transferred to the wall via a succession of ES and CX collisions.[12]

[11] At sufficiently high densities, at which the neutral–neutral collisional mean free path becomes comparable to the gradient lengths in the SOL, this technique must be supplemented by neutral–neutral scattering, or by a hybrid fluid/Monte-Carlo approach.

[12] As before, quantitative treatment requires Monte-Carlo simulations of *neutral* trajectories.

On average, a hydrogen atom emits less than one photon before being ionized. Emission occurs via radiative de-excitation involving H_α, H_β, H_γ, ... transitions. The number of ionizations per photon, known as a *Johnson–Hinov factor* and denoted as S/XB, is close to unity for $T_e \sim$ 1–3 eV and increases with increasing electron temperature; as a result, typical energy radiated per ionization event varies from ~30 eV for $T_e >$ 20 eV to ~150 eV for $T_e \sim$ 2 eV. Together with the ionization potential, radiative de-excitation and charge exchange, provide the dominant cooling process (energy sink) in purely hydrogenic plasmas, while in actual tokamak discharges line radiation from intrinsic impurities, especially from low-Z elements such as C, N and Ne, whose radiative potential $L_Z(T_e)$ peaks below 100 eV, Fig. 7.4, usually dominates the overall radiative plasma cooling.[13] At low to moderate densities, the mean free path for photon re-absorption exceeds the width of the SOL, $\lambda_{h\nu} > \lambda_n$, so that the radiated energy is deposited directly on the vessel wall, while at sufficiently high density the neutral gas becomes opaque to selected photons, which are reabsorbed before striking the wall, causing photonic excitation followed by photon emission, and so on.[14]

Finally, the dissociation of an H_2 molecule, whose energy spectrum is a function of the inter-nuclear distance, electronic configuration, and molecular vibration and rotation, produces two atoms, which share ~3–4 eV of liberated energy; these *Franck–Condon* atoms are much more energetic than the thermal (<0.1 eV) molecules desorbed from the vessel wall, and thus provide a source of warm neutrals which can heat the cold ($T_i <$ 2 eV) plasma regions and thus slow down the rate of volumetric recombination.

7.1.2.2 Plasma–impurity interactions

Impurity ($Z > 1$) atomic reactions generally involve more than one orbital electron, complicating the quantum mechanics, and different charge states, X_Z^{j+}, quickly making the reactant set $\mathbf{Y}_+$ intractably large. The situation is simplified somewhat by the fact that ionization and recombination normally occur in a stepwise manner, $X_Z^{(j-1)+} \leftrightarrow X_Z^{j+} \leftrightarrow X_Z^{(j+1)+}$, and so on, with each charge state (except for the fully stripped nucleus) able to emit line radiation following excitation by electron, ion or photon impact. The radiated power density of all the charge states of a given impurity atom per background electron, is known as its *radiative potential*, $L_Z \equiv \sum P_{rad}\left(X_Z^{j+}\right)/n_e n_Z$. Assuming a uniform electron density in thermal equilibrium with the impurity species, so-called *coronal equilibrium*

[13] In particular, impurity radiation greatly facilitates divertor plasma detachment.
[14] Once again, quantitative treatment requires Monte-Carlo simulation of *photon* trajectories.

on account of its prevalence in the solar corona, L_Z can be calculated for each impurity species (of atomic charge Z) as a function of the electron temperature, Fig. 7.4. One finds that $L_Z(T_e)$ increases gradually with Z and peaks at some critical T_e^* which likewise increases with Z.[15] In a non-uniform plasma, $L_Z(T_e)$ is modified by the transport of impurities, which permits a charge state X_Z^{j+} to sample a range of temperatures before being either ionized to $X_Z^{(j+1)+}$ or recombined to $X_Z^{(j-1)+}$; as for the case of neutrals or photons, impurity transport can be tackled by Monte-Carlo techniques, i.e. by following a large number of independent impurity trajectories, or by solving multi-fluid transport equations, one for each charge state X_Z^{j+}.

7.1.2.3 Source and sink terms for plasma fluid equations

In the context of edge plasma modelling, we are interested in the plasma–neutral interactions as sources of mass, momentum and energy for the plasma fluid equations. Based on the above discussion, we can construct these dominant *neutral sources* as follows,

$$I_0 = m_i(I_{iz} - I_{rec}), \qquad \mathbf{I}_1 = m_i(\mathbf{V}_0 I_{iz} - \mathbf{V}_i I_{rec}) - m_i(\mathbf{V}_0 - \mathbf{V}_i)I_{CX},$$
$$I_{2s} = \tfrac{1}{2}m_s\left(\left\langle v_0^2\right\rangle I_{iz} - \left\langle v_s^2\right\rangle I_{rec}\right) + I_{2s}^{(n)} + I_{2s}^{(Z)}, \tag{7.23}$$

where I_{iz}, I_{rec} and I_{CX} are the local ionization, recombination and charge exchange rates, respectively, $\mathbf{V}_i$ and $\mathbf{V}_0$ are the ion and neutral flow velocities, v_s and v_0 are the velocities of species s and of neutrals (at birth), and $I_{2s}^{(n)} = \sum E_j n_s n_n \langle\sigma v\rangle_j$ and $I_{2s}^{(Z)} = \sum E_j n_s n_Z \langle\sigma v\rangle_j$ are the energy sources for species s due to plasma–neutral and plasma–impurity interactions.

In the above, $I_{2e}^{(n)}$ contains electron energy loss due to radiative de-excitation, $-\sum E_k I_k$, atomic ionization, $-13.6\,\text{eV}\times I_{iz}$, fast dissociation, $-10.4\,\text{eV}\times I_{fd}$, slow dissociation, $-5.0\,\text{eV}\times I_{sd}$, dissociative molecular ionization, $-8.0\,\text{eV}\times I_{dmiz}$, as well as electron energy gain due to three-body recombination, $+13.6\,\text{eV}\times I_{rec}$.[16] Similarly, $I_{2i}^{(n)}$ contains ion energy gain due to atomic ionization, $f_H E_H I_{iz}$, where $f_H = n_H/(n_H + n_{H2})$ is the atomic fraction of the neutral density, molecular ionization, $(1 - f_H)E_{H2}I_{iz}$, with $E_{H2} \sim 3$–$4\,\text{eV}$ which is the Franck–Condon dissociation energy, recombination, $-E_i I_{rec}$, with $E_i = \frac{3}{2}T_i$, and charge exchange, $-\left(E_i - E_H + \frac{1}{2}m_i V_i^2\right) I_{CX}$, where $I_{CX} = n_i n_H \langle\sigma v\rangle_{CX}$. The effect of the various energy terms on the edge plasma solution will be discussed in the following sections.

[15] This maximum in $L_Z(T_e)$ gives rise to a range of radiative instabilities in tokamak edge plasmas, most notably, the so-called MARFE (*Multifaceted Axis-symmetric Radiation from the Edge*) instability which is linked to the density limit of the discharge, terminating in a disruption.

[16] Clearly, two-body recombination transfers 13.6 eV directly to the photon and not the plasma.

7.1.3 SOL geometry: limiter, divertor and ergodic SOL

As we already noted at the opening of the chapter, the exhaust properties of a magnetized plasma are determined largely by (i) the *scrape-off layer* (SOL), defined as the region of *open* field lines beyond the last closed flux surface (LCFS), $r > r_\wr \equiv r_{LCFS}$, and (ii) by the plasma *edge*, defined as the region of *closed* field lines inside of, and yet in close vicinity to, the LCFS. The magnetic field lines are open in the sense that they penetrate a solid surface, with $\nabla \cdot \mathbf{B} = 0$ satisfied by closure of the magnetic field lines within the solid structure. The transition between the *core* and *edge* regions can be defined in terms of the dimensionless parameters introduced in Section 6.3. For simplicity, the edge region may be approximated as the outermost 10% of the confined plasma adjacent to the LCFS, i.e. $0.9 < (r - r_\wr)/a < 1$.

This topological distinction between the core and edge regions on the one hand, and the SOL region on the other, has profound consequences: since every field line in the SOL is in direct contact with some solid surface, the SOL plasma is always subject to parallel losses to solid surfaces and hence is home to all the physics discussed in Sections 7.1.1 and 7.1.2.

There are three distinct scrape-off layer geometries, which reflect different plasma exhaust strategies:[17] (i) the *limiter SOL*, formed by any solid object (known as a *limiter*) protruding furthest away from the vessel wall, and thus defining the LCFS, i.e. $r_\wr = r_{lim}$; (ii) the *divertor SOL*, formed by shaping the equilibrium poloidal field using additional current carrying coils to create a magnetic *separatrix* and a poloidal field null (*X-point*) and thereby *diverting* the SOL plasma into a divertor volume and onto specially designed *target plates*, see Fig. 7.5. Here we focus exclusively on the *poloidal* divertor, which is by far the most common approach. The equilibrium flux surfaces, defined as the surfaces of constant *poloidal* magnetic flux, are modified by *toroidal* coil currents in such a way as to create one, or more, points of vanishing *poloidal* field ($\mathbf{B}_P = 0$), called the X-point(s); the *separatrix* flux surface passes through the X-point(s) and intersects the solid targets which can be some distance away and therefore not in direct contact with the edge plasma. By modulating the coil current, the separatrix may be periodically swept across the target surface, producing a more uniform heat load profile. Finally, (iii) the *ergodic SOL*, formed by actively perturbing the poloidal field, $\mathbf{B}_P$, with additional coils until the magnetic flux surfaces break up into chaotic volumes.[18] In all three cases, the plasma exhaust is directed to selected

[17] Each of these have been explored experimentally on many devices (Stangeby, 2000).

[18] Recalling our discussion in Section 3.1, a more appropriate name would be *chaotic SOL*. The transition between the edge and SOL regions in the ergodic SOL is gradual, rather than abrupt, i.e. there is no clearly defined LCFS, but rather a 3D transition layer in which the connection lengths to the vessel wall, $L_\parallel(r, \theta)$, decrease progressively, but not monotonically, with radius.

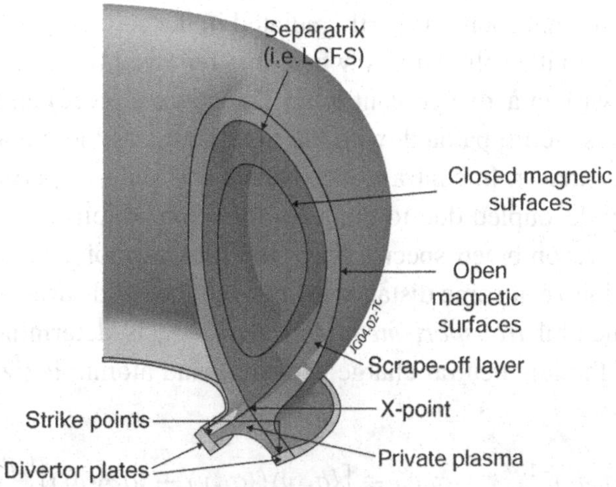

Fig. 7.5. Schematic of magnetic flux surfaces in a tokamak equilibrium with a poloidal X-point, illustrating the resulting divertor-SOL geometry.

Fig. 7.6. Composite picture of the inside of JET, in the absence (left) and presence (right) of a plasma discharge. Note the limiter and divertor tiles and their interaction with the plasma, as indicated by the recycling light. © EFDA-JET.

plasma facing components (PFCs), which are made from load resistant materials, e.g. CFC or W, and can be actively cooled to remove the deposited power. As a result, the plasma particle and power loads are restricted to a small fraction of the vessel wall, thus shielding the remainder from undesirable plasma–surface interaction, see Fig. 7.6.

In all three cases the source of energy reaching the targets can be firmly located inside the LCFS, where it is deposited by plasma heating or generated by fusion reactions. This thermal energy is partly radiated in the core/edge/SOL plasma with the rest conveyed to the solid targets, where it is removed from the system, e.g. by heat exchange with an active coolant loop. The sources (core) and sinks (targets) of energy are thus well separated, with the SOL acting as the medium/conduit for power flow and exhaust. In contrast, the sources and sinks of particles in the SOL cannot be easily decoupled due to the recombination of ions and electrons at the target tiles. The recombined species leave the tiles as molecules, dissociate into atoms and then ionize at some distance away from the solid surface. This distance, known as the neutral *transport mean free path*, λ_0, is determined by the relative strength of the ion–neutral charge exchange and atomic ionization processes (Stangeby, 2000),

$$\lambda_0 = (\sigma_0 n_e)^{-1}, \qquad \sigma_0 v_{ti} \equiv \left[\langle\sigma_{iz} v\rangle(\langle\sigma_{iz} v\rangle + \langle\sigma_{CX} v\rangle)\right]^{1/2}, \tag{7.24}$$

where σ_0 is the average neutral transport cross-section.

Differences in the distribution of particle (ion–electron) sources, and the related plasma recycling patterns, are key factors distinguishing plasma dynamics in the three SOL geometries, e.g. the limiter SOL is relatively *simple*, being fuelled by radial particle transport from the core region, the divertor SOL is generally more *complex*, both geometrically and physically, being dominated by plasma recycling in the divertor volume, and the ergodic SOL is characterized by strong radial convection in the edge region. Below we consider each SOL geometry in turn, examining both its merits and shortcomings in achieving the exhaust criteria outlined in Chapter 1.

7.1.3.1 Limiter SOL

We begin, chronologically, with the limiter SOL, which is the earliest and simplest method of shielding the main chamber from plasma wall fluxes. There are four types of limiters: discrete (mushroom), vertical (rail), toroidal (belt) and poloidal, again listed roughly in chronological order.

By definition, a limiter defines the LCFS and so is always in close proximity to the edge plasma. This proximity leads to a strong influx of both fuel and impurity neutrals into the edge, which ionize and distribute roughly uniformly on each closed flux surface. They return to the SOL via ambipolar radial transport[19] and flow towards the limiter targets along the magnetic field lines; the resulting flow increases linearly from the stagnation point located roughly halfway between the bounding limiter plates to super-sonic velocities in front of the limiters as required

[19] Radial transport, which is driven by thermodynamic gradients and magnetic curvature, and thus exhibits a strongly ballooning character, is generally dominated by edge-SOL turbulence.

by the *Bohm criterion*, recall Section 7.1.1. The closed cycle, involving a recirculation of plasma species and their recombination at some plasma–solid interface, is known as *plasma recycling*. Aside from the obvious effect of providing a local plasma source, I_0, recycling also cools and slows down the plasma via the $\mathbf{I}_1$ and I_2 neutral source terms, respectively, (7.23), in regions of plasma–neutral interaction. This applies to both *SOL* and *edge* recycling,[20] the latter being the dominant recycling process in the limiter SOL, in which case most neutrals are ionized in the edge plasma. As such, there is little, if any, SOL recycling in the limiter SOL,[21] and hence little radiative cooling/slowing down of the SOL plasma. As a result, the limiter SOL is typically hot and tenuous, leading to high heat loads on limiter targets, high erosion yields (due to high plasma temperature) and poor pumping (due to low neutral density).[22]

When the resulting target heat loads exceed the melting and/or sublimation limits, they can lead to a massive influx of impurities into the plasma, with detrimental effects on the plasma performance, even terminating the discharge via a radiative collapse. However, even below these limits, the physical and chemical erosion of the limiter tiles by plasma ions results in an influx of impurity atoms into the edge plasma, and thus to local cooling and dilution. These processes degrade tokamak performance (reduce P_α and Q_α) by increasing Z_{eff}, increasing radiation from the edge/core regions, P_{rad}. In addition, edge plasma recycling produces a flux of energetic neutrals onto the vessel walls via charge exchange collisions between recycling hydrogenic neutrals and edge plasma ions, again resulting in physical sputtering and impurity influx of vessel material into the edge/SOL plasma.

In summary, the limiter approach is simple, robust and cheap, but suffers from several important disadvantages which limit its usefulness for power exhaust in a fusion reactor. Specifically, physical sputtering of the limiter targets and its impact on plasma purity is the main reason why limiters are supplemented by divertors in most of today's tokamaks, including ITER.

7.1.3.2 Divertor SOL

The divertor SOL concept emerged in direct response to the problems associated with the limiter SOL, as outlined above. As the name suggest, the *divertor* SOL involves diverting the SOL plasma flow away from the LCFS and onto solid targets by creating one or more poloidal field nulls (X-points) in the equilibrium poloidal

[20] The distinction between SOL and edge recycling is made on the basis of the ionization region.
[21] That is, SOL ionization followed by parallel flow to the limiter targets.
[22] It is also barely collisionless, with $\nu_e^* = L_\parallel/\lambda_{ei}$ typically of order ten or less, so that parallel transport obeys the sheath limited regime, see Section 7.1.4.5.

magnetic field.[23] This has the effect of increasing plasma recycling in the near-target SOL, producing a colder and denser near-target plasma than in the limiter geometry, and hence: (i) increasing the near-target plasma compression, pumping efficiency and particle exhaust,[24] (ii) reducing material erosion and impurity sources, and (iii) reducing the penetration of both fuel and impurity neutrals into the edge plasma,[25] thus improving core plasma purity (lower Z_{eff}), and ultimately, reactor performance (higher Q_{DT}). Moreover, the divertor SOL geometry has two important consequences, namely the appearance of (i) divertor plasma detachment at sufficiently high plasma densities, see Section 7.1.4.5, and of (ii) the high confinement regime (H-mode) at sufficiently high power densities, Section 7.3. It is these two, largely unforseen and highly beneficial, phenomena which are the ultimate cause of the dominance of the divertor SOL geometry in modern-day tokamaks.[26] Below we consider some of its other advantages, the main disadvantages being cost, size and complexity.

Whereas in the limiter SOL, the safety factor, $q(r)$, and magnetic shear, $(r/q)q'(r)$, profiles are continuous across the LCFS, in the divertor SOL, where $\mathbf{B}_P$ vanishes at the X-point, they become singular at the magnetic separatrix, $r = r_{\wr}$. As a result, the SOL flux surfaces expand *radially* in the vicinity of the X-point to conserve the poloidal magnetic flux,[27] and the attack angle of field lines on the divertor target, $\theta_{\perp}$, decreases roughly linearly as one approaches the intersection of the separatrix with the target, also known as the *magnetic strike point*; both these effects represent an increase in the plasma wetted area and hence a reduction of the plasma loads, i.e. fluxes normal to the solid surface, on the divertor *target*, t. Since *upstream*, u, SOL radial profiles of n_e, T_e and T_i decay roughly exponentially with radius, see Section 7.1.4.3, one finds that as the X-point gradually approaches the target, the plasma load profile is reduced at the strike point, $r = r_{\wr}$, and peaks at some radial location in the SOL, $r > r_{\wr}$, which depends on the poloidal field and the poloidal shape of the target. Obviously, the plasma wetted area may be increased further by reducing the poloidal angle between the target tile and the magnetic field,[28] and/or by sweeping the magnetic strike point over the tile surface. The net SOL broadening between the outer mid-plane and the target, measured by the ratio of distances (one radial, r, the other along tile, s) between two flux surfaces, $FX = ds/dr$, is known as the *effective flux expansion*. It is a useful measure of

[23] The different divertor SOLs are commonly classified as single null (one X-point) or double null (2 X-points) types. The former are further divided into lower and upper single null.
[24] Compression of neutrals in the divertor volume increases their cryogenic pumping rate.
[25] The influx of impurities is attenuated by ionization of impurity neutrals in the SOL and by impurity–plasma friction sweeping the impurities back towards the target.
[26] A good review of experimental divertor physics may be found in Pitcher and Stangeby (1997).
[27] This broadening, typically by a factor of 3–5, is commonly known as *poloidal flux expansion*.
[28] Minimum angle being determined by the magnetic pre-sheath conditions, see Section 7.1.1.

the net increase in the plasma wetted area due to a combined effect of poloidal flux expansion and poloidal tile shaping. As a result, the footprint of the plasma loads on the divertor targets is typically several times broader than the radial profiles of upstream plasma quantities.

In Section 7.1.3.1 we saw that the limiter SOL is characterized by intimate contact between the LCFS and the limiter tiles and hence by the absence of local (near-target) recycling. By contrast, in the *divertor* SOL, in which the poloidal distance between the divertor tiles and the X-point is typically larger than λ_0, plasma recycling is typically localized to the divertor volume and, as a result, only influences the plasma flow in the vicinity of the divertor tiles. As the *upstream* density, n_u, increases, the large particle source in the divertor augments the parallel plasma flow to the targets, Γ_t, an effect known as *flow amplification*; this flow increases as the square of the upstream density, $\Gamma_t \propto n_u^2$, while the *target* density increases as its cube, $n_t \propto n_u^3$.[29] In the recycling region, the strong convective flow of energy reduces the parallel temperature gradients, while excitation/de-excitation + ionization of neutral hydrogen effectively cools the divertor plasma. At sufficiently low temperatures (typically below 5 eV) the loss of parallel plasma momentum due to charge exchange with background neutrals reduces the plasma flux to the divertor tiles, even as the upstream density is increased, an effect known as *plasma flux roll-over*. It corresponds to *plasma detachment* from the divertor plates and the *detached regime* of SOL transport, $\Gamma_t \propto n_u^{-\alpha}$, $n_t \propto n_u^{-\beta}$ where $\alpha, \beta > 0$, see Section 7.1.4.5. At still lower temperatures (typically below 1 eV) and sufficiently high densities, plasma detachment is further enhanced by volumetric recombination, which can extinguish the plasma flame entirely, preventing any plasma–surface contact.

The effectiveness with which the divertor structure traps neutral particles, an essential requirement for effective pumping of both hydrogenic and impurity species, is quantified in terms of the fraction of neutrals which ionize outside the divertor volume and is commonly referred to as divertor *closure*.[30] Although a complicated function of both divertor target geometry, magnetic geometry and divertor plasma parameters, it can be roughly approximated as the geometrical *closure*, i.e. the angle subtended by the separatrix and the divertor targets, with the strike point as the pivot location. Not surprisingly, sub-divertor neutral pressures and neutral compression (ratio of divertor to main chamber pressures) both increase with increasing divertor closure. As the divertor plasma detaches, the ionization front moves away from the target, effectively reducing the closure of the divertor structure and thus degrading its ability to trap neutral

[29] That is, SOL transport follows the conduction limited regime, see Section 7.1.4.5.
[30] A *closed* divertor trapping a larger fraction of neutrals than an *open* divertor.

particles, which is manifest in the roll-over of neutral compression in an upstream density scan.

In summary, the divertor SOL addresses the power exhaust problem by distancing the solid targets away from the LCFS, allowing local recycling to produce a dense, cold near-target plasma, and reducing the heat loads by a combination of field shaping, target design and plasma detachment.

7.1.3.3 Ergodic SOL

The final SOL concept deserving our attention is the so-called ergodic SOL, in which the magnetic flux surfaces in the edge region are destroyed by external magnetic perturbations. As a result, the edge plasma is subject to increased radial transport (radial component of parallel losses along the perturbed field) degrading the ability of the edge field to magnetically confine the plasma. This is typified in the so-called magnetic pump-out effect, referring to the reduction of the edge plasma density with the *Chirikov* parameter, i.e. a measure of the degree of magnetic island overlap.[31]

The absence of a clearly defined LCFS and a spaghetti-like mixture of field lines with different connection lengths to the vessel wall, $L_{\parallel}(r, \theta)$, translates into complicated 2D heat load patterns on the 'divertor' targets. While dynamic sweeping of the magnetic strike points can distribute the resulting 'hot spots' over a larger area, this does not fully compensate for loss of the beneficial effect of near-target recycling. Thus, the ergodic SOL is used to best effect when either combined with a standard divertor geometry, or when the plasma occupies only a small fraction of the vacuum vessel, thus leaving space for shaping long connection lengths to the solid targets.[32]

Despite these limitations, the ergodic SOL approach, namely the active control of the magnetic field structure, has one clear advantage. As we will see in Section 7.3, this ability proves highly valuable for controlling the edge instabilities (ELMs) which periodically destroy the edge transport barrier following a transition to the H-mode regime. Currently, it is this feature which is the dominant driver for development of ergodic SOL geometries.

Finally, we note that ITER will be equipped with limiters for the start-up/ramp-down phase, a lower single null divertor for the main heating phase and ergodization coils for ELM control. It will thus make combined use of all three SOL geometries to address different plasma exhaust problems, an approach which may prove even more important in a fusion reactor.

[31] The energy confinement is less adversely affected, due to a number of compensating factors.
[32] Indeed, the ergodic SOL with multiple X-points is the standard SOL geometry in stellarators.

7.1.4 SOL equilibrium, stability and transport

Having introduced the geometry of the tokamak boundary plasma, we next examine its equilibrium, stability and transport properties.

7.1.4.1 SOL equilibrium and stability

Since the SOL plasma is quite rarefied compared to core and edge regions, it is often neglected, or rather approximated as a vacuum envelope, when considering the global plasma stability against fast, MHD motions. This omission, while clearly reasonable, is rarely taken to its logical conclusion, namely that the SOL plasma is *not* in MHD equilibrium! This is not to say that nested magnetic flux surfaces are necessarily destroyed in the open field line region, but rather that their origin lies almost exclusively in currents external to the SOL – the toroidal field being generated due to external poloidal coils, the poloidal field due to external toroidal coils and toroidal currents *inside* the LCFS, including any edge localized bootstrap current – such that the SOL plasma carries only a small toroidal current in response to the toroidal inductive electric field. The cause of this behaviour is two fold: (i) the SOL plasma is cold (1–100 eV) and thus electrically resistive;[33] and (ii) the presence of open field lines which terminate at either end with electrostatic sheaths at the solid targets, likewise impedes the flow of charge.

The large effective SOL resistivity, $\eta_\parallel = \eta_\parallel^c + \eta_\parallel^{sh}$, defined as the sum of collisional and 'sheath' resistivities, invalidates the ideal MHD Ohm's law, $\mathbf{E}' = \mathbf{E} + \mathbf{V} \times \mathbf{B} = \eta_\parallel \mathbf{J} = 0$, (2.244), along with the *frozen-in* property of the magnetic flux and the Grad–Shafranov equation of two-dimensional MHD equilibrium. Consequently, the SOL cannot be in equilibrium in the sense of ideal MHD, because the parallel return currents required to balance the thermal and Lorentz forces, $\nabla p \approx \mathbf{J} \times \mathbf{B}$, are impeded by the large $\eta_\parallel$. The SOL plasma is thus inherently unstable to a range of dynamical instabilities, foremost among which is the interchange (R–T) instability. For the same reason, the resulting dynamics are unable to follow the fast, *MHD ordering*, recall (2.41), and are governed instead by the slower, *drift ordering*, (2.42). We will return to this theme in the context of edge-SOL turbulence, Section 7.2, which as we'll see is dominated by electrostatic, $\mathbf{E} \times \mathbf{B}$ motions.

7.1.4.2 SOL transport

Particles crossing the LCFS carry mass, momentum and energy from the edge into the SOL plasma, which in the absence of volumetric sinks are deposited on the

33 Low T_e implies high electrical resistivity, $\eta_\parallel^c \propto 1/\sigma_e \propto n/\nu_{ei} \propto T_e^{-3/2}$, see (5.154).

solid targets. The radial thickness of the SOL, or more accurately, the decay lengths of density, temperature, pressure, etc.,

$$\lambda_n \equiv n/\nabla_\perp n, \qquad \lambda_{T_s} \equiv T_s/\nabla_\perp T_s, \qquad \lambda_{p_s} \equiv p_s/\nabla_\perp p_s, \tag{7.25}$$

and hence the plasma loads on divertor or limiter tiles, are determined by competition between transport processes in the poloidal and radial directions, specifically between parallel and perpendicular flows. Despite the axis-symmetry inherent in the tokamak concept[34] and its primary consequence, namely the existence of toroidal, nested flux surfaces, one should not lose sight of the 3D nature of magnetized plasma transport, particularly in the region of open magnetic field lines which define the scrape-off layer. Hence, we discuss SOL plasma transport separately in the three magnetic directions: parallel ($\|$), diamagnetic ($\wedge$) and radial ($\perp$).

7.1.4.3 Radial SOL transport

Collisional transport theory predicts that in a (laminar) confined plasma, the heat flows of species s in these directions are ordered by (5.151), or

$$q_{\|s} : q_{\wedge s} : q_{\perp s} \approx 1 : \nu_s/\Omega_s : (\nu_s/\Omega_s)^2, \qquad \nu_s/\Omega_s \sim \delta_s \nu_s^*, \tag{7.26}$$

where $\Omega_s = e_s B/m_s$ is the gyro-frequency and ν_s the collisional frequency of species s; flows of particles, $\mathbf{\Gamma}_s = n_s \mathbf{V}_s$, are ordered similar to that of electron heat, $\mathbf{q}_e$. Since $\nu_s/\Omega_s \sim \delta_s \nu_s^* \ll 1$ is one of the conditions defining a *magnetized* plasma, the other being $\delta_s \ll 1$, one expects to find a strong scale separation in the magnitudes of the parallel, diamagnetic and radial flows, with $V_\| \gg V_\wedge \gg V_\perp$. This fact underlies the thin, elongated, *boundary layer* aspect of the SOL, whose radial extent is determined by competition between *poloidal* (parallel and diamagnetic) and *radial* fluxes.

The radial decay lengths of the parallel particle and heat flows, $\Gamma_\|$ and $q_{\|s}$, defined as $\lambda_\Gamma \equiv \Gamma_\|/\nabla_\perp \Gamma_\|$ and $\lambda_{q_s} \equiv q_{\|s}/\nabla_\perp q_{\|s}$, may be estimated by neglecting volumetric sources and equating the divergences of $\|$ and $\perp$ flows,

$$\nabla \cdot \mathbf{\Gamma} \approx 0 \Rightarrow \lambda_\Gamma/L_\| \sim \Gamma_\perp/\Gamma_\| \sim (\nu_e/\Omega_e)^2 \ll 1, \tag{7.27}$$

$$\nabla \cdot \mathbf{q}_e \approx 0 \Rightarrow \lambda_{q_e}/L_\| \sim q_{\perp e}/q_{\| e} \sim (\nu_e/\Omega_e)^2 \ll 1, \tag{7.28}$$

where $L_\|$ is half the magnetic connection length between the solid targets. The corresponding e-folding lengths for particle density, temperature and pressure may be shown to be related to λ_Γ and λ_{q_a} as follows,

$$\lambda_\Gamma^{-1} = \lambda_n^{-1} + (2\lambda_T)^{-1}, \qquad \lambda_q^{-1} = \lambda_\Gamma^{-1} + \lambda_T^{-1} = \lambda_n^{-1} + 1.5\lambda_T^{-1} \tag{7.29}$$

[34] We forget for the moment about the toroidal field asymmetry due to a finite number of coils, localized heating by RF antennas and neutral beams, localized cooling by gas and pellet injection, and localized recycling from poloidal limiters and other non-uniformities of the PFCs.

and $\lambda_{p_a}^{-1} = \lambda_n^{-1} + \lambda_{T_a}^{-1}$. The estimates (7.27)–(7.28) are highly idealized, since unlike poloidal fluxes, which are typically well understood, radial fluxes in the SOL are governed by turbulent advection and can exceed the classical values by two–three orders of magnitude; we will return to this point in Section 7.2. As a result, the SOL is generally much broader than predicted by (7.27)–(7.28), with typical values of $\lambda_n \sim \lambda_{T_e} \sim \lambda_{p_e} \sim 0.01a$.

It is conventional to divide the SOL into two radial regions: the *near-SOL*, defined as the region of high plasma potential, $\varphi \sim 3T_{e\wr}$, strong radial electric fields, $E_\perp \approx \nabla_\perp \varphi \sim 3T_{e\wr}/\lambda_{T_e}$ and poloidal flow-shear, $\nabla_\perp V_E \sim E_\perp / B\lambda_{T_e}$, which extends $\sim\lambda_q$ beyond the LCFS, and the *far-SOL*, which refers to the remaining region of the SOL extending towards the vessel wall. The near-SOL is of particular importance to power deposition on divertor/limiter tiles, while the far-SOL determines the degree of plasma interaction with the main chamber wall. The practical implications are clear: since radial fluxes decay radially away from the LCFS,[35] the divertor/limiter SOL guards the vessel wall from contact with the hot edge plasma provided that $r_{wall} - r_\wr \gg \lambda_\Gamma, \lambda_q$,

$$\Gamma_\parallel(r) = \Gamma_{\parallel\wr} \exp[-(r - r_\wr)/\lambda_\Gamma(r)], \qquad q_{\parallel s}(r) = q_{\parallel s\wr} \exp[-(r - r_\wr)/\lambda_q(r)],$$

effectively localizing the plasma–surface interactions to target tiles.

If the net radial fluxes of particles and heat are expressed in terms of effective radial velocities, $V_{\perp n} = \Gamma_\perp / n$ and $V_{\perp T} = q_\perp / \frac{3}{2} nT$, or diffusivities, $D_\perp = \Gamma_\perp / \nabla n$ and $\chi_\perp = q_\perp / n\nabla T$, the e-folding lengths become,

$$\lambda_\Gamma \approx V_{\perp n} \tau_{\parallel n}, \qquad \lambda_n \lambda_\Gamma \approx D_\perp \tau_{\parallel n}, \qquad \tau_{\parallel n} \approx L_\parallel / c_S \tag{7.30}$$

$$\lambda_q \approx V_{\perp T} \tau_{\parallel T}, \qquad \lambda_T \lambda_q \approx \chi_\perp \tau_{\parallel T}, \qquad \tau_{\parallel T} \approx \tfrac{3}{2} L_\parallel^2 / \chi_{\parallel e}, \tag{7.31}$$

were $\tau_{\parallel n}$ and $\tau_{\parallel T}$ are the parallel loss times for density and temperature, $c_S = [(ZT_e + T_i)/m_i]^{1/2}$ is the plasma sound speed and $\chi_{\parallel e} = \kappa_{\parallel e}/n_e$ is the parallel electron heat diffusivity, which is larger than $\chi_{\parallel i} = \kappa_{\parallel i}/n_i$ by the square root of the mass ratio, see (5.164)–(5.165).[36] Under typical Ohmic and L-mode conditions, $V_{\perp n}$ and $D_\perp$ are inferred to be in the range ~10–100 m/s and ~0.1–1 m²/s, respectively (Lipschultz *et al.*, 2007).

Simple models of the SOL suggest that all the above lengths are roughly linearly proportional, $\lambda_n \sim \lambda_\Gamma \sim \lambda_T \sim \lambda_p \sim \lambda_q$, so that

$$\lambda_n \sim \lambda_\Gamma \sim V_{\perp n} \tau_{\parallel n} \sim \sqrt{\tau_{\parallel n} D_\perp}, \qquad \lambda_T \sim \lambda_q \sim V_{\perp T} \tau_{\parallel T} \sim \sqrt{\tau_{\parallel T} \chi_\perp}. \tag{7.32}$$

[35] This is the origin of the label scrape-off layer, i.e. the plasma is scraped-off by parallel losses.

[36] Note that (7.31) presupposes the conduction limited regime, $10 < \nu_e^* < 100$, see below, i.e. that parallel energy transport is dominated by electron conduction; in the sheath limited regime, $\nu_e^* < 10$, parallel convection dominates and $\tau_{\parallel T} \approx \tau_{\parallel n}$.

In all the above expressions, the volumetric sources in the SOL were assumed to be negligible, which is reasonable for the *limiter SOL*, but not for the *divertor SOL*. The influence of volumetric sources of particles, I_0, and heat, I_2, may be estimated by including these in (7.27)–(7.28). For simplicity, let us consider the particle conservation equation, $\nabla \cdot \mathbf{\Gamma} \approx I_0 \approx I_{iz} - I_{rec}$, where I_{iz} and I_{rec} are the ionization and recombination densities (typically, $I_0 \approx I_{iz} \gg I_{rec}$), which may be re-written as

$$\lambda_\Gamma = \frac{\Gamma_\perp}{I_{iz} + \Gamma_\parallel / L_\parallel^{iz}}, \qquad L_\parallel^{iz} \equiv \int_0^{L_\parallel} I_{iz} l \mathrm{d}l \Big/ \int_0^{L_\parallel} I_{iz} \mathrm{d}l. \tag{7.33}$$

This expression differs from (7.27) in two respects: (i) the connection length, $L_\parallel$, is replaced by the ionization-weighted, parallel distance away from the target, $L_\parallel^{iz}$, which corresponds to the beginning of the recycling region; since $L_\parallel^{iz} < L_\parallel$, the parallel density removal time is thereby reduced, $\tau_{\parallel n}^{iz} = L_\parallel^{iz}/c_S < \tau_{\parallel n}$; and (ii) an additional (source) term appears in the denominator, and tends to reduce λ_Γ when $I_0 = I_{iz} - I_{rec}$ is positive and comparable to $\Gamma_\parallel / L_\parallel^{iz} \approx n/\tau_{\parallel n}^{iz}$. Evidently, both these effects act to increase the radial density gradient (reduce $\lambda_n \sim \lambda_\Gamma$). This is most clearly observed by re-writing the advective/diffusive expressions for λ_n and λ_Γ, (7.30), in terms of (7.33),

$$\lambda_\Gamma \approx \frac{V_{\perp n} \tau_{\parallel n}^{iz}}{1 + \alpha_{iz}}, \qquad \lambda_n \lambda_\Gamma \approx \frac{D_\perp \tau_{\parallel n}^{iz}}{1 + \alpha_{iz}}, \qquad \tau_{\parallel n}^{iz} \approx \frac{L_\parallel^{iz}}{c_S}, \qquad \alpha_{iz} = \frac{S_{iz} \tau_{\parallel n}^{iz}}{n}. \tag{7.34}$$

These divertor SOL expressions reduce to the limiter SOL results, (7.30), when $S_{iz} \ll n/\tau_{\parallel n}^{iz}$, in which case $\alpha_{iz} \to 0$, $L_\parallel^{iz} \to L_\parallel$ and $\tau_{\parallel n}^{iz} \to \tau_{\parallel n}$.

Similar expressions may be derived for λ_{T_s} and λ_{q_s}, starting with the steady-state energy conservation equation, $\nabla \cdot \mathbf{Q}_s \approx I_2 \approx I_{heat} - I_{rad}$, where I_{heat} and I_{rad} are the volumetric densities of heating and radiation. Since I_{rad} is typically larger than I_{heat} in the SOL, I_2 is generally negative and represents a net sink of heat. Thus, in contrast to the particle sources in the SOL, which are generally positive, and thus tend to reduce λ_Γ and λ_n, energy sources are typically negative, and thus tend to increase λ_{q_s} and λ_{T_s}, i.e. to broaden the temperature profiles.

7.1.4.4 Diamagnetic SOL transport

In the presence of strong radial gradients, as is the case in the near-SOL, the magnetization condition, $\delta_s = \rho_s / L_\perp \ll 1$, and the associated scale separation (7.26), are less pronounced, such that diamagnetic flows, $\Gamma_\wedge$ and $q_{\wedge s}$, can become comparable to parallel flows, $\Gamma_\parallel$ and $q_{\parallel s}$, and must be retained when calculating the net transport within the flux surface, i.e. the net poloidal flows,

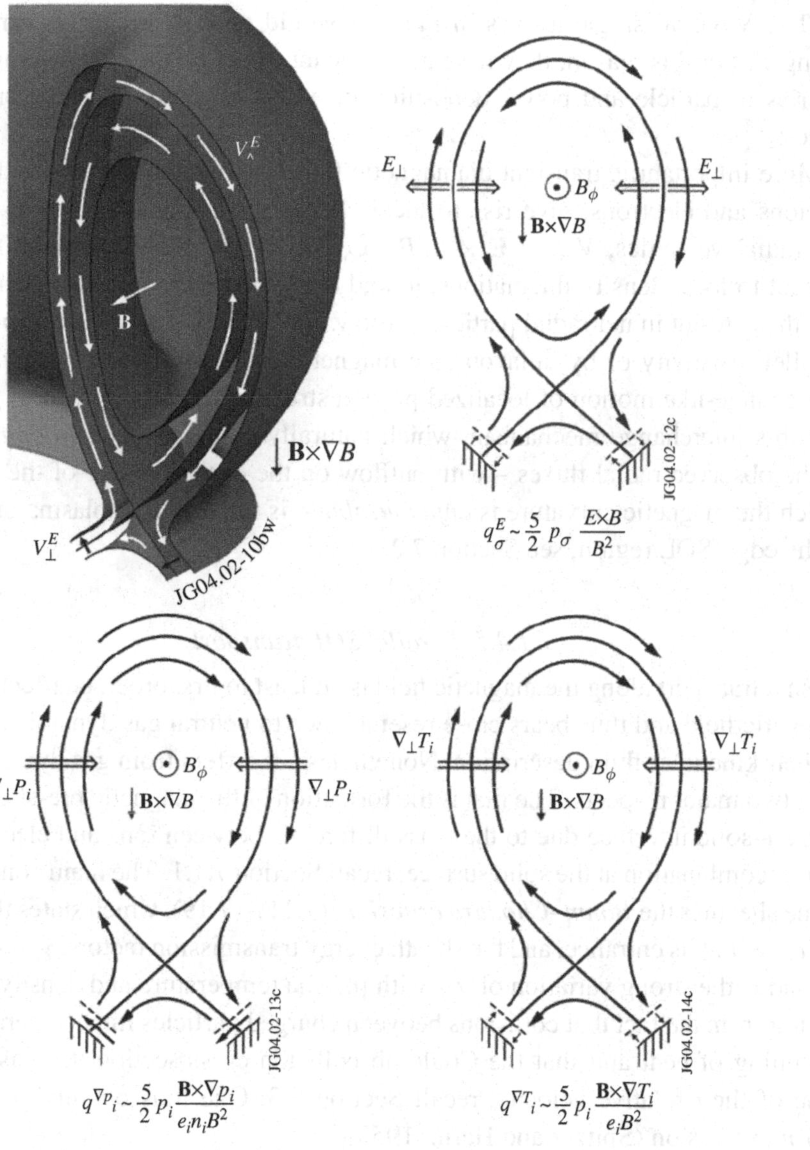

Fig. 7.7. Poloidal projections of electric and diamagnetic drift related flows of particles and heat in the edge and SOL plasmas: the $E \times B$ flow (upper frames), the ion diamagnetic convective flow (lower left) and conductive flow (lower right). *Source*: Fundamenski *et al.* (2005).

Γ_θ and $q_{\theta s}$, should replace $\Gamma_\parallel$ and $q_{\parallel s}$ in the estimate of the SOL thickness (7.27)–(7.28). The poloidal projection of the leading diamagnetic flows in a divertor SOL configuration is shown schematically in Fig. 7.7, with a normal direction of the toroidal magnetic field ($\mathbf{B} \times \nabla B$ towards the divertor). Since diamagnetic flows depend on the direction of the magnetic field (or rather the direction of $\mathbf{B} \times \nabla B$), all single arrows in Fig. 7.7 would reverse when this direction is changed. For this reason, they have a significant effect on the time-average asymmetries in particle and power deposition on inner vs. outer divertor (or limiter) plates.

More importantly, transient diamagnetic flows, which have opposite directions for ions and electrons, give rise to local charge separation and thus to fluctuating radial velocities, $\widetilde{V}_\perp = \widetilde{\mathbf{E}} \times \mathbf{b}/B \cdot \hat{\mathbf{e}}_\wedge = \widetilde{E}_\wedge/B$. When phase-shifted with respect to local density fluctuations, $\widetilde{n}$, and combined with a radial density gradient, these result in net radial particle outflow. Such phase shift can be produced by parallel resistivity or by unfavourable magnetic curvature, both of which lead to interchange-like motion of localized plasma structures. It is now widely accepted that this interchange mechanism, which naturally explains the *ballooning* nature of the observed radial fluxes – with outflow on the *outboard* side of the torus, on which the magnetic curvature is *unfavourable* – is the origin of plasma turbulence in the edge-SOL region, see Section 7.2.

7.1.4.5 Parallel SOL transport

Plasma transport along the magnetic field is, at least to first order, unaffected by the magnetic field and thus bears close resemblance to neutral gas dynamics, whether in their kinetic or fluid description. Nonetheless, it differs from gas dynamics in at least two major respects. The first is the formation of the magnetic pre-sheath at the plasma–solid interface due to the mass difference between ions and electrons and their recombination at the solid surface, recall Section 7.1.1. The main consequence of the sheath is the *Bohm–Chodura criterion*, (7.11)–(7.19), which states that $M_\parallel = V_\parallel/c_S \geq 1$ at its entrance, and the sheath energy transmission factors, γ_i and γ_e. The second is the strong variation of $\chi_{\parallel e}$ with plasma temperature and density, (5.165), arising from the fact that collisions between charged particles impede parallel free-streaming of heat and that the Coulomb collision cross-section decreases as the cube of their relative velocity, recall Section 5.3. One thus obtains the *Spitzer–Härm* expression (Spitzer and Härm, 1953),

$$\chi_{\parallel e}^{SH} = r_3 v_{te}^2/\nu_e = r_3 v_{te}\lambda_e \propto T_e^{5/2}/n_e \propto L_\parallel v_{te}/\nu_e^*, \tag{7.35}$$

where v_{te} is the electron thermal velocity, $\lambda_e = v_{te}/\nu_e$ is the electron–ion collisional mean free path and $\nu_e^* = L_\parallel/\lambda_e$ is the electron collisionality, see (5.165); it is only valid under collisional conditions, $\nu_e^* \gg 1$.[37]

[37] As $\nu_e^* \to 1$, Spitzer–Härm conduction reduces to parallel free-streaming of electron heat at some fraction of v_{te}, $q_{\parallel e} = \alpha_e v_{te} n T_e$, where α_e is known as the flux limiting factor. Heat flux limiting in SOL parallel transport was reviewed in Fundamenski (2005).

Returning to the parallel loss times of density and temperature, (7.30)–(7.31), which are defined as $\mathrm{d}_t\mathcal{A} \equiv -\mathcal{A}/\tau_{\mathcal{A}}$ and represent the temporal decay of $\mathcal{A}$ due to parallel losses, we may now write these more accurately as

$$\tau_{\parallel n} = L_\parallel/M_\parallel c_S, \qquad \tau_{\parallel T_e} = \tfrac{3}{2}L_\parallel^2/\chi_{\parallel e}, \qquad \tau_{\parallel T_i} = \tfrac{3}{2}L_\parallel^2/\chi_{\parallel i}, \tag{7.36}$$

where $\chi_{\parallel e} = \chi_{\parallel e}^{SH}/\left(1 + r_3/\nu_e^*\alpha_e\right)$ is the flux limited electron heat diffusivity and $\chi_{\parallel i} = \chi_{\parallel i}^{SH}/\left(1 + 3.9/\nu_i^*\alpha_i\right)$ is the flux limited ion heat diffusivity. The ratio of 'convective' and 'conductive' losses can be approximated as

$$\frac{\tau_{\parallel n}}{\tau_{\parallel T_e}} \sim \frac{2}{3}\frac{L_\parallel/c_S}{L_\parallel^2/\chi_{\parallel e}} \sim \frac{2}{3}\frac{\sqrt{m_i/m_e}}{\nu_{r_3}^* + r_3/\alpha_e} \sim \frac{20\sqrt{A}}{\nu_e^*/32 + 1/\alpha_e}, \tag{7.37}$$

where the last expression was evaluated for a hydrogenic plasma ($r_3 \approx 3.2$) assuming a free-streaming heat flux limit of $\alpha_e \approx 0.3$. A is the atomic number of the ion, $A = m_i/m_p$. For high collisionality $\left(\nu_e^* \gg 10\right)$ one finds that $\tau_{\parallel n}/\tau_{\parallel T_e} \to 64\sqrt{A}/\nu_e^* < 1$, so that convective loss dominate, while under low collisionality $\left(\nu_e^* \ll 10\right)$, $\tau_{\parallel n}/\tau_{\parallel T_e} \to 64\sqrt{A}$, so that conductive losses are dominant. The transition between the two regimes occurs for $\nu_e^* \sim 64\sqrt{A} - 10$, for which $\tau_{\parallel n}/\tau_{\parallel T_e} \sim 1$, so that the convective and conductive losses are comparable. Finally, the decay time of energy (pressure) can be expressed in terms of (7.36) as

$$\tau_{\parallel p_s}^{-1} = \tau_{\parallel n}^{-1} + \tau_{\parallel T_s}^{-1} = \tfrac{5}{3}\left(\tau_{\parallel n}^{-1} + \tau_{\parallel \chi_s}^{-1}\right), \qquad \tau_{\parallel \chi_s} \equiv \tfrac{5}{2}L_\parallel^2/\chi_{\parallel s}. \tag{7.38}$$

Combining the above results with conservation equations of particles, momentum and energy, (2.163)–(2.165), and retaining only the dominant transport channels, namely mass and momentum convection and electron thermal conduction, yields the simplest description of parallel SOL transport, the so-called *two-point model*, which relates upstream and target quantities, denoted by the subscripts u and t, along a single flux tube.[38] Energy transport is simplified by adding the two energy equations, adopting a classical estimate for $\kappa_{\parallel e} = n_e\chi_{\parallel e} = \kappa_{e0}T_e^{5/2}$, where $\kappa_{e0} \sim 2000$ W/eVm, and using $\gamma = \gamma_i + \gamma_e \sim 7$–$8$ as target boundary conditions. We thus find,

$$\nabla_\parallel(nV_\parallel) = I_0, \qquad \nabla_\parallel\left(p + nmV_\parallel^2\right) = I_1, \qquad \nabla_\parallel\left(\kappa_{e0}T_e^{5/2}\nabla_\parallel T_e\right) = I_2, \tag{7.39}$$

which can be integrated along $\mathbf{B}$ from the target ($l = 0$) to some distance l,

$$\Gamma_\parallel = \Gamma_{\parallel t} + \textstyle\int_0^l I_0\mathrm{d}l, \qquad \left(1 + M_\parallel^2\right)p = \left(1 + M_{\parallel t}^2\right)p_t + \textstyle\int_0^l I_1\mathrm{d}l, \tag{7.40}$$

$$T_e^{7/2} = T_{et}^{7/2} + \kappa_{e0}^{-1}\textstyle\int_0^l\int_0^{l'} I_2\mathrm{d}l'\mathrm{d}l. \tag{7.41}$$

Choosing $l = L_\parallel$, the last integral gives the total energy into the flux tube, which in the *absence* of volumetric losses must equal the energy flow to the target,

[38] Typically, the magnetic field is assumed uniform, i.e. flux expansion is neglected.

$q_{\|t} = \gamma p_t M_{\|t} c_{St}$, where $M_{\|t} \sim 1$ and $\gamma \sim 7$–8. Assuming $T_e \approx T_i \approx T$,[39] we find the simple *two-point model* relating n_t, T_t, n_u, T_u,

$$\left(1 + M_{\|t}^2\right) n_t T_t = \left(1 + M_{\|u}^2\right) n_u T_u, \qquad T_u^{7/2} - T_t^{7/2} = \tfrac{7}{4} q_{\|t} L_{\|}/\kappa_{e0}. \qquad (7.42)$$

Finally, since the flow is expected to be quasi-stagnant at the upstream location, the left equation is often approximated as $2n_t T_t \approx n_u T_u$.[40]

Subjecting this model to systematic analysis, with $L_{\|}$, $q_{\|} = q_{\|t}$ and n_u treated as independent variables, and n_t, T_t, T_u as dependent variables,[41] reveals two regimes of parallel transport in the SOL, which appear for different values of upstream collisionality, $\nu_u^* \propto q R n_u / T_u^2$ (Stangeby, 2000):[42]

(i) at low collisionality $\left(\nu_u^* < 10\right)$, one finds the *sheath limited*, or *low recycling*, regime in which efficient parallel heat conduction ensures that both temperature and total pressure are roughly uniform along the magnetic field, $T_u \approx T_t$, $2p_t \approx p_u$, and parallel power (out)flow is dominated by the electrostatic sheath. This yields a linear relation between densities and flows, $\Gamma_{\|t} \propto \Gamma_{0t} \propto n_t \propto n_u$, where Γ_{0t} is the flux of neutrals to the target, and a weaker than linear decrease of temperature with density, $T_u \approx T_t \propto n_u^{-2/3}$.

(ii) at moderate collisionality, $\nu_u^* \sim 10$ to $\sqrt{m_i/m_e}$, one obtains the *conduction limited*, or *high recycling*, regime in which total pressure remains uniform but parallel heat conduction is sufficiently reduced to support significant parallel temperature and density gradients $\left(T_t^{7/2} \ll T_u^{7/2},\ 2p_t \approx p_u\right)$,[43] thus cooling and compressing the target plasma compared to the upstream SOL. Including these conditions into (7.42) yields the following scalings,

$$T_u \propto (q_{\|} L_{\|})^{2/7}, \qquad T_t \propto (q_{\|}/p_u)^2 \propto q_{\|}^{10/7} L_{\|}^{-4/7} n_u^{-2} \qquad (7.43)$$

$$n_t \propto n_u T_u / T_t \propto q_{\|}^{-8/7} L_{\|}^{6/7} n_u^3, \qquad \Gamma_{\|t} \approx n_t c_{St} \propto q_{\|}^{-3/7} L_{\|}^{4/7} n_u^2. \qquad (7.44)$$

We first note that due to the strong temperature dependence of $\kappa_{\|e}$, T_u is only weakly sensitive to $q_{\|} L_{\|}$ and to the $I_2(l)$ profile; similarly, due to the omission of parallel convection, it is independent of n_u and the $\Gamma_{\|}(l)$ and $I_0(l)$

39 In reality, the ions are usually much warmer than electrons, $T_{iu}/T_{eu} \sim 2$, since the ion parallel heat conduction is much less efficient, i.e. $\chi_{\|e}/\chi_{\|i} \sim 60$ and $T_{iu}/T_{eu} \propto (\kappa_{0e}/\kappa_{0i})^{2/7} \sim 3$.

40 Note that (7.42) assumes that the energy enters the flux tube uniformly, $I_2 = const$. If all the power entered at $l = L_{\|}$, than the factor 7/4 is replaced by 7/2.

41 This reflects the experimental situation where we control the magnetic field, hence $L_{\|}$, the heating power, hence $q_{\|}$, and the rate of fuelling, hence n_u.

42 Since ν_u^* depends on both density and temperature, the three SOL regimes can be traversed by varying either of these quantities. In keeping with the two-point model, we will consider the gradual increase n_u for constant input power – a common experimental situation.

43 In reality, these gradients only develop in the region upstream of the near-target recycling flow, while the recycling region itself is dominated by parallel convection.

profiles. In contrast, all target quantities are very sensitive to n_u, with T_t and $\Gamma_{\parallel t}$ decreasing/increasing as n_u^2, and n_t and Γ_{0t} increasing as n_u^3.[44]

At still higher collisionality, $\nu_u^* > \sqrt{m_i/m_e}$, we expect the target plasma to cool sufficiently to allow volumetric losses to become significant. We can estimate the effect of volumetric momentum loss (plasma–neutral scattering), volumetric power loss (line radiation[45]) and convective power flow by introducing three new dimensionless factors, $0 < f_1 < 1$, $0 < f_2 < 1$, $0 < f_\chi < 1$, respectively, defined by $q_{\parallel t} = f_2 q_\parallel = \gamma p_t M_t c_{St}$ and

$$\left(1 + M_{\parallel t}^2\right) n_t T_t = f_1 \left(1 + M_{\parallel u}^2\right) n_u T_u, \qquad T_u^{7/2} - T_t^{7/2} = \tfrac{7}{4} f_\chi q_\parallel L_\parallel / \kappa_{e0}, \quad (7.45)$$

thus forming a *modified two-point model.* When these factors are set equal to unity, we recover the original two-point model, (7.42), and its scalings, (7.43). For smaller values, $T_u, n_t, T_t, \Gamma_{\parallel t}$ scale with f_1, f_2, f_χ as

$$T_u \propto f_\chi^{2/7}, \quad T_t \propto f_1^{-2} f_\chi^{-4/7} f_2^2, \quad n_t \propto f_1^3 f_2^{-2} f_\chi^{6/7}, \quad \Gamma_{\parallel t} \propto f_1^2 f_2^{-1} f_\chi^{2/7}.$$

Consistent with (7.43), we find that T_u is independent of volumetric losses (f_1, f_2) and is only weakly sensitive to convection (f_χ), while $n_t, T_t, \Gamma_{\parallel t}$ are highly sensitive to f_1 and f_2, and moderately sensitive to f_χ.

The modified two-point model (7.45) can be used to estimate the ratio of upstream and target temperatures in deuterium plasmas as,

$$(T_u/T_t)^{1/2}[1 - (T_u/T_t)^{-7/2}] \approx (\nu_u^*/20)\, f_\chi^{3/7} f_1/f_2, \qquad (7.46)$$

which approaches unity when the right-hand side tends towards zero and increases as $T_u/T_t \approx \left(\nu_u^*/20\right)^2 f_\chi^{6/7} f_1^2/f_2^2$ when it is much larger than one.[46] This result explains the gradual transition between the sheath-limited, conduction-limited and CX-limited (*detached*) regimes with increasing values of ν_u^*:

(i) for $\nu_u^* \ll 30$, $\kappa_{\parallel e}$ is so large that even small $\nabla_\parallel T_e$ is sufficient to carry all the power along the flux tube, while atomic interactions are ineffective at removing either momentum or energy from the plasma, so that $f_1, f_2, f_\chi \sim 1$;
(ii) as ν_u^* approaches $\sim$30, $\kappa_{\parallel e}$ is progressively reduced, requiring larger $\nabla_\parallel T_e$ to carry the same power, while at the same time, volumetric energy losses begin to appear, reducing f_2 below unity. At still higher ν_u^*, the near-target recycling flow begins to dominate the power balance, i.e. the associated convection carries most of the power, reducing $\nabla_\parallel T_e$ in this region. Upstream conditions are thus separated into a convectively limited

[44] There is direct experimental evidence supporting these predictions, e.g. the transition from $T_{eu} \propto n_u^{-2/3}$ at low density to $T_{eu} \propto$ const at high density is well documented.
[45] This includes impurity line radiation, which is an important catalyst for divertor detachment.
[46] This is the origin of the transitional value of $\nu_u^* \sim 10$ cited above, at which $T_u/T_t \sim 2$.

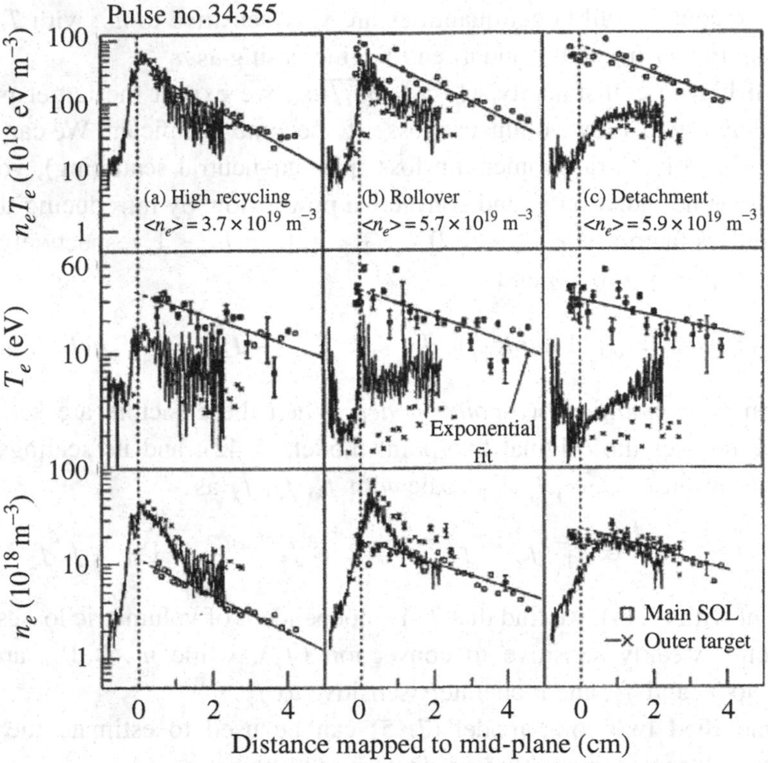

Fig. 7.8. Radial profiles of electron pressure (top), temperature (middle) and density (bottom) in the upstream SOL and at the outer divertor target in JET L-mode discharges (MkI divertor). The three columns represent a density scan by additional gas fuelling: (left) high recycling, conduction limited regime, (middle) the plasma flux 'rollover', and (right) partially detached regime. *Source*: Loarte *et al.* (1998).

(transonic, near-target) region, and a conductively limited (near-stagnant, upstream) region;[47] and, finally,

(iii) as ν_u^* increases above $\sqrt{m_i/m_e}$, and T_e drops below ~5 eV, the rate of ionization drops sufficiently so that a typical neutral experiences several charge exchange collisions before being ionized. This process removes significant energy and momentum from the plasma flow ($f_1, f_2 \ll 1$), producing a net pressure drop along the flux tube ($T_t \ll T_u, 2p_t \ll p_u$), and so reducing n_t and $\Gamma_{\parallel t}$ provided f_1 *decreases* faster than linearly with n_u, i.e. $f_1 \propto n_u^{-\alpha}$ with $\alpha > 1$; the point $\alpha = 1$ defines the plasma flux rollover.[48] This results in *detachment* of the plasma 'flame', or more accurately of its radiation and ionization fronts, from the divertor targets, see Fig. 7.8.

[47] This behaviour is typically referred to as the high recycling regime, emphasizing the role of convection ($f_\chi \ll 1$), although the dividing line between the regimes is somewhat arbitrary.

[48] While both n_t and $\Gamma_{\parallel t}$ decrease with n_u, the neutral flux, Γ_{0t}, continues to increase; this translates into an increase of the divertor neutral density and hydrogenic line radiation.

Once volumetric recombination becomes significant (below $T_{et} \approx 1$ eV in D), it effectively forms a *gas target*, reducing n_t and $\Gamma_{\parallel t}$ towards zero, and thus suppressing all plasma–surface contact, much as a candle flame is extinguished by the cold gas into which it flows.[49] At this point, the non-linearities in the volumetric sources (I_0, I_1, I_2) begin to dominate the plasma dynamics, producing a thermal instability whereby the radiation and ionization fronts rapidly propagate upstream, until they reach the vicinity of the X-point. The result is a region of dense, cold and highly radiative plasma, commonly known as an X-point MARFE (*Multifaceted Axis-symmetric Radiation from the Edge*). The MARFE instability is driven by the inverted temperature dependence of line-radiative power, $L_Z(T_e)$ for mid-Z elements, e.g. C, O, N, Ne, see Section 7.1.2.2. It saturates at sufficiently low T_e, when parallel heat flow can no longer replenish the radiative losses. Aside from cooling the edge plasma, and thus degrading reactor performance, it is a precursor to the wall MARFE, which in turn is linked to the density limit of the discharge, terminating in a disruption.

Three different approaches are commonly used to quantify the *degree of detachment*:[50] (i) the ratio of the actual $\Gamma_{\parallel t}$ to that predicted by the two-point model $(\Gamma_{\parallel t} \propto n_u^2)$ based on some reference point in the *attached* (high-recycling) regime. This approach is typically used when only target data are available, and rests on the tenuous assumption, recently challenged by experimental results, that n_u scales linearly with core average plasma density, $\overline{n}_e$; (ii) the ratio of upstream and target total pressures, $\approx p_u/2p_t$, which measures the parallel momentum losses, i.e. the physical mechanism at the heart of detachment, a preferred approach when reliable upstream measurements are available;[51] and (iii) the position of the plasma density peak, $l_n/L_\parallel$, and/or the ratio of peak to target densities, n_{max}/n_t. However, one should not lose sight of the only true measure of detachment, namely the reduction of $\Gamma_{\parallel t}$ with increasing n_u, i.e. the plasma flux roll-over.

As was already said in Section 7.1, divertor *closure* facilitates plasma detachment by increasing the neutral target density (neutral compression) for the same upstream power and density, e.g. it is easier to achieve on vertical compared to horizontal targets. This underlines the role of neutral penetration, λ_0, (7.24), in the detachment dynamics: as the near-target plasma cools, λ_0 increases, pushing the ionization front further away from the target, and so extending the recycling region in which parallel heat convection dominates and $\nabla_\parallel T_e$ is small, i.e. $T_e(l)$ is relatively flat.

[49] Of course, neutral–surface interaction at the divertor target is not suppressed; indeed, divertor neutral pressure increases during the onset of detachment, later saturating and/or decreasing as the ionization front moves away from the target.

[50] Consistent with the larger average power flow into the outer divertor volume for normal field direction ($B \times \nabla B$ pointing towards the X-point), see Section 7.1.4.4, the inner target typically detaches earlier, i.e. at lower n_u, than the outer target.

[51] Somewhat arbitrarily one might consider the divertor SOL plasma detached when $p_u/2p_t > 3$.

Detachment is currently viewed, along with the concept of an edge radiative mantle, as an essential part of resolving the power exhaust problems in fusion reactors. It is beneficial in both (i) reducing the steady-state heat loads, by transferring part of the power entering the divertor from the plasma to the neutrals, and hence distributing it over a larger PFC area, and (ii) cooling the divertor plasma and hence reducing the impurity source due to physical sputtering and the impurity influx to the core plasma. The second item is especially important if the divertor targets are made from high-Z materials, e.g. tungsten,[52] which has a physical sputtering threshold of $\sim$200 eV for deuterium. Since ion impact energy is roughly $2T_i + 3ZT_e$, it means that W sputtering by D vanishes when $T_e \sim T_i < 40$ eV; for higher Z projectiles, a further reduction of T_e is required, e.g. for $T_e \sim T_i < 6$ eV for $Z = 10$, corresponding to $T_e < 5$ eV needed for detachment.

Since, as we will see in Section 7.3.2, most of the power entering the SOL during the inter-ELM phase is deposited on the outer target, it follows that the inner divertor leg typically detaches at a lower upstream density than the outer leg. Indeed, partial detachment of the outer divertor leg can be viewed as one of the main exhaust challenges for a tokamak fusion reactor. The term *partial* is chosen carefully, since *complete* detachment of *both* divertor legs leads to a radiation limited divertor regime, accompanied by the movement of the radiation front from the divertor volume into the X-point region, i.e. by the formation of an X-point MARFE. This results in strong cooling of the edge plasma and a reduction of pedestal pressure by roughly a factor of 2 in ELMy H-mode, and since $W_{ped} \sim W/3$, to a $\sim$15–20% drop in the normalized energy confinement, $H_{98} = \tau_E/\tau_{E98y}$.[53]

7.1.5 SOL modelling approaches

'Entia non sunt multiplicanda praeter necessitatem.'
Lex parsimoniae

'Everything should be made as simple as possible, but not simpler.'
Attributed to Albert Einstein (c. 1930)

In other words, the explanation of any phenomenon should make as few assumptions as possible, eliminating those that make no difference in the observable predictions of a hypothesis or theory, i.e. all other things being equal, the simplest solution is the best. This thought, known as *Ockam's razor*, should be the guiding light of any modelling effort.

[52] Tungsten is envisioned for the D–T phase of ITER and is the most likely PFC material for DEMO.

[53] On JET, outer target detachment is typically accompanied by transition from Type-I to Type-III ELMy H-mode, which appears to be determined by a critical temperature and/or resistivity, rather than a critical collisionality, see Section 7.3.1.3.

Despite reproducing the basic trends, the simple models of SOL transport developed above, i.e. the radial decay lengths (7.30)–(7.31) and the modified two-point model (7.45), are clearly too crude to quantitatively interpret real experiments and/or predict future outcomes. For this purpose, the following '*laminar*' modelling approach is usually adopted:[54]

(i) the axis-symmetric magnetic equilibrium is first reconstructed by solving the Grad–Shafranov equation constrained by in-vessel magnetic coil signals. The equilibrium field includes the edge, separatrix, X-point(s) and the SOL region, extending all the way to the vessel wall, although any currents that may flow in the SOL are typically neglected in the calculation;

(ii) the equilibrium poloidal field in the edge-SOL region is then mapped onto a structured numerical mesh/grid of arbitrary resolution, which typically includes ~100 poloidal and ~30 radial grid points,[55] and extends radially over $0.8a \sim r \sim r_{lim}$. The plasma core, $r/a < 0.8$, and the limiter shadow, $r > r_{lim}$, are normally excluded from the numerical grid;

(iii) the plasma fluid equations (2.163)–(2.165) for electrons, hydrogen ions and impurity ions,[56] are then averaged in time and in the toroidal angle to yield 2D evolution equations for the *mean* quantities, $\langle \mathcal{A} \rangle = \overline{\mathcal{A}}$. These are then discretized and evolved numerically on the 2D magnetic mesh, subject to the Bohm–Chodura boundary conditions at the solid surfaces, recall Section 7.1.1.[57] Radial transport is treated by a diffusive approximation relating mean radial flows to mean radial gradients,

$$\overline{\Gamma}_\perp = \langle n V_\perp \rangle = D_\perp \nabla_\perp \overline{n}, \qquad \overline{\pi}_\perp = \mu_\perp \nabla_\perp \overline{\mathbf{V}}, \qquad \overline{q}_\perp = \kappa_\perp \nabla_\perp \overline{T}, \tag{7.47}$$

with $D_\perp$, $\mu_\perp$ and $\kappa_\perp$ varied to match, or rather minimize the error with, experimentally measured radial profiles of plasma quantities in the edge and SOL regions. Finally, the volumetric neutral sources are calculated using a Monte-Carlo (M-C) technique, i.e. by launching a large number of neutrals (injected from gas valves or recycling from the solid targets) and following their trajectories as they interact with electrons, ions and other neutrals, which populate the magnetic grid. The multi-fluid plasma and Monte-Carlo neutrals are then coupled iteratively with a typical time step of $\sim L_\parallel / V_A \sim 1\ \mu$s, i.e. much shorter than the SOL equilibration time of $\sim L_\parallel / c_S \sim 1$ ms. The solution is then converged until steady-state for given radial flows of particles, momentum and power across the inner most grid (flux) surface.

[54] The label 'laminar' refers to the neglect of turbulent fluctuations.
[55] These typical grids are too crude to resolve turbulent plasma dynamics, which occur on the ion gyro-radius ρ_i, or rather the sound speed gyro-radius, ρ_S scale, recall Section 6.3.1.
[56] Each impurity ion charge state is described by a separate continuity and momentum equation, but a common energy equation is typically employed for all ion species.
[57] The magnetic field is not evolved in time, i.e. the equilibrium is assumed stationary and stable. All plasma dynamics involving $\mathbf{B}_P$ perturbations, e.g. reduced MHD, are thus suppressed.

The laminar modelling approach, as implemented in several code packages (the plasma fluid codes include B2 (Schneider *et al.*, 2000), EDGE2D (Simonini *et al.*, 1994) and UEDGE (Porter *et al.*, 1996), TECXY (Zagórski *et al.*, 2007), etc.; the most common M-C neutral transport code is EIRENE (Reiter, 1992); the most widely used combination of these, B2/EIRENE, labelled as SOLPS, was used to design the ITER divertor), quantifies the effect of magnetic flux expansion and divertor closure on neutral penetration, and is thus able to capture the main features of SOL plasma–neutral dynamics, including such vital phenomena as divertor detachment and X-point MARFE formation, see Section 7.1.4.2. Nonetheless, it has so far failed to reproduce all the diagnostically available signals *simultaneously*, even under relatively simple Ohmic, attached conditions. The three notable examples of major discrepancies between laminar modelling and experiment are (i) the radial electric field in the near-SOL (Chankin *et al.*, 2007), (ii) the parallel flow in the upstream SOL (Coster *et al.*, 2007) and (iii) in–out asymmetry in divertor detachment (Wischmeier *et al.*, 2007).[58] Finally, the above approach is of course unable to address such issues as the L–H transition, ELM filament transport and/or 3D (ergodic SOL) effects, see Section 7.3.1.3.

What is the reason for these discrepancies? While this question remains under active investigation, there are several plausible explanations:

(i) As we will see in Section 7.2, the SOL plasma is nearly always highly turbulent, with the RMS values of relative thermodynamic fluctuations increasing with radius and approaching unity in the far-SOL, i.e. $n_{rms}/\overline{n} \sim 1$ where $n_{rms}^2 = \langle \widetilde{n}^2 \rangle$. Moreover, since they originate in radial advection of turbulent filaments, the density, velocity and temperature fluctuations are closely correlated, e.g. $\langle \widetilde{n}\widetilde{T} \rangle \propto n_{rms} T_{rms}$. Consequently, the mean-field approximation on which the laminar approach is based, and which relies on the smallness of relative fluctuations, introduces significant errors in the description of plasma dynamics, e.g. by approximating $\langle nT \rangle = \overline{n}\overline{T} + \langle \widetilde{n}\widetilde{T} \rangle$ as $\overline{n}\overline{T}$, one picks up an error of $\sim n_{rms} T_{rms}$, which is comparable to $\overline{n}\overline{T}$. The same is true for all non-linear combinations of n, $\mathbf{V}$ and T, with the relative error increasing with the order of the moment. Hence, it is comparatively large for parallel heat conduction, $\kappa_{\parallel} \propto T^{5/2}$ and even larger for thermally activated plasma–neutral processes, $\sigma_{iz} \propto \exp(-E_{iz}/T)$, etc.

It is worth stressing that these effects are purely a remnant of the time averaging of original plasma fluid equations. Thus even if these offered an accurate description of plasma dynamics at all times, their mean-field form would not be accurate due to the non-linearities omitted in the averaging. To estimate these errors, let us assume a simplified time variation,

$$\{n = 1, T = 1\}, \quad 0 < t < 9, \qquad \{n = a_n > 1, T = a_T > 1\}, \quad 9 < t < 10,$$

[58] The last item is particularly worrying for fusion reactor design. Indeed, our predictive capability relating to divertor detachment is at present highly limited.

Table 7.2. *Estimate of selected errors in laminar SOL modelling.*

a_n, a_T	$\sqrt{\langle\widetilde{T}^2\rangle}/\langle T\rangle$	$\langle nT\rangle/\langle n\rangle\langle T\rangle$	$\langle T^{5/2}\rangle/\langle T\rangle^{5/2}$	$\langle e^{-10/T}\rangle/e^{-10/\langle T\rangle}$	$\langle e^{-3/T}\rangle/e^{-3/\langle T\rangle}$
1.5	0.14	1.02	1.04	2.3	1.02
3	0.5	1.25	1.56	15.0	0.99
6	1	2	3.52	14.9	0.79
21	2	5	13.0	1.74	0.36

and so on, in a cyclical sequence (all quantities in arbitrary units). The resulting errors in various mean-field quantities are then easily computed as a function of a_n and a_T, which in turn can be related to the relative RMS fluctuations, see Table 7.2. As expected, we find that for nominal fluctuations levels in the SOL, the error in pressure is of order unity, while those in $\kappa_\parallel \propto T^{5/2}$ and $\sigma_{iz} \propto \exp(-E_{iz}/T)$ can be much larger than one.[59]

(ii) The turbulent closure scheme typically used in the laminar modelling is none other than the Prandtl–Boussinesq mixing length approximation with $D_\perp \sim D_{tb}$, $\nu_\perp \sim \nu_{tb}$, $\chi_\perp \sim \chi_{tb}$; the turbulent diffusivities are typically treated as constant parameters. In contrast, measurements and simulations indicate radial transport is dominated by advection of turbulent filaments originating at the separatrix, so that local radial fluxes are *not* related to the local gradients (Naulin, 2007), i.e. even when viewed instantaneously, SOL turbulence is simply not a diffusive process!

(iii) The collisional, Spitzer–Härm (S–H) closure for the parallel heat flux (and viscosity) becomes invalid in regions of sharp parallel temperature gradients, e.g. in the ionization/radiation front. Moreover, even the heat flux limit corrections become inadequate, since the heat flux can actually exceed the S–H value as hot electrons stream into a cold plasma region (Fundamenski, 2005).[60] These findings call into question the validity of the isotropic plasma fluid approach (n, $\mathbf{V}$, T) and require either higher-order moments and/or the gyro-fluid (gyro-tropic) formulation, i.e. $T_\parallel$, $T_\perp$, $q_\parallel$, $q_\perp$, etc.

(iv) Finally, other possible sources of error include neutral, photon and impurity transport effects. For instance, under dense divertor conditions, neutrals become fluid-like and most likely turbulent, while photon transport becomes opaque in certain lines, e.g. Lyman alpha.

All the above limitations call into question the validity of the laminar modelling approach and argue for the development of global edge turbulence codes, which can follow total, rather than mean and/or fluctuating quantities, see Section 7.2.[61]

59 The same arguments apply for ELMs, which represent even larger SOL perturbations.

60 Such *non-local heat transport* has recently been shown to yield parallel variation of the flux limiting factors from <0.1 to $\gg 100$, a hopelessly intractable result (Tskhakaya *et al.*, 2008)!

61 This is further reinforced by the fact that SOL transport is three- rather than two-dimensional, and the assumed axis-symmetry is broken by the presence of turbulent plasma filaments.

Nonetheless, one could, and perhaps should, attempt to salvage laminar modelling with the help of the turbulent closure CFD techniques developed for HD turbulence, recall Section 6.1.2, specifically the appropriately modified $\mathcal{K}-\varepsilon$ and *Reynold's stress* models.[62] At the very least, radial diffusivities should be corrected for the measured level of relative fluctuations, along the lines of Table 7.2, and should be made dependent upon local plasma and field parameters (Kirnev *et al.*, 2007).

7.2 L-mode power exhaust: edge-SOL turbulence

As we saw in Section 7.1.4.2, parallel SOL transport is generally well understood and can be described to a fair accuracy by either the Fokker–Planck or Braginskii equations (with kinetic corrections for non-local heat flow) in the fluid and kinetic approximation, respectively. In contrast, perpendicular (radial) transport is generally governed by turbulent advection, whose mechanism and dynamics has for a long time remained elusive, acquiring the somewhat fatalistic label '*anomalous*', i.e. abnormal, irregular, not understood. Hence, our ability to predict the steady-state plasma loads on PFCs (divertor or limiter tiles) reflects the degree to which we can understand and model edge-SOL turbulence. It is therefore encouraging to find that significant progress has been made in recent years in this domain, with the governing mechanisms having been identified and predictive capability coming into view. Let us briefly summarize the emerging physical picture.

It is now widely recognized that SOL turbulence originates in the edge region, where it is driven by a combination of drift-wave, or rather drift-Alfvén wave, and interchange dynamics, recall Section 6.3.1. These reduce to resistive-MHD (resistive ballooning) dynamics at sufficiently high collisionality, and to ideal MHD (ideal ballooning) dynamics at sufficiently high pressure gradient, i.e. there is a continuity between the mechanisms determining edge plasma (in)stability and transport. Since both drift- and Alfvén waves are strongly damped on open field lines, turbulent motions in the SOL are dominated by (largely electrostatic) interchange dynamics, supplemented by parallel losses and sheath dissipation at the plasma–solid interface. Overall, the free energy driving edge-SOL turbulence is provided by radial pressure gradients, which build up together with poloidal $\mathbf{E} \times \mathbf{B}$ flow shear during relatively long, quiescent periods. Effectively, the large radial electric fields in the near-SOL and the associated poloidal velocity shear

62 It is left as an exercise, and as a challenge, to the interested reader to extend these techniques to the plasma fluid equations, (2.163)–(2.165), with Braginskii transport coefficients.

(zonal flow), play an active and self-regulating role in edge-SOL turbulence by de-correlating (ripping apart) coherent structures, thus reducing radial turbulent fluxes.[63] This build-up is sporadically interrupted by formation and ejection of isolated, field aligned structures, variously known as plasma *blobs* or *filaments*, which advect mass, momentum and energy into the far-SOL, while draining to the divertor or limiter tiles.[64] This highly intermittent nature of SOL transport results in strong fluctuations in far-SOL quantities, which for most of the time represent a cool and rarefied plasma background, occasionally punctuated by relatively dense and hot plasma filaments. In the SOL, the thermodynamic quantities of the plasma filament represent a strong perturbation to the ambient values, so that plasma filaments are isolated, field-aligned plasmoids, which are driven largely by their own field and pressure gradients, and not by the ambient gradients in these quantities. When averaged over all scales, this radial advection is the origin of the 'anomalous' radial transport in the SOL.

Below we investigate edge-SOL turbulence in more detail by briefly reviewing experimental observations in tokamaks, Section 7.2.1, and comparing these with numerical simulations, Section 7.2.2.

7.2.1 Experimental observations

'The facts, the facts, Hastings, they are the stones that make up the road upon which we travel.'

Hercule Poirot, A. Christie (c. 1930)

There are several comprehensive reviews of edge-SOL turbulence measurements in fusion devices, one of the most recent and approachable being Zweben *et al.* (2007). Generally, SOL measurements on virtually all tokamaks indicate intense electrostatic activity, with very small levels of magnetic flutter, i.e. while the relative fluctuations of density ($\widetilde{n}/n$), temperature ($\widetilde{T}/T$) and electric potential ($e\widetilde{\varphi}/T_e$) are found to increase with radius, and approach, or even exceed, unity in the far-SOL, and to exhibit bursty, intermittent time signatures, the magnetic fluctuations are always much smaller, typically $\widetilde{B}_\perp/B < 10^{-4}$, and poorly correlated with $\widetilde{n}/n$, so that $\langle\widetilde{n}\widetilde{B}_\perp\rangle/nB \ll \widetilde{n}^{rms}\widetilde{B}_\perp^{rms}/nB \sim 10^{-4}$. The radial flows associated with electric and magnetic field fluctuations, $\widetilde{E}_\wedge$ and $\widetilde{B}_\perp$, respectively,[65] are

63 This mechanism has been demonstrated in numerical simulations, see Section 7.2.2.

64 The filaments are of course formed in the edge region due to the inverse energy cascade typical of 2D turbulence, see Section 6.1.3. Having crossed the LCFS, they continue to move in the drift ($\perp$–$\wedge$) plane in the inherently unstable SOL region, recall Section 7.1.4.1.

65 This should not be confused with radial fluxes due to perturbed field lines in the ergodic SOL, which can be the dominant radial transport channel.

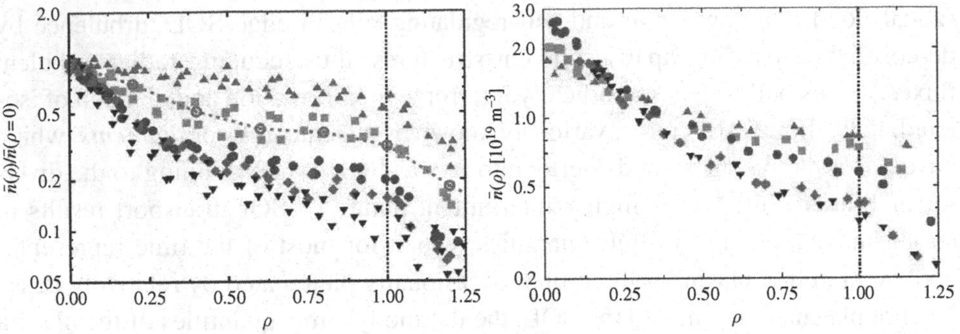

Fig. 7.9. Radial average plasma density profiles in the SOL ($\rho = 0$ at LCFS, $\rho = 1$ at first limited flux surface) as measured by an outboard reciprocating Langmuir probe at TCV. The left frame shows a density (gas fuelling) scan, while the right shows an plasma current scan. Both scans indicate a tendency for SOL density profiles to broaden with increasing collisionality, with ▼ representing the lowest and ▲ the highest ν_u^*. The dotted line are results of an edge-SOL electrostatic turbulence simulation, see Section 7.2.2. *Source*: Garcia *et al.* (2007a,b).

$$\Gamma_\perp^E \equiv \langle n v_\perp \rangle = \frac{\langle \widetilde{n}\widetilde{E}_\wedge \rangle}{B}, \quad Q_{\perp s}^E \equiv \frac{3}{2}\langle n T_s v_\perp \rangle \approx \frac{3}{2} T_s \Gamma_\perp^E + \frac{3}{2} n \frac{\langle \widetilde{T}_s \widetilde{E}_\wedge \rangle}{B}, \tag{7.48}$$

$$\Gamma_\perp^B \approx \frac{\langle \widetilde{n} c_S \widetilde{B}_\perp \rangle}{B}, \quad Q_{\perp e}^B \approx \frac{5}{2} T_e \Gamma_\perp^B + q_{\parallel e} \left\langle \frac{\widetilde{B}_\perp^2}{B^2} \right\rangle \frac{L_\parallel}{\lambda_{T_e}}, \tag{7.49}$$

where $Q_{\perp s}$ is written as the sum of the convective and conductive terms. Consistent with the above fluctuation and correlation levels, the net radial flow is dominated by $\mathbf{E} \times \mathbf{B}$ advection, i.e. $\Gamma_\perp^E \gg \Gamma_\perp^B$, $Q_\perp^E \gg Q_\perp^B$.

From the wide range of available data, the recent measurements on the *Tokamak de Configuration Variable* (TCV) offer perhaps the clearest insight yet into the nature of SOL turbulence. The experiments, described in Garcia *et al.* (2006b, 2007a,b,c), consist of two separate scans in otherwise identical L-mode plasmas: the first in the plasma density, the second in the plasma current, both of which (independently) vary the upstream separatrix collisionality, ν_u^*. In agreement with previous observations, e.g. LaBombard (2003), the SOL mean density profile (measured near the outer mid-plane with a fast-scanning, reciprocating Langmuir probe) was found to broaden with increasing ν_u^*, especially in the far-SOL, Fig. 7.9, and to be characterized by large density fluctuations, increasing with radius from ~0.25 in the near-SOL to ~0.75 in the far-SOL, Fig. 7.10. The radial plasma flow, $\Gamma_\perp$, was also found to increase with collisionality, Fig. 7.11, while the effective radial velocity, defined as $V_\perp = \Gamma_\perp/\overline{n}_e$, increased gradually with radius, approaching

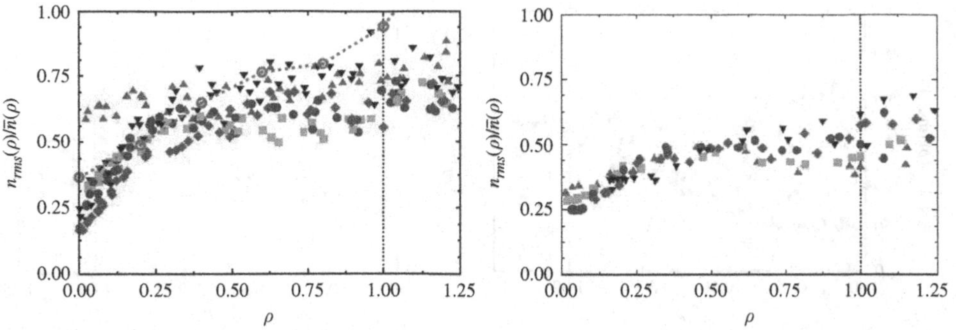

Fig. 7.10. Same as Fig. 7.9 except showing the RMS density fluctuation level, normalized by the mean value. *Source*: Garcia *et al.* (2007a,b).

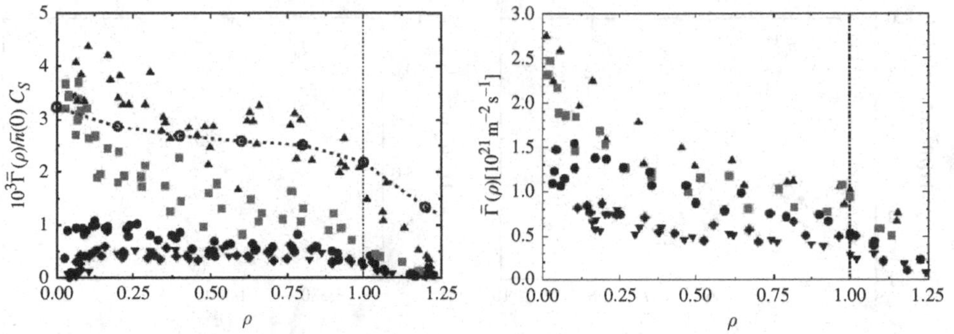

Fig. 7.11. Same as Fig. 7.9 except showing the mean radial flow due to combined density and potential fluctuations. *Source*: Garcia *et al.* (2007a,b).

a flat profile at high ν_u^*, Fig. 7.12;[66] as a result, the radial flux to the wall, as measured by the flow at the first limited flux surface, increased roughly as the square of the upstream density, and decreased inversely with the plasma current, $\Gamma(\rho = 1) \propto \bar{n}_e^{1.8} I_T^{-1}$, Fig. 7.13.

The probability distribution function (PDF) of both density and radial velocity fluctuations in the far-SOL was found to be universal and highly intermittent, Fig. 7.14. Finally, the temporal pulse shape of the plasma filaments recorded by the probe in the far-SOL revealed a sharp leading front and a long trailing wake, Fig. 7.15, suggesting a highly dispersive radial motion.[67] Such filaments have been

[66] A similar increase of $V_\perp$ with radius was observed on many tokamaks (Lipschultz *et al.*, 2007).

[67] The filament velocity is somewhat higher, typically in the range 0.5–2 km/s, or ~1–10% of the plasma sound speed evaluated at the upstream separatrix temperature, $T_u \sim$ 25–100 eV (the values reflect measurements on various machines). Recalling our earlier discussion, Section 2.2.1, we may conclude that SOL turbulence is dominated by drift-ordered dynamics.

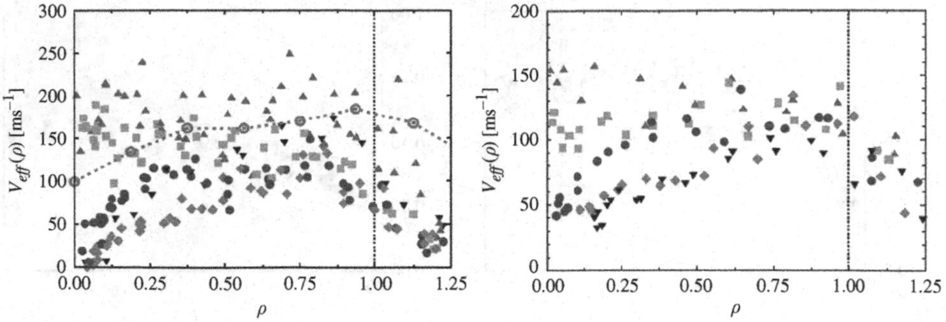

Fig. 7.12. Same as Fig. 7.9 except showing the effect radial velocity of the turbulent flow. *Source*: Garcia *et al.* (2007a,b).

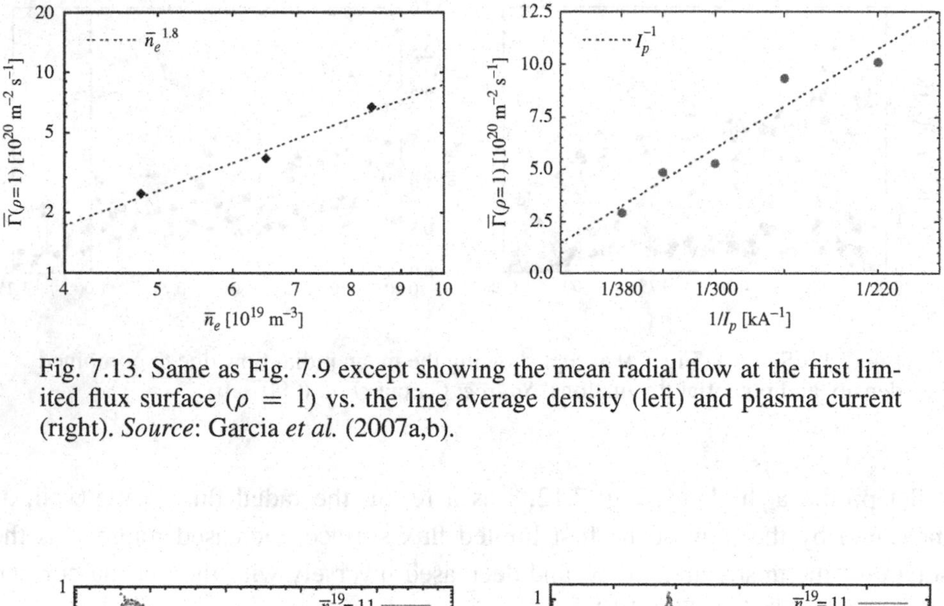

Fig. 7.13. Same as Fig. 7.9 except showing the mean radial flow at the first limited flux surface ($\rho = 1$) vs. the line average density (left) and plasma current (right). *Source*: Garcia *et al.* (2007a,b).

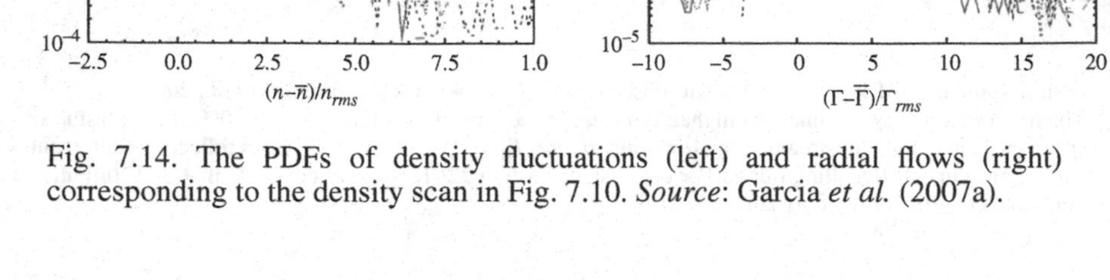

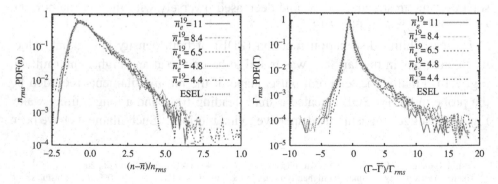

Fig. 7.14. The PDFs of density fluctuations (left) and radial flows (right) corresponding to the density scan in Fig. 7.10. *Source*: Garcia *et al.* (2007a).

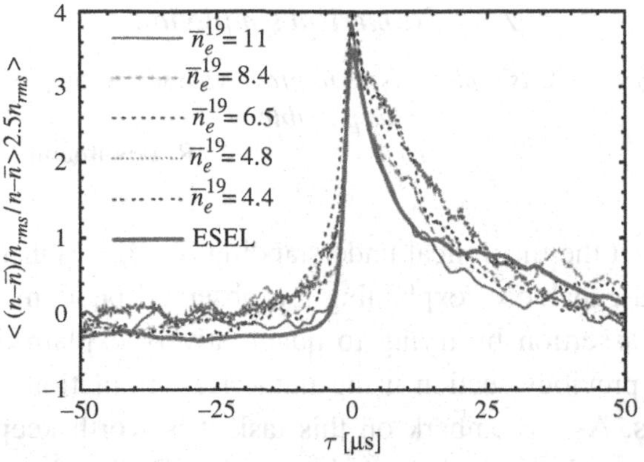

Fig. 7.15. The temporal pulse shape of a density perturbation in the SOL corresponding to the density scan in Fig. 7.10. *Source*: Garcia *et al.* (2007a).

observed by many other diagnostic techniques, the most spectacular being fast camera imaging on MAST (Kirk *et al.*, 2004) and NSTX (Zweben *et al.*, 2007) spherical tokamaks.

It is interesting to observe a link between the increase of radial fluxes with ν_u^* and the maximum density, known as the *density limit*, achievable in a tokamak discharge. Although the detailed scaling of the density limit is still under debate, it can be approximated to a fair accuracy by the *Greenwald density* (Greenwald, 2002),

$$n_{GW} = I_T/\pi a^2 \approx I_T/A_p. \tag{7.50}$$

It is suggestive that $n_{GW}/I_T \sim n_u L_\parallel \sim T_u^2 \nu_u^*$ defines some critical collisionality for a given edge temperature. Since the (turbulent) radial transport in the SOL increases with ν_u^*, this suggests some link between main chamber recycling and the density limit, as observed in LaBombard *et al.* (2005).

Finally, one might expect the upstream SOL plasma to be virtually stagnant due to the localized plasma recycling in the divertor volume, irrespective of ν_u^*, i.e. of the SOL transport regime. In reality, strong parallel flows, with Mach number of order unity, are observed over much of the upstream SOL when the $\mathbf{B} \times \nabla B$ direction points towards the X-point.[68] These parallel SOL flows are believed to originate due to a combination of Pfirsch–Schlüter return flows and ballooning return flows, the latter caused by the turbulent outflow of plasma on the low field side of the torus, see below.

[68] The flows are smaller and point towards the outer target when the $\mathbf{B} \times \nabla B$ direction is reversed.

7.2.2 Numerical simulations

'Your first task as a physicist is to prove yourself wrong as quickly as possible.'

R. Feynmann (c. 1960)

We said earlier that the theoretical understanding of edge-SOL turbulence is relatively mature and capable of explaining the observed phenomenology. It is now time to test this assertion by trying to quantitatively explain the measurements presented in the previous section using numerical simulations based on various drift-fluid models. As we embark on this task it is worth keeping in mind that comparison between theoretical predictions and reality is often a humbling experience, as it has proven many times in the history of fusion research. Indeed, most experimental *discoveries* in this field were not anticipated by theory, but occurred accidently; what is even more sobering, the crucial ones, e.g. the L–H transition, have not been comprehensively explained even after decades of intense research. In short, theory constantly struggles to catch up with empirical fact, making a healthy partnership of the two an essential component of successful scientific research.

7.2.2.1 Interchange motions of plasma filaments

The radial motion of plasma filaments under interchange (curvature + pressure) forcing is governed by the *generalized* vorticity equation, (4.48),[69] which represents the radial momentum balance and/or the quasi-neutrality (charge conservation) condition, $\nabla\cdot\mathbf{J}=0$, with $\mathbf{J}=\mathbf{J}_p+\mathbf{J}_*+\mathbf{J}_\parallel$. In the thin layer approximation, it becomes the *reduced* vorticity equation, derived previously in the context of both flute-reduced MHD, Section 4.2.5, and drift-fluid, Section 6.3.1, approximations. Since we are now interested in plasma filament motion on *open* field lines, we need to modify these results by one additional consideration, namely, the divergence of the parallel current due to parallel losses to the divertor/limiter tiles, Section 7.1.1.

This is most easily done by averaging $\nabla\cdot\mathbf{J}=0$ over a single flux tube stretching between the two solid targets assuming a constant magnetic field,

$$\langle\mathcal{A}\rangle=\int_{-L_\parallel}^{L_\parallel}\mathcal{A}B^{-1}\mathrm{d}l\Big/\int_{-L_\parallel}^{L_\parallel}B^{-1}\mathrm{d}l\approx\int_{-L_\parallel}^{L_\parallel}\mathcal{A}\mathrm{d}l/2L_\parallel. \tag{7.51}$$

[69] Recall that this equation is valid irrespective of the plasma ordering scheme.

Using the sheath-limited expression for the parallel current, (7.14), or its linearized form, (7.15), the average divergence of $J_\parallel$ may be estimated as[70]

$$\langle \nabla_\parallel J_\parallel \rangle \approx \frac{J_\parallel}{L_\parallel} \approx \frac{e n_{e\prec} c_S}{L_\parallel}\left[1-\exp\left(-\frac{e\widehat{\varphi}}{T_e}\right)\right] \approx \frac{e^2 n_{e\prec} c_S}{T_e L_\parallel}\widehat{\varphi}. \tag{7.52}$$

Its impact on SOL plasma dynamics is best understood in terms of the energy balance, $\int \widehat{\varphi}\nabla\cdot\mathbf{J}d\mathbf{x} = 0$, integrated over some region of interest,[71]

$$\int \widehat{\varphi}\nabla\cdot\mathbf{J}_p d\mathbf{x} = -\int \widehat{\varphi}\nabla\cdot\mathbf{J}_* d\mathbf{x} - \int \widehat{\varphi}\nabla\cdot\mathbf{J}_\parallel d\mathbf{x}. \tag{7.53}$$

The first, polarization term gives the change in the perpendicular kinetic energy due to local time variation and transport effects,

$$\int \widehat{\varphi}\nabla\cdot\mathbf{J}_p d\mathbf{x} = \int [\partial_t \mathcal{K} + \nabla\cdot(\mathbf{V}_E\mathcal{K})]d\mathbf{x} = \partial_t \int \mathcal{K}d\mathbf{x} + \int \mathcal{K}\mathbf{V}_E\cdot d\mathbf{S}, \tag{7.54}$$

where $\mathcal{K} = \frac{1}{2}\rho V_E^2$ is the perpendicular kinetic energy density; the second, diamagnetic term gives the interchange drive along with a surface integral,

$$\begin{aligned} -\int \widehat{\varphi}\nabla\cdot\mathbf{J}_* d\mathbf{x} &= \int p\nabla\cdot\mathbf{V}_E d\mathbf{x} - \int \nabla\cdot(\widehat{\varphi}\mathbf{J}_*)d\mathbf{x}, \\ &= -2\int\left[p\mathbf{V}_E\cdot\boldsymbol{\kappa} + \frac{1}{2}\beta\mathbf{J}\cdot\nabla\widetilde{\varphi}\right]d\mathbf{x} - \int \widehat{\varphi}\mathbf{J}_*\cdot d\mathbf{S}, \end{aligned} \tag{7.55}$$

where the beta term is typically negligible in the SOL. The interchange term above is positive, and hence destabilizing, in the region of bad curvature on the outboard side of the torus. Finally, the third, sheath-current term,[72]

$$-\int \widehat{\varphi}\langle\nabla_\parallel J_\parallel\rangle d\mathbf{x} \approx -\int \frac{p_e c_S}{L_\parallel}\frac{e\widehat{\varphi}}{T_e}\left[1-\exp\left(-\frac{e\widehat{\varphi}}{T_e}\right)\right]d\mathbf{x} \approx -\int \frac{p_e c_S}{L_\parallel}\left(\frac{e\widehat{\varphi}}{T_e}\right)^2 d\mathbf{x},$$

is either negative or zero and is hence always stabilizing, i.e. provides a sink for the perpendicular kinetic energy of the plasma filaments;[73] note that the above expression is simply the *sheath dissipation* introduced in (7.16). Finally, it is instructive to combine all terms in a conservative form,

$$\partial_t \int \mathcal{K}d\mathbf{S}_\parallel + \int (\mathcal{K}\mathbf{V}_E + \widehat{\varphi}\mathbf{J}_*)\cdot\frac{d\mathbf{S}_\perp}{2L_\parallel} \approx -\int\left[2p\mathbf{V}_E\cdot\boldsymbol{\kappa} + \frac{p_e c_S}{L_\parallel}\left(\frac{e\widehat{\varphi}}{T_e}\right)^2\right]d\mathbf{S}_\parallel.$$

70 Here we assume symmetry between the inner, i, and outer, o, targets, so that $J_{\parallel i} = -J_{\parallel o}$, where a positive sign indicates current flow towards the outer target, i.e. we assume that there is net flow of charge of $2J_{\parallel o}$ out of the flux tube and no recirculating current between the two targets. More generally, we should write $\langle\nabla_\parallel J_\parallel\rangle \approx J_{\parallel ave}/L_\parallel$, where $J_{\parallel ave} = (J_{\parallel o} - J_{\parallel i})/2$. This symmetry allows us to define $\varphi_0 = 0$ so that $\varphi_{sh} = \varphi_\prec$ and $\widehat{\varphi}_{sh} = \widehat{\varphi} = \varphi_\prec - \varphi_{\prec fl}$.

71 The infinitesimal elements are given by $d\mathbf{x} = dl\,d\mathbf{S}_\parallel$, $d\mathbf{S} = 2d\mathbf{S}_\parallel + d\mathbf{S}_\perp$, $dS_\parallel = \pi dl_\perp^2$.

72 The same result can of course be reached by performing an integration by parts on $\widehat{\varphi}\nabla_\parallel J_\parallel$.

73 That is, it removes part of the energy which would otherwise be transferred into filament motion.

Choosing $dl_\perp$ much larger than the perpendicular size of an isolated filament, the second term on the left-hand side vanishes, leaving

$$\partial_t \int \mathcal{K} d\mathbf{S}_\| \approx -\int \left[2p\mathbf{V}_E \cdot \boldsymbol{\kappa} + \frac{p_e c_S}{L_\|} \left(\frac{e\widehat{\varphi}}{T_e} \right)^2 \right] d\mathbf{S}_\|. \tag{7.56}$$

Including the effects of sheath, as well as viscous, dissipation in the reduced vorticity equation, and approximating the evolution of n and p by an advection-diffusion equation for a single thermodynamic variable,[74] we find

$$d_t \Omega_E + (2/\rho R)\nabla_\wedge p \approx \mu_\perp \nabla_\perp^2 \Omega_E + (e^2 n c_S / L_\| T_e)\widehat{\varphi}, \quad d_t p = \kappa_\perp \nabla_\perp^2 p,$$

where $d_t = \partial_t + \mathbf{V}_E \cdot \nabla$ is the advective derivative, and $\mu_\perp$ and $\kappa_\perp$ are the collisional diffusivities of momentum and energy. Normalizing the perpendicular distance by the characteristic perpendicular size of the filament, ℓ, the time by the ideal interchange growth time, $1/\gamma$, and the pressure, p, which measures a perturbation from a background value, p_0, by some nominal variation, Δp, we find the following dimensionless equation set,

$$d_t \Omega_E + \nabla_\wedge p \approx \widehat{\mu}_\perp \nabla_\perp^2 \Omega_E + \Lambda \widehat{\varphi}, \qquad d_t p = \widehat{\kappa}_\perp \nabla_\perp^2 p, \tag{7.57}$$

where $d_t = \partial_t + \mathbf{b} \times \nabla\widehat{\varphi} \cdot \nabla$. The interchange growth rate is given by

$$\gamma = \left[(g/\ell)(\Delta p/p_0) \right]^{1/2}, \qquad g = 2c_S^2/R, \qquad c_S^2 = p_0/\rho, \tag{7.58}$$

where g is the effective gravity and c_S is the plasma sound speed. The dissipative terms are characterized by the Rayleigh and Prandtl numbers,

$$\mathrm{Ra} = (\widehat{\kappa}_\perp \widehat{\mu}_\perp)^{-1} = (g\ell^3/\kappa_\perp \mu_\perp)(\Delta p/p_0), \qquad \mathrm{Pr} = \widehat{\mu}_\perp/\widehat{\kappa}_\perp = \mu_\perp/\kappa_\perp, \tag{7.59}$$

which measure the ratio of buoyancy to viscosity, and viscosity to heat diffusion, respectively, and the sheath dissipation coefficient,

$$\Lambda = 2c_S\ell^3/\gamma L_\| \rho_S^2 = (2\ell R)^{1/2}(\ell/\rho_S)^2, \qquad \rho_S = c_S/\Omega_i. \tag{7.60}$$

For large Ra and small Λ, the right-hand side in (7.57) is negligible, leaving the only dimensionally allowable scaling for the transverse velocity,

$$M_\perp^{int} = V_\perp^{int}/c_S = (2\ell\Delta p/R p_0)^{1/2}. \tag{7.61}$$

In this ideal interchange scaling, the radial Mach number increases as the square root of the cross-field filament size normalized by the radius of curvature, ℓ/R,

74 Of course, by doing so we neglect any parallel losses of particles and energy from the filament to the solid targets. We will return to consider these losses in the following section.

and the pressure perturbation relative to some background value, $\Delta p/p_0$.[75] In other words, provided dissipation forces are small, we expect spatially larger and more intense perturbations to travel faster relative to the plasma sound speed, and since parallel density losses are also characterized by c_S, recall (7.32), to penetrate further into the SOL,

$$\lambda_n \sim V_\perp \tau_{\parallel n} \sim V_\perp L_\parallel / c_S, \qquad \lambda_n / L_\parallel \sim V_\perp / c_S \sim M_\perp. \tag{7.62}$$

As we shall see presently, the inclusion of viscous and sheath dissipative effects leads to a reduction of $M_\perp$ below the ideal interchange value (7.61), with a corresponding reduction in λ_n. However, before looking at the numerical solutions of (7.57), it is instructive to study the effects of dissipation by taking a spectral decomposition (Fourier transform) of this equation,

$$\mathrm{d}_t \widehat{\varphi}_{\mathbf{k}} - \mathrm{i}\left(k_\wedge / k_\perp^2\right) p_{\mathbf{k}} \approx -\left[\widehat{\mu}_\perp k_\perp^2 + \Lambda / k_\perp^2\right] \widehat{\varphi}_{\mathbf{k}}, \tag{7.63}$$

which reveals the major difference between viscous and sheath dissipation, namely, the fact that the former is most effective at small spatial scales (large $k_\perp$), and the latter at large spatial scales (small $k_\perp$).

As a result, we expect the morphology of the plasma filaments evolving in the two limiting dissipation regimes to differ substantially. Such differences have indeed been observed in numerical solutions of (7.57), starting from an initially stationary, Gaussian filament, $p(\mathbf{x}, t = 0) = \exp\left(-\frac{1}{2}|\mathbf{x}|^2\right)$. The results, as described in Garcia *et al.* (2006a), indicate that: (i) for $\mathrm{Ra} \gg 1$, $\mathrm{Pr} \sim 1$ and $\Lambda \ll 1$, the filament evolves into a mushroom-like shape, typical to all R–T instabilities, consisting of two counter-rotation lobes and an increasing accumulation of pressure at the leading tip, Fig. 7.16; (ii) for $\mathrm{Ra} \sim 100$ and $\Lambda \ll 1$, it takes on a quasi-circular, kidney shape, with the degree of broadening dependent on the Prandtl number, Fig. 7.17; and (iii) for $\mathrm{Ra} \gg 1$, $\mathrm{Pr} \sim 1$ and $\Lambda \geq 1$, it forms a much thinner, arc-like structure, Fig. 7.18. In the ideal case, the centre-of-mass of the filament accelerates in a few interchange times to a maximum velocity which is comparable to, but less than, $V_\perp^{int}$ (7.61), and then decelerates gradually as the filament breaks up into smaller structures. In the sheath dissipative case, $\Lambda \geq 1$, this maximum velocity, and hence the penetration depth of the filament, are substantially reduced, e.g. by roughly a factor of 4 for $\Lambda = 1$.[76] In both cases,

[75] For typical large tokamak values, $n_e \sim 10^{19}\,\mathrm{m}^{-3}$, $T_e \sim T_i/2 \sim 50\,\mathrm{eV}$, $R \sim 2\,\mathrm{m}$, $B \sim 2\,\mathrm{T}$, $q \sim 3$, and cross-field filament sizes and amplitudes of $\ell/R \sim 1\%$, $\Delta p/p_0 \sim 1$, one finds $c_S \sim 80\,\mathrm{km/s}$, $\rho_S \sim 1\,\mathrm{mm}$, $g \sim 7 \times 10^9\,\mathrm{m/s^2}$, $1/\gamma \sim 1.6\,\mu\mathrm{s}$, $M_\perp \sim 0.14$, $V_\perp \sim 10\,\mathrm{km/s}$, which is substantially larger than the typical values of 0.5–2 km/s found in the experiments.

[76] In earlier theories, the interchange term was assumed to be non-linear in p, which allows an analytical solution in the ideal limit of $\mathrm{Ra} \gg 1$, namely, $V_\perp = V_\perp^{int}/\Lambda$. This solution is clearly incorrect for $\Lambda \ll 1$, for which it predicts $V_\perp \gg V_\perp^{int}$, but is also inaccurate for $\Lambda \sim 1$.

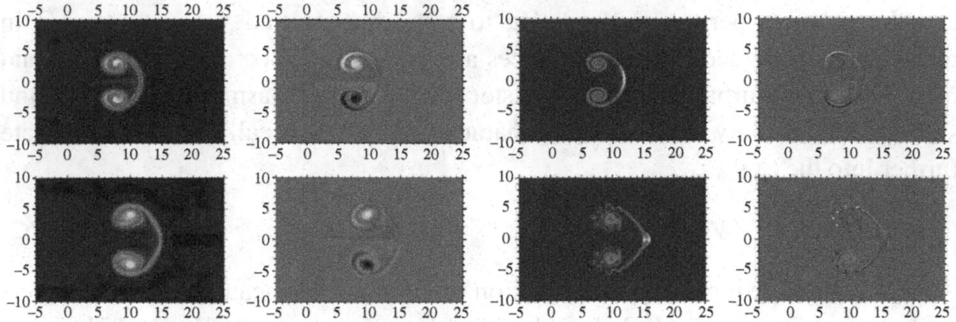

Fig. 7.16. Filament pressure and vorticity profiles at $t = 15$ (top) and 20 (bottom) for $\mathrm{Pr} = 1$, $\Lambda = 0$ and $\mathrm{Ra} = 10^4$ (left) and 10^6 (right). Reused with permission from Garcia *et al.* (2006a), ©AIP.

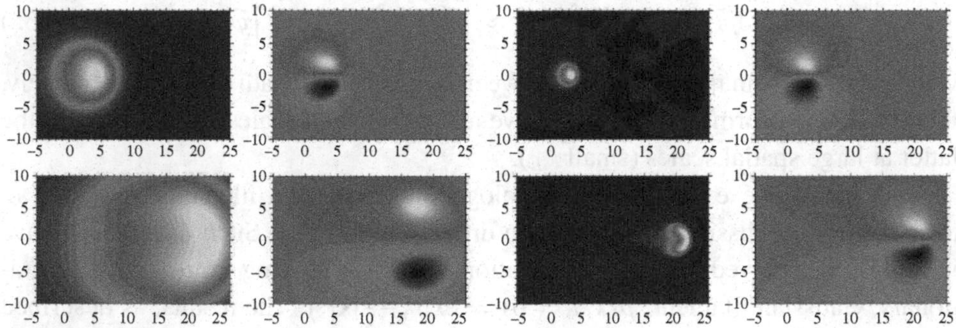

Fig. 7.17. Filament pressure and vorticity profiles at $t = 10$ (top) and 50 (bottom) for Ra = 100, $\Lambda = 0$ and Pr = 0.04 (left) and 400 (right). Reused with permission from Garcia *et al.* (2006a), ©AIP.

the filament travels a distance many times its initial size after ~50 interchange times.

The evolution (in time and space) of the pressure along the symmetry line in the default case ($\mathrm{Ra} = 10^4$, $\mathrm{Pr} = 1$, $\Lambda = 0$, Fig. 7.16) is shown in Fig. 7.19. Note the formation of the leading front and trailing wake, similar to experimental observations, Fig. 7.15.[77] Since parallel losses have been neglected in the model, the pressure in Fig. 7.19 decays due to filament broadening in the drift plane. The evolution of the maximum pressure along the symmetry line for several values of Ra and Pr (all with $\Lambda = 0$) is shown in Fig. 7.20, with the corresponding radial velocities and radial diffusivity shown in Figs. 7.21 and 7.22; these two quantities express filament motion in terms of its two lowest order moments, i.e. its centroid

[77] It is also similar to the break up of vortices in incompressible, two dimensional fluids, recall Section 6.1.3. Since different parts of the filament travel at different radial velocities the filament wave-packet disperses or broadens in the radial direction.

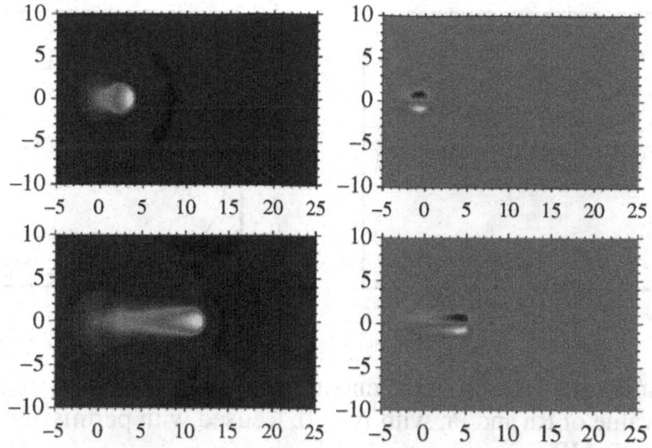

Fig. 7.18. Filament pressure and vorticity profiles at $t = 10$ (top) and 50 (bottom) for Pr = 1, Ra = 10^4 and $\Lambda = 1$. Reused with permission from Garcia *et al.* (2006a), ©AIP.

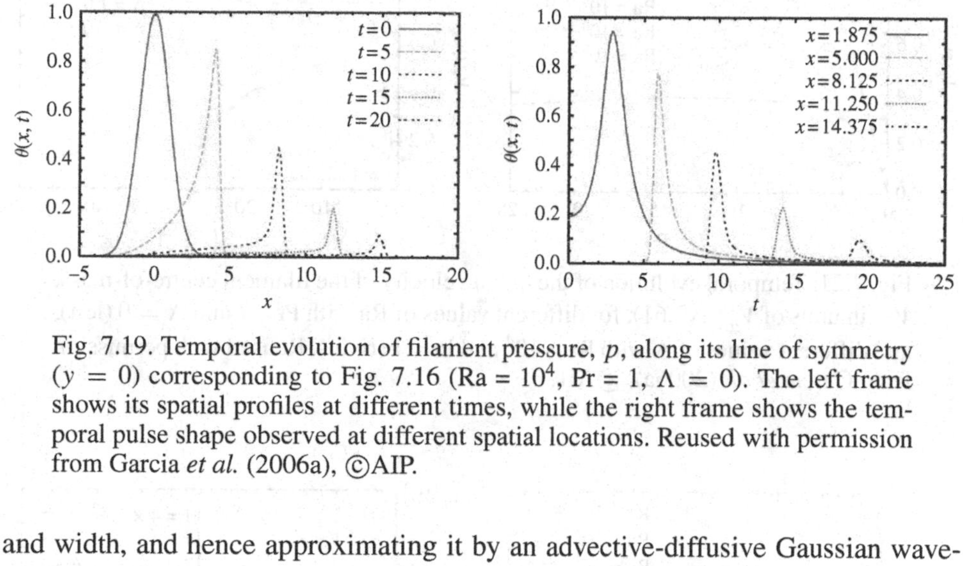

Fig. 7.19. Temporal evolution of filament pressure, p, along its line of symmetry ($y = 0$) corresponding to Fig. 7.16 (Ra = 10^4, Pr = 1, $\Lambda = 0$). The left frame shows its spatial profiles at different times, while the right frame shows the temporal pulse shape observed at different spatial locations. Reused with permission from Garcia *et al.* (2006a), ©AIP.

and width, and hence approximating it by an advective-diffusive Gaussian wave-packet,

$$p(\mathbf{x}, t) = (\pi D_\perp t)^{-1/2} \exp[-(r - V_\perp t)^2/D_\perp t], \qquad \int_{-\infty}^{\infty} p d\mathbf{x} = \text{const.} \quad (7.64)$$

Once again, the filament accelerates to a maximum velocity which is some fraction of the ideal interchange velocity, $V_\perp^{int}$, given by (7.61), and then decelerates as the pressure gradients which drive its motion are reduced.

The variation of the maximum radial velocity, in units of $V_\perp^{int}$, with Ra, Pr and Λ is shown in Fig. 7.23. For a sufficiently high Rayleigh number, when viscous

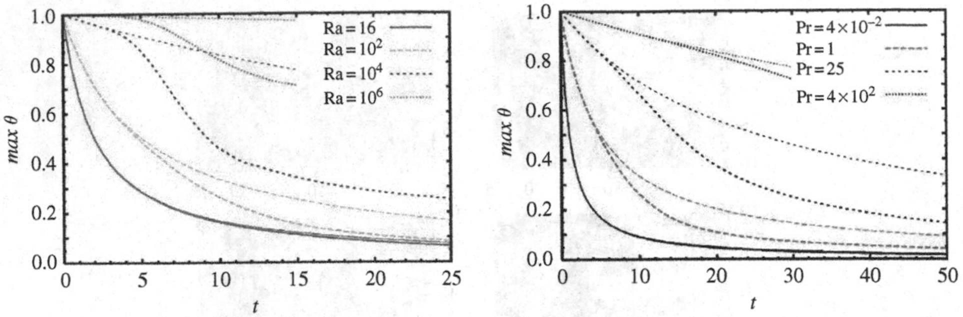

Fig. 7.20. Temporal evolution of the maximum filament pressure, p_{max}, with time for different value of Ra and Pr, with $\Lambda = 0$. Reused with permission from Garcia *et al.* (2006a), ©AIP.

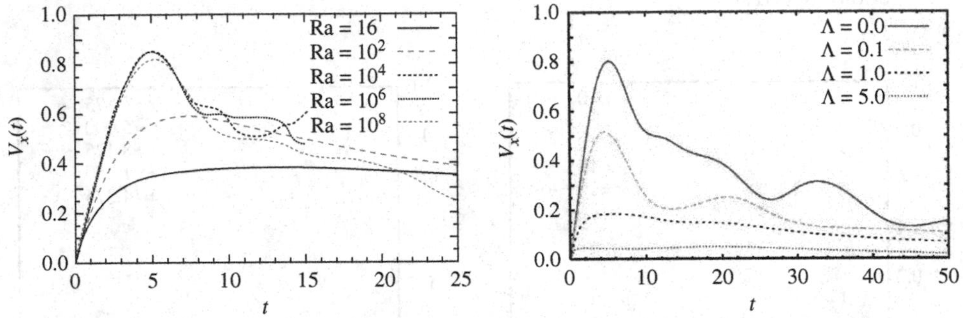

Fig. 7.21. Temporal evolution of the radial velocity of the filament centre-of-mass, $V_\perp$, in units of $V_\perp^{int}$, (7.61), for different values of Ra with Pr = 1 and $\Lambda = 0$ (left), and different values of Λ with Ra = 10^4 and Pr = 1 (right). Reused with permission from Garcia *et al.* (2006a), ©AIP.

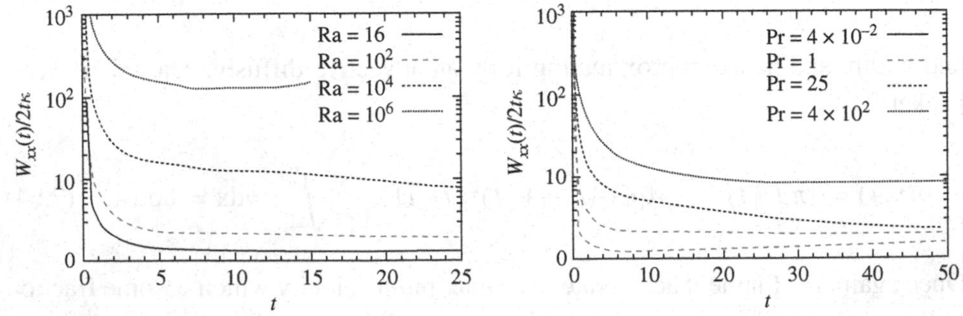

Fig. 7.22. Temporal evolution of the radial diffusivity, $D_\perp = W_{\perp\perp}/2t\kappa$, for different values of Ra with Pr = 1 and $\Lambda = 0$ (left), and different values of Pr with Ra = 10^4 and $\Lambda = 0$ (right). Reused with permission from Garcia *et al.* (2006a), ©AIP.

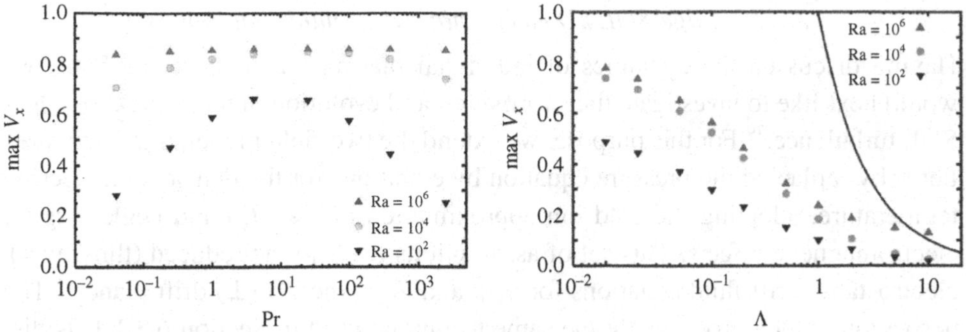

Fig. 7.23. Temporal evolution of the maximum radial velocity of the filament centre-of-mass, $V_\perp$, in units of $V_\perp^{int}$, (7.61), with Ra, Pr and Λ. Reused with permission from Garcia *et al.* (2006a), ©AIP.

effects are negligible, $V_\perp^{max}$ converges to $\sim 0.8 V_\perp^{int}$, and shows little dependence on Pr, but decreases substantially with increasing Λ. In short, $V_\perp^{max} / V_\perp^{int}$ is only weakly dependent on collisional dissipation, but strongly dependent on sheath dissipation.[78] Aside from these variations, the radial Mach number scales with the interchange velocity, (7.61), so that larger and more intense filaments penetrate further into the SOL.

In the presence of parallel losses, (7.64) is modified to read

$$p(\mathbf{x}, t) = (\pi D_\perp t)^{-1/2} \exp[-(r - V_\perp t)^2 / D_\perp t - t/\tau_{\parallel p}], \qquad \int_{-\infty}^{\infty} p d\mathbf{x} = e^{-t/\tau_{\parallel p}},$$

where the first term in the exponential represents the radial translation and broadening and the second term the parallel energy losses. Treating the density and pressure separately, their average values evolve as

$$d_t \overbrace{n} \approx -\overbrace{n} / \tau_{\parallel n}, \qquad d_t \overbrace{T_s} \approx -\overbrace{T_s} / \tau_{\parallel T_s}, \tag{7.65}$$

$$\overbrace{n}(t) \approx \overbrace{n}(0) e^{-t/\tau_{\parallel n}}, \qquad \overbrace{T_s}(t) \approx \overbrace{T_s}(0) e^{-t/\tau_{\parallel T_s}}, \tag{7.66}$$

where the over-brace denotes an average over the cross-section of the filament. Since $\tau_{\parallel n} > \tau_{\parallel T_i} > \tau_{\parallel T_e}$, see (7.36), we expect the filament to cool faster than it rarefies, and for electrons to cool faster than the ions. We will revisit the topic of plasma filament evolution in Section 7.3.1.

[78] It does not scale as $1/\Lambda$, shown by the solid line in Fig. 7.23, as predicted by earlier theories.

7.2.2.2 Edge-SOL 2D electrostatic drift-fluid simulations

Having discussed the dynamics of individual plasma filaments in the SOL, we would next like to investigate their formation and evolution in the context of edge-SOL turbulence.[79] For this purpose, we extend the two-field model, (7.57), to three fields by replacing the pressure equation by equations for the density and electron temperature, adopting the cold ion approximation ($T_i \ll T_e$) and neglecting all electromagnetic effects. This set of assumptions leads to the reduced (thin layer), electrostatic, drift-fluid equations for n, φ and T_e in the 2D ($\perp$) drift plane.[80] The derivation, which proceeds by the same technique used in Section 6.3.1.1, is discussed in detail in Garcia *et al.* (2005). It yields the following set of evolution equations for n, φ and T_e,

$$d_t n + n\mathcal{K}(\varphi) - \mathcal{K}(nT_e) = S_n, \tag{7.67}$$

$$d_t T_e + \tfrac{2}{3}T_e\mathcal{K}(\varphi) - \tfrac{7}{3}T_e\mathcal{K}(T_e) - \tfrac{2}{3}\left(T_e^2/n\right)\mathcal{K}(n) = S_{T_e}, \tag{7.68}$$

$$d_t\Omega_E - \mathcal{K}(nT_e) = S_\Omega, \tag{7.69}$$

where $\Omega_E = \nabla_\perp^2\varphi$ is the electric drift vorticity, d_t is the advective derivative with respect to $\mathbf{V}_\perp = \mathbf{V}_E$, and $\mathcal{K}$ is the (toroidal) curvature operator,

$$d_t = \partial_t + \mathbf{V}_E \cdot \nabla\varphi = \partial_t + B^{-1}\mathbf{b} \times \nabla\varphi \cdot \nabla, \tag{7.70}$$

$$B^{-1} \approx 1 + R/R_0 = 1 + (a + \rho_{S0}x)/R_0, \tag{7.71}$$

$$\mathcal{K} = -(\rho_{S0}^2/R_0)\nabla_\wedge = -(\rho_{S0}/R_0)\partial_y. \tag{7.72}$$

All the quantities in the above equations, with the exception of the minor and major radii, a and R_0, and the (hybrid) thermal gyro-radius, $\rho_{S0} = c_{Se0}/\Omega_{i0}$, are dimensionless and expressed in the *Bohm normalized* form, (6.214), with the temporal and spatial scales normalized by $\Omega_{i0} = eB_0/m_i$ and ρ_{S0},

$$e\varphi/T_{e0} \to \varphi, \qquad t\Omega_{i0} \to t, \qquad \mathbf{x}/\rho_{S0} \to \mathbf{x}, \tag{7.73}$$

$$n/n_0 \to n, \qquad T_e/T_{e0} \to T_e, \qquad B/B_0 \to B. \tag{7.74}$$

In the above, $c_{Se0} = (ZT_{e0}/m_i)^{1/2}$ is the cold ion plasma sound speed, B_0 is the magnitude of the local magnetic field strength and the zero subscript indicates nominal (dimensional) values.[81] Finally, the distances in the radial ($\perp$) and diamagnetic

[79] Due to the formation of filaments, which generate strong fluctuations in SOL plasma quantities, the corresponding dynamics is not amenable to the mean-field approach, and requires instead a fully non-linear treatment, i.e. following full quantities, rather than only their fluctuations.

[80] Note that by integrating over the parallel direction, all drift-wave dynamics, which are the dominant processes in edge plasma turbulence, are effectively removed. Nonetheless, the model is appropriate to electrostatic plasma dynamics in the SOL region, where drift-waves are quickly damped by the sheath dissipation, leaving the interchange forcing to dominate the dynamics.

[81] This is typically chosen at the separatrix location on the outboard mid-plane of the torus.

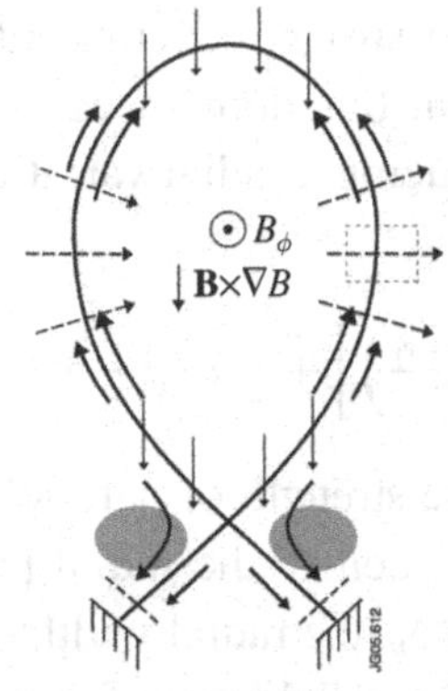

Fig. 7.24. Schematic of neoclassical, Pfirsch–Schlüter transport on closed and open flux surfaces, for $\mathbf{B} \times \nabla B$ direction towards the X-point (vertically down). The thin arrows represent downward ion magnetic drifts, the thick arrows the parallel return flows and the dashed arrows the resulting radial transport. The shaded region illustrates the area of strong plasma–neutral interaction and thus increased recycling in the divertor plasma. The location of the magnetic pre-sheath (not to scale) is indicated by the dashed lines adjacent to the divertor targets. Finally, the dotted box represents the poloidal projection of the computational domain (not to scale) used in ESEL simulations. *Source*: Fundamenski *et al.* (2007b).

($\wedge$) directions in the 2D drift plane are described by the dimensionless co-ordinates x and y.

The S_α terms (with $\alpha = n, \Omega, T_e$) on the right-hand side of (7.67)–(7.69) represent dissipation as a result of perpendicular collisional diffusion, $D_{\perp\alpha}\nabla_\perp^2$, and parallel losses to the divertor targets, $\alpha/\tau_{\|\alpha}$,

$$S_n = D_{\perp n}\nabla_\perp^2 n - n/\tau_{\|n}, \tag{7.75}$$

$$S_{T_e} = D_{\perp T_e}\nabla_\perp^2 T_e - T_e/\tau_{\|T_e}, \tag{7.76}$$

$$S_\Omega = D_{\perp\Omega}\nabla_\perp^2 \Omega - \Omega/\tau_{\|\Omega}, \tag{7.77}$$

where $D_{\perp a}$ are the perpendicular diffusivities of particles, momentum and electron heat and $\tau_{\|a}$ are the corresponding parallel loss times.[82] These six parameters are both necessary and sufficient to close the dynamical system (7.67)–(7.69). The equations are then discretized in two spatial dimensions, x and y, with the grid size comparable to ρ_{S0} and the box size having side lengths of hundreds of ρ_{S0}. In keeping with the thin layer approximation for the curvature operator, the simulation domain is limited to the neighbourhood of the outer mid-plane, where

[82] In the Bohm normalization, the units of $D_{\perp\alpha}$ and $\tau_{\|\alpha}$ are $\rho_{S0}^2\Omega_{i0}$ and $1/\Omega_{i0}$, respectively. Since we are following a rigorous theory of neoclassical transport on open field lines, the latter must be constructed with some care from the collisional expressions on closed field lines, see Fig. 7.24.

the interchange instabilities are strongest.[83] One of the advantages of the parallel loss time formulation is that this domain can easily span both the edge and SOL regions, simply by introducing a radial variation to the parallel loss times in (7.75)–(7.76), e.g.

$$\tau_{\|\alpha}^{-1} = \tfrac{1}{2}\sigma_{\alpha l}\left[1 + \tanh\left(\tfrac{x-x_l}{\delta_l}\right)\right] + \tfrac{1}{2}(\sigma_{\alpha w} - \sigma_{\alpha l})\left[1 + \tanh\left(\tfrac{x-x_w}{\delta_w}\right)\right], \quad (7.78)$$

where $\sigma_{\alpha l}$ and $\sigma_{\alpha w}$ represent the strength of parallel losses in the SOL and limiter shadow, respectively, x_l and x_w denote the radial positions of the LCFS and first limited flux surface, and δ_l and δ_w the radial widths over which the parallel losses are activated.[84] The absence of parallel losses for $x < x_l - 2\delta_l$ and their presence for $x > x_l + 2\delta_l$, with $\tau_{\|\alpha} = 1/\sigma_\alpha$, simulate the regions of closed (edge) and open (SOL) field lines, respectively.

The resulting model is frequently referred to as ESEL, to emphasize its treatment of *edge-SOL electrostatic* turbulence.[85] Despite its simplicity, it has been remarkably successful at reproducing a wide range of experimental measurements of edge-SOL turbulence, e.g. most of the observations discussed in Section 7.2.1, including the radial profile of the density, temperature and radial flows, the PDFs of relative fluctuations, the degree of intermittency, and finally, the pressure pulse due to the filament in the far-SOL, see Figs. 7.9–7.15. The level of agreement is impressive, considering the fully a priori nature of the model,[86] and provides strong evidence for the dominance of interchange dynamics in SOL turbulence. This is further confirmed by the morphology of the plasma structures, e.g. as observed by fast camera imaging on NSTX (Zweben *et al.*, 2004) and obtained in ESEL simulations – a snapshot of the typical density and vorticity profiles in the x–y plane during the ejection of a plasma filament (blob) is shown in Fig. 7.25 – both of which are characterized by mushroom-like structures followed by thin tails, characteristic of the R–T instability.

Nonetheless, the model does not contain the effect of sheath dissipation and is hence only relevant to the highly collisional, conduction-limited regime, as is indeed the case for the simulation shown in Figs. 7.9–7.15. It is therefore not able to address the observed collisionality dependence in the experiments. However, a likely explanation for this dependence may be formulated based on the filament

[83] The inner mid-plane is of course stable to interchange dynamics.

[84] Typically $\delta_l = \delta_w = 1$, with the same radial profile shape used for all three fields.

[85] It should be stressed that the ESEL model/simulations owe a large debt to the work of Sarazin and Ghendrih (1998) and Beyer *et al.* (1999) and that they are part of a larger edge-SOL drift-fluid modelling effort, e.g. Xu *et al.* (2002); Myra *et al.* (2006, 2008); Russell *et al.* (2007).

[86] Additional information, beyond that contained in the reduced fluid model, is clearly required to determine the magnitude and the functional form of the collisional and parallel damping terms, $D_{\perp\alpha}$ and σ_α. These may be derived from first principles based on a judicious combination of neoclassical and classical collisional transport coefficients in the perpendicular and parallel directions, respectively, see Fundamenski *et al.* (2007a).

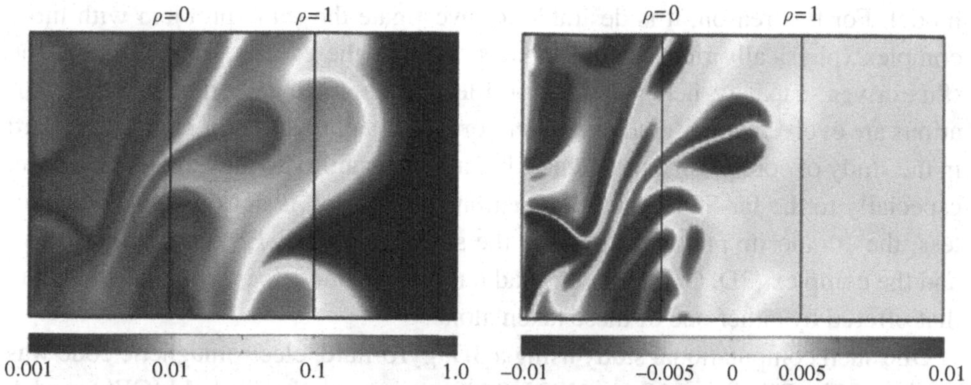

Fig. 7.25. Snapshots of two-dimensional distribution of the density (left) and potential (right) in the edge, SOL and limiter shadow regions near the outer mid-plane, as predicted by an ESEL simulation for typical TCV conditions, i.e. corresponding to the dotted lines in Figs. 7.9–7.12. *Source*: Garcia *et al.* (2006b).

propagation study, specifically Fig. 7.23: At low ν_u^* (in the sheath-limited regime), the sheath energy dissipation tends to reduce the interchange drive and slow down radial filament motion. As ν_u^* increases, plasma filaments become electrically isolated from the sheath, making the interchange drive more effective, thus increasing the effective radial velocity and broadening the far-SOL density profile. This mechanism could explains why the above ESEL model has overestimated, by roughly a factor of 3, the density width on JET, whose SOL is far less collisional than that on TCV (Fundamenski *et al.*, 2007a). However, a numerical validation is clearly necessary to test the above hypothesis.

Finally, the ESEL model offers a simple explanation of the 'ballooning' component, i.e. the part independent of the $\mathbf{B} \times \nabla B$ direction, of the measured parallel SOL flow, which can be explained in terms of the over-pressurization of individual SOL flux tubes as they are perturbed by the radially moving plasma filament. This explanation, as put forth in Fundamenski *et al.* (2007a), has been confirmed by 3D simulations in limiter geometry, e.g. see Tamain *et al.* (2009). As a result we next turn to the discussion of three-dimensional simulations.

7.2.2.3 Edge-SOL 3D electromagnetic gyro-fluid simulations

Aside from the perennial virtue of simplicity, the 2D ESEL model discussed above has the additional benefit of being 'flux driven' or 'global', i.e. of following total quantities, rather than only their deviations away from mean (profile) values, which as we saw in Section 7.2.1 can approach unity in the SOL. However, this very simplicity is also its main shortcoming. Indeed, were it not for the strong support from experimental data, it would call into question the verisimilitude of such a sparse model. For this reason, it is desirable to investigate the same problem with more complex (physically richer) fluid models, even at the expense of sacrificing the 'flux driven' capacity, i.e. even if the profiles remain fixed and only plasma fluctuations are evolved. This 'gradient driven' or 'local' approach is, of course, standard in the study of core plasma turbulence, but its applicability to the edge-SOL region, especially to the far-SOL where fluctuations are largest, is not obvious. Nonetheless, the composite picture offered by the simple (2D, DF, ES) and 'flux driven', and the complex (3D, GF, EM) yet 'gradient driven', models is more valuable than that offered by either one of these taken alone.

One such computational study using a 3D, gyro-fluid, electromagnetic code was carried out by Ribeiro and Scott (2005, 2008), first using the six-field (GF6) model, then using the more complete 14-field (GF14) model, both of which were described in Section 6.3.1. Before discussing the findings of this study, let us begin with an extended quote from Ribeiro and Scott (2005), which nicely summarizes the differences between plasma turbulence in the closed (edge) and open (SOL) field line regions:[87]

'In gradient driven turbulence in magnetized plasmas, the evolution of the system is dictated by the balance between the linear and nonlinear processes. The latter include an inverse energy cascade (for $\mathbf{E} \times \mathbf{B}$ drift energy, $\widetilde{\varphi}$) due to the vorticity nonlinearity and a direct energy cascade (for the thermal energy, $\widetilde{p}_e$) due to the $\mathbf{E} \times \mathbf{B}$ advection of the pressure. The former involve the adiabatic response and the interchange forcing, both controlling the energy transfer between electrostatic potential, $\widetilde{\varphi}$, and the electron pressure, $\widetilde{p}_e$. The source of free energy is given by the advection down the background pressure gradient, and the sink provided by resistive dissipation, acting on the parallel current, $\widetilde{J}_\parallel$ ...

In the tokamak SOL, the field lines end on the divertor plates or the limiters, leading to linearized forms of the sheath dissipation, (7.16). The field lines are therefore not closed, and the periodicity constraint, (4.136), does not apply. Hence, the *convective cell* ($k_\parallel = 0$) modes, finite-sized eddies in the perpendicular plane which are perfectly aligned to the magnetic field ($k_\parallel = 0$ with $k_\perp \neq 0$), are allowed to exist. These are distinguished by the fact that they do not involve parallel dynamics. This allows pure *interchange* (κ) dynamics to control the energetic phenomena at all perpendicular wavelengths in the spectrum. In contrast, the situation on closed field lines (edge), with no $k_\parallel = 0$ modes, is that the parallel electron dynamics acts on all available degrees of freedom, keeping $\widetilde{\varphi}$ coupled to $\widetilde{p}_e$ through parallel forces mediated by $\widetilde{J}_\parallel$. This is called *adiabatic coupling*, since it also involves the compressibility of the electron flow along the magnetic field, specifically $\nabla_\parallel \widetilde{J}_\parallel$, which must be balanced by a compressible polarization current in the ions. At the nonlinear level, this type of dynamics is referred to as *drift wave* (DW) turbulence, and with finite resistivity allowing an imbalance in the parallel electric and pressure forces on the electrons, as *collisional drift wave* (CDW) turbulence. This parallel force imbalance, necessary for free energy access and transport, is referred to as *nonadiabatic electron*

[87] The introductory section in Ribeiro and Scott (2008) offers an equally good summary. To assist the reader, notation and references to equations in the present text have been added.

dynamics and can be quantified using mode structure diagnostics. CDW turbulence persists in toroidal geometry even with toroidal compressibility, leading to the possibility of κ dynamics. This situation also persists when, due to the finite pressure, the electron response is electromagnetic, leading to *drift Alfvén* (DA) dynamics. However, in the SOL the adiabatic response does not apply to the $k_\parallel = 0$ modes, and large scale κ dynamics becomes dominant.

The basic character of low frequency drift dynamics is for the very longest parallel wavelengths, and medium range in the perpendicular wavelengths between profile scale length, $L_\perp \approx L_p$, and the drift scale, $\rho_S = c_S/\Omega_i$, to be responsible for the energetics. On closed flux surfaces, the $k_\parallel = 0$ mode is absent and the energetics resides in a set of modes with the average $k_\parallel$ between the lowest allowed value and one to two times $1/qR$. On open flux surfaces, the $k_\parallel = 0$ mode is available, so that energy placed into finite-$k_\parallel$ modes tends to be passed into the convective cells through nonlinear wave–wave coupling (inverse energy cascade). Indeed, there is experimental evidence that in the SOL the observed values of $k_\parallel$ are statistically indistinguishable from zero. Therefore, the fact that $k_\parallel = 0$ modes are allowed in one situation but not in the other can be expected to lead to significant qualitative differences in the basic results, mode structure and transport.'

The key point to take away from the above passage is the impact of magnetic topology (field line periodicity in the edge, and its absence in the SOL) on the physical mechanisms governing plasma turbulence in the two regions. Indeed, one may paraphrase this point as follows: the presence of the Debye sheath at the divertor/limiter tiles suppresses the drift-wave energy coupling, leaving the interchange coupling to dominate the dynamics (Naulin *et al.*, 2004). It is this fact, borne out by detailed simulations including both mechanisms, which justifies the use of the simplified, interchange driven drift-fluid model discussed in the previous section, and largely explains its success in matching the experimental data.

As with the DF4 and DF6 simulations discussed in Section 6.3.1, the GF6 and GF14 models were phrased in terms of field aligned co-ordinates,

$$x = r - a, \quad y_k = [q(\theta - \theta_k) - \zeta]a/q_a = \xi a/q_a, \quad s = q_a R_0 \theta, \tag{7.79}$$

where θ and ζ are the 2π-periodic poloidal and toroidal angles. Hence, the s co-ordinate follows the field line, the y co-ordinate acts as a field line label and x is a radial co-ordinate. Since the Jacobian of the above system is unity, (7.79) are an instance of Hamada co-ordinates, see Section 3.2.1.

The simulations were performed for a set of conditions corresponding to the L-mode edge of a typical medium sized tokamak,[88] c.f. (6.190)–(6.191),

$$\hat{\epsilon} \approx 18\,350, \quad \omega_B = 0.05, \quad \hat{s} = 1, \quad \tau_i = 1, \tag{7.80}$$

$$\hat{\beta} = 2, \quad \hat{\mu} = 5, \quad \hat{\nu} = 3, \quad \hat{C} = 0.51\hat{\mu}\hat{\nu} = 2.55\hat{\nu} = 7.65, \tag{7.81}$$

[88] For example, Asdex Upgrade: $n_e = 4.5 \times 10^{19}\,\mathrm{m}^{-3}$, $T_e = T_i = 80\,\mathrm{eV}$, $B_0 = 2.5\,\mathrm{T}$, $R_0 = 1.65\,\mathrm{m}$, $a = 0.5\,\mathrm{m}$, $q = 3$, $L_\perp = L_p = 3.65\,\mathrm{cm}$, and single-ion species with $A = 2$, $Z = 1$.

which we expect to exhibit both EM/DA ($\hat{\beta} > 1$) and CDW ($\hat{C} > 1$) character; it is worth recalling the significance of these parameters, Section 6.3.1:

$$\hat{\epsilon} \gg 1 \quad \Rightarrow \quad \text{flute-reduced ordering,} \tag{7.82}$$

$$\hat{\mu} > 1 \quad \Rightarrow \quad \text{electron inertia important,} \tag{7.83}$$

$$\hat{\beta} > 1 \quad \Rightarrow \quad \text{induction important,} \tag{7.84}$$

$$\hat{C} > 1 \quad \Rightarrow \quad \text{collisions important,} \tag{7.85}$$

where $\hat{\mu} > 1$ can be said to define the *edge*, as opposed to the *core*, region. For the same set of parameters,[89] (7.80)–(7.81), simulations were performed with different boundary conditions at the inner mid-plane ($s = \pm\pi$): (i) *periodic*, corresponding to the closed field line (edge) region; (ii) Debye *sheath limited*, corresponding to the open field line (SOL) region;[90] and (iii) *vanishing boundary flow*, corresponding to the fictional (unphysical) case of open field lines without an electrostatic sheath, which, by comparison with (ii), measures the impact of the sheath on SOL turbulence.

The background profiles are not fixed, as in a standard local simulation, but rather folded into the fluctuation variables, whose time averages are now finite and give the mean quantities, i.e. $\widetilde{n}_e$ is really $n_e = \overline{n}_e + \widetilde{n}_e$. Since these are evolved using the fluctuation equations, the implicit assumption is that $\widetilde{n}_e/\overline{n}_e \ll 1$, which as we saw earlier can be violated in the SOL, especially the far-SOL region. Moreover, the thin layer approximation is invoked, yet the profiles vary significantly over the radial extent of the grid. Finally, the simulation parameters (7.80)–(7.81) are held fixed across the domain, whereas in reality they would depend on the mean quantities, which vary in space and time. For these reasons, the model can be said to be semi-global, in that it evolves the background quantities, but does so in a way which is partially inconsistent with its own assumptions.

The flux-surface average, or 'zonal', density profile is controlled by means of a localized source (sink) on the left (right) boundary of the domain, which control the mean radial particle flow through the system, and since radial transport is self-organizing, control the mean radial density gradient. The resulting profiles in the three cases are shown in Fig. 7.26, in which the flattening of the density profile from (i) to (iii) reflects the increase in the radial turbulent flow.[91] To compare with a fully local computation, one could use these profiles to calculate $L_n = 1/|\nabla \ln n_{e0}|$ and $\omega_n = L_\perp/L_n$, which would decrease progressively from (i) to (iii), e.g. case

89 As a result, the simulations do not constitute a comparison between plasma turbulence in the edge and SOL regions of the same discharge, in which these parameters would differ substantially between the two regions, but only a study of the effect of poloidal boundary conditions.

90 Or rather, to the *sheath limited* regime of parallel SOL transport, recall Section 7.1.4.5.

91 The small peak near the left boundary is caused by the above-mentioned source there.

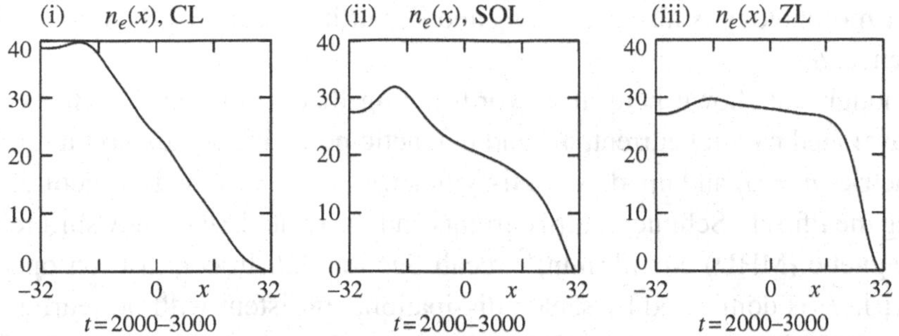

Fig. 7.26. Time- and flux-surface (zonal) average radial density profiles in the edge (i), SOL (ii) and SOL without sheath (iii). *Source*: Ribeiro and Scott (2005).

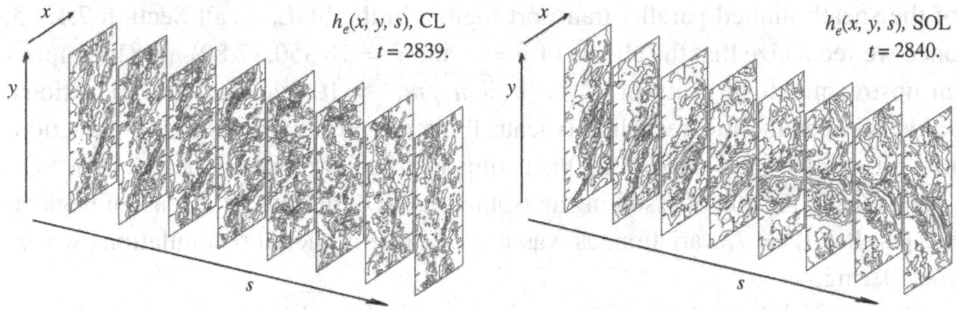

Fig. 7.27. The left frame shows the snapshot of the 3D distribution of $\widetilde{h}_e = \widetilde{n}_e - \widetilde{\varphi}$ in the edge region (closed field lines, periodic boundary conditions at the inner mid-plane, $s = \pm\pi$), while the right frame shows a similar snapshot for the same simulation parameters but with SOL boundary conditions (Debye sheath at $s = \pm\pi$). The s, y and x co-ordinates measure distances in the $\|$, $\wedge$ and $\perp$ directions; the entire computational domain is shown. *Source*: Ribeiro and Scott (2005).

(i) corresponds roughly to $\omega_n \approx 1/2$. Indeed, a slow feedback process, on the transport time scale, is used to adjust the source/sink to attain this nominal value.

Let us next examine the 3D distribution of $\widetilde{h}_e = \widetilde{n}_e - \widetilde{\varphi}$ in cases (i) and (ii) at the time of fully developed turbulence, when transport equilibrium has been attained, Fig. 7.27. Comparing the two plots reveals the fundamental difference between edge and SOL turbulence, i.e. the presence of large-scale, field-aligned, convective cell ($k_\| \ll 1/qR$) modes in the latter and their absence from the former, confirming our expectations. This is directly observed by following individual structures in subsequent s frames and noting that those with finite perpendicular extent lose their identity before a full turn around the torus, so that $k_\| \sim 1/qR$;[92]

[92] In the closed field line region, $k_\| = (m - qn)/qR$ cannot vanish on irrational flux surfaces.

similar morphology is also seen in $\widetilde{n}_e$ and $\widetilde{\varphi}$, but it is most clearly visible in their difference, $\widetilde{h}_e$.

Although not shown here, it is worth noting that on closed field lines (i) the time-averaged parallel current, $\widetilde{J}_\parallel$, and magnetic potential, $\widetilde{A}_\parallel$, exhibit a toroidally symmetric ($n = 0$) and up–down anti-symmetric ($m = \pm 1$) mode structure, representing the Pfirsch–Schlüter return currents and the related Shafranov shift required for magnetic (MHD) equilibrium,[93] recall Section 3.2. In contrast, on open field lines (ii), $\widetilde{J}_\parallel$ is dominated by sheath dissipation, consistent with our earlier claim that the SOL plasma in not in MHD equilibrium. Moreover, in the latter case $\widetilde{n}_e$ and $\widetilde{\varphi}$ are coupled by the linear sheath boundary condition, (7.16), and are roughly constant along the field line between the two sheaths, so that both quantities become effective flux surface labels. This result is also consistent with our earlier discussion of the sheath-limited parallel transport regime in the SOL, recall Section 7.1.4.5, once we recognize that the choice of $\hat{\nu} = 3$ and $\hat{\epsilon} = 18\,350$, (7.80)–(7.81), implies an upstream collisionality of $\nu_u^* \approx \hat{\nu}\sqrt{\hat{\epsilon} m_e/m_i} \approx 10$, which is the transitional value between the low recycling (sheath-limited) and high recycling (conduction-limited) regimes, see page 314. One is thus justified in omitting the effect of SOL plasma recycling and of assuming an isothermal SOL plasma, or at least of neglecting parallel $T_e = T_i$ variation, as was indeed done in the GF6 simulations we are considering.

The poloidal dependence of the average squared amplitudes of $\widetilde{\varphi}$, $\widetilde{p}_e$ and $\widetilde{h}_e$, as well as the corresponding radial $\mathbf{E} \times \mathbf{B}$ outflow of particles are shown in Fig. 7.28. The latter, along with $\widetilde{p}_e$ and $\widetilde{h}_e$, are less ballooned in the SOL (ii) than in the edge (i), due to the appearance of the $k_\parallel \approx 0$ modes, and become virtually uniform in (iii). The relative amplitude of the three terms is also significant, with $\langle \widetilde{p}_e^2 \rangle$ comparable to, and even locally larger than, $\langle \widetilde{\varphi}^2 \rangle$, in (i), becoming much smaller everywhere in (ii), and negligibly small in (iii); indeed, averaged over the whole domain one finds that their ratio is roughly inverted between the edge and SOL regions, with $\langle \widetilde{p}_e^2 \rangle / \langle \widetilde{\varphi}^2 \rangle \sim 4$ in (i), ~ 0.25 in (ii) and $\ll 1$ in (iii). In all cases, the magnetic energy, $\langle \widetilde{A}_\parallel^2 \rangle$ (not shown here), is more than two orders of magnitude smaller than either the thermal, $\langle \widetilde{p}_e^2 \rangle$, or drift, $\langle \widetilde{\varphi}^2 \rangle$, energies.[94] The resulting radial heat transport, averaged over s and t, is much larger in (ii) than in (i), with $Q_e = \langle \widetilde{p}_e \widetilde{V}_E^x \rangle \approx 11$ and ≈ 2.5 in the two cases; moreover, when sheath dissipation is artificially suppressed (iii), radial transport increases further to $Q_e \approx 15.5$, as the interchange drive is then wholly unchecked and drives the outward motion of

[93] Since $\widetilde{J}_\parallel$ and $\widetilde{A}_\parallel$ involve the mean quantities in this semi-global model, the MHD (or neoclassical) equilibrium becomes part of the turbulent dynamics, as shown in Fig. 6.2, i.e. it can be said to be 'dynamic' with flux surfaces fluctuating about their mean values.

[94] This smallness of the magnetic energy and the related magnetic flutter transport is the implied meaning of the label 'electrostatic transport' often associated with $E \times B$ turbulence.

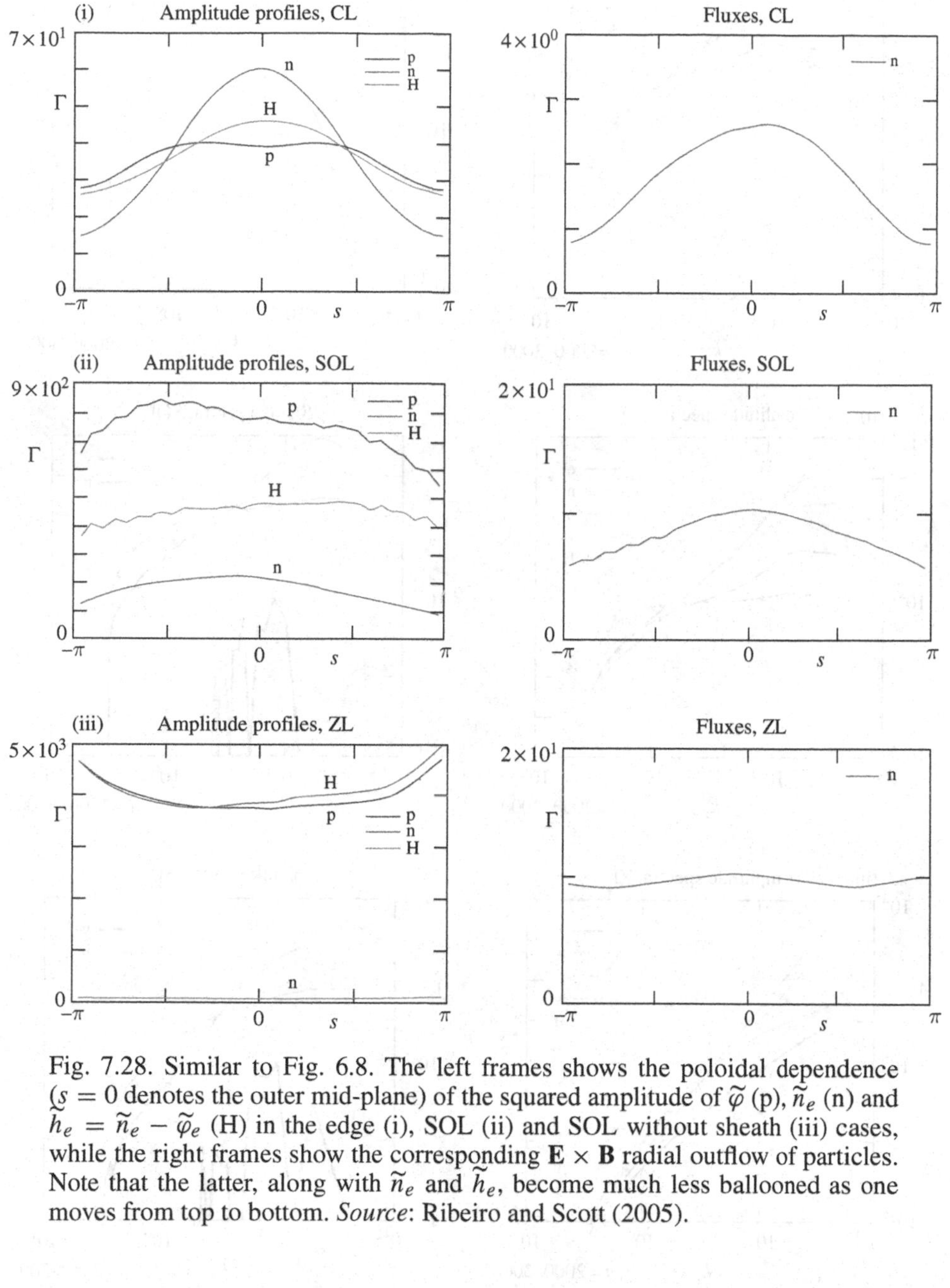

Fig. 7.28. Similar to Fig. 6.8. The left frames shows the poloidal dependence ($s = 0$ denotes the outer mid-plane) of the squared amplitude of $\widetilde{\varphi}$ (p), $\widetilde{n}_e$ (n) and $\widetilde{h}_e = \widetilde{n}_e - \widetilde{\varphi}_e$ (H) in the edge (i), SOL (ii) and SOL without sheath (iii) cases, while the right frames show the corresponding $\mathbf{E} \times \mathbf{B}$ radial outflow of particles. Note that the latter, along with $\widetilde{n}_e$ and $\widetilde{h}_e$, become much less ballooned as one moves from top to bottom. *Source*: Ribeiro and Scott (2005).

$k_\parallel \approx 0$ structures to maximum effect; these findings are broadly consistent with our earlier discussion in Sections 7.2.2.1 and 7.2.2.2.

The $k_\wedge$ spectra of the squared amplitude of $\widetilde{p}_e$, $\widetilde{\varphi}$, $\widetilde{h}_e$, $\widetilde{\mathbf{E} \times \mathbf{B}}$ and $\widetilde{\Omega}$ in (i) and (ii), as well as the corresponding sources and sinks, are shown in Fig. 7.29. In all

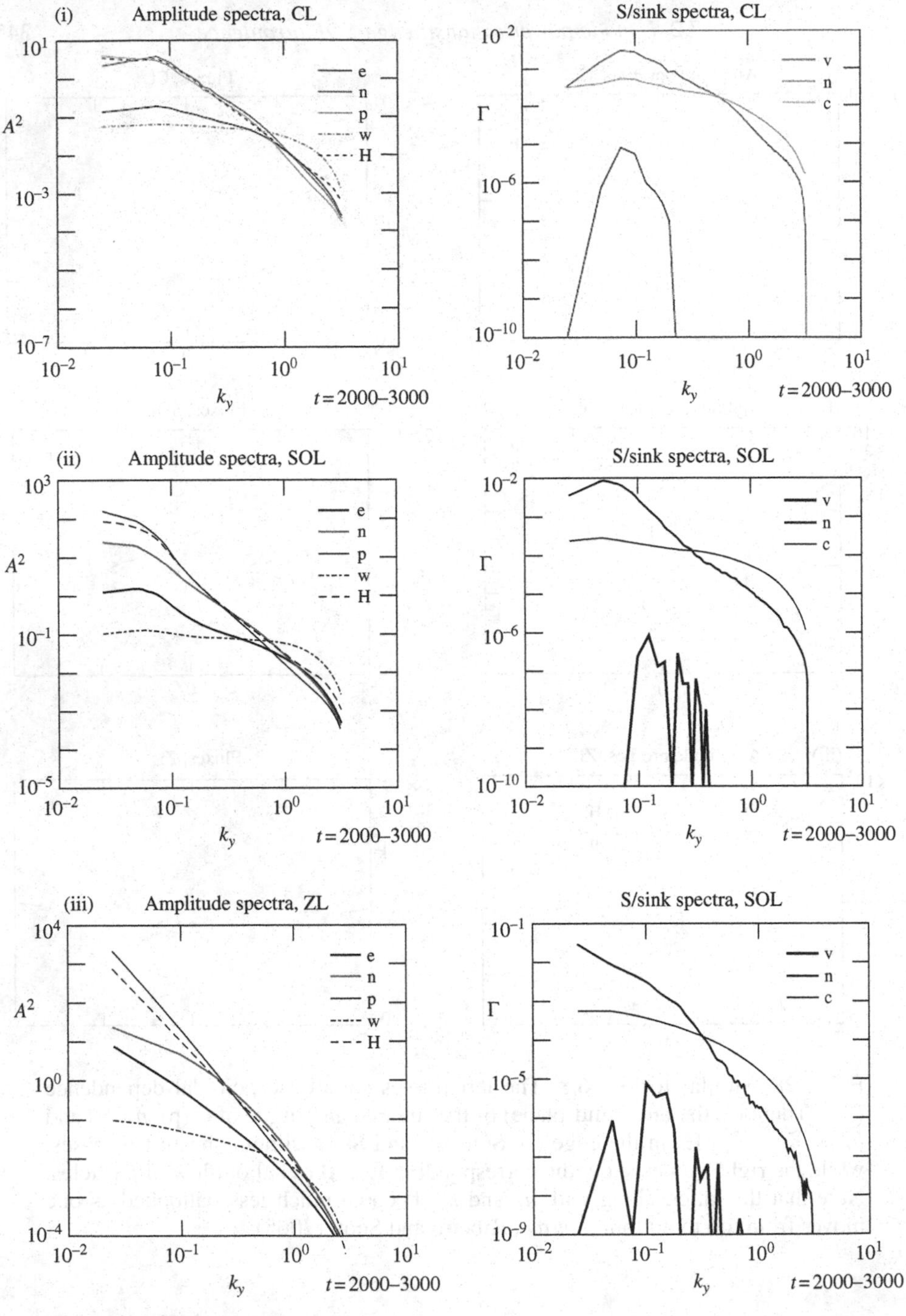

Fig. 7.29. Similar to Figs. 6.4–6.5. The left frames shows the $k_\wedge$ spectra of the squared amplitude of the $\widetilde{\mathbf{E}} \times \mathbf{B}$ energy (e), $\widetilde{\varphi}$ (p), $\widetilde{n}_e$ (n), $\widetilde{\Omega}$ (w) and $\widetilde{h}_e$ (H) in the edge (i), SOL (ii) and SOL without sheath (iii) cases. The right frames show the corresponding sources and sinks in the three cases: the magnetic flutter drive (v), $\widetilde{\mathbf{E}} \times \mathbf{B}$ gradient drive (n) and collisional/sheath dissipation (c); note the cross-over at $k_\wedge \rho_S \sim 0.3$–0.5. *Source*: Ribeiro and Scott (2005).

cases, the magnetic flutter drive is two orders of magnitude smaller than the $\widetilde{\mathbf{E}} \times \mathbf{B}$ drive, and is thus practically negligible. At large scales (small $k_\wedge$) the latter always exceeds the dissipation, while at small scales ($\rho_S k_\wedge > 0.3$–0.5) dissipation always dominates, i.e. the dissipative scale is found as two-to-three ρ_S in both (i) and (ii).[95]

In the edge (i), the $\widetilde{p}_e$, $\widetilde{\varphi}$ and $\widetilde{h}_e$ amplitude spectra are closely coupled due to the electron adiabatic response, whereas in the SOL (ii) this coupling is broken predominantly at large scales (small $k_\wedge$), at which $\widetilde{p}_e$ and $\widetilde{\varphi}$ amplitudes begin to separate, with the latter becoming dominant;[96] this separation is even more pronounced without sheath dissipation (iii), when the slope of the $\widetilde{\varphi}$ amplitude spectrum develops a robust inertial range and remains steep up to the largest scale, $L_\wedge \sim L_\perp$.

The reason for the above phenomena is the efficient suppression of the parallel electron dynamics in the emergent $k_\parallel \approx 0$ modes on open field lines, and hence of the adiabatic electron energy coupling and drift-wave dynamics, leaving the interchange coupling unchallenged. This leads to a dual energy cascade: the *inverse* cascade of the $\mathbf{E} \times \mathbf{B}$ drift energy, $\widetilde{\varphi}$, via the vorticity non-linearity and a *direct* cascade via $\mathbf{E} \times \mathbf{B}$ advection of the pressure, $\widetilde{p}_e$. The former is of course the familiar feature of all quasi-2D HD and MHD flows leading to the formation of large structures. Since this cascade is more effective in (ii), not being counteracted by transfer into drift-waves, it is no surprise that one finds the appearance of larger structures in the SOL compared to the edge regions;[97] in the case of the SOL these structures, i.e. the convective cells ($k_\parallel \approx 0$ modes), are just the turbulent filaments which we analyzed in Section 7.2.2.1.

Finally, the cross-correlation between $\widetilde{p}_e$ and $\widetilde{\varphi}$, as well as the $k_\wedge$ spectra of their phase shift are shown in Fig. 7.30. Recalling the analysis of DW and κ dynamics in Section 6.3.1, it is easy to recognize (in both spectral diagnostics) the characteristic signature of the former in (i) and of the latter in (ii); the interchange character is even more pronounced in (iii) where all correlation between $\widetilde{p}_e$ and $\widetilde{\varphi}$ disappears and their phase shift peaks at $\pi/2$ at the largest possible scale, $L_\wedge \sim L_\perp$. It is interesting to note that the reduction of this phase shift at this scale in (ii) is a consequence of sheath dissipation, which is most effective at large scales.

In a subsequent study (Ribeiro and Scott, 2008), the above simulations were extended in three important aspects: first, the GF6 model was replaced by the more complete GF14 model, thus allowing the study of temperature gradient, as well

[95] Here it is worth recalling that sheath dissipation is most effective at large scales, while collisional dissipation is largest at small scales, recall Section 7.2.2.1. The above values of the dissipative scale are largely consistent with the analytic estimate in Fundamenski *et al.* (2005).

[96] At small scales ($\rho_S k_\wedge > 0.3$–0.5), the $\widetilde{n}_e$ amplitude is generally larger.

[97] Aside from direct inspection of Fig. 7.27, this is also confirmed by comparison of auto-correlation lengths in the y direction, which are indeed found to be roughly twice as large in the SOL, $\lambda_\wedge \approx 20\rho_S$, compared to the edge, $\lambda_\wedge \approx 10\rho_S$.

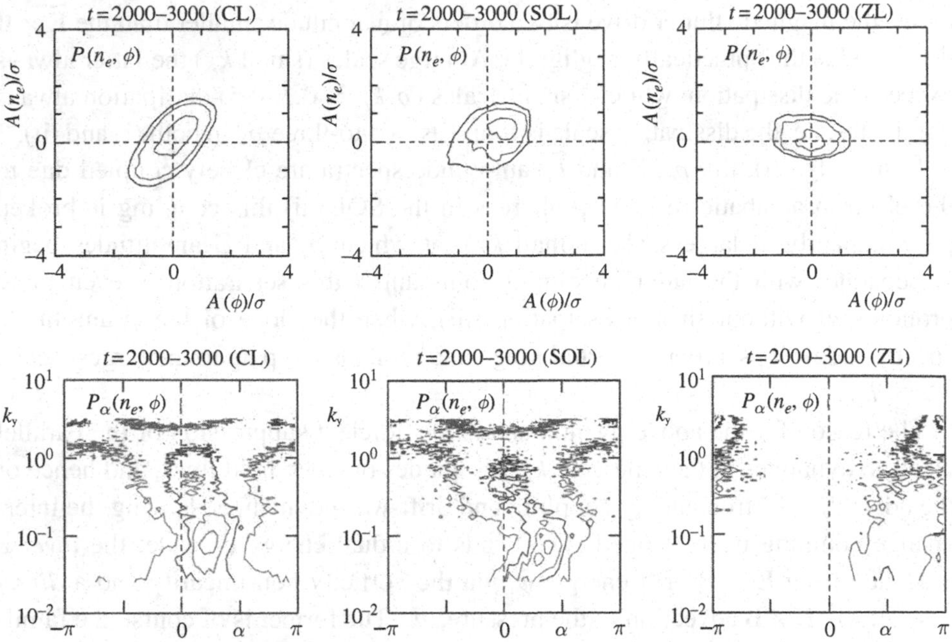

Fig. 7.30. Similar to Figs. 6.4-6.5. The upper frames show the correlation between $\widetilde{p}_e = \widetilde{n}_e$ and $\widetilde{\varphi}$ for the edge (left), SOL (middle) and SOL without sheath (right) cases in a GF6 simulation. The lower frames show the corresponding $k_\wedge$ spectra of the phase shift between $\widetilde{p}_e$ and $\widetilde{\varphi}$. *Source*: Ribeiro and Scott (2005).

as density gradient, driven turbulence; second, the edge and SOL regions were evolved simultaneously, by constructing a computational domain with two radial regions: the inner, edge region of closed field lines, and the outer, SOL region of open field lines, see Fig. 7.31; and third, by varying the location and number of 'limiters' in the SOL, i.e. the locations at which the sheath boundary conditions are imposed. Whereas in the earlier study the limiter was always placed at the inner mid-plane, labelled as *inner single null* (ISN), here four distinction single null locations were introduced, see Fig. 7.31: ISN (inner, $s = \pm\pi$), OSN (outer, $s = 0$), USN (upper, $s = \pi/2$), LSN (lower, $s = -\pi/2$), as well as an combination of upper and lower limiters, denoted as *double null* (DN) in the figure.[98]

Parameters used in the GF14 simulations were the same as those listed in (7.80)–(7.81), except $\omega_B = 0.046$, $\hat{\beta} = 1.5$ and $\hat{C} = 2.7$.[99] The same semi-global approach is adopted, with additional sources for perpendicular and parallel

[98] LSN, USN and DN are also common in the divertor SOL geometry, although here the complexity of the X-point magnetic null is clearly not addressed, e.g. see Xu *et al.* (2000).

[99] In terms of physical quantities, these are the same as before, see footnote on page 341, except $n_e = 2.2 \times 10^{19}\,\mathrm{m}^{-3}$, $B_0 = 2.2\,\mathrm{T}$, $q = 3.1$, $L_\perp = 3.8\,\mathrm{cm}$.

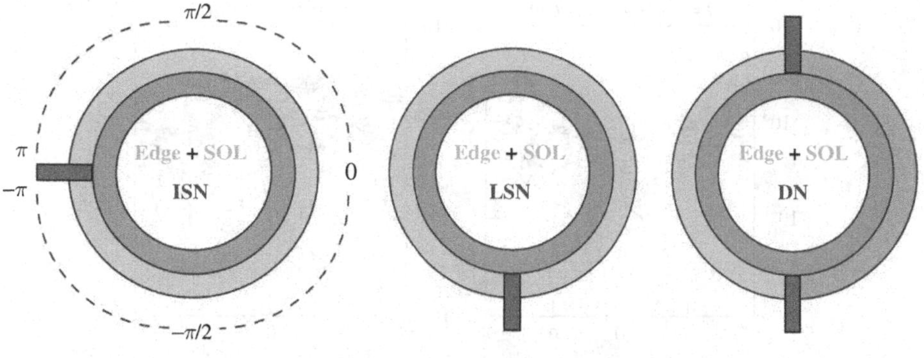

Fig. 7.31. Schematic of computational domain used in the GF14 simulations, showing the edge and SOL regions and the location of the 'limiters' at which the sheath boundary conditions were imposed. The major axis is on the left and the direction of B_T is out of the page, so that $\mathbf{B} \times \nabla B$ points down. *Source*: Ribeiro and Scott (2008).

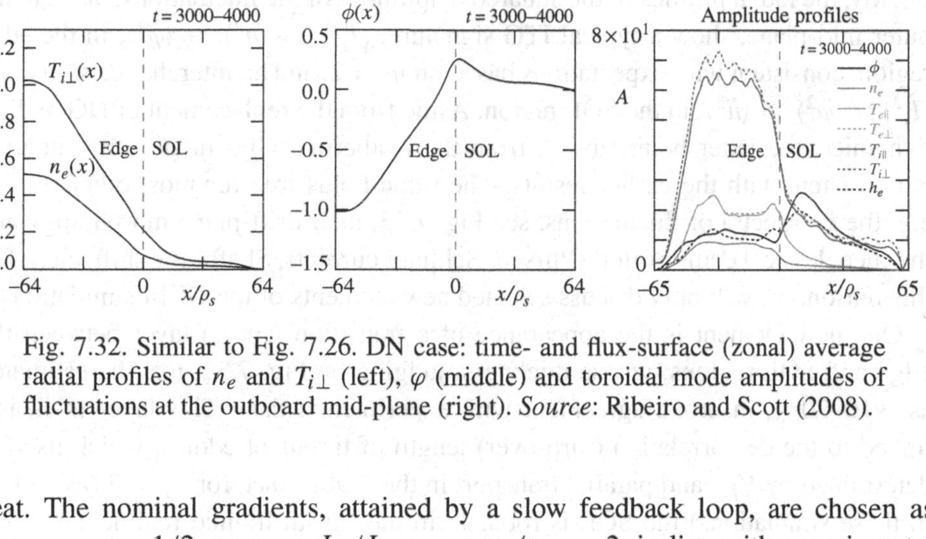

Fig. 7.32. Similar to Fig. 7.26. DN case: time- and flux-surface (zonal) average radial profiles of n_e and $T_{i\perp}$ (left), φ (middle) and toroidal mode amplitudes of fluctuations at the outboard mid-plane (right). *Source*: Ribeiro and Scott (2008).

heat. The nominal gradients, attained by a slow feedback loop, are chosen as $2\omega_n = \omega_T = 1/2$, or $\eta_i = L_n/L_{Ti} = \tau_i\omega_T/\omega_n = 2$, in line with experimental values;[100] in view of Section 6.3.1, one thus expects ITG dominated turbulence in the edge region.

The resulting radial profiles of the mean zonal quantities, shown in Fig. 7.32 for the DN case, indicate a smooth transition of both n_e and $T_{i\perp}$ profiles across the LCFS, with a gradual flattening in the far-SOL, suggesting a substantial increase in radial transport. Also shown in Fig. 7.32 is the variation in the electric potential, indicating a V_E^y which is constant for most of the edge region and changes direction

[100] Recall the definitions: $\omega_n = L_\perp/L_n$, $\omega_T = L_\perp/L_{Ti}$, $L_n = 1/|\nabla \ln n_{e0}|$, $L_{Ti} = 1/|\nabla \ln T_{i0}|$.

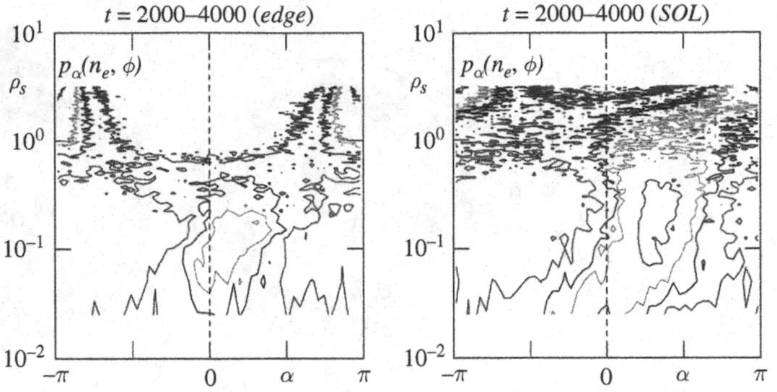

Fig. 7.33. Similar to Fig. 7.30. LSN case: the $k_\wedge$ spectra (note the y-axis should read $k_\wedge$, not ρ_s) of the phase shift between $\widetilde{p}_e$ and $\widetilde{\varphi}$ in the edge (left) and SOL (right) regions at the outer mid-plane. *Source*: Ribeiro and Scott (2008).

in the SOL, setting up a localized shear layer as shown schematically in Fig. 7.7. Finally, the radial profiles of the squared amplitudes of the fluctuations, taken at the outer mid-plane, show a typical ITG signature, $\langle \widetilde{T}_i^2 \rangle \gg \langle \widetilde{n}^2 \rangle \sim \langle \widetilde{\varphi}^2 \rangle$, in the edge region, consistent with expectations based on $\eta_i = 2$, and an interchange signature, $\langle \widetilde{T}_i^2 \rangle \sim \langle \widetilde{\varphi}^2 \rangle \gg \langle \widetilde{n}^2 \rangle$, in the SOL region. Aside from the replacement of DG by ITG dynamics, the latter being absent from the isothermal GF6 model, the situation is consistent with the earlier results. The same holds true for most other results, e.g. the $k_\wedge$ spectra of fluctuations, see Fig. 7.33, their drift-plane morphology and the 'neoclassical equilibrium' (Pfirsch–Schlüter currents, Shafranov shift, etc.). For this reason, we will only discuss selected new elements of the GF14 simulations.

One new element is the appearance of a transition (shear) layer between the edge and SOL regions, whose thickness, judging by Fig. 7.32, may be estimated as $\sim$10–20ρ_S, in broad agreement with experimental data. This layer is clearly linked to the de-correlation (turn-over) length of turbulent eddies, which itself is determined by $V'_{E\wedge}$ and parallel transport in the SOL. Since for $\nu_u^* \sim 3$ (as chosen in these simulations) the SOL is robustly in the sheath limited regime, such that parallel conduction in the SOL plays little role in power exhaust, we expect the near-SOL power width to be determined directly by the upstream gradients in the transition (shear) layer between edge and SOL, as seen in the left frame of Fig. 7.32. It is also worth noting that, because of the dual energy cascade, the plasma energy resides preferentially in the large scales, above $1/k_\wedge \sim 10\rho_S$ in the edge and in the largest available scales, here $1/k_\wedge \sim 40\rho_S$, in the SOL.

Another new element is the poloidal mode structure of the ITG mode, Fig. 7.34, which is seen to peak near the outer top of the vessel ($\pi/4 < s < \pi/2$) in the edge and, in a diminished form, towards the outer bottom ($-\pi/2 < s < -\pi/4$) in the SOL. This up–down asymmetry is cause by the 'neoclassical equilibrium',

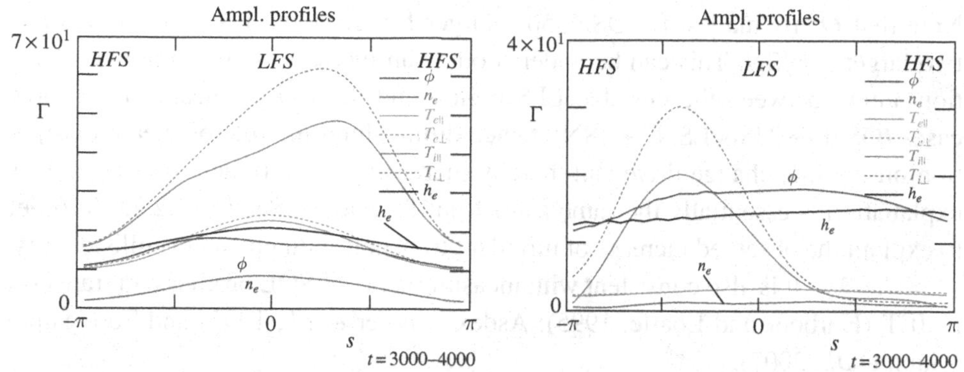

Fig. 7.34. Similar to Fig. 7.28. ISN case: poloidal dependence of the squared amplitude of fluctuations in the edge (left) and SOL (right). Note the peak of T_i towards the outer top ($\pi/4 < s < \pi/2$) in the edge and the outer bottom ($-\pi/2 < s < -\pi/4$) in the SOL. *Source*: Ribeiro and Scott (2008).

recall Section 5.3.3, which for the given $\mathbf{B} \times \nabla B$ direction causes $T_{i\perp}$ to be larger at the bottom and n_e to be larger at the top of the torus (for an otherwise up–down symmetric ISN case), allowing the pressure to remain roughly constant. This translates into steeper radial T_i gradient near the top, and hence increased ITG activity. The situation is reversed in the SOL due to a difference in the mean T_i profiles in the SOL, which yield steeper gradients near the bottom of the torus. These effects can be quantified using the familiar, *return flow* approach, extended to the gyro-fluid equations, see Ribeiro and Scott (2008).

In the GF14 model, the radial energy outflow may be calculated as

$$Q_s/\tau_s = \frac{3}{2}\left[\widetilde{n}_s \nabla_\wedge(\Gamma_1\widetilde{\varphi}) + \widetilde{T}_{\perp i}\nabla_\wedge(\Gamma_2\widetilde{\varphi})\right] + \left(\frac{1}{2}\widetilde{T}_{\parallel i} + \widetilde{T}_{\perp i}\right)\nabla_\wedge(\Gamma_1\widetilde{\varphi}) + (\widetilde{n}_i + 2\widetilde{T}_{\perp i})\nabla_\wedge(\Gamma_2\widetilde{\varphi}), \quad (7.86)$$

where for electrons $\Gamma_1\widetilde{\varphi} = \widetilde{\varphi}$ and $\Gamma_2\widetilde{\varphi} = 0$. If such gyro-radius corrections are neglected, one finds the usual expression $Q_s \approx \frac{3}{2}\widetilde{p}_s\widetilde{V}_E^x$, where $\widetilde{V}_E^x = \nabla_\wedge\widetilde{\varphi}$ is the radial $\mathbf{E} \times \mathbf{B}$ velocity and $\widetilde{p}_s = \frac{1}{3}\widetilde{p}_{\parallel i} + \frac{2}{3}\widetilde{p}_{\perp i}$ is the isotropic pressure. Defining the radial outflow across the LCFS as the average of (7.86) over $-L_x/4 < x < L_x/4$, the energy outflow in the five limiter combinations was found as (with comparable values for ions and electrons, $Q_e \approx Q_i \approx Q_s$),[101]

$$Q_s^{OSN} \approx 0.7, \quad Q_s^{USN} \approx 1.0, \quad Q_s^{LSN} \approx 1.1, \quad Q_s^{ISN} \approx 1.3, \quad Q_s^{DN} \approx 2.0.$$

[101] The fact that $Q_e \sim Q_i$ despite $\widetilde{T}_e \ll \widetilde{T}_i$ can be explained by the close coupling between $\widetilde{T}_e$ and $\widetilde{V}_E$ via the parallel adiabatic response. In effect, the electrons are more efficient energy carriers in the radial direction than the ions, which are only weakly coupled to $\widetilde{V}_E$.

Note that Q_s is smallest for OSN, 50% larger for USN and LSN, and a factor of two larger for ISN. This can be understood as an increase in the parallel connection length between the unstable LFS region and the Debye sheath in the three cases (OSN → USN/LSN → ISN), hence diminishing the role of sheath dissipation on the interchange drive and thereby increasing the turbulent transport. This explanation is essentially the same as that put forward in Section 7.2.2.2 in order to explain the observed increase of turbulent transport with upstream collisionality, $\nu_u^* = L_\parallel/\lambda_{ei}$. It is also consistent with measurements of SOL profiles and transport in JET (Harbour and Loarte, 1995), Asdex (Endler *et al.*, 1995) and Tore-Supra (Gunn *et al.*, 2007).

Finally, the fact that energy outflow is roughly twice as large in the DN case as in both USN and/or LSN cases is a direct consequence of the separation of the inner and outer SOL regions, thus preventing the interchange stable HFS region from exerting a stabilizing influence on the interchange unstable LFS region. As a result, the level of fluctuations decreases/increases strongly in the inner/outer SOL, with the increase on the LFS more than compensating for the decrease on the HFS. This fact is illustrated in Fig. 7.35, which shows the electron density in the drift ($y-x$ or $\wedge-\perp$) plane for a combination of DN and LSN cases. Evidently, the removal of the upper limiter has a dramatic effect on the HFS (inboard) SOL, which in the DN case is nearly quiescent, and in the LSN case is filled in by turbulence 'spilling

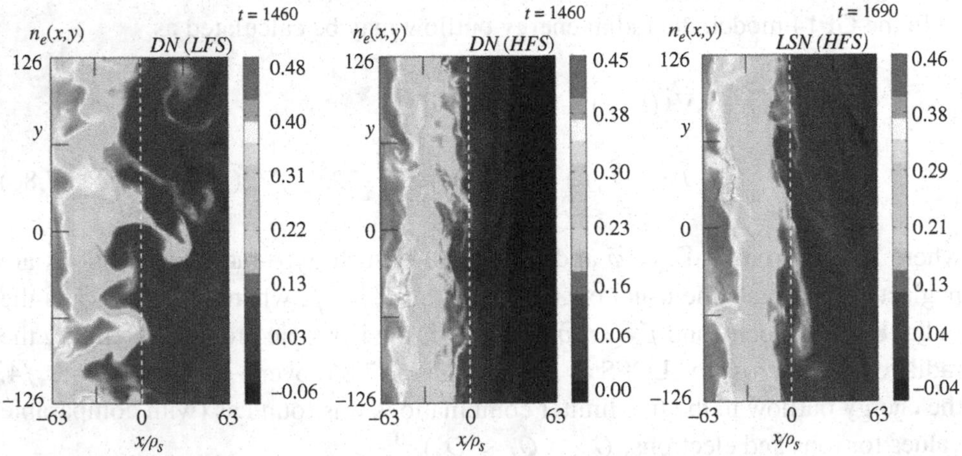

Fig. 7.35. Similar to Fig. 7.25 and Fig. 7.27. Snapshots of total electron density in the drift plane (recall that y acts as the toroidal angle) for: DN taken at outer mid-plane (LFS), DN taken at inner mid-plane (HFS) and LSN taken at inner mid-plane (HFS). Both edge and SOL regions are shown with LCFS indicated by the dashed line. Comparison of the latter two shown the turbulence 'spilling over' from the LFS to the HFS in the LSN case. *Source*: Ribeiro and Scott (2008).

over' from the LFS (outboard) SOL. As a result, one would expect significant SOL flow at the top of the vessel, due to the ballooning outflow on the LFS and the acoustic free-streaming towards the HFS. Such flows have indeed been observed on many machines, e.g. $M_{\parallel} \sim 0.5$ at the top of the torus on JET (Erents *et al.*, 2004) and $M_{\parallel} \sim 1$ on the inboard side on Alcator C-mod (LaBombard, 2003). Their $\mathbf{B} \times \nabla B$ *independent* component can indeed be explained by turbulence simulations (Fundamenski *et al.*, 2007a; Tamain *et al.*, 2009), with the $\mathbf{B} \times \nabla B$ *dependent* component reflecting the Pfirsch–Schlüter return flows required for the 'neoclassical equilibrium' discussed above and illustrated in Fig. 7.24.

To conclude the section, it is fair to say that our understanding of edge-SOL plasma transport, as encapsulated by both DF and GF models (treating the GK models as work in progress), is sufficiently advanced to abandon the 'anomalous' label. Indeed, the fluid models offer a route to a quantitative relation between flows across the LCFS and loads on main chamber PFCs in fusion plasmas, and hence to a prediction of power exhaust limits in a fusion reactor. Unfortunately, this route only applies to the turbulence dominated, low confinement (L-mode) regime. The more attractive, high confinement (H-mode) regime is further complicated by the appearance of the *edge transport barrier* (ETB) and its periodic destruction by *edge localized modes* (ELMs).

7.3 H-mode power exhaust: edge localized modes (ELMs)

The formation of the ETB and the related H-mode regime are still not fully understood and remain the chief source of uncertainty in virtually all aspects of tokamak plasma behaviour, including power exhaust. Therefore, somewhat ironically, it is the absence, rather then presence of edge-SOL turbulence, which may now be properly described as being anomalous.

Below we divide the discussion of H-mode power exhaust into formation, evolution and relaxation of the ETB, exhaust in between and during ELMs, and various techniques to reduce the inter-ELM and ELM heat loads.

7.3.1 Edge transport barrier

'Richie Mo stared through the window panes of his Chicago office. The rain was coming down hard. It was Wednesday. Anything could happen.'

T. Eich (2008)

7.3.1.1 ETB formation: the L–H transition

As the heating power in an X-point (divertor) tokamak discharge is increased, at some point the radial transport in the edge plasma undergoes a sudden bifurcation, in which the turbulence (fluctuation) level and the associated radial transport are strongly reduced.[102] This reduction in radial plasma flows for a given radial gradient only occurs in the outermost shell of the plasma, typically for $0.95 < r/a < 1$, which is hence known as the *edge transport barrier* (ETB). Since the presence of the ETB increases the plasma stored energy, W, and hence the energy confinement time, $\tau_E = W/P_{heat}$, the plasma is said to enter a *high confinement* (H-mode) regime.[103]

The low-to-high confinement (L–H) transition occurs when the power crossing the LCFS exceeds some threshold value $P_{\text{L–H}}$, or the radial energy flow exceeds $Q_\perp^{\text{L–H}}$, which scale roughly as (Doyle *et al.*, 2007)

$$P_{\text{L–H}} \propto \langle n_e \rangle_a^{0.72} B_T^{0.8} A_p^{0.94}, \qquad Q_\perp^{\text{L–H}} \propto \langle n_e \rangle_a^{0.72} B_T^{0.8}, \tag{7.87}$$

where $A_p \approx 4\pi^2 R\kappa a$ is the area of the LCFS and $\langle n_e \rangle_a$ is the average electron density inside the LCFS. It should be stressed that the dataset behind (7.87) contains a large scatter which most likely reflects a number of hidden variables, e.g. $P_{\text{L–H}}$ depends on ion species, poloidal plasma shape (triangularity, X-point height), topology (single vs. double null), toroidal rotation, etc. Recalling (2.253), replacing the volume average quantities with edge quantities,[104] we find that, roughly,

$$P_{\text{L–H}}/A_p \equiv Q_\perp^{\text{L–H}} \equiv p_a V_\perp^{\text{L–H}} \propto n_a V_{Aa}^{0.8}, \tag{7.88}$$

which indicates that the threshold energy flow per particle, $Q_\perp^{\text{L–H}}/n_a$, increases roughly linearly with the edge Alfvén speed, V_{Aa}, and/or that the corresponding Alfvénic Mach number $V_\perp^{\text{L–H}}/V_{Aa}$ decreases as $1/T_a \propto \eta_{\|a}^{2/3}$,

$$V_\perp^{\text{L–H}} \equiv Q_\perp^{\text{L–H}}/p_a \propto 1/\beta V_{Aa} \propto 1/\sqrt{\beta_a} V_{Sa}, \tag{7.89}$$

$$V_\perp^{\text{L–H}}/V_{Aa} \propto T_a^{-1} \propto \eta_{\|a}^{2/3}, \quad Q_\perp^{\text{L–H}}/n_a V_{Aa} \approx T_{\text{L–H}}. \tag{7.90}$$

Despite the appearance of these basic parameters of plasma physics,[105] with density and field roughly combining into the Alfvén speed, the theoretical understanding of the L–H transition remains elusive and the bifurcation is yet to be quantitatively reproduced by either analytic theories (Connor and Wilson, 2000) or turbulence codes (Scott, 2007). Nonetheless, there are some basic physical processes which are known to play a role in the transition: (i) it is known that edge turbulence is reduced due to radial gradients in the $\mathbf{E}\times\mathbf{B}$ plasma flow, V_E', so-called

[102] The effective mass, $D_\perp \equiv \Gamma_\perp/\nabla_\perp n$, and heat, $\chi_\perp \equiv q_\perp/n\nabla_\perp T$, diffusivities are reduced by roughly an order of magnitude down to (ion) neoclassical levels, e.g. from ~1 to ~0.1 m²/s.

[103] Although not reported in Wagner *et al.* (1982), it must have been Wednesday when the plasma density suddenly doubled, and the first ever ETB was observed in the Asdex control room.

[104] Since the L–H transition is local to the edge plasma; however, by doing this the corresponding density exponent in (7.87) may change slightly if $n_a/\langle n \rangle_a \neq$ const.

[105] Here $\langle T \rangle_a$ may indicate the role of either the plasma resistivity, $\eta_{\|a} \propto T_a^{-3/2}$, or plasma–neutral interactions, since PNI cross-sections are likewise a strong function of electron temperature.

flow shear – the appearance of the flow shear, as inferred from the measured radial electric field, coincides with that of the ETB, and is hence one of the unmistakable signatures of the H-mode – which breaks up (de-correlates) the turbulent eddies, thus reducing the effective mixing length and the radial transport; (ii) it is also known that the flow shear is at least partially generated spontaneously by the turbulence, i.e. it is an aspect of *zonal flow* generation, although it is not clear how and why it is generated to the levels needed to affect the transition; nonetheless, perhaps the leading theory of L–H transition is based on zonal flow generation by drift-Alfvén wave turbulence (Guzdar *et al.*, 2001a); (iii) in view of the axis-symmetric toroidal equilibrium expression for the plasma velocity, (5.237), it is known that the flow shear is related to the toroidal rotation, $\omega_s(\psi)$, via the radial electric field, $E_\perp = -\varphi'$; (iv) thus, one can infer that the flow shear, and hence the L–H transition itself, depend on all factors affecting charge continuity (collisional resistivity, ion orbit losses due to the X-point, etc.) and/or radial momentum transport (collisional viscosity, toroidal momentum sources in the core, parallel flows in the SOL, fuelling location, etc.); and (v) comparison between limiter and divertor plasmas reveals that the edge magnetic shear in the X-point geometry facilitates ETB formation and extends its lifetime, before it is destroyed by the ELM. More generally, it is observed that the L–H transition is affected by the magnetic geometry (divertor shape, triangularity) and topology (magnetic balance between single vs. double nulls), see Martin *et al.* (2008).

In short, as in some murder mysteries, we know who did it, namely the quasi-neutrality condition, $\nabla \cdot \mathbf{J} = 0$, but not quite how. What is clear is that the drift-fluid and gyro-fluid models developed in Section 6.3 and applied to edge-SOL turbulence in Section 7.2.2 have so far been unable to capture the L–H transition. By deduction, at least one piece of physics missing from these models, e.g. kinetic effects, magnetic X-point, plasma rotation, magnetic equilibrium evolution, atomic/molecular physics, etc., must play a vital role in ETB formation. We are thus faced with the sobering thought that the understanding of the mechanisms governing edge-SOL turbulence is a necessary, but not a sufficient, condition for understanding its suppression.

7.3.1.2 ETB evolution: the H-mode pedestal

Following the L–H transition, the reduction of the radial diffusivity and the resulting steepening of the radial pressure gradients within the ETB tend to build up a so-called *H-mode pedestal*, Fig. 7.36.[106] This pedestal, whose top is defined as

[106] One thus speaks of a density, temperature and pressure pedestal, with the quantities at the pedestal top denoted as n_{ped}, T_{ped} and p_{ped}, respectively. The pedestal width and height are related by the radial gradient within the ETB, e.g. $\nabla_\perp p \approx (p_{ped} - p_{sep})/\Delta_{p,ped} \approx p_{ped}/\Delta_{p,ped}$.

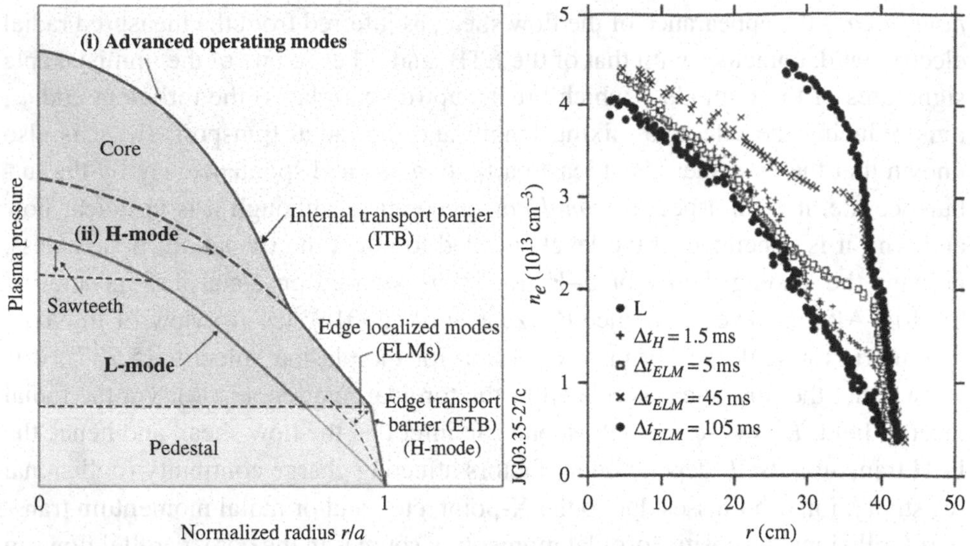

Fig. 7.36. The left frame illustrates the radial pressure profile in L-mode, H-mode and Advanced Tokamak regimes, showing the location of the edge (ETB) and internal (ITB) transport barriers, as well as the relaxation of the core profile by sawteeth, and the H-mode pedestal by ELMs. © EFDA-JET. The right shows the original observations of ETB formation on Asdex (Wagner *et al.*, 1982). Reproduced from Connor and Wilson (2000).

the inflection point between the core and ETB radial gradients, builds up gradually on the edge transport time scale (typically tens of ms); since the radial gradients in the ETB remain roughly constant during this evolution, the pedestal height increases linearly with its width, i.e. by extension of the inner boundary (broadening) of the ETB region.[107] The growth rate decreases steadily with time, tending towards some asymptotic value, and is eventually terminated by an ELM, see Section 7.3.1.3.

On average, the formation of the pedestal significantly increases the edge pressure, $p_{ped} \gg p_{sep}$, and since the pressure gradient in the core remains roughly constant, it also raises the central pressure, $p(0)$ and the fusion power, P_{fus}.[108] The saturated (pre-ELM) value of the pedestal pressure is a function of the ELM regime, with highest p_{ped} achieved for so-called Type-I ELMs and smaller (by roughly a factor of two) p_{ped} for Type-III ELMs. Defining the stored energy associated with the pedestal as $W_{ped} = p_{ped}\mathcal{V}_a = \frac{3}{2}n_{ped}(T_{i,ped} + T_{e,ped})\mathcal{V}_a$,

[107] This in an instance of the inverse of the more familiar problem of turbulence spreading, e.g. as in internal flow in a pipe, and may be termed *quiescence spreading*.

[108] Typically, it roughly doubles the volume averaged plasma pressure compared to the L-mode level, $\langle p\rangle_a^H \sim 2\langle p\rangle_a^L$, with a similar increase in the global energy confinement time, $\tau_E^H \sim 2\tau_E^L$.

where $\mathcal{V}_a$ is the plasma volume, one typically finds $W_{ped} \sim \frac{2}{3} W_{core}$ and $W_{ped} \sim 0.4 W^H \sim 0.8 W^L$ in Type-I ELMy H-mode and $W_{ped} \sim \frac{1}{3} W_{core}$ in Type-III ELMy H-mode.

We next turn to the evolution cycle of the ETB, also known as the *ELM cycle* since the growth of the ETB is necessarily terminated by an ELM, focusing for the time being on Type-I ELMs which lead to higher global energy confinement, see Section 7.3.1.3. Assuming the ELM to be an ideal MHD instability, and recalling Section 4.2, we expect the (in)stability of the H-mode pedestal (against ELMs) to be determined by a combination of edge current driven (external kink/peeling, Section 4.2.7) and edge pressure driven (ballooning, Section 4.2.8) modes. Indeed, the *peeling–ballooning* (P–B) model (Connor *et al.* 1998; Connor and Hastie 1999; Wilson *et al.* 1999; Snyder *et al.* 2002; Snyder *et al.* 2004) in which both these effects are retained for finite-n modes, see Figs. 7.38 and 7.41, has been very successful at describing the pedestal stability boundary against Type-I ELMs. However, it does not generally apply to other ELM types, e.g. Type-III ELMs occur when the pedestal is ideal ballooning stable, see Fig. 7.39.[109] Consequently, the Type-I ELM cycle is well described in terms of the evolution of the two quantities governing the pedestal (in)stability, namely the edge pressure gradient, $\nabla_\perp p$, and the edge current, J_a; the former is typically replaced by the ideal ballooning parameter, α, (4.156), while the latter is normalized by the average plasma current, to yield $j = J_a/\langle J \rangle_a$, see page 126. Note that j is related to the magnetic shear as $s_a = \mathrm{d}\ln q/\mathrm{d}\ln r = 2(1 - j)$, i.e. the edge current has a destabilizing and the magnetic shear a stabilizing effect.

On this basis, the Type-I ELM cycle can be divided into three stages, Fig. 7.37: (i) the *pressure growth* stage, in which α increases at roughly constant j towards the ideal ballooning limit (set by $n = \infty$ ballooning modes) on the edge radial transport time scale; (ii) the *current growth* stage, in which j increases at roughly constant α towards the peeling limit (set by low-n external kink modes) on the edge resistive time scale;[110] and (iii) the *ELM crash*, in which both α and j rapidly decay during the ELM (due to medium-n, $n \sim$ 5–20, P–B modes) on the edge parallel transport time scale, Section 7.3.1.3. The time between ELMs is determined by the rate at which the pedestal is rebuilt, which in turn

109 Moreover, the success of the P–B theory does not rule out a more comprehensive theory, containing P–B as its subset, e.g. a combination of peeling and tearing modes, i.e. resistive MHD effects, may offer a more realistic description of the Type-I ELM cycle (Huysmans, 2005b). As always, a DHD theory would offer the most accurate fluid description.

110 The origin of the edge current growth during this stage is a combination of the reduced resistivity due to higher edge temperature and increased edge bootstrap current, (4.206), driven by the pressure gradient in the ETB region. More generally, the latter also depends on the density and ion/electron temperature gradients and the collisionality (Sauter *et al.*, 1999).

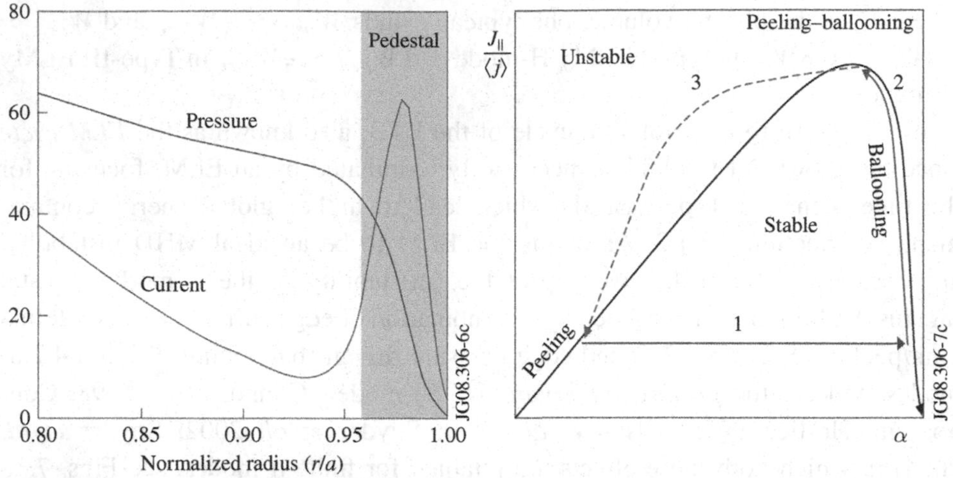

Fig. 7.37. Schematic of the ELM cycle in the ideal MHD ($j-\alpha$) stability diagram (left) and radial profiles of edge pressure and (bootstrap) current (right). Adapted from Snyder *et al.* (2002, 2004) and Snyder and Wilson (2003).

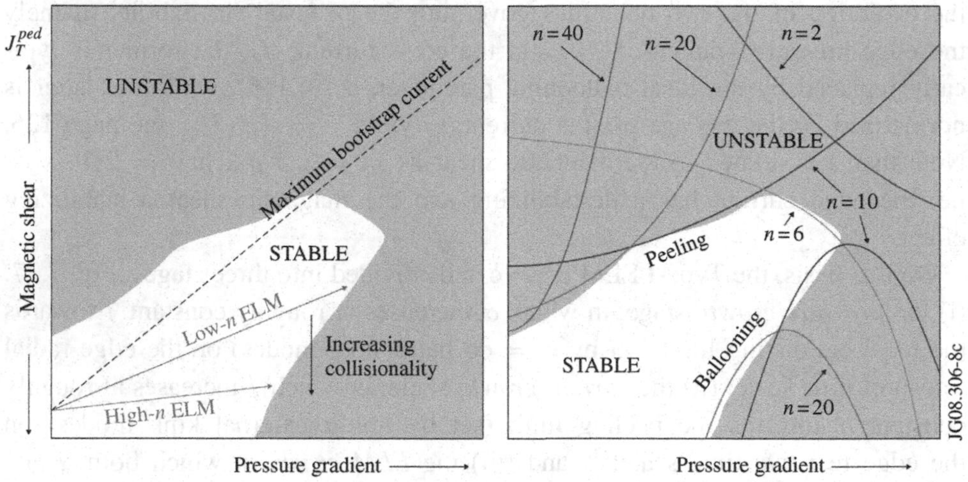

Fig. 7.38. Schematic of pedestal plasma (in)stability on an ideal MHD ($j-\alpha$) diagram; the instability region is indicated in grey, along with pedestal evolution as a function of collisionality (left) and the stability boundaries of selected toroidal mode number modes (right). Adapted from Snyder *et al.* (2004).

depends on both sources and radial transport of particles, momentum and energy in the edge plasma by a combination of collisional and turbulent processes, e.g. inside the ETB the radial ion heat diffusion is dominated by neoclassical conduction, but electron heat and mass diffusion are governed by residual levels of turbulence.

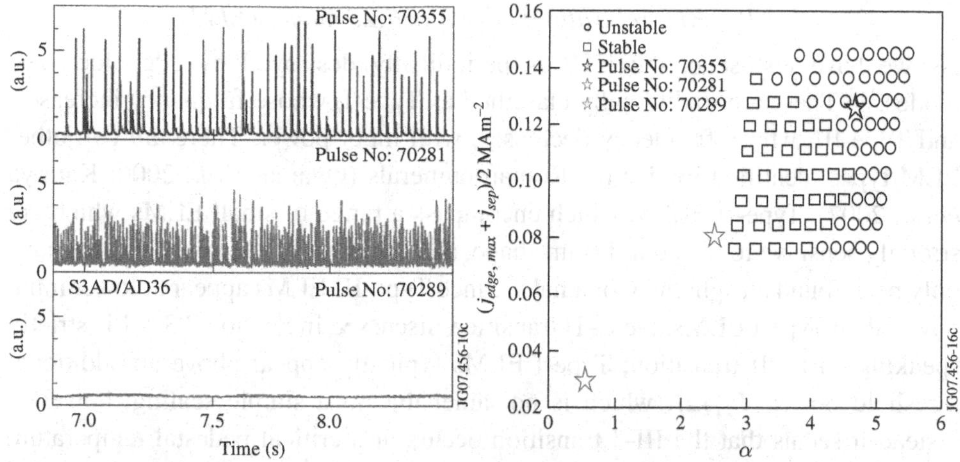

Fig. 7.39. The left frame shows typical time traces of recycling, D_α, light in the divertor in JET during L-mode (70289), Type-III ELMy H-mode (70281) and Type-I ELMy H-mode (70355), with the spikes caused by plasma ejected by the ELMs, see Section 7.3.1.3. The right frame locates the pre-ELM edge pedestal plasma of these three discharges on an ideal MHD ($j-\alpha$) stability diagram, showing that the Type-I ELMy H-mode edge plasma is marginally stable to medium-n ideal MHD (peeling-ballooning) modes, i.e. sits close to the stability boundary, while Type-III and L-mode profiles are ideal MHD stable. *Source*: Beurskens *et al.* (2008).

In summary, as with the L–H transition, there is now a loose understanding of the underlying mechanisms of ETB dynamics, but a lack of quantitative models capable of replicating real experiments. The most acute problem is posed not by the ETB gradients, which can be at least partly explained in terms of global stability requirements, but by the ETB width, specifically, its saturated (pre-ELM) value. Most models, which typically infer this width from the velocity shear region, in which the $E \times B$ shearing rate exceeds the linear growth rate, are unable to explain the experimental data.[111] The ultimate problem is of course the same as that of the L–H transition, but is compounded further by the radial quiescence spreading (of turbulence suppression). Hence, until the L–H transition itself can be captured numerically, there appears to be little hope of predicting the ETB width evolution. In the meantime, one has to rely on empirical scalings, i.e. dimensionless scalings derived from experimental data, which indicate that $\Delta_{ped}/a \sim 2.5$–5% across a range of machines and conditions, with an indication of a square-root scaling on poloidal beta and a weak scaling on the magnetization parameter, $\Delta_p \propto \beta_p^{0.5} \rho_*^{0.2}$.

[111] The situation is best for the density pedestal which has been convincingly linked to the neutral inflow from the SOL, and worst for the ion and electron temperature pedestals.

7.3.1.3 ETB relaxation: edge localized modes (ELMs)

As we have just seen, the ETB is periodically destroyed by edge localized modes.[112] These are traditionally classified as Type-I, whose frequency increases, and Type-III, whose frequency decreases, with input power. There are two other ELM types identified by distinct Roman numerals (Oyama *et al.*, 2006; Kamiya *et al.*, 2007): Type-II ELMs, which encompass a range of small ELMs which are strongly sensitive to the equilibrium shape, and Type-V ELMs, which to date have only been found on spherical tokamaks. Since Type-III ELMs appear at lower input power than Type-I ELMs, the L–H transition discussed in Section 7.3.1.1 is strictly speaking an L–III transition; Type-I ELMs typically appear above an additional threshold power, P_{III-I}, which is not amenable to a simple scaling, however; instead it seems that the III–I transition occurs at a critical pedestal temperature and/or parallel resistivity.[113] The two characteristic features of this secondary, III–I transition are a strong increase in the pedestal pressure, by roughly a factor of two, and a pronounced drop in the ELM frequency, by roughly a factor of ten, so that in general, Type-I ELMs are much less frequent and hence much larger, in both energy and extent, than Type-III ELMs.

As can be gathered from the ideal MHD stability diagram, Fig. 7.39, a Type-I ELM is most likely dominated by an ideal MHD (P–B) instability, while a Type-III ELM, being ideal MHD (P–B) stable, must involve resistive MHD and/or DHD instabilities. Thus, a Type-I ELM occurs when the pressure gradient and edge current exceed the P–B stability limits, while a Type-III ELM occurs above a corresponding limit for DHD (resistive MHD) stability. In either case, the ETB is (mostly) destroyed by the ELM, Fig. 7.36, leading to a (partial) collapse of the pedestal pressure, Δp. This can be divided into a density drop, Δn, associated with losses of edge particles, known as 'convective' losses, and a temperature drop, ΔT, associated with cooling of edge particles, known as 'conductive' losses (Leonard *et al.*, 2006). The relative pressure drop, $\Delta p/p$, or $\Delta W_{ELM}/W_{ped}$ as it is more commonly expressed, is found to decrease strongly with the pedestal collisionality (Loarte *et al.*, 2002, 2004). Moreover, this reduction is associated primarily with the reduction of conductive losses, $\Delta T/T \sim \Delta W/W_{ped}$, so that sufficiently small ELMs are dominated by convective losses (Loarte *et al.*, 2004; Leonard *et al.*, 2006), see Fig. 7.40.

The Type-I ELM itself, as distinct from its cycle, can also be roughly divided into three stages (Fundamenski *et al.*, 2006), as illustrated in Fig. 7.42: (i) the

[112] See Section 7.4 for suggested references concerning ELM observations in tokamaks.

[113] It should be stressed that the appearance of Type-I ELMs does not rule out the intermittent re-appearance of Type-III ELMs, which can co-exist in a mixed (I-III) ELM regime. Indeed, 'clean' Type-I ELMs are only observed when the power crossing the LCFS significantly exceeds the III-I threshold. This behaviour is most likely due to the statistical variability of ELMs.

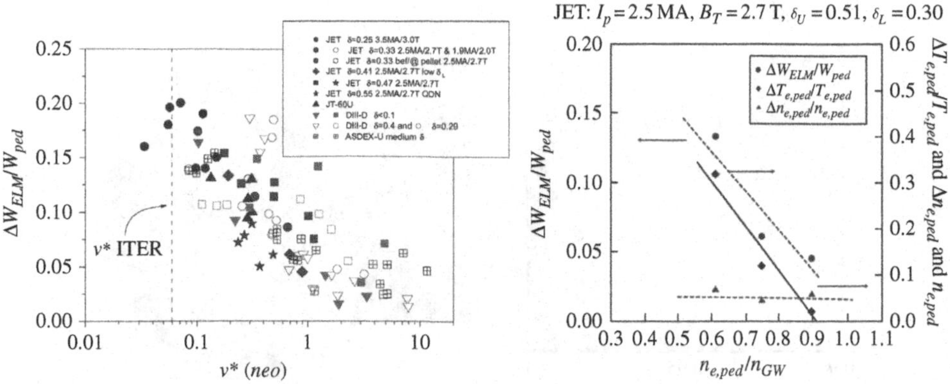

Fig. 7.40. Relative ELM 'size', i.e. the relative energy drop, $\Delta W_{ELM}/W_{ped}$, vs. the neoclassical pedestal collisionality on a number of tokamaks (left) and vs. the Greenwald fraction, $f_{GW} = n_e/n_{GW}$, on JET (right). The right frame also shows the break down of this energy drop into convective, $\Delta n/n$, and conductive, $\Delta T/T$, components. *Sources*: (Loarte *et al.*, 2003; Leonard *et al.*, 2006).

growth stage, or onset of the ELM, in which the instability grows linearly[114] forming a finite number of flute-like 'ripples', or weak perturbations, in pedestal quantities ($\widetilde{p}/p_{ped} \ll 1$, $k_\parallel \ll k_\perp$), including the poloidal magnetic field ($\widetilde{B}_\perp/B_{ped} \ll 1$), which are characterized by a strongly ballooning character; this stage is accompanied by increased magnetic fluctuations, often described as *MHD activity*;[115] (ii) the *saturation stage*, in which these develop into a comparable number of filaments ($\widetilde{p}/p_{ped} \sim 1$, $k_\parallel \ll k_\perp$) during the non-linear growth of the instability; this growth of filamentary structures may be accompanied by formation of magnetic structures, or, more generally, by some form of re-arrangement of the poloidal magnetic field, e.g. reconnection in the X-point region or ergodization in the ETB region – this stage is the beginning of transport and parallel losses, potentially of reconnection and/or of ergodization of the edge magnetic field; and (iii) the *exhaust stage* in which the filaments move (accelerate then decelerate) outward, driven mainly by the interchange (pressure + curvature) force, while draining to the divertor targets, recall Section 7.2.2.1.[116]

While the above picture of a Type-I ELM is far from perfect due to the lack of clear scale separation between the three stages,[117] it is nonetheless a useful

[114] For Type-I ELMs, P–B modes with $n \sim 5$–20 have highest linear growth rates.

[115] The term *magnetic activity* would be more accurate, since magnetic fluctuations do not imply the MHD ordering, unless $V_\perp/V_S \sim 1$, recall Section 2.1.2.

[116] The filamentary nature of ELMs is now well documented, see Section 7.4.

[117] For instance, not all filaments are formed, or indeed, begin to move radially, at the same time, nor are the exhaust processes exclusive to the 'exhaust' stage, as parallel losses usually become active during the formation of the filaments, i.e. during the 'saturation' stage.

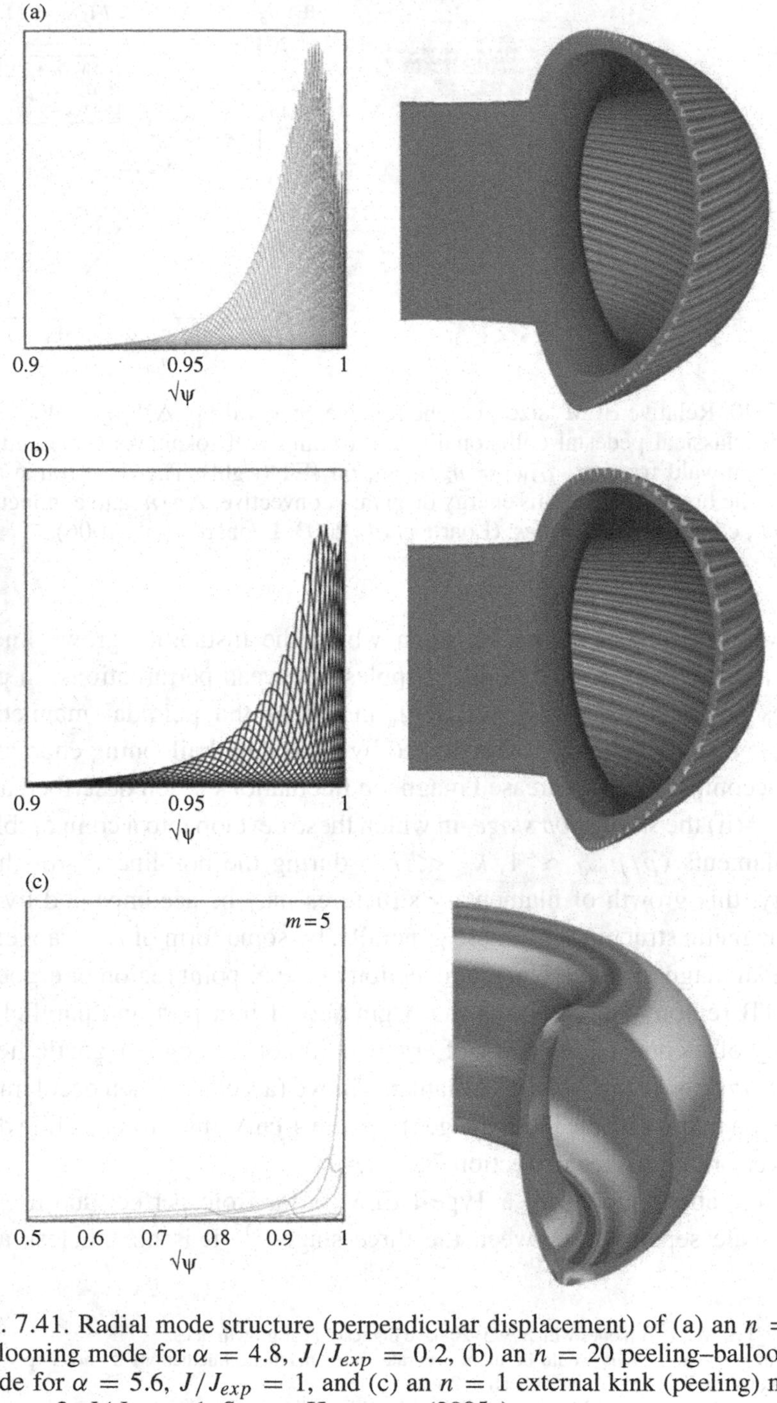

Fig. 7.41. Radial mode structure (perpendicular displacement) of (a) an $n = 32$ ballooning mode for $\alpha = 4.8$, $J/J_{exp} = 0.2$, (b) an $n = 20$ peeling–ballooning mode for $\alpha = 5.6$, $J/J_{exp} = 1$, and (c) an $n = 1$ external kink (peeling) mode for $\alpha = 2$, $J/J_{exp} = 1$. *Source*: Huysmans (2005a).

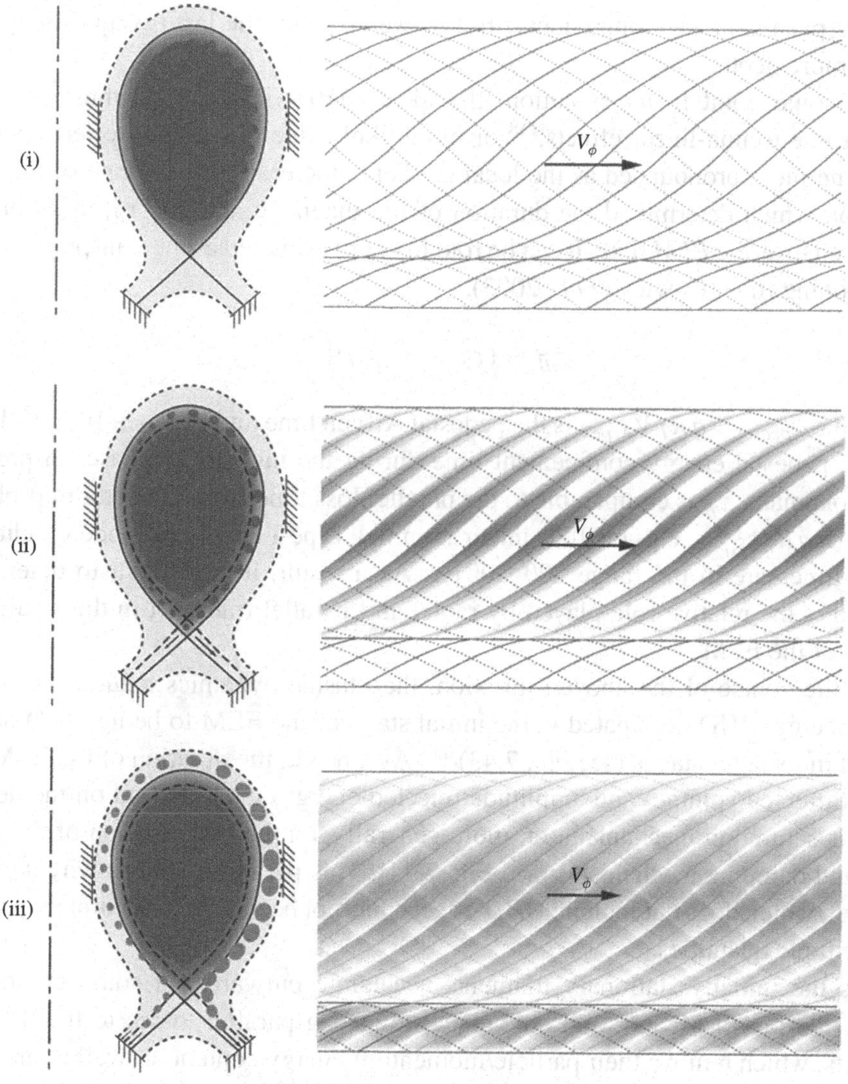

Fig. 7.42. Schematic illustration of the three stages (growth, saturation and exhaust) of ELM evolution. *Source*: Fundamenski *et al.* (2006).

framework within which to discuss ELM evolution. Specifically, it has the virtue of distinguishing between the ELM as a coherent, global (although edge-localized), *linear* eigenmode, and the individual ELM filaments, which are the *non-linear* consequences of this instability in the scrape-off layer. The three stages correspond, respectively, to the three key questions relating to ELM dynamics: (i) when does an ELM occur; (ii) how long does it last and how many particles/how much energy does it eject into the SOL; and (iii) how are the ejected particles/energy distributed on the plasma facing components? Below we consider the last two questions in a little more depth.

At some point in its evolution, the ideal MHD (P–B) mode must saturate, either due to non-linear effects,[118] or more likely, due to transport effects, which become more pronounced as the local gradients increase; it is the rate of this saturation which determined the duration of the magnetic activity, $\tau_{\tilde{B}}$, traditionally referred to as the *ELM time*. It can be roughly approximated as the transport-limited ballooning time (Cowley *et al.*, 2003),

$$\tau_{\tilde{B}} \sim (\tau_{A,ped}^2 \tau_{E,ped})^{1/3}, \tag{7.91}$$

where $\tau_{A,ped} \sim \pi q R / V_{A,ped}$ is the pedestal Alfvén time and $\tau_{E,ped} \sim W_{ped} / \partial_t W_{ped}$ is the pedestal energy confinement time during the inter-ELM phase. In present day tokamaks, $\tau_{\tilde{B}}$ is comparable to the parallel loss time from the outer mid-plane, $\tau_{\|} \sim \pi q R / c_{S,ped}$, e.g. on JET, under nominal Type-I ELMy H-mode conditions both times are in the range 200–300 μs. As a result, it is difficult to determine which is the relative role played by radial and parallel transport in the saturation stage of the ELM.

In the course of the above saturation, the plasma dynamics naturally changes from being MHD dominated in the initial stage of the ELM to being DHD dominated in its latter stages (see Fig. 7.43).[119] As a result, the duration of the ELM, its size (affected volume) and amplitude (ejected energy) must depend on the details of this saturation mechanism, or, more generally, on the combination of the driving and damping mechanisms (see Fig. 7.44). This perhaps explains why the P–B model, which does not include DHD effects, has not been able to accurately predict these three quantities.

As the initially stationary filaments accelerate outwards, at some point they enter the pre-ELM SOL and are subject to strong parallel losses to the divertor targets, which remove their particle/momentum/energy content, as well as any current density, initially associated with the pedestal plasma, and later advected by the filaments. Indeed, the pedestal plasma as a whole may begin to experience parallel losses much earlier than this, i.e. during the saturation stage of the ELM, due to changes in magnetic structure, which during the instability is clearly not in equilibrium. For instance, at some point between their formation and crossing the pre-ELM separatrix, the ELM filaments may be opened by a magnetic reconnection

118 Although these can also lead to increased, even explosive, growth of the instability into filamentary structures (Cowley *et al.*, 2003; Wilson and Cowley, 2004). However, such explosions are generally reduced with the inclusion of dissipative effects.

119 Such gradual transition from MHD to drift-wave dynamics during an ELM, including the formation of filaments, their acceleration, break-up and slowing down, has been recently demonstrated using a gyro-fluid (GF14) model, Section 6.3.1, Fig. 7.43 (Kendl and Scott, 2009).

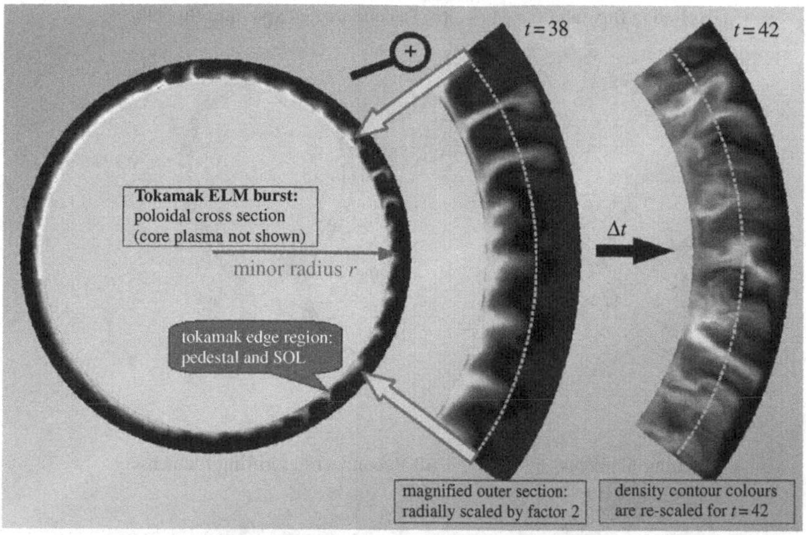

Fig. 7.43. Poloidal density contours during an ELM 'blow off' simulation using a gyro-fluid (GF14) turbulence code, see Section 7.2.2.3, in which the initial pressure gradient was set above the ideal ballooning limit. The left frame is taken at the maximum of the linear growth rate and the right frame is taken $\sim 15\,\mu s$ later. *Source*: Kendl and Scott (2009).

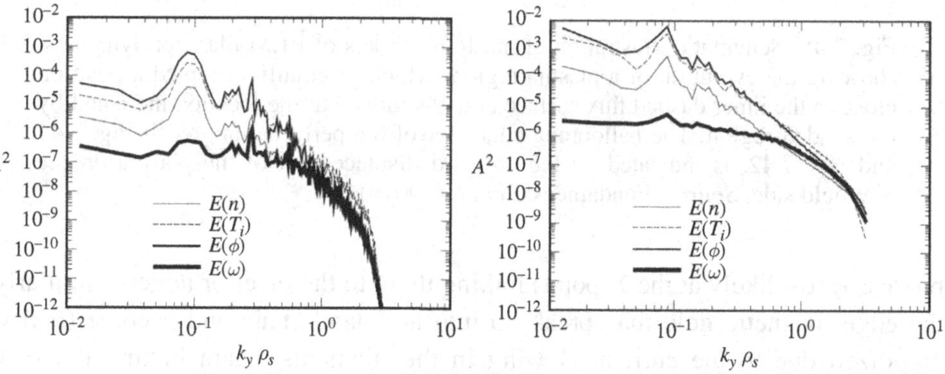

Fig. 7.44. Evolution of $k_\wedge$ spectra of squared fluctuation amplitudes of density, $\widetilde{n}$, ion temperature, $\widetilde{T}_i$, electric potential, $\widetilde{\varphi}$ and parallel vorticity, $\widetilde{\Omega}_\parallel$ during the ELM 'blow off': $t = 38$ (left) and $t = 42$ (right). Note the dominant peak at $k_\wedge \rho_S \sim 0.1$, corresponding to a toroidal mode number of $n \sim 10$, due to the dominant ideal ballooning (MHD) and ion temperature gradient (ITG) modes, and the rapid dual cascade towards both large and small scales only $15\,\mu s$ after the peak of the linear growth phase. *Source*: Kendl and Scott (2009).

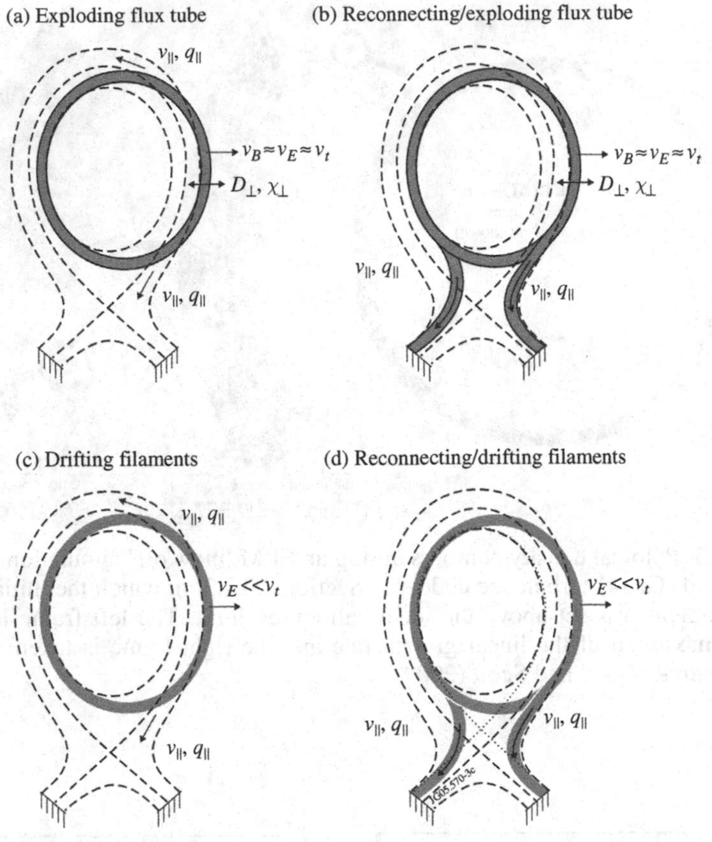

Fig. 7.45. Schematic illustration of the four models of ELM filament dynamics, showing the evolution of a plasma region, which originally occupied a position close to the inner dashed flux surface and has moved to the location indicated by the shaded region. The ballooning character of the perturbation, recall Fig. 7.41 and Fig. 7.42, is indicated by the outward displacement of the plasma on the low-field side. *Source*: Fundamenski *et al.* (2007b).

process (most likely at the X-point) linking them to the divertor targets. Similarly, the edge magnetic field may break up into an island chain and become (partly) ergodized due to the currents flowing in the filaments, again linking the ETB region directly to the divertor targets, albeit with a wide range of parallel connection lengths. Finally, the degree to which the original magnetic flux tube moves with the plasma filament[120] is still under debate. Taking these uncertainties into account one may construct a number of models of the exhaust stage of the ELM, as shown in Fig. 7.45. Reality lies somewhere between these four limiting cases, although a comparison with tokamak data suggests that models (c) and (d) may be

120 Since these are identically coupled in the MHD ordering and largely decoupled in the drift ordering limits, a gradual transition between the two is the most likely outcome.

the most appropriate for predicting ELM wall loads; we will return to this point in Section 7.3.3.2. Based on these arguments, one may conclude that the distribution of particles/current/energy on plasma facing components, and hence the energy split between the divertor and limiter tiles, is determined by competition between radial plasma motion, filamentary or not, and parallel (poloidal) losses. Since the latter almost always dominate, as they do in the L-mode and inter-ELM SOL, one finds that most of the energy released by the ELM is deposited on the divertor targets, in agreement with experiments.

7.3.2 Power exhaust in between ELMs

Since the duration of the ELM is typically much shorter than the mean inter-ELM period, one may divide H-mode power exhaust into two distinct phases: the relatively long and quiescent inter-ELM phase in which the edge pedestal is slowly being rebuilt, and the relatively short, yet intense, ELM phase, in which particles and energy are ejected into the SOL in a quick burst. Despite the nearly two orders of magnitude difference between the typical duration of these two phases, the amount of power exhausted in each phase is comparable, i.e. the short duration of the ELMs is compensated by their large energy density. Indeed, one typically finds that the inter-ELM phase dominates the overall power exhaust, with roughly two-thirds of the plasma power removed between, and one third during, ELMs.[121] Below we consider these two phases separately, beginning with power exhaust in between ELMs, and returning to the effects of ELMs in the next section.

Consistent with the reduction of radial transport in the ETB region, radial SOL transport in the inter-ELM phase is likewise reduced, as observed by the steepening of the near-SOL profiles by roughly a factor of two, although this effect becomes progressively less pronounced at higher upstream collisionality, ν_u^*; invariably, the profiles begin to broaden with radius, i.e. towards the far-SOL, with the degree of broadening again increasing with ν_u^*.[122] Since the radial gradients vary almost continously across the LCFS, it is tempting to interpret this result as an extension of the ETB region a short distance across the LCFS into the near-SOL, in which case the inter-ELM radial transport in the near-SOL should be governed by similar processes as in the ETB region. This observation leads us to make two hypotheses: (i) since the radial power flow across the ETB is dominated by the ion channel, with the ion heat diffusivity close to neoclassical levels, one may speculate that inter-ELM power exhaust, i.e. near-SOL energy transport, should likewise be dominated by ion heat conduction, or at least, if some residual level of turbulence is present,

[121] As a result, the time-averaged plasma loads on divertor PFCs, which determine their steady-state temperature, depend largely on inter-ELM power exhaust.

[122] At high ν_u^*, the far-SOL profiles in ELMy H-mode and L-mode discharges begin to converge.

that the latter should provide a lower limit to the effective radial heat flow;[123] and (ii) since the ETB pressure profiles are marginally stable with respect to ideal ballooning modes, recall Section 7.3.1.2, and the pressure profile across the LCFS is continuous, it suggests that the radial transport across the LCFS is such as to maintain this gradient at marginal stability, i.e. when the ballooning limit is briefly exceeded, a DHD instability quickly develops, which increases the radial energy transport, thus restoring the pressure profile to marginal stability. The two hypotheses are certainly not equivalent, but neither are they contradictory. Rather they constitute complementary approaches to attacking the inter-ELM power exhaust problem. It is therefore encouraging that both find substantial support in dedicated experimental data.

7.3.2.1 Extension of ETB into the near-SOL: (neo)classical transport

The first hypothesis is supported by multi-fluid, laminar edge plasma modelling of JET ELMy H-mode discharges (Fundamenski *et al.*, 2003; Kallenbach *et al.*, 2004), which found that the best match to the measured SOL profiles required a reduction of $D_\perp$ and $\chi_\perp$ in the near-SOL region down to ion neoclassical levels.[124] This hypothesis was explored in a series of dedicated studies on JET, in which the time-averaged heat loads on divertor tiles were measured in ELMy H-mode plasmas for a range of toroidal field, B_T, plasma current, I_T, neutral beam heating power, ion drift direction, $\mathbf{B} \times \nabla B$, and ion species, D vs. He (Fundamenski *et al.*, 2005, 2006). Different diagnostics and measurement methods were used to derive the target heat load profiles: (a) infra-red (IR) thermography, which measures the total heat load on the target, but, crucially, relies on sufficient resolution and the absence of loose deposits on the target surface; (b) Langmuir probes (LP) mounted in divertor targets, which deliver information on the electron temperature and ion flux impinging on the target, but are not always consistent with optical measurements of density and temperature; (c) thermocouples (TC) mounted on the divertor tiles, which can be used to infer a time-averaged heat load profile by shifting the plasma strike vertically, either on a shot-to-shot basis (Matthews *et al.*, 2003), or slowly, during a single shot (Riccardo *et al.*, 2001). The first approach depends on the identity of the discharges, whereas the latter involves a numerical deconvolution of the profile using a finite-element modelling of the thermal response of the target material and comparison with the time-dependent heat pulse.

123 The role of collisional effects in edge-SOL turbulence was discussed briefly in Section 7.2.2. A more comprehensive discussion may be found in Fundamenski *et al.* (2007a).

124 Additionally, this modelling suggested a presence of a radially inward velocity in the ETB region and an outward in the near-SOL region (Kallenbach *et al.*, 2004).

The use of time-averaged, outer target profiles can be partly justified by the fact that outer target power widths during ELMs are comparable to those during the inter-ELM phases, i.e. ELMs do not markedly broaden the profiles,[125] while contributing only a third to the total power. Moreover, as we will see in Section 7.3.3, ELMs deposit most of their energy on the inner target, and thus contribute little to the outer target heat load.[126] In contrast, as is evident from Fig. 7.46, the time-averaged power is deposited mainly on the outer targer, with an inner:outer asymmetry of ~1 : 2–2.5, where it is concentrated in a narrower profile, with a typical peak heat load asymmetry of ~1 : 5.[127] The former asymmetry is largely accounted for by (i) the larger surface area of the outboard surface, (ii) the Shafranov shift compressing the flux surfaces on the outboard side of the torus, and (iii) the ballooning nature of edge-SOL turbulence; the latter asymmetry, nor indeed the inferred inter-ELM power asymmetry which may be as larger as ~1 : 5 cannot be accounted for by these effects alone. Since the ELM power is preferentially deposited on the inner target, with an asymmetry of ~2 : 1, see Section 7.3.3, the inter-ELM power asymmetry is roughly ~1 : 5. Hence, detachment of the outer divertor leg is the main challenge of inter-ELM power exhaust. Since both asymmetries increase (decrease) with input power for normal (reversed) field direction, i.e. are sensitive to the $\mathbf{B} \times \nabla B$ direction, this suggests a poloidal contribution from guiding centre (mainly $\mathbf{E} \times \mathbf{B}$) drifts, recall Fig. 7.7, which increases in magnitude with the magnetization parameter, δ_i, recall (2.208).

Comparison of the IR, TC and LP heat load profiles for a typical ELMy H-mode plasma on JET is shown in Fig. 7.46. Under natural density conditions, i.e. with low gas fuelling, the integral power width at the outer target, which receives most of the steady-state power, was found to scale as

$$\lambda_q \equiv \int q_{div} \mathrm{d}s / q_{div}^{max} \propto A^\alpha Z^\beta B_T^{-1.03} q_{95}^{0.6} P_{div}^{-0.41} n_u^{0.25}, \quad \alpha + \beta \approx 1.04, \qquad (7.92)$$

i.e. to decrease with $B_T, I_T \propto q_{95}^{-1}$ and power to the (outer) target, P_{div}, while increasing with n_u and a combination of A and Z.[128] In contrast, a multi-machine scaling (Loarte *et al.*, 2002) of the integral power width mapped to the outer mid-plane, using data from JT-60, Asdex Upgrade (DIV I), and DIII-D shows a positive dependence on the power to the divertor; the clear tendency for an increase

125 This point is still under investigation. There are some indications that the degree of broadening increases with ELM size, approaching unity for small ELMs.
126 Nonetheless, the time-averaged heat loads must be treated as an approximation to the inter-ELM ones, introducing a degree of uncertainty to both the results and the conclusions.
127 This is why we have emphasized the outer target profiles in our discussion.
128 The mass, A, and charge, Z, scaling cannot be decoupled in the JET experiments since they are obtained from a comparison of D and He plasmas for which $A/Z = 2$.

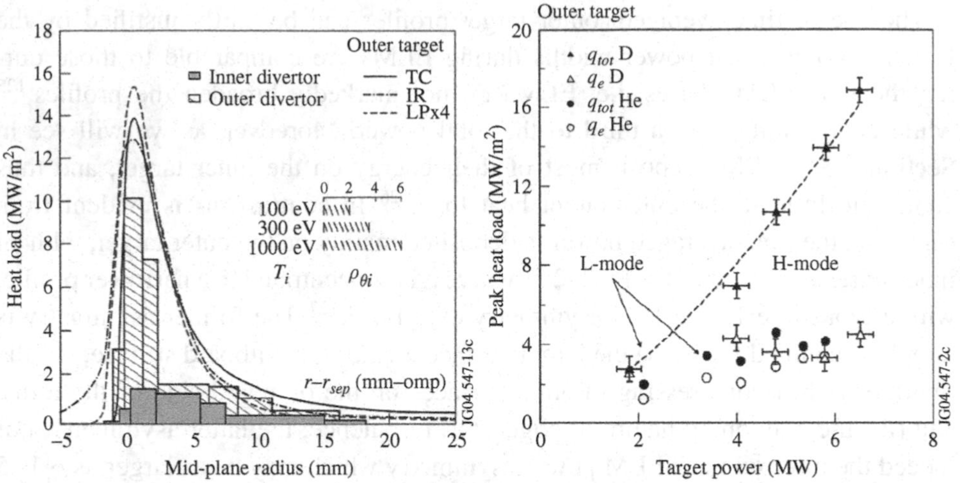

Fig. 7.46. The left frame shows time-averaged power deposition profiles at the outer target for a typical ELMy H-mode discharge on JET measured by four different diagnostic techniques; note that the scale of the electron heat flux (LP) is four times smaller. For comparison the poloidal gyro-radius at the outer mid-plane is shown for three values of the ion energy, c.f. $T_{i,ped} \sim 1$ keV. The right frame shows corresponding total (TC) and electron (LP) peak heat loads for D and He plasmas as a function of power entering the SOL; note that the excess power becomes more pronounced in D for high power H-modes. *Source*: Fundamenski *et al.* (2005).

in the integral power width with input power is revealed for JT-60 and DIII-D, (Loarte *et al.*, 2002, Fig. 7.5).

Concentrating on the dedicated JET experiments, the decrease of λ_q, and hence an increase in the peak heat load, Fig. 7.46, (7.92), was linked to a narrowing of the near-SOL profile, while the far-SOL profile remained largely unaffected. Comparison of various diagnostic techniques revealed that the increased heat load was primarily due to more energetic (hotter) ions, and that the ion heat load was closely correlated with a narrowing of the near-SOL power profiles.[129] Indeed, the smallest λ_q, mapped to the outer mid-plane, was comparable to the ion poloidal gyro-radius at the mid-pedestal temperature, Fig. 7.46. The energy of D^+ ions striking the outer target was inferred as ~300 eV, or $\nu_i^* \sim 1$, for the highest heating power, consistent with CXRS measurement of separatrix ion temperature. In general, the peak heat load correlated well with the inferred ion collisionality, see Fig. 7.47, and increased strongly, relative to the inter-ELM LP value, for $\nu_i^* < 5$. This broadening with ν_u^*, or n_u, was also observed in additionally fuelled ELMy H-mode discharges,

[129] It should be stressed that the contribution of supra-thermal ions and electrons to the divertor load profiles remains largely unresolved and constitutes an active area of research.

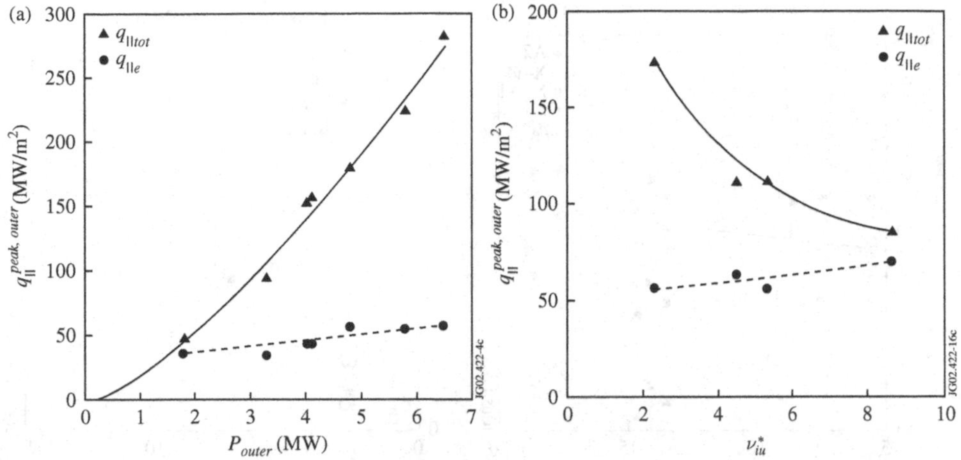

Fig. 7.47. Similar to Fig. 7.46: (a) peak parallel power densities (total and electron) at the outer target vs. power reaching the target at the 'natural', or unfuelled, H-mode plasma density; (b) the same quantities vs. upstream SOL collisionality at constant input power (12 MW) and different levels of gas fuelling; note the decrease of $q_{\parallel}^{\rm peak}$ (increase of λ_q) with increasing ν_{iu}^*. *Source*: Matthews *et al.* (2003).

although in this case the peak heat load decreased nearly quadratically with the upstream density, i.e. the near-SOL profile broadened as $1/q_{div}^{peak} \propto \lambda_q \propto \nu_u^{*1.8}$, Fig. 7.47. This is consistent with an increase in both the parallel temperature gradient and the radial convective velocity with ν_u^*, Section 7.1.4.2, suggesting an increase in SOL turbulence with density.

The measured λ_q scaling, (7.92), was compared with two dozen models of radial heat diffusivity in the SOL, as compiled by Connor and Hastie (1999), with the SOL transport analysis initially based on the two-point model estimates and simple relations derived in Section 7.1.4.2 (Fundamenski *et al.*, 2005), and later repeated using multi-fluid, laminar edge-SOL simulations (Kirnev *et al.*, 2007). In both cases, the best agreement between theory and experiment was obtained with collisional ion heat conduction. Comparison between the various (neo)classical predictions and the measured power widths,[130] shown in Fig. 7.48, suggests that *collisional* ion orbit loss (IOL) is the most likely (neo)classical process to explain the observed profiles.[131]

130 Here the prefix *neo* appears in brackets since neoclassical theory assumes closed flux surfaces and hence is invalid in the open field line (SOL) region. In the absence of a rigorous neoclassical theory of the SOL, the term (neo)classical transport is used to denote some variant of collisional transport in the SOL, bounded by classical and neoclassical expressions.

131 The term *ion orbit loss* refers to the opening-up and loss to divertor tiles of either passing or trapped ion orbits in the edge region due to the vanishing poloidal field at the X-point.

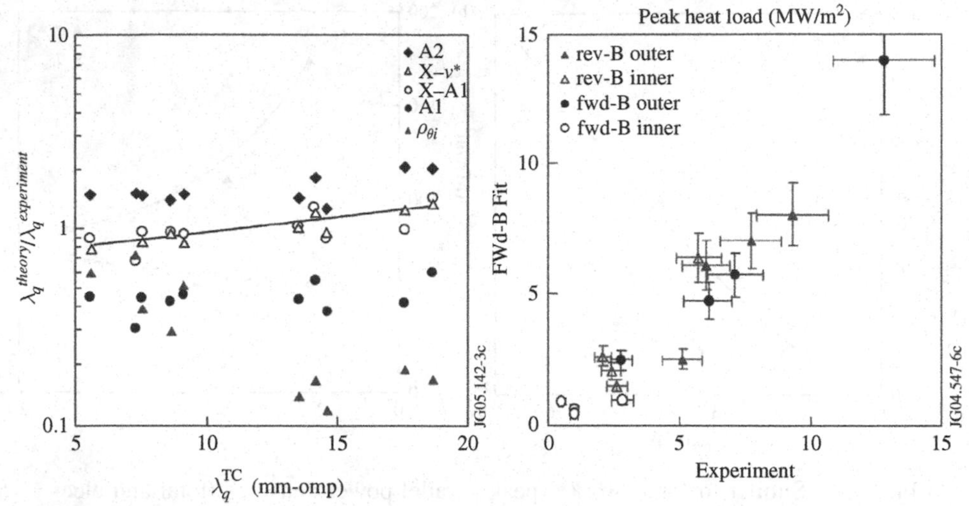

Fig. 7.48. The left frame shows the ratio of theoretical and experimental power widths, λ_q, in JET normal (forward) field experiments. The former include (neo)classical and classical ion conduction, **A2** and **A1**, two collisionally modified ion orbit loss estimates, $\mathbf{X} - \nu^*$ (IOL–coll) and $\mathbf{X} - \mathbf{A1}$ (IOL–neo), and the poloidal gyro-radius, $\rho_{\theta i}$. The right frame compares outer target, peak heat loads, with the best fit to normal (forward) field data plotted vs. experimental values, for both field directions and both targets. *Source*: Fundamenski *et al.* (2005).

Collisionless, or *direct* IOL, which was postulated to explain the observations of narrow power profiles in earlier studies, was discounted based on the results of reversed field (upward $\mathbf{B} \times \nabla B$) discharges together with Monte-Carlo simulations of ion orbit loss using an orbit following code (ASCOT) coupled to a realistic magnetic equilibrium and plasma background. While the code predicted ion orbit loss profiles to be strongly affected by field reversal, no such effect was observed in the experiment, Fig. 7.48. The footprint of direct IOL on the outer divertor target, was calculated based on ASCOT simulations over a wide range of R, B_T, q_{95}, A and Z,

$$\lambda_q^{\text{IOL}} \propto A^{0.35} Z^{-0.8} B_T^{-0.89} q_{95}^{0.88} T_{i,ped}^{0.39} n_{e,ped}^{-0.08} R^{0.8}, \tag{7.93}$$

which was used to define two coll–IOL widths (Fundamenski *et al.*, 2005),

$$\lambda_q^{\text{IOL-coll}} = \lambda_q^{\text{IOL}} \nu_i^{*1/2}, \qquad \lambda_q^{\text{IOL-neo}} = 2.4\zeta\lambda_q^{\text{neo}} + (1-\zeta)\lambda_q^{\text{IOL}}, \tag{7.94}$$

where $0 < \zeta = \nu_i^*/(1+\nu_i^*) < 1$ is a parameter measuring the degree of upstream ion collisionality. Both expressions agree, within the experimental error, with the measured power widths, Fig. 7.48.

Since (7.92) has been derived under natural density, attached divertor conditions, the above conclusions, as well as the expressions (7.94), cannot be easily extended to detached divertor operation; in that case, target heat load profiles, which become strongly affected by recycling neutrals and impurity radiation in the divertor, can become decoupled from SOL transport properties. At best, they represent the heat flux profiles at the entrance to the divertor volume, roughly at the X-point location.

7.3.2.2 *Marginal stability of near-SOL radial pressure gradient*

The second hypothesis proposed on page 368 has been recently explored on Alcator C-mod by examining the electron density and temperature profiles in the edge and SOL regions over a wide range of plasma densities, currents and magnetic fields in Ohmic L-mode discharges and for selected Ohmic H-mode discharges (LaBombard *et al.*, 2005, 2008). In all plasmas, the radial gradients were found to be steepest, i.e. the radial profile of $L_{p_e} = 1/\nabla_\perp \ln p_e$ to have a local minimum, in the near-SOL ($r - r_{sep} \approx 2\,\text{mm}$). To compare with predictions of DHD theory, the measured L_{p_e} have been expressed in terms of the *ideal ballooning parameter*, α_{MHD}, and the *diamagnetic parameter*, α_d, representing the ratio of the drift frequency, ω_{*n}, and the resistive ballooning rate. These can be related to the inductivity, $\hat{\beta}$, and the collision parameter, $\hat{C}$, of Section 6.3.2, as

$$\alpha_{MHD} = (4L_{p_e}/R)\hat{\beta}, \qquad \alpha_d = (R/2L_n)^{1/4}(R/2L_{p_e})^{1/2}/4\pi\hat{C}^{1/2}. \tag{7.95}$$

The results for a series of Ohmic L-mode discharges are shown in Fig. 7.49. The measured gradient lengths in the near-SOL are well ordered by α_d, and hence increase with $\hat{C} \propto \alpha_d^{-2}$. Moreover, at fixed values of $\hat{C}$, the near-SOL L_{p_e} decrease with I_T^2, such that α_{MHD} remains constant. As a result, they occupy a triangular region in the two-parameter phase space formed by α_{MHD} and α_d, with the largest α_{MHD} (corresponding to the ideal ballooning limit, $\alpha_{MHD} \sim 1$) achieved for an intermediate value of $\alpha_d \sim 0.3$, and decreasing for both larger and smaller values of α_d.

The interpretation put forward by LaBombard *et al.* (2005) is that radial particle and heat transport are strong functions of α_{MHD} and α_d, increasing rapidly when a certain threshold in the $\alpha_{MHD} - \alpha_d$ space is exceeded. This leads to marginal stability of the near-SOL profiles against DHD instabilities driving edge-SOL turbulence, i.e. when these boundaries are exceeded the radial transport increases relaxing the pressure gradients below the marginal stability boundary. The two disallowed regions correspond roughly to the L-mode density limit (high $\hat{C}$, low α_d) and a natural density boundary (low $\hat{C}$, high α_d). In other words, these observations are consistent with the idea that the operational space of the edge plasma, including boundaries associated with the tokamak density limit(s), are controlled by DHD turbulence, e.g. as α_d decreases below $\sim$0.3, radial heat convection increases

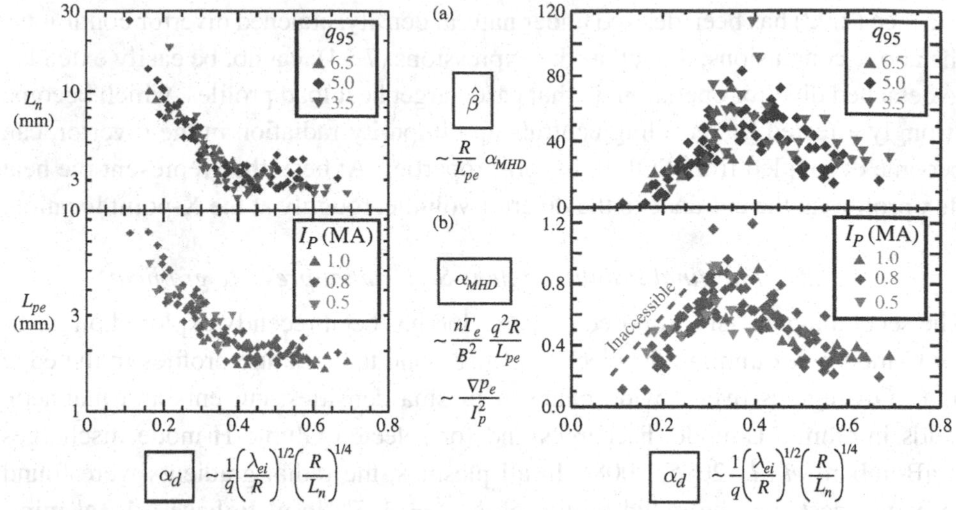

Fig. 7.49. The left frame shows the variation of radial gradient lengths of n_e and p_e in the near-SOL of Ohmic L-mode discharges on Alcator C-mod vs. the diamagnetic parameter, α_d. The right frame shows L_{p_e} normalized in terms of $\hat{\beta}$ and α_{MHD}. *Source*: LaBombard *et al.* (2005).

sharply and competes with parallel heat conduction along open field lines, making the high plasma density regions of $\alpha_{MHD} - \alpha_d$ space energetically inaccessible. The above threshold boundaries, including the low collisionality, *L–H* threshold boundary, which can be estimated as $\alpha_{MHD}\alpha_d^2 = 0.15$, consistent with the theory of Guzdar *et al.* (2001a,b) are also illustrated in Fig. 7.50, along with the radial pressure profiles in the SOL.

It is interesting to note that since the mechanisms determining radial energy transport in the near-SOL were not known at the time of the ITER design, the marginal stability hypothesis of near-SOL pressure profiles (with respect to the ideal ballooning limit, $\alpha_{MHD} \sim 1$) was used as the starting ansatz for multi-fluid, laminar edge-SOL plasma modelling of ITER exhaust (Kukushkin *et al.*, 2001). With hindsight, this assumption appears relatively well founded, and it is encouraging that the predictions of divertor power deposition profiles based on ideas of Sections 7.3.2.1 and 7.3.2.2 agree reasonably well; both approaches predict a power width of ~5 mm, mapped to the outer mid-plane, see Section 8.1.

7.3.2.3 Combination of collisional and turbulent transport in the near-SOL

Let us leave the topic of inter-ELM radial transport by examining the compatibility between the two hypotheses, or more generally, between the co-existence of collisional and turbulent transport in the near-SOL. The key to reconciling this

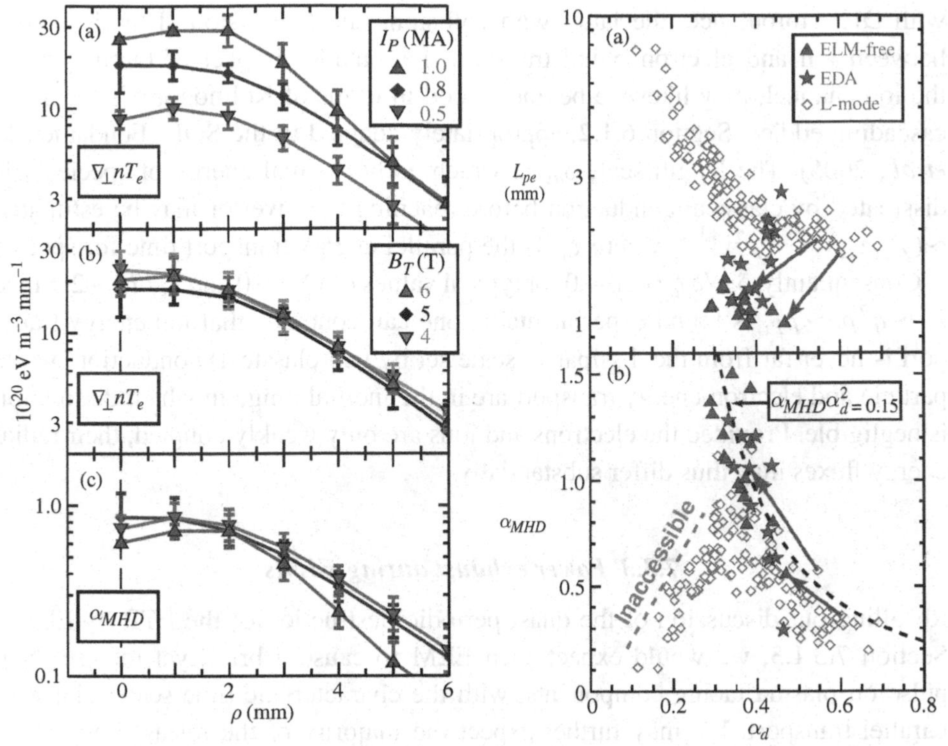

Fig. 7.50. The left frame shows the radial profiles of $\nabla_\perp p_e$ in the SOL in several Alcator C-mod Ohmic L-mode discharges, constituting a current and field scan. The right frame shows L_{p_e} and its normalization in terms of α_{MHD} as a function of the diamagnetic parameter, α_d. *Source*: LaBombard *et al.* (2005).

apparent dilemma is the concept of the dissipative scale, see Section 6.1.2, which represents the scale at which turbulent energy can be effectively converted into heat via collisional viscosity.[132] Based on the gyro-fluid edge-SOL simulations reported in Ribeiro and Scott (2008), see Section 7.2.2.3, this scale has been estimated as several ρ_S, $\delta_d \sim (3\text{–}5)\rho_S$, the transition between the edge and SOL regions as $\sim(10\text{–}20)\rho_S$, and the extent of the near-SOL as $\lambda_q \sim 10\rho_S \sim q^2\rho_S$. Thus even in the absence of the ETB, the width of the near-SOL is comparable to the dissipative scale, $\lambda_q/\delta_d \sim 2$, both of which scale with the sound speed gyro-radius, ρ_S.

Notwithstanding the uncertainty involved in relating the target load profiles to the SOL energy transport, the results of Fig. 7.46 suggest different mechanisms involved in electron and ion transport in the SOL, the former likely linked

[132] A comparable scale can be defined for dissipation of thermal energy by collisional heat diffusivity, etc.

with DHD turbulence, the latter with collisional heat diffusion. This difference between ion and electron radial transport, i.e. smaller impact of turbulence on the ion channel, may likewise be understood in terms of Kolmogorov's theory of cascading eddies, Section 6.1.2, appropriately adapted to the SOL (Fundamenski *et al.*, 2005). The length scale δ_{ds} at which eddy thermal energy of species s is dissipated by classical conduction before reaching the divertor may be estimated $\sim(\chi_{\perp s}^{cl}\tau_{\parallel s})^{1/2} \sim \rho_s^2 \nu_s^{*1/2}$, where $\tau_{\parallel s}$ is the parallel energy transport time for species s. Consequently, $\delta_{di}/\delta_{de} \sim$ 10–20 for typical values of $\nu_u^* \sim 10$ and $T_i/T_e \sim 2$. Since $\lambda_q \sim q^2\rho_i \sim \rho_{Pi}$ is found experimentally, one can conclude that ion energy transport is never far from the dissipative scale set by neo(classical) conduction, while particle and electron energy transport are in the inertial range in which dissipation is negligible. Provided the electrons and ions are only weakly coupled, their radial energy fluxes may thus differ substantially.

7.3.3 Power exhaust during ELMs

Recalling our discussion of the quasi-periodic destruction of the ETB by ELMs, Section 7.3.1.3, we would expect each ELM to cause a brief, yet intense, heat pulse on plasma facing components, with the characteristic time scale related to parallel transport. We may further expect the majority of the released energy to be deposited on the divertor tiles, with only a small fraction reaching the main chamber wall, i.e. the limiter tiles. Finally, we may expect the resulting heat load pattern, on both divertor and limiter tiles, to exhibit complicated fine structure characteristic of (i) filamentation of the edge plasma, and/or (ii) rearrangement of the edge magnetic field due to reconnection, ergodization, etc. It is therefore reassuring that these expectations are broadly supported by experimental data.[133]

Consequently, it is helpful to divide the discussion of power exhaust during the ELM into the associated heat loads on divertor and limiter tiles. In the former case, the key issues include the inner:outer energy asymmetry, the temporal pulse shape and the degree of broadening with respect to the inter-ELM heat load profile, in the latter case, the fraction of ELM energy reaching the limiters, the degree of localization (peaking) of the heat loads on the limiter tiles and the mean energy of ions and electrons at the limiter location. This division has an additional basis in the physics of parallel transport: since the pedestal plasma is typically collisionless, $\nu_{ped}^* < 1$, the divertor heat loads are likely dominated by kinetic effects, and since collisionality increases as the filaments cool in the SOL, ELM filament dynamics, and hence limiter heat loads, may be tractable, by the much simpler, fluid description.

[133] Type-I ELM observations are reviewed in Leonard *et al.* (2006); Fundamenski *et al.* (2007b).

7.3.3.1 ELM heat loads on divertor tiles

As with the inter-ELM power flow, the energy released during the ELM is conveyed primarily to the divertor targets. For small-to-moderate ELMs, one observes little broadening of the near-SOL heat load profile compared to the inter-ELM phase, i.e. the integral power width, λ_q, remains roughly unchanged. In contrast, further into the SOL, one observes a number a striations in the heat load pattern, whose shape and number is consistent with those expected from ELM filaments moving radially into the pre-ELM SOL magnetic field, see Fig. 7.51 (Eich *et al.*, 2005a). While this point is still under investigation, early indications suggest that the relative amplitude of these striations increases with ELM size, and consequently, so does the ELM heat load profile broadening. Interestingly, the absence of fine structure in the near-SOL heat load profile can be explained by the strong magnetic shear in the X-point region, which tends to smear out distinct filamentary structures and thus produce a roughly toroidally uniform pattern.

The temporal shape of the ELM heat pulse, as measured by IR thermography on both AUG and JET, is characterized by a steep rise and a gradual decay, on the parallel ion transport time, Figs. 7.52 and 7.53. Although half the energy is initially stored in the electrons ($p_e \sim p_i$), this energy cannot flow to the divertor on the parallel electron time scale due to the quasi-neutrality constraint. In effect, the electrons, which move much faster than ions, quickly set up a strong repelling potential which limits further electron outflow to the level determined by the ion

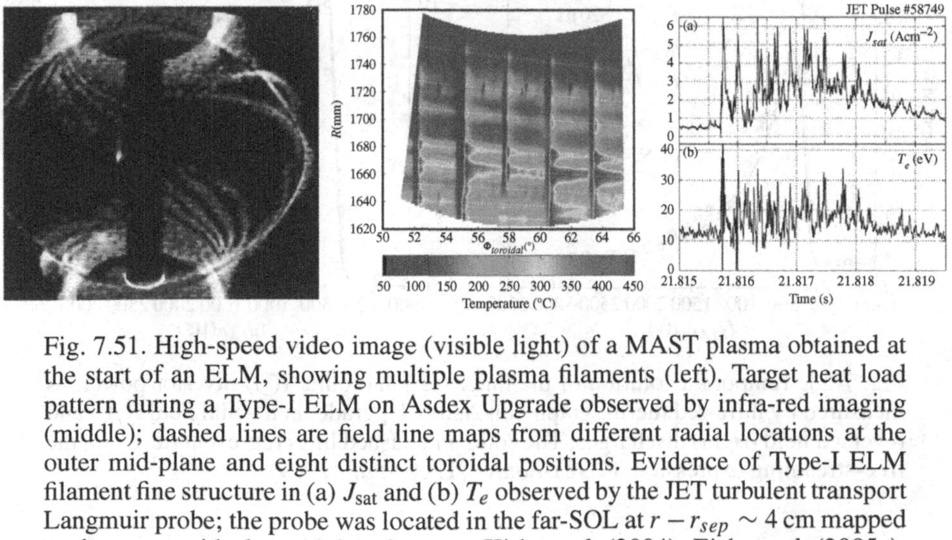

Fig. 7.51. High-speed video image (visible light) of a MAST plasma obtained at the start of an ELM, showing multiple plasma filaments (left). Target heat load pattern during a Type-I ELM on Asdex Upgrade observed by infra-red imaging (middle); dashed lines are field line maps from different radial locations at the outer mid-plane and eight distinct toroidal positions. Evidence of Type-I ELM filament fine structure in (a) J_{sat} and (b) T_e observed by the JET turbulent transport Langmuir probe; the probe was located in the far-SOL at $r - r_{sep} \sim 4$ cm mapped to the outer mid-plane (right). *Sources*: Kirk *et al.* (2004); Eich *et al.* (2005a); Pitts *et al.* (2006); Silva *et al.* (2005).

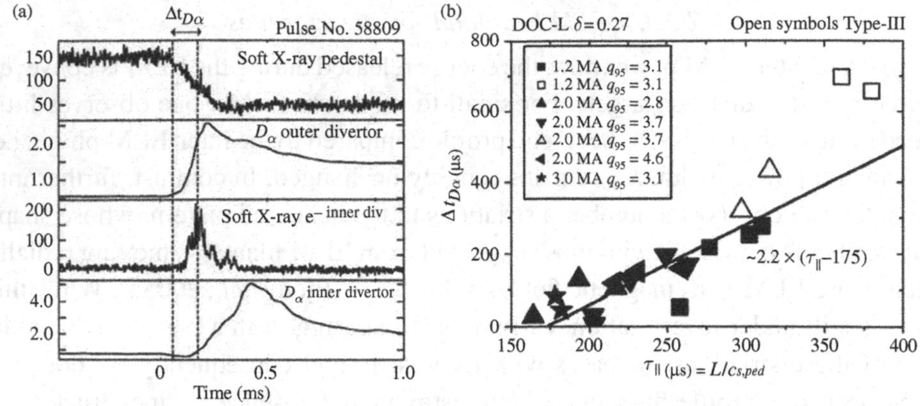

Fig. 7.52. Time traces of soft X-ray emission from the pedestal region (indicating reduction in pedestal T_e) and from the inner divertor (evidence of energetic electron impact), as well as D_α emission from inner and outer divertor targets, during a Type-I ELM in JET (left). The delay between the start of the ELM and the (rise of) D_α emission at the outer targets, which increase (roughly linearly) with the parallel transit time at the pedestal sound speed, $\tau_\parallel = L_\parallel / c_{s,ped}$, consistent with the outer mid-plane origin of the ELM pulse (right). Reused with permission from Loarte *et al.* (2004), ©AIP.

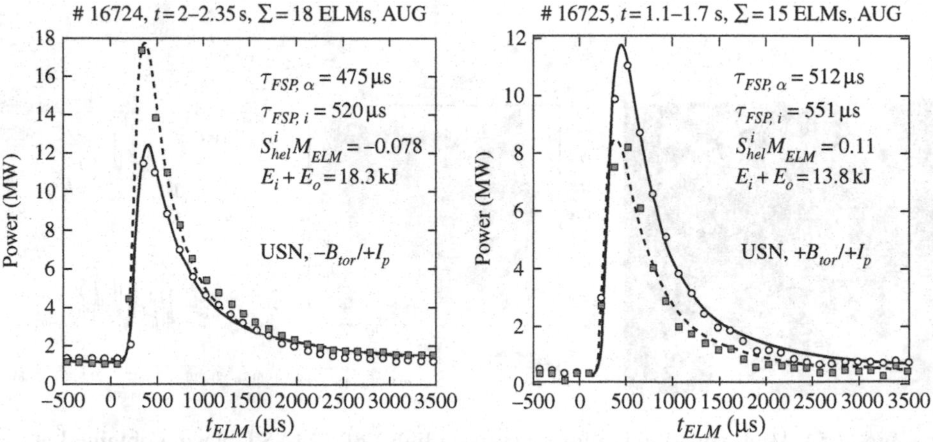

Fig. 7.53. Temporal evolution of the inner (■) and outer (○) divertor power, as measured by infra-red thermography on Asdex Upgrade, in normal field (left) and reversed field (right) discharges. The solid and dashed lines represent fits using the free-streaming expression, (7.98). *Source*: Eich *et al.* (2009).

outflow.[134] This picture is largely confirmed by the finding that the temporal pulse shape of the ELM heat load, at both targets, is in good agreement with that expected due to free-streaming of ions from the ETB region characterized by an initially Maxwellian velocity distribution at mid-pedestal ion temperature,

$$\Gamma_{ELM}^{outer}(t)\tau_{\|ELM}^{outer}/N_{ELM}^{outer} = x^2 e^{-x^2}/\sqrt{\pi}, \tag{7.96}$$

$$P_{ELM}^{outer}(t)\tau_{\|ELM}^{outer}/W_{ELM}^{outer} = \tfrac{2}{3}(1+x^2)x^2 e^{-x^2}/\sqrt{\pi}, \tag{7.97}$$

see Fundamenski *et al.* (2006), where $x = v_{cr}/v_{Ti} = \ell_{\|}^{outer}/tv_{Ti} = \tau_{\|ELM}^{outer}/t$ is a normalized critical velocity and $\tau_{\|ELM}^{outer} = \ell_{\|}^{outer}/v_{Ti} \approx \ell_{\|}^{outer}/c_S$ is the characteristic loss time to the outer target (an analogous expression exists for the inner target). While the above pulse shape assumes a Dirac delta function impulse, it remains roughly unchanged in shape (with the substitution $\tau_\| \to \tau_\| + \tau_{\tilde{B}}$) when convoluted with a top hot impulse with a characteristic time of $\tau_{\tilde{B}}$. Likewise, spatial convolution with a range of different $\ell_\|$ values gives a similar temporal broadening by a factor of $(\ell_{max} - \ell_{min})/v_{Ti}$. Note that the power pulse peaks earlier $\left(x \approx 1.3,\ t/\tau_{\|ELM}^{outer} \approx 0.77\right)$, than the particle pulse $(x \approx 1, t/\tau_{\|ELM}^{outer} \approx 1)$, due to the fact that faster particles, which arrive sooner, have a higher kinetic energy.

Clearly, (7.96) assumes the absence of inter-particle collisions and hence is only applicable under low collisionality conditions, ν_{ped}^*.[135] It is interesting to note that the best agreement is found when $\tau_\| - \tau_{\tilde{B}}$, where $\tau_{\tilde{B}} \sim 200\,\mu s$ is the time of magnetic fluctuations denoting the saturation stage of the ELM, is substantially longer than the parallel ion transit time from the outer mid-plane to the outer target, $\tau_\| \sim L_\|/v_{Ti}$, where $L_\| \sim \pi q R$ is the corresponding connection length. This implies either an increase of the effective parallel length, $\ell_\|^{outer} \sim 2L_\|$, or a reduction of the effective velocity of the free-streaming ions, v_{Ti}. Here the former option seems far more likely, since the ions lost from the ETB originate from various poloidal locations, including the inner mid-plane. This suggests either (i) the extension of $\ell_\|$ from the outer target via the outer mid-plane to the inner mid-plane region, i.e. $\ell_\|^{outer} \sim 2.5L_\|$, or (ii) a degree of ergodization of the ETB region creating a range of $\ell_\|^{outer}$ values, with the average close to $2L_\|$. At present it is difficult to distinguish between these two cases on the basis of experimental data.

The free-streaming expression (7.96) is in fair agreement with kinetic modelling of the ELM pulse using 1D particle-in-cell codes, which solve the Fokker–Planck equation along an SOL flux tube during the ELM transient, e.g. Tskhakaya *et al.* (2009). The results show that the initial part of the transient is well represented by

134 This process may be viewed as a transient Debye sheath formation.

135 However, recall that sufficiently supra-thermal particles are always collisionless due to the strong increase of the collisional mean free path with the particle velocity, $\lambda \sim v^4$. By the same argument, sufficiently sub-thermal particles are always collisional.

(7.96), with a small correction (enhancement) on the parallel electron time scale, $\sim 2.5L_{\parallel}/v_{Te}$, due to the arrival of the thermal electrons. However, in line with the quasi-neutrality argument, the electron pulse does not lead to an earlier peak in the particle or power deposition, but rather results in the increase of the electron temperature and the sheath potential,[136] e.g. the electron sheath heat transmission coefficients increase transiently from the pre-ELM value of $\gamma_e \sim 5$ by roughly an order of magnitude. Integrating in time over the ELM pulse, the codes find that the majority of the power is carried and deposited by the ions, which are accelerated by the transient electron potential along the flux tube.

In contrast to the time-averaged power, which in the normal field direction ($\mathbf{B} \times \nabla B$ towards the X-point) is deposited mostly on the outer target, the ELM energy appears to be deposited mostly on the inner target with an inner:outer asymmetry of $\sim 2:1$ (Eich *et al.*, 2007; Pitts *et al.*, 2007b). One plausible explanation for this opposite asymmetry which emerges based on a number of recent studies may be phrased thus: it is clear that the radial electric field in the edge and SOL regions points in opposite directions, due to an electric potential well near the separatrix. With the normal magnetic field direction, the electric drift in the SOL region increases the convective energy flow to the outer target, and that in the edge region to the inner target, while parallel flows tend to convect energy equally towards both targets. However, the parallel flow of ions on circulating orbits, which has an effective poloidal flow component, would tend to increase the energy deposited on the target facing towards this flow, since the parallel ion momentum would be roughly preserved as the ELM filaments connect to the divertor targets. This would suggest a link between net poloidal plasma rotation as the ultimate origin of the inner:outer ELM energy asymmetry (Eich *et al.*, 2009).

This hypothesis is given further credence by modifying (7.96) to include the above flows (Eich *et al.*, 2009; Fundamenski, 2009),

$$\Gamma^{outer}_{ELM}(t)\tau^{outer}_{\parallel ELM}/N^{outer}_{ELM} = x^2 e^{-(x-M)^2}/\sqrt{\pi}, \tag{7.98}$$

$$P^{outer}_{ELM}(t)\tau^{outer}_{\parallel ELM}/W^{outer}_{ELM} = \tfrac{2}{3}(1+x^2)x^2 e^{-(x-M)^2}/\sqrt{\pi}, \tag{7.99}$$

where $M = V/v_{Ti}$ is the Mach number of the parallel projection of the poloidal ion flow in the filament – clearly it is the net poloidal flow which contributes to the inner:outer asymmetry – with positive value indicating flow towards the outer target. In the presence of both $\mathbf{E} \times \mathbf{B}$ and parallel flows, when $V_P \approx V_E + (B_P/B_T)V_{\parallel}$, its parallel projection becomes,

$$V = V_P(B_T/B_P) \approx V_E(B_T/B_P) + V_{\parallel}. \tag{7.100}$$

[136] Such increases have been measured using Langmuir probes at the divertor tiles.

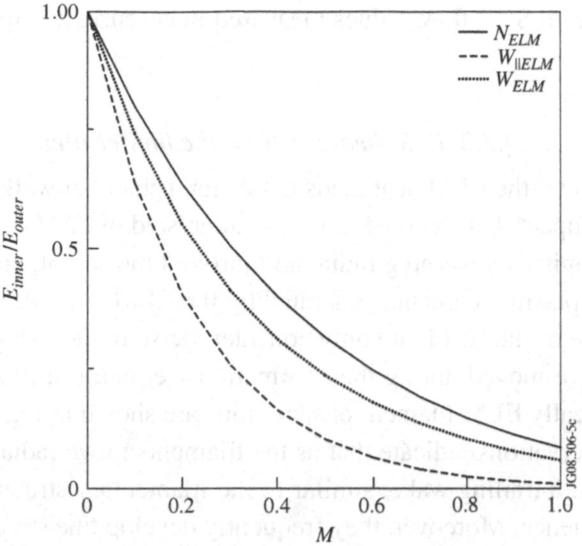

Fig. 7.54. ELM particle and energy asymmetries vs. the parallel Mach number.

Clearly, in the free-streaming approximation the time-integrated particle and energy loads on the divertor are symmetric for $M = 0$, with half the particle and energy content deposited on the outer target and the other half on the inner target. This symmetry is preserved even when the connection lengths to the two targets differ, $\ell_{\|}^{in} \neq \ell_{\|}^{out}$, since in the absence of collisions particles cannot change their initial velocities and hence terminate their trajectories on the divertor target towards which they were initially moving.[137] For a non-zero Mach number, these may be calculated as, see Fig. 7.54,

$$N_{ELM}^{outer}/N_{ELM} = W_{\perp ELM}^{outer}/W_{\perp ELM} = \tfrac{1}{2}\,[1+\mathrm{erf}(M)]\,, \tag{7.101}$$

$$W_{\|ELM}^{outer}/W_{\|ELM} = \tfrac{1}{2}[1+\mathrm{erf}(M)] + [M/(1+2M^2)]\mathrm{e}^{-M^2}/\sqrt{\pi}, \tag{7.102}$$

$$\begin{aligned} W_{ELM}^{outer}/W_{ELM} &= \tfrac{2}{3}W_{\perp ELM}^{outer}/W_{\perp ELM} + \tfrac{1}{3}W_{\|ELM}^{outer}/W_{\|ELM}, \\ &= \tfrac{1}{2}\,[1+\mathrm{erf}(M)] + [M/(1+2M^2)]\mathrm{e}^{-M^2}/3\sqrt{\pi}, \end{aligned} \tag{7.103}$$

where $N_{ELM}^{outer} = 1 - N_{ELM}^{inner}$ and $W_{ELM}^{outer} = 1 - W_{ELM}^{inner}$ are the number of particles and the amount of energy deposited on the outer target, from which the inner:outer asymmetries are easily obtained. Hence, only a modest Mach number towards the inner target, $M \sim 0.2$, is needed to account for the observed inner:outer energy asymmetry of $\sim 2:1$. Such Mach numbers are consistent with toroidal rotation,

137 In contrast, the time scales of deposition, and hence the peak loads are a function of $\ell_{\|}$.

as well as upstream SOL flow, values measured in co-current neutral beam heated plasmas.[138]

7.3.3.2 ELM heat loads on the limiter tiles

Let us next consider the ELM heat loads to the main chamber wall as a result of the ELM filament impact. In this context, we are interested in *ELM filament dynamics*, i.e. in the mechanisms governing radial and toroidal motion of, and parallel transport within, the plasma filaments generated by the ELM. Hence, we will focus on the exhaust stage of the ELM and only consider the saturation stage in as much as the filaments have moved during their formation, i.e. before full saturation of the instability. Typically ELM filament observations are shown in Fig. 7.51.

Tokamak observations indicate that as the filaments move radially they develop a steep front and a trailing wake, similar to the filamentary structures ('blobs') in edge-SOL turbulence. Moreover, they frequently develop fine structure, i.e. appear to stretch and disintegrate into ever smaller filaments. Initially, each filament rotates toroidally at some fraction of the pedestal toroidal velocity, while accelerating radially, so that viewed from a fixed location, the combined effect of toroidal rotation and radial dispersion is observed as a succession of arriving filaments, whose intensity generally decreases with time.[139] The duration of the observed wave-train of spikes corresponds to the duration of such a trailing wake, and can be a factor of ten longer than the length of magnetic activity. Finally, since electrons are much more mobile than the ions, the former are more effectively cooled. As a result, filaments become more collisional as they travel radially, with ion and electron energies becoming increasingly coupled. This increase in the degree of collisionality and the ion–electron coupling suggest the use of fluid description to treat the evolution of the ELM filaments in the far-SOL.

Such a model can be constructed starting from (7.65)–(7.66) with the parallel loss times given by (7.36)–(7.38). Suppressing the brace notation and introducing the ion–electron collisional energy exchange, one may write

$$\partial_t n \approx -n/\tau_{\parallel n}, \tag{7.104}$$

$$\partial_t T_e \approx -T_e/\tau_{\parallel T_e} - (T_e - T_i)/\tau_{ie}^{\varepsilon}, \tag{7.105}$$

$$\partial_t T_i \approx -T_i/\tau_{\parallel T_i} + (T_e - T_i)/\tau_{ie}^{\varepsilon}, \tag{7.106}$$

where we replaced d_t by ∂_t, thus implicitly adopting a frame of reference moving with the filament and introduced the ion–electron equilibration time $\tau_{ie}^{\varepsilon} = 1/\nu_{ie}^{\varepsilon}$,

138 The idea is further supported by the observations that with counter-current heating, when M is negative, one finds most of the ELM energy on the outer target, Fig. 7.53 (Eich *et al.*, 2009).

139 In general, filaments decelerate toroidally, but can both accelerate and decelerate radially.

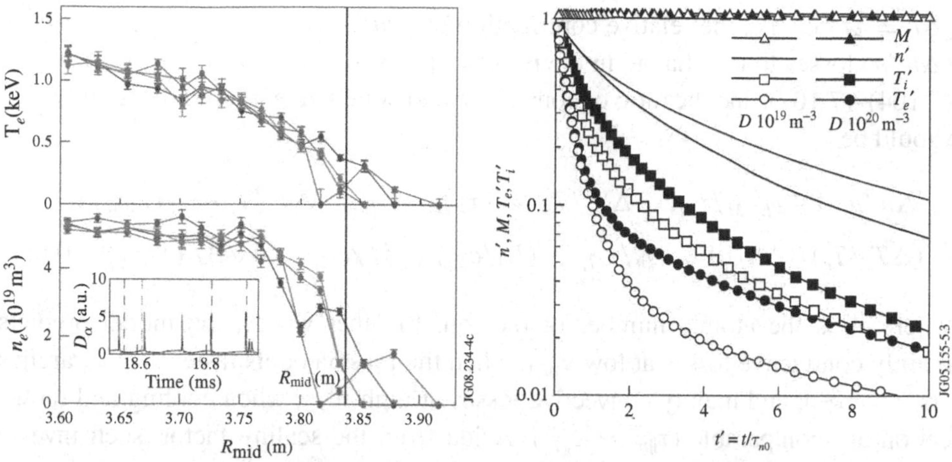

Fig. 7.55. The left frame shows the radial n_e and T_e profiles in the edge-SOL region, measured using high resolution Thompson scattering at JET, various times during a Type-I ELM cycle. The left frame shows the typical solution of the parallel loss model equations, (7.104)–(7.106), for JET relevant parameters, at two different values of the initial density. *Sources*: Fundamenski *et al.* (2006); Beurksens et al. (2009).

recall (5.111) and (5.113). The sum of the last two equations,

$$2\partial_t T \approx -T_e/\tau_{\|T_e} - T_i/\tau_{\|T_i}, \qquad T = p/n = (T_e + T_i)/2, \qquad (7.107)$$

indicates that the average energy content per particle is reduced only due to parallel losses, i.e. is independent of τ_{ie}^{ε} (as it must be since the latter simply transfers energy between the two species). The typical solution of these equations is shown in Fig. 7.55, which may be compared with a similar evolution in the free-streaming picture, see Fundamenski *et al.* (2006). Initially, the electrons cool faster than the ions, until the filament becomes sufficiently collisional for the energy exchange to begin to equilibrate the two temperatures. As a result the ratio T_i/T_e increases from unity to some value (~3–5) before again decreasing asymptotically towards unity. Finally, the rate of rarefaction due to particle losses is slower than both cooling processes.

The above set of kinetic and fluid equations form the basis of the so-called *parallel loss model* of ELM exhaust (Fundamenski *et al.*, 2006). Generally, the model describes the radial motion of the pedestal plasma as an effective filament moving with some mean radial velocity, which must be obtained from experiment, and subject to parallel losses to the divertor tiles.[140] One general prediction of this model is the relative change of pedestal temperature and density during the ELM

[140] Typically, the filament density and temperature are evolved using the fluid equations. Although these are typically not valid in the initial phase of the ELM (recall the discussion of parallel free-streaming), they become appropriate to the latter, collisional phase of the ELM.

($\Delta t = \tau_{ELM}$), i.e. the relative contribution of *conductive* ($\Delta T/T$) and *convective* ($\Delta n/n$) losses to the change in the pedestal pressure ($\Delta p/p$), which according to (7.104)–(7.107), and the ratio of convective and conductive times derived in (7.37), should be

$$\Delta n/n \approx -\tau_{ELM}/\tau_{\|n}, \quad \Delta T_e/T_e \approx -\tau_{ELM}/\tau_{\|T_e}, \quad \Delta T_i/T_i \approx -\tau_{ELM}/\tau_{\|T_i},$$
$$(\Delta T_e/T_e)/(\Delta n/n) \approx \tau_{\|n}/\tau_{\|T_e} \sim (L_\|/c_S)/\left(L_\|^2/\chi_{\|e}\right) \sim 64\sqrt{A}/\left(\nu^*_{ped} + 10\right),$$

where A is the atomic number of the ion. In other words, the model predicts mainly conductive losses at low ν^*_{ped}, when the plasma cools faster than it rarefies ($\tau_{\|n} \gg \tau_{\|T_e}$), and mainly convective losses at high ν^*_{ped}, when cooling and rarefication are comparable ($\tau_{\|n} \sim \tau_{\|T_e}$). Aside from the scaling factor, such inverse scaling of $(\Delta T_e/T_e)/(\Delta n/n)$ with ν^*_{ped} is indeed observed in tokamak experiments. Moreover, since ν^*_{ped} is typically varied by a change in pedestal density, the model explains why $\Delta n/n$ remains roughly independent of ν^*_{ped} and why $\Delta T_e/T_e$ decreases roughly as $1/\nu^*_{ped}$. It also helps to explain why $\Delta W_{ELM}/W_{ped}$ decreases with increasing ν^*_{ped} and hence why large (small) ELM are dominantly (largely) conductive (convective).

The parallel loss model, combined with the measured SOL-averaged radial filament velocities of $V_\perp^{ELM} \sim 600$ m/s (Fundamenski, 2005), has been successful at reproducing a range of ELM filament measurements on JET (Fundamenski *et al.*, 2007c), including the radial e-folding lengths of density, electron temperatures and energy, $\lambda_n^{ELM} \sim 12$ cm, $\lambda_{T_e}^{ELM} \sim 3$ cm, and $\lambda_W^{ELM} \sim 3.5$ cm inferred from dedicated outer gap-scan experiments for medium sized ($\Delta W/W_{ped} \sim 12\%$) Type-I ELMs, and the far-SOL ELM ion energies, measured using a retarding field analyzer probe head on fast scanning assembly, which indicate that $\lambda_{T_i}^{ELM} \sim 8$ cm (Pitts *et al.*, 2007b). Recently, the model has also been compared against infra-red measurements of ELM filament heat loads on the outboard limiters for discharges with well characterized pedestal density and temperature profiles. Starting from the inter-ELM pedestal profiles, Fig. 7.55, obtained using high-resolution Thompson scattering, the radial evolution of the fraction of ELM released energy reaching a given radial location, $W'(r) = W(r)/W(0)$ where $W(0) = W_{ELM}$, has been calculated for different initial filament positions (pedestal-top, mid-pedestal and separatrix) and radial velocities (600 m/s and 1200 m/s), see Fig. 7.56. With the default model assumptions (initial position at mid-pedestal, $V_{\perp,ELM} \sim 600$ m/s), the model predicts 9% of the ELM energy reaching the wall, in fair agreement with the measured value.

Another important observation, against which the model may be tested is the direct measurement of the pedestal plasma n_e and T_e evolution and formation of filaments during an ELM. Such measurements on JET, indicate that for nominal Type-I ELM conditions ($\Delta W_{ELM} \sim 100$ kJ), the inferred energy density in

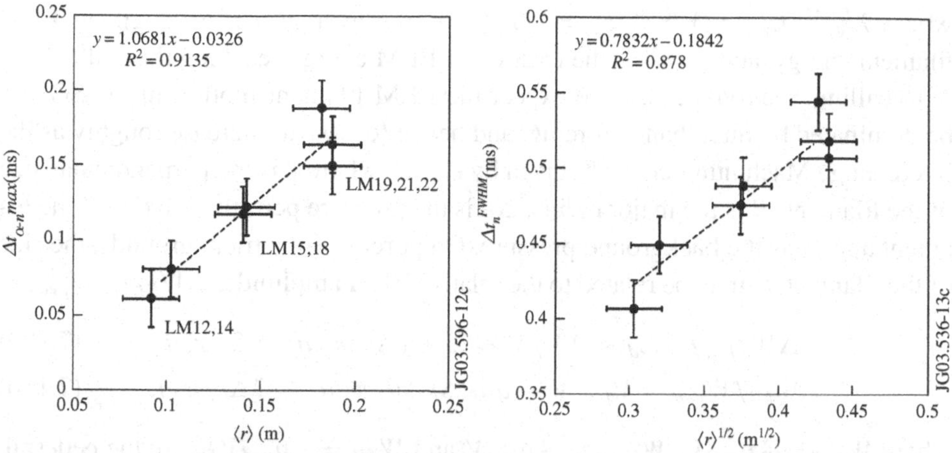

Fig. 7.56. The left frame shows the average time delay, $\Delta t_{\alpha-n}^{max} = t_n^{max} - t_\alpha^{max}$, of the peak plasma flux at the limit probe during an ELM (with respect to the start of the ELM) for each probe as a function of the average probe radius, $\langle r \rangle$, indicating a radial velocity of the wave-train of ELM filaments, i.e. of the effective or average ELM filament, of ~900 m/s in the limiter shadow. The right frame shows the full width at half maximum (FWHM) of the plasma flux for each probe vs. the square root of the average probe radius, indicating a diffusive broadening of the wave-train of ELM filaments in the limiter shadow. *Source*: Fundamenski *et al.* (2004a).

a single filament observed at ~5–7 cm from the pedestal top, i.e. 1–3 cm in the SOL, is roughly 3 kJ.[141] If ten such filaments are present ($n_0 \sim 10$), then their energy content at $r - r_{ped} \sim$ 5–7 cm is 30 kJ, or ~30% of the ELM. This is in fair agreement with model predictions (with the same assumptions as above), Fig. 7.55, which predicts ~20–30% of the released ELM energy at that radial location.

Crucially, the fraction of ELM energy observed at the limiters ($\Delta r \sim 5$ cm) increases with relative ELM size, e.g. from ~5% for $\Delta W_{ELM} \sim 250$ kJ ($\Delta W_{ELM}/W_{ped} \sim 0.09$) to ~10% for $\Delta W_{ELM} \sim 500$ kJ ($\Delta W_{ELM}/W_{ped} \sim 0.18$) (Pitts *et al.*, 2009); similar conclusions have been drawn earlier based on the deficit of the ELM energy reaching the divertor targets (Loarte *et al.*, 2007; Pitts *et al.*, 2007a). This observation suggests that smaller ELM filaments travel slower, i.e. have a lower radial Mach number, since for dominant convective losses, as appropriate to the far-SOL, one finds

$$f_{wall}^{ELM} = W_{wall}^{ELM}/\Delta W_{ELM} \approx \exp\left[-(0.5\Delta_{ped} + \Delta_{SOL})/\lambda_W^{ELM}\right], \qquad (7.108)$$

[141] Here the radial size of the filament was measured as $l_\perp \sim 5$ cm, the poloidal is taken as $l_\wedge \sim 3l_\perp \sim 15$ cm, a uniform distribution of n_e, T_e and $T_i \sim 2T_e$ is assumed along its length of $2Rq_{95} \sim 50$ (Beurskens *et al.*, 2008).

where $\lambda_W^{ELM}/L_\parallel \approx V_\perp^{ELM}/c_S = M_\perp^{ELM}$ is the average decay length of ELM filament energy, and f_{wall}^{ELM} is the fraction of ELM energy reaching the wall.

Recalling Section 7.2.2.1, we expect the ELM filament motion in the SOL to be dominated by interchange forcing, and hence for $M_\perp$ to increase roughly as the interchange Mach number, $M_\perp^{int}$, given by (7.61), where ℓ is the perpendicular size of the filament, R is the major radius, Δp is the pressure perturbation due to the filament and p_0 is the background pressure. On purely geometrical grounds, the size of the filament, ℓ, may be related to the relative ELM amplitude, $\Delta W_{ELM}/W_{ped}$, as

$$\Delta W_{ELM}/W_{ped} \approx \Delta V/V \approx 2\pi\kappa a\Delta r/\pi\kappa a^2 \approx 2\Delta r/a, \tag{7.109}$$

$$W_{fil}/W_{ped} \approx V_{fil}/V \approx qa\kappa\Delta r\Delta\theta/\pi\kappa a^2 \approx 2\Delta r/n_0 a, \tag{7.110}$$

where $W_{ped} = \frac{3}{2}p_{ped}V$, $W_{ELM} = \frac{3}{2}p_{ped}V$ and $W_{fil} = \frac{3}{2}p_{ped}V_{fil}$ are the pedestal, ELM and ELM filament energies, respectively; ΔV, V and V_{fil} are the plasma, ELM-affected and filament volumes; Δr and a are the ELM-affected and plasma minor radii; $m_0 = 2\pi/\Delta\theta$ and $n_0 = 2\pi/\Delta\phi$ are the initial poloidal and toroidal mode numbers of the ELM; $q = m_0/n_0$ is the safety factor; and κ is the plasma elongation (typically $q \sim 3$ and $\kappa \sim 1.5$). Since (7.61) was derived for a circular, Gaussian filament shape, $\exp(-(|\mathbf{x}|/\ell)^2)$, one is justified in treating the size scaling as a free parameter,

$$\ell \approx \ell_\perp^\gamma \ell_\wedge^{1-\gamma} \approx (\Delta r)^\gamma (a\Delta\theta)^{1-\gamma} \tag{7.111}$$

where $\gamma = 0.5$ corresponds to equal scaling with radial and poloidal widths and $\gamma = 0$ to a purely poloidal scaling.[142] Making use of an empirical result from JET, where it was found that the quasi-toroidal mode number at the main chamber wall, n_w, decreases with the relative ELM size, $n_w \sim 6/(\Delta W_{ELM}/W_{ped})$, the two widths can be estimated as[143]

$$\ell_\perp/a \approx c_\perp\Delta r/a \approx 3c_\perp/n_w \approx c_\perp/n_0, \tag{7.112}$$

$$\ell_\wedge/a \approx c_\wedge a\kappa\Delta\theta/a \approx 2\pi\kappa c_\wedge/n_0 q, \tag{7.113}$$

$$\ell_\wedge/\ell_\perp \approx 2\pi\kappa c_\wedge/qc_\perp \approx 3c_\wedge/c_\perp \sim 3. \tag{7.114}$$

Here we assumed $c_\perp$ and $c_\wedge$ to be comparable, so that the initial filament aspect ratio, $\ell_\perp/\ell_\wedge$, is independent of n_0 and hence of $\Delta W_{ELM}/W_{ped}$. This result, namely $\ell_\wedge/\ell_\perp \sim 3$, is confirmed by direct measurements of $\ell_\perp$ and $\ell_\wedge$ on several tokamaks (Kirk *et al.*, 2007).[144] It is noteworthy that JET infra-red measurements of ELM filament heat loads indicate that the filament aspect ratio changes to

142 Since the poloidal gradient is the main driving term, the latter will be our default ansatz.

143 The relative width, $\delta\theta/\Delta\theta$, was roughly independent of ELM size, ranging from ~0.6 ± 0.2 at the upper dump plates and ~0.8 ± 0.2 at the outer limiters (Jakubowski *et al.*, 2009).

144 It appears that the aspect ratio of the ELM filaments is the same as that of the plasma itself!

$\ell_\wedge/\ell_\perp \approx 2\pi\kappa c_\wedge/3qc_\perp \approx c_\wedge/c_\perp \sim 1$ by the time they make contact with the wall, i.e. the filaments appear to change from being poloidally elongated to roughly circular as they travel through the SOL.

Combining the above results, one finds a general expression for the interchange scaling of the radial Mach number of ELM filaments,

$$M_\perp^{int} \sim \sqrt{\frac{2l_\perp^\gamma l_\wedge^{1-\gamma}}{a/\epsilon_a}\frac{p_{ped}}{p_{sep}}} \sim \sqrt{\frac{2\epsilon_a}{n_0}\left(\frac{2\pi}{q}\right)^{1-\gamma}\frac{p_{ped}}{p_{sep}}} \sim \sqrt{\epsilon_a\left(\frac{2\pi}{q}\right)^{1-\gamma}\left(\frac{\Delta W}{W_{ped}}\right)^{1+\delta}},$$

where $\epsilon_a = a/R$; for default values $\gamma = \delta = 0$ this yields

$$M_\perp^{int} \sim \sqrt{(2\pi\epsilon/q)(\Delta W_{ELM}/W_{ped})}. \tag{7.115}$$

Note that the scaling with ELM amplitude is independent of γ due to the observed constancy of the initial filament aspect ratio. This prediction compares favourably with exponents inferred from measurements on JET

$$M_\perp \propto (\Delta W_{ELM}/W_{ped})^\alpha, \tag{7.116}$$

where $\alpha \sim 0.3$–0.4 based on turbulence transport probes (Silva *et al.*, 2009), $\alpha \sim 0.3$–0.5 based on the divertor energy deficit (Fundamenski *et al.*, 2007a; Kirk *et al.*, 2007; Pitts *et al.*, 2007a), and $\alpha \sim 0.4$ based on direct infra-red measurements of heat loads on the outboard limiters (Fundamenski, 2009; Pitts *et al.*, 2009).[145] In Section 8.1, we will assume $\alpha \sim 0.4$, which is consistent with all three measurements, when extrapolating to ITER.

The above results suggest that (7.115) slightly overestimates the exponent α unless the ratio of pedestal and separatrix pressures decreases weakly with ELM amplitude ($\delta \sim -0.2$); at present this ratio has not been accurately measured and we assume that $|\delta| \sim 0$. More likely, sheath dissipative effects modify the interchange scaling with ELM amplitude, reducing α below $0.5 + \delta$ (Fundamenski, 2009). The predictions of $M_\perp^{int}$ increasing with ϵ_a and decreasing with q are yet to be tested, although the former is consistent with results from MAST ($\epsilon_a \sim 1$) in which higher $M_\perp^{int}$ are measured than in conventional tokamaks ($\epsilon_a \sim 1/3$) (Kirk *et al.*, 2008). Finally, (7.115) can also explain why Type-III ELMs deposit far less energy to the wall in terms of the drop in pedestal pressure in the I–III transition, i.e. p_{ped}/p_{sep} is smaller by roughly a factor of two in Type-III compared to Type-I ELMs.

[145] This last value is derived from the data in Pitts *et al.* (2009), by noting that $f_{ELM}^{wall} \sim 5$ and 10% for $\langle\Delta W_{ELM}\rangle \sim 250$ and 500 kJ, and/or $\Delta W_{ELM}/W_{ped} \sim 0.09$ and 0.18, which yields $\lambda_{W,ELM}$ of 23 mm and 30 mm, respectively, and using (7.108) and (7.116) to find $\alpha \sim 0.4$.

7.3.4 Power exhaust control techniques

In present day tokamaks, the parallel heat flux in the SOL is of the order of a few hundred MW/m^2 and is expected to be $\sim$1 GW/m^2 in ITER. Projected onto the divertor tiles, the resulting heat load has to be reduced to $\sim$10 MW/m^2 (up to 20 MW/m^2 during transients), the technological feasible maximum for actively cooled structures.[146] The peak heat load may be reduced by maximizing the area of power deposition either by geometric measures, such as target tilting and flux expansion, or by profile broadening due to increased radial transport in the SOL. This reduction must be sufficient so as to achieve partially detached divertor conditions.

Similar reduction is also necessary for the energy transients due to ELMs, which cannot exceed $\sim$0.5 MJ/m^2 in $\sim$250 μs, above which the surface of PFCs is adversely affected by either micro-cracking (tungsten) or fibre erosion (carbon fibre composites). Moreover, since large ELMs cause the most damage, the average ELM size must be well below the threshold value. Hence, effective techniques for controlling the average ELM size are required.

7.3.4.1 Inter-ELM heat load control techniques

In Section 7.1.4.5, we observed that the peak divertor heat load may be reduced by increasing the upstream collisionality, $\nu_u^* \propto L_\parallel n_u / T_u^2$, which has two distinct beneficial effects: (a) it reduces the parallel electron heat conduction, thus allowing larger parallel temperature gradients; and (b) it increases the radial velocity of turbulent structures, thus broadening the SOL pressure profile. Both effects combine to cool and compress the divertor plasma and hence facilitate its eventual detachment from the divertor tiles.

In line with its definition, there are three ways to achieve the desired increase in ν_u^*: (i) increasing n_u via additional gas fuelling; (ii) decreasing T_u via injecting extrinsic, radiative impurities; and (iii) increasing $L_\parallel$ by increasing the edge safety factor, q_{95}, or elongating the divertor legs. The first strategy is ultimately constrained by the tokamak density limit at $n_u \sim n_{GW}$. The second strategy, in which the edge plasma is cooled by extrinsic impurities has the adverse effect of reducing the pedestal pressure and the associated stored energy, W_{ped}. Both strategies have a detrimental effect on the pedestal stored energy and the global energy confinement, i.e. $H_{98} = \tau_E / \tau_E^{98}$ decreases when either the Greenwald fraction, $f_{GW} = n / n_{GW}$, or the radiative fraction, $f_{rad} = P_{rad} / P_{heat}$ exceed values of $\sim$0.7–0.8. It should be added that a significant reduction of the peak heat load can

[146] One can of course conceive of advanced divertor target designs including liquid metals, e.g. lithium, or pebble beds, which could have significantly larger heat loading capabilities.

be achieved by optimizing the divertor geometry in respect to radiation capability (Herrmann, 2002; Takenaga *et al.*, 2003). The additional hazard associated with impurity seeding is the tendency of high-Z impurities to accumulate in the centre of the plasma, with a detrimental effect on core fuel density and fusion power, under the influence of purely collisional transport, recall (5.189) and Section 5.3.3.3. Since, to a fair approximation, this situation is achieved within the internal transport barrier (ITB), it is not surprising to find central peaking of high-Z impurity densities in discharges with strong ITBs (Takenaga *et al.*, 2003). This problem is particularly acute for non-inductive, advanced tokamak (AT) scenarios, which rely on the ITB to recover the stored energy lost by operating at higher q_{95} (lower I_T), than the baseline scenario; typically $q_{95} \sim 5$ in AT vs. $q_{95} \sim 3$ in inductive scenarios. Although impurity transport is one of the least understood processes in fusion plasmas, involving a combination of turbulent and collisional processes,[147] an empirical scaling of $Z_{eff} - 1$ is found to increase roughly linearly with the radiative heat flux and decrease roughly as the square of line averaged electron density (Matthews *et al.*, 1999), i.e.

$$Z_{eff} - 1 \propto P_{rad}/A_p\overline{n}_e^2, \qquad A_p = \pi\kappa a^2. \tag{7.117}$$

This indicates that operating close to the density limit is doubly beneficial to maximizing the fusion power: first, via maximizing the core fuel density, and second, via minimizing its dilution by impurity species.

The third strategy, namely increasing the connection length, is the most benign from the viewpoint of plasma performance, provided it can be achieved without increasing q_{95} (decreasing I_T), although it comes at the price of expensive reactor volume (which must be dedicated to the elongated divertor). There are a number of possible variants of this approach, e.g. Kotschenreuther *et al.* (2007), including increased flux expansion, multiple X-points and ergodization of the divertor magnetic field; in all cases, the ultimate aim is to maximize the plasma wetted area and divertor volume, thus lowering the plasma heat loads for the same upstream conditions. Despite lack of experimental confirmation, such designs offer a promising path to increasing the power handling of fusion reactors, with an added benefit of maximizing the distance between the burning plasma and the load bearing surfaces.

7.3.4.2 ELM heat load control techniques

Finally, let us examine the techniques for reducing the ELM heat loads. We may begin by asking whether most of the ELM energy can be radiated in the divertor

[147] While attempts have been made at predicting the impurity concentration in the core plasma by measuring the 'screening' of the SOL plasma to quantified extrinsic impurity sources, these have been restricted to low-Z impurity species, specifically carbon (Strachan *et al.*, 2004, 2008).

volume before striking the targets? Unfortunately, for present day divertor designs, the answer appears to be 'no'. For instance, dedicated experiments on JET have shown that the divertor plasma can at most radiate a small fraction (~10–20%) of the energy released during a typical, Type-I ELM above 50 kJ, and that this fraction only approaches unity for sufficiently small ELMs, e.g. Type-III ELMs below 20 kJ.[148] This result may be explained in terms of the limited amount of energy that can be radiated in the divertor on the time scale of the ELM (Maddison *et al.*, 2003; Rapp *et al.*, 2004; Monier-Garbet *et al.*, 2005). One may hope that divertor detachment would facilitate such radiative shielding of the divertor but again this would be false. The ELM heat pulse effectively 'burns through' the neutral gas buffer and transiently reattaches the plasma, so that under Type-I H-mode conditions detachment is only maintained during the inter-ELM phase; again, this is not true for the small, Type-III ELMs whose size is below some threshold.

In the above discussion care should be taken not to confuse radiative buffering with ELM-induced radiation. The former, beneficial effect, consists in radiation of the ELM energy *before* the ELM heat pulse strikes the target, the latter, generally adverse effect, represents the radiation spike often observed *after* the ELM heat load strikes the target. This spike can be comparable to the energy released by the ELM (Pitts *et al.*, 2007a; Huber *et al.*, 2009), and is simply a consequence of the influx of impurities released from the targets by the ELM plasma load. It can lead to strong cooling of the X-point region, which for sufficiently large ELMs can induce a back transition to the L-mode regime.

The problem of ELM-induced impurity inflows is particularly acute for tungsten PFCs, since the ELM pulse is characterized by hot ions (at keV energies), including any impurity ions present in the pedestal region, which can lead to significant physical sputtering of the tungsten tiles. This process has been recently observed at AUG, where the inter-ELM W influx from the outer divertor is strongly reduced as divertor plasma is cooled below ~10 eV (Dux *et al.*, 2009). With the outer divertor detached, the average W influx is dominated by ELMs and the transient W influx increases with ELM energy. Significantly, both the inter-ELM and ELM influxes are dominated by impurity sputtering. It was found that argon seeding decreased erosion *between* ELMs (by cooling the plasma below the threshold temperature) but increased W erosion *during* ELMs (since the ELM ions were always energetic enough to sputter W atoms). The optimum seeding rate (smallest W influx) is thus determined by competition between erosion by argon ions and cooling by argon radiation.[149]

[148] These findings have been broadly reproduced by multi-fluid edge-SOL simulations.
[149] More generally, it favours the use of low-Z radiative impurities such as neon or nitrogen.

Is there then no way to reduce the ELM heat load by means of extrinsic, radiative impurities? We have already touched on the answer in the form of the Type-III ELMs, which represent one of the three leading methods to significantly reduce the ELM size, the others being pellet injection and magnetic field perturbation. Recalling Section 7.3.1.2, we know that the ELM frequency can be increased (and the ELM size can be reduced) by an order of magnitude factor by cooling the pedestal plasma and thus affecting a transition from Type-I, to Type-III, ELMs.[150] At present, Type-III ELMy H-mode is the only scenario which has been demonstrated to be compatible with all ITER exhaust requirements,

$$f_{rad} \sim 0.75, \qquad f_{GW} \sim 0.85, \qquad q_{95} \sim 3, \qquad Z_{eff} < 2, \tag{7.118}$$

outer divertor detachment, and most notably, $\Delta W_{ELM}/W_{ped} < 0.01$, (Rapp *et al.*, 2004, 2005), although at a 'penalty' of a ~50% drop in pedestal pressure/temperature and hence a ~15–20% drop in H_{98}. If applied to ITER this solution would clearly have an adverse effect on the fusion gain, recall (1.34), with the reduction variously estimated as between ~30% and ~50% (Rapp *et al.*, 2009).[151]

The second method of reducing the ELM size consists of increasing the Type-I ELM frequency by injection of small hydrogen (deuterium) ice pellets at a frequency at least 50% greater than the natural ELM frequency (Lang *et al.*, 2006). As the pellet ablates in the edge region it perturbs the background plasma, triggering some type of edge localized instability: under L-mode conditions, it leads to increased MHD activity, while under H-mode conditions it simply triggers an ELM. The strength of this perturbation (effectiveness of the ELM triggering) appears to be correlated only to the penetration depth of the pellet, so that, provided the pellet is sufficiently large, it can trigger a Type-I ELM at any point in the ELM cycle! The physics of this triggering is still poorly understood due to a number of factors: (i) the local (non-axis-symmetric) nature of the perturbation; (ii) details of the ablation of the ice pellet leading to uncertainties in the source pattern; (iii) impact of these sources on the plasma background; and (iv) uncertainties in the dynamics of the pellet generated plasmoid. Recalling Section 7.3.1.3, however, it is safe to assume that the pellet-induced perturbation pushes the local pressure gradient above the (ideal) ballooning threshold, thus triggering a local ballooning instability.

The technique of triggering ELMs by pellet injection, known as *pellet pacing* of ELMs, has been extensively tested in AUG, where f_{ELM} has been increased by up to a factor of two, albeit at the reduction of the plasma stored energy and energy

150 Whereas the Type-I ELM stability boundary is determined by ideal MHD (peeling–ballooning) modes, Type-III ELMs are ideal MHD stable and are thus controlled by resistive MHD.

151 It should be noted however that $Q \sim 10$ could be recovered were it possible to increase the plasma current from 15 to 17 MA (i.e. by ~15%).

confinement, H_{98}, by roughly 10–20%. This reduction, however, which stems from increased convective losses, that is, from the additional plasma fuelling provided by the pellets, simply indicates that the pellets used in the experiments were too large for the task, i.e. that the effects of plasma fuelling and ELM pacing could not be decoupled on AUG.[152] Nonetheless, there are two major questions to be resolved regarding this ELM control technique, namely: can pellet ELM pacing be demonstrated above $f_{GW} \sim 0.85$ and can pellet injection produce a strong enough perturbation on ITER to trigger an edge instability (Type-I ELM) without triggering a core instability, e.g. a neoclassical tearing mode?

The third common technique of controlling the ELM size, and indeed, under certain conditions of suppressing ELMs altogether, consists of distorting, and to a large extent ergodizing (chaotizing), the edge magnetic field by applying a *resonant magnetic perturbation* (RMP). The resulting field, which can be described using the canonical representation, (3.23)–(3.28), is produced via a parametric resonance of the RMP with the natural 'frequency' of the magnetic field lines circumnavigating the torus, i.e. the resonance relies on the matching of the poloidal/toroidal mode numbers (m/n) of the perturbation to the edge safety factor, q_{95}. The RMP technique, which has been developed primarily on DIII-D with a toroidal mode number of $n = 3$ (Evans *et al.*, 2006, 2008), works most effectively with in-vessel coils, containing a significant poloidal harmonic, thus ensuring that the RMP is localized to the edge region. One of its hallmarks, related to the associated magnetic ergodization (chaotization), is the reduction of the edge plasma density due to increased parallel losses to the divertor targets.[153] This effect, known as *magnetic pump-out*, was first observed with ergodic divertors, see Ghendrih *et al.* (1996); Ghendrih (1999). Significantly, and somewhat surprisingly in view of basic parallel loss physics, (7.65)–(7.66), it represents edge plasma rarefication, $|\Delta n/n| \sim 1/3-1/2$, but not cooling, $|\Delta T/T| \ll 1$, i.e. the edge temperature remains roughly constant! As a result the pedestal density and pressure are reduced by ~15–30%, while the energy confinement is roughly preserved (Fenstermacher *et al.*, 2007; Evans *et al.*, 2008).

The mechanism responsible for mitigation and/or suppression of ELMs by RMP has yet to be unambiguously identified, although a number of candidate theories have been proposed so far (Tokar *et al.*, 2007, 2008; Evans *et al.*, 2008). All that can be said with confidence at present is that (i) ELMs are suppressed due to the increased radial transport caused by 'magnetic flutter' introduced by the RMP field (thus preventing the pressure gradient reaching the ideal ballooning limit)

152 This issue should be addressed shortly in the planned experiments on JET, where the two effects of fuelling and pacing become easier to decouple.

153 Although the effect generally occurs only under collisional edge plasma conditions, $\nu^*_{ped} < 1$.

and (ii) ELM frequency is modified by the same mechanism, which changes both the pedestal (in)stability conditions and the time to rebuild the pedestal during the ELM cycle.

Despite its overall attractiveness, there are a number of hurdles to be cleared by the RMP technique: (i) it remains to be seen whether this density drop can be recovered (to the required level of $f_{GW} \sim 0.85$) by increased particle fuelling, either by gas puffing or pellet injection, without loss in energy confinement, i.e. an effective reduction of W_{ped} or H_{98} due to the edge cooling associated with increased re-fuelling[154] and (ii) it is not clear whether ELM suppression could be maintained in the presence of pellet fuelling (as required for ITER), since large, 'fuelling' pellets generally trigger an ELM anywhere in the ELM cycle, even when the pedestal pressure gradient is ideal MHD stable. Nonetheless, the technique offers a very promising route of actively controlling edge pressure gradients and ELM heat loads, although clearly more experiments are needed to answer the above questions.

Finally, it is worth mentioning that a significant reduction of the ELM size (increase of ELM frequency) has also been observed with external perturbations to the toroidal field. These include both low ($n = 1, 2$) and high ($n = 16$) toroidal mode numbers, the former generated with error field correction coils (EFCCs) (Liang *et al.*, 2007), the latter by varying the toroidal field ripple (TFR) (Saibene *et al.*, 2007), e.g. when the ripple was increased from ~0.1% to ~1% in JET, the relative ELM size, for a given pedestal collisionality, was reduced by a factor of two (this change was related to smaller conductive losses, i.e. mainly convective ELMs, and was much less pronounced at higher density, i.e. for $f_{GW} \sim 0.85$). With both techniques, the pedestal density and pressure was reduced due to, what appears to be, a 'magnetic pump-out' effect. In the case of TFR, which is inherent in any tokamak and is expected to have a nominal value at the outer mid-plane of ~0.3% in ITER (compared with ~0.08% on JET), this loss of pedestal pressure was not recovered by increased fuelling, which merely led to increased cooling of the pedestal and so a reduction of H_{98}.[155]

In summary, on the basis of our present knowledge (experiment) and understanding (theory) of the plasma exhaust processes, it appears that this cost, (1.33), is negligible for existing tokamaks with carbon PFCs, in which PFC limits are rarely reached and impurity seeding is not a necessary condition, but could be significant (of order ~0.3–0.5 or more) for ITER and DEMO with metal PFCs. The dominant contribution to this reduction is the requirement of small ELMs ($\Delta W_{ELM}/W_{ped} < 0.01$), which entails a reduction of the pedestal pressure

[154] So far, the experience with the ergodic divertor in Tore-Supra, and with error field correction coil and toroidal field ripple experiments, see below, suggests otherwise.

[155] Similar results were found in the case of EFCC (Jachmich *et al.*, 2009), with the maximum achievable density reduced from $f_{GW} \sim 0.95$ to ~0.8 and H_{98} reduced by ~10–20%.

by ~0.3–0.5 and of the energy confinement by ~0.1–0.15. Although active ELM control by pellet injection and magnetic perturbations hold much promise and are topics of intense research in tokamak programmes worldwide, it remains to be seen whether these methods offer a smaller performance cost than the more conventional method of Type-III ELMy H-mode. At the moment, the latter appears to be the only plasma scenario compatible with all ITER exhaust criteria, including the heat load limits on PFCs. Indeed, since significant radiative impurity seeding is necessary in ITER to ensure detached divertor operation (in order to minimize W sputtering), and the exact mechanism governing the Type-I to Type-III transition is poorly understood, one may yet find that the transition to Type-III ELMy H-mode becomes unavoidable in ITER at the levels of radiation ($f_{rad} > 0.75$) required by the steady-state PFC limit of 10 MW/m^2. Finally, and perhaps most worryingly of all, is the unknown effect of the large transient W influxes after the ELM, on the edge transport barrier and ELM dynamics – a process which could become self-limiting if the W inflow destroys the ETB by effecting an H to L back transition. Indeed, the very access and maintenance of the H-mode appears problematic, as the required $P_{SOL}/P_{LH} \sim 2$ may exceed the steady-state PFC limit if sufficient energy ($> 50\%$) cannot be radiated in the divertor volume.

7.4 Further reading

The indispensible reference for virtually all aspects of SOL and divertor physics, with the exception of turbulence and ELMs, is the monumental compendium by Stangeby (2000). Complementary texts are given below: *Scrape-off layer and divertor physics*: Stangeby and McCracken (1990); Pitcher and Stangeby (1997); Loarte (2001); Fundamenski (2008). *L-mode and edge-SOL turbulence*: Endler (1999); Endler *et al.* (2005); Naulin (2007); Scott (2007); Zweben *et al.* (2007). *Edge-SOL modelling*: Schneider *et al.* (2006); Coster *et al.* (2007). *L–H transition*: Ryter and the H-mode Threshold Database Group (2002); Martin *et al.* (2008). *H-mode and edge localized modes*: Zohm (1996a,b); Connor (1998); Connor *et al.* (1998); Connor and Wilson (2000); Suttrop (2000); Bécoulet *et al.* (2003); Hermann (2003); Bécoulet *et al.* (2005); Eich *et al.* (2005b); Huysmans (2005a); Leonard *et al.* (2006); Oyama *et al.* (2006); Wilson *et al.* (2006); Fundamenski *et al.* (2007b); Kamiya *et al.* (2007); Kirk *et al.* (2008). *ELM filaments*: Endler (1999); Gonçalves *et al.* (2003); Kirk *et al.* (2004); Fundamenski *et al.* (2004a); Boedo *et al.* (2005); Eich *et al.* (2005a); Rudakov *et al.* (2005); Silva *et al.* (2005).

8

Outlook: power exhaust in fusion reactors

'Put out the light and then, put out the light.'
Othello, W. Shakespeare (c. 1603)

The ultimate aim of our discussion throughout Chapters 1 to 7 was the development of predictive capability regarding power exhaust in magnetically confined fusion reactors. It is therefore fitting to close this discussion with a brief examination of heat load predictions in two specific tokamak reactors: the next step burning plasma experiment, ITER, Section 8.1 and Fig. 8.1, and a demonstration fusion power plant, DEMO, Section 8.2, and finally commercial fusion reactors, Section 8.3.

8.1 ITER

'In any crisis, the best choice is to do the right thing. The second best is to do the wrong thing. The worst is to do nothing at all.'
Theodor Roosevelt (c. 1910)

Let us begin by considering the global power balance in ITER, assuming 40 MW of auxiliary heating power and $Q_\alpha = 2$. Since 80% of the D–T fusion power is released in the form of neutrons, this leaves $P_{core} = 40 + 80 = 120$ MW transferred to the plasma. Of this, let us assume that some substantial fraction, say $f_{rad}^{core} \sim 0.5$, is radiated in the core of the plasma by bremsstrahlung, synchrotron and line radiation, so that only $P_{SOL} \sim 60$ MW crosses the separatrix. In fact, this fraction should be determined by the requirement that the remaining power, $P_{edge} = \left(1 - f_{rad}^{core}\right) P_{core}$, which reaches the edge region is sufficient for accessing and maintaining the desired Type-I ELMy H-mode, with $P_{edge}/P_{L-H} > 1$ being essential and > 1.5 being desirable; the L–H threshold power in ITER ($I_T \sim 15$ MA, $B_T \sim 5$ T) at the desired operating density, $f_{GW} \sim 0.85$, may be estimated using (7.87) by extrapolating from typical Type-I ELMy H-mode

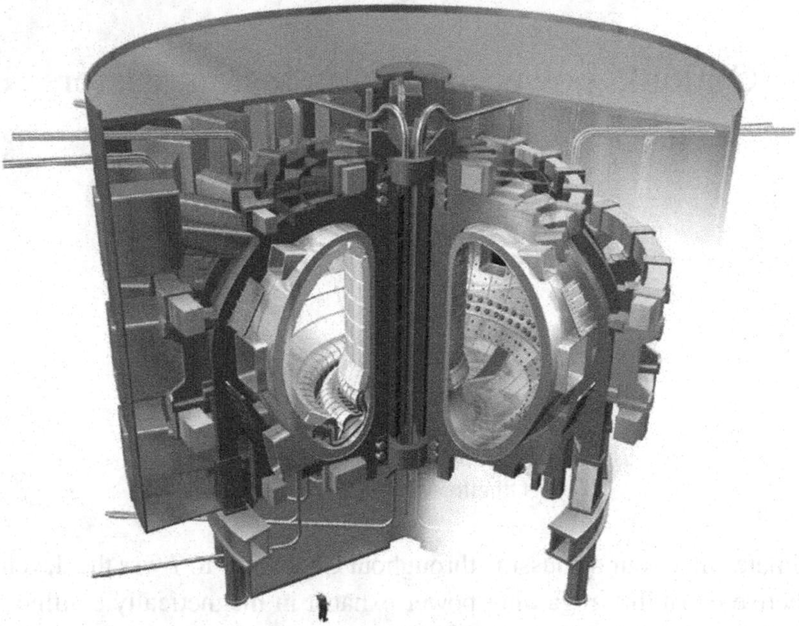

Fig. 8.1. Cut-away drawing of the ITER experiment (artist's impression). Source: © 2008 Eric Verdult, www.kennisinbeeld.nl.

conditions on JET, for which $P_{L-H} \sim 5\,\mathrm{MW}$ for $f_{GW} \sim 0.5$, $I_T \sim 2.5\,\mathrm{MA}$, $B_T \sim 2.5\,\mathrm{T}$, to find

$$P_{L-H}^{ITER}/P_{L-H}^{JET} \sim (f_G f_I/f_A)^{0.64} f_T^{0.78} f_A^{0.94} \sim (1.7 \times 6)^{0.64} 2^{0.78} 4^{0.3} \sim 12.5,$$

where f_G, f_I, f_T and f_A are the ratios of f_{GW}, I_P, B_T and A_p between ITER and JET. One thus finds an rough estimate of $P_{L-H}^{ITER} \sim 60\,\mathrm{MW}$.[1]

Of the remaining 60 MW, we expect ~2/3 to be exhausted between ELMs, and ~1/3 by the ELMs themselves. Assuming an inner:outer asymmetry of 1:2 in the average power deposited on the divertor, and of 1:1 in the ELM energy deposition,[2] recall Section 7.3.1.3, we expect the power to the inner and outer divertor legs between and during the ELMs to be

$$\left(P_{inner}^{inter}, P_{inner}^{ELM}\right) \sim (10, 10)\,\mathrm{MW}, \qquad \left(P_{outer}^{inter}, P_{outer}^{ELM}\right) \sim (30, 10)\,\mathrm{MW}. \quad (8.1)$$

Estimating the respective plasma wetted areas as $A = 2\pi R_{div} \lambda_q F_{eff}$, where R_{div} is the local major radius, λ_q is the integral width of the average deposited power

[1] This value is in fair agreement, within assumed error bars, with more detailed extrapolations.

[2] Here we assume that the ELM energy asymmetry is determined by edge plasma rotation, recall Section 7.3.3.1, and that this rotation will be much smaller on ITER than on present devices.

Table 8.1. *Nominal reactor parameters for JET, ITER and DEMO.*

	R_0/a (m)	q_a	B_0 (T)	I_T (MA)	n_{GW} (m^{-3})	P_α (MW)	Q_α
JET	3 / 1	3	2.5	2.5	10^{20}	2	0.1
ITER	6 / 2	3	5	15	1.5×10^{20}	80	2
DEMO	9 / 3	3	7.5	30	1.5×10^{20}	1000	20

mapped to the outer mid-plane, and F_{eff} is the effective flux expansion (ratio of distances along the target and outer mid-plane), we find

$$A_{in} \sim 2 \times 5\,\mathrm{m} \times 5\,\mathrm{mm} \times 10 \sim 1.2\,\mathrm{m}^2, \quad A_{out} \sim 2 \times 6\,\mathrm{m} \times 5\,\mathrm{mm} \times 10 \sim 1.5\,\mathrm{m}^2.$$

Here we estimated λ_q using the empirical scaling, (7.92), with P_{SOL} replaced by P_{SOL}/R^2, and two limiting cases of parallel SOL transport (7.31): (i) ion convection and (ii) electron conduction,

$$\lambda_q^{conv} \propto B_T^{-1}(P_{SOL}/R^2)^{-0.5} n_u^{0.25} (Rq_{95})^{0.5} \propto B_T^{-1} P_{SOL}^{-0.5} n_u^{0.25} q_{95}^{0.5} R^{1.5}, \quad (8.2)$$

$$\lambda_q^{cond} \propto B_T^{-1}(P_{SOL}/R^2)^{-0.5} n_u^{0.25} Rq_{95} \propto B_T^{-1} P_{SOL}^{-0.5} n_u^{0.25} q_{95} R^2. \quad (8.3)$$

Alternatively, one could use the simple power width estimate, $\lambda_q \sim \sqrt{\chi_\perp \tau_\parallel}$, (7.32), with $\chi_\perp$ given by the ion (neo)classical expression and $\tau_\parallel$ by either the convective or conductive loss time. Both approaches yield an estimate of the ITER power width (at the entrances into the ITER divertor) as ~3.5–4.5 mm, mapped to the outer mid-plane (Fundamenski *et al.*, 2004b, 2005). Since these values are obtained based on natural density H-modes, they could be viewed as a lower limit; on the other hand, they correspond to ν_u^* values which are comparable to those expected in ITER.[3] Finally, since the ITER divertor is much more closed than the divertor on JET – and indeed most of today's tokamaks – it has a capacity to broaden the power profile further. These values are consistent with those found by assuming marginal stability of the pedestal plasma pressure profile against ideal ballooning modes, combined with a multi-fluid laminar plasma edge modelling, which yields $\lambda_q \sim 5$ mm on the divertor targets (Kukushkin *et al.*, 2001).[4]

The maximum average and transient heat loads for divertor PFCs (irrespective of the target material) have been imposed as ~10 MW/m^2 and 0.5 MJ/m^2 in 250 μs

[3] Taking the lowest value of AUG, DIII-D, JT-60U data from the multi-machine scaling (Loarte *et al.*, 2002, Fig. 5) and the AUG scaling for DIV II from IR-measurements, a lower limit for the power deposition width in ITER of 5 mm is likewise predicted.

[4] It should be stressed however that such an approach only constrains the pressure profile, not the parallel heat flux profile.

to peak, respectively, which translate into maximum inter-ELM power deposition and maximum ELM energy deposition of,[5]

$$\left(P_{inner}^{inter}, W_{inner}^{ELM}\right)^{PFC} \sim (12\,\text{MW}, 0.6\,\text{MJ}), \tag{8.4}$$

$$\left(P_{outer}^{inter}, W_{outer}^{ELM}\right)^{PFC} \sim (15\,\text{MW}, 0.75\,\text{MJ}), \tag{8.5}$$

This implies that of the 40 MW reaching the outer divertor volume, at least 25 MW has to be removed by volumetric losses (charge exchange, elastic scattering and line radiation) in the divertor itself. The ELM time scale is too short for these processes to be useful, recall Section 7.3.4, with volumetric losses during the ELM in ITER estimated at less than 1 MJ. Hence, these losses must take place during the inter-ELM phase, requiring an effective radiation (volumetric loss) fraction during this phases of $f_{rad}^{outer} \sim 25/30 = 0.83$. Repeating this analysis for the inner target, we find a comparable inter-ELM radiative (volumetric loss) fraction of $f_{rad}^{inner} \sim 8/10 = 0.8$.

Similarly, the PFC limits on the maximum transient energy load imply a *maximum* ELM size and a *minimum* ELM frequency of

$$\Delta W_{ELM}^{max} \sim 2 \times 0.6\,\text{MJ} \sim 1.2\,\text{MJ} \tag{8.6}$$

$$f_{ELM}^{min} \sim \max(10\,\text{MW}/0.6\,\text{MJ}, 10\,\text{MW}/0.75\,\text{MJ}) \sim 17\,\text{Hz}, \tag{8.7}$$

or less than 1% of the pedestal stored energy, $W_{ped} \sim \frac{1}{3}W \sim 133\,\text{MJ}$. The best estimates of core energy transport predict that the attainment of a fusion gain factor of $Q_\alpha = 2$ in ITER requires a pedestal temperature of $T_{i,ped} \sim 2T_{e,ped} \sim 4\,\text{keV}$, assuming an average Greenwald density, $f_{GW} = 1$ (Doyle *et al.*, 2007). Based on the collisionality dependence of the relative ELM size, recall Section 7.3.1.3, the corresponding collisionality, $\nu_{ped}^* \sim 0.06$, would predict a relative ELM size and frequency in ITER of roughly

$$\Delta W_{ELM}^{nat}/W_{ped} \sim 0.2, \quad \Delta W_{ELM}^{nat} \sim 25\,\text{MJ}, \quad f_{ELM}^{nat} \sim 20\,\text{MW}/25\,\text{MJ} \sim 0.8\,\text{Hz}.$$

The temporal heat pulse due to an ELM on the divertor tiles should follow the free-streaming expression, (7.98), with a parallel convective time,[6]

$$\tau_\parallel \sim \tau_{\parallel ELM}^{outer} \sim 2.5\pi q R/c_{S,ped} \sim 260\,\mu\text{s}, \tag{8.8}$$

[5] Here we assume that there is no broadening of the power deposition profile during an ELM. This assumption is generally supported by experimental measurements, e.g. Eich *et al.* (2009), specifically for relatively small ELMs, which we would like to address here.

[6] Note that the parallel convective time is comparable on ITER and JET, since the increase in size is roughly compensated by the increase in the pedestal plasma sound speed: $\tau_\parallel^{ITER}/\tau_\parallel^{JET} \sim f_R/\sqrt{f_T} \sim 2/\sqrt{4} \sim 1$, where f_R and f_T are the ratios of R and T_{ped} on ITER and JET.

convoluted by the temporal duration of the parallel losses from the pedestal. The latter may be estimated as some fraction of the magnetic fluctuation time, $\tau_{\tilde{B}}$, (7.91), which may be evaluated under different assumptions concerning τ_E (Fundamenski *et al.*, 2006, Appendix) to yield $\tau_{\tilde{B}} \sim 300-700\,\mu s$. One is thus left with a wide range of predictions regarding the effective parallel loss time, ranging from $\tau_{\parallel}$, in the case where parallel losses dominate the saturation process, to $\tau_{\parallel} + \tau_{\tilde{B}}$, in the case where they play merely a passive role in the dynamics, simply draining the energy to the targets. The peak heat load could be reached after anywhere from $0.77\tau_{\parallel} \sim 200\,\mu s$ (as in the specifications for the 0.5 MJ/m^2 limit) to 750 μs (a factor of three longer than in the plasma gun tests). Taking the more conservative value, we may conclude that the transient PFC limits impose a reduction of the ELM size (increase of the ELM frequency) by a factor of roughly 20 with respect to their natural values,

$$\Delta W_{ELM}^{max} / \Delta W_{ELM}^{nat} \sim f_{ELM}^{min} / f_{ELM}^{nat} \sim 20. \tag{8.9}$$

There are of course some caveats in this requirement, e.g. the heat pulse shape used in the plasma gun tests differs somewhat from that expected for the ITER ELM as does the ratio of actual plasma pressures. However, while these discrepancies may relax the PFC limit by some factor of order unity, this is likely to be offset by the statistical variations in the ELM size.[7]

According to our analysis in Section 7.3.3.2, the added benefit of this reduction of the ELM size is a substantial reduction in the fraction of the ELM energy reaching the main chamber wall. Using the modified interchange scaling, (7.116) with an empirically obtained exponent, $\alpha \sim 0.4$,

$$M_{\perp}^{ITER} / M_{\perp}^{JET} \sim [(\Delta W / W_{ped})^{ITER} / (\Delta W / W_{ped})^{JET}]^{0.4}, \tag{8.10}$$

and inserting $M_{\perp}^{JET} \sim 1.6 \times 10^{-3}$ (obtained from the measured 600 m/s with c_S evaluated at $T_{ped} \sim 1600\,\text{eV}$) and $(\Delta W / W_{ped})^{JET} \sim 0.12$ one obtains

$$V_{\perp}^{ITER}[\text{m/s}] \sim 600 \left(T_{ped}^{ITER}[\text{keV}]/1.5 \right) [(\Delta W / W_{ped})^{ITER} / 0.12]^{0.4}. \tag{8.11}$$

We combine this expression with the parallel loss model as described in Section 7.3.3.2 and set $\Delta_{ped}/2 = 2.5\,\text{cm}$, $\Delta_{SOL} = 5\,\text{cm}$ to the upper dump plate to get the average transit time of $t = (\Delta_{ped}/2 + \Delta_{SOL})/V_{\perp}^{ITER}$. The solution indicates that $W' \sim 0.25$ and 0.04 reaching the upper baffle for natural ($\Delta W_{ELM} \sim 20$ MJ, $\Delta W_{ELM}/W_{ped} \sim 0.13$) and mitigated ($\Delta W_{ELM} \sim 1$ MJ, $\Delta W_{ELM}/W_{ped} \sim 0.0066$) ELMs, respectively.[8] On the basis of this analysis we

[7] Recent studies indicate that the PDF of ΔW_{ELM} can be intermittent (Moulton *et al.*, 2010).
[8] However, it is worth remembering a substantial error bar entailed in these predictions.

may conclude that the mitigated ELMs would (on average) deliver only a small fraction of their energy to the main chamber wall. Provided the associated heat loads, corresponding to ~40 kJ/ELM or ~800 kW (with $f_{ELM} \sim 20$ Hz) can be tolerated by the limiter tiles, the maximum ELM size on ITER would be determined by the heat load limits on divertor, rather than main chamber, PFCs.[9]

As we have seen in Section 7.3.4, it is presently not clear whether the required reduction in the ELM size, (8.9), can be achieved without adversely affecting the pedestal pressure, i.e. whether the desired combination of

$$f_{rad}^{core} < 0.5, \qquad f_{rad}^{inner}, f_{rad}^{outer} > 0.8, \qquad f_{ELM} > 20\,\text{Hz}, \tag{8.12}$$

can be achieved while maintaining $T_{ped} \sim 4$ keV, $f_{GW} \sim 0.85$ and $H_{98} \sim 1$, which are estimated to be required to achieve $Q_\alpha \sim 2$. One could, of course, repeat the above analysis in the opposite direction, using the specified PFC limits, (8.4)–(8.5), to estimate the resulting fusion gain, Q_α, for a given set of $f_{rad}^{core}, f_{rad}^{inner}, f_{rad}^{outer}, f_{ELM}, \ldots$, or, more generally, for a given set of modelling assumptions. Such calculations can be performed with various degrees of sophistication ranging from the simple power balance to a fully integrated core/edge/SOL transport model (Telesca *et al.*, 2007).

It is important to recognize that the actual interdependence of the above factors cannot be predicted at present, nor is it likely to be possible on the time scale of ITER construction. Instead, it will only become evident from actual ITER discharges, and indeed, only once ELMy H-mode can be robustly accessed. Therefore, the power exhaust aspects of ITER discharges should be viewed as being as much a part of the ITER experiment as the burning plasma or fusion technology issues.

Having said this, our understanding of power exhaust can be progressed given continued experimental and theoretical investigations. On the experimental front, tokamak operation with metal walls as in Alcator C-Mod (Mo), AUG (W) and the upcoming JET ITER-like wall (Be,W) is highly valuable. On the theoretical front, essential tools are intergrated simulations linking the three tokamak regions (core, edge, SOL) with PFC elements, which must be compared and benchmarked against experimental data to offer predictive capability for ITER. Moreover, it would be highly desirable to further benchmark these tools in the hydrogen phase of ITER, i.e. before the nuclear phase of the experiment, and to extrapolate to the deuterium phase based on modelling of hydrogen and deuterium plasmas in existing tokamaks.

[9] This does not exclude the possibility of occasional local damage due to arrival, in quick succession, of very large ELM filaments (Moulton *et al.*, 2010).

8.2 DEMO

'Fusion will be ready when society needs it.'
A. Artsimovitch (c. 1960)

The brief analysis of power exhaust in ITER, as outlined above, could be repeated for any particular design of a demonstration fusion power plant, e.g. the various conceptual designs considered in the ARIES (Najmabadi *et al.*, 1991) and/or EFDA studies (Maisonnier *et al.*, 2005).[10] Without entering into a detailed discussion, one can identify a number of critical power exhaust issues facing any DEMO reactor:

(i) The transient heat load problem (due to ELMs and disruptions), which is already severe in ITER, becomes practically insurmountable in DEMO; indeed, even the normally negligible transients caused by Type-III ELMs and/or (L-mode) turbulent filaments become a cause for concern. Therefore, unless results from ITER indicate otherwise, unmitigated Type-I ELMy H-mode cannot at present be contemplated as a viable scenario for DEMO.

(ii) In order to prevent a transition into Type-I ELMy H-mode, strong core radiation, primarily via bremsstrahlung and synchrotron radiation becomes necessary. This radiation requires a high-Z impurity, such as tungsten, which radiates preferentially in the centre of the plasma. Moreover, tungsten is also one of the best candidates for a PFC material. The cycle of physical erosion at the divertor, ingress into the plasma, accumulation in the core, decrease of the plasma burn, followed by reduction of physical erosion, etc. offers a natural feedback process, which is likely to determine the operating cycle for a given input power and fuelling rate.[11] The control and stabilization of this non-linear cycle offers a major challenge to fusion reactor operation.

(iii) Depending on the divertor design, high radiative fractions may also become necessary in order to handle the time-average heat load. As we have seen in the ITER analysis, λ_q is a weak function of R, so that the plasma wetted area increases linearly with size. As a result, the factor P_{SOL}/R may be used as a very rough scaling factor for the peak heat loads on the divertor tiles. Since for a constant fusion power density, P_{SOL} increases as R^3, we expect P_{SOL}/R to increase as $R^2\left(1 - f_{rad}^{cor}\right)$ in a family of geometrically similar fusion reactors. Hence, to compensate for the increase in size, it is necessary to reduce the power entering the SOL by increasing f_{rad}^{cor} in excess of 0.9.

In summary, the power exhaust limits in DEMO may be constructed as follows. Since we expect $Q_\alpha = P_\alpha/P_{heat} \gg 1$, the power deposited in the core is roughly equal to the fusion heating power,

[10] In one such recent analysis the same multi-fluid, laminar approach used to design ITER was applied to a tokamak DEMO with a major radius of roughly 9 m (Pacher *et al.*, 2007).

[11] This process, albeit without the fusion burn element, has been observed on long pulse, metal wall devices, e.g. LHD, and has been described as *plasma breathing* due to the long time scales involved in the collapse of the core pressure profile and inhaling/exhaling of impurities into/from the core plasma.

$$P_{core} = P_\alpha + P_{heat} = P_\alpha \left(1 + Q_\alpha^{-1}\right) \approx P_\alpha, \tag{8.13}$$

$$P_\alpha \approx C_\alpha n_D n_T \langle \sigma v \rangle_{DT} \mathcal{V} \approx C_\alpha p_D p_T \mathcal{V} \approx C_\alpha C_\mathcal{V} \beta^2 B_T^4 R^3, \tag{8.14}$$

where $\mathcal{V} \approx C_\mathcal{V} R^3$ is the plasma volume and $C_\mathcal{V} = 2\pi^2 \varepsilon^2 \kappa$ and C_α are constants. The power to the divertor can be expressed in terms of the core and SOL radiative fractions, $f_{rad}^{core} \equiv P_{rad}^{core}/P_{core}$ and $f_{rad}^{SOL} \equiv P_{rad}^{SOL}/P_{SOL}$,

$$P_{div} \approx P_\alpha \left(1 - f_{rad}^{core}\right)\left(1 - f_{rad}^{SOL}\right) \approx P_\alpha (1 - f_{rad}), \tag{8.15}$$

$$P_{div}^{PFC} \approx 4\pi R \lambda_q F_{eff} q_{div}^{PFC} \approx C_{PFC} R, \tag{8.16}$$

where $f_{rad} = P_{rad}/P_{fus}$ and we used $\lambda_q \sim 5$ mm, $F_{eff} \sim 10$ and $q_{div}^{PFC} \sim 10\,\text{MW/m}^2$ to estimate C_{PFC} as $\sim$60 MW/m. Requiring $P_{div} \leq P_{div}^{PFC}$ and inserting (8.14)–(8.16), we find a limit on the product $\beta B_T^2 R$,

$$\beta B_T^2 R \leq \sqrt{\frac{C_{PFC}/C_\alpha C_\mathcal{V}}{1 - f_{rad}}}, \tag{8.17}$$

which increases with the square root of C_{PFC}, and decreases with the square root of C_α, $C_\mathcal{V}$ and $1 - f_{rad}$. Hence for a given combination of fusion fuel, toroidal field, reactor size and geometry, the minimum radiative fraction imposed by the PFC limits is given by (8.17). In addition, unless active ELM suppression can be demonstrated, the minimum *core* radiative fraction is determined by the requirement of avoiding the H-mode, $P_{SOL} < P_{L-H}$,

$$\begin{aligned} f_{rad}^{core} &\geq 1 - P_{L-H}/P_\alpha \approx 1 - C_{L-H} n_{GW}^{0.65} B_T^{0.8} A_p / C_\alpha C_\mathcal{V} \beta^2 B_T^4 R^3, \\ &\approx 1 - C_{rad}^{core} q^{-0.65} \beta^{-2} B_T^{-2.55} R^{-1.65}, \end{aligned} \tag{8.18}$$

where $C_{rad}^{core} \sim C_{L-H}/C_\alpha C_\mathcal{V}$. In the course of the derivation, we used (7.87) for P_{L-H} and assumed that $\overline{n}_e$ is given by the Greenwald density, (7.50)

$$n_{GW} \approx \frac{I_T}{\pi a^2} \sim \frac{2\pi a \epsilon B_T/\mu_0 q}{\pi a^2} \sim \frac{2\kappa}{qR} \frac{B_T}{\mu_0}. \tag{8.19}$$

Above we used Ampere's law to relate the plasma current and the poloidal field, $\mu_0 I_P \sim 2\pi a \kappa B_P$, the definition of the safety factor, $q \approx \epsilon B_T/B_P$, to relate the poloidal and toroidal fields, so that

$$\begin{aligned} P_{L-H} &\approx C_{L-H} n_{GW}^{0.65} B_T^{0.8} A_p, \\ &\sim C_{L-H} \pi \kappa (2/\mu_0)^{0.65} B_T^{1.45} (qR)^{-0.65} (\epsilon R)^2, \end{aligned} \tag{8.20}$$

Combining (8.17) and (8.18), one finds the minimum *divertor* radiative fraction, which is again determined by the PFC heat load limits,

$$f_{rad}^{div} \geq 1 - C_{rad}^{div} \left(1 - f_{rad}^{core}\right)^{-1} (\beta B_T^2 R)^{-2},$$
$$\approx 1 - \left(C_{rad}^{div} / C_{rad}^{core}\right) q^{0.65} B_T^{-1.55} R^{-0.35} \tag{8.21}$$

where $C_{rad}^{div} = C_{PFC}/C_{DT}C_V$. Note that although both f_{rad}^{core} and f_{rad}^{div} tend towards unity as $R \to \infty$, the requirement of avoiding the L–H transition implies that the former dependence is much stronger.[12] Alternatively, if small ELMs may be tolerated, then (8.20) and (8.21) could be replaced by a corresponding threshold for access to Type-I ELMs, P_{I-III}. The above exhaust limits must of course be combined self-consistently with criteria of plasma ignition and stability, as outlined in Section 1.1, in order to determine the extent of the operating window of a given reactor design.

8.3 PROTO and beyond

'Insanity is doing the same thing over and over again while expecting a different outcome.'

A. Einstein (1935)

Looking beyond DEMO to prototype power plants (PROTO) and commercial fusion reactors, one can envisage not only advanced magnetic confinement schemes, such as the stellarator concept, see Fig. 3.1, which would offer significant benefits for steady-state operation (no need for current drive to maintain the poloidal field) and machine safety (absence of disruptions), but likewise more sophisticated power exhaust strategies based on direct (non-thermodynamic) energy conversion (Miley, 1976). These may include (i) hard X-ray (bremsstrahlung) capture in a low- to medium-Z first wall, (ii) synchrotron radiation capture and/or reflection from a metallic wall, and (iii) direct kinetic-to-electrical energy conversion via charged particle deceleration in a plasma exhaust stream. The last strategy becomes particularly attractive when combined with open-ended magnetic geometries, such as field reversed configurations (FRC), which can operate at plasma β approaching unity and which are naturally embedded in a thick, elongated scrape-off layer. When combined with low-neutron fusion fuel cycles, such as $D + He^3 \to p + He^4$, (1.4), in which up to 98% of the fusion energy is released in the form of charged particles and hence deposited/captured by the plasma column, such direct energy conversion offers a far more efficient

[12] Similarly, $f_{rad}^{core}, f_{rad}^{div} \to 1$ as $B_T \to \infty$, with a weaker difference between the two scalings.

means of exploiting the fusion burn process.[13] Here the on going ARIES studies (Najmabadi *et al.*, 1991, 1992; El-Guebaly, 1992) provide an interesting insight into the far-reaching possibilities of fusion energy.

8.4 Further reading

Nuclear fission reactors: Lamarsh (1983). *Nuclear fusion reactors*: Dolan (1982). *ITER*: ITER Physics Basis Editors *et al.* (1999); Shimada *et al.* (2007). *ITER power exhaust modelling*: Igitkhanov *et al.* (1998); Kukushkin *et al.* (2002, 2003); Kukushkin and Pacher (2002); Pacher *et al.* (2005, 2007, 2008). *DEMO:* Najmabadi *et al.* (1991, 1992); El-Guebaly (1992); Maisonnier *et al.* (2005, 2006). *D–He^3 fusion*: Fundamenski and Harms (1996). *Direct energy conversion*: Miley (1976).

[13] For a review of physics and technology of D–He^3 fusion, see Fundamenski and Harms (1996).

Appendix A
Maxwellian distribution

In thermal equilibrium $f_s(\mathbf{x}, \mathbf{v})$ relaxes to an isotropic distribution $f_s(v)$ with a vanishing flow velocity and uniform density and temperature. Its velocity dependence is given by the Maxwellian distribution f_{Ms} defined by

$$f_{Ms}(\mathbf{v}) = (\sqrt{\pi}v_{Ts})^{-3}n_s \mathrm{e}^{-v^2/v_{Ts}^2}, \qquad v_{Ts} \equiv \sqrt{2}v_{ts} = (2T_s/m_s)^{1/2}. \tag{A.1}$$

In the presence of a gentle drive away from equilibrium (whose strength is measured by ϖ), such as a local source of particles, momentum or energy, $f_s(\mathbf{x}, \mathbf{v})$ approaches a family of Maxwellian distributions, whose moments $n_s(\mathbf{x})$, $\mathbf{V}_s(\mathbf{x})$ and $T_s(\mathbf{x})$ change gradually with position. Hence $f_s(\mathbf{x}, \mathbf{v}) = f_{Ms}(\mathbf{x}, \mathbf{w}) + O(\varpi)$ where $\mathbf{w}(\mathbf{x}) = \mathbf{v} - \mathbf{V}_s(\mathbf{x})$ is the relative velocity and

$$f_{Ms}(\mathbf{x}, \mathbf{w}) = (\sqrt{\pi}v_{Ts})^{-3}n_s \mathrm{e}^{-w^2/v_{Ts}^2}. \tag{A.2}$$

Since the shifted Maxwellian is isotropic in the fluid frame, $\mathbf{V}_s$, its viscous stress, $\boldsymbol{\pi}_s$, and the heat flow, $\mathbf{q}_s$, vanish exactly, as do all moments related to $\mathbf{w}$-space anisotropy. Since $f_s(\mathbf{v}) = f_{Ms}(\mathbf{w}) + O(\varpi)$, we find

$$\boldsymbol{\pi}_s(f_s) \sim O(\varpi^2), \quad \mathbf{q}_s(f_s) \sim O(\varpi). \tag{A.3}$$

An approximation in which all $O(\varpi)$ corrections are neglected is aptly known as *local thermodynamic equilibrium* (LTE), since the entropy density, s_s,

$$s_s \equiv -\int f_s \ln f_s \mathrm{d}\mathbf{v} = n_s \langle \ln f_s \rangle_{f_s}, \tag{A.4}$$

remains constant. To demonstrate this, we evaluate (A.4) using (A.2),

$$s_s = n_s \ln\left(T_s^{3/2} n_s^{-1}\right) = \tfrac{3}{2} n_s \ln\left(p_s n_s^{-5/3}\right). \tag{A.5}$$

Hence $\mathrm{d}_t s_s = 0$ implies adiabatic expansion, $p_s n_s^{-5/3} = \text{const}$. The evolution equation for s_s follows directly from (2.169)–(2.171),

$$p_s \mathrm{d}_{t,s}(s_s/n_s) + \boldsymbol{\pi}_s : \nabla\mathbf{V}_s + \nabla\cdot\mathbf{q}_s = \mathcal{Q}_s + \tfrac{1}{2}i_{s2} - \tfrac{5}{2}T_s i_{s0}, \tag{A.6}$$

and indicates that entropy is conserved in the absence of collisions, viscosity and heat flow. The above can be restated in divergence form as

$$\partial_t s_s + \nabla \cdot \mathbf{s}_s = \Theta_s, \qquad \mathbf{s}_s = s_s \mathbf{V}_s + \mathbf{q}_s / T_s \tag{A.7}$$

$$\Theta_s T_s \equiv \mathcal{Q}_s - \boldsymbol{\pi}_s : \nabla \mathbf{V}_s - (\mathbf{q}_s / T_s) \cdot \nabla T_s + \tfrac{1}{2} i_{s2} - \tfrac{5}{2} T_s i_{s0}, \tag{A.8}$$

where $\mathbf{s}_s$ is the entropy flow and Θ_s is the entropy production rate. We note that entropy is conveyed by both advection and heat flow and is generated due to collisional dissipation, viscous heating and thermal diffusion. In the absence of these processes, the flow evolves isentropically.

In terms of GC-variables, $(\mathcal{E}, \mu, \gamma)$, (A.2) becomes

$$f_{Ms}(\mathbf{x}, \mathbf{v}) = (\sqrt{\pi} v_{Ts})^{-3} n_s \exp\left(- \frac{\mathcal{E} - e_s \varphi}{T_s} \right). \tag{A.9}$$

Note the absence of any μ or γ dependence, as expected for an isotropic distribution. The spatial dependence now also enters $f_{Ms}(\mathbf{x}, \mathbf{v})$ via the potential $\varphi(\mathbf{x})$ as is evident from the gradient of (A.9),

$$\nabla \ln f_{Ms} = \nabla \ln n_s + \left(\frac{\mathcal{E} - e_s \varphi}{T_s} - \frac{3}{2} \right) \nabla \ln T_s + \frac{e_s \nabla \varphi}{T_s}. \tag{A.10}$$

Velocity-space averages involving (A.9) are best calculated by introducing the normalized speed, $x = v / v_{Ts}$, and the pitch angle cosine, $\xi \equiv v_\parallel / v = \cos \alpha$, in terms of which $(\mathcal{E} - e_s \varphi)/T_s = x^2$ and $(v_\perp / v_{Ts})^2 = x^2 (1 - \xi)^2$. With these definitions, the $\mathbf{v}$-space average reduces to a double integral,

$$\langle \mathcal{A} \rangle_{f_{Ms}} = \frac{2}{\sqrt{\pi}} \int_{-1}^{1} \int_{0}^{\infty} \mathcal{A} x^2 \mathrm{e}^{-x^2} \mathrm{d}x \mathrm{d}\xi, \tag{A.11}$$

whose evaluation is elementary for polynomial and exponential functions (Arfken and Weber, 2001).

In a magnetized plasma, the velocity distributions of charged particles are in general non-isotropic, with distinct temperatures in the parallel and perpendicular directions, $T_{\parallel s} \neq T_{\perp s}$, defined by (2.194). In approaching LTE, they tend to relax to a bi-Maxwellian distribution f_{2Ms},

$$f_{2Ms}(v_\parallel, v_\perp) = (\sqrt{\pi} v_{Ts})^{-3} n_s \exp\left(-v_\perp^2 / v_{T\perp s}^2 - v_\parallel^2 / v_{T\parallel s}^2 \right), \tag{A.12}$$

where $v_{T\parallel s}^2 \equiv 2T_{\parallel s}/m_s$, $\frac{1}{2} v_{T\perp s}^2 \equiv 2T_{\perp s}/m_s$, $v_{Ts}^2 = v_{T\parallel s}^2 + v_{T\perp s}^2$ and the particle velocity $\mathbf{v}$ is written in gyro-tropic vector notation (2.17).

Appendix B
Curvilinear co-ordinates

Any vector field $\mathbf{A}(\mathbf{x})$, with $\mathbf{x}$ being the position vector, can be represented in two different bases derived from a set of spatial co-ordinates (z^1, z^2, z^3),

$$\mathbf{A}(\mathbf{x}) = A^i \mathbf{e}_i = A_i \mathbf{e}^i, \quad A^i = \mathbf{A} \cdot \mathbf{e}^i, \quad A_i = \mathbf{A} \cdot \mathbf{e}_i, \tag{B.1}$$

where $\mathbf{e}_i(\mathbf{x})$ and $\mathbf{e}^i(\mathbf{x})$ are the *co-variant* and *contra-variant* basis vectors,

$$\mathbf{e}_i = \partial_{z^i} \mathbf{x}, \qquad \mathbf{e}^i = \partial_{\mathbf{x}} z^i = \nabla z^i, \tag{B.2}$$

and $A_i(\mathbf{x})$ and $A^i(\mathbf{x})$ are the co- and contra-variant components of $\mathbf{A}$; summation over repeated indices is implied. To understand the origin of this terminology, it is helpful to visualize the (constant) co-ordinate surfaces, $z^1 = c^1$, $z^2 = c^2$ and $z^3 = c^3$. The intersection of any two surfaces defines the (constant) co-ordinate curves, e.g. the z^1-curve is defined by $z^2 = c^2$ and $z^3 = c^3$. In this picture, the co-variant basis vectors, $\mathbf{e}_i = \partial_{z^i} \mathbf{x}$, are tangent (parallel) to the co-ordinate curves, while the contra-variant basis vectors, $\mathbf{e}^i = \nabla z^i$, are normal (perpendicular) to the co-ordinate surfaces. As a result, the *co-variant* components A_i vary *with* ∇z^i, while the *contra-variant* components A^i vary *against* ∇z^i. The co- and contra-variant bases are generally distinct, being identical only for rectilinear co-ordinates, e.g. (x, y, z). They are mutually reciprocal in the sense that $\mathbf{e}^i \cdot \mathbf{e}_j = \delta^i_j$ and

$$\mathbf{e}^i = \frac{\mathbf{e}_j \times \mathbf{e}_k}{\mathbf{e}_i \cdot (\mathbf{e}_j \times \mathbf{e}_k)} = \frac{\mathbf{e}_j \times \mathbf{e}_k}{J}, \quad \mathbf{e}_i = \frac{\mathbf{e}^j \times \mathbf{e}^k}{\mathbf{e}^i \cdot (\mathbf{e}^j \times \mathbf{e}^k)} = J(\mathbf{e}^j \times \mathbf{e}^k).$$

The *Jacobian* of the transformation, J, is conveniently expressed as $J = \sqrt{g}$, where $g \equiv \det \mathbf{g}$ is the determinant of the *metric tensor*, $\mathbf{g} \equiv [g_{ij}]$,

$$A_i = A^j \mathbf{e}_j \cdot \mathbf{e}_i \equiv g_{ij} A^j \quad \Rightarrow \quad g_{ij} \equiv \mathbf{e}_i \cdot \mathbf{e}_j, \tag{B.3}$$

$$A^i = A_j \mathbf{e}^j \cdot \mathbf{e}^i \equiv g^{ij} A_j \quad \Rightarrow \quad g^{ij} \equiv \mathbf{e}^i \cdot \mathbf{e}^j. \tag{B.4}$$

Here $g_{ij} = g_{ji}$ are the co-variant and $g^{ij} = g^{ji}$ the contra-variant *metric coefficients*, satisfying $g_{ij}g^{jk} = \delta_i^k$. With these definitions, the co- and contra-variant basis vectors and vector components are related by

$$\mathbf{e}_i = g_{ij}\mathbf{e}^j, \quad \mathbf{e}^i = g^{ij}\mathbf{e}_j, \quad A_i = g_{ij}A^j, \quad A^i = g^{ij}A_j. \tag{B.5}$$

It is also useful to define the *unit basis vectors*, $\hat{\mathbf{e}}_i$ and *scale factors*, h_i,

$$\hat{\mathbf{e}}_i \equiv \mathbf{e}_i/h_i, \quad h_i = |\mathbf{e}_i| = \sqrt{g_{ii}}, \quad g_{ij} = h_i h_j \hat{\mathbf{e}}_i \cdot \hat{\mathbf{e}}_j, \tag{B.6}$$

with similar definitions for $\hat{\mathbf{e}}^i$ and scale factors h^i. When the basis vectors are orthogonal, $\mathbf{e}_i \cdot \mathbf{e}_j = 0$ for $i \neq j$, the off-diagonal metric coefficients, g_{ij}, vanish, such that $[g_{ij}]$ can be represented by the three scale factors h_i and the Jacobian reduces to $\jmath = \sqrt{g} = h_1h_2h_3$. The differential arc length $\mathrm{d}l^i = |\mathrm{d}\mathbf{x}^i|$, differential area element, $\mathrm{d}S^i = |\mathrm{d}\mathbf{x}^j \times \mathrm{d}\mathbf{x}^k|$ and differential volume element, $\mathrm{d}V = \mathrm{d}\mathbf{x}^1 \cdot \mathrm{d}\mathbf{x}^2 \times \mathrm{d}\mathbf{x}^3$ can then be evaluated as,

$$\mathrm{d}l^i = h_i \mathrm{d}z^i, \quad \mathrm{d}S^i = h_j h_k \mathrm{d}z^j \mathrm{d}z^k, \quad \mathrm{d}V = h_1h_2h_3 \mathrm{d}z^1 \mathrm{d}z^2 \mathrm{d}z^3, \tag{B.7}$$

where no summation over repeated indices is implied.

The scalar (dot) product of two vectors $\mathbf{A}$ and $\mathbf{B}$ is easily found as,

$$\mathbf{A} \cdot \mathbf{B} = A_i B^i = A^i B_i = g_{ij} A^i B^j = g^{ij} A_i B_j, \tag{B.8}$$

while the vector (cross) product can be evaluated as

$$(\mathbf{A} \times \mathbf{B})_k = \jmath \varepsilon_{ijk} A^i B^j, \quad (\mathbf{A} \times \mathbf{B})^k = \jmath^{-1} \varepsilon_{ijk} A_i B_j, \tag{B.9}$$

where ε_{ijk} are the Levi–Civita symbols (equal to $+1$ when ijk form even permutations of 123, to -1 for odd permutations, and to 0 otherwise).

Differential calculus in curvilinear co-ordinates takes advantage of the fact that the gradient operator is most naturally written in contra-variant form, allowing a simple expression for the gradient of scalar $\mathcal{A}$,

$$\nabla = \nabla z^k \partial_{z^k} = \mathbf{e}^k \partial_{z^k}, \quad \mathcal{A}_{,k} \equiv (\nabla \mathcal{A})_k = \partial_{z^k} \mathcal{A}. \tag{B.10}$$

This is generally known as the *co-variant derivative* and is typically denoted by a comma subscript. The co-variant derivative of a vector $\mathbf{A}$ consists of two terms, representing changes of A^j and $\mathbf{e}_i$ with the co-ordinates z^k,

$$A^j_{,k} \equiv \left(\partial_{z^k}\mathbf{A}\right)^j = \partial_{z^k} A^j + \Gamma^j_{ik} A^i, \quad A_{j,k} \equiv \left(\partial_{z^k}\mathbf{A}\right)_j = \partial_{z^k} A_j - \Gamma^i_{jk} A_i, \tag{B.11}$$

where the Christoffel symbols, Γ^i_{jk}, depend only on the metric tensor,

$$\Gamma^j_{ik} \equiv \mathbf{e}_j \cdot \partial_{z^k} \mathbf{e}_i = \frac{1}{2} g^{jn} \left(\partial_{z^k} g_{ni} + \partial_{z^i} g_{nk} - \partial_{z^n} g_{ik}\right). \tag{B.12}$$

Since Γ^i_{jk} measure the spatial variation of the basis vectors, they must vanish in rectilinear co-ordinates where these are independent of position, i.e. $\mathbf{e}_i(\mathbf{x}) = \mathbf{e}^i(\mathbf{x}) =$ const. Although Christoffel symbols offer the most general representation of any co-variant derivative, they are rather cumbersome, even tedious, to work with. Fortunately, Γ^i_{jk} typically reduce to much simpler combinations of g_{ij} and g^{ij} when evaluating the most common differential operators. For instance, the divergence and curl of **A** reduce to

$$\nabla \cdot \mathbf{A} = \jmath^{-1} \partial_{z^i} (\jmath A^i), \qquad \nabla \times \mathbf{A} = \jmath^{-1} \varepsilon_{ijk} \partial_{z^i} A_j \mathbf{e}_k, \tag{B.13}$$

which, in orthogonal basis, can be further simplified to yield

$$\nabla \cdot \mathbf{A} = \jmath^{-1} \left[\partial_{z^1} \left(h_2 h_3 \mathcal{A}^1 \right) + \cdots \right], \qquad \jmath = h_1 h_2 h_3, \tag{B.14}$$

$$\nabla \times \mathbf{A} = \frac{1}{h_2 h_3} \left[\partial_{z^2} (h_3 \mathcal{A}_3) - \partial_{z^3} (h_2 \mathcal{A}_2) \right] \hat{\mathbf{e}}_1 + \cdots . \tag{B.15}$$

Here the continuation implies similar terms pertaining to co-ordinates z^2 and z^3, obtained by symmetric permutation of the 123-indices. For reference, we write down two more frequently encountered expressions,

$$\nabla^2 A = \jmath^{-1} \left[\partial_{z^1} \left(\frac{h_2 h_3}{h_1} \partial_{z^1} A \right) + \cdots \right], \tag{B.16}$$

$$\begin{aligned} \mathbf{A} \cdot \nabla \mathbf{B} = \Bigg[& \frac{A_1}{h_1} \left(\partial_{z^1} B_1 + \frac{B_2}{h_2} \partial_{z^2} h_1 + \frac{B_3}{h_3} \partial_{z^3} h_1 \right) \\ & + \frac{A_2}{h_2} \left(\partial_{z^2} B_1 - \frac{B_2}{h_1} \partial_{z^1} h_2 \right) + \frac{A_3}{h_3} \left(\partial_{z^3} B_1 - \frac{B_3}{h_1} \partial_{z^1} h_3 \right) \Bigg] \hat{\mathbf{e}}_1 + \cdots . \end{aligned} \tag{B.17}$$

Evaluating (B.14)–(B.17) in Cartesian (x, y, z), cylindrical (r, θ, z) and spherical (r, θ, ϕ) co-ordinates, for which the scale factors are $h_1 = h_2 = h_3 = 1$, $h_1 = h_2/r = h_3 = 1$ and $h_1 = h_2/r = h_3/r \sin\theta = 1$, respectively, we recover the familiar differential operators of vector calculus. For orthogonal toroidal co-ordinates (r, θ, ζ) we find $h_r = 1$, $h_\theta = r$ and $h_\zeta = R$.

For more details on curvilinear co-ordinates and tensor calculus, the reader is referred to Synge and Schild (1949), Aris (1962) and Lovelock and Rund (1989). Application to (magnetic) flux co-ordinates is comprehensively discussed in D'haeseleer *et al.* (1991).

References

Alfven, H. and Falthammar, C.-G. (1963). *Cosmical Electrodynamics*, 2nd edn. Oxford University Press.

Anderson, D. A., Tannehill, J. C. and Pletcher, R. H. (1984). *Computational Fluid Mechanics and Heat Transfer*. McGraw-Hill.

Arfken, G. B. and Weber, H. J. (2001). *Mathematical Methods for Physicists*, 5th edn. Harcourt.

Aris, R. (1962). *Vectors, Tensors and the Basic Equations of Fluid Mechanics*. Dover Publishing.

Arnold, V. I. (1989). *Mathematical Methods of Classical Mechanics*, 2nd edn. Springer-Verlag.

Ashby, M. and Jones, D. (2005). *Engineering Materials, Vols. I–II*, 3rd edn. Elsevier.

Balescu, R. (1960). Irreversible processes in ionized gases. *Physics of Fluids* **3**, 52–63.

Balescu, R. (1988a). *Transport Processes in Plasmas, 1: Classical Transport*. North Holland.

Balescu, R. (1988b). *Transport Processes in Plasmas, 2: Neoclassical Transport*. North Holland.

Balescu, R. (2008). *Aspects of Anomalous Transport in Plasmas*. Institute of Physics Publishing.

Batchelor, G. (1967). *An Introduction to Fluid Dynamics*. Cambridge University Press.

Batchelor, G. K. (1953). *The Theory of Homogeneous Turbulence*. Cambridge University Press.

Batchelor, G. K. (1969). Computation of the energy spectrum in homogeneous two-dimensional turbulence. *Physics of Fluids* **12**, 233.

Batemann, G. (1978). *MHD Instabilities*. MIT Press.

Bécoulet, M., Huysmans, G., Sarazin, Y. *et al.* (2003). Edge localized mode physics and operational aspects in tokamaks. *Plasma Physics and Controlled Fusion* **45**, 12A (Dec.), A93–A113.

Bécoulet, M., Huysmans, G., Thomas, P. *et al.* (2005). Edge localized modes control: experiment and theory. *Journal of Nuclear Materials* **337**, 677–683.

Beer, M. A. and Hammett, G. W. (1996). Toroidal gyrofluid equations for simulations of tokamak turbulence. *Physics of Plasmas* **3**, 4046–4064.

Benedict, M. *et al.* (1981). *Nuclear Chemical Engineering*. McGraw-Hill.

Beurksens, M. *et al.* (2009). *Nuclear Fusion*, in press.

Beurskens, M. N. A., Arnoux, G., Brezinsek, A. S. *et al.* (2008). Pedestal and ELM response to impurity seeding in JET advanced scenario plasmas. *Nuclear Fusion* **48**, 9 (Sept.), 095004.

Beyer, P., Sarazin, Y., Garbet, X., Ghendrih, P. and Benkadda, S. (1999). 2D and 3D boundary turbulence studies. *Plasma Physics and Controlled Fusion* **41**, A757–A769.

Bhatnagar, P. L., Gross, E. P. and Krook, M. (1954). A model for collision processes in gases. I. Small amplitude processes in charged and neutral one-component systems. *Physical Review* **94**, 511–525.

Bird, R., Stewart, W. and Lightfoot, E. (1960). *Transport Phenomena*. John Wiley and Sons.

Birdsall, C. K. and Langdon, A. B. (1991). *Plasma Physics via Computer Simulation*. Institute of Physics Publishing.

Biskamp, D. (1993). *Nonlinear Magnetohydrodynamics*. Cambridge University Press.

Biskamp, D. (2000). *Magnetic Reconnection in Plasmas*. Cambridge University Press.

Biskamp, D. (2003). *Magnetohydrodynamics Turbulence*. Cambridge University Press.

Boas, M. (1983). *Mathematical Methods in the Physical Science*, 2nd edn. John Wiley and Sons.

Boedo, J. A., Rudakov, D. L., Hollmann, E. *et al.* (2005). Edge-localized mode dynamics and transport in the scrape-off layer of the DIII-D tokamak. *Physics of Plasmas* **12**, 7 (July), 072516.

Bogoliubov, N. N. and Mitropolski U. (1961). *Asymptotic Methods in the Theory of Nonlinear Oscillations*. Gordon and Breach.

Boozer, A. H. (1984). Time-dependent drift Hamiltonian. *Physics of Fluids* **27**, 2441–2445.

Boyd, T. J. M. and Sanderson, J. J. (2003). *The Physics of Plasmas*. Cambridge University Press.

Braginskii, S. I. (1965). Transport Processes in a Plasma. *Reviews of Plasma Physics* **1**, 205.

Budynas, R. (1977). *Advanced Strength and Applied Stress Analysis*. McGraw-Hill.

Callen, H. (1985). *Thermodynamics and an Introduction to Thermostatistics*, 2nd edn. John Wiley and Sons.

Catto, P. J., Tang, W. M. and Baldwin, D. E. (1981). Generalized gyrokinetics. *Plasma Physics* **23**, 639–650.

Chandrasekhar, S. (1941). The time of relaxation of stellar systems. *Astrophysics Journal* **93**, 285.

Chandrasekhar, S. (1942). *Principles of Stellar Dynamics*. University of Chicago Press.

Chandrasekhar, S. (1943). Stochastic problems in physics and astronomy. *Reviews of Modern Physics* **15**, 1–89.

Chang, C. S. and Ku S. (2006). Particle simulation of neoclassical transport in the plasma edge. *Contributions to Plasma Physics* **46**, 496–503.

Chang, C. S. and Ku, S. (2008). Spontaneous rotation sources in a quiescent tokamak edge plasma. *Physics of Plasmas* **15**, 6 (June), 062510.

Chang, C. S., Klasky, S., Cummings, J. *et al.* (2008). Toward a first-principles integrated simulation of tokamak edge plasmas. *Journal of Physics Conference Series* **125**, 1 (July), 012042.

Chang, C. S., Ku, S. and Weitzner, H. (2004). Numerical study of neoclassical plasma pedestal in a tokamak geometry. *Physics of Plasmas* **11**, 2649–2667.

Chankin, A. V., Coster, D. P., Asakura, N. *et al.* (2007). Discrepancy between modelled and measured radial electric fields in the scrape-off layer of divertor tokamaks: a challenge for 2D fluid codes? *Nuclear Fusion* **47**, 479–489.

Chapman, S. (1916). On the law of distribution of molecular velocities, and on the theory of viscosity and thermal conduction, in a non-uniform simple monatomic gas. *Royal Society of London Philosophical Transactions Series A* **216**, 279–348.

Chapman, S. and Cowling, T. G. (1939). *The Mathematical Theory of Non-Uniform Gases*. Cambridge University Press.

Chen, F. F. (1984). *Introduction to Plasma Physics and Controlled Fusion*, 2nd edn. Plenum Press.

Chew, G. F., Goldberger, M. L. and Low, F. E. (1956). The Boltzmann equation and the one-fluid hydromagnetic equations in the absence of particle collisions. *Royal Society of London Proceedings Series A* **236**, 112–118.

Chirikov, B. V. (1959). The passage of a nonlinear oscillating system through resonance. *Soviet Physics Doklady* **4**, 390.

Chodura, R. (1982). Plasma-wall transition in an oblique magnetic field. *Physics of Fluids* **25**, 1628–1633.

Choudhuri, A. R. (1998). *The Physics of Fluids and Plasmas*. Cambridge University Press.

Claassen, H. A., Gerhauser, H., Rogister, A. and Yarim C. (2000). Neoclassical theory of rotation and electric field in high collisionality plasmas with steep gradients. *Physics of Plasmas* **7**, 3699–3706.

Cohen, R. H. and Xu, X. Q. (2008). Progress in kinetic simulation of edge plasmas. *Contributions to Plasma Physics* **48**, 212–223.

Cohen-Tannoudji, C. *et al.* (1977). *Quantum Mechanics, Vols. I–II*. John Wiley and Sons.

Cohen-Tannoudji, C. *et al.* (1992). *Atom-Photon Interactions*. John Wiley and Sons.

Connor, J. W. (1998). Review article: A review of models for ELMs. *Plasma Physics and Controlled Fusion* **40**, 191–213.

Connor, J. W. and Hastie, R. J. (1999). Collisionless and resistive ballooning stability. *Physics of Plasmas* **6**, 4260–4264.

Connor, J. W. and Taylor, J. B. (1987). Ballooning modes or Fourier modes in a toroidal plasma? *Physics of Fluids*. **30**(10), 3180–3185.

Connor, J. W. and Wilson, H. R. (2000). Review article: A review of theories of the L-H transition. *Plasma Physics and Controlled Fusion* **42**, 1.

Connor, J. W., Hastie, R. J., Wilson, H. R. and Miller, R. L. (1998). Magnetohydrodynamic stability of tokamak edge plasmas. *Physics of Plasmas* **5**, 2687–2700.

Corrsin, S. (1961). Turbulent flow. *American Science* **49**, 3.

Coster, D. P., Bonnin, X., Mutzke, A., Schneider, R. and Warrier, M. (2007). Integrated modelling of the edge plasma and plasma facing components. *Journal of Nuclear Materials* **363**, 136–139.

Cowley, S. C., Wilson, H., Hurricane, O. and Fong, B. (2003). Explosive instabilities: from solar flares to edge localized modes in tokamaks. *Plasma Physics and Controlled Fusion* **45**, 12A (Dec.), A31–A38.

Davidson, P. A. (2004). *Turbulence: An Introduction for Scientists and Engineers*. Oxford University Press.

Dendy, R. (ed.) (1993). *Plasma Physics: An Introductory Course*. Cambridge University Press.

D'haeseleer, W. D., Hitchon, W. N. G., Callen, J. D. and Shohet, J. L. (1991). *Flux Coordinates and Magnetic Field Structure*. Springer-Verlag.

Diamond, P. H., Itoh, S.-I., Itoh, K. and Hahm, T. S. (2005). Zonal flows in plasmas – a review. *Plasma Physics and Controlled Fusion* **47**, 5, R35–R161.

Dolan, T. J. (1982). *Fusion Research: Principles, Experiments and Technology, Vols. I–III*. Pergamon Press.

Dorf, R. (1989). *Introduction to Electric Circuits*. John Wiley and Sons.

Dorland, W. and Hammett, G. W. (1993). Gyrofluid turbulence models with kinetic effects. *Physics of Fluids B* **5**, 812–835.

Doyle, E. J., Houlberg, W. A., Kamada, Y. *et al.* (2007). Chapter 2: Plasma confinement and transport. *Nuclear Fusion* **47**, 18.

Dubrulle, B. (1994). Intermittency in fully developed turbulence: log–Poisson statistics and generalized scale covariance. *Physical Review Letters* **73**, 959–962.

Dux, R., Bobkov, A., Herrmann, A. *et al.* (2009). Plasma–wall interaction and plasma behaviour in the non-boronised all tungsten ASDEX Upgrade. *Journal of Nuclear Materials* **390–391**, 858–863.

Eckstein, W. (1997). Physical sputtering and reflection processes in plasma–wall interactions. *Journal of Nuclear Materials* **248**, 1–8.

Eich, T., Andrew, P., Herrmann, A. *et al.* (2007). ELM resolved energy distribution studies in the JET MKII Gas-Box divertor using infra-red thermography. *Plasma Physics and Controlled Fusion* **49**, 573–604.

Eich, T., Herrmann A., Neuhauser, J. *et al.* (2005a). Type-I ELM substructure on the divertor target plates in ASDEX Upgrade. *Plasma Physics and Controlled Fusion* **47**, 815–842.

Eich, T., Herrmann, A., Pautasso, G. *et al.* (2005b). Power deposition onto plasma facing components in poloidal divertor tokamaks during type-I ELMs and disruptions. *Journal of Nuclear Materials* **337**, 669–676.

Eich, T., Kallenbach, A., Fundamenski, W. *et al.* (2009). On the asymmetries of ELM divertor power deposition in JET and ASDEX Upgrade. *Journal of Nuclear Materials* **390–391**, 760–763.

El-Guebaly, L. A. (1992). Shielding aspects of D-3He fusion power reactors. *Fusion Technology* **22**, 124.

Endler, M. (1999). Turbulent SOL transport in stellarators and tokamaks. *Journal of Nuclear Materials* **266**, 84–90.

Endler, M., García-Cortés, I., Hidalgo, C., Matthews, G. F., ASDEX Team and JET Team (2005). The fine structure of ELMs in the scrape-off layer. *Plasma Physics and Controlled Fusion* **47**, 219–240.

Endler, M., Niedermeyer, H., Giannone, L. *et al.* (1995). Measurements and modelling of electrostatic fluctuations in the scrape-off layer of ASDEX. *Nuclear Fusion* **35**, 1307–1339.

Enskog, D. (1917). *Kinetische Theorie der Vorgaenge in maessig verduennten Gasen. I. Allgemeiner Teil*. Almquist Wiksells Boktryckeri.

Erents, S. K., Pitts, R. A., Fundamenski, W., Gunn, J. P. and Matthews, G. F. (2004). A comparison of experimental measurements and code results to determine flows in the JET SOL. *Plasma Physics and Controlled Fusion* **46**, 1757–1780.

Evans, T. E., Burrell, K. H., Fenstermacher, M. E. *et al.* (2006). The physics of edge resonant magnetic perturbations in hot tokamak plasmas. *Physics of Plasmas* **13**, 5 (May), 056121.

Evans, T. E., Fenstermacher, M. E., Moyer, R. A. *et al.* (2008). RMP ELM suppression in DIII-D plasmas with ITER similar shapes and collisionalities. *Nuclear Fusion* **48**, 2 (Feb.), 024002.

Falchetto, G. L., Scott, B. D., Angelino, P. *et al.* (2008). The European turbulence code benchmarking effort: turbulence driven by thermal gradients in magnetically

confined plasmas. *Plasma Physics and Controlled Fusion* **50**, 12 (Dec.), 124015.
Fenstermacher, M. E., Evans, T. E., Moyer, R. A. *et al.* (2007). Pedestal, SOL and divertor plasma properties in DIII-D RMP ELM-suppressed discharges at ITER relevant edge collisionality. *Journal of Nuclear Materials* **363**, 476–483.
Feynman, R., Leighton, R. and Sands, M. (1963). *The Feynman Lectures on Physics, Vols. I–III*. Addison-Wesley.
Freidberg, J. P. (1991). *Ideal Magnetohydrodynamics*. Plenum Press.
Freidberg, J. P. and Haas, F. A. (1973). Kink instabilities in a high-β tokamak. *Physics of Fluids* **16**, 1909–1916.
Frisch, U. (1995). *Turbulence*. Cambridge University Press.
Frisch, U., Pouquet, A., Leorat, J. and Mazure A. (1975). Possibility of an inverse cascade of magnetic helicity in magnetohydrodynamic turbulence. *Journal of Fluid Mechanics* **68**, 769–778.
Fundamenski, W. (2005). Topical review: Parallel heat flux limits in the tokamak scrape-off layer. *Plasma Physics and Controlled Fusion* **47**, 163.
Fundamenski, W. (2008). Chapter 6: Scrape-off layer transport on JET. *Fusion Science and Technology* **53**, 1023–1065.
Fundamenski, W. (2009). Power and particle exhaust in tokamaks: integration of plasma scenarios with plasma facing materials and components. *Journal of Nuclear Materials* **390–391**, 10–19.
Fundamenski, W. and Harms, A. A. (1996). Evolution and status of D-He3 fusion: a critical review. *Fusion Technology* **29**, 313–349.
Fundamenski, W., Garcia, O. E., Naulin, V. *et al.* (2007a). Dissipative processes in interchange driven scrape-off layer turbulence. *Nuclear Fusion* **47**, 417–433.
Fundamenski, W., Naulin, V., Neukirch, T., Garcia, O. E. and Rasmussen, J. J. (2007b). Topical review: On the relationship between ELM filaments and solar flares. *Plasma Physics and Controlled Fusion* **49**, 43.
Fundamenski, W., Pitts R. and JET EFDA contributors (2007c). ELM-wall interaction on JET and ITER. *Journal of Nuclear Materials* **363**, 319–324.
Fundamenski, W., Pitts, R. A. and JET EFDA contributors (2006). A model of ELM filament energy evolution due to parallel losses. *Plasma Physics and Controlled Fusion* **48**, 109–156.
Fundamenski, W., Pitts, R. A., Matthews, G. F., Riccardo, V., Sipilä, S. and JET EFDA contributors (2005). ELM-averaged power exhaust on JET. *Nuclear Fusion* **45**, 950–975.
Fundamenski, W., Sailer, W. and JET EFDA contributors (2004a). Radial propagation of Type-I ELMs on JET. *Plasma Physics and Controlled Fusion* **46**, 233–259.
Fundamenski, W., Sipila, S. *et al.* (2003). Narrow power profiles seen at JET and their relation to ion orbit losses. *Journal of Nuclear Materials* **313**, 787–795.
Fundamenski, W., Sipilä, S. and JET EFDA contributors (2004b). Boundary plasma energy transport in JET ELMy H-modes. *Nuclear Fusion* **44**, 20–32.
Furth, H. P., Killeen, J. and Rosenbluth, M. N. (1963). Finite-resistivity instabilities of a sheet pinch. *Physics of Fluids* **6**, 459–484.
Galeev, A. A. and Sagdeev, R. Z. (1979). Theory of neoclassical diffusion. *Reviews of Plasma Physics* **7**, 257.
Galeev, A. A. and Sudan, R. N. (eds.) (1983). *Basic Plasma Physics, Vols. I–II*. North Holland.
Garbet, X. (2006). Introduction to turbulent transport in fusion plasmas. *Comptes Rendus Physique* **7**, 573–583.

Garbet, X. and Waltz, R. E. (1998). Heat flux driven ion turbulence. *Physics of Plasmas* **5**, 2836–2845.

Garbet, X., Sarazin, Y., Beyer, P. *et al.* (1999). Flux driven turbulence in tokamaks. *Nuclear Fusion* **39**, 2063–2068.

Garcia, O. E., Bian, N. H. and Fundamenski, W. (2006a). Radial interchange motions of plasma filaments. *Physics of Plasmas* **13**, 8 (Aug.), 082309.

Garcia, O. E., Horacek, J., Pitts, R. A. *et al.* (2006b). Letter to the editor: Interchange turbulence in the TCV scrape-off layer. *Plasma Physics and Controlled Fusion* **48**, L1–L10.

Garcia, O. E., Horacek, J., Pitts, R. A. *et al.* (2007a). Fluctuations and transport in the TCV scrape-off layer. *Nuclear Fusion* **47**, 667–676.

Garcia, O. E., Naulin, V., Nielsen, A. H. and Juul Rasmussen J. (2005). Turbulence and intermittent transport at the boundary of magnetized plasmas. *Physics of Plasmas* **12**, 6 (June), 062309.

Garcia, O. E., Pitts, R. A., Horacek, J. *et al.* (2007b). Collisionality dependent transport in TCV SOL plasmas. *Plasma Physics and Controlled Fusion* **49**, 47.

Garcia, O. E., Pitts, R. A., Horacek, J. *et al.* (2007c). Turbulent transport in the TCV SOL. *Journal of Nuclear Materials* **363**, 575–580.

Ghendrih, P. (1999). Comparison of ergodic and axisymmetric divertors. *Journal of Nuclear Materials* **266**, 189–196.

Ghendrih, P., Grosman, A. and Capes, H. (1996). Review article: Theoretical and experimental investigations of stochastic boundaries in tokamaks. *Plasma Physics and Controlled Fusion* **38**, 1653–1724.

Gill, R. D. (ed.) (1981). *Plasma Physics and Nuclear Fusion Research.* Academic Press.

Goedbloed, H. and Poedts, S. (2004). *Principles of Magnetohydrodynamics.* Cambridge University Press.

Goldstein, H. (1980). *Classical Mechanics*, 2nd edn. Addison-Welsey.

Goldston, R. J. and Rutherford, P. H. (1995). *Plasma Physics.* Institute of Physics Publishing.

Gombosi, T. I. (1994). *Gaskinetic theory.* Cambridge University Press.

Gonçalves, B., Hidalgo, C., Pedrosa, M. A. *et al.* (2003). Edge localized modes and fluctuations in the JET SOL region. *Plasma Physics and Controlled Fusion* **45**, 1627–1635.

Grad, H. (1949). On the kinetic theory of rarified gases. *Communications in Pure and Applied Mathematics* **2**, 331.

Greenwald, M. (2002). Topical review: Density limits in toroidal plasmas. *Plasma Physics and Controlled Fusion* **44**, 27.

Gunn, J. P., Boucher, C., Dionne, M. *et al.* (2007). Evidence for a poloidally localized enhancement of radial transport in the scrape-off layer of the Tore Supra tokamak. *Journal of Nuclear Materials* **363**, 484–490.

Guzdar, P. N., Kleva, R. G., Das, A. and Kaw, P. K. (2001a). Zonal flow and field generation by finite beta drift waves and kinetic drift-Alfvén waves. *Physics of Plasmas* **8**, 3907–3912.

Guzdar, P. N., Kleva, R. G., Das, A. and Kaw, P. K. (2001b). Zonal flow and zonal magnetic field generation by finite β drift waves: a theory for low to high transitions in tokamaks. *Physical Review Letters* **87**, 1 (July), 015001.

Hahm, T. S. (1988). Nonlinear gyrokinetic equations for tokamak microturbulence. *Physics of Fluids* **31**, 2670–2673.

Haken, H. and Wolf, H. (1987). *Atomic and Qunatum Physics.* Springer-Verlag.

Hammett, G. W., Beer, M. A., Dorland, W., Cowley, S. C. and Smith S. A. (1993). Developments in the gyrofluid approach to tokamak turbulence simulations. *Plasma Physics and Controlled Fusion* **35**, 973–985.

Hanjalic K. (2002). One-point closure models for buoyancy-driven turbulent flows. *Annual Review of Fluid Mechanics* **34**, 321–347.

Harbour, P. J. and Loarte, A. (1995). The scrape-off layer in a finite-aspect-ratio torus: the influence of limiter position. *Nuclear Fusion* **35**, 759–772.

Hasegawa, A. (1975). *Plasma Instabilities and Nonlinear Effects*. Springer-Verlag.

Hazeltine, R. D. (1973). Recursive derivation of drift-kinetic equation. *Plasma Physics* **15,** 1, 77–80.

Hazeltine, R. D. and Meiss, J. D. (1992). *Plasma Confinement*. Addison-Wesley.

Hazeltine, R. D. and Waelbroeck, F. L. (2004). *The Framework of Plasma Physics*. Westview Press.

Heikkinen, J. A., Henriksson, S., Janhunen, S., Kiviniemi, T. P. and Ogando, F. (2006). Gyrokinetic simulation of particle and heat transport in the presence of wide orbits and strong profile variations in the edge plasma. *Contributions to Plasma Physics* **46**, 490–495.

Heikkinen, J. A., Janhunen, S. J., Kiviniemi, T. P. and Ogando, F. (2008). Full f gyrokinetic method for particle simulation of tokamak transport. *Journal of Computational Physics* **227**, 5582–5609.

Helander, P. and Sigmar, D. J. (2002). *Collisional Transport in Magnetized Plasmas*. Cambridge University Press.

Henriksson, S. V., Janhunen, S. J., Kiviniemi, T. P. and Heikkinen, J. A. (2006). Global spectral investigation of plasma turbulence in gyrokinetic simulations. *Physics of Plasmas* **13**, 7 (July), 072303.

Herrmann, A. (2002). Overview on stationary and transient divertor heat loads. *Plasma Physics and Controlled Fusion* **44**, 883–903.

Herrmann, A. (2003). Stationary and transient divertor heat flux profiles and extrapolation to ITER. *Journal of Nuclear Materials* **313**, 759–767.

Hinton, F. L. (1983). Collisional transport in plasma. In *Handbook of Plasma Physics*, M. N. Rosenbluth and R. Z. Sagdeev (eds.), North Holland, 147.

Hinton, F. L. and Hazeltine, R. D. (1976). Theory of plasma transport in toroidal confinement systems. *Reviews of Modern Physics* **48**, 239–308.

Hirsch, C. (1990a). *Numerical Computation of Internal and External Flows. Vol. 1 – The Fundamentals of Computational Fluid Dynamics*. John Wiley and Sons.

Hirsch, C. (1990b). *Numerical Computation of Internal and External Flows. Vol. 2 – Computational Methods for Inviscid and Viscous Flows*. John Wiley and Sons.

Hirschman, S. P. and Sigmar, D. J. (1981). Neoclassical transport of impurities in tokamak plasmas. *Nuclear Fusion* **21**, 1079.

Hora, H. (1981). *Physics of Laser Driven Plasmas*. John Wiley and Sons.

Huang, K. (1987). *Statistical Mechanics*, 2nd edn. John Wiley and Sons.

Huba, J. D. (2006). *NRL Plasma Formulary*. Naval Research Laboratory, Washington, DC.

Huber, A., Pitts, R., Loarte, A., *et al.* (2009). Plasma radiation distribution and radiation loads onto the vessel during transient events in JET. *Journal of Nuclear Materials*, in press.

Hutchinson, I. (2002). *Principles of Plasma Diagnostics*, 2nd edn. Cambridge University Press.

Huysmans, G. T. A. (2005a). ELMs: MHD instabilities at the transport barrier. *Plasma Physics and Controlled Fusion* **47**, B165–B178.

Huysmans, G. T. A. (2005b). External kink (peeling) modes in x-point geometry. *Plasma Physics and Controlled Fusion* **47**, 2107–2121.

Ichimaru, S. (1973). *Basic Principles of Plasma Physics*. W. A. Benjamin.

Ichimaru, S. (1992). *Statistical Plasma Physics*. Addison-Wesley.

Igitkhanov, Y., Janeschitz, G., Pacher, G. W. *et al.* (1998). Edge parameter operational space and trajectories for ITER. *Plasma Physics and Controlled Fusion* **40**, 837–844.

Iroshnikov, P. S. (1964). Turbulence of a conducting fluid in a strong magnetic field. *Soviet Astronomy* **7**, 566.

ITER Physics Basis Editors, ITER Physics Expert Group Chairs, ITER Joint Central Team and Physics Integration Unit (1999). Chapter 1: Overview and summary. *Nuclear Fusion* **39**, 2137–2174.

Itoh, K., Itoh, S.-I. and Fukuyama, A. (2000). *Transport and Structural Formation in Plasmas*. Institute of Physics Publishing.

Jachmich, S., Liang, Y., Arnoux, G. *et al.* (2009). ELM Filament interaction with the main JET chamber. *Journal of Nuclear Materials* **390–391**, 781–784.

Jackson, J. D. (1975). *Classical Electrodynamics*, 2nd edn. John Wiley and Sons.

Jakubowski, M., Fundamenski, W., Arnoux, G. *et al.* (2009). ELM filament interactions with the JET main chamber. *Journal of Nuclear Materials*, in press.

Janev, R. K. (ed.) (1995a). *Atomic and Molecular Processes in Fusion Edge Plasmas*, Plenum.

Janev, R. K. (1995b). Basic properties of fusion edge plasmas and role of atomic and molecular processes. In *Atomic and Molecular Processes in Fusion Edge Plasmas*, R. K. Janev (ed.) Plenum.

Janev, R. K. *et al.* (1988). *Elementary Processes in Hydrogen-Helium Plasmas*. Springer-Verlag.

Janhunen, S. J., Ogando, F., Heikkinen, J. A., Kiviniemi, T. P. and Leerink, S. (2007). Collisional dynamics of Er in turbulent plasmas in toroidal geometry. *Nuclear Fusion* **47**, 875–879.

Kadomtsev, B. B. (1965). *Plasma Turbulence*. Academic Press.

Kallenbach, A., Andrew, Y., Beurskens, M. *et al.* (2004). EDGE2D modelling of edge profiles obtained in JET diagnostic optimized configuration. *Plasma Physics and Controlled Fusion* **46**, 431–446.

Kamiya, K., Asakura, N., Boedo, J. *et al.* (2007). Edge localized modes: recent experimental findings and related issues. *Plasma Physics and Controlled Fusion* **49**, 43.

Kendl, A. and Scott, B. D. (2005). Shear flow reduction by the geodesic transfer mechanism in tokamak edge turbulence. *Physics of Plasmas* **12**, 6 (June), 064506.

Kendl, A. and Scott, B. D. (2009). Submitted to *Physics of Plasmas*.

Kirk, A., Asakura, N., Boedo, J. A. *et al.* (2008). Comparison of the spatial and temporal structure of type-I ELMs. *Journal of Physics Conference Series 123*, 1 (July), 012011.

Kirk, A., Counsell, G. F., Cunningham, G. *et al.* (2007). Evolution of the pedestal on MAST and the implications for ELM power loadings. *Plasma Physics and Controlled Fusion* **49**, 1259–1275.

Kirk, A., Counsell, G. F., Wilson, H. R. *et al.* (2004). ELM characteristics in MAST. *Plasma Physics and Controlled Fusion* **46**, 551–572.

Kirnev, G., Fundamenski, W. and Corrigan, G. (2007). Modelling of ELM-averaged power exhaust on JET using the EDGE2D code with variable transport coefficients. *Plasma Physics and Controlled Fusion* **49**, 689–701.

Kittel, C. (1995). *Introduction to Solid State Physics*, 7th edn. Freeman and Company.

Kolmogorov, A. (1941). The local structure of turbulence in incompressible viscous fluid for very large Reynolds' numbers. *Akademiia Nauk SSSR Doklady* **30**, 301–305.

Kolmogorov, A. N. (1962). A refinement of previous hypotheses concerning the local structure of turbulence in a viscous incompressible fluid at high Reynolds number. *Journal of Fluid Mechanics* **13**, 82–85.

Kotschenreuther, M., Valanju, P. M., Mahajan, S. M. and Wiley, J. C. (2007). On heat loading, novel divertors, and fusion reactors. *Physics of Plasmas* 14, 7 (July), 072502.

Kraichnan, R. H. (1965). Inertial-range spectrum of hydromagnetic turbulence. *Physics of Fluids* **8**, 1385–1387.

Kraichnan, R. H. (1967). Inertial ranges in two-dimensional turbulence. *Physics of Fluids* **10**, 1417–1423.

Krall, N. A. and Trivelpiece, A. W. (1973). *Principles of Plasma Physics.* McGraw-Hill.

Krane, K. S. (1988). *Introductory Nuclear Physics.* John Wiley and Sons.

Kruer, W. L. (1988). *The Physics of Laser Plasma Interactions.* Addison-Welsey.

Kukushkin, A. S. and Pacher, H. D. (2002). Divertor modelling and extrapolation to reactor conditions. *Plasma Physics and Controlled Fusion* **44**, 931–943.

Kukushkin, A., Janeschitz, G., Loarte, A. *et al.* (2001). Critical issues in divertor optimisation for ITER FEAT. *Journal of Nuclear Materials* **290**, 887–891.

Kukushkin, A. S., Pacher, H. D., Janeschitz, G. *et al.* (2002). Basic divertor operation in ITER-FEAT. *Nuclear Fusion* **42**, 187–191.

Kukushkin, A. S., Pacher, H. D., Pacher, G. W. *et al.* (2003). Scaling laws for edge plasma parameters in ITER from two-dimensional edge modelling. *Nuclear Fusion* **43**, 716–723.

LaBombard, B. (2003). Toroidal rotation as an explanation for plasma flow observations in the Alcator C-Mod scrape-off layer. *Journal of Nuclear Materials* **313**, 995–999.

LaBombard, B., Hughes, J., Mossessian, D. *et al.* (2005). Evidence for electromagnetic fluid drift turbulence controlling the edge plasma state in the Alcator C-Mod tokamak. *Nuclear Fusion 45,* **12**, 1658–1675.

LaBombard, B., Hughes, J. W., Smick, N. *et al.* (2008). Critical gradients and plasma flows in the edge plasma of Alcator C-Mod. *Physics of Plasmas* **15**, 5 (May), 056106.

Lamarsh, J. R. (1983). *Introduction to Nuclear Engineering.* Addison-Wesley.

Lamb, H. (1932). *Hydrodynamics.* New York: Dover.

Landau, L. D. (1936). *Phys. Z. Sowjetunion* **10**, 154.

Landau, L. D. (1944). On the problem of turbulence. *CR (Doklady) Acad. Sci. URSS.*

Landau, L. D. (1946). On the vibrations of the electronic plasma. *Journal of Plasma Physics (USSR)* **10**, 25.

Landau, L. D. and Lifschitz E. M. (1960). *Course of Theoretical Physics, Vols. I–X.* Pergamon Press.

Lang, P. T., Gal, K., Hobirk, J. *et al.* (2006). Investigations on the ELM cycle by local 3D perturbation experiments. *Plasma Physics and Controlled Fusion* **48**, 5A (May), A141–A148.

Lawson, J. D. (1957). *Proceedings of the Physical Society B* **70**, 6.

Lee, W. W. (1983). Gyrokinetic approach in particle simulation. *Physics of Fluids* **26**, 556–562.

Leerink, S., Heikkinen, J. A., Janhunen, S. J., Kiviniemi, T. P., Nora, M. and Ogando, F. (2008). Gyrokinetic full f analysis of electric field dynamics and poloidal velocity in the FT2-tokamak configuration. *Plasma Physics Reports* **34**, 716–719.

Lenard, A. (1960). On Bogoliubov's kinetic equation for a spatially homogeneous plasma. *Annals of Physics* **10**, 390–400.

Leonard, A. W., Asakura, N., Boedo, J. A. *et al.* (2006). Survey of Type I ELM dynamics measurements. *Plasma Physics and Controlled Fusion* **48,** A149–A162.

Liang, Y., Koslowski, H. R., Thomas, P. R. *et al.* (2007). Active control of type-I edge localized modes on JET. *Plasma Physics and Controlled Fusion* **49,** 581.

Lifschitz, E. M. and Pitaevskii, L. P. (1981). *Physical Kinetics*. Pergamon Press.

Linke, J., Escourbiac, F., Mazul, I. V. *et al.* (2007). High heat flux testing of plasma facing materials and components. Status and perspectives for ITER related activities. *Journal of Nuclear Materials* **367**, 1422–1431.

Lipschultz, B., Bonnin, X., Counsell, G. *et al.* (2007). Plasma–surface interaction, scrape-off layer and divertor physics: implications for ITER. *Nuclear Fusion* **47**, 1189–1205.

Littlejohn, R. G. (1983). Variational principles of guiding centre motion. *Journal of Plasma Physics* **29**, 111–125.

Loarte, A. (2001). Review article: Effects of divertor geometry on tokamak plasmas. *Plasma Physics and Controlled Fusion* **43**, 183.

Loarte, A., Becoulet, M., Saibene, G. *et al.* (2002). Characteristics and scaling of energy and particle losses during Type I ELMs in JET H-modes. *Plasma Physics and Controlled Fusion* **44**, 1815–1844.

Loarte, A., Monk, R. D., Martín-Solís, J. R. *et al.* (1998). Plasma detachment in JET Mark I divertor experiments. *Nuclear Fusion* **38**, 331–371.

Loarte, A., Saibene, G., Sartori, R. *et al.* (2003). Characteristics of type I ELM energy and particle losses in existing devices and their extrapolation to ITER. *Plasma Physics and Controlled Fusion* **45**, 1549–1569.

Loarte, A., Saibene, G., Sartori, R. *et al.* (2004). Characterisation of pedestal parameters and edge localized mode energy losses in the Joint European Torus and predictions for the International Thermonuclear Experimental Reactor. *Physics of Plasmas* **11**, 2668–2678.

Loarte, A., Saibene, G., Sartori, R. *et al.* (2007). Transient heat loads in current fusion experiments, extrapolation to ITER and consequences for its operation. *Physica Scripta Volume T* **128**, 222–228.

Lovelock, D. and Rund, H. (1989). *Tensors, Differential Forms and Variational Principles*. Dover Publishing.

Maddison, G. P., Brix, M., Budny, R. *et al.* (2003). Impurity-seeded plasma experiments on JET. *Nuclear Fusion* **43**, 49–62.

Maisonnier, D., Cook, I., Pierre, S. *et al.* (2005). The European power plant conceptual study. *Fusion Engineering and Design* **75–79**, 1173–1179.

Maisonnier, D., Cook, I., Pierre S. *et al.* (2006). DEMO and fusion power plant conceptual studies in Europe. *Fusion Engineering and Design* **81**, 1123–1130.

Martin, Y., Takizuka, T. *et al.* (2008). Power requirement for accessing the H-mode in ITER. *Journal of Physics Conference Series* **123**, 1 (July), 012033.

Matthews, G. F., Andrew, P., Eich, T. *et al.* (2003). Steady-state and transient power handling in JET. *Nuclear Fusion* **43**, 999–1005.

Matthews, G. F., Balet, B., Cordey, J. G. *et al.* (1999). Studies in JET divertors of varied geometry. II: Impurity seeded plasmas. *Nuclear Fusion* **39**, 19–40.

McComb, W. D. (1992). *The Physics of Fluid Turbulence*. Oxford University Press.

Mikhailovskii, A. B. (1974). *Theory of Plasma Instabilities, 1: Instabilities in a Homogeneous Plasma*. Institute of Physics Publishing.

Mikhailovskii, A. B. (1992). *Electronmagnetic Instabilities in an Inhomogeneous Plasma.* Institute of Physics Publishing.
Mikhailovskii, A. B. (1998). *Instabilities in a Confined Plasma.* Institute of Physics Publishing.
Miley, G. H. (1976). *Fusion Energy Convesion.* American Nuclear Society.
Milne-Thomson, L. (1968). *Theoretical Hydrodynamics.* New York: Dover.
Miyamoto, K. (1989). *Plasma Physics for Nuclear Fusion*, revised edn. MIT Press.
Monier-Garbet, P., Andrew, P., Belo, P. *et al.* (2005). Impurity-seeded ELMy H-modes in JET, with high density and reduced heat load. *Nuclear Fusion* **45**, 1404–1410.
Monin, A. S. and Yaglom, A. M. (1971). *Statistical Fluid Mechanics: Mechanics of Turbulence.* MIT Press.
Morse, P. M. and Feschbach, H. (1953). *Methods of Theoretical Physics.* McGraw-Hill.
Motz, H. (1979). *The Physics of Laser Fusion.* Academic Press.
Moulton, D., Fundamenski, W. *et al.* (2010). Submitted to *Plasma Physics and Controlled Fusion.*
Müller, W.-C., Biskamp, D. and Grappin, R. (2003). Statistical anisotropy of magnetohydrodynamic turbulence. *Physical Review E* **67**, 6 (June), 066302.
Myra, J. R., Russell, D. A. and D'Ippolito, D. A. (2006). Collisionality and magnetic geometry effects on tokamak edge turbulent transport. I. A two-region model with application to blobs. *Physics of Plasmas* **13**, 11 (Nov.), 112502.
Myra, J. R., Russell, D. A. and D'Ippolito, D. A. (2008). Transport of perpendicular edge momentum by drift-interchange turbulence and blobs. *Physics of Plasmas* **15**, 3 (Mar.), 032304.
Najmabadi, F., Conn, R. and the ARIES team (1991). The ARIES-I tokamak reactor study. *Fusion Technology* **90**, 253.
Najmabadi, F., Conn, R. and the ARIES team (1992). The ARIES-II and ARIES-IV second stability reactors. *Fusion Technology* **21**, 1721–1728.
Naulin, V. (2007). Turbulent transport and the plasma edge. *Journal of Nuclear Materials* **363**, 24–31.
Naulin, V., Garcia, O. E., Nielsen, A. H. and Rasmussen, J. J. (2004). Statistical properties of transport in plasma turbulence. *Physics Letters A* **321**, 355–365.
Nave, M. F. F. and Wesson, J. A. (1990). Mode locking in tokamaks. *Nucl. Fusion* **30**, 2572.
Nicholson, D. R. (1983). *Introduction to Plasma Theory.* John Wiley and Sons.
Nishikawa, K. and Wakatani, M. (1994). *Plasma Physics*, 2nd edn. Springer-Verlag.
Obukhov, A. M. (1962). Some specific features of atmospheric turbulence. *Journal of Geophysical Research* **67**, 3011.
Oyama, N., Gohil, P., Horton, L. D. *et al.* (2006). Pedestal conditions for small ELM regimes in tokamaks. *Plasma Physics and Controlled Fusion* **48**, 5A (May), A171–A181.
Pacher, G. W., Pacher, H. D., Janeschitz, G. *et al.* (2005). Modelling of ITER improved H-mode operation with the integrated core pedestal SOL model. *Nuclear Fusion* **45**, 581–587.
Pacher, G. W., Pacher, H. D., Janeschitz, G. and Kukushkin, A. S. (2008). ITER operation window determined from mutually consistent core-SOL-divertor simulations: definition and application. *Nuclear Fusion* **48**, 10 (Oct.), 105003.
Pacher, H. D., Kukushkin, A. S., Pacher, G. W. *et al.* (2007). Effect of the tokamak size in edge transport modelling and implications for DEMO. *Journal of Nuclear Materials* **363**, 400–406.

Padmanabhan, T. (2000). *Theoretical Astrophysics, Vols. I–III*. Cambridge University Press.

Pauling, L. (1960). *The Nature of the Chemical Bond*. Cornell University Press.

Pauling, L. (1970). *General Chemistry*. Dover Publishing.

Penrose, O. (1960). Electrostatic instabilities of a uniform non-Maxwellian plasma. *Physics of Fluids* **3**, 258–265.

Pitcher, C. S. and Stangeby, P. C. (1997). Review article: Experimental divertor physics. *Plasma Physics and Controlled Fusion* **39**, 779–930.

Pitts, R., Arnoux, G., Brezinsek, S. *et al.* (2009). The impact of large ELMs on JET. *Journal of Nuclear Materials*, in press.

Pitts, R. A., Andrew, P., Arnoux, G. *et al.* (2007a). ELM transport in the JET scrape-off layer. *Nuclear Fusion* **47**, 1437–1448.

Pitts, R. A., Fundamenski, W., Erents, S. K. *et al.* (2006). Far SOL ELM ion energies in JET. *Nuclear Fusion* **46**, 82–98.

Pitts, R. A., Horacek, J., Fundamenski, W. *et al.* (2007b). Parallel SOL flow on TCV. *Journal of Nuclear Materials* **363**, 505–510.

Pope, S. B. (2000). *Turbulent Flows*. Cambridge University Press.

Popov, E. (1978). *Mechanics of Materials*, 2nd edn. Prentice-Hall.

Porter, G. D., Allen, S. L., Brown, M. *et al.* (1996). Simulation of experimentally achieved DIII-D detached plasmas using the UEDGE code. *Physics of Plasmas* **3**, 1967–1975.

Prandtl, L. (1925). Berichtuber Untersuchungen zur ausgebildeten Turbulenz. *Zeitschritt für Angewandte Mathematic and Mechanik*.

Rapp, J., de Baar, M., Fundamenski, W. *et al.* (2009). Highly radiating type-III ELMy H-mode with bw plasma core pollution. *Journal of Nuclear Materials* **390–391**, 238–241.

Rapp, J., Matthews, G. F., Monier-Garbet, P. *et al.* (2005). Strongly radiating type-III ELMy H-mode in JET an integrated scenario for ITER. *Journal of Nuclear Materials* **337**, 826–830.

Rapp, J., Monier-Garbet, P., Matthews, G. F. *et al.* (2004). Reduction of divertor heat load in JET ELMy H-modes using impurity seeding techniques. *Nuclear Fusion* **44**, 312–319.

Rayleigh, L. (1912). On the propagation of waves through a stratified medium, with special reference to the question of reflection. *Royal Society of London Proceedings Series A* **86**, 207–226.

Reif, F. (1965). *Fundamentals of Statistical and Thermal Physics*. McGraw-Hill.

Reiser, D. and Scott, B. (2005). Electromagnetic fluid drift turbulence in static ergodic magnetic fields. *Physics of Plasmas* **12**, 12 (Dec.), 122308.

Reiter, D. (1992). The EIRENE code: version: Jan. 92; users manual. Forschungszentrum Juelich, Zentralbibliothek, available at: www.eirene.de.

Ribeiro, T. T. and Scott, B. (2005). Tokamak turbulence computations on closed and open magnetic flux surfaces. *Plasma Physics and Controlled Fusion* **47**, 1657–1679.

Ribeiro, T. T. and Scott, B. (2008). Gyrofluid turbulence studies of the effect of the poloidal position of an axisymmetric Debye sheath. *Plasma Physics and Controlled Fusion* **50**, 5 (May), 055007.

Riccardo, V., Fundamenski, W. and Matthews, G. F. (2001). Reconstruction of power deposition profiles using JET MkIIGB thermocouple data for ELMy H-mode plasmas. *Plasma Physics and Controlled Fusion* **43**, 881–906.

Richardson, L. F. (1922). *Weather Prediction by Numerical Process*. Cambridge University Press.

Riemann, K. U. and Meyer, P. (1996). Comment on 'Bohm criterion for the collisional sheath' [*Physics of Plasmas* **3**, 1459 (1996)]. *Physics of Plasmas* **3**, 4751–4753.

Rosenbluth, M. N. and Rostoker, N. (1959). Theoretical structure of plasma equations. *Physics of Fluids* **2**, 23–30.

Rosenbluth, M. N., MacDonald, W. M. and Judd, D. L. (1957). Fokker–Planck equation for an inverse-square force. *Physical Review* **107**, 1–6.

Ross, S. (1985). *Introduction to Probability Models*, 4th edn. Academic Press.

Roth, J., Preuss, R., Bohmeyer, W. *et al.* (2004). Letter: Flux dependence of carbon chemical erosion by deuterium ions. *Nuclear Fusion* **44**, L21–L25.

Rudakov, D. L., Boedo, J. A., Moyer, R. A. *et al.* (2005). Far SOL transport and main wall plasma interaction in DIII-D. *Nuclear Fusion* **45**, 1589–1599.

Russell, D. A., Myra, J. R., D'Ippolito, D. A. (2007). Collisionality and magnetic geometry effects on tokamak edge turbulent transport. II. Many-blob turbulence in the two-region model. *Physics of Plasmas* **14**, 10 (Oct.), 102307.

Rutherford, P. H. (1973). Nonlinear growth of the tearing mode. *Physics of Fluids* **16**, 1903–1908.

Ryter, F. and the H-mode Threshold Database Group (2002). Progress of the international H-mode power threshold database activity. *Plasma Physics and Controlled Fusion* **44**, A415–A421.

Sagdeev, R. Z. and Galeev, A. A. (1969). *Nonlinear Plasma Theory*. W. A. Benjamin.

Saha, M. N. (1921). On a physical theory of stellar spectra. *Royal Society of London Proceedings Series A* **99**, 135–153.

Saibene, G., Oyama, N., Lönnroth, J. *et al.* (2007). The H-mode pedestal, ELMs and TF ripple effects in JT-60U / JET dimensionless identity experiments. *Nuclear Fusion* **47**, 969–983.

Sarazin, Y. and Ghendrih, P. (1998). Intermittent particle transport in two-dimensional edge turbulence. *Physics of Plasmas* **5**, 4214–4228.

Sauter, O., Angioni, C. and Lin-Liu, Y. R. (1999). Neoclassical conductivity and bootstrap current formulas for general axisymmetric equilibria and arbitrary collisionality regime. *Physics of Plasmas* **6**, 2834–2839.

Sauter, O., La Haye, R. J., Chang, Z. *et al.* (1997). Beta limits in long-pulse tokamak discharges. *Physics of Plasmas* **4**, 1654–1664.

Schmidt, G. (1979). *Physics of High Temperature Plasmas*. Academic Press.

Schneider, R., Bonnin, X., Borrass, K. *et al.* (2006). Plasma edge physics with B2-Eirene. *Contributions to Plasma Physics* **46**, 3–191.

Schneider, R., Coster, D., Braams, B. *et al.* (2000). B2-solps5.0: SOL transport code with drifts and currents. *Contributions to Plasma Physics* **40**, 328–333.

Scott, B. (1997). Three-dimensional computation of drift Alfvén turbulence. *Plasma Physics and Controlled Fusion* **39**, 1635–1668.

Scott, B. (1998). Computation of warm-ion drift Alfv́en turbulence. *Contributions in Plasma Physics* **38**, 171–176.

Scott, B. (2000). ExB shear flows and electromagnetic gyrofluid turbulence. *Physics of Plasmas* **7**, 1845–1856.

Scott, B. (2001). *Low Frequency Fluid Drift Turbulence in Magnetized Plasmas*. Max-Planck-Institut für Plasmaphysik Report.

Scott, B. D. (2002). The nonlinear drift wave instability and its role in tokamak edge turbulence. *New Journal of Physics* **4**, 52.

Scott, B. (2003a). The character of transport caused by $E \times B$ drift turbulence. *Physics of Plasmas* **10**, 4, 963–976.

Scott, B. D. (2003b). Computation of electromagnetic turbulence and anomalous transport mechanisms in tokamak plasmas. *Plasma Physics and Controlled Fusion* **45**, 26 (Dec.), A385–A398.

Scott, B. D. (2005a). Drift wave versus interchange turbulence in tokamak geometry: linear versus nonlinear mode structure. *Physics of Plasmas* **12**, 6 (June), 062314.

Scott, B. D. (2005b). Dynamical alignment in three species tokamak edge turbulence. *Physics of Plasmas* **12**, 8 (Aug.), 082305.

Scott, B. D. (2005c). Free-energy conservation in local gyrofluid models. *Physics of Plasmas* **12**, 10 (Oct.), 102307.

Scott, B. D. (2005d). GEM – an energy conserving electromagnetic gyrofluid model. *ArXiv Physics e-prints*.

Scott, B. D. (2006). Computation of turbulence in magnetically confined plasmas. *Plasma Physics and Controlled Fusion* **48**, B277–B293.

Scott, B. D. (2007). Tokamak edge turbulence: background theory and computation. *Plasma Physics and Controlled Fusion* **49**, 25.

Shafranov, V. D. (1966). Plasma equilibrium in a magnetic field. *Reviews of Plasma Physics* **2**, 103.

She, Z.-S. and Leveque, E. (1994). Universal scaling laws in fully developed turbulence. *Physical Review Letters* **72**, 336–339.

Shimada, M., Campbell, D. J., Mukhovatov, V. *et al.* (2007). Chapter 1: Overview and summary. *Nuclear Fusion* **47**, 1.

Shkarofsky, I. P. *et al.* (1966). *The Particle Kinetics of Plasmas*. Addison-Wesley.

Silva, C., Gonçalves, B., Hidalgo, C. *et al.* (2005). Determination of the particle and energy fluxes in the JET far SOL during ELMs using the reciprocating probe diagnostic. *Journal of Nuclear Materials* **337**, 722–726.

Silva, C., Gonçalves, B., Hidalgo, C. *et al.* (2009). ELM transport in the JET far SOL. *Journal of Nuclear Materials*, in press.

Simonini, R., Corrigan, G., Radford, G., Spence, J. and Taroni, A. (1994). Models and numerics in the multi-fluid 2-D edge plasma code EDGE2D/U. *Contributions to Plasma Physics* **34**, 368–373.

Sitenko, A. and Malnev, V. (1995). *Plasma Physics Theory*. Chapman and Hall.

Smirnov, V. I. (1964). *A Course of Higher Mathematics, Vols. I–VI*. Pergamon Press.

Snyder, P. B. and Hammett, G. W. (2001). Electromagnetic effects on plasma microturbulence and transport. *Physics of Plasmas* **8**, 744–749.

Snyder, P. B. and Wilson, H. R. (2003). Ideal magnetohydrodynamic constraints on the pedestal temperature in tokamaks. *Plasma Physics and Controlled Fusion* **45**, 1671–1687.

Snyder, P. B., Wilson, H. R., Ferron, J. R. *et al.* (2002). Edge localized modes and the pedestal: a model based on coupled peeling-ballooning modes. *Physics of Plasmas* **9**, 2037–2043.

Snyder, P. B., Wilson, H. R., Ferron, J. R. *et al.* (2004). ELMs and constraints on the H-mode pedestal: peeling ballooning stability calculation and comparison with experiment. *Nuclear Fusion* **44**, 320–328.

Spitzer, L. (1940). The stability of isolated clusters. *Monthly Notes of the Royal Astronomical Society* **100**, 396.

Spitzer, L. (1956). *Physics of Fully Ionised Gases*. Interscience.

Spitzer, L. and Härm, R. (1953). Transport phenomena in a completely ionized gas. *Physical Review* **89**, 977–981.

Stangeby, P. and McCracken, G. (1990). Plasma boundary phenomena in tokamaks. *Nuclear Fusion* **30**, 1225.

Stangeby, P. C. (2000). *The Plasma Boundary of Magnetic Fusion Devices*. Institute of Physics Publishing.
Stix, T. H. (1992). *Waves in Plasmas*. American Institute of Physics.
Strachan, J. D., Corrigan, G., Kallenbach, A. *et al.* (2004). Diverted tokamak carbon screening: scaling with machine size and consequences for core contamination. *Nuclear Fusion* **44**, 772–787.
Strachan, J. D., Likonen, J., Coad, P. *et al.* (2008). Modelling of carbon migration during JET carbon-13 injection experiments. *Nuclear Fusion* **48**, 10 (Oct.), 105002.
Strauss, H. R. (1976). Nonlinear, three-dimensional magnetohydrodynamics of noncircular tokamaks. *Physics of Fluids* **19**, 134–140.
Streeter, V., Wylie, E. and Bedford, K. (1962). *Fluid Mechanics*, 3rd edn. McGraw-Hill.
Suttrop, W. (2000). The physics of large and small edge localized modes. *Plasma Physics and Controlled Fusion* **42**, 5A (May), A1–A14.
Swanson, D. G. (2003). *Plasma Waves*, 2nd edn. Institute of Physics Publishing.
Synge, J. L. and Schild, A. (1949). *Tensor Calculus*. Dover Publishing.
Taguchi, M. (1988). Ion thermal conductivity and ion distribution function in the banana regime. *Plasma Physics and Controlled Fusion* **30**, 1897–1904.
Tajima, T. (1989). *Computational Plasma Physics*. Addison-Wesley.
Takenaga, H., Higashijima, S., Oyama, N. *et al.* (2003). Relationship between particle and heat transport in JT-60U plasmas with internal transport barrier. *Nuclear Fusion* **43**, 1235–1245.
Takizuka, T. (2003). Two-dimensional particle simulation of the flow control in SOL and divertor plasmas. *Journal of Nuclear Materials* **313**, 1331–1334.
Tamain, P., Ghendrih, E., Tsitrone, Y. (2009). 3D modelling of edge parallel flow asymmetries. *Journal of Nuclear Materials* **390–391**, 347–350.
Telesca, G., Zagórski, R., Kalupin, D., Stankiewicz, R., Tokar, M. Z. and Van Oost, G. (2007). Modelling of radiative power exhaust by sputtered and seeded impurities in fusion reactors with carbon and molybdenum target plates. *Nuclear Fusion* **47**, 1625–1633.
Tennekes, H. and Lumley, J. L. (1972). *A First Course in Turbulence*. MIT Press.
Tokar, M. Z., Evans, T. E., Gupta, A. *et al.* (2008). Modelling of pedestal transport during ELM suppression by external magnetic field perturbations. *Nuclear Fusion* **48**, 2 (Feb.), 024006.
Tokar, M. Z., Evans, T. E., Gupta, A., Singh, R., Kaw, P. and Wolf, R. C. (2007). Mechanisms of edge-localized-mode mitigation by external-magnetic-field perturbations. *Physical Review Letters* **98**, 9 (Mar.), 095001.
Troyon, F., Gruber, R., Saurenmann, H., Semenzato, S. and Succi, S. (1984). MHD-limits to plasma confinement. *Plasma Physics and Controlled Fusion* **26**, 209–215.
Trubnikov, B. A. (1958). Plasma radiation in a magnetic field. *Soviet Physics Doklady* **3**, 136.
Tskhakaya, D., Pitts, R., Fundamenski, W. *et al.* (2009). Kinetic simulations of the parallel transport in the JET scrape-off layer. *Journal of Nuclear Materials*, in press.
Tskhakaya, D., Subba, F., Bonnin, X. *et al.* (2008). On kinetic effects during parallel transport in the SOL. *Contributions to Plasma Physics* **48**, 89–93.
Van de Vegte, J. (1990). *Feedback Control Systems*, 2nd edn. Prentice-Hall.
Wagner, F., Becker, G., Behringer, K. *et al.* (1982). Regime of improved confinement and high beta in neutral-beam-heated divertor discharges of the ASDEX tokamak. *Physical Review Letters* **49**, 1408–1412.

Ware, A. A. and Haas, F. A. (1966). Stability of a circular toroidal plasma under average magnetic well conditions. *Physics of Fluids* **9**, 956–964.

Weiland, J. (2000). *Collective Modes in Inhomogeneous Plasma.* Institute of Physics Press.

Wesson, J. (2004). *Tokamaks*, 3rd edn. Oxford University Press.

Whyte, R. B. (1989). *Theory of Tokamak Plasmas.* North-Holland.

Wilson, H. R. and Cowley, S. C. (2004). Theory for explosive ideal magnetohydrodynamic instabilities in plasmas. *Physical Review Letters* **92**, 17 (April), 175006.

Wilson, H. R., Connor, J. W., Field, A. R. *et al.* (1999). Ideal magnetohydrodynamic stability of the tokamak high-confinement-mode edge region. *Physics of Plasmas* **6**, 1925–1934.

Wilson, H. R., Cowley, S. C., Kirk, A. and Snyder, P. B. (2006). Magneto-hydrodynamic stability of the H-mode transport barrier as a model for edge localized modes: an overview. *Plasma Physics and Controlled Fusion* **48**, 5A (May), A71–A84.

Wischmeier, M., Kallenbach, A., Chankin, A. V. *et al.* (2007). High recycling outer divertor regimes after type-I ELMs at high density in ASDEX Upgrade. *Journal of Nuclear Materials* **363**, 448–452.

Wood, B. (1982). *Applications of Thermodynamics*, 2nd edn. Waveland Press.

Xu, X., Xiong, Z., Dorr, M. *et al.* (2007). Edge gyrokinetic theory and continuum simulations. *Nuclear Fusion* **47**, 8, 809–816.

Xu, X. Q., Cohen, R. H., Rognlien, T. D. and Myra, J. R. (2000). Low-to-high confinement transition simulations in divertor geometry. *Physics of Plasmas* **7**, 1951–1958.

Xu, X. Q., Nevins, W. M., Cohen, R. H., Myra, J. R. and Snyder, P. B. (2002). Dynamical simulations of boundary plasma turbulence in divertor geometry. *New Journal of Physics* **4**, 53.

Yaglom, A. M. (1966). The influence of fluctuations in energy dissipation on the shape of turbulence characteristics in the inertial interval. *Soviet Physics Doklady* **11**, 26.

Yoshizawa, A., Itoh, S.-I. and Itoh, K. (2003). *Plasma and Fluid Turbulence.* Institute of Physics Publishing.

Zagórski, R., Gunn, J. P. and Nanobashvili, I. (2007). Numerical investigations of edge plasma flows in the Tore Supra tokamak. *Plasma Physics and Controlled Fusion* **49**, 97.

Zawaideh, E., Kim, N. S. and Najmabadi, F. (1988). Generalized parallel heat transport equations in collisional to weakly collisional plasmas. *Physics of Fluids* **31**, 3280–3285.

Zawaideh, E., Najmabadi, F. and Conn, R. W. (1986). Generalized fluid equations for parallel transport in collisional to weakly collisional plasmas. *Physics of Fluids* **29**, 463–474.

Zohm, H. (1996a). Review article: Edge localized modes (ELMs). *Plasma Physics and Controlled Fusion* **38**, 105–128.

Zohm, H. (1996b). The physics of edge localized modes (ELMs) and their role in power and particle exhaust. *Plasma Physics and Controlled Fusion* **38**, 1213–1223.

Zweben, S. J., Boedo, J. A., Grulke, O. *et al.* (2007). Edge turbulence measurements in toroidal fusion devices. *Plasma Physics and Controlled Fusion* **49**, 1.

Zweben, S. J., Maqueda, R. J., Stotler, D. P. *et al.* (2004). High-speed imaging of edge turbulence in NSTX. *Nuclear Fusion* **44**, 134–153.

Zwillinger, D. (1989). *Handbook of Differential Equations.* Academic Press.

Index

Printed in the United States
By Bookmasters